Flow Measurement

D. W. Spitzer, Editor

Flow Measurement

D. W. Spitzer, Editor

Practical Guides
for Measurement and Control

Instrument Society of America

FLOW MEASUREMENT

Printed in the United States of America

INSTRUMENT SOCIETY OF AMERICA
67 Alexander Drive
P.O. Box 12277
Research Triangle Park, NC 27709

Library of Congress Cataloging in Publication Data

Flow Measurement: practical guides for measurement and control / D. W. Spitzer, editor.
p. cm.
ISBN 1-55617-334-2
1. Fluid dynamic measurements. 2. Flow meters. I. Spitzer, David W. II. Instrument Society of America.
TA357.5.M43F57 1991 91-31239
681'.2—dc20 CIP

About This Series

This volume is part of the Practical Guides Series developed and published by the Instrument Society of America (ISA).

The Practical Guides were conceived because of a shortage of published material in the field of measurement and control that bridges the gap between theory and actual industrial practice. Many books in the field have catered to the needs of technical students, who need to be oriented to basic control theory and concepts, or college-level readers, who are interested in engineering mainly from a classroom perspective. There are handbooks for practicing engineers that cover measurement and control, but these handbooks often devote only a chapter or two to topics that merit more attention. Within the Practical Guides Series, separate volumes address each of the important topics and give them comprehensive, book-length treatments. Each book in the series can be understood and used by technical students, sales engineers, sales personnel, and managers, and relied upon by those who have "real-life" industrial concerns such as correct application, safety, installation, and maintenance.

Another unique feature of the Practical Guides is the stress placed on the actual experience of measurement and control practitioners. The Practical Guides are overseen by three Series Editors and one Series Technical Editor, who have extensive experience in measurement and control. The Series Editors guide the Volume Editors, who have been selected for their specific expertise in the volume topics and who bring together numerous Contributing Writers with even more specialized knowledge.

The Practical Guides capture the hard-earned experience of the writers and, by employing examples and recording anecdotal observations, make that experience as applicable for the reader as possible. Case studies, either hypothetical or based on real case histories, are used to illustrate typical situations and show how good planning and practical applications made the difference between success and failure. Some of this information has never been documented before.

This volume is designed to be at home in a library, in a classroom, or on the plant floor. Thc comfortable reading style, large pages, and frequent illustrations will contribute to ease of use. The page design uses graphics to "call out" some of the major points of the text, such as crucial safety checks and important examples. Each Practical Guide gathers widely scattered information in a single text, with bibliographies directing the reader to other sources.

Providing editorial guidance for the Practical Guides Series are some of the most distinguished names in the field of measurement and control.

Paul W. Murrill, Ph. D., Series Editor

Paul W. Murrill has authored or co-authored ten textbooks and over 70 articles on process control, computers, and mathematical models. Formerly Chancellor of Louisiana State University, he currently serves as Special Advisor to the Chairman of the Board of the Gulf States Utilities, after having served as chairman and CEO of the company for five years.

Thomas M. Stout, Ph. D., Series Editor

Thomas M. Stout is the author or co-author of more than 125 technical papers and holds four patents as co-inventor of computer and process controls. Dr. Stout is a pioneer in the application of computers in process control, particularly in petroleum refining.

Robert H. Zielske, Series Editor

Robert H. Zielske is Chief Instrument Engineer for Georgia Pacific Corporation, where he oversees design, testing, installation, and start-up of new facilities and major expansions. During his 40-year career, he has worked in design and applications engineering, marketing and sales, and training.

John W. Bernard, Series Technical Editor

John W. Bernard has worked on the leading edge of industrial automation systems for 35 years. He has extensive experience in process management and research. Among his many achievements is installing the first digital computers for direct process control. Bernard has written numerous papers on computer control and systems engineering and has been deeply involved in standards activities.

Call for Participation

The major purpose of this series is to collect in one volume most of the existing practical knowledge about specific measurement and control subjects.

Additional material for inclusion in subsequent updates of this volume is most welcome.

If you wish to contribute any helpful hints, case studies, or other material, please contact:

Manager, Publication Services
Instrument Society of America
P.O. Box 12277
Research Triangle Park, NC 27709
(919) 549-8411

Table of Contents

Preface

Virtually every technical book contains an introduction that expounds upon the importance and applicability of its content. In reality, this section is rarely read because this information adds little or nothing to the overall technical content of the text.

Often, that which is not presented can be a book's most valuable lesson.

Consider giving a person clear verbal directions on how to get between two places, understanding that the route may not be the shortest. Teach that same person how to read a map, and the best route to get anywhere in the world can be determined.

The content of this book is no more than numerous verbal directions that supply answers to questions. Learning to read the map entails intimate knowledge of the process and the laws that govern it.

Most individuals succumb to pressure and are content to find the quick solution to a perceived problem. This approach typically yields mediocre results and may even camouflage the real problem.

Even though most individuals are capable of thought, the process of defining the real problem before finding a solution occurs far too infrequently. The real problem and a good solution thereto must be determined before one can decide whether the contents of this book are truly useful.

To apply information effectively, one must think.

David W. Spitzer
Spring Hill, NY
August 1991

About the Volume Editor

David W. Spitzer, author of ISA's *Industrial Flow Measurement and Variable Speed Drives: Principles and Applications*, is Lead Utility and Instrument Engineer at Nepera, Inc., where he is responsible for the technical support and overall direction in the electrical, instrument, and utility areas on a plant-wide basis, as well as assisting in the preparation and execution of projects, systems, and long-range planning. With degrees in engineering from the University of Connecticut and the University of Illinois, Mr. Spitzer teaches courses in flow measurement and variable speed drive technology under the aegis of both the Instrument Society of America and the Association of Energy Engineers. His numerous papers on flowmeter technology have appeared in conference proceedings and technical journals.

Mr. Spitzer serves on the American Society of Mechanical Engineers' Committee on Measurement of Fluid Flow in Closed Conduits. His other interests include the design and implementation of energy saving projects, distributed control systems and programmable controllers, and control valves and other final control elements.

Contributors

Biographies of each of the following contributors are included at the end of their respective chapter(s).

William C. Baker
Teledyne Hastings-Raydist

R. Gary Barnes
Brooks Instrument Division

Mead Bradner
Consultant

Alvin E. Brown
Consultant

Thomas H. Burgess
Fischer & Porter Company

William S. Buzzard
Fischer & Porter Company

Theron A. Carman, Jr.
Nepera Chemical Company

Charles E. Fees
Fischer & Porter Company

William C. Gotthardt
Endress & Hauser

Douglas M. Grant
ISCO, Inc.

Zaki D. Husain
Daniel Flow Products, Inc.

Thomas M. Kegel
Consultant

Frederick A. Kilpatrick
U. S. Geological Survey

John G. Kopp
Flow Measurement and Marketing Associates

Michael A. Lucas
Accurate Metering Systems, Inc.

Wade M. Mattar
The Foxboro Company

George E. Mattingly
National Institute of Standards and Technology

Raymond C. Mills, Jr.
Fischer & Porter Company

Paul D. Oliver
Flow Dynamics, Inc.

James R. Reusch
Micro Motion, Inc.

M. Joseph Sergesketter

Lee Smith
Micro Motion, Inc.

David W. Spitzer
Copperhill and Pointer, Inc.

E. Loy Upp
Daniel Industries, Inc.

Mason P. Wilson, Jr.
University of Rhode Island

1

Why Measure Flow?

The audience for a book on the subject of flow measurement is relatively limited; however, flow measurement is a common part of everyday life. For centuries, domestic drinking water bills have been based upon flowmeter measurements, and in ancient times flow measurement techniques were used in water clocks. More recently, natural gas, fuel oil, gasoline, and potable water flowmeters have been used to bill domestic customers.

While similar, industrial flowmeters exhibit some subtle differences when compared to their utility counterparts, both measure flow. However, most utility flowmeters totalize for internal billing and efficiency purposes, while most industrial flowmeters measure flow rate for control, alarm, and indication purposes. Utility flowmeters are optimized for many installations of a given application, while the more expensive industrial flowmeters are of a more universal design that must be carefully applied to each application.

Due to the vast number and types of fluids (such as liquids, gases, saturated steam, superheated steam, wet steam, slurries, corrosives, and abrasives) that must be measured in industrial processes and exposed to a myriad of operating and ambient conditions (such as high and low pressures and temperatures), selection of an industrial flowmeter can be a formidable challenge.

The magnitude of the challenge is illustrated by experience, whereby most flowmeters will "work," but do not "work accurately." It is not uncommon to encounter situations where the focus is on installing and maintaining a flowmeter to be operational without even a passing thought to whether the measurement is correct.

Why should there be a concern about measurement accuracy? The answer to this question lies in the answer to the question posed at the onset: Why measure flow? The importance of flow measurement accuracy reflects the importance placed upon the primary purpose of the flow measurement.

Initially, in industry there was little interest in measuring flow, and flow measurements were not considered. As time went on, the primary concern was whether or not there was flow, and rudimentary flow measurement was considered. More recently, interest is focused on obtaining reasonably accurate flow measurements to monitor and improve process efficiency, quality, and safety. Investments are made in flowmeters with the expectation of receiving a return that exceeds the costs associated with the installation. Flowmeters that "work" but do not "work accurately" may not achieve their potential return.

When To Measure Flow

Flow measurement technology has flourished because it is an economical means to obtain primary information that is necessary to generate an invoice for billing purposes and/or to control, alarm, or indicate a process condition.

Billing

Flow measurement is applied to measure quantities of fluid that change ownership when it is not practical to more accurately perform the measurement using a more direct method, such as directly measuring the fluid weight or fluid volume. When the flow measurement is used for billing purposes (custody transfer), the flow measurement should be as accurate as possible, given the economic value of the product passing through the flowmeter.

When significant quantities of a high value fluid pass through a flowmeter, the cost of purchasing, installing, operating, and maintaining the most accurate (and often the most expensive) flowmeter is small relative to the value of the product. Selection of the most accurate flowmeter possible is dictated in these applications.

In other applications, when large quantities of a low value product are measured for billing purposes, the dollar value of the billed flow may be high enough to warrant good measurement accuracy.

Inexpensive flowmeters with good measurement accuracies for the service are usually applied to the measurement of small quantities of low value fluids.

Control

Flow measurement is applied when process considerations indicate that the flow of a given fluid should be controlled and the user is willing to support the costs involved.

Flow measurement accuracy requirements are dependent upon process considerations. Experience has shown that accuracy requirements based upon process considerations are usually more stringent than the actual flowmeter performance, even though the flowmeter "works." Measurement inaccuracy is usually a result of inappropriate flowmeter selection, unanticipated operating conditions, improper calibration, improper maintenance, and/or incorrect installation.

The inherent accuracy and maintenance requirements for control purposes are not as stringent as those required for billing. However, flowmeter accuracy requirements should reflect not only the economic value of the flowing fluid, but also the economic effect of flow measurement errors on the process that affect product quality. It is not uncommon to install a highly accurate flowmeter for a low value fluid stream when that fluid has a high impact on final product quality.

Indication

Flow measurement is applied when process considerations show that an indication of the flow rate of a given fluid would be helpful and the user is willing to support the costs involved. A flowmeter used for indication should usually be selected to provide an accuracy similar to that of a flowmeter used for control.

Alarm

Flow measurement is applied when process considerations show that an abnormal flow rate of a given fluid should be alarmed and the user is willing to support the costs involved. Due to the subjective nature of the alarm settings, flowmeter accuracy requirements are usually not stringent.

Flowmeters

Flow measurement technology has evolved rapidly in recent decades. Some measurement techniques have survived, while others have fallen by the wayside or have never been commercially developed. Physical phenomena discovered centuries ago have been the starting point for many viable flowmeter designs. In recent years, technical development, namely in fluid mechanics, optics, acoustics, electromagnetism, and electronics, have resulted not only in improved sensor and electronic designs but also in new flowmeter concepts.

This technology "explosion" has enabled flowmeters to handle many more applications than could have been imagined centuries ago. Flow measurements encompass applications that range from capillary blood flow to flow over spillways, flows of gases, plasmas, pseudo-plastics, solids, and corrosives, to name a few.

Effective flowmeter selection requires a thorough understanding of flowmeter technology in addition to a practical knowledge of the process and the fluid being measured. The difficulty in bringing these two facets of flow measurement to bear on a practical application is challenging even to experienced engineers, technicians, and sales personnel.

Why Measure Flow?

Flow is measured accurately because one cannot afford *not* to measure accurately because the competitive environment dictates that one does not have the *luxury* of producing wastefully. Someone is always watching; if not the plant manager, then the vice-president of finance; if not the plant neighbors, then the EPA; and, of course, there are the laws of physics to keep everyone honest.

This book provides technical methods, flowmeter designs, clues, guides, and direction to measure flow accurately.

2

Historical Perspective

Millions of years ago a small tribe of men found that the area in which they were living produced less and less vegetation and game each year to feed upon. One particular year, the situation became almost unbearable and Ogu, the leader of this small group, gathered his clan together and announced that he thought they should move out of the area to find better hunting and foraging grounds. He knew that the journey would be very difficult because they had to traverse a vast area of barren land. They gathered as many supplies as they could and set off for better lands.

After several days of traveling and being without water, the tribe became somewhat edgy, and Ogu became very concerned. He knew that without water the tribe could last only for a few more days, so he sent some of his most trusted and able warriors ahead to search for water. A few days later one of the warriors returned but did not seem too overjoyed, so the tribe thought that he failed to find water. However, the warrior had found water but not in sufficient quantity that the whole tribe would be sated, and he knew that the stronger ones would make sure that they got their water even at the expense of the weaker members of the tribe. Their animal instinct to survive far outweighed their commitment to community survival, so the warrior knew that he must tell of his find only to Ogu.

Ogu was puzzled; he never had to face this problem in the past. He knew that the survival of the clan depended on the survival of the weaker members, the women and the children, as well as the stronger members. There was no easy solution, but he knew that they had to set out in the direction of this new find.

At last, as Ogu and the tribe reached the top of a knoll, he looked down and saw a very small oasis—so small that it supported only one palm tree. Suddenly Ogu knew the answer. He knew he had to limit each member to a small quantity of water. He finally got the courage to call his clan together and explain what had to be done and why it had to be done. At the end of his speech, Ogu decreed that each member should have only five handfuls of water so that all could survive.

Whether this story is true or mere conjecture, it could have happened and possibly did happen in much the same way as it was told. This is the same problem one faces in trying to write the history of flow measurement: we have some idea concerning the history of hydraulics and fluid mechanics but little is known about actual flow measurement. It wasn't a very popular subject then as it isn't a very popular topic today. Flow measurement had to be accomplished to achieve some of the successes in hydraulics in the past, but little was written about it. The layman today thinks very little about it even though he purchases his natural gas, domestic water, and gasoline with this technology. In the story above, the handfuls of water could have been the first time that primitive man measured a defined volume of liquid. Only in the years to come was the making of pottery discovered, and wine and water were sold and traded according to liquid volume.

This little story also points out another aspect of flow measurement: it is a common and natural thing to do. The velocity of a moving air mass or the swiftness of a stream are topics often cited in the normal conversation of the day. For example, this morning when I woke up the wind was howling, or was it screeching? We would almost all agree that howling is not as bad as screeching. Why? Experience tells us that as the wind velocity increases, the pitch of the sound associated with it also increases. Whether or not we know the reason for this naturally occurring phenomenon, we use the same principle to estimate wind velocity as the modern vortex shedding flowmeter uses to measure flow in a closed conduit. In modern times, your child, who is sitting in the back seat of the car with his hand out the window, might say, "Dad, go faster, I can still hold my hand out the window." Could this be the same principle used to measure flow in target meters and pitot tubes?

In early history the Chinese, the Egyptians, and the Romans showed a great degree of understanding of hydraulics. The Egyptians, with their massive irrigation ditches and the weirs they used to distribute the flow to the fields, had to have some knowledge of flow measurement; otherwise, why did they use weirs to distribute the flow? In some cases when irrigation was required in fields higher than the source they were required to use bailers, as shown in Figure 2-1. I'm quite sure these bailers were slaves and had to bail a certain number per day—a precursor of modern-day displacement meters.

Since man always tries to make work easier or more efficient, along came the first screw pump (probably invented by one of the Egyptian bailers) shown in Figure 2-2. Historians normally attribute the invention of the screw pump to Archimedes, but some evidence exists that the Egyptians developed it first.

The Chinese needed to have some knowledge of the amount of flow in their rivers to design their reservoirs and massive canal systems. Today the same systems are used in the same way to prevent floods as they were in 1500 B.C. The Romans, in order to design their famous aqueducts, needed to know something about flow measurement since they supplied water equivalent to the amount a modern city of 700,000 uses today. The individual Roman consumer was taxed according to the size of the nozzle installed in his house. Even in those times, the people who knew the system could cheat on it. This was abetted by the fact that there was no standardized system for measuring the size of a pipe. The size of a pipe measuring five digits in one system might be actually larger than the size of

(a) IN ANCIENT EGYPT

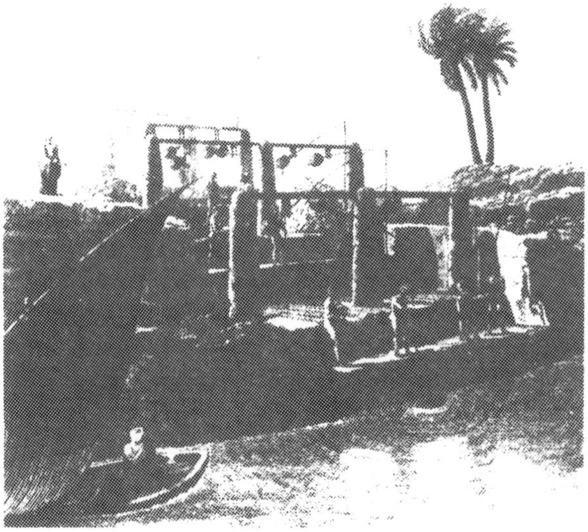

(b) IN 19th-CENTURY EGYPT

Figure 2-1. Egyptian Water Bailers

a five-digit pipe in another system. If the suppliers wanted to cheat a particular customer, they would give him the system that had the smaller size, and they and their friends could tap into the aqueduct for free without the officials knowing of any shortage.

Unfortunately, the Romans didn't have the necessary understanding for accurate flow measurement, nor did they know which variables were important. They didn't understand that pressure affected the flow through a nozzle. Therefore, there were no specifications regarding the height location of the individual spout (adjutis) in the aqueduct; however, surprisingly, they did know that installing the pipe perpendicular to the aqueduct was essential. They fully understood that if it faced upstream, more flow would go through the tap, and, conversely, if it faced downstream, less flow would occur. Figure 2-3 illustrates the proper installations and shows what happens if the nozzle is installed improperly.

The Romans sometimes needed to pump water, as the Egyptians did. Their adaptation of the Archimedean screw pump is shown in Figure 2-4. It bears a striking resemblance to the screw pumps used in the food industry today. The reason for mentioning the screw pump here is that it is the first time that blades in the form of a screw were used to move a fluid from one place to another. It requires mechanical work input. The Romans did not realize that they could use the same screw to take work out of the fluid in the form of a turbine. Surprisingly, we must wait until the Industrial Revolution for this invention.

In order to measure flow rate or velocity it is mandatory that some concept of time be introduced. Early man measured time in terms of the number of days, moons, or seasons—a rather crude approximation for the degree of accuracy required to measure flow rate today, except in rare circumstances. The next refinement to the measurement of time took the form of shadow clocks or sundials. These relied on the position of the sun and the shadow it cast on a fixed object. These instruments didn't work very effectively during inclement weather and were impossible to use at night. Thus, the early experimental fluid mechanist could not have been a workaholic because he always had to quit at sunset and had many holidays.

About 1400 B.C., the Egyptians invented a water clock (see Figure 2-5). Their first attempt was to fill up a bowl of water that had a small hole in the bottom and calibrate it by observing the number of bowls required for a full day's operation. Unfortunately they found that it took many more than twice the number of half bowls to last a full day and deduced that the height of the water in the bowl was

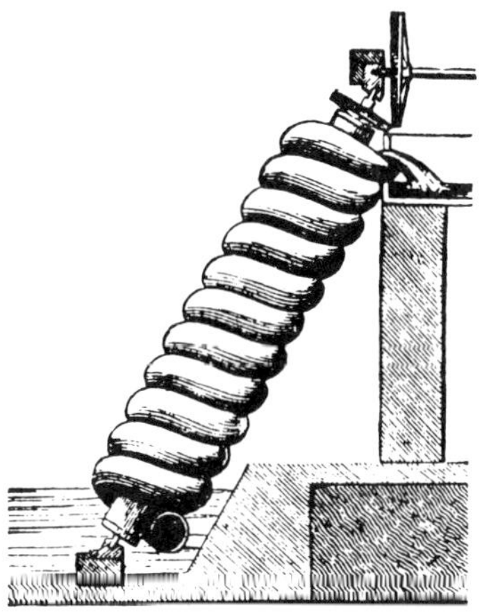

Figure 2-2. Egyptian Form of Archimedean Screw

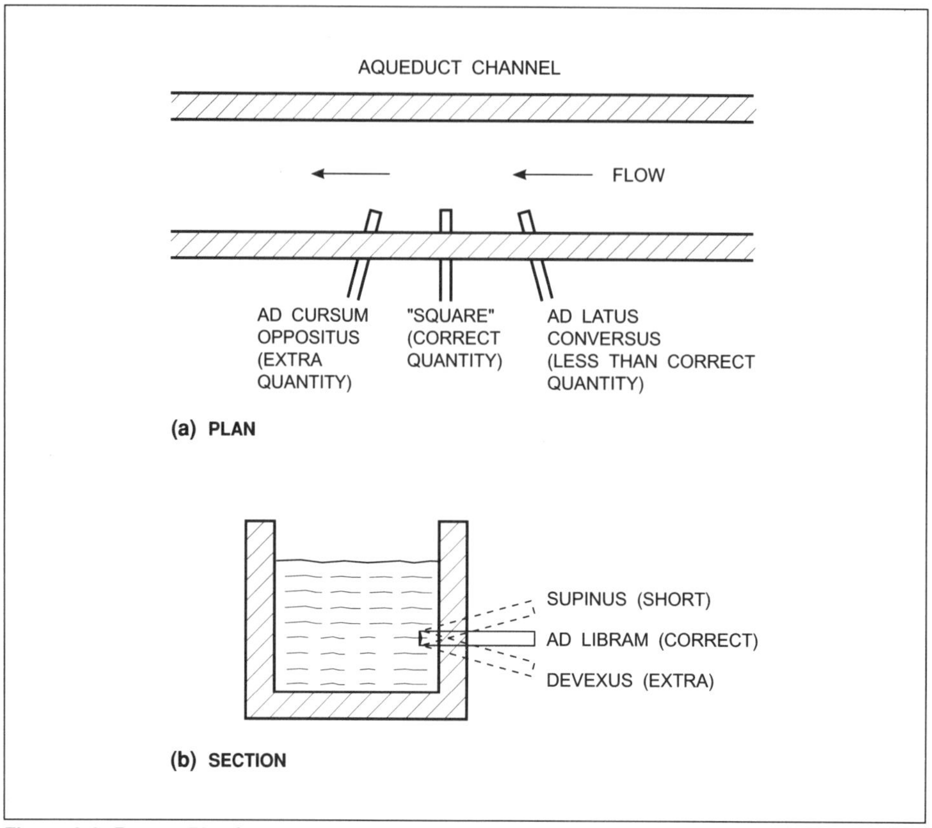

Figure 2-3. Roman Pipe Installation

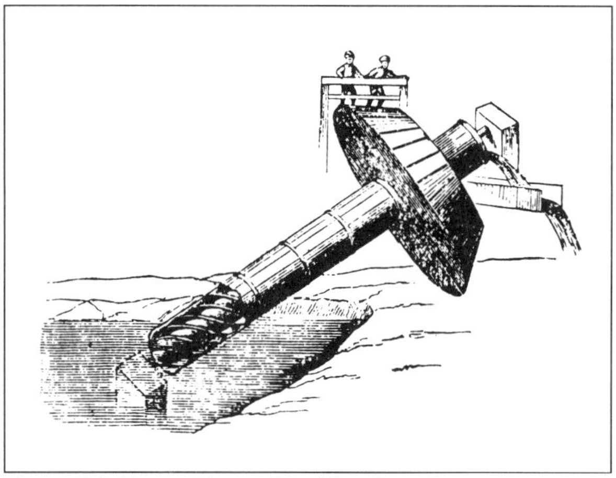

Figure 2-4. Roman Form of Archimedean Screw

Figure 2-5. Clepsydra or Water Clock

an important factor in the rate at which water flows out of the bowl. They found that a system of two bowls, one above the other, worked much better. The upper bowl, having the small hole in the bottom, was maintained full of water by providing sufficient water that it continued to overflow; time was measured by counting the number of lower level bowls being filled. Refinements to this time could be easily achieved by replacing the bowls with smaller bowls or by using cylindrical bowls having marks to indicate the level of water in the bowl. This was the first indication that the height of a water column was an important parameter that affected the rate of flow. Prior to this, the early hydraulist knew only that the reservoir had to be at a higher level than the irrigation canals.

Hero of Alexandria (*ca* 150 B.C.) picked up this idea and clearly understood that flow rate was a function of area and velocity. He stated the modern concept of the continuity equation—that the flow of an incompressible fluid is equal to the cross-sectional area times the velocity, i.e., $Q = A \times \bar{v}$. Of course he didn't say it quite the same way; he stated, "Given a spring (a source of water), to determine its flow, that is, the quantity of water it delivers...

"Now it is necessary to block in all the water of the spring so that none of it runs off at any point, and to construct a lead pipe of rectangular cross section (at that time they didn't realize that lead pipes could cause sterility). Care should be taken to make the dimensions of the pipe considerably greater than those of the stream of water. The pipe should then be inserted at a place such that the water in the spring will flow out through it... Now the water that flows through the pipe will cover a portion of the cross section of the pipe at its mouth. Let this portion be, for example, 2 digits. Now suppose that the width of the opening of the pipe is 6 digits. $6 \times 2 = 12$ digits. Thus, the flow of the spring is 12 digits. It is to be noted that in order to know how much water the spring supplies it does not suffice to find the area of the cross section of the flow, which in this case we say is 12 digits. It is necessary also to find the speed of flow, for the swifter is the flow, the more water the spring supplies, and the slower it is, the less. One should, therefore, dig a reservoir under the stream, note with the help of a sundial how much water flows into the reservoir in a given time, and thus calculate how much will flow in a day. It is, therefore, unnecessary to measure the area of the cross section of the stream. For the amount of water delivered will be clear from the measure of time."[Ref. 1]

It is surprising that Hero had such a remarkable understanding of how to measure flow rate. Furthermore, his description of a using a reservoir and a sundial is a clear forerunner of the modern calibration stand.

In summary, the contributions of the ancients to flow measurement include the invention of the clock, the screw pump, the use of a weir, and Hero's method for determining the flow rate in a channel. Archimedes established his buoyancy laws, but these added little to our understanding of how to measure flow rate.

From these early times to about the year 1500, very little was done in the area of hydraulics or fluid mechanics.

Renaissance to Modern Times

The modern era of fluid mechanics started with such giants as Galileo, da Vinci, Newton, and many others. So widespread and diverse were the contributions of these early investigators that to treat each investigator thoroughly would detract from the focus of the history of flow measurement. The approach taken herein, therefore, is to look at each type of meter used today and trace its origin as far back as possible, giving due credit where possible. It must be realized that there was very little demand for flowmeters throughout history except in recent times, and most of the principles used to measure flow today were the result of independent investigations in the general field of fluid mechanics—not investigations directed towards the development of a particular flowmeter. Another limitation to the development of flowmeters was the development of instrumentation, which, until recently, lagged far behind our knowledge of fluid mechanics. With the advent of the microprocessors, transistors, and a variety of sensors, etc., the instrumentation field now allows one to detect fluid processes that only a short time ago seemed impossible and enables new types of flowmeters to enter the market.

Open Channel Flow

The measurement of flow in canals, aqueducts, irrigation ditches, streams, and rivers falls under this category. The Roman aqueducts were primarily open channel flow systems. At that time they knew how to make pipe from lead sheets and fired clays but did not perfect the art of successfully joining pipes together. Consequently, the few pipe systems they constructed were not very successful. Open channel flows, from the time of the Romans to the present, play a major role in providing water supplies to major cities. One of the largest water systems in the world, Los Angeles, currently receives its water through an elaborate system of open channels.

It should be noted that Hero, as mentioned above, knew how to calculate the flow in a conduit and essentially stated that the volume flow rate is equal to the product of the cross-sectional area and the velocity of the stream. Unfortunately, he did not know how to measure the velocity.

It wasn't until the early 18th century that an Italian astronomer, physics and mathematics professor developed the first meaningful expression for the flow rate over a weir. Marquis Giovanni Poleni (1683-1761) is credited with the following equation:

$$Q = \tfrac{2}{3} hb\sqrt{f} \tag{2-1}$$

where h is the depth and b is the breadth of the channel and f represents a velocity function that later becomes $2gh$. Inclusion of this term in the above equation results in what is now called the Poleni Equation.

Leonardo da Vinci (1452-1519), among a myriad of other topics, was also interested in hydraulics. His works are collected in a nine-part treatise called "Del moto e misura dell'acqua." In this work he discusses floating bodies, water waves, eddies, the movement of water, and a host of other topics. Leonardo probably would have had a much greater impact on the field of hydrodynamics had he not kept his findings and sketches to himself. Only after his death was the above treatise found by his relatives. Besides his interest in hydraulics, architecture, art, etc., his brilliant mind found time to be a chief engineer overlooking a vast system of canals and harbors in northern Italy during the years 1502 to 1503; he held a similar type of position in France some time later. A sketch of flow over a contracted weir by Leonardo is shown in Figure 2-6.

During this same period, one of Poleni's contemporaries, Henri de Pitot (1695-1771), invented what is today referred to as the Pitot tube (see Figure 2-7). It consists of two parallel tubes, one of which has a right angle bend at the end. This tube is then aligned with the flow, and the difference in the pressure between the two tubes can then be shown to be a function of the stream velocity. Pitot would later claim that the idea was so simple and natural that the moment he thought of it he ran immediately to the river with a glass tube to experiment. His first experiment verified his predictions, and he would later admit that it was hard to imagine how such a simple invention could escape so many skilled people who had worked on the motion of water. It is interesting to note that Pitot was one of those youths who showed very little promise until his early twenties, when he became an avid student of mathematics and the physical sciences. He studied in Paris under Reaumur and returned to his native province of Languedoc in 1740, where he was named superintendent of the Canal du Midi and remained for the rest of his life. In this capacity he built bridges and aqueducts and improved the flood protection of his district. The Pitot tube became one of the first and most reliable velocity measurement instruments during this period. Had Hero had such an instrument, it is doubtful that he would have recommended diverting the stream to measure its flow rate. This instrument gave the fluid mechanist the opportunity to determine the flow rate in any stream, river, or aqueduct according to the following equation:

$$Q = A\bar{v} \tag{2-2}$$

where:

Q = volume flow rate
A = cross-sectional area of flow
$\bar{v}$ = average flow velocity

The Pitot tube continues to be a very popular instrument for measuring velocity, especially in the aircraft and maritime industries, where it is the prime instrument for indicating speed.

In the early nineteenth century another Italian hydraulician, Giorgio Bidone (1781-1839), investigated flow rate over a weir. Bidone was a professor at the University of Turin who also investigated the contraction and "percussion" of water jets. He is sometimes credited with the discovery of the hydraulic jump. Most of his works of interest were published between 1820 and 1826.

Pierre Louis Georges du Buat (1734-1809) amplified Poleni's analysis of the discharge over a weir and obtained a discharge coefficient of 0.65. A French engineer and aristocrat, du Buat later acquired the title of Count. He believed there was a stagnant layer near solid boundaries that the remainder of the fluid would glide over. If this were the precursor of the modern concept of the boundary layer, he failed to mention viscosity as one of the variables.

Around this same time Reinhard Woltman (1757-1837) invented a current meter having a vane and counter to measure river flow; the vane rotated in the water. This can be considered the precursor of the modern day turbine meter (see Figure 2-8).

In 1845, Julius Weisbach (1806-1871) further analyzed the flow over weirs and developed the equation that, except for minor corrections, is the one presently used to calculate flow rate. He studied in Freiburg, Vienna, and Göttingen and later became a teacher of mathematics at the school of mines in Freiburg. It is interesting to note that Gaspard Gustave de Coriolis (1792-1843) also worked on the analysis of flow over weirs. Coriolis is better known for his discovery of the force that now bears his name and which is now used to describe a certain type of mass flowmeter. Michael Faraday (1791-1867) was also interested in open channel

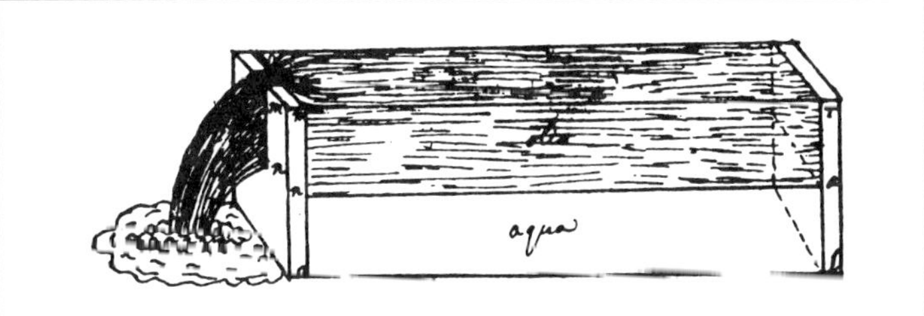

Figure 2-6. Sketch by Leonardo of Flow over a Contracted Weir

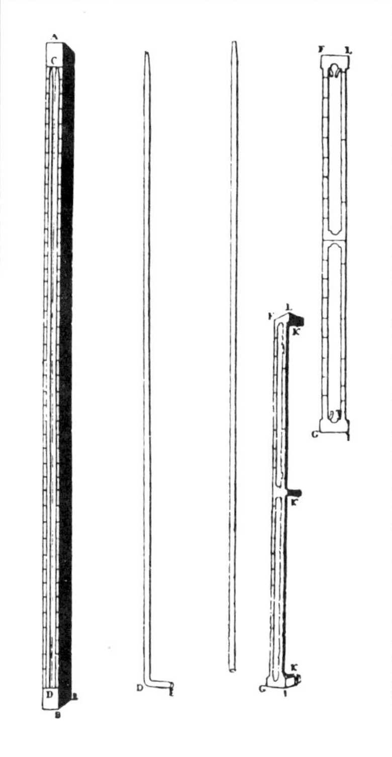

Figure 2-7. Details of Original Pitot Tube

flow; he had tried to measure the flow in the Thames using an electromagnetic method. He had hoped the earth's magnetic field would be sufficient to induce an electric field that he could measure with two large metal plates. Unfortunately, his instrumentation wasn't sensitive enough to pick up the induced field, but he must be credited with conceiving the first modern day electromagnetic flow-meter.

Differential Pressure Measuring Devices

This section includes all the devices that measure flow using differential pressure to indicate the flow rate, such as the orifice plate, the Venturi tube, the elbow meter, and so on.

ORIFICE PLATES

Flow through an orifice has probably been the most studied. In its most simple version, the device consists of a small hole in a container through which flow occurs. It was an integral part of the water clock and has fascinated scientists, artists, and philosophers for centuries. The artist or sculptor used it to create beautiful fountains, while the scientist studied patterns of its efflux from early times until rather recently. Evangelista Torrecelli (1608-1684) was the first to generalize the projectile of water as it issued from an orifice and the first to recognize that the flow was a function of the height of the liquid column above the water. You might remember him for being the inventor of the barometer; the space above the mercury barometer is often referred to as the Torrecelli vacuum. He had the fortunate experience of meeting Galileo and was his constant companion for the last three months of Galileo's life. He later occupied Galileo's chair of mathematics at Florence. He also was the first to present an expression for virtual work.

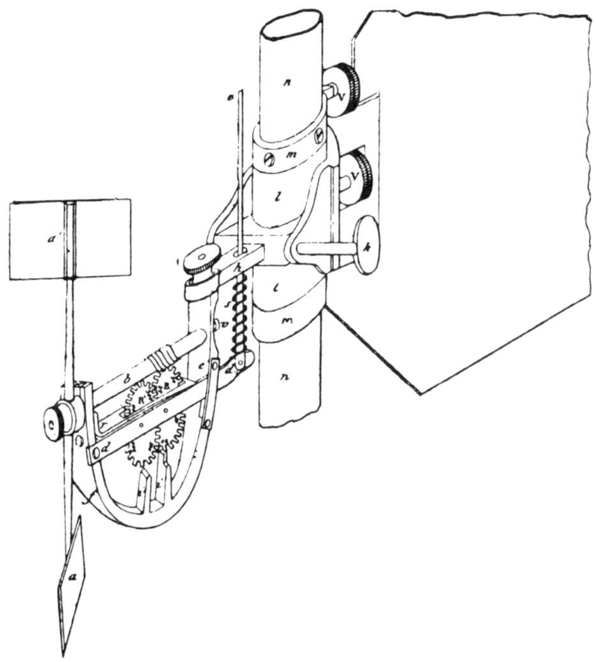

Figure 2-8. Woltman Current Meter

Isaac Newton (1642-1727), primarily noted for his works in mechanics and his theory of gravity, was also interested in hydraulics and gave us the first glimpse that shear stresses existed in fluids. He was the first to define what is called the vena contracta and introduced the contraction coefficient. Figure 2-9 shows Newton's view of flow through an orifice.

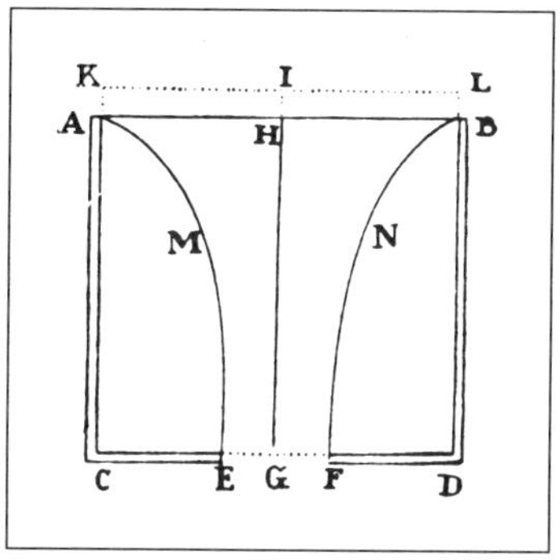

Figure 2-9. Newton's Original Concept of Orifice Discharge

Poleni, mentioned above for his work on weirs, also worked on the flow through orifices and showed the contraction coefficient to be 0.62, a great improvement over Newton's value of $1/\sqrt{2}$.

Jean Charles Borda (1733-1790) was also interested in the flow through orifices and showed the velocity as $\sqrt{2gh}$, where h was the head of fluid. Borda was also the first to introduce into the analysis of fluid flow the concept of the streamline. Prior to his time all the analysis of flow was done by looking at slices through the flow. He was educated at La Fleche, the scene of Descartes' early training. He began his career, as so many did at that time, as a military engineer, and it was in his early twenties that most of his work on hydraulics was done.

Du Buat (1734-1809), also mentioned above, studied losses at the abrupt corner of the orifice and showed the entrance loss coefficient to be 0.51, which is the presently accepted value.

No one is sure when the first orifice plate was used to measure flow in a pipe for the purposes of custody transfer, yet it is very feasible that it could have been accomplished, to some degree, around this period. However, one important link was missing—one that is required for the practical use of the orifice plate as a meter—that is, a differential pressure recording device connected in some way to a timing device. The first of these chart recorders appeared on the market in the late 1800s. It allowed both the orifice and any differential pressure measuring device to be used as a flowmeter.

One of the first patents on orifices was by Max Gehre in 1896. The first practical orifice meter appears to have arrived on the market in 1909 and was used to measure steam. Meanwhile, the burgeoning petroleum and gas industries began to use the orifice for several reasons but mostly because it was easily standardized and rather maintenance-free. This increased usage of orifice meters prompted three major engineering organizations to investigate the orifice coefficients more precisely; they were the American Petroleum Institute (API), the American Gas Association (AGA), and the American Society of Engineers (ASME).

Much of the work was done in the 1930s between two organizations, Ohio State University and the National Bureau of Standards. At Ohio State, Samuel Reid Beitler did much of the experimental work, and Edgar Buckingham and Howard Bean at the National Bureau of Standards did the correlation of the experimental results and developed equations that were still in use through the 1980s.

In the 1960s and early 1970s Rodger Birtwell Dowdell showed that there were many discrepancies between the known coefficients and the more recent experimental data. He later corroborated with Mason P. Wilson, Jr. in an attempt to more accurately correlate and analyze the flow through orifices. As an outgrowth of these studies, one showed (Wilson and Teyssandier) that the vena contracta does not occur at the location of the minimum pressure and that all the tap location data and coefficients should be correlated along a pressure profile. This discovery prompted many investigators to enter the field, developing much more accurate experimental data and new correlations that make the modern day orifice meter a highly accurate flowmeter. The most recent correlations of the orifice coefficients are available through ISO (International Standards Organization) or the American Society of Mechanical Engineers (ASME).

VENTURI AND OTHER FLOW UBES

Giovanni Battista Venturi (1746-1822), an Italian physicist, lived in the same era as Woltman (the first person to use rotating vanes as a current meter). Venturi was born in Reggio and at the age of 23 was ordained a priest. He taught geometry and philosophy at Modena University and was later appointed state engineer on projects involving rivers and drainage. He did his early work in Italy on the same equipment Poleni used. He was sent to Paris to study, and it was there that he wrote his dissertation on what he was best known for: "Recherches Experimentales sur le Principe de la Communication Lateral du Mouvement dans les Fluids Applique a l'Explication de Differens Phenomenes Hydraulics." It was in this work that he demonstrated the separation of flow downstream of mouthpieces (nozzles) to show that a gradual expansion of the flow eliminated the separation and, in doing so, regained most of the head from the nozzle. He was the first to propose a reduced depth of high speed flow downstream of a spillway and just prior to the hydraulic jump in order to increase the head (height) of the stream. The familiar Venturi tube or meter consists of a short nozzle, followed by a gradual expansion as shown in Figure 2-10. The flow through the tube can be determined from the pressure drop across the nozzle and the geometry of the nozzle.

While in Paris, Venturi became great friends of General Bonaparte. When Bonaparte conquered Italy he made Venturi a Professor at Pavia University. As indicated from various sources, Venturi was a very gifted individual and had a reputation for being very witty.

His work didn't receive its full impact until some 90 years later when Clemens Herschel (1842-1930), at the age of 56, published his most popular paper, "The Venturi Water Meter," in 1898. For this invention he received the Elliot Cresson medal from the Franklin Institute. Both Boston and Vienna, Austria, claim to be his birthplace, and his education was truly international. Herschel studied at Harvard, Karlsruhe, and Paris. His primary interest was civil engineering, but the influence of Francis (inventor of the Francis turbine) led him to a career in hydraulics. Herschel was president of both the Boston Society of Engineers and the American Society of Civil Engineers. He worked for Allis-Chalmers, served as a consultant to Niagara Power, and was a chief engineer for a New Jersey water company. While at Holyoke he was faced with the problem of how to charge the local power companies for the amount of water they used. It was this

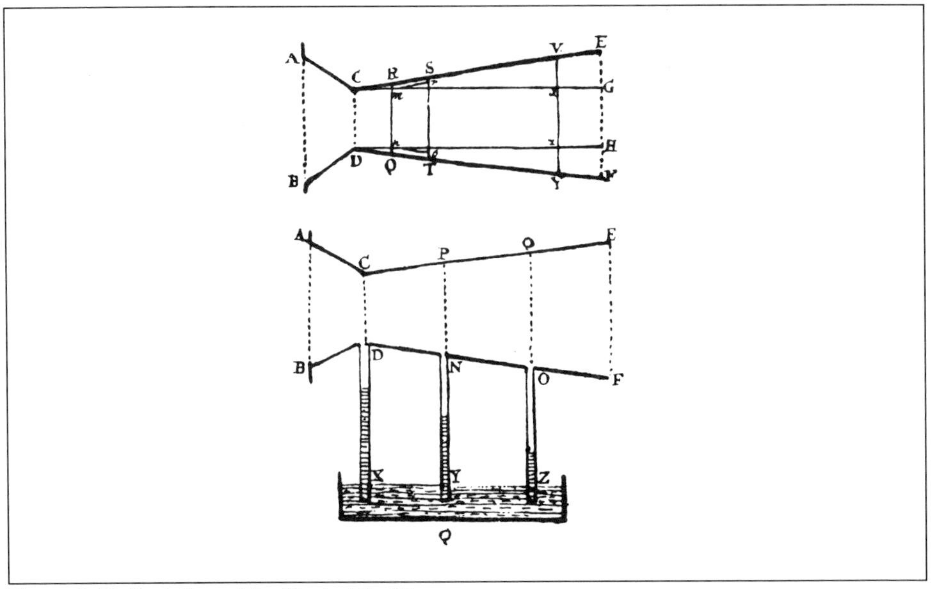

Figure 2-10. Evolution of the Venturi Profile

that prompted him to investigate the Venturi tube experimentally. Surprisingly, he did not use the Bernoulli principle to determine the meter coefficient until he had already correlated the results.

Positive Displacement Meters

The origins of positive displacement meters date back to the early 1800s and closely parallel the development of the gas and water industry. The first gas meter was invented by Samuel Clegg in 1815. The gas industry in the U.S.A. started around 1817 and somewhat earlier in Europe. The first water meter was invented around 1820. Clegg's meter was a water-sealed rotating drum meter, quite often found in chemistry and physics laboratories. It wasn't until 1843 that William Richards invented the diaphragm-type gas meter with sliding vanes that is very similar to the domestic gas meter of today. Some of the early water meters operated on the same principle and could not operate at any substantial pressure.

Between 1820 and the turn of the century the meter industry developed into the industry that we know today. It was driven by the development of the gas and water industries and was influenced by the petroleum and chemical industries to a lesser extent. The type of thinking that existed throughout this time is best illustrated by quoting from a brochure distributed by the National Meter Co. in 1908.

> Fifty years ago a water meter was practically unknown as an article of trade, and, moreover, it is the one thing to which the user did not take kindly. The average consumer who has not been enlightened will oppose its use, on his pipe at least, even if he may think it is a good thing for his neighbor. For this reason the introducers of water meters have had the combined opposition of all who should have been their best friends. This is an instance in which the demand for an article was negative. Yet, in spite of this fact, the instrument has made for itself a permanent place in the trade of today.
>
> Such a victory speaks in the highest terms both for the unfailing devotion and pluck of the early promoters and the worth of the instrument itself, as well as the equity of the meter system of selling water.
>
> Twenty-five years ago there was only one superintendent of water works in the United States who dared, in convention, to state without qualification that he was in favor of metering every tap. Today (1908) the superintendents who are opposed to meters are lonely in their isolation.
>
> The National Meter Company, the pioneer in the perfection of water meters and in the missionary work of introducing their general use, has for over thirty-eight years occupied a position of pre-eminence in this line and takes pleasure in presenting to you in the following pages some idea of the extent of this industry and a complete description of its products.
>
> The development of the water meter to its present state of perfection has been a long line of the survival of the fittest device. There have been many proposed devices for measuring water, and about everything has been tried that the imagination of man can conceive, but the exacting requirements of actual service have eliminated the great majority of these makes, leaving only a few successful meters on the market today.

This statement accurately describes that period of flowmeter history, from 1850 to 1900. The number of patents assigned to flowmeters was greater than to any other field of engineering, and only a few types remained on the market at the turn of the century. For a water meter to gain acceptance in the field it was expected to remain in service a number of years, a significant reason for so few to survive. Cost, maintenance, testing, and certification also played important roles in the success of a particular meter. So important were these items that they prompted Mr. Thomas Hawksley to state in 1882 that increase of cost—which included, in addition to the prime cost of the instrument itself, the expenses of fixing and repairs, replacement every third year, testings from time to time, renewals, and official inspection and registration—was so monstrous that he did not think "we should ever in this country come to be supplied with water for domestic consumption by means of meters."[Ref. 1]

A few rather remarkable meters during this time are worthy of mention. One particular meter, invented by C. W. Siemens, was of the variable area type, which consisted of a hinged gate, the deflection of which indicated the rate of flow. The deflection angle has to be integrated with respect to time in order for this device to become a meter. This was accomplished by adopting a mechanical wind-up clock and a twin cone variable speed device. Winding up the clock became cumbersome, and Siemens and Kennedy conceived a device for extracting the motive power for the clock from the fluid, as shown in Figure 2-11. It wasn't long after they built this twin rotor turbine that they discovered it functioned as a meter, the revolutions of which could be mechanically integrated to give totalized flow. It became the first of a long line of turbine meters.

The nutating disc type of meter in common use today as the domestic water meter was invented in 1830 when James and Edward Dakwyne were granted a patent for a hydraulic pump acting on the same principle; however, it wasn't until 1850 that this principle was used in a meter by Bryan. A number of others improved its design, and around the turn of the century the disc began to be made of hard rubber. The combination of hard rubber sliding on brass greatly extended the life of the meter, so it became practical for many water companies to use this meter to measure domestic water use. Many of these same meters are in use today, and the design remained untouched until the late 1950s, when the brass meter body and brass chamber were replaced by composites and plastics (see Figure 2-12).

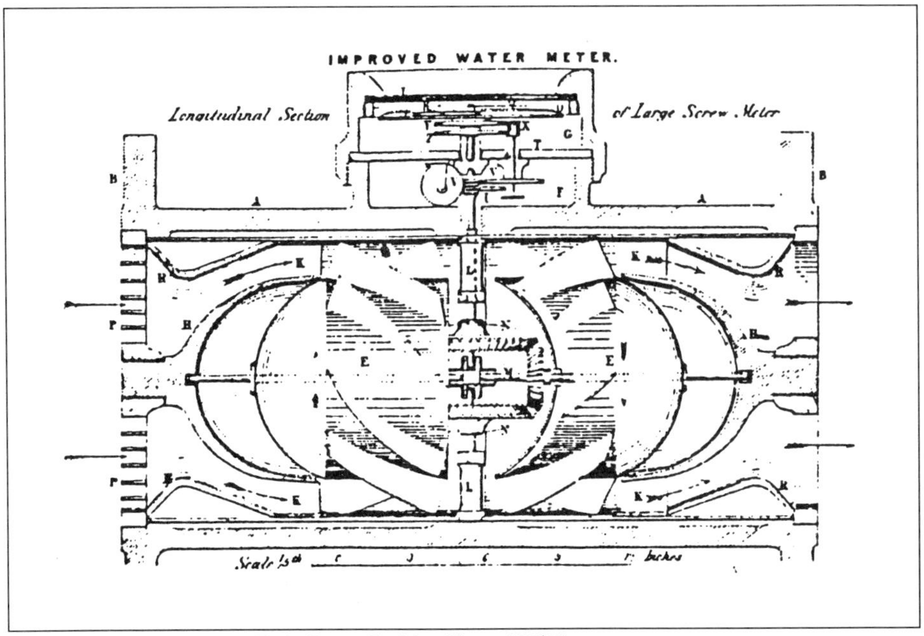

Figure 2-11. The Siemens Twin Rotor Turbine Meter (1850)

During the early and mid 1800s a number of piston type meters were introduced, some with opposing pistons, but most of these didn't stand up very well under the rigors of the field. In the late 1800s the rotary piston meter was invented and is still a common meter in the market today, especially in the food and diary industry where the designs are such that they are easily disassembled and cleaned.

Turbine Meters

The original principle of using a vane to impart motion to a fluid dates all the way back to Archimedes' screw pump. However, it wasn't until 1790 that Woltman developed the first known meter to work on the principle of using vanes geared to the fluid in a device used to measure velocity. Strange as it might seem, a horizontal turbine wheel, very similar to the modern day turbine, used water power to grind flour in ancient Greece. There is evidence that the Greeks obtained it from the barbarians. This type of water wheel was handed down through the centuries, and it appears that the Irish used one as early as the 4th century when King Cormac macArt sent for a skilled man from "over the wide sea" to build a mill (see Figure 2-13(b)). A horizontal mill, circa 1588, designed much like the impulse turbine of Pelton (1829-1908), is shown in Figure 2-13(a). One can easily make the comparison with the Pelton wheel (see Figure 2-13(c)). At least one person believes that Pelton didn't get his idea from the Greeks or the Irish, as shown in Figure 2-14. It appears that philosophers from antiquity up to almost the present ignored what the farmers were doing.

Siemens accidently developed the first turbine meter as we know it today (see Figure 2-11). However, the turbine meter did not gain popular acceptance until Potter in 1938 developed a pick-off sensor that utilized a magnet and rotating conductor to produce a pulsed emf, which was easily counted to produce totalized flow. These first practical turbine meters were used exclusively to measure liquid flows and were extensively employed in the petroleum industry to measure the

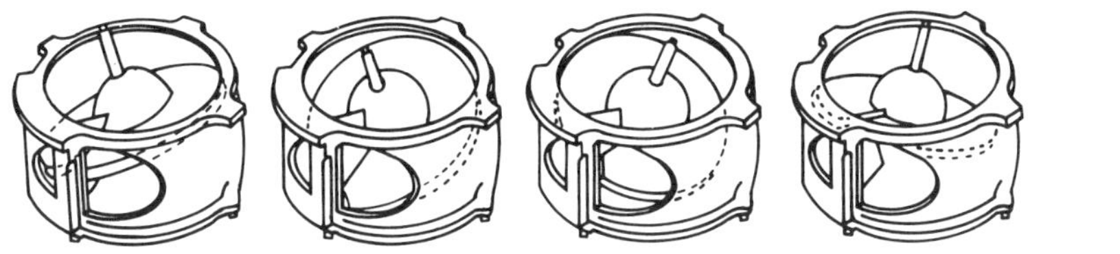

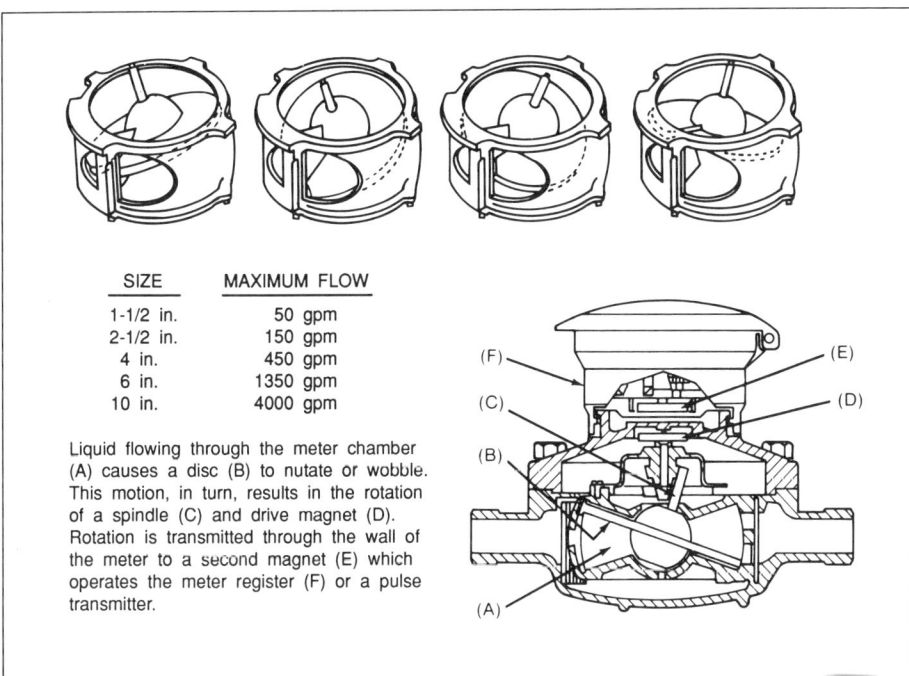

SIZE	MAXIMUM FLOW
1-1/2 in.	50 gpm
2-1/2 in.	150 gpm
4 in.	450 gpm
6 in.	1350 gpm
10 in.	4000 gpm

Liquid flowing through the meter chamber (A) causes a disc (B) to nutate or wobble. This motion, in turn, results in the rotation of a spindle (C) and drive magnet (D). Rotation is transmitted through the wall of the meter to a second magnet (E) which operates the meter register (F) or a pulse transmitter.

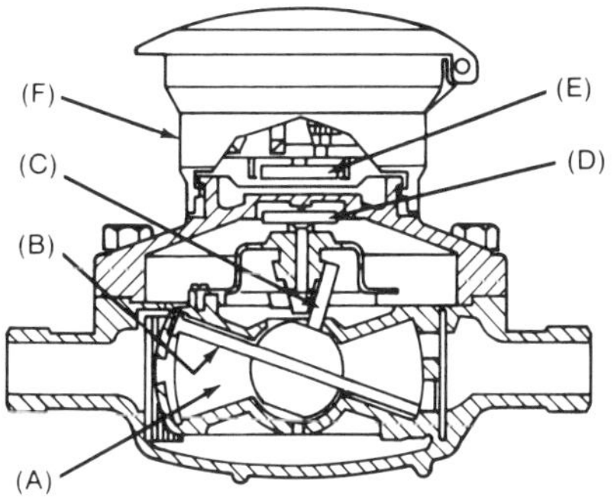

Figure 2-12. Nutating Disc Operation
(Courtesy of Badger Meter, Inc.)

(a) A HORIZONTAL MILL FROM
LE DIVERSE ET ARTIFICIOSE MACHINE
BY AGOSTINO RAMELLI, 1588

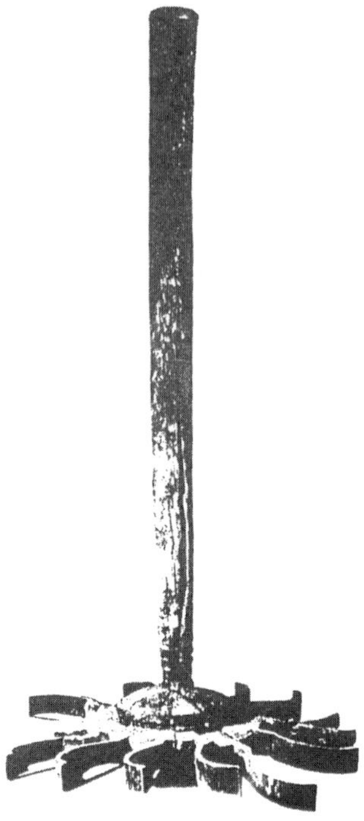

(b) THE MOYCRAIG TIRL,
BELFAST MUSEUM

(c) SOUTHERN CALIFORNIA EDISON, BIG CREEK 2A.
8 1/2-IN.-DIAMETER JET IMPULSE BUCKETS AND DISK
IN PROCESS OF BEING REAMED. 56,000 HP,
2200 FT HEAD, 300 RPM.
1948. *(ALLIS-CHALMERS MFG. CO.)*

Figure 2-13. Horizontal Mill

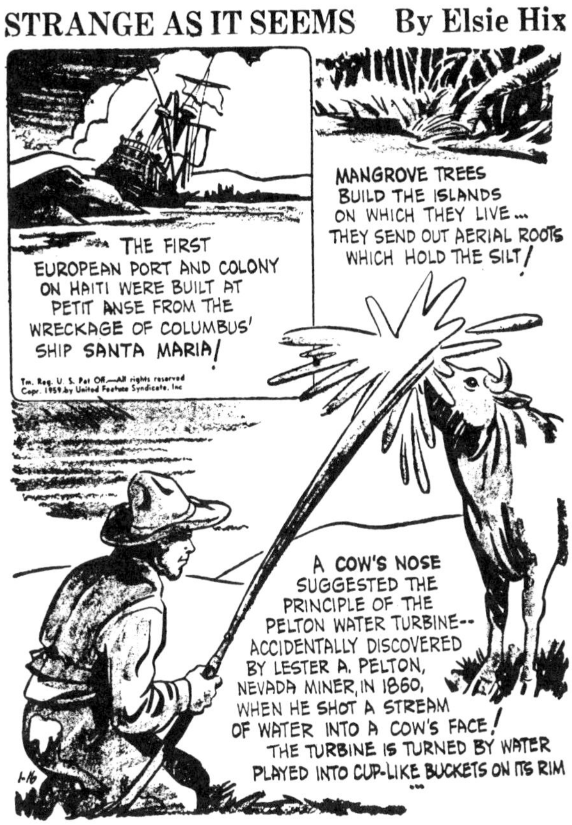

Figure 2-14. Pelton's Revelation

flow of hydrocarbons, most of which were natural lubricants and ideally suited to the turbine meter.

Not until 1953 was the first gas turbine meter introduced that had the same basic characteristics of the present day gas turbine meter. In 1963, Lee et al., of Rockwell introduced to the gas industry a much improved gas turbine meter. It took approximately 10 years of testing for this meter to be widely accepted in that industry where it now enjoys the reputation of being a highly accurate and reliable meter. In 1981, the American Gas Association (AGA) published "Measurement of Fuel Gas by Turbine Meters," AGA Report #7. In 1982, Rockwell introduced a self-correcting and self-checking (auto-adjust) turbine meter developed by Winston F. Z. Lee (see Figure 2-15). This new feature allowed the meter to be remotely inspected to determine whether or not it was functioning properly—a unique feature since most of these meters are employed in gas transmission lines and in places that are often remote.

Oscillatory Meters

The term "oscillatory meters" includes what today are known as the vortex and fluidic flowmeters and refers to any flowmeter that utilizes naturally occurring oscillations whose frequency varies with flow velocity. It is the shedding of vortices from tree limbs that makes the wind "howl" and the flag wave. However, there is no evidence that vortices were examined in any detail until Leonardo da Vinci sketched them downstream of a bluff body in 1513 (see Figure 2-16). This is surprising, since it is such an easily observable, naturally occurring phenomenon. It seems almost impossible not to observe the vortex shedding from a paddle when canoeing. Nevertheless, it appears that this phenomenon slipped by many of the best thinkers in history. Hermann Ludwig Ferdinand von Helmholtz (1821-1894), a native of Potsdam and a physiology professor at Konigsberg, Bonn, and Heidelberg, developed the theory of fluid rotation and vortex motion. The first comprehensive study of vortex shedding from bluff bodies was conducted by V. Strouhal who published his results in the *Ann. Phys. and Chemie* in 1878. He showed that the rate of shedding of vortices was proportional to the free stream velocity.

The shedding of vortices from blunt bodies was also investigated extensively by Theodor von Karman (1881-1963), and he applied for and was rewarded a patent for a flowmeter using this principle. von Karman was the son of a profes-

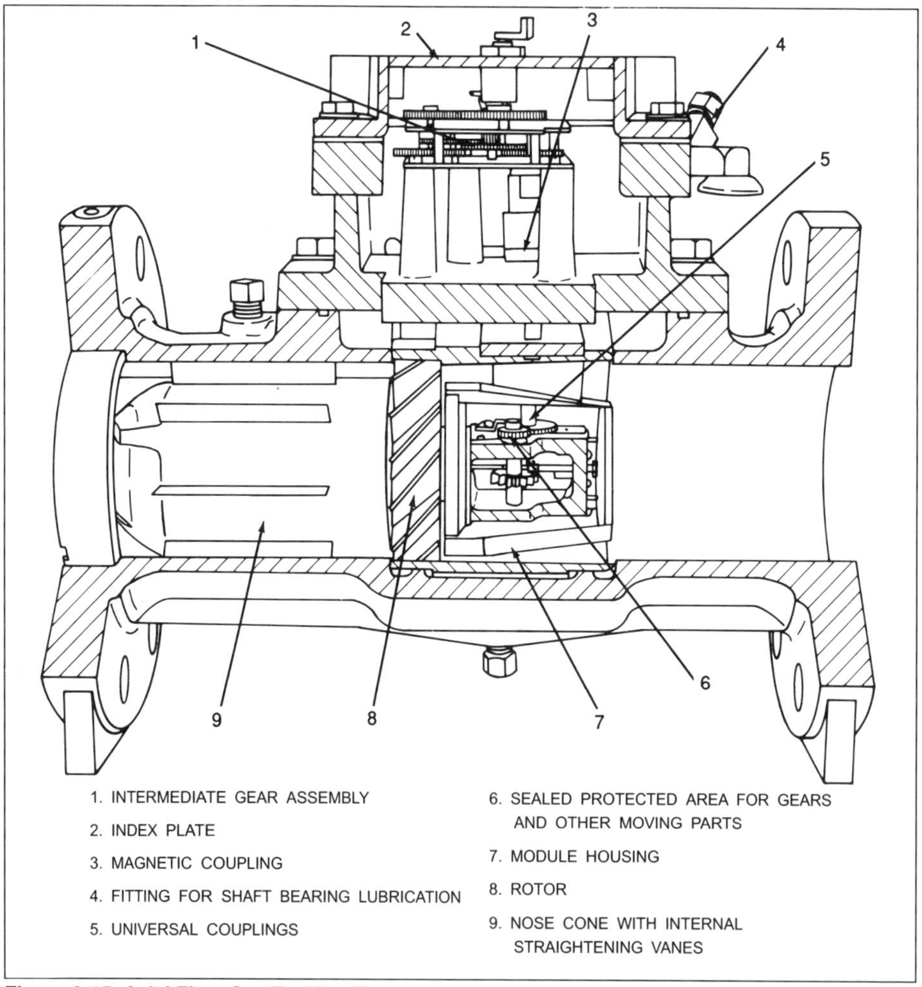

Figure 2-15. Axial Flow Gas Turbine Flowmeter
(Courtesy of Rockwell International)

Figure 2-16. Eddy Formation in Zones of Separation

sor at the University of Budapest. He graduated from the Royal Polytechnic Institute of Budapest with highest honors in mechanical engineering and went on to obtain his doctorate at Göttingen, where he later held a teaching post. It was at Göttingen that the famous Kaiser Wilhelm Institut für Stromungsforschung, under the direction of Ludwig Prandtl, graduated so many famous fluid mechanists, among them von Karman, Blasius, Betz, Tollmien, Shiller, Tietjens, and Ackeret. In 1912, von Karman was appointed director of an aeronautical laboratory at the Polytechnic Institute of Aachen, and in 1930 he accepted a similar post at the California Institute of Technology.

In 1954, Roshko patented a vortex shedding-type flowmeter that consisted of a pivoted rod that oscillated with a frequency proportional to the velocity. Bird in 1959 developed a pivoted bluff body having a flag on its downstream side that also oscillated at a frequency proportional to flow velocity.

Not until 1958 did a vortex shedding flowmeter, developed by Alan E. Rodely, have a limited commercial success. It consisted of a blunt body that occupied approximately one third of the pipe cross-sectional area. In the early designs, thermistor sensors located on the upstream side of the blunt body were used to sense the vortex motion. In the field, the protective covering for the thermistors quickly eroded away, causing the meter to fail. Rodely's meter was very successful in some applications, but in others it failed to live up to expectations. At times the meter would suffer "dead zones" when the meter would not oscillate at all. A series of new sensors and sensor locations were tried and received moderate success. Today, a host of other companies have picked up Rodely's work and have developed the meter so that it presently enjoys a place in the flowmeter market (see Figure 2-17).

The first flowmeter using fluidic principles was reported by Mason P. Wilson, Jr. and Southall in 1969 in ASME Paper 69-WA/FM-3, "Experimental Investigation of a Fluidic Volume Flowmeter." It works on the principle of attached jets or the Coanda effect. The use of this principle to produce pneumatic digital logic devices was born at the U.S. Army's Harry Diamond Laboratories in 1960. A diagram of a fluidic meter is shown in Figure 2-18 and operates as follows: A two-dimensional jet leaves the power nozzle and attaches to the side wall because of the Coanda effect and forms a separation bubble. The jet remains attached to one wall until it is switched to the other side by flow through the appropriate control nozzle. As flow is introduced into the separation bubble, the bubble grows, causing the attachment point to move along the wall and finally causing the jet to switch to the other wall. The first commercial flowmeter using this principle was marketed in 1972.

Ultrasonic Meters

This section includes all flowmeters that rely on the acoustic velocity in the flowing medium as a primary indicator of the fluid velocity. The simplest and easiest to explain is that using the Doppler principle to indicate the velocity of flow. The Doppler effect is the reason a train whistle has a high-pitched tone as it approaches and a low-pitched tone as it leaves (see Figure 2-19). For the stationary sound source it is seen that the sound waves are all at the same distance from the source in all directions, and, consequently, the frequency or "pitch" is the same throughout the field. For the case of the traveling source, the sound waves are compressed in the direction of travel by an amount equal to the difference between the sound velocity and the speed of the object and, hence, have an apparent higher pitch. The reverse is true once the source passes.

Although this method for measuring flow rate has been known for over seventy years, only in recent years has it been commercially successful, primarily because of the problems relating to instrumentation required to pick up the signal and separate it from other noise sources. The greatest number of patents filed for ultrasonic flowmeters was in the period 1960-1970, and a number of these found a place in the market.

Mass Flowmeters

Probably more patents are filed on mass flowmeters than any other kind. In principle and in practice almost all meters are used to measure mass flow. The use of orifices to measure gas flow rate requires a multitude of equations and corrections to reduce the measurement to standard cubic feet. To obtain the mass transferred, all that is required is to multiply the standard cubic feet by the gas density of a standard cubic foot of the gas. One measures the volume flow rate of liquids assuming that the density remains constant. Although most flowmeters imply mass flow measurement, few truly measure the inertia or mass of the fluid. The first mass flowmeter that measured the change in momentum and, thereby, measured the mass flow rate appears to have been by Y. T. Li and S. Y. Lee, who in 1953 described a Coriolis mass flowmeter. This meter consisted of two turbine wheels, each differently geared to the fluid and attached to each other by a helical spring. The torque, measured by the angular deflection of the spring, was proportional to the mass flow rate (see Figure 2-20).

A year later, V. A. Orlando and F. B. Jennings of the Aircraft Instrument Engineering Division of General Electric Company reported on a similar meter that became quite successful. It was used to measure the mass flow rate of aircraft

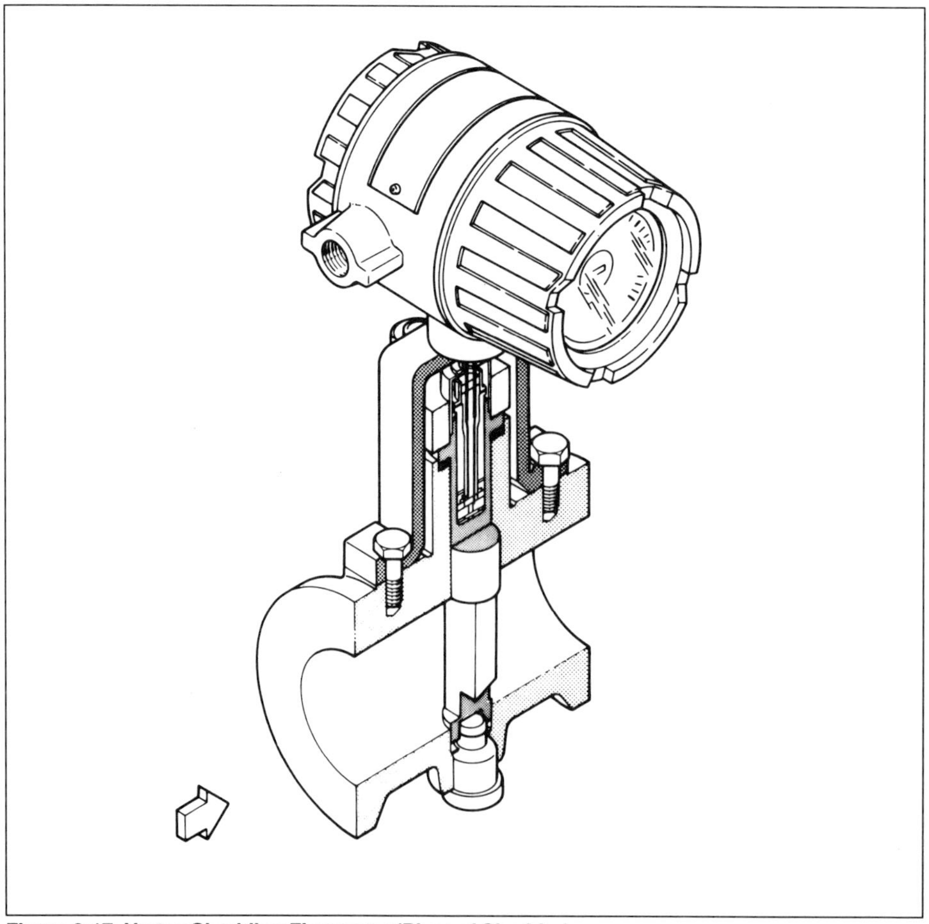

Figure 2-17. Vortex Shedding Flowmeter (Pivot of Shedder)
(Courtesy of Yokogawa Corporation of America)

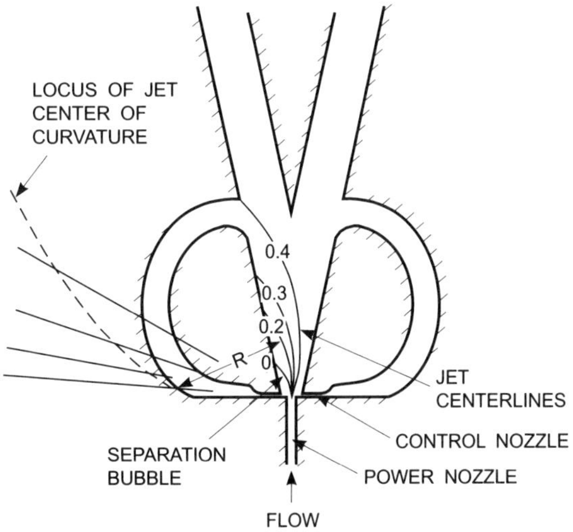

(a) FLUIDIC FEEDBACK OSCILLATOR

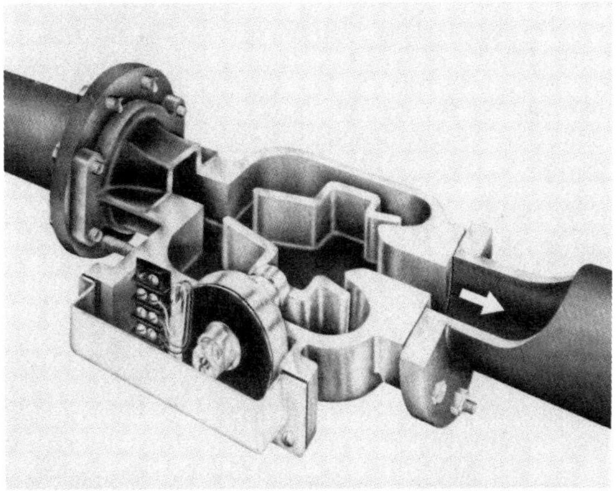

(b) FLUIDIC FLOWMETER CONSTRUCTION
(Courtesy of Moore Products Company)

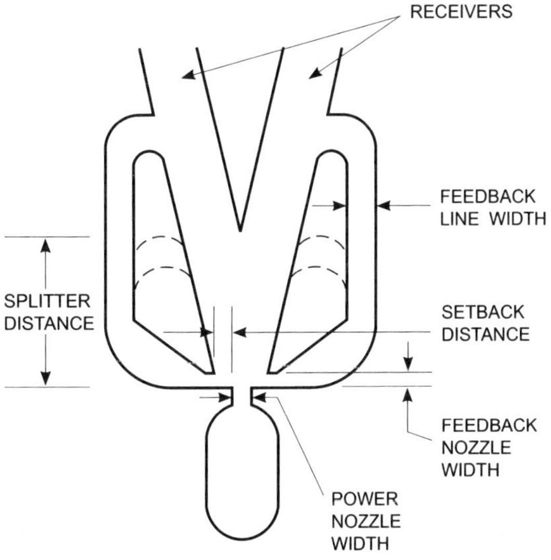

(c) JET ATTACHMENT AND SWITCHING

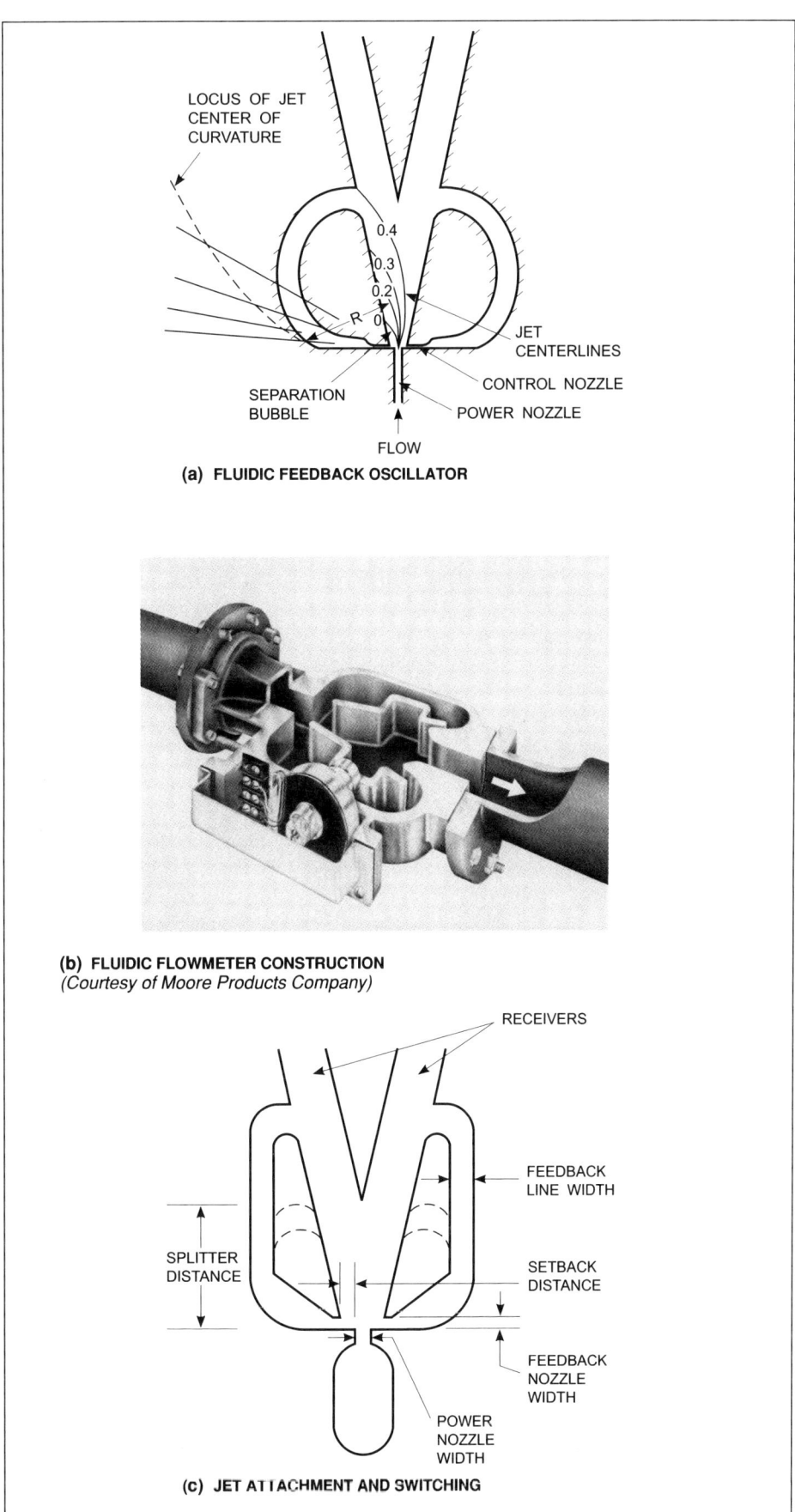

Figure 2-18. Fluidic Flowmeter

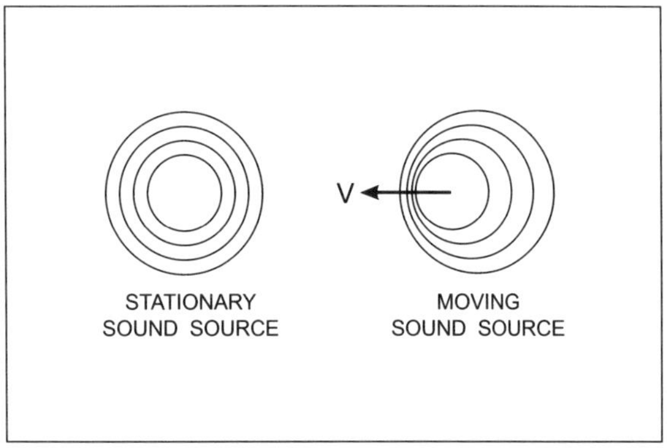

Figure 2-19. Doppler Effect

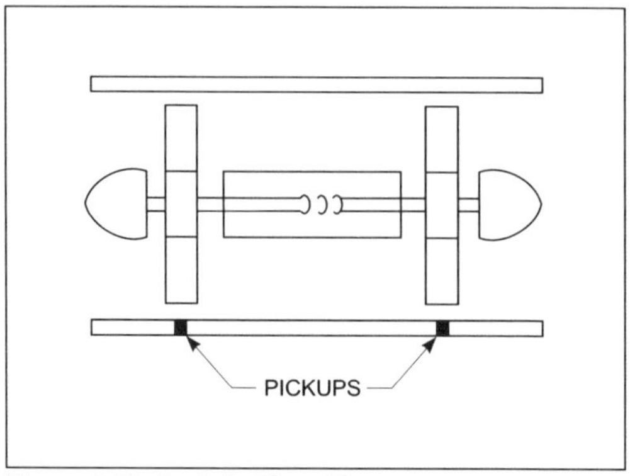

Figure 2-20. Coriolis Mass Flowmeter

fuel and consisted of a constantly rotating cylindrical cage that imparted a constant angular velocity to the fluid. Downstream was located a stationary cage that was kept from rotating by a helical spring. As in the meter by Li and Lee, the deflection of the spring was proportional to the mass flow rate. A number of other mass flowmeters were invented during this period, among them a gyroscopic meter that consisted of rotating a loop of pipe as shown in Figure 2-21. The problems associated with this mass flowmeter were immense. M. M. Decker in 1960 found that it was necessary only to vibrate the loop and that the torque required to vibrate the loop was proportional to the mass flow rate. This was the predecessor of the modern day mass flowmeter. In 1979, Micromotion introduced a somewhat similar flowmeter. Since then, a host of other manufacturers have entered the market (see Figure 2-22).

Electromagnetic Flowmeters

Faraday's experiment to measure the flow in the Thames, mentioned in the earlier portion of this chapter, was truly the first attempt to measure flow using induced emf as an indicator of flow rate. The problem of measuring small emf's and separating the signal from background noise was the chief stumbling block in getting this type of flowmeter on the market. In 1917 Smith and Slepian patented a device based on this principle for measuring the speed of a ship. In 1932 J. Williams applied this principle to measure the flow in pipes. The first commercial

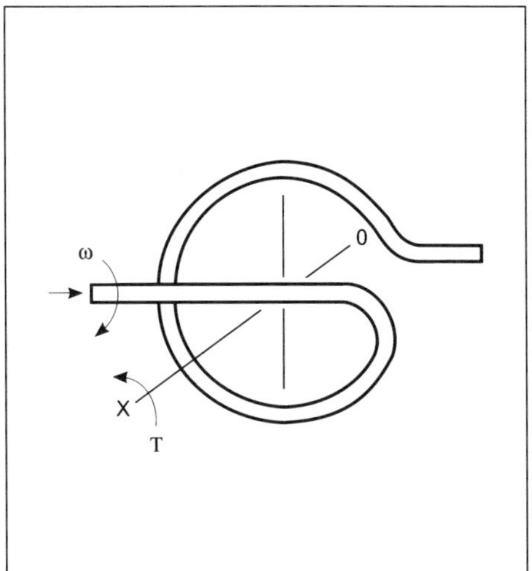

Figure 2-21. Gyroscopic Mass Flowmeter

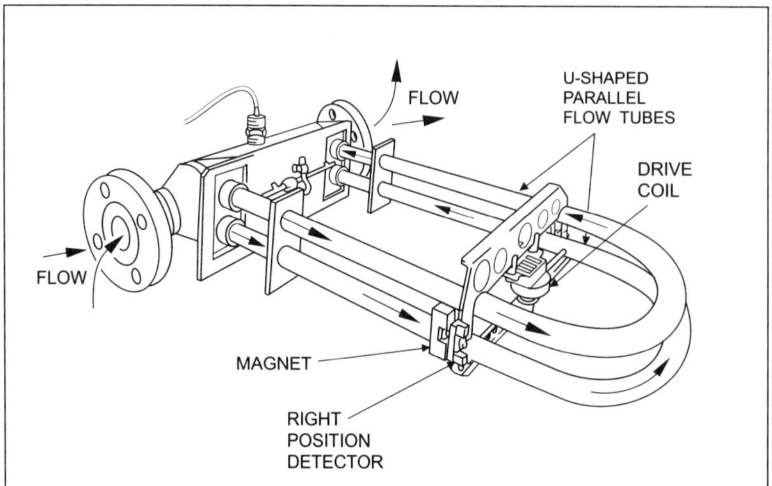

Figure 2-22. Mass Flowmeter
(Courtesy of Micromotion, Inc.)

electromagnetic (em) flowmeter was marketed by the Tobinmeter company in Holland in 1952. The Foxboro Company acquired the rights to this meter and marketed their own version in 1954. Since that time many companies throughout the world have developed variations of this basic meter. In 1962, J. A. Shercliffe wrote a book on the design and analysis of em flowmeters, which has since become a classic in the field. Rummel, Ketelson, and M. K. Bevir in 1968 expanded the analysis. Today the em flowmeter enjoys a comfortable place in the market, especially in the wastewater field where it is uniquely suited because it is a non-obtrusive meter.

References

1. Medlock, R. S., "Flow Measurement Special Issue," *Measurement Control*, June 1986.
2. Kinghorn, F.C., *Measurement Control*, Vol. 21, October 1988.
3. Rouse, H., and Ince, S., *History of Hydraulics*, Iowa Institute of Hydraulic Research, 1957.
4. Rouse, H., *Hydraulics in the United States*, 1776-1976, Iowa Institute of Hydraulics, 1976.
5. Landels, J. G., *Engineering in the Ancient World*, University of California Press, 1978.
6. Bean, H. S., "Formulation of Equations for Orifice Coefficients," ASME Paper No. 70-WA/FM-2.
7. *History of Orifice Meters and the Calibration, Construction, and Operation of Orifices for Metering*, Rept. Joint AGA, ASME Committee on Orifice Coeff., 1935.

Acknowledgments

It is almost impossible to give an accurate appraisal of all those who have contributed to this chapter. My own experience in this field and the people I have been associated with have all made an impact. More recently, my communication with W. F. Z. Lee, formerly of the Rockwell Corporation, has been very helpful.

About the Author

Dr. Mason P. Wilson, Jr., graduated from SUNY, Albany, with a B.S. in physics and completed his master's and doctorate degrees in mechanical engineering at the University of Connecticut. He served in the U.S. Navy as a weatherman in 1951-53. His industrial experience spanned a term of thirteen years, beginning with his flow measurement experience at Neptune Research Laboratories of Neptune Meter Co. In 1968 he joined the faculty at the University of Rhode Island where he now holds the rank of Professor. The holder of three patents, Dr. Wilson has done sponsored research for the American Gas Association Pipeline Research Committee, the United States Departments of Energy and Agriculture, the Office of Naval Research, and the United States Air Force Cambridge Research Laboratory. He has done much consulting and research in the area of flow measurement and has produced more than 75 publications. He is a member of the honor societies of Pi Tau Sigma, Sigma Xi, and Phi Kappa Phi and is a recipient of the American Gas Association's Award of Merit.

3

Physical Properties of Fluids

In a discussion of flow measurement it is important to become familiar with the various fluid properties, their nomenclature, and their units of measurement. Depending on the type of flowmeter, these properties can have a major effect on the resulting system design and performance.

Fluid properties discussed in this chapter are:

Temperature	Vapor Pressure
Pressure	Boiling Point
Density	Electrical Conductivity
Specific Gravity	Sonic Conductivity
Viscosity	Specific Heat
Liquid State	Velocity
Gaseous State	Flashing/Cavitating

Those listed on the left are of fundamental importance, while those on the right are important in certain flowmeters or in certain situations.

The basic flow equation is: $Q = A \times v$, where Q is volumetric flow rate, A is the cross-sectional area of flow, and v is the flowing velocity. The equation applies in all cases. If flow is in a pipe, the cross-sectional area can be found in numerous piping handbooks.

Units of Measurement

Units used to describe fluid properties are usually a combination of the SI system and the English system. In some countries that formerly used the metric system, a combination of the SI and the old metric system is used. Most manufacturers in the United States publish their data in both SI and the English system.

Keep a unit conversion table handy.

The SI system (Système Internationale d'Unités) is a modernized and improved version of the metric system. It is proposed as an eventual standard throughout the world.

The old metric system, sometimes called the "c.g.s." system, has been used for many years. Centimeters, grams, and degrees Celsius are some of its units.

The English system had formerly been used in most English-speaking countries. Inches, pounds, and degrees Fahrenheit are examples of its units. Some of these countries have now switched partly or totally to the SI system.

Fluid Temperature

A scientific way to describe temperature is to say that a body at a certain temperature and in contact with another body at a lower temperature will experience a flow of heat from the former to the latter. This is not a very practical definition; therefore, suffice it to say that temperature is a measure of hotness or coldness.

Kelvin and Celsius Scales

In the SI system temperature is expressed as absolute temperature in kelvins (K). Absolute scales start at zero, representing the lowest theoretical temperature possible. The Celsius scale, based on the behavior of water at atmospheric pressure, is also used. Freezing is 0°C and boiling is 100°C. A temperature in degrees Celsius is 273.15 less than the same temperature expressed in kelvins, and a Celsius degree is the same size as a kelvin.

The proper name for degrees C is "degrees Celsius." Often one will hear it called "degrees Centigrade." This is actually a French word, and it is completely interchangeable with "Celsius."

Rankine and Fahrenheit Scales

The English system of temperature units is similar to the SI system except that the size of its degree is smaller. It takes 1.8 of these degrees to equal one SI system degree. In the English system the absolute scale is Rankine and the convenient scale is Fahrenheit. The Rankine scale starts with the lowest possible temperature as zero, exactly as in the Kelvin scale. The Fahrenheit scale is not related to water behavior but seems to relate to weather conditions. Zero degrees Fahrenheit is about as cold as it gets, and 100°F is about as hot as it gets in many parts of the world. As it turns out, water freezes at 32°F and boils at 212°F. A temperature expressed in degrees F is 359.67 less than when expressed in kelvins. A Fahrenheit degree and a Rankine degree are the same size.

The Reaumur Scale

The only other temperature scale in practical use today is the Reaumur scale. It is rare and limited to use in some breweries. The Reaumur degree (°Rea) is 1.25 times as big as the Celsius degree. Water freezes at 0 degrees Reaumur and boils at 80 degrees Reaumur.

Figure 3-1 shows the relationship between the various temperature scales, and some useful equivalents are given below:

$$K = {}^\circ C + 273.15$$

$$^\circ R = {}^\circ F + 459.67$$

$$^\circ C = \frac{(^\circ F + 32)}{1.8}$$

$$^\circ F = 1.8\,(^\circ C) + 32$$

$$^\circ R = 1.8\,(K)$$

$$K = \frac{(^\circ R)}{1.8}$$

It is important to understand that most calculations for gas flow utilize absolute temperature in the equations. Signals representing °F or °C can be used but must be conditioned to give the corresponding absolute temperature values.

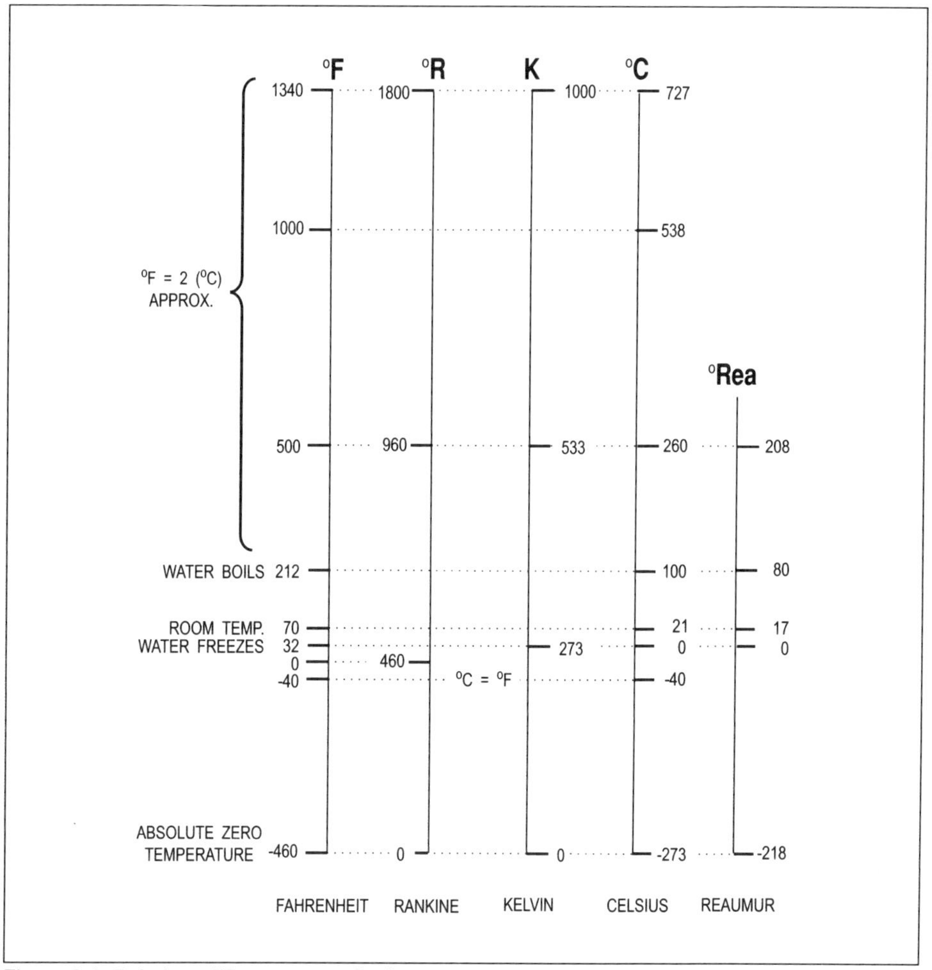

Figure 3-1. Relation of Temperature Scales

Fluid Pressure

Pressure is defined as force divided by the area over which the force is evenly distributed. Since liquids and gases are fluid, they exert pressure on their containers evenly. The pressure of the flowing fluid becomes a very useful criterion when defining the performance of a flowmetering system.

The SI system of units measures pressure in newtons (force) per square meter (area), which has the special name "pascal" (Pa). The pascal turns out to be so small that it is common to use kilopascals or megapascals, (kPa or MPa), which are 1000 and 1,000,000 pascals, respectively. In countries that formerly used the metric system, there is a tendency to remain with the older units of bars or kilograms/centimeter^2. In the United States the English units for pressure, lb/in.^{2} gage (psig) and lb/in.^{2} absolute (psia) have remained very common.

Three distinct types of pressure must be described by the units, as follows:

(1) *Absolute pressure* is the actual pressure of the fluid with respect to a perfect vacuum, regardless of the atmospheric pressure on the outside of the container.

(2) *Gage pressure* is the fluid pressure with respect to the atmospheric pressure outside its container.

(3) *Differential pressure* is the difference between two pressures. (Notice that gage pressure is actually a differential pressure between fluid pressure and atmospheric pressure.)

The SI system does not distinguish between absolute and gage pressure; therefore, it is always necessary to spell it out as "kPa abs," "kPa gage," and perhaps "kPa diff." The same is true when using units of bars or kg/cm^2. In the English system, the abbreviations "psig," "psia," and "psid" clearly describe the pressure type intended.

The SI system differentiates between force units and mass units (newtons and kilograms). In the English and the old metric systems the same name is used for force as for mass, (pounds and kilograms, respectively). This leads to some confusion and requires the inclusion of a unit converter, g_c, which is numerically equal to the acceleration of gravity at sea level.

$$F = mg_c$$

Yet another unit for pressure is the height of a column of water or mercury that would create the designated pressure at the base of the column. This unit is very commonly used for differential pressure. Typical units are "inches of water," "millimeters of mercury," etc. It is easily seen that the units of measurement for pressure are not clear and straightforward. The following equivalents, along with Figure 3-2, will help.

An absolute pressure transmitter is actually a differential pressure transmitter with one port completely evacuated and sealed. Therefore, it is more expensive than a gage pressure transmitter. The latter can be substituted for the former when measuring at high enough pressure so that variation in atmospheric pressure is negligible. If the pressure is above about 50 psig, changes in atmospheric pressure will usually have no significant effect on a gas flow computing system.

1 psi = 6.895 kPa

1 kPa = 0.1450 psi

1 bar = 100 kPa

1 bar = 14.50 psi

1 MPa = 145.0 psi

1 psi = 27.73 inches of water

1 psi = 2.310 feet of water

1 kPa = 7.5 mm of water

1 kPa = 4.019 inches of water

At sea level the atmospheric pressure will vary with the weather conditions. It is approximately 14.7 ±0.5 psia (101 ±3 kPa abs). If gage pressure is less than atmospheric, it is sometimes reported as "vacuum." This can be misleading because it means "the amount below atmospheric" but is shown with a positive sign rather than a negative one. Modern usage avoids "vacuum" as a unit of pressure. Instead, it should simply be called a "negative gage pressure"; for example, "- 3 psig," not "3 psi vacuum."

Fluid Density

Density is defined as the mass of the fluid per unit volume ($\rho = m/V$). In the English system, density is usually expressed in pounds per cubic foot, where the pounds represent mass rather than force. In the SI system density is often expressed either in kilograms per cubic meter or kilograms per liter. Several equivalence formulas are:

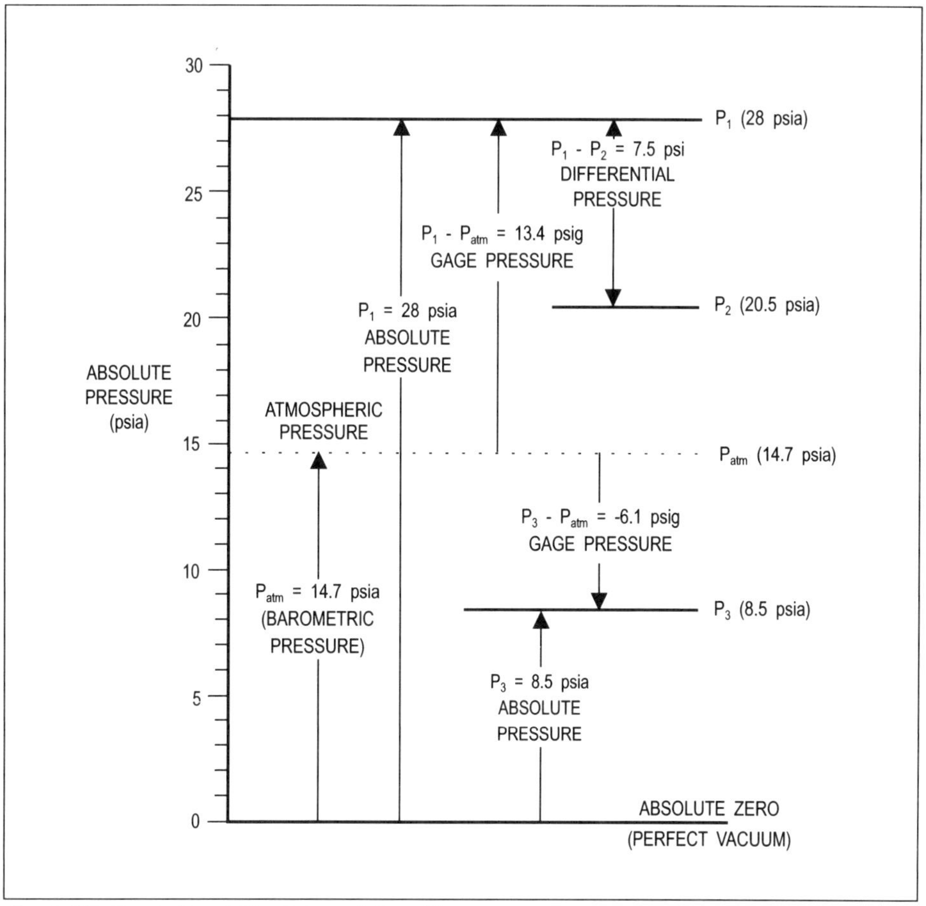

Figure 3-2. Examples of Absolute and Gage Pressure

$$1 \text{ lb/ft}^3 = 16.026 \text{ kg/m}^3$$

$$1 \text{ lb/ft}^3 = 0.016026 \text{ kg/l}$$

$$1 \text{ kg/l} = 0.0624 \text{ lb/ft}^3$$

Density of Liquids

The density of a liquid can vary with temperature and pressure. In nearly all cases the effect of pressure is so small that it can be ignored completely. Temperature, however, does have a significant effect on density. In general, liquids expand as temperature increases, and, therefore, the density decreases. Much data is available to give the density of various liquids at different temperatures. Sometimes it is in tabular form, as in Table 3-1. This makes it very easy to read the values.

In other cases there is only a listing of equations for the volumetric expansion of the liquid. To use this data, the density must be known at one temperature. Then, by applying the volumetric expansion coefficient values in proper ratio, the density at another temperature can be calculated.

Often graphical density data is available. Figure 3-3 is an example. These graphs are sometimes difficult to read accurately but can often serve as useful information. Figure 3-3 shows densities for a variety of liquids. Notice that over large portions of the graph the lines are quite straight. Therefore, over a designated range of temperature, it is quite possible to write a simple equation that expresses density as a linear function of temperature. For example, the equation for 65°Brix (Bx) sucrose solution over the range of 20°C to 60°C is:

Table 3-1. Example of Tabular Density Data

Barium Chloride Solution in Water ($BaCl_2$)						
%$BaCl_2$	**0°C**	**20°C**	**40°C**	**60°C**	**80°C**	**100°C**
2	1.0181	1.0159	1.0096	1.0004	0.9890	0.9755
4	1.0368	1.0341	1.0275	1.0181	1.0066	0.9931
8	1.0760	1.0721	1.0648	1.0551	1.0434	1.0299
12	1.1178	1.1128	1.1047	1.0948	1.0827	1.0692
16	1.1627	1.1564	1.1478	1.1373	1.1249	1.1113
20	1.2105	1.2031	1.1938	1.1828	1.1702	1.1563
24	———	1.2531	1.2430	1.2316	1.2186	1.2045
26	———	1.2793	1.2688	1.2571	1.2440	1.2798

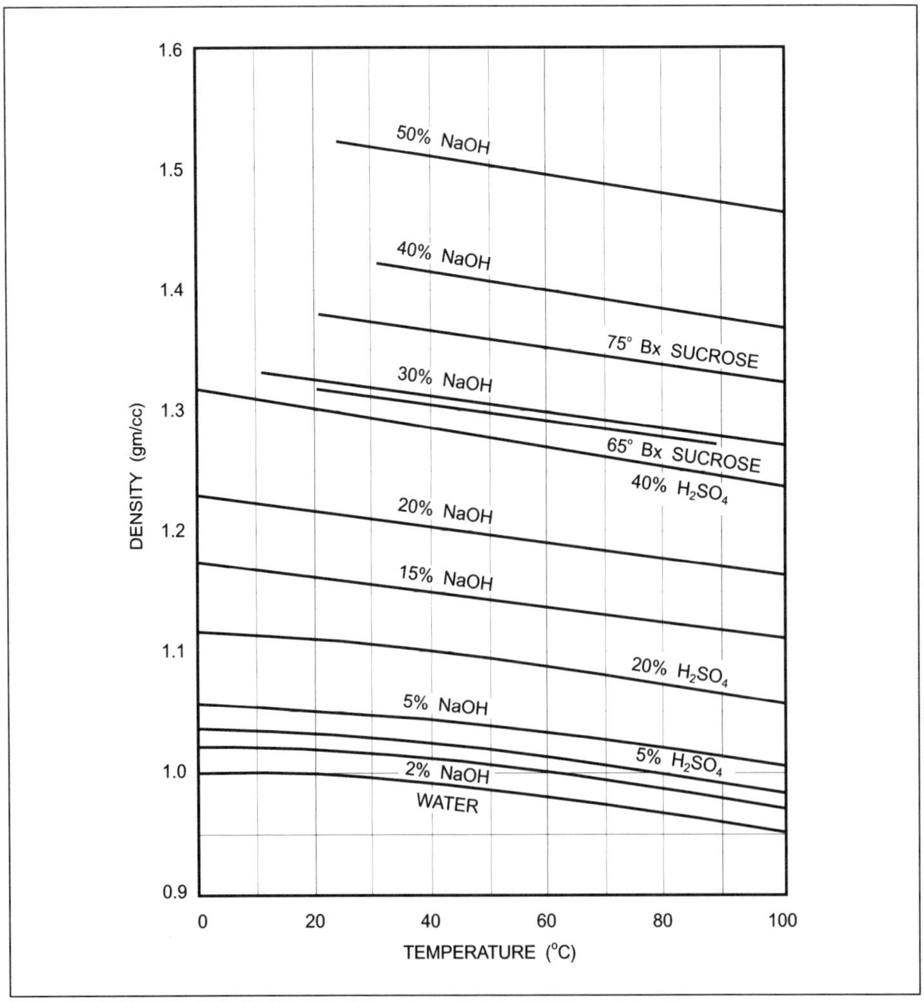

Figure 3-3. Density Data in Graphic Form
(Courtesy of Fischer & Porter)

$$\text{Density in kg/l} = 1.33 - 0.058°\text{C} \quad (3\text{-}1)$$

The Brix scale is a hydrometer scale for sugar solutions in which its readings at a specified temperature represent percentages by weight of sugar in the solution.

Specific Volume

Some data provide specific volume rather than density. This term is volume/mass and is merely the reciprocal of density. Tables and graphs for properties of steam show specific volume of steam rather than density. English system units are ft^3/lb, and SI units are usually m^3/kg.

Density can be derived from the volumetric expansion equation. For example: The density of 100% sulfuric acid at 20°C is 1.8305 kg/liter. What is its density at 50°C? The equation is found to be:

$$V_t = V_0\,[1 + A(t) + B(t)^2]$$

where:

V_t = **volume at t°C**
V_0 = **volume at 0°C**
A = **0.5758^{-3}**
B = **−0.432^{-6}**

Solving:

$$\rho 50°\text{C} = \rho 20°\text{C}\,\frac{(1 + A(20) + B(20)^2}{1 + A(50) + B(50)^2} = 1.8013 \text{ kg/l}$$

Density of Gases

The density of gases can vary greatly with changes in both temperature and pressure. The Ideal Gas Law incorporates Charles' and Boyle's Laws and also the concept of molecular weight. Charles' Law states that the density of a gas at constant temperature is directly proportional to its absolute pressure. Boyle's Law states that the density of a gas at constant pressure is inversely proportional to its absolute temperature.

Also it is known that the density of different gases at the same temperature and pressure varies in direct proportion to their molecular weights. The Ideal Gas Law is:

$$PV = nRT \quad (3\text{-}2)$$

where:

P = absolute pressure
V = volume
n = the mass/molecular weight (i.e., number of moles)
R = the Universal Gas Constant
T = absolute temperature

Since density is mass/volume:

$$\text{Gas Density} = \frac{(P)(\text{Mol. Wt.})}{(R)(T)} \tag{3-3}$$

The Universal Gas Constant, R, is an experimentally determined number. Values for R depend upon the units chosen for pressure, temperature, and density. Table 3-2 represents some of the values that might be required.

Gas Compressibility Factor

Unfortunately, the Ideal Gas Law is accurate only at relatively low pressures and high temperatures. To account for the deviation from the ideal situation, another factor is included. It is called the Gas Compressibility Factor, or Z-factor. This correction factor is dependent on pressure and temperature for each gas considered. The True Gas Law, or the Non-Ideal Gas Law, becomes:

$$PV = ZnRT \tag{3-4}$$

and density becomes:

$$\text{True Density} = \frac{(P)(\text{Mol. Wt.})}{(Z)(R)(T)} \tag{3-5}$$

The Ideal Gas Law is only a crude approximation. Be sure to include the *Z*-factor in your gas density calculations.

The Z-factor should always be considered when calculating a gas density. It is equal to unity at low pressures and high temperatures but can deviate widely from unity under other conditions. Some gases are not ideal even when at atmospheric pressure and ambient temperature.

There are some published tabulations of Z-factor for specific gases, but the most common method to find Z is to use the normalized compressibility charts. These charts offer a remarkable correlation of Z-factor with the pressure and temperature of gases normalized with respect to their critical values. Critical pressure and critical temperature are the conditions above which liquid cannot exist. Critical values for a number of gases are listed in Table 3-0. Graphs in Figure 3-4, Figure 3-5, Figure 3-6, and Figure 3-7 give Z-factors from values of reduced pressure and temperature. Accuracy is better than 1% for most gases. In some cases the critical values in Table 3-0 have been slightly modified so that the Z-charts are more accurate.

The normalized or "reduced" pressure and temperature values are calculated as:

$$T_r = \frac{T}{T_c} \tag{3-6}$$

$$P_r = \frac{P}{P_c} \tag{3-7}$$

Table 3-2. Values of the Universal Gas Constant, R

Units				Value of R
Mass	Press.	Vol.	Temp.	
lb	psia	ft^3	Deg. R	10.73
lb	psfa	ft^3	Deg. R	1544
kg	kPa abs	m^3	K	8.314
kg	kPa abs	liter	K	8314
kg	kg/cm^3	liter	K	84.78
kg	bars	liter	K	83.14

Table 3-3. Critical Values for Some Gases
(Courtesy of Fischer & Porter)

Gas	Mol. Wt.	Formula	T_c (°F)	P_c (psia)	T_c (°C)	P_c (kPa)
Acetic Acid	60	CH_3COOH	1071	840	595	5792
Acetylene	26	C_2H_2	556	911	309	6280
Air	(29)	———	239	547	133	3770
Ammonia	17	NH_3	730	1640	405	11310
Argon	40	A	272	705	151	4860
Benzene	78	C_6H_6	1011	702	562	4840
Butane	58	C_4H_{10}	765	551	425	3800
Carbon Dioxide	44	CO_2	548	1072	304	7390
Carbon Monoxide	28	CO	239	507	133	3500
Carbon Tetrachloride	154	CCl_4	1001	661	556	4560
Chlorine	71	Cl_2	751	1118	417	7709
Cyclohexane	84	C_6H_{12}	997	594	554	4100
Decane	142	$C_{10}H_{22}$	1115	312	619	2150
Ethane	30	C_2H_6	550	708	305	4880
Ethanol	46	C_2H_5OH	929	927	516	6390
Ethyl Chloride	64.5	C_2H_5Cl	829	764	460	5270
Ethyl Ether	74	$C_4H_{10}O$	839	522	466	3600
Ethylene	28	C_2H_4	509	748	283	5160
Helium*	4	He	(24)	(151)	(13.3)	(1050)
Heptane	100	C_7H_{16}	972	377	540	2600
Hexane	86	C_6H_{14}	914	436	508	3010
Hydrogen*	2	H_2	(74)	(306)	(41)	(2110)
Hydrogen Chloride	36.5	HCl	584	1200	324	8270
Hydrogen Cyanide	27	HCN	822	735	457	5070
Methane	16	CH_4	343	673	191	4640
Methanol	32	CH_3OH	924	1450	513	10000
Methyl Chloride	50.5	CH_3Cl	749	967	416	6670
Neon*	20	Ne	(95)	(498)	(52)	(3430)
Nitric Oxide	30	NO	323	955	179	6590
Nitrogen	28	N_2	227	492	126	3390
Nonane	128	C_9H_{20}	1072	336	596	2320
Octane	114	C_8H_{18}	1025	362	569	2500
Oxygen	32	O_2	278	730	154	5030
Pentane	72	C_5H_{12}	847	486	470	3350
Propane	44	C_3H_8	666	617	370	4250
Propanol	76	C_3H_7OH	914	779	508	5370
Propylene	42	C_3H_6	658	662	365	4562
Sulfur Dioxide	64	SO_2	775	1142	430	7870
Sulfur Trioxide	80	SO_3	885	1228	491	8470
Toluene	92	C_6H_7CH	1069	612	594	4220
Water	18	H_2O	1165	3206	647	22100

*Pseudo-critical values shown.

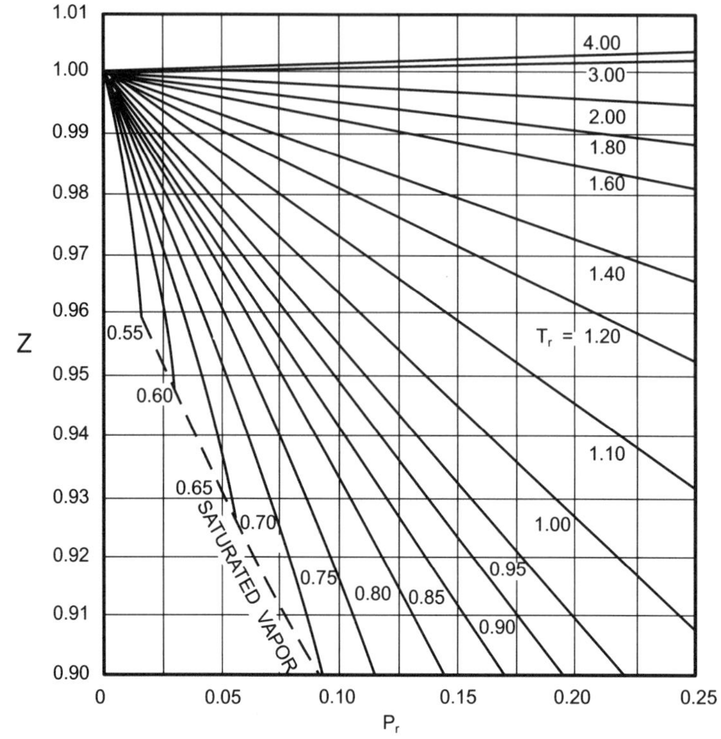

Figure 3-4. Compressibility Chart (Low Range)
(Courtesy of Fischer & Porter)

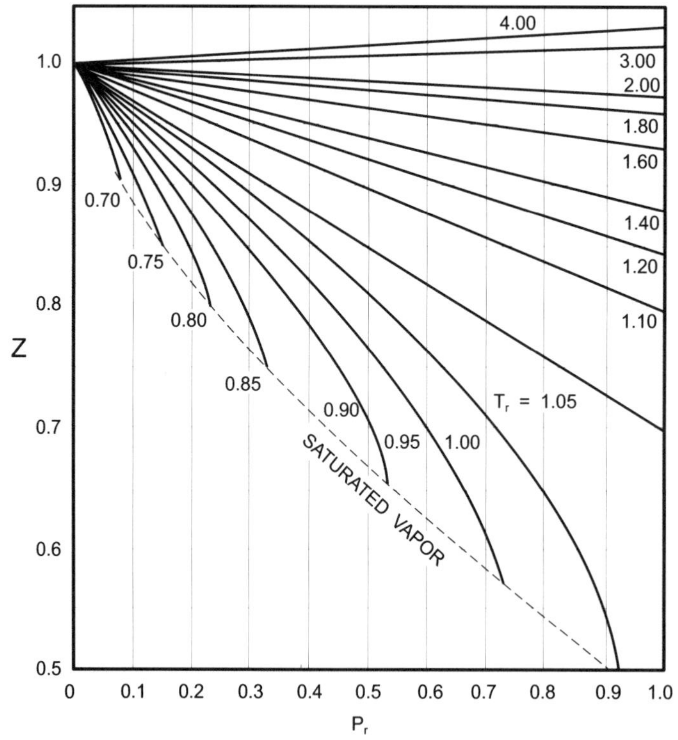

Figure 3-5. Compressibility Chart (Mid-Range)
(Courtesy of Fischer & Porter)

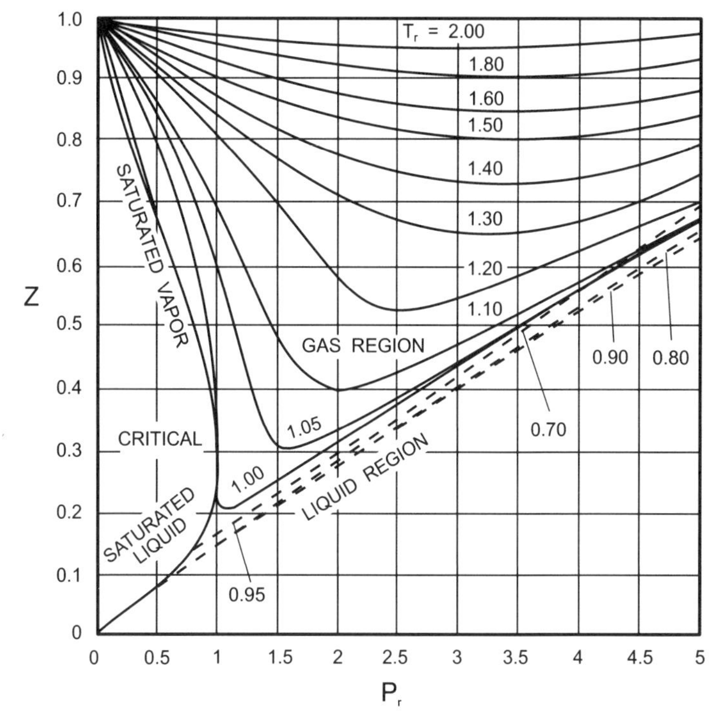

Figure 3-6. Compressibility Chart (High Range)
(Courtesy of Fischer & Porter)

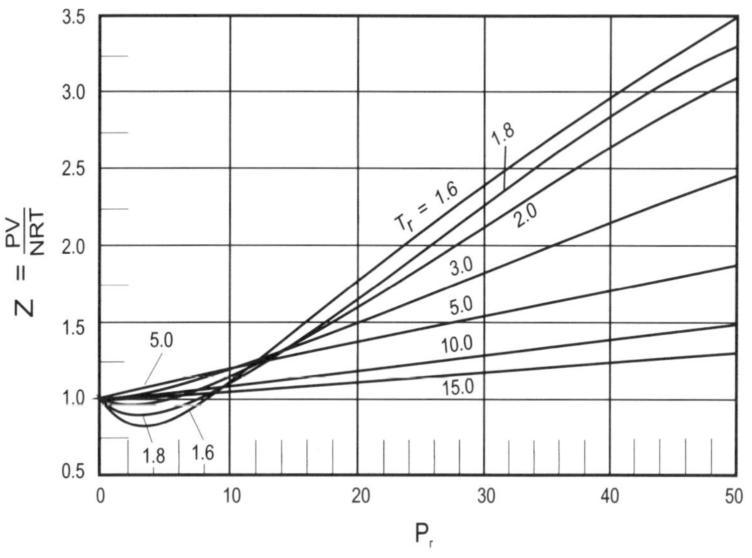

Figure 3-7. Compressibility Chart (Very High Range)
(Courtesy of Fischer & Porter)

Data for the density of air are rather scarce. Fortunately, air acts very nearly as an ideal gas if operating below 1400 kPa abs and above 0°C (204 psia and 32°F). Use the Ideal Gas Law with an approximate molecular weight of 29 for air.

where:

T_r = reduced temperature
P_r = reduced pressure
T = absolute temperature
P = absolute pressure
T_c = critical temperature
P_c = critical pressure

T and T_C and also P and P_C must be in the same units.

When accurate tables or graphs of density at different operating conditions exist, it is not necessary to use the Z-charts and the Ideal Gas Law. For example, very accurate values for specific volumes of steam are presented in tables of the thermodynamic properties of steam.

In the case of natural gas and other hydrocarbon mixtures high in methane, the American Gas Association has published Z-factors based on five criteria: pressure, temperature, specific gravity, % carbon dioxide, and % nitrogen. This report is called AGA Report No. 3. More recently, AGA has published a very comprehensive method (AGA Report No.8) to solve for the Z-factor of hydrocarbon-based gases with as many as 20 different components in varying proportions.

Specific Gravity

Specific gravity has two very different meanings, one for liquids and another for gases. When applied to liquids, it is the ratio of the density of the fluid at one specified temperature to the density of water at another specified temperature. As such, specific gravity has no dimensional units and should always be accompanied by two temperature values. Often, the water reference temperature is omitted, and the reference temperature is assumed to be either 4°C or 60°F. Luckily, the density of water is practically constant between those two values, so the omission usually presents no problem. On the other hand, the fluid reference temperature should always be included when using the term "specific gravity" for liquids.

Specific gravity is sometimes used to represent the density of liquids. This is unfortunate because the two temperatures are often omitted. The implied meaning probably is that the specific gravity is at flowmetering temperature and is referred to water at 4°C, but one can never be certain. Specific gravity is a poor substitute for density.

The specific gravity of a gas is a constant value regardless of its pressure or temperature.

The specific gravity of a gas is the ratio of the density of the gas to the density of air, both being measured at "standard" pressure and temperature. When stating a gas specific gravity, the standard temperature is usually omitted. Standard conditions are usually near ambient pressure and temperature, and the ratio of densities is essentially unaffected, regardless of the ambient conditions chosen. Air at ambient conditions acts as an ideal gas, as do many other gases. For these gases, the specific gravity becomes the ratio of their molecular weight to that of air. If the gas is not ideal at ambient conditions, this is no longer valid, but only slight differences will be found.

One interesting simplification arises because the density of water at 4°C is exactly 1.000 kg/l. This means that when specific gravity is referred to water at 4°C it is numerically equal to the liquid density in units of kg/l at its specified temperature.

There are many other "relative" units similar to specific gravity. They are used more to describe the concentration of various solutions than to describe their densities; examples are: degrees Baumé, degrees Brix, degrees API proof. Figure 3-8 shows how they relate to specific gravity.

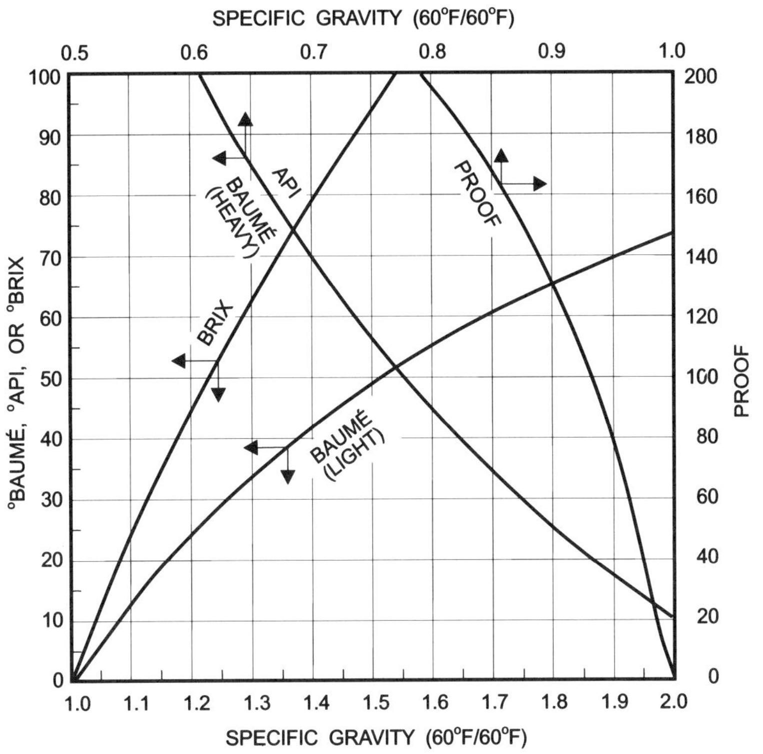

$$°Bé = 145 - \frac{145}{G_{60°/60°}}$$

$$°Bé = \frac{140}{G_{60°/60°}} - 130$$

$$°API = \frac{141.5}{G_{60°/60°}} - 131.5$$

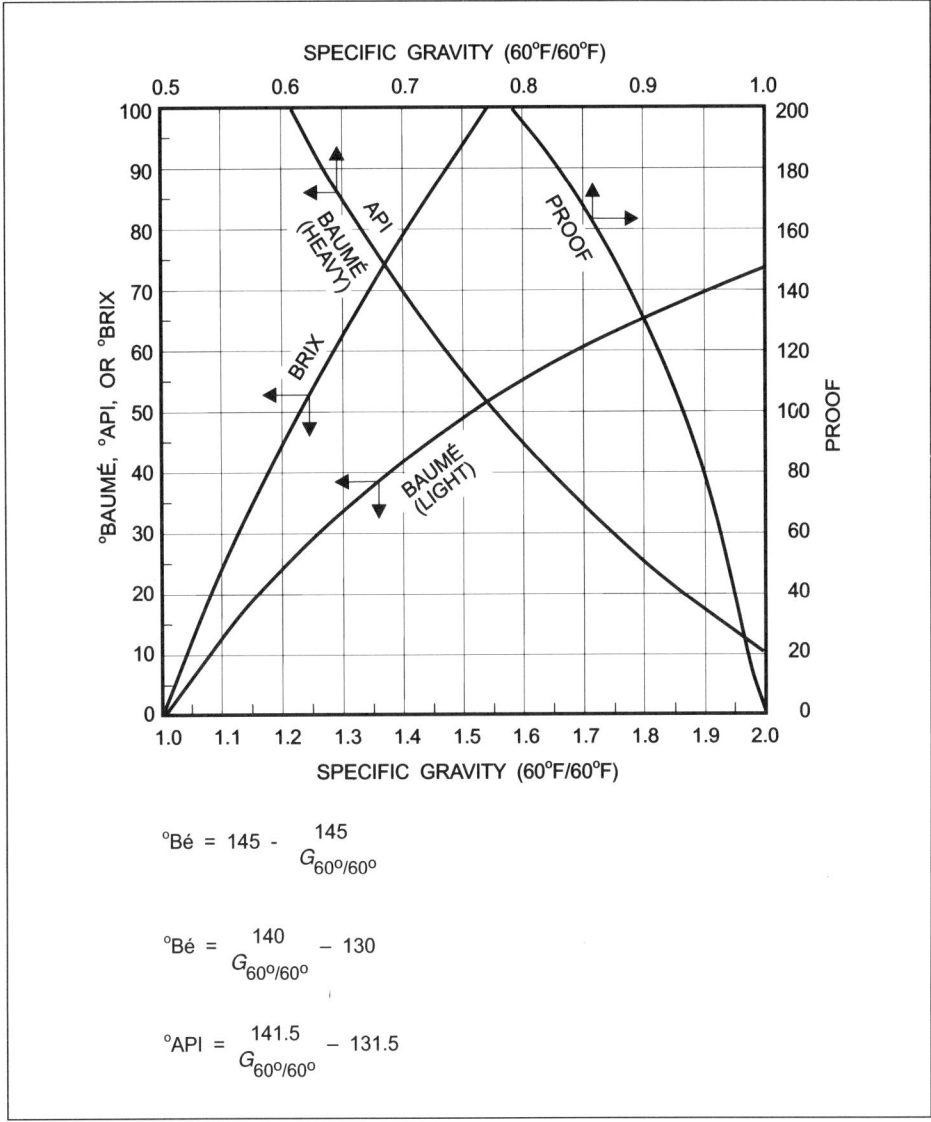

Figure 3-8. Relation of °Baumé, °API, and Proof to Specific Gravity of Liquids
(Courtesy of Fischer & Porter)

Fluid Viscosity

Viscosity is the internal resistance of a fluid to flow. Water has low viscosity, while syrup has high viscosity. When syrup is cold it is very much more viscous than when it is hot. Viscosity is highly temperature dependent.

Absolute Viscosity and Kinematic Viscosity

There are two ways to measure the viscosity of a fluid. One is to rotate a cup or disk in the fluid and measure the rotational speed along with the torque required to keep it rotating. The density of the fluid doesn't play any part in the measurement. Viscosity is the torque divided by the speed and divided by the contacting surface area. When measured this way, the viscosity is called "absolute viscosity." Its units are centipoise, pascal-seconds, or lb/ft-sec.

Another way to measure viscosity is to allow a certain volume of fluid to drain by gravity out of a container through a capillary tube or some other restriction.

The time to drain is directly related to viscosity and is often reported in "seconds." Since the flow is by gravity, the force causing the flow depends on the density of the fluid. Viscosity measured this way is called "kinematic viscosity." Its units are centistokes, m^2/sec, or ft^2/sec. As suspected, centipoise (cP) equals centistokes (cSt) multiplied by density. Some equivalents are:

$$\text{cP} = (\text{cSt})(\text{density in kg/l})$$

$$\text{lb/ft-sec} = 0.000672(\text{cP})$$

$$\text{Pa-sec} = (\text{m}^2/\text{sec})(\text{density in kg/l})$$

Figure 3-9 shows the relationship between many of the kinematic viscosity scales.

Gas viscosities are treated exactly as are liquid viscosities, but in general the absolute viscosity of a gas is much lower than that of a liquid.

Temperature has a very great effect on the viscosity of a fluid. For liquids, increasing temperature will lower viscosity, but, for gases, increasing temperature increases the viscosity. Figure 3-10 presents graphs that allow estimation of viscosity of a liquid at one temperature if viscosity is known for that liquid at some other temperature. Table 3-4 and Table 3-5 give equations and constants for the calculation of viscosity values for a number of specific liquids and gases.

Many flowmetering phenomena actually depend on kinematic viscosity, but it is common to show absolute viscosity in the formulas and include density in the formula.

The Influence of Viscosity on Flowmeter Performance

"Re" is the general symbol for Reynolds number and applies to any flow configuration. When flow is in a pipe with circular cross section, the symbol is R_D. The subscript refers to the inside pipe diameter.

Flowmeters that operate on hydraulic principles will usually operate very well regardless of viscosity, so long as that viscosity is below some limit. The limit itself is found to change with the nature of the flowing stream. It has been established that the influence of viscosity can be related to a criterion called Reynolds number (Re).

This is a dimensionless number defined for flow in a pipe as:

$$Re_D = \frac{(D)(\bar{v})}{(\text{kinetic viscosity})} = \frac{4(\text{mass flow})}{(\pi)(D)(\text{abs. viscosity})} \tag{3-8}$$

where:

D = pipe diameter
$\bar{v}$ = average flowing velocity
π = 3.1416

Several more useful forms of the equation for Reynolds number are:

$$Re_D = \frac{6.316(\text{lb/hr})}{(D)(\text{cP})} = \frac{6.316(\text{ft}^3/\text{hr})(\text{lb/ft}^3)}{(D)(\text{cP})} = \frac{3160(\text{gpm})(\text{SG})}{(D)(\text{cP})} \tag{3-9}$$

where D is in inches.

$$Re_D = \frac{\text{kg/h}}{2827(\text{D})(\text{Pa-s})} = \frac{(\text{m}^3)(\text{kg/m}^3)}{2827(D)(\text{Pa-s})} = \frac{1.272(\text{l/s})(\text{kg/l})}{(D)(\text{Pa-s})} \tag{3-10}$$

where D is in meters.

Reynolds number can be thought of as the ratio between the inertial forces and the viscous forces in a flowing stream. In a pipe, the flow is "laminar" when Re_D is less than 2000, and when Re_D is greater than 4000 the flow becomes "turbulent." The kinetic energy is different in identical flowing streams operating in laminal rather than turbulent flow. A similar effect would be expected in the performance of flowmeters, and it turns out to be true. To predict this behavior, the Reynolds number can be evaluated and correlated to the variation in flowmeter performance.

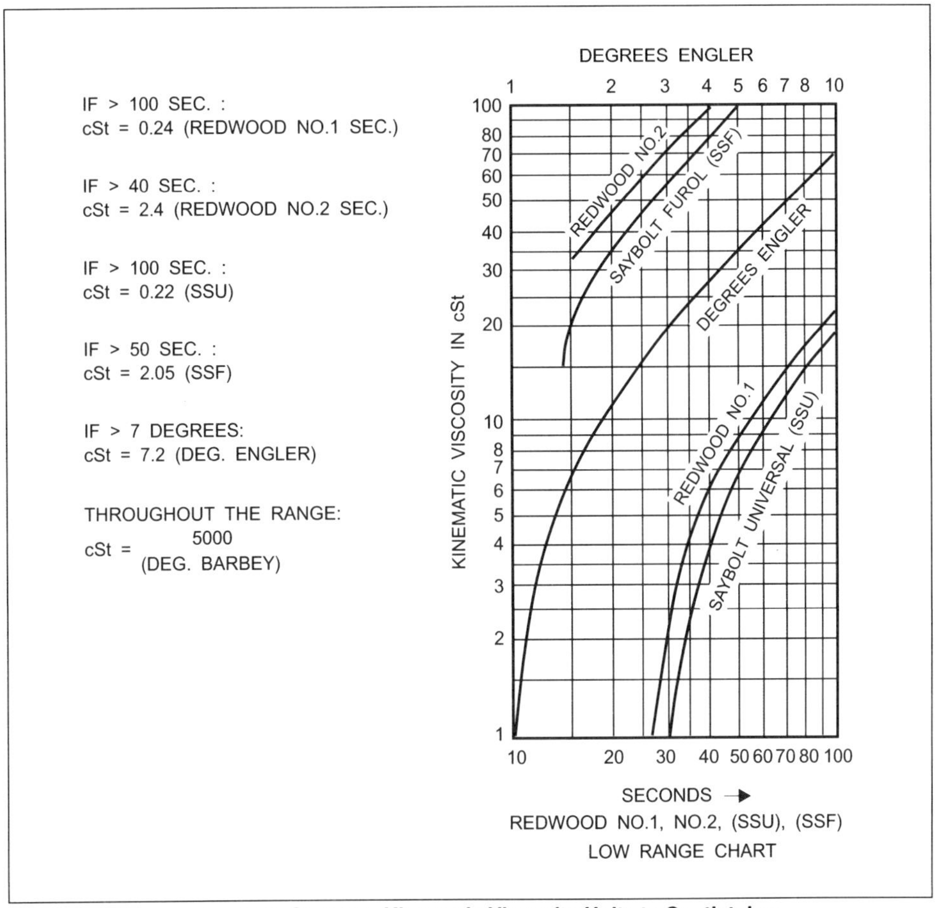

Figure 3-9. Conversion from Common Kinematic Viscosity Units to Centistokes
(Courtesy of Fischer & Porter)

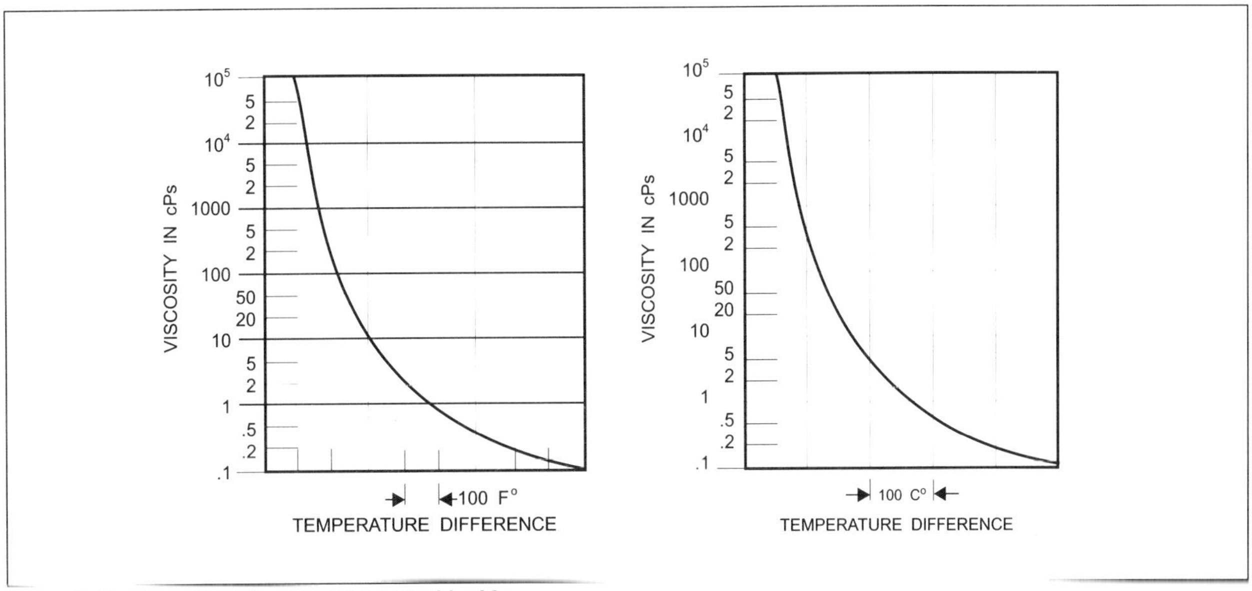

Figure 3-10. Viscosity Estimating Charts for Liquids
(From Perry, *Chemical Engineering Handbook.*)

Table 3-4. Viscosities of Liquids

$$cP = \frac{K}{\left(\frac{^\circ C + 273}{100}\right)^c}$$

Name of Liquid	K	c	Name of Liquid	K	c
Acetaldehyde	3.31	2.47	p-Chlorotoluene	42.1	3.56
Acetic Acid, 100%	140.2	4.34	m-Cresol	1.113×10^9	16.40
Acetic Acid, 70%	2610	6.39	Cyclohexane	764	6.12
Acetic Anhydride	67.8	3.94	Cyclohexanol	1.337×10^9	15.5
Acetone, 100%	7.45	2.85	Dibromoethane	122	3.94
Acetone, 35%	9660	7.96	Dichloroethane	42.5	3.62
Acetonitrile	8.17	2.91	Dichloromethane	9.546	2.8
Acrylic Acid	112.9	4.21	Diethyl Ketone	19.8	3.43
Allyl Alcohol	717	5.78	Diethyl Oxalate	548	5.14
Allyl Bromide	12.2	2.91	Diethylene Glycol	1.236×10^7	11.69
Allyl Iodide	22.9	3.14	Dimethyl Oxalate	165	4.21
Ammonia, 100%	8.67	4.00	Diphenyl	348	4.41
Ammonia, 26%	717	5.87	Dipropyl Ether	21.3	3.62
Anyl Acetate	125	4.55	Dipropyl Oxalate	1380	5.70
Amyl Alcohol	38200	8.39	Ethyl Acetate	17.1	3.31
Aniline	19100	7.74	Ethyl Acrylate	42.3	3.94
Anisole	104.2	4.21	Ethyl Alcohol, 100%	485	5.54
Arsenic Trichloride	41.0	3.20	Ethyl Alcohol, 95%	1050	6.12
Benzene	53.8	4.07	Ethyl Alcohol, 40%	94400	9.59
Brine, 25% $CaCl_2$	67700	9.46	Ethyl Benzene	37.2	3.62
Brine, 25% NaCl	1190	5.78	Ethyl Bromide	8.76	2.85
Bromine	26.6	3.02	2, Ethyl Butyl Acrylate	277	4.99
Bromotoluene	2.11	0.359	Ethyl Chloride	5.06	2.96
Butyl Acetate	63.2	4.21	Ethyl Ether	5.28	2.85
i-Butyl Acetate	38.4	3.68	Ethyl Formate	11.0	3.02
n-Butyl Acrylate	162	4.77	Ethyl Iodide	11.6	2.74
Butyl Alcohol	7330	7.23	Ethyl Propionate	27.4	3.62
i-Butyl Bromide	35.5	3.68	Ethyl Propyl Ether	9.58	3.14
i-Butyl Chloride	25.1	3.68	Ethyl Sulfide	15.5	3.25
i-Butyl Iodide	41.2	3.56	Ethylene Bromide	222	4.48
Butyric Acid	175	4.34	Ethylene Chloride	60.2	3.94
Carbon Dioxide	11.7	4.70	Ethylene Glycol	1.38×10^6	10.24
Carbon Disulfide	3.33	2.02	2, Ethyl Hexyl Acrylate	2770	6.85
Carbon Tetrachloride	71.8	3.94	Ethylidene Chloride	19.6	3.37
Chlorobenzene	83.6	4.21	Fluorobenzene	21.8	3.31
Chloroform	13.6	2.91	Formic Acid	626	5.38
Chlorosulfonic Acid	658	4.99	Freon-11	10.92	2.91
o-Chlorotoluene	0.602	3.75	Freon-12	1.73	1.68
m-Chlorotoluene	42.1	3.56	Freon-22	1.24	1.50

Table 3-4. Viscosities of Liquids (continued)

$$cP = \frac{K}{\left(\frac{^{\circ}C + 273}{100}\right)^{c}}$$

Name of Liquid	K	c	Name of Liquid	K	c
Freon-113	59.4	4.07	Phenol	164000	9.09
Glycerol, 100%	1.588×10^{11}	17.60	Phosphorus Tribromide	66.5	3.25
Glycerol, 50%	1.193×10^{5}	9.09	Phosphorus Trichloride	5.62	1.97
Heptane	11.7	3.08	Prpionic Acid	76.9	3.87
Hexane	6.41	2.74	Propyl Acetate	31.6	3.68
Hldrochloric Acid, 31.5%	114	3.75	Propyl Alcohol	3560	6.75
Iodobenzene	116	3.87	Propyl Bromide	11.5	2.85
Isobutyl Alcohol	58900	8.85	i-Propyl Bromide	13.6	3.08
Isobutyric Acid	135	4.27	Propyl Chloride	8.32	2.91
Isopropyl Alcohol	8680	7.64	i-Propyl Chloride	10.3	3.20
Kerosene	1280	5.78	Propyl Formate	28.1	3.68
Linseed Oil	366000	8.39	Propylene Glycol	6.345×10^{8}	15.10
Mercury	4.63	0.982	Propyl Iodide	21.1	3.08
Methanol, 100%	53.3	4.14	i-Propyl Iodide	25.3	3.31
Methanol, 90%	74.2	4.21	Sodium	8.42	1.87
Methanol, 40%	12400	8.06	Sodium Hydroxide, 50%	1.013×10^{9}	14.90
Methyl Acetate	10.6	3.02	Stannic Chloride	39.1	3.43
Methyl Acrylate	28.9	3.75	Succinonitrile	3340	5.87
Methal i-Butyrate	48.8	4.21	Sulfur Dioxide	4.97	2.47
Methyl n-Butyrate	29.5	3.62	Sulfuric Acid, 110%	600000	8.74
Methyl Chloride	3.07	2.58	Sulfuric Acid, 100%	107000	7.85
Methyl Ethyl Ketone	13.7	3.20	Sulfuric Acid, 98%	411000	8.97
Methylene Chloride	12.4	3.08	Sulfuric Acid, 60%	3380	5.78
Methyl Formate	9.35	3.02	Sulfuryl Chloride	12.6	2.47
Methyl Iodine	12.3	2.96	Tetrachloroethane	222	4.48
Methyl Propionate	19.0	3.43	Tetrachloroethylene	24.3	3.02
Methyl Propyl Ketone	12.7	2.96	Thiophene	33.8	3.62
Methyl Sulfide	4.16	2.42	Titanium Tetrachloride	19.9	2.91
Naphthalene	20500	8.00	Toluene	21.8	3.31
Nitric Acid, 95%	76.9	3.87	Trichloroethylene	11.3	2.69
Nitric Acid, 60%	743	5.30	Triethylene Glycol	2.32×10^{7}	12.17
Nitrobenzene	747	5.46	Turpentine	261	4.77
Nirogen Dioxide	26.0	3.81	Vinyl Acetate	13.4	3.14
Nirtotoluene	632	5.14	Vinyl Toluene	35.7	3.50
Octane	20.2	3.31	Water	538	5.78
Octyl Alcohol	278,000	9.46	o-Xylene	34.2	3.43
Pentachloroethane	725	5.22	m-Xylene	19.8	3.20
Pentane	4.17	2.63	p-Xylene	21.0	3.20

Table 3-5. Viscosities of Gases

$$cP = K\left(\frac{^{\circ}C + 273}{10{,}000}\right)^{c}$$

Name of Gas	K	c	Name of Gas	K	c
Acetic Acid	0.399	1.114	Freon-113	0.100	0.642
Acetone	0.208	0.935	Helium	0.211	0.685
Acetulene	0.181	0.816	Hexane	0.209	0.997
Air	0.194	0.674	Hydrogen	0.089	0.652
Ammonia	0.353	1.006	$3H_2 + 1N_2$	0.139	0.652
Argon	0.293	0.731	Hydrogen Bromide	0.502	0.949
Benzene	0.250	0.992	Hydrogen Chloride	0.397	0.949
Bromine	0.401	0.935	Hydrogen Cyanide	0.181	0.816
Butene	0.200	0.894	Hydrogen Iodide	0.480	0.921
Butylene	0.208	0.935	Hydrogen Sulfide	0.402	0.977
Carbon Dioxide	0.298	0.854	Iodine	0.354	0.921
Carbon Disulfide	0.421	1.067	Mercury	4.525	1.563
Carbon Monoxide	0.194	0.674	Methane	0.185	0.803
Chlorine	0.354	0.921	Methyl Alcohol	0.324	0.992
Chloroform	0.277	0.935	Nitric Oxide	0.211	0.685
Cyanogen	0.233	0.894	Nitrogen	0.224	0.720
Cyclohexane	0.167	0.894	Nitrosyl Chloride	0.504	1.067
Ethane	0.225	0.907	Nitrous Oxide	0.410	0.949
Ethyl Acetate	0.250	0.992	Oxygen	0.219	0.674
Ethyl Chloride	0.324	0.992	Pentane	0.467	1.230
Ethyl Alcohol	0.210	0.894	Propane	0.153	0.829
Ethyl Ether	0.208	0.935	Propyl Alcohol	0.266	1.006
Ethylene	0.206	0.854	Propylene	0.218	0.921
Flourine	1.453	1.179	Sulfur Dioxide	0.241	0.841
Freon-11	0.139	0.720	Toluene	0.220	0.977
Freon-12	0.129	0.663	2, 3, 3-Trimethyl Butane	0.129	0.854
Freon-21	0.133	0.697	Water	0.422	1.067
Freon-22	0.222	0.779	Xenon	0.513	0.881

If the Reynolds number is very high for all operating conditions of a flowmeter, the measurement will probably show no variation from the theoretical performance. As Reynolds number becomes lower, predictable deviations will occur. Since viscosity is in the denominator of the equation for Reynolds number, combinations of high viscosity and low velocity lead to low Re_D and deviations from constant flowmeter performance. For flowmeters influenced by viscosity, it is common to prepare graphs of flowmeter performance vs. Re_D. A typical graph is shown in Figure 3-11.

For turbine meters and vortex meters that have a frequency output signal, a more convenient ratio is often used instead of Reynolds number. This number is the frequency divided by the kinematic viscosity, and it is applied only to individual meters of designated size. Its units are usually Hz/cSt. The two numbers are quite similar. Frequency relates closely to velocity, and both have viscosity in the denominator of their equation. The diameter appears only in Reynolds number. The Hz/cSt number is especially useful if large variations occur.

Newtonian and Non-Newtonian Fluids

Some fluids exhibit different apparent viscosities under different conditions. Actually, the definition of viscosity implies that there is a fixed ratio between shear stress and shear rate in the fluid. Most homogeneous fluids follow this relationship. These are called "Newtonian" fluids. At a fixed temperature they have a certain viscosity that has the same value when measured at any shear rate.

"Non-Newtonian" fluids show different viscosities at different shear rates. Many non-Newtonian fluids are actually slurries of suspended particles in a liquid. Paint, clay slips, and wet sand are examples of fluids that exhibit this strange behavior. Figure 3-12 is a graph of shear stress vs. shear rate. In the case of a rotational viscosimeter, it would be rpm vs. torque. Newtonian fluids show a straight line through the origin with a slope equal to the viscosity. Paint should act as a low viscosity material when being sheared by the action of a paint brush, but then it should become more firm and stick to the wall without running after the brush moves on. The particles and pigment in the paint are chosen to produce a product that has a higher slope (apparent viscosity) at low shear rate than at high shear rate. Clay slips usually should have similar characteristics if being applied as a coating. Kaolinite clays do this, but montmorillonite clays and wet sand behave in just the opposite way. As shown in Table 3-6, these non-Newtonian fluids are called plastic, pseudoplastic, or dilatant fluids. Another term, thixotropic, is used to describe fluids whose apparent viscosity changes with time as well as shear rate.

The effect of non-Newtonian fluid properties on the performance of flowmeters is a very complex subject, with little experimental data. It is recommended that such metering installations somehow be calibrated in place. Often the results will show that the metering system is indeed practical and only slightly affected by these properties.

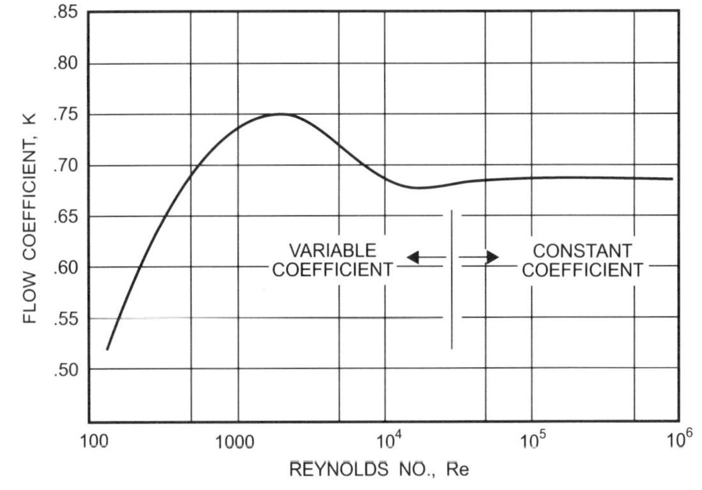

Figure 3-11. Effect of Reynolds Number on Differential Pressure Meter Performance

Vapor Pressure and Boiling Point of Liquids

Vapor Pressure

The vapor pressure of a liquid, at a designated temperature, is the absolute pressure at which the liquid would boil if its pressure were lowered to that point. The vapor pressure of water at 212°F is 14.7 psia. That is why water boils away into steam at 212°F when the atmospheric pressure is 14.7 psia. High on a mountain top water might boil at 200°F. The vapor pressure of water at 200°F is 11.5 psia, and, therefore, the barometric pressure up on that mountain must have been 11.5 psia.

Vapor pressure can be very important in a flowmetering application if the liquid enters the meter at a pressure near its vapor pressure. Any pressure reduction in the meter could cause the liquid to flash partly to a gas. This can completely destroy the performance of many types of flowmeters. If the gas pressure rises above the vapor pressure, the gas bubbles will implode, causing cavitation to occur. Cavitation can destroy many types of flowmeters.

Always check to be sure that the requirements are met for minimum inlet line pressure based on the vapor pressure of the liquid at the highest temperature expected in the application. These limits are published by the flowmeter manufacturers.

A common problem arises when trying to measure steam condensate if it is not subcooled or, if not, pumped to increase its pressure. Vapor pressure does not vary in proportion to temperature. Instead, it typically changes only slightly at lower temperatures and then increases much more drastically at relatively higher temperatures.

Boiling Point

The boiling point of a liquid is a property directly related to vapor pressure. It is simply the temperature at which a liquid would boil at a designated pressure.

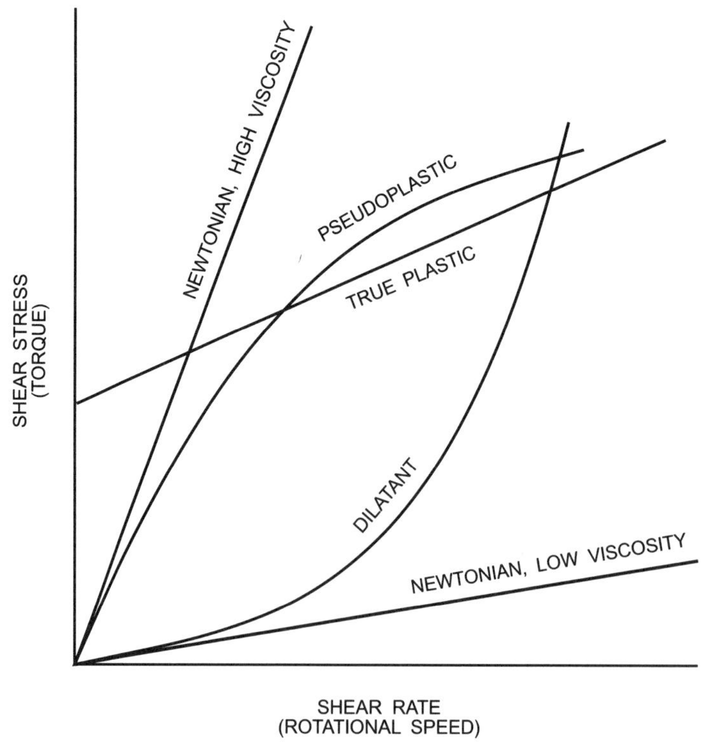

Figure 3-12. Behavior of Newtonian and Non-Newtonian Fluids

Table 3-6. Specific Heats of Gases at Atmospheric Pressure

$$C_p\text{(Heat Capacity)} = A(e)^{°C/Y}$$

where: e = 2.7183

Name of Gas	Range, °C	A	Y	Name of Gas	Range, °C	A	Y
Acetylene	0 - 200	0.410	929	Hydrogen	0 - 600	3.45	78200
Acetylene	200 - 400	0.440	1500	Hydrogen	600 - 1400	3.25	7336
Acetylene	400 - 1400	0.520	4200	Hydrogen Bromide	0 - 1400	0.085	7336
Air	0 - 1400	0.248	8020	Hydrogen Chloride	0 - 1400	0.188	8020
Ammonia	0 - 600	0.505	1470	Hydrogen Fluoride	0 - 1400	0.335	8410
Ammonia	600 - 1400	0.620	3210	Hydrogen Iodide	0 - 1400	0.053	6020
Carbon Dioxide	0 - 400	0.205	1334	Hydrogen Sulfide	0 - 700	0.240	2350
Carbon Dioxide	400 - 1400	0.255	5250	Hydrogen Sulfide	700 - 1400	0.275	4780
Carbon Monoxide	0 - 1400	0.248	6620	Methane	0 - 300	0.520	770
Chlorine	0 - 200	0.113	2740	Methane	300 - 700	0.590	1110
Chlorine	200 - 1400	0.120	27600	Methane	700 - 1400	0.860	2950
Ethane	0 - 200	0.390	469	Nitric Oxide	0 - 700	0.230	4500
Ethane	200 - 600	0.480	912	Nitric Oxide	700 - 1400	0.253	12200
Ethane	600 - 1400	0.750	2850	Nitrogen	0 - 1400	0.248	6500
Ethylene	0 - 200	0.350	470	Oxygen	0 - 500	0.230	4500
Ethylene	200 - 600	0.440	1060	Oxygen	500 - 1400	0.244	12160
Ethylene	600 - 1400	0.650	3360	Sulfur	300 - 1400	0.132	24100
Freon-11	0 - 150	0.128	709	Sulfur Dioxide	0 - 400	0.152	1650
Freon-21	0 - 150	0.141	790	Sulfur Dioxide	400 - 1400	0.181	8200
Freon-22	0 - 150	0.148	705	Water	0 - 1400	0.440	3750

Tabulations of boiling point for various liquids at atmospheric pressure are commonly published. They are of little help in flowmeter design because they apply only at one pressure.

Electrical Conductivity

This property is of importance when considering magnetic flowmeters. In most designs, the meter is able to detect flow only if the liquid is capable of conducting electricity to a small extent. Organic liquids such as oils and hydrocarbons are almost perfect insulators with no significant conductivity. In most cases they cannot be measured by magnetic meters. The minimum conductivity is always specified by the manufacturer, and it is very important to be sure the liquid is sufficiently conductive. A spot check for conductivity sometimes indicates that it is high enough, but it is quite possible that, at times, the material changes and the meter can fail to operate.

When conductivity is borderline, be sure to make a number of tests at different times — not just once or twice.

The dimensional units for conductivity are usually given as micromhos/cm, micromhos/cm^3, or siemens (S). Surprisingly, all these mean the same thing; they refer to the conductance between two plates, one cm^2 in area and one centimeter apart, with the liquid in between them.

When choosing a magnetic meter to avoid flashing in a liquid with a low vapor pressure, don't forget to check its conductivity.

Sometimes data is presented as "resistivity" rather than "conductivity." The two terms are merely reciprocals of each other, and the units of resistivity are usually ohms/cm^3.

Sonic Conductivity

Ultrasonic flowmeters depend on the conduction of sound through the liquid being measured. When using the Doppler-type meters, the sound wave must bounce off particles in the liquid, but it is not necessary that a sound wave traverse the entire diameter of the pipe. When using time-of-flight meters, the liquid must be relatively free of suspended material to allow passage of sound from one side of the pipe to the other. This unique property of liquids is not easily expressed quantitatively. Experience and specific testing of liquids can lead to the decision of meter suitability.

Specific Heat and Ratio of Specific Heats

The specific heat of fluids is important when computing heat flow from a mass flow measurement and differential temperature. The equation is:

$$Q = mC_p \Delta T \quad (3\text{-}11)$$

where:

Q = heat flow rate
m = mass flow rate
C_p = specific heat
ΔT = temperature difference (inlet and outlet of a heater, for example)

Specific heat can be considered as the amount of energy required to increase the temperature of one unit of mass of a material by one degree. Common units are calories/gram-°C, joules/gram-°C, or Btu/pound-°F.

Table 3-6 gives an equation for specific heat and constants for a number of gases. Table 3-7 does the same for specific heat of liquids.

For gases, it is sometimes convenient to use "molal heat capacity," which is specific heat normalized to one molecular weight rather than actual mass of the gas.

The ratio of specific heats for steam ranges from 1.3 to 3.0 at different pressures and temperatures. Steam is a very non-ideal gas and the C_p/C_v ratio should not be used. The actual isentropic exponent varies only slightly from 1.3.

Liquids are essentially incompressible and have only one form of specific heat, C_p. Gases have two forms of specific heat: C_p measured at constant pressure, and C_v measured at constant volume. The ratio of C_p/C_v is quite important when designing differential pressure flowmeters for gas flow.

Differential pressure meters use an equation based on a velocity change. Velocities are inversely proportional to the inlet cross-sectional area and the restriction throat area:

$$q = A_1 v_1 = A_2 v_2$$

where:

q = volumetric flow rate
A_1 and A_2 = cross-sectional of areas of inlet and throat
v_1 and v_2 = velocities at inlet and throat

The equation is true for liquids, but gases will expand due to lower pressure at the throat. Therefore, a correction factor, Y, is included in gas flow equations. This is called the Gas Expansion Factor, and it depends on line pressure, differential pressure, meter geometry, and the isentropic exponent for the particular gas at operating conditions. For ideal gases, the isentropic exponent is equal to the ratio of specific heats, C_p/C_v. Figure 3-13 is a graph of C_p/C_v values for common gases. If the gas does not act as an ideal gas, specific thermo-dynamic tables or graphs must be used to determine the isentropic exponent.

Table 3-7. Specific Heats of Liquids (Heat Capacity)

$$C_p = A + \frac{^{\circ}C}{K}$$

Name of Gas	Range, ˚C	A	K	Name of Gas	Range, ˚C	A	K
Acetic Acid 100%	0 — 80	0.465	1100	Ethyl Iodide	0 — 100	0.168	620
Acetone	20 — 50	0.511	1320	Ethylene Glycol	-40 — 200	0.551	866
Ammonia	-20 — 50	1.120	833	Freon 11	-20 — 70	0.210	5610
Amyl Alcohol	-50 — 25	0.527	665	Freon 12	-40 — 15	0.240	1510
Amyl Acetate	0 — 100	0.458	1490	Freon 21	-20 — 70	0.250	5350
Aniline	0 — 130	0.467	1110	Freon 22	-20 — 60	0.285	1340
Benzene	10 — 80	0.384	960	Freon 113	-20 — 70	0.215	4320
Benzyl Alcohol	-20 — 30	0.440	700	Glycerol	-40 — 20	0.543	805
Benzyl Chloride	-30 — 30	0.339	3130	Heptane	0 — 60	0.475	714
Brine, 25% $CaCl_2$	-40 — 20	0.670	1250	Hexane	-60 — 20	0.520	1560
Brine, 25% NaCl	-40 — 20	0.810	3800	Hydrochloric Acid, 30%	20 — 100	0.570	554
Butyl Alcohol	0 — 100	0.501	356	Isoamyl Alcohol	10 — 100	0.485	354
Carbon Disulfide	-100 — 25	0.235	52500	Isobutyl Alcohol	0 — 100	0.502	293
Carbon Tetrachloride	10 — 60	0.195	2830	Isopropyl Alcohol	-20 — 50	0.580	329
Chlorobenzene	0 — 100	0.303	1350	Isopropyl Ether	-60 — 20	0.494	1400
Chloroform	0 — 50	0.227	3380	Methyl Alcohol	-40 — 20	0.575	995
Decane	-80 — 25	0.412	1850	Methyl Chloride	-60 — 20	0.376	2480
Dichloroethane	-30 — 60	0.293	2300	Naphthalene	90 — 200	0.356	1340
Dichloromethane	-40 — 50	0.284	4010	Nitrobenzene	0 — 100	0.350	1580
Diphenyl	50 — 120	0.370	1490	Nonane	-50 — 25	0.512	1840
Diphenylmethane	30 — 100	0.365	833	Octane	-50 — 25	0.510	1510
Diphenyl Oxide	0 — 200	0.356	1000	Perchloroethylene	-30 — 140	0.195	2870
Dowtherm A	0 — 200	0.356	1000	Propyl Alcohol	-20 — 100	0.530	427
Ethyl Acetate	-50 — 25	0.451	3040	Pyridine	-50 — 25	0.393	1650
Ethyl Alcohol, 100%	30 — 60	0.488	276	Sulfuric Acid, 98%	10 — 45	0.338	2390
Ethyl Alcohol, 95%	20 — 80	0.545	262	Sulfur Dioxide	-20 — 100	0.320	1240
Ethyl Alcohol, 50%	20 — 80	0.858	476	Toluene	0 — 60	0.382	959
Ethyl Benzene	0 — 100	0.401	850	Water	10 — 200	1.000	14290
Ethyl Bromide	5 — 25	0.217	87700	Xylene, Ortho	0 — 100	0.390	1300
Ethyl Chloride	-30 — 40	0.370	1844	Xylene, Meta	0 — 100	0.380	1280
Ethyl Ether	-100 — 25	0.545	1690	Xylene, Para	0 — 100	0.370	1120

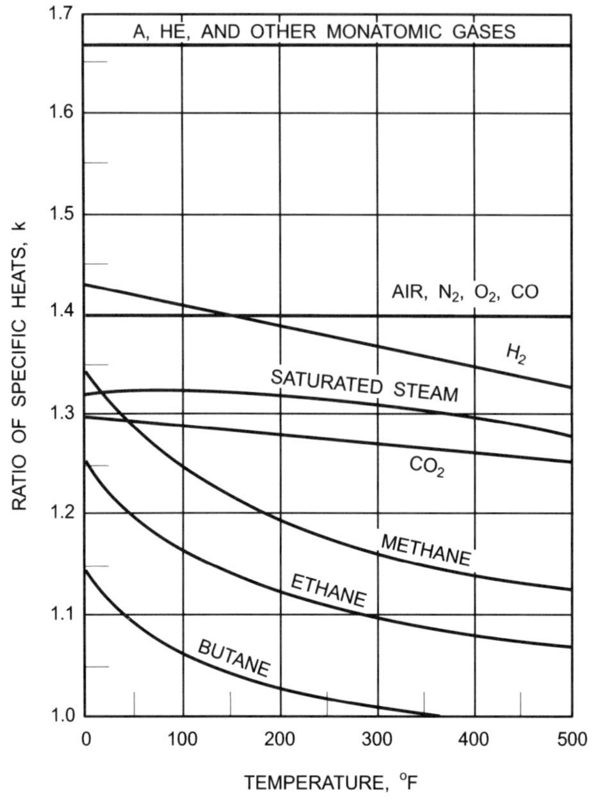

Figure 3-13. Ratio of Specific Heats, *k*, (C_p/C_v) versus Temperature

About the Author

Bill Buzzard served with the 42nd Infantry Division in Europe from 1944 to 1946. After earning his degree in Chemical Engineering from the Massachusetts Institute of Technology, he spent 5 years in pulp mill operation for Westvaco, Inc. He was with Fischer & Porter Co. for the next 37 years as Product Manager, Application Engineer, Project Manager, Chief Technologist, and Manager of the Application Department before his retirement in 1990.

A Charter Member of the Delaware Valley ISA Section, Mr. Buzzard has presented numerous technical papers and ISA training sessions. He is the author of the *Fischer & Porter Flowmetering Orifice Sizing Handbook.*

4

Fundamentals of Flow Measurement

Flowmetering is measuring matter in motion. An understanding of flow measurement is based on an understanding of the medium and the characteristics of its motion. Since with flow there is motion, the measurement of flow is a dynamic measurement. In fact, flow measurement signals are described as "noisy," which is a recognition of the dynamic, changing nature of flow. A quality measurement is required under these conditions and can be made with an understanding of the flowing medium, how that medium moves in piping, and the effect of the piping on that motion.

Matter

In simplest terms, flowmetering is the measurement of matter in motion. Matter occurs in three forms (phases): solid, liquid, and gas. Flow measurement is usually the measurement of moving matter in the liquid phase or the gas phase. Collectively, liquids and gases are known as fluids. However, it is not uncommon to have flow of matter in more than one phase:

(1) Solids in liquid

(2) Gases in liquids

(3) Solids in gas

(4) Liquids in gas

(5) Solids and gases in liquid

(6) Solids and liquids in gas

Mixed Phases of Matter

Each multi-phase medium presents its own set of problems to being successfully measured while in motion. The interest in the solid phase is only as it occurs with either the liquid phase, as where the liquid is used as a vehicle to convey solids (called slurries), or as it occurs with the gas phase, as in the pneumatic conveying of solids. This interest extends to the size of the solid particles, their distribution in the flow stream, and their mass relative to the fluid vehicle, that is, whether the solid particles tend to float or sink in the flow stream. This interest extends further to the nature of the solid particles, whether they are soft and amorphous, hard and sharp-edged, or fibrous and the length of the fibers, etc. Further, there is interest in solids beyond those fluids in which the liquid or gas is used to convey solids from place to place. There is interest in those flow streams where solids are a very small percent by volume of the total stream flow, in what usually are called "dirty streams."

Newtonian and Non-Newtonian Fluids

Solids exhibit internal molecular resistance to externally applied shear. It is most easily recognized in metals; in fact, it is a characteristic of all matter, including fluids. This has particular significance when matter in motion is examined. In fluids, when the resistance to shear is linear and goes through zero, the physical property is called viscosity, and the fluid is called Newtonian (see Figure 4-1). Viscosity, usually expressed in poise or equivalent units, is absolute viscosity (μ). Kinematic viscosity (ν) is absolute viscosity divided by density and is expressed in stokes or equivalent units.

$$\nu(\text{cS}) = \frac{\mu(\text{cP})}{\rho(\text{kg/m}^3)} \qquad (4\text{-}1)$$

With liquids, increasing temperature usually results in lower viscosity. The change in viscosity with temperature is the Viscosity Index. For gases, an increase in temperature increases the kinematic viscosity—just the opposite effect from liquids.

Fluids that do not exhibit a linear resistance to shear that goes through zero are called non-Newtonian (see Figure 4-1). When the relationship between shear rate and the rate of deformation is not linear, these non-Newtonian fluids are called power-law fluids, which are in two subgroups: dilatant and pseudoplastic. There are other non-Newtonian fluids that require a given shear rate before deformation begins, so the curves do not go through zero. These are classified as Bingham plastic fluids. Additionally, there are other fluids that have time-dependent characteristics. Thixotropic fluids have a decrease in apparent viscosity[1] with time. Rheopectic fluids, up to some limit, have an increase in apparent viscosity with time. Finally, some fluids, normally considered Newtonian, do not immediately return to a condition of zero shear rate when the stress is removed; these are viscoelastic fluids.

Temperature and Pressure Effects on Fluids

Knowing the state of matter and its characteristics is the first step in making a good measurement and assuring reliable, consistent operation of the flowmeter.

Temperature and pressure have their individual effects on the state of matter. At a given pressure, progressively adding energy to a solid first increases its temperature, then at a given constant temperature there is a conversion of state from solid to liquid phase (see Figure 4-2). The amount of heat necessary to make the conversion is the latent heat of liquefaction. Once all the solid is converted to liquid, the addition of heat will raise the temperature of the liquid until again a new temperature is reached where there is no further temperature increase. At this temperature, with the addition of heat there is a conversion of the liquid phase to the gaseous phase. The amount of heat necessary to make the conversion is the latent heat of vaporization. At this temperature liquid and gaseous phases coexist. The gas is called saturated; it is at this temperature that the gas can condense. Once all the liquid is converted to gas, further addition of heat increases the temperature of the gas; it is now termed superheated (see Figure 4-2). For each liquid temperature there is a corresponding vapor pressure. If the pressure is reduced below the vapor pressure at a given temperature, there will be a change of state from liquid to gas phase. This phenomenon can occur in flow-

1 Apparent viscosity is the ratio of shear stress to shear rate for non-Newtonian fluids. The ratio changes with stress, and for time-dependent non-Newtonians, the ratio changes as well with time.

meters that involve a pressure loss in making the flow measurement, particularly of high vapor pressure liquids.

Lowering the local pressure in the meter below the vapor pressure of the liquid will cause a change of state to gas. With this change of state there is a huge increase in volume. It is not uncommon for gas to occupy 50 to 200 times its volume as a liquid. This dramatic increase in volume with change of state from liquid to gas can have dire effects on the ability of the flowmeter to make the measurement and can, in some cases, lead to catastrophic failure of the flowmeter.

Matter in Motion

The characteristics of matter in motion, fluid flow specifically, is derived from:

(1) the Equation of Continuity,

(2) Bernoulli's Theorem, and

(3) the work of Osborne Reynolds.

The Equation of Continuity

The volume rate of flow (Q) passing a point is equal to the normal cross section (A) at that point times the average velocity across that area ($\bar{v}$) (see Figure 4-3). If there is constant volume flow rate for a given area change (change in pipe size), there is an inverse change in average velocity. This is the Equation of Continuity (see Figure 4-4); it is based on average velocity at the cross-sectional area.

Stated another way, volumetric flow rate through piping can be calculated by multiplying the area of the pipe, at a location, by the average velocity at that location. This is the basis of operation for all velocity-type measurement flowmeters. They are based on detecting an average velocity. For a known and fixed cross-sectional area, that is, no coating and no corrosion, volumetric flow rate is proportional to average velocity. The proportionality constant is the cross-sectional area at the point of measurement, which, incidentally, is usually ascertained by actual fluid calibration.

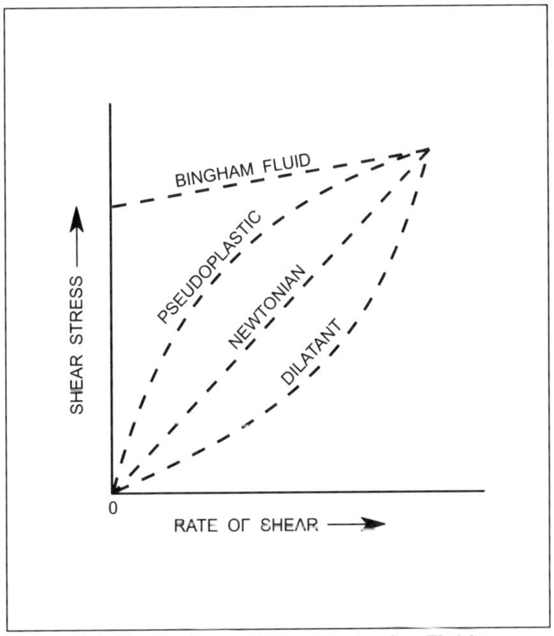

Figure 4-1. Newtonian and Non-Newtonian Fluids

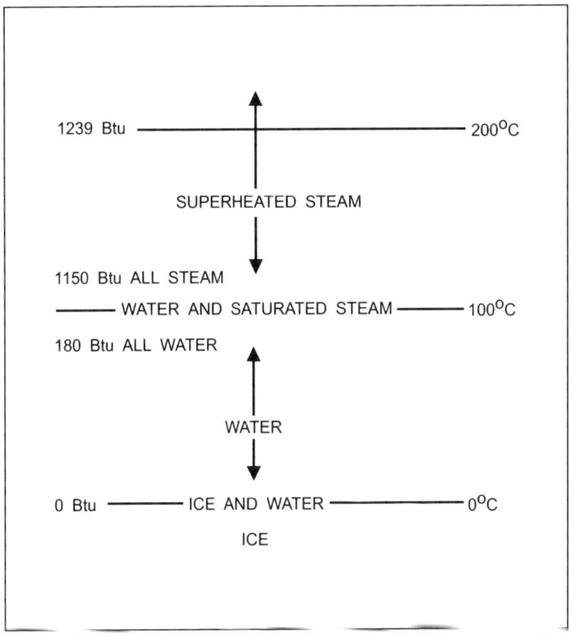

Figure 4-2. States of Water at Atmospheric Pressure and Various Temperatures

Bernoulli's Theorem

The Equation of Continuity shows the relationship between area and average velocity for a given volumetric flow rate. The work of Daniel Bernoulli and others in the 18th century recognized that a velocity change has associated with it a conversion of energy (head) from velocity head (kinetic energy) to pressure head (potential energy) (see Figure 4-5). In other words, if total energy is constant (energy is neither added nor removed), a change in velocity will result in an inverse change in pressure. For a reduction in area upstream of the reduction, there is a given velocity and a corresponding pressure; downstream of the reduction, there is increased flow velocity and a corresponding lower pressure. It is this characteristic of flow that is used in all variable differential pressure types of flowmeters, including orifices, Venturis, flow tubes, flow nozzles, Pitots, and on and on.

Reynolds Number: Flow Profile and Fluid Properties

In the late 19th century (1883 specifically), further insight into the nature of flow was provided by Osborne Reynolds in a group of classic experiments, shown in Figure 4-6. He injected dye into a flow stream and observed significant changes in how the dye moved. At low flow velocities (and higher viscosities) he observed that the dye traced a straight line from the point of injection; he called this condition direct flow (the upper pipe of Figure 4-6). We call it laminar flow because the fluid is moving as if it were composed of laminations or plates. As he increased the velocity a condition was reached where the dye started out as a straight line but began to wander, indicating instability (the middle pipe in Figure 4-6). As Reynolds further increased the velocity he observed that there was immediate dispersion of the dye throughout the flow stream; he called this sinuous flow (the lower pipe in Figure 4-6). Today this is called turbulent flow.

Reynolds' work showed that in laminar flow viscous forces (internal friction of the fluid) dominated fluid behavior, and an analysis of the local velocities across the pipe define a parabolic profile with a velocity at the pipe centerline that is twice the mean velocity (see Figure 4-7). In the turbulent regime, inertial forces (dynamic forces) dominate fluid behavior, evidenced by lateral and transverse forces that cause the immediate dispersion of the dye in the flow steam. These lateral forces cause intimate mixing of the dye into the flow stream and result in a

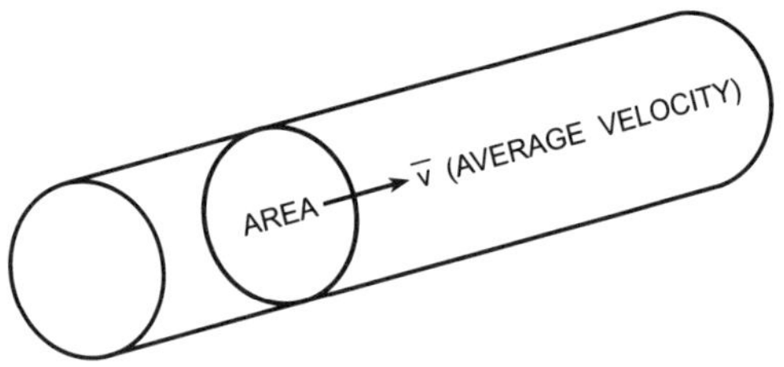

VOLUME FLOW RATE = AREA x AVERAGE VELOCITY

$$Q = A \times \bar{v}$$

$$\frac{(\text{FEET})^3}{\text{TIME}} = (\text{FEET})^2 \times \frac{\text{FEET}}{\text{TIME}}$$

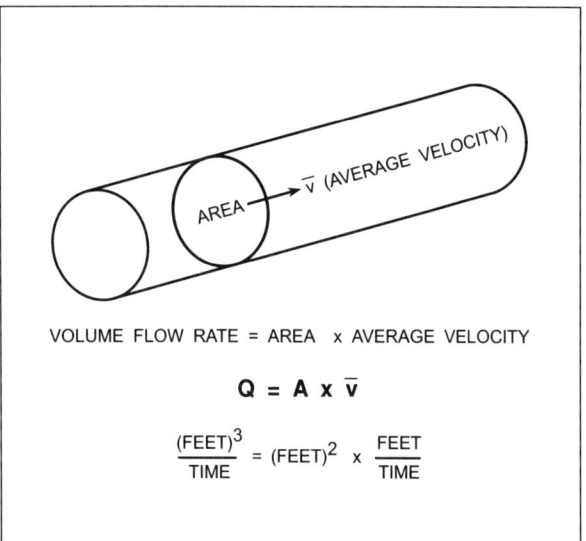

Figure 4-3. The Volume Flow Rate Relationship to Area and Linear Velocity

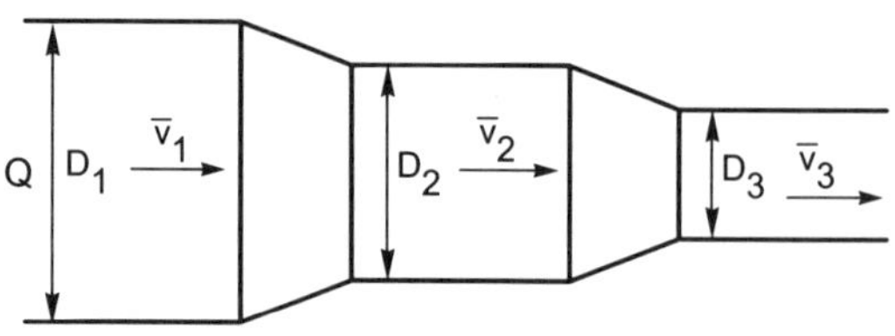

$$Q = \frac{\pi\,(D1)^2}{4}\,\bar{v}_1 = \frac{\pi\,(D2)^2}{4}\,\bar{v}_2 = \frac{\pi\,(D3)^2}{4}\,\bar{v}_3 \ldots$$

$$= A_1\bar{v}_1 = A_2\bar{v}_2 = A_3\bar{v}_3$$

and $\bar{v}_1 < \bar{v}_2 < \bar{v}_3 \ldots$

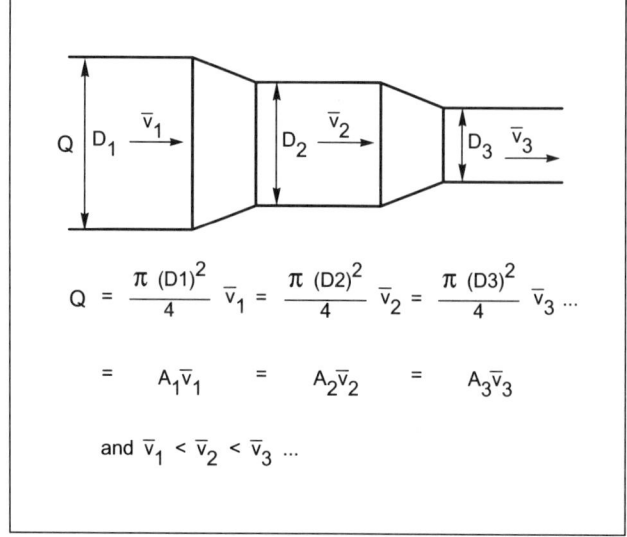

Figure 4-4. The Equation of Continuity

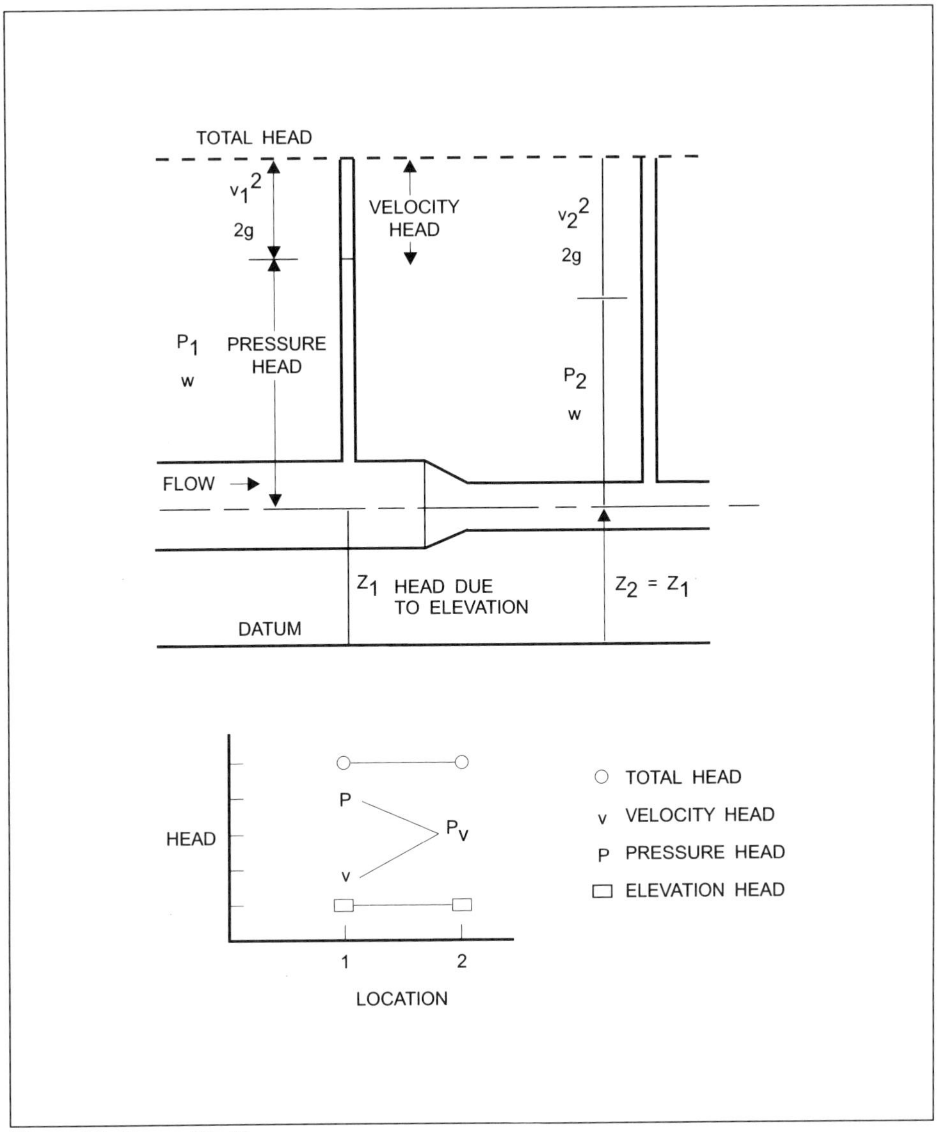

Figure 4-5. Bernoulli's Theorem

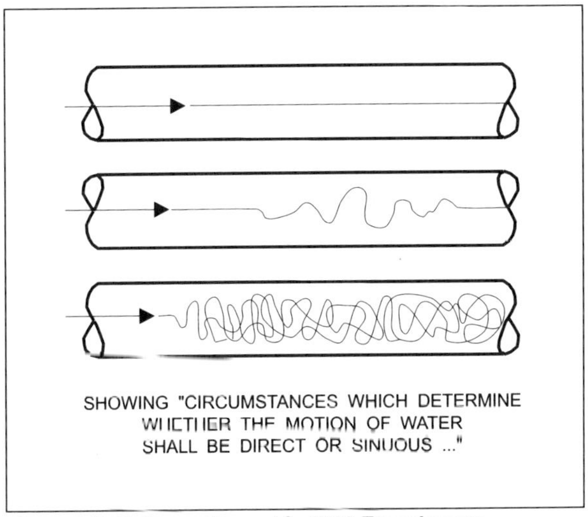

Figure 4-6. Osborne Reynolds 1883 Experiments

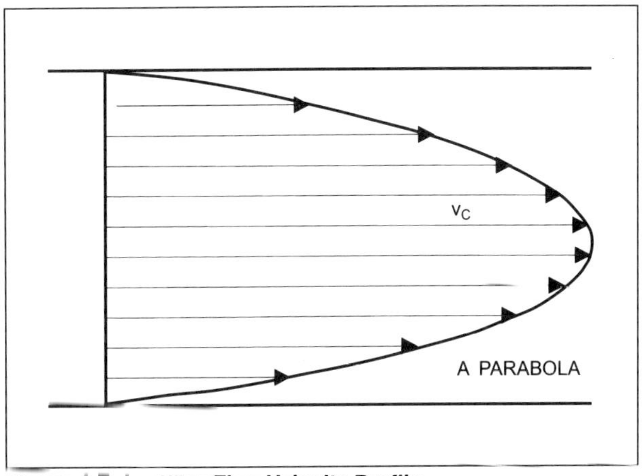

Figure 4-7. Laminar Flow Velocity Profile

profile that has relatively small changes in local velocity across the pipe area a given distance away from the pipe wall (see Figure 4-8). The distance from the pipe wall (boundary layer thickness) is a function of fluid properties and the condition (roughness) of the pipe wall. The profile is squared up, but not square, as idealized in some texts (see Figure 4-8). The degree to which the profile is squared up changes with changing fluid velocity and viscosity (see Figure 4-9). There are two definite regimes, laminar flow and turbulent flow, with a zone of instability in between that is known as transition. In this zone, flow may behave as laminar flow or turbulent flow and may "jump" at random from one condition to the other, usually caused by obscure perturbations in the flow stream. Most flowmeters have poor repeatability if operated in this unstable zone.

Reynolds did not stop at this point in his experimental work, and his analysis of the tests yielded a tool that has added to our understanding of how fluid flows. He provided a dimensionless number, the Reynolds number (Re), which defines in which regime the flow is operating: laminar, turbulent, or transition.

$$\text{Re} = \frac{\rho v D}{\mu}$$

$$\text{Re}_\text{D} = \frac{3160 Q_{\text{gpm}} \text{G}}{\mu_{\text{cP}} D_{\text{in}}}$$

where:

Re = Reynolds number
Re_D = pipe Reynolds number
ρ = fluid denstiy
D = a dimension
v = average fluid velocity
μ = absolute fluid viscosity
Q_{gpm} = volume flow rate, gallons per minute
G = liquid specific gravity referred water at 4°C
D_{in} = inside pipe diameter, inches
μ_{cP} = fluid viscosity, centipoise

Flow Profile and Piping Effects

The Reynolds number defines whether the flow is laminar (viscous forces dominate) or is turbulent (dynamic forces dominate). When the Reynolds number

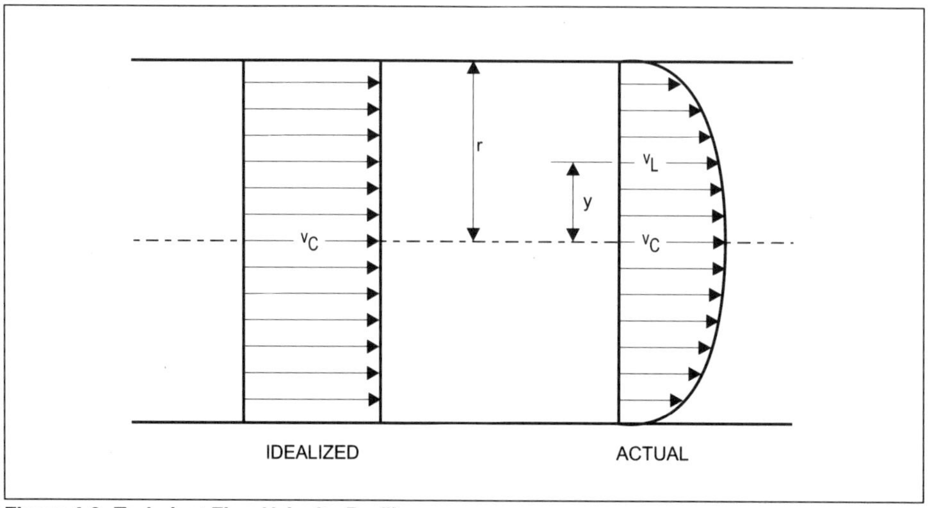

Figure 4-8. Turbulent Flow Velocity Profile

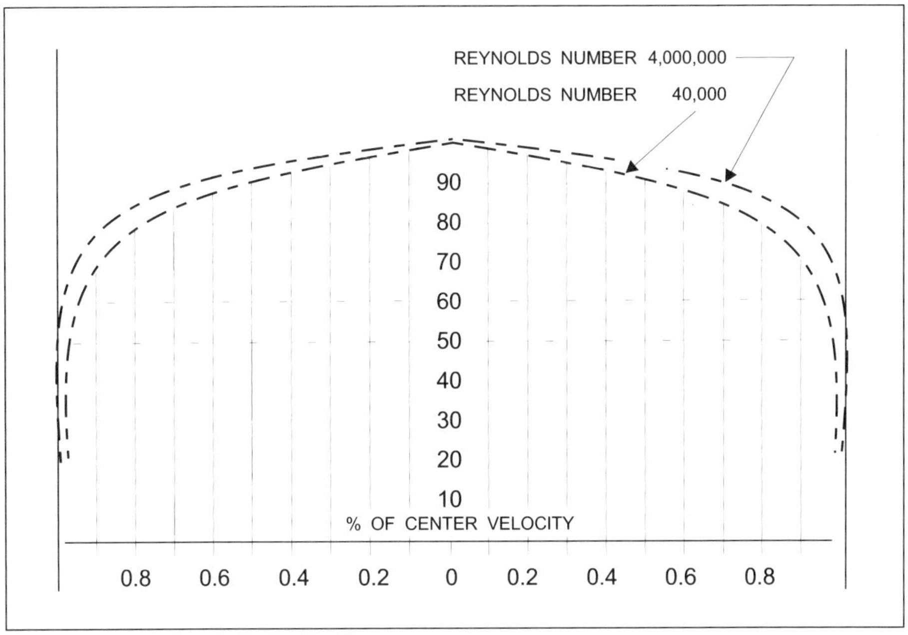

Figure 4-9. Turbulent Velocity Profiles

is 4000 or greater, the profile is squared up and flow is turbulent; when Reynolds number is 2000 and less, the profile is parabolic and flow is laminar. Between Reynolds numbers of 2000 and 4000 is the transition zone. This analysis is based on symmetrical profiles; that is, the profile is identical regardless of which transverse plane cuts through the pipe centerline. Only those parameters that determine Reynolds number, fluid velocity, a dimension such as inside pipe diameter, and kinematic viscosity determine the flow profile. However, flow profiles are symmetrical only if special care is taken. As flow moves through piping, the profiles become distorted or are asymmetrical (lack symmetry). A single 90 degree elbow in piping distorts flow profile, as in Figure 4-10. As flow moves through the elbow, flow accelerates around the outside of the bend and slows down on the inside of the bend. The profile is distorted with a high velocity zone occurring at other than the pipe centerline. If the flow is turbulent, the dynamic forces cause mixing of the flow, the profile is gradually restored and, if there are no other disturbances, returns to a fully developed turbulent profile (which is a preferred profile for the most commonly used flowmeters).

If two 90 degree elbows are close-coupled[2] but in different planes (see Figure 4-11), the distorted profile introduced by the first elbow is further distorted as it passes through the second elbow; flow leaving that elbow is spinning, that is, at one location the profile is continually changing with the high velocity zone having a planetary motion. This is a very persistent phenomenon, and it takes longer pipe length for the flow profile to return to a fully developed turbulent profile. Piping geometry can result in distorted profiles, spinning (swirling) flows, and jets, and in some cases the flow can actually be tumbling as it moves through the pipe. These motions are unfavorable for most flowmeters. It is usual to install most flowmeters an adequate distance from

2 Separation of the two 90 degree elbows by more than 3 diameters of straight pipe results in a flow pattern from the second elbow like that from a single elbow.

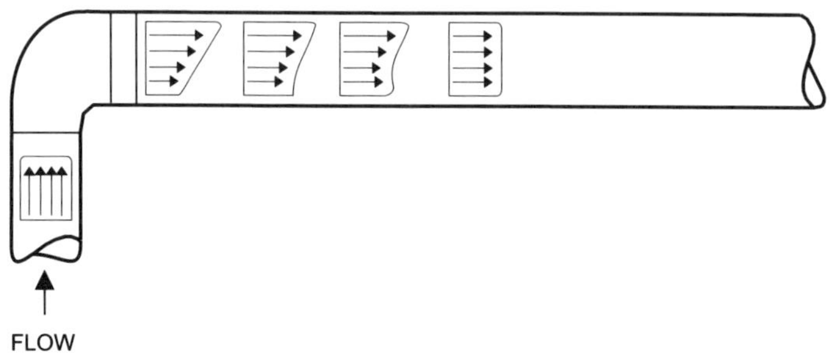

Figure 4-10. Effect of a Single Piping Elbow on Flow Profile

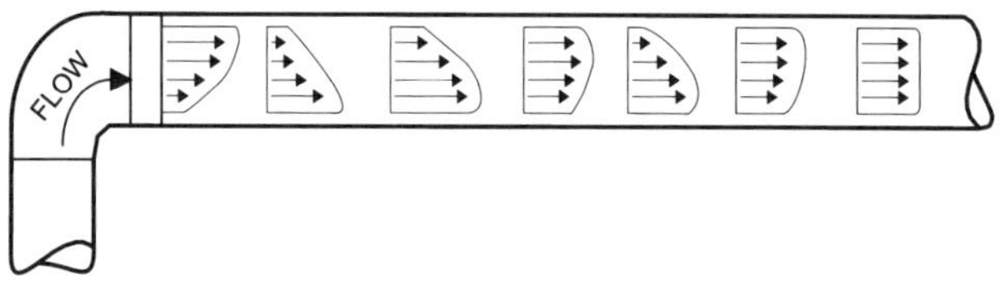

Figure 4-11. The Effect on Flow Profile of Two 90° Elbows Close-Coupled and in Different Planes

disturbances so that the mixing in the flow returns the profile to the preferred fully developed turbulent profile. This is why there are required lengths of straight upstream piping before most flowmeters. For most flowmeters there are also requirements for lengths of straight downstream piping between the flowmeter and a disturbance. Disturbances such as a control valve should be located downstream of the flowmeter. Some downstream disturbances can be so upsetting to the flow that the effect can move opposite to the direction of flow, get back into the meter, and effect the measurement. Figure 4-12 is a graphical representation of the ASME recommendations for upstream and downstream piping for orifices or flow nozzles after a specific flow disturbance. Similar graphs are prepared for other types of upstream piping (flow disturbances) and for other types of flowmeters.

Flow Conditioning: Controlling Flow Profile

The most common forms of flow conditioners—flow straighteners—are not recommended for distorted flow profiles, such as from a single elbow, where, in fact, they may be a detriment to getting a squared up profile because they isolate elements of flow from lateral mixing.

Ideally, the flow profile is defined by Reynolds number and the condition (roughness) of the inside wall of the piping. Actually, the profile is modified by the piping configuration and can be restored ultimately by the mixing action of the fluid as it moves. The profile can also be restored by flow conditioners (see Figure 4-13), but they must be applied with discretion.

The proprietary flow conditioners (Sprenkle™, Zanker™, Mitsubishi™, and VORTAB™) are all effective in eliminating grossly distorted flow profiles, jets, and persistent swirl but have individual head loss, which may be a factor (see Table 4-1).

How fluid moves in pipes does have an effect on most flowmeters. Those that are Reynolds number (Re) dependent are given in Table 4-2.

A general performance curve for these meters is shown in Figure 4-14. There is a zone at higher Reynolds number (Re) where the curve is almost linear; that is, there is essentially constant meter coefficient. It is this zone

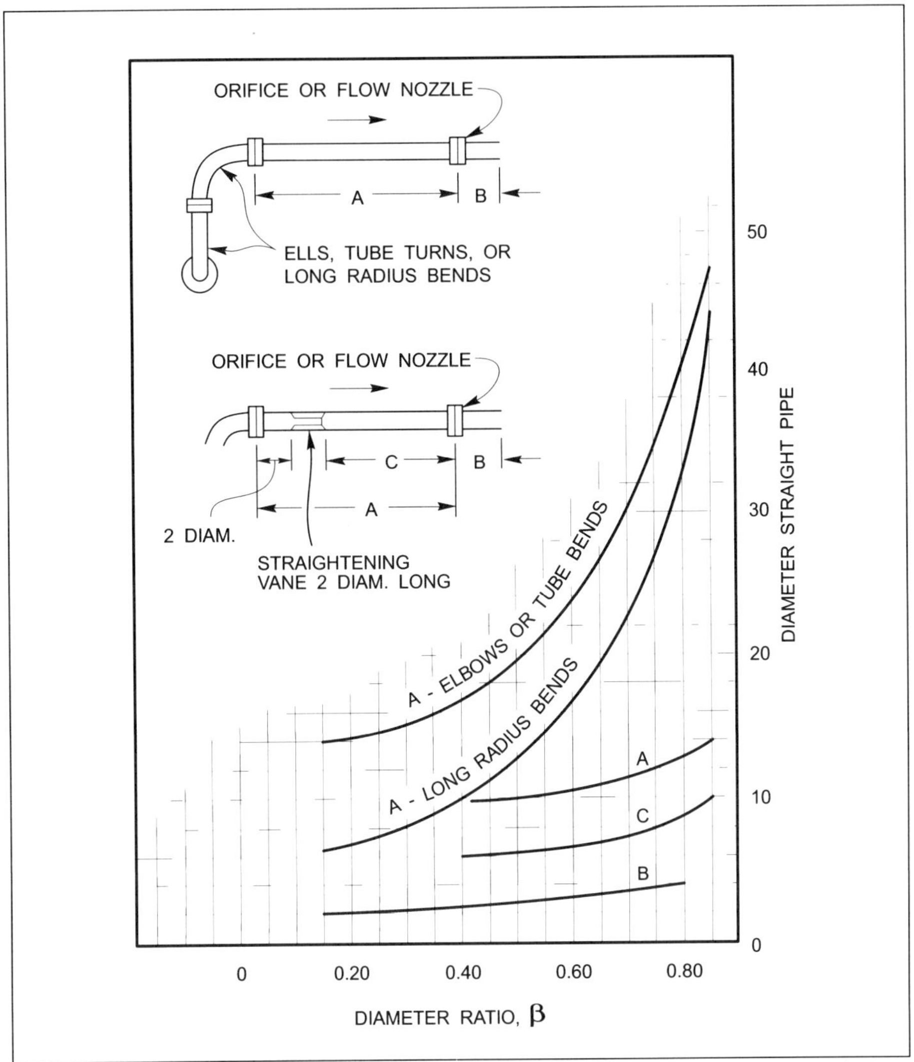

Figure 4-12. Upstream and Downstream Piping Requirements for Orifice or Flow Nozzle for a Specific Upstream Disturbance
(From ASME, *Fluid Meters*, 1971.)

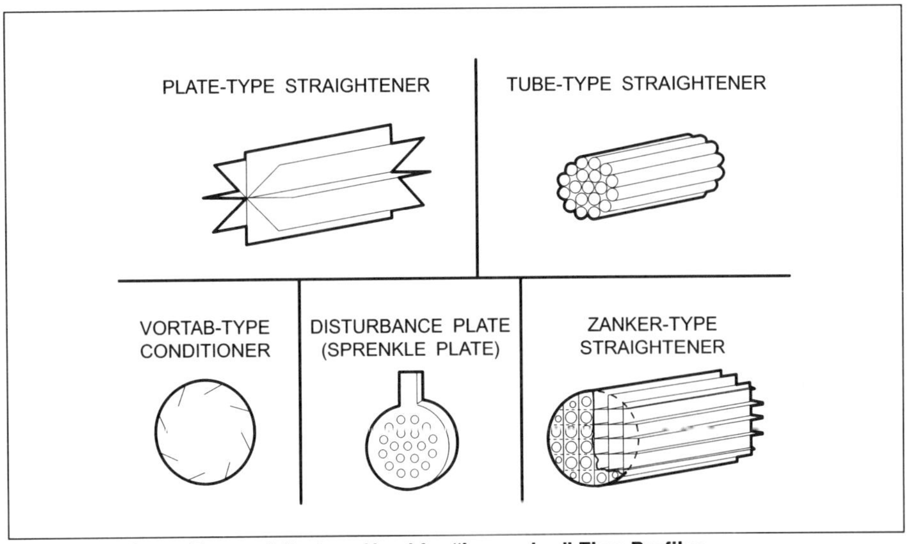

Figure 4-13. Some Types of Devices Used for "Improving" Flow Profiles

Table 4-1. Relative Head Loss of Proprietary Flow Conditioners

Flow Conditioner	Relative Head Loss (Four Highest)
Mitsubishi	2
Sprenkle	4
VORTAB	1
Zanker	3

Table 4-2. Reynolds Number Dependent Flowmeters

Averaging Pitot	Segmental Orifice
Concentric Orifice	Sonic—Doppler
Eccentric Orifice	Sonic—Time-of-Flight
Flow Nozzle	Turbine
Flow Tube	Variable Area
Fluidic	Venturi
Pitot	Vortex Shedding
Positive Displacement*	

*Positive displacement-type flowmeters are included here even though technicallly they are not Reynolds number dependent. They are, however, viscosity dependent. Viscosity is the fluid property specified to provide an adequate fluid seal between fixed and moving parts. Postive displacement meters, which provide excellent measurement at higher viscosities, may not provide the same quality of measurement at lower viscosities.

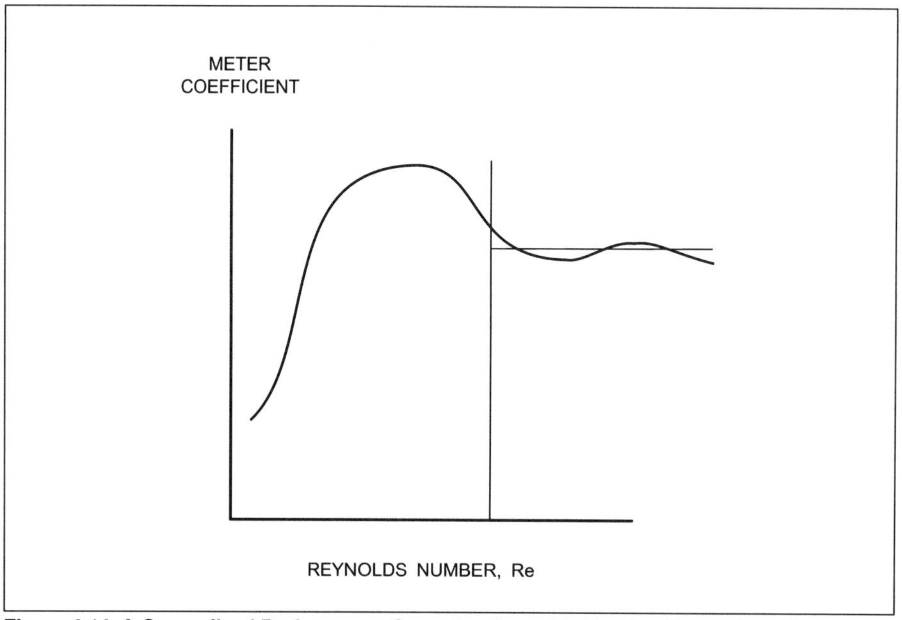

Figure 4-14. A Generalized Performance Curve for Reynolds Number Dependent Flowmeters

where these meters are generally applied. At lower Reynolds number a hump is developed (the viscous hump) that covers transition and laminar flow regions. Performance in these zones changes with both flow rate and viscosity change. The orifice meter specific curves of Figure 4-15 have a change in slope with increasing beta ratio (the ratio of the diameter of the orifice to the inside diameter of the pipe). Table 4-3 is an example that is derived from work by William Buzzard [Ref. 1].

The flow coefficient for orifice meters is Reynolds number dependent. It is recommended that the coefficient be checked at both maximum and minimum Reynolds number and the variation in coefficient be calculated.

Figure 4-16 shows typical curves for other selected Reynolds number dependent flowmeters. The meters that are not dependent on Reynolds number are Coriolis, mass, magnetic, and thermal types.

Some flowmeters do not require straight lengths of pipe before and after the meter; these are the Coriolis, mass, positive displacement, and variable area.

Measurement

It is revealing to look at fluid flow measurement on the basis of information desired from that measurement (see Table 4-4), and to look also at the units of flow that are traditionally used (see Table 4-5).

A comparison and consolidation of these two tables shows that information desired from flow measurement and the flow units most frequently used fall into two groups—mass or volume information. Separating meters into these two groups reveals that only three types of meters fit:

MASS—Coriolis and thermal types

VOLUME—Positive displacement type

A look at the other types of meters is required.

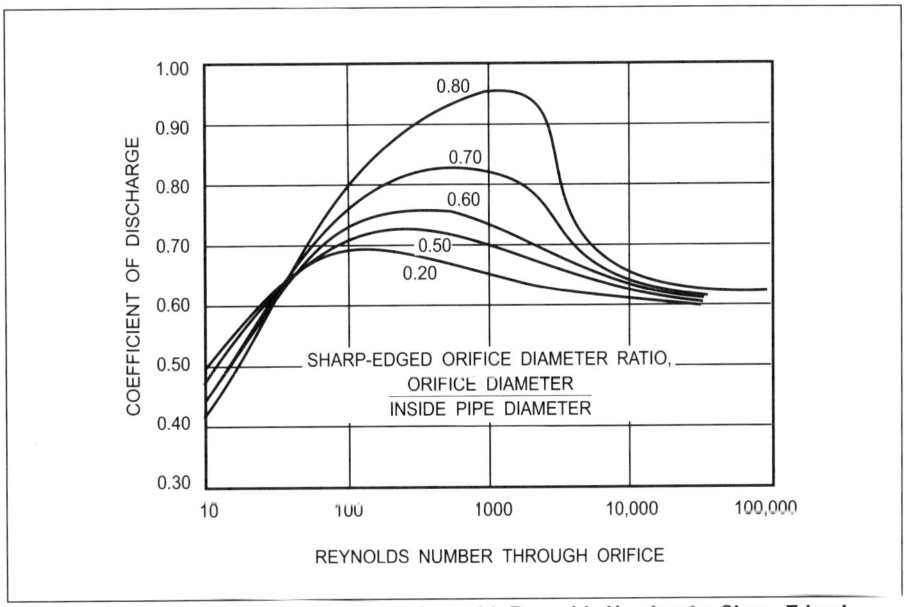

Figure 4-15. Variation of Discharge Coefficient with Reynolds Number for Sharp-Edged Orifices

Table 4-3. Flow Coefficient Variations with Flow Rate for Orifice Meters

ΔP RANGE (in. H_2O)	APPROXIMATE DIAMETER RATIO		VISC. OF LIQUID, cSt	PIPE SIZES WITH MAXIMUM FLOWS FOR RANGES OF APPROXIMATELY 0-4 ft/sec and (0-8 ft/sec) PIPE VELOCITY									
				2 in.		3 in.		4 in.		6 in.		8 in.	
				40 gpm	(80 gpm)	90 gpm	(180 gpm)	150 gpm	(300 gpm)	350 gpm	(700 gpm)	620 gpm	(1240 gpm)
0-20	β=0.72	—	0.5	2.3%	—	1.9%	—	1.6%	—	1.3%	—	1.2%	—
			1.0	3.7%	—	2.5%	—	2.3%	—	2.1%	—	1.8%	—
			5.0	6.7%	—	5.2%	—	4.6%	—	4.1%	—	3.4%	—
0-50	β=0.61	—	0.5	1.7%	—	1.2%	—	1.1%	—	1.07%	—	0.9%	—
			1.0	2.3%	—	1.8%	—	1.6%	—	1.3%	—	1.1%	—
			5.0	5.0%	—	3.5%	—	3.3%	—	2.9%	—	2.3%	—
0-100	β=0.52	(β=0.70)	0.5	1.3%	(1.6%)	1.0%	(1.2%)	0.9%	(1.0%)	0.7%	(1.0%)	0.6%	(0.9%)
			1.0	1.7%	(2.3%)	1.3%	(1.8%)	1.1%	(1.7%)	0.9%	(1.5%)	0.8%	(1.2%)
			5.0	3.1%	(4.4%)	2.8%	(3.8%)	2.6%	(3.6%)	2.1%	(2.9%)	1.6%	(2.4%)
0-200	β=0.45	(β=0.63)	0.5	0.8%	(1.3%)	0.7%	(0.9%)	0.7%	(0.9%)	0.7%	(0.9%)	0.5%	(0.7%)
			1.0	1.4%	(1.9%)	1.1%	(1.2%)	1.0%	(1.1%)	0.9%	(1.1%)	0.7%	(1.0%)
			5.0	2.3%	(4.0%)	2.2%	(3.1%)	2.0%	(2.7%)	1.8%	(2.3%)	1.5%	(1.8%)

Figures shown are the percent variation of K for a turndown of 5:1 (100% to 20% of max. flow).
As an estimate:
For 3:1 turndown, multiply figures by 0.8
For 10:1 turndown, multiply figures by 1.6

Table 4-4. Classifications of Fluid Flow Measurement

Mass Flow
Volume Flow
Inferred Mass or Volume Flow - Constant Density
Computed Mass or Volume Flow - Measured Density
Flow of Dissolved/Suspended Solids - Inferred, Constant Concentration - Computed, Measured Concentration

Table 4-5. Units of Flow Measurement Commonly Used

Liquids	Gases	Steam
Actual Volume	Actual Volume	Gravimetric (Mass)
Gravimetric (Mass)	Gravimetric (Mass)	
Volume @ "Base" Temperature	Volume at Standard Conditions	

Velocity Detectors Used as Volume Flowmeters

A number of meters are velocity detectors but are used as volumetric flowmeters. This is valid as long as their cross-sectional area at the point of measurement is known and constant; that is, there is no coating of the meter interior and no corrosion. Meters in this group include fluidic, magnetic, turbine, ultrasonic Doppler, ultrasonic time-of-flight, and vortex shedding. In all cases except magnetic, a change in cross-sectional area (assumed to be known by

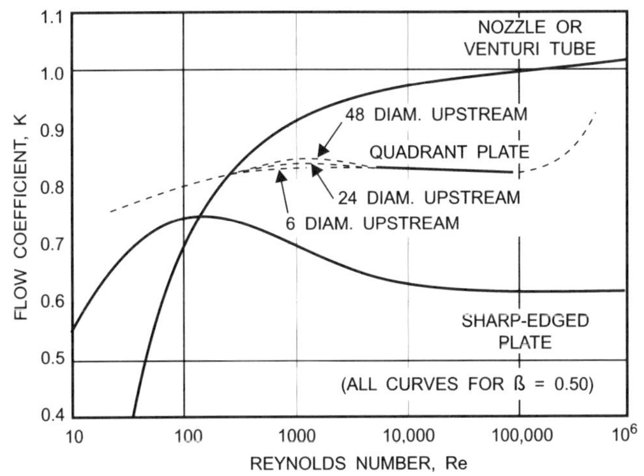

Figure 4-16. Typical Performance Curves for Selected Variable Differential Pressure Flowmeters, Beta Ratio of 0.50

calibration or measurement and assumed to be constant) will result in an inverse change in signal. For example, if there is coating on the inside of the pipe, the flow area decreases, and, for a given flow rate, the velocity increases, which means a higher signal for the same flow rate (the Equation of Continuity). For magnetic flowmeters, if the coating has the same electrical conductivity as the liquid (which may or may not be the case), the meter's ability to average velocity contributions from across the profile will sense the coating as zero velocity and provide the correct signal for volumetric flow rate.

Flowmeters That Cannot Be Categorized as Mass or Volume Types

Some flowmeters do not fit into mass, volume, or velocity types; they are differential pressure types. The output signal is dependent on conversion of head, so both differential pressure and the density of the fluid must be known.

This category includes all forms of orifice meters, Venturis, flow nozzles, flow tubes, Pitots, averaging Pitots, target types, elbows, and all the proprietary forms such as the Wedge™, V-cone™, etc. All of these differential pressure-type flowmeters are, more explicitly, *variable* differential pressure flowmeters. There are also *constant* differential pressure flowmeters, which in their most common industrial form are called variable area flowmeters. Regardless of whether the meters are variable or constant differential pressure types, it is fundamental to their successful application that the fluid density be known.

Inferential Mass Flow Measurement

Since only a few types of mass meters are available, and not all applications can be handled by them, other types of meters have traditionally been used to infer mass flow rate (see Figure 4-17). The measurement of volume multiplied by density has had limited use even though that method is independent of the fluid and the composition of mixtures. The most common method used to infer mass flow rate is to multiply flow rate by density that is inferred from pressure and/or temperature measurement (see Figure 4-18). This method applies to one fluid or a mixture only, since the pressure/temperature/density relationship applies to one fluid composition or mixture.

Perhaps the most common form of inferred mass flow measurement is units of standard volumes, normal volumes, volumes at base conditions, or volumes

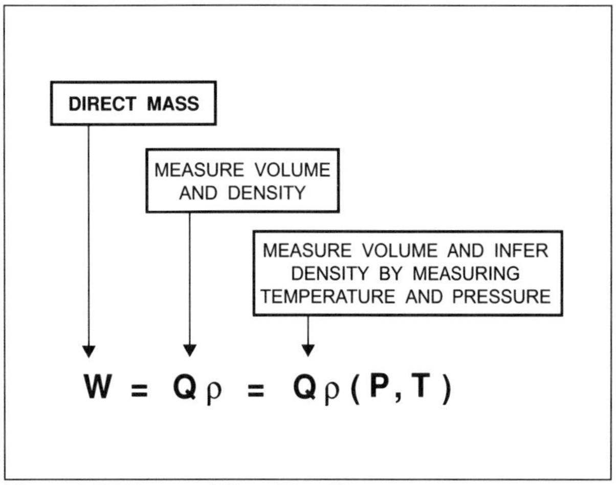

Figure 4-17. Forms of Mass Flow Measurement

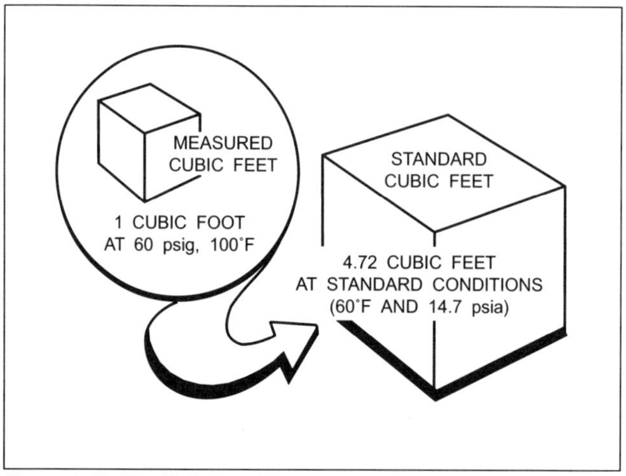

Figure 4-18. The Relationship between Measured Volume (Volume at Operation Conditions) and Volumes at a Condition Called "Standard"

at reference conditions. Although volume is used in the terminology, these are inferential mass measurements. In each case the required measurement is a volume at "standard," "normal," "base," or "reference" conditions. These are specific pressure and temperature conditions; hence, density is inferred. Although these conditions approximate atmospheric pressure and ambient temperature, they vary widely (see Table 4-6).

There is a problem: there is neither a universal "standard" temperature nor a universal "standard" pressure! There are many conventions, and these vary by locale and industry and even within given industries. Before using units of standard, normal, base, or reference volumes, determine what are the "standard" temperature and "standard" pressure.

Significant errors may result if the incorrect standard is used. For example, if normal conditions of 14.2 psia and 15°C were used when the correct ones were 14.696 psia and 15°C, the calculated normal volume would be in error +3.375% of the volume at 14.696 psia. This speaks to the issue that the traditional meters that are used are cpable only of measuring volume at operating conditions. Standard, normal, base, and reference conditions are computed values (see Figure 4-18).

The ratio of operating pressure to standard pressure and the ratio of standard temperature to operating temperature use the Ideal Gas Law to calculate standard volumes or the other flow units (see Figure 4-19). The temperature and pressure ratios infer the ratio of the fluid density at the operating condition to the fluid density at standard conditions.

In these relationships pressure must be expressed as absolute pressure, that is, gage pressure plus atmospheric pressure. Most often gage pressure transmitters are used and a fixed value for atmospheric pressure is added to obtain the absolute pressure. At lower operating pressures the variation in the pressure of the atmosphere can add unacceptable error for some applications. The user should consider using absolute pressure transmitters when the operating pressure is 200 psig or less.

Temperature must also be on the absolute scale, so for the Fahrenheit scale add 460°F, for the Celsius scale add 273°C. Figure 4-19 is fine for ideal gases, but most applications warrant a check of the compressibility of the gases, especially where accurate measurement is needed. Compressibility is the deviation of the gas from the Ideal Gas Law. This law predicts an exact inverse relation-

Table 4-6. Some of the More Commonly Used Standard Pressures and Standard Temperatures (May Be Used in Any Combination)

Standard Pressure, psia	Standard Temperature
14.2	0°C
14.4	15°C
14.696	32°F
14.73	59°F
15.05	00°F
	70°F

ship of volume change with absolute pressure change and a direct change in volume with absolute temperature change. Most gases have a large enough deviation from an ideal gas to have an effect on the accuracy of computation of standard volumes. The calculation should include ratio Z_s/Z_o where Z is compressibility at standard conditions and operating conditions, respectively. Rather than leave this as rote, remember that mass is identical whether at operating conditions or at "standard" conditions.

$$W_o = W_s$$

where:

W_o = the mass at operating conditions
W_s = the mass at standard conditions

Since volume times density equals mass: $Q_o\rho_o = Q_s\rho_s$ = mass flow, where $Q_o\rho_o$ and $Q_s\rho_s$ are volume rate of flow and fluid density at operating conditions and standard conditions, respectively. The volumetric rate of flow at standard conditions is the required information, so solving for Q_s yields $Q_s = Q_o\frac{\rho_o}{\rho_s}$.

The basic equation for the density of an ideal gas is: $\rho = \frac{PM}{RT}$ and for an imperfect gas: $\rho = \frac{PM}{RTZ}$

where:

P = pressure
M = molecular weight
R = the universal gas constant
T = temperature
Z = compressibility

Substituting this in $Q_s = Q_o\frac{\rho_o}{\rho_s}$ is $Q_s = Q_o\frac{P_oM}{RT_oZ_o} \times \frac{RT_sZ_s}{P_sM}$ which reduces to $Q_s = Q_o\frac{P_o}{P_s} \times \frac{T_s}{T_o} \times \frac{Z_s}{Z_o}$.

Density of saturated steam is defined by measuring either pressure or temperature. For superheated steam both temperature and pressure must be measured to define steam density.

The measurement of steam is always in mass units (see Table 4-5). For saturated steam, that is, when all the liquid has just been converted to vapor with no liquid carryover, one temperature defines the condition for the given pressure.

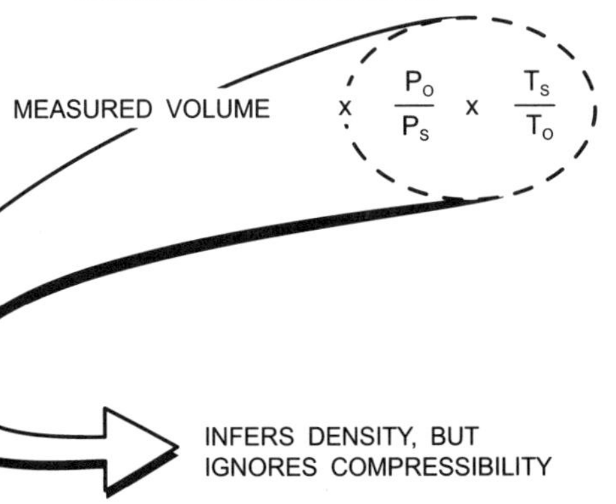

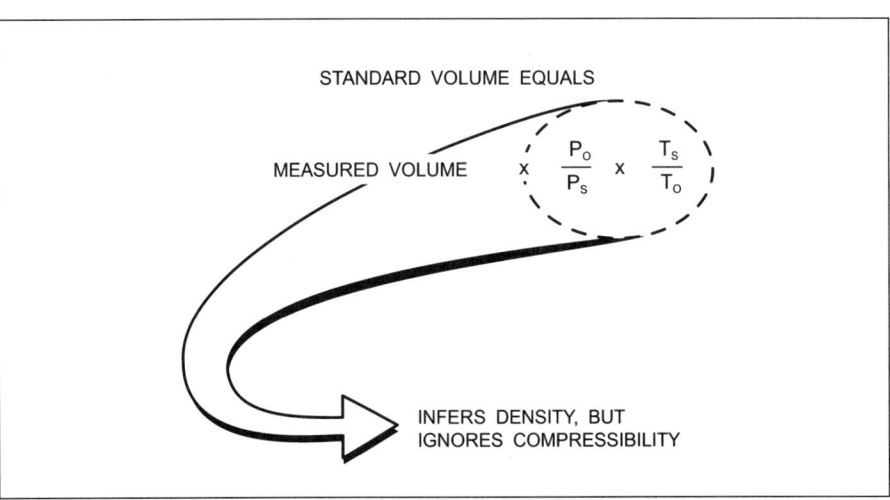

Figure 4-19. Calculating Standard Volumes from Temperature and Pressure Ratios

The Quality of the Measurement: Accuracy

The ability of the flowmeter to make the measurement is expressed in a composite statement called accuracy.

Flowmeters have accuracy statements expressed as:

(1) percent or actual flow rate, or

(2) percent of full scale flow (or percent of meter capacity), or

(3) percent of maximum differential pressure.

The equation for each is:

$$(1)\quad \%\text{ of Rate Accuracy} = \pm \frac{\text{Flow Uncertainty}}{\text{Instantaneous Flow Rate}} \times 100$$

This statement applies to turbine meters, pulsed DC magnetic meters, Coriolis meters, vortex meters, flow integrators, and some special meters or instruments that have logarithmic flow scale characteristics.

$$(2)\quad \%\text{ of Full Scale Accuracy} = \pm \frac{\text{Flow Uncertainty}}{\text{Full Scale Flow Rate}} \times 100$$

This type of statement of accuracy is most suited to describing primary meters such as rotameters, AC-excited magnetic flowmeters, and components such as linear flow recorders, square root extractors, and linear transmitters.

$$(3)\quad \%\text{ Maximum dP Accuracy} = \pm \frac{\text{dP Uncertainty}}{\text{Maximum dP}} \times 100$$

where dP is differential pressure. This type of accuracy statement usually applies to differential pressure flow transmitters and recorders having square root graduations on the chart or scale.

If performance is stated as ± 1%, that is an incomplete statement. The uncertainty may be ± 1%, but the reference (the denominator of the fraction) has not been stated. In that case, assume the worst.

Accuracy statements of percent of actual flow rate mean that the measurement uncertainty is within the given percent value within the stated range for the flowmeter. All other types of statements mean there is a *fixed quantity uncertainty* in the measurement regardless of the flow rate.

For example, a flowmeter with an operating range of 10 to 100 gpm, which has a measurement uncertainty of ± 1 gpm, has that uncertainty at all flow rates from 10 through 100 gpm. To simplify the performance statement, the accuracy claim for the meter is given as ± 1% of maximum flow, that is:

$$\frac{\textbf{Flow uncertainty} \times \textbf{100}}{\textbf{Maximum flow rate}} = \pm \frac{\textbf{1 gpm} \times \textbf{100}}{\textbf{100 gpm}} = \pm \textbf{1\% of maximum flow rate}$$

At 100 gpm the percent of rate accuracy is also ± 1% of rate:

$$\frac{\textbf{Flow uncertainty} \times \textbf{100}}{\textbf{Instantaneous flow rate}} = \pm \frac{\textbf{1 gpm} \times \textbf{100}}{\textbf{100 gpm}} = \pm \textbf{1\% of rate}$$

At 50 gpm the performance statement is still ± 1% of maximum flow rate, but at 50 gpm the percent of rate accuracy is:

$$\frac{\textbf{Flow uncertainty} \times \textbf{100}}{\textbf{Instantaneous flow rate}} = \pm \frac{\textbf{1 gpm} \times \textbf{100}}{\textbf{50 gpm}} = \pm \textbf{2\% of rate}$$

(continued)

(continued)

Likewise at 10 gpm, this ± 1% of maximum flowmeter has ± 10% of instantaneous flow rate accuracy:

$$\frac{\textbf{Flow uncertainty} \times 100}{\textbf{Instantaneous flow rate}} = \pm \frac{\textbf{1 gpm} \times 100}{\textbf{10 gpm}} = \pm 10\% \textbf{ of rate}$$

This is shown graphically in Figure 4-20.

The message:

(1) There is no difference for numerically equal accuracy statements of percent of maximum flow (or equivalent statements) and percent of rate accuracy when the meters are operated at maximum flow rate. Either type of flowmeter could be chosen.

(2) However, if the meter to be selected is to operate over a broad flow range, for numerically equal accuracy statements the percent of rate flowmeters are the meters of choice for best accuracy over the full operating range.

(3) Accuracy statements that are not numerically equal can be evaluated graphically as in Figure 4-20.

Transducing Steps and System Accuracy

In evaluating the quality of the measurement made, do not ignore the effect of all the transducing steps that are made. It is essential that each of the individual elements have their performance stated in the same way—preferably,

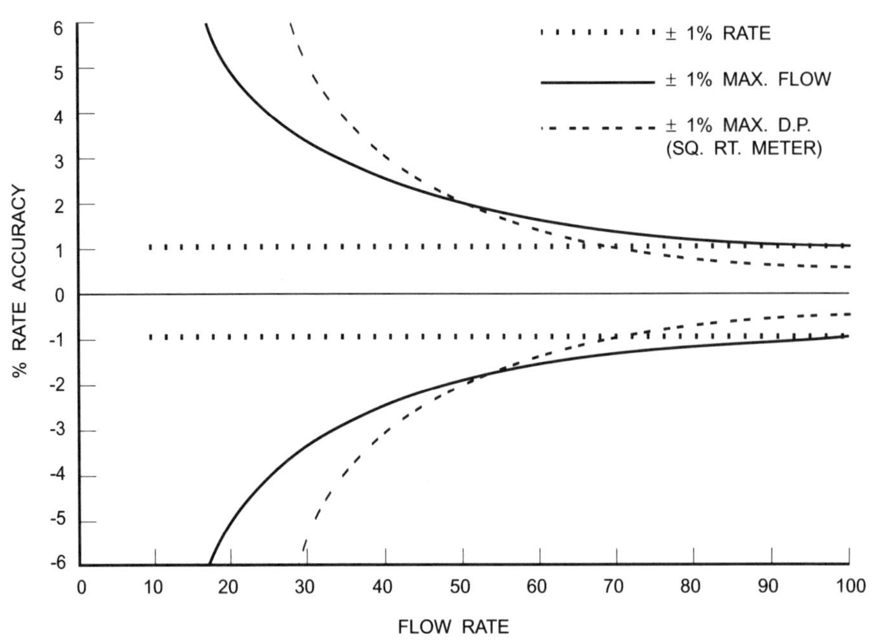

Figure 4-20. Graphical Comparision of Three Numerically Equal Accuracy Statements

as a percent of rate. The statistical analysis of the performance leaves the probability that a particular measurement will fall beyond the expected computed accuracy. However, this approach gives a realistic, not overly pessimistic view of the expected overall system performance. For example, what is called an orifice meter is, in reality, a small system. One hypothetical configuration would consist of:

(1) the orifice plate (a percent of rate device),

(2) a differential pressure transmitter (a percent of maximum differential pressure device),

(3) a square root extractor (a percent of maximum flow device), and

(4) a linear recorder (a percent of maximum flow device).

To define the performance of the flowmeter system, the effect of each element (the plate, the transmitter, the square root extractor, and the recorder) must be combined. To estimate the accuracy of such systems, the performance of each element at a selected flow rate is calculated on the same basis, that is, percent of rate. The usual method is to combine the accuracy of the individual elements statistically by calculating the square root of the weighted sum of their squares. This method takes into account that the various errors will probably be neither all positive nor all negative at one time, the resulting uncertainty should not reflect the worst case of arithmetic addition of errors, and not all of the various errors affect the measurement by the same amount.

$$\text{System Accuracy} = \sqrt{x_1(\text{Accuracy}_1)^2 + x_2(\text{Accuracy}_2)^2 + x_3(\text{Accuracy}_3)^2 + \ldots}$$

where x_i are the sensitivity coefficients of the various errors.

System Needs

The quality of measurement that is required has to be considered in the light of how the measurement is to be used. For many processes, flow variations are minor. Reproducibility (repeatability over a long time) is the required characteristic for this circumstance. However, do not neglect to review what quality of measurement is needed during start-up, upset, or shutdown conditions. Although these conditions may occur infrequently, in some processes there may be a real requirement for a good quality of measurement at other than normal conditions.

Using that portion of the meter performance curve that matches the expected process flow conditions can improve the quality of flowmeter performance, but it is often overlooked.

For many processes, measurement over varying flow conditions is a reality. Flowmeters with essentially constant coefficients (for example, pulses per volume for digital meters) have been the most easy to apply. This characteristic, called linearity, is ± 1% of instantaneous flow rate or better for the more modern flowmeters. Do not overlook that this performance may be improved by using a coefficient for the operating range rather than the coefficient selected by the manufacturer for the meter range (see Figure 4-21). It is necessary to have a calibration curve for the flowmeter to make this determination. Curves are available from most manufacturers for a nominal extra charge when ordered at the time the flowmeter is purchased.

Sampling Measurement

All discussion up to this point has been based on measuring the whole body of flow. However, some meters infer the whole body of flow by measuring only a portion of the flow; some forms are shown in Figure 4-22. Regardless of the form, sampling flowmeters are velocity profile dependent, even though they are generally used at high Reynolds number. They may be excellent

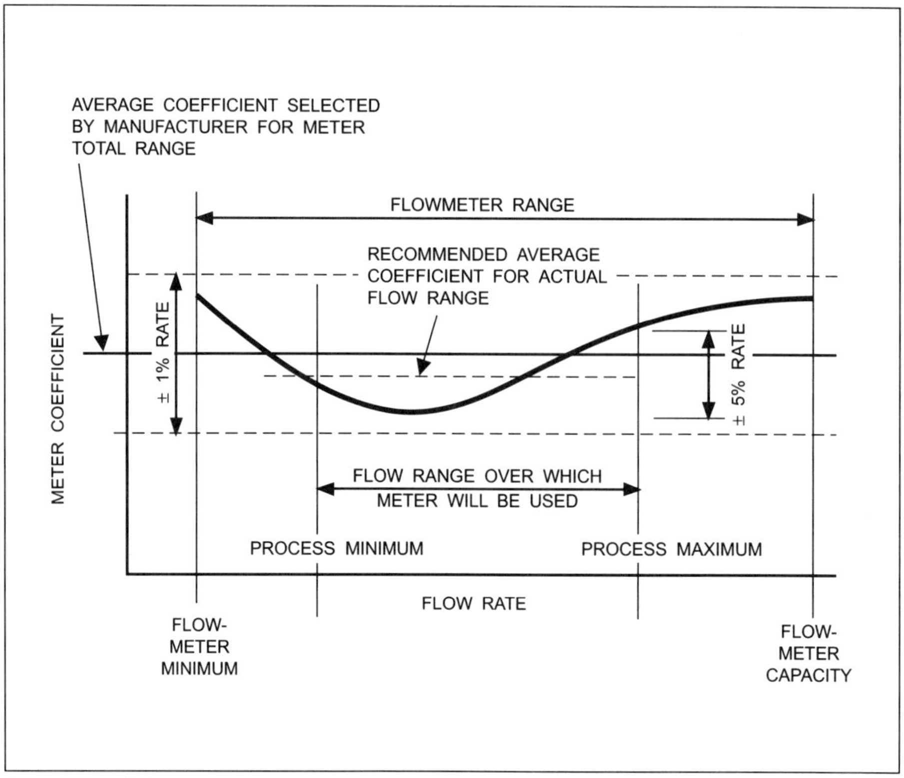

Figure 4-21. Selecting an Average Meter Coefficient for Actual Flow Range

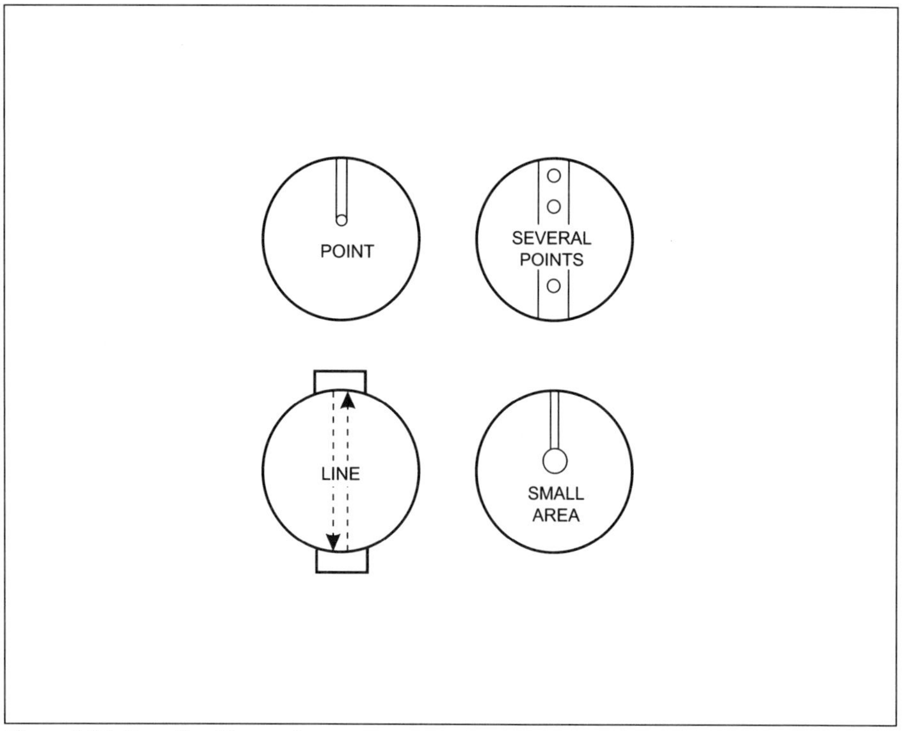

Figure 4-22. Sampling Flowmeters

transducers, but their accuracy statements apply only to transducing a local velocity to an output signal. In the case of Doppler ultrasonic flowmeters, what is sensed is the velocity (a component of velocity) of particles, bubbles, or other discontinuities in the flow. It falls to the user to find the relationship between bubble or particle velocity and the local velocity.

For all types of sampling meters, it also falls to the user to find the relationship of the local velocity to the required average velocity of the fluid (see Figure 4-23). In order to calculate flow rate, the cross-sectional area must be known, and that may not be easy to determine since these meters are usually installed in existing pipe. The flow area uncertainty can be considerable; for example, for 6-inch schedule 40 pipe, the difference between nominal flow area that appears in the literature and the maximum flow area allowed by manufacturing tolerances of the pipe is $\pm 4\frac{1}{2}\%$ *of the nominal area.*

Summary

Making a good measurement of flow rate is assured when adequate attention is given to fundamentals. It means understanding the nature and behavior of the fluids to be measured and a recognition that the measurement is made under dynamic conditions. It also means understanding how fluids move in pipes and the importance of flow profile to most measurements. It means understanding what is being measured and what is required to convert that information to the information that is required. It is this added care and attention to fundamentals that will assure that the actual results are as good as the expected results.

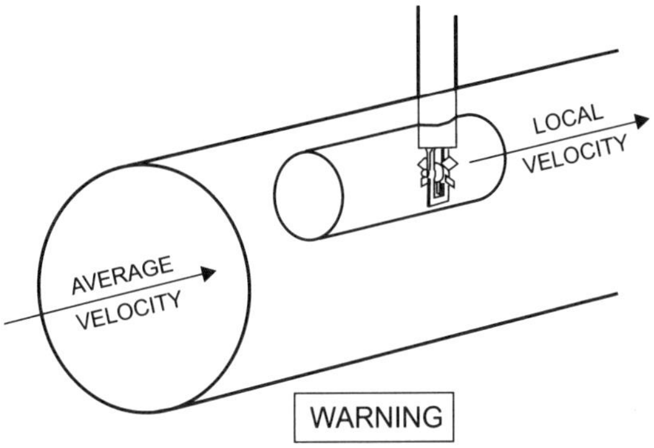

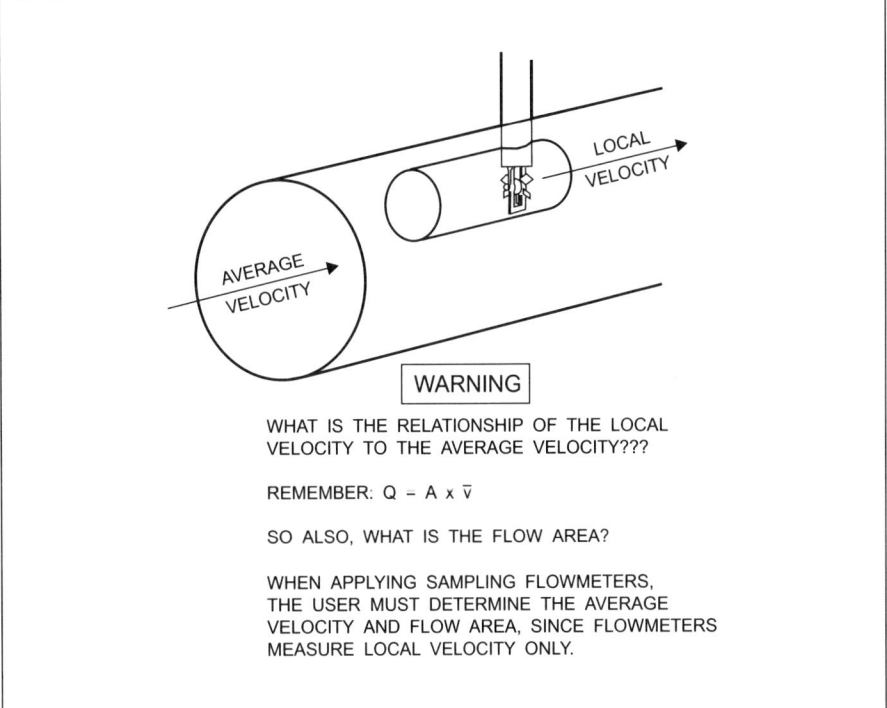

Figure 4-23. Sampling Flowmeter Installation

Bibliography

1. Buzzard, W. S., "Predicting Flow Rate System Accuracy," Instrument Society of America Symposium, Newark, DE, 1979.
2. DeCarlo, J. P., *Fundamentals of Flow Measurement*, Instrument Society of America, 1984.
3. Miller, R. W., *Flow Measurement Engineering Handbook*, Second Edition, McGraw-Hill Publishing Company, 1989.
4. Spitzer, D. W., *Industrial Flow Measurement*, Instrument Society of America, 1990.
5. American Society of Mechanical Engineers, *Fluid Meters*, 6th Edition, 1971.

About the Author

John G. Kopp is a graduate of Drexel University with a Bachelor of Science degree in Mechanical Engineering and is a licensed Professional Engineer in Pennsylvania.

He began his career as an application engineer for flow products and associated equipment including magnetic flowmeters, oscillatory-type flowmeters, and turbine flowmeters and has been a product manager for various flow products and product marketing manager for magnetic flowmeters and vortex shedding flowmeters.

A contributor to *Instrument Engineer's Handbook* (Chilton), Mr. Kopp is now a consultant for product planning, recommending content, enhancements, or extensions for flow products including magnetic, target, turbine, and vortex shedding flowmeters.

5

Linearization, Compensation, and Totalization of Flow Signals

With few exceptions, the purpose of a flowmeter is to generate a signal that is directly proportional to flow rate. Certain types of flow signals are not necessarily proportional to flow, i.e., they are nonlinear. Furthermore, if fluid properties such as density, viscosity, and temperature are not constant, the signals from many flowmeters become incorrect and no longer represent true flow.

"Linearization" of a flow signal is the characterization of a nonlinear signal to form a linear signal.

"Compensation" of a flow signal is the mathematical combining of the flowmeter output signal with other signals representing the fluid properties in such a way as to form a true flow signal.

Often one or both of these functions are required to attain a true linear flow signal.

Linearization

Many types of flowmeters produce a linear signal, meaning that the magnitude of the signal is directly proportional to the flow rate of fluid passing through the meter. This group includes turbine meters, magnetic meters, positive displacement meters, rotameters, and others. Figure 5-1 shows the output signal vs. flow rate characteristic of these meters.

Certain types of flowmeters produce nonlinear signals. The most common of these are the differential pressure meters in which the output signal varies as the square of the flow rate. Figure 5-2 shows this relationship. Notice that a change in flow at low rates produces a very small signal change. The same change in flow at high rates produces a much greater signal change. For example, a 0 to 10% flow gives 0 to 1% signal change, whereas 90 to 100% flow gives 81 to 100% signal change. Open channel meters generate signals that are exponential curves. For rectangular weirs and many flumes the exponent is approximately 1.5, and for V-notch weirs the exponent is 2.5. Figure 5-3 illustrates some nonlinear flow signal characteristics.

Flowmeters that normally produce linear or squared signals are sometimes used to measure fluids with high viscosity. The output signal, although repeatable, may not adhere to the theoretical linear or squared characteristic. These signals must also be considered nonlinear. Linearization is the conversion of a nonlinear signal to a linear one. It can be performed by a separate transducer, by components in the receiving instrument, or by components located in the flow transmitter itself (see Figure 5-4).

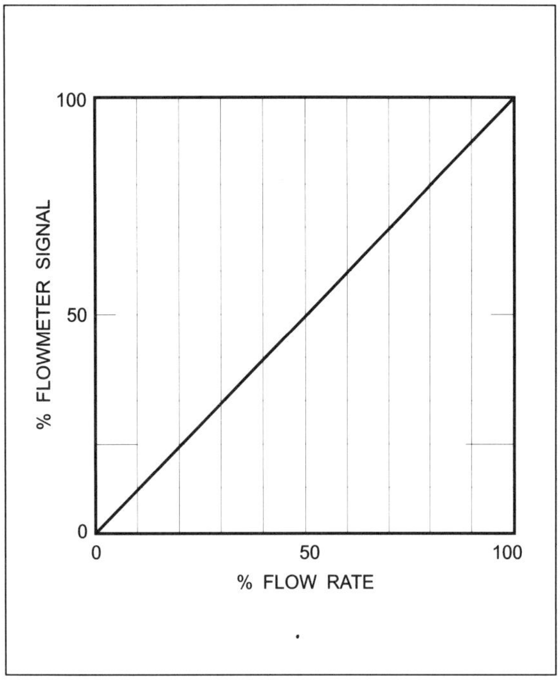

Figure 5-1. Linear Flow Signal

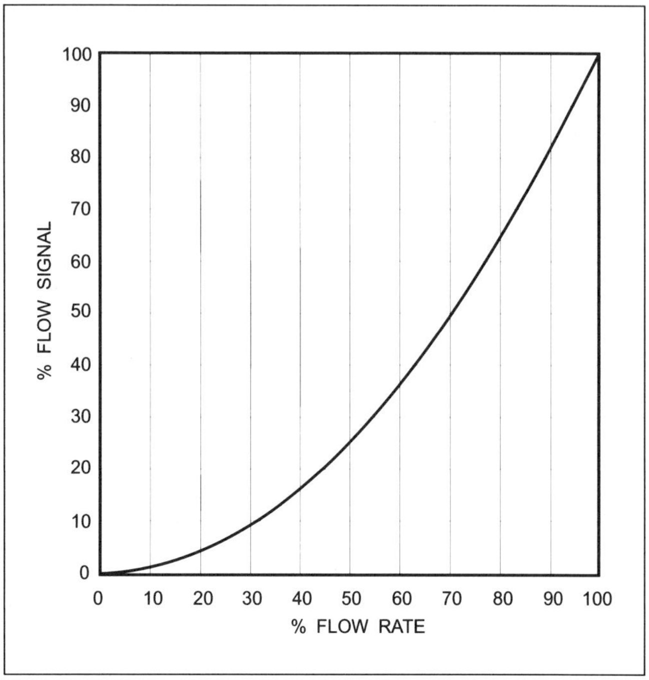

Figure 5-2. Squared Flow Signal (Sometimes Called "Square Root" Signal

A simple linearization method can be used for mechanical indicators and recorders that display the flow by a moving pointer on a fixed scale. The scales can have graduations unevenly spaced so that they read directly in flow. Some such scales are shown in Figure 5-5. Nonlinear scales have been widely used in the past, but linear scales for indicators and recorders are generally preferred.

Square Root Extraction

A separate transducer called a square root extractor is commonly used to receive a squared signal and retransmit a linear flow signal. Figure 5-6 shows the input/output relation of a square root extractor. Notice that it is just the opposite of the curve shown in Figure 5-2.

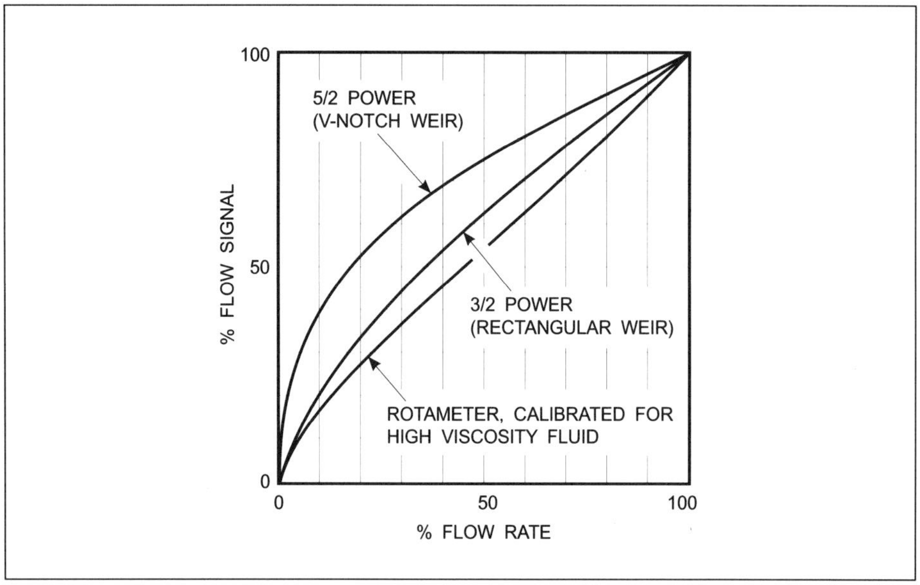

Figure 5-3. Examples of Nonlinear Signals

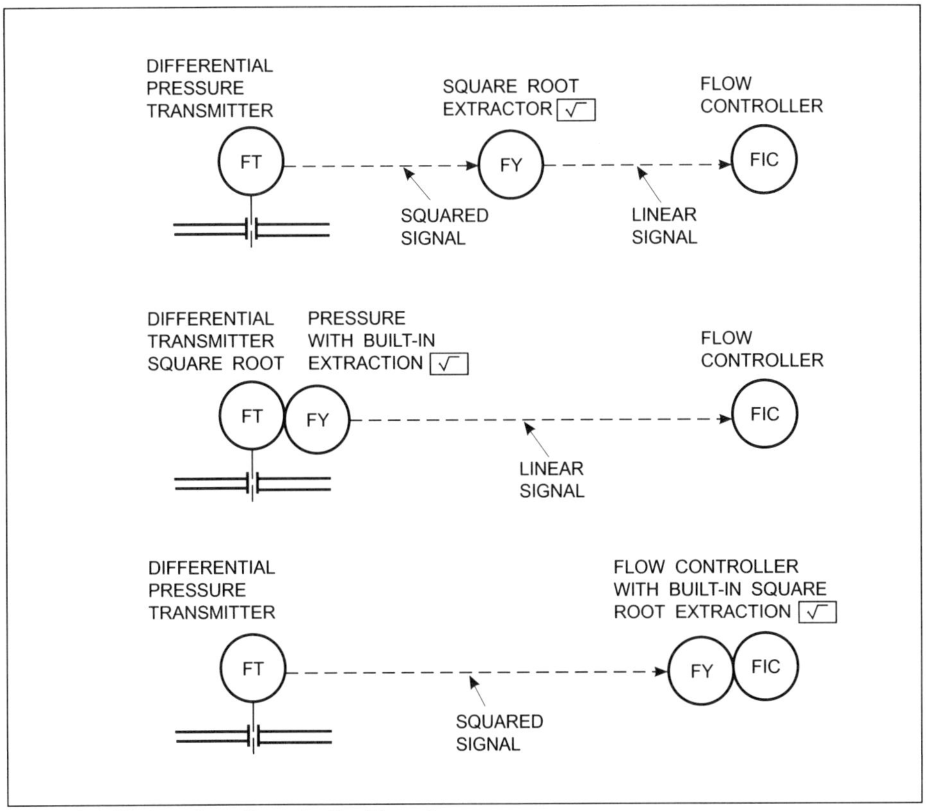

Figure 5-4. Various Locations for the Linearizer

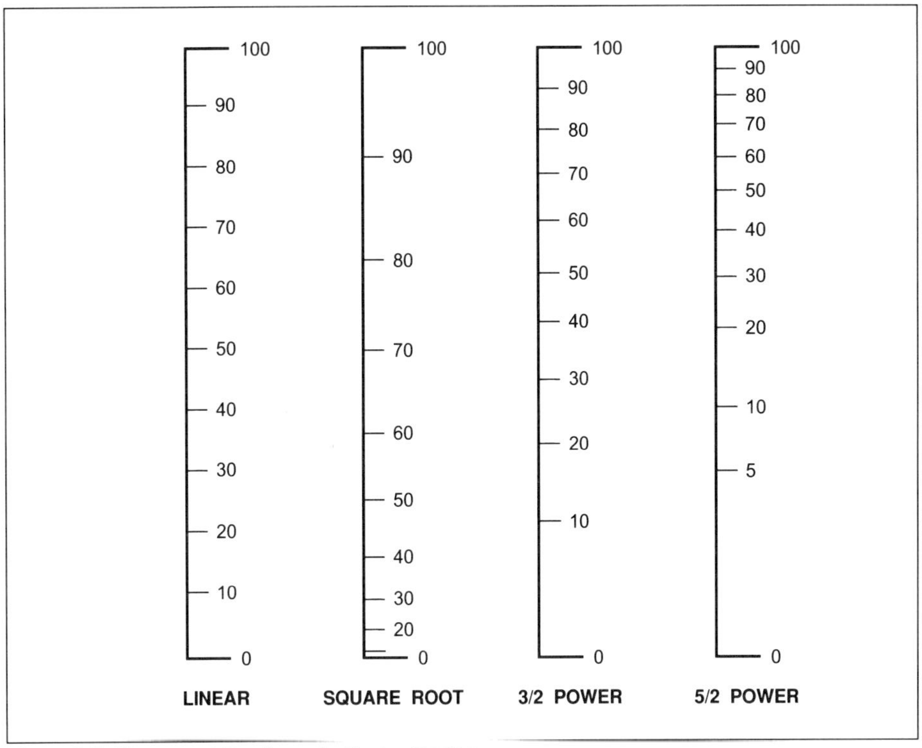

Figure 5-5. Linear and Nonlinear Indicator Scales

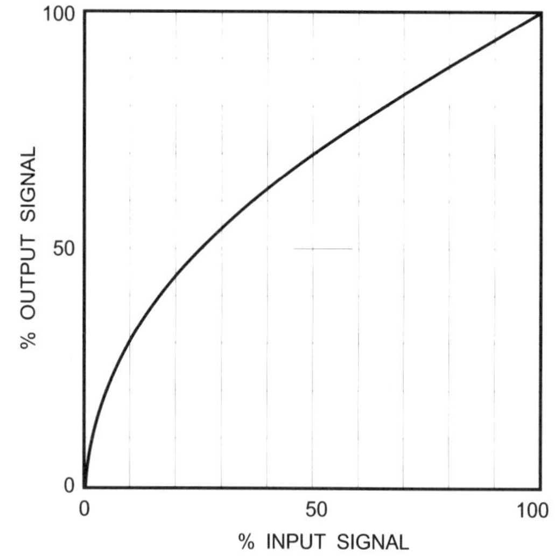

Figure 5-6. Square Root Extractor Characteristics

Signals from differential pressure flowmeters are called "squared signals" in this text. Often, however, they will be called "square root signals." Although this seems to be wrong, the intent is to describe the same characteristic. Try to avoid arguments over this matter.

The low gain of the differential pressure flowmeter at low flows requires very high gain in the square root extractor. Consequently, any error or noise in the flowmeter signal can cause large errors or fluttering in the output when at low flow rate or when flow is stopped completely. Most square root extractors include a built-in "dropout" feature that artificially forces the output to zero whenever the flow falls to approximately 7% of full scale (input = 0.5% of differential pressure). This greatly improves the performance at very low signal levels and guarantees a firm zero when flow is stopped. Figure 5-7 shows the discontinuity introduced by the dropout feature in a square root extractor.

A digital receiver accepts an input signal and immediately converts it to a number that represents the magnitude of the signal. In most designs this number includes optional square root extraction, zero dropout, and a scaling factor to allow

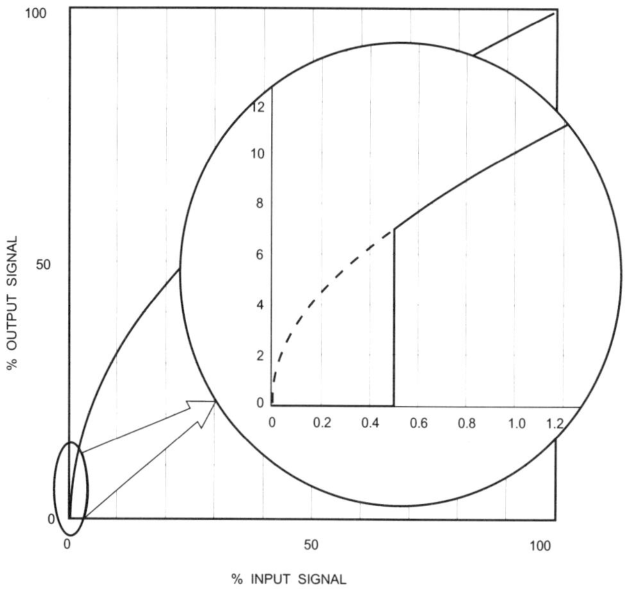

Figure 5-7. Square Root Extractor Dropout

reading in actual engineering units such as gallons per minute rather than merely "0 to 100%" of full scale. Therefore, a separate square root extractor is not required. The square root function is performed very accurately, and engineering units are immediately available for display, recording, totalizing, or other requirements. Care must be taken that an exact squared signal is generated by the flowmeter. Otherwise, further linearization must be performed to get a true linear flow signal.

Linearization of Other Nonlinear Flow Signals

Many flowmeters produce signals that are neither linear nor squared. Open channel meters have an exponential relationship between flow and signal, but other flow signals can exhibit almost any imaginable nonlinear characteristic. Often, meters are laboratory calibrated. If so, extremely accurate data are obtained, and the signal usually turns out to be somewhat nonlinear.

An analog "function generator" or "characterizer" can be used to modify the input signal and to retransmit a linear flow signal. The usual mechanism in these devices can be described graphically as locating a number of straight line segments so that they approximate the nonlinear function required. Figure 5-8 shows a 5-segment approximation of a curve. The break points must first be established and each segment is calibrated as required. There are obvious limits to the accuracy of function generators of this type, and they can be tedious to calibrate. Nevertheless, they can often be used advantageously.

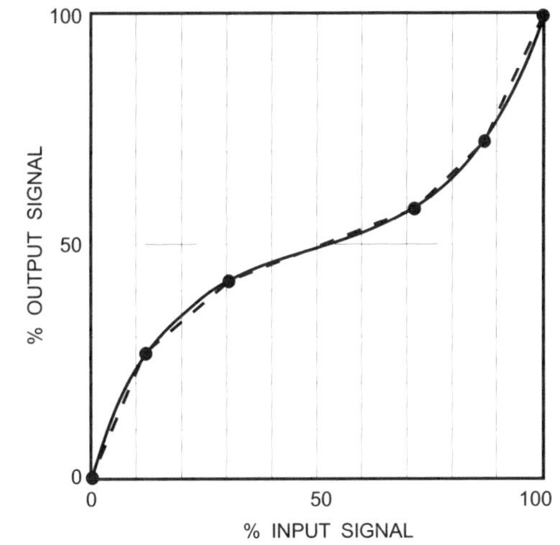

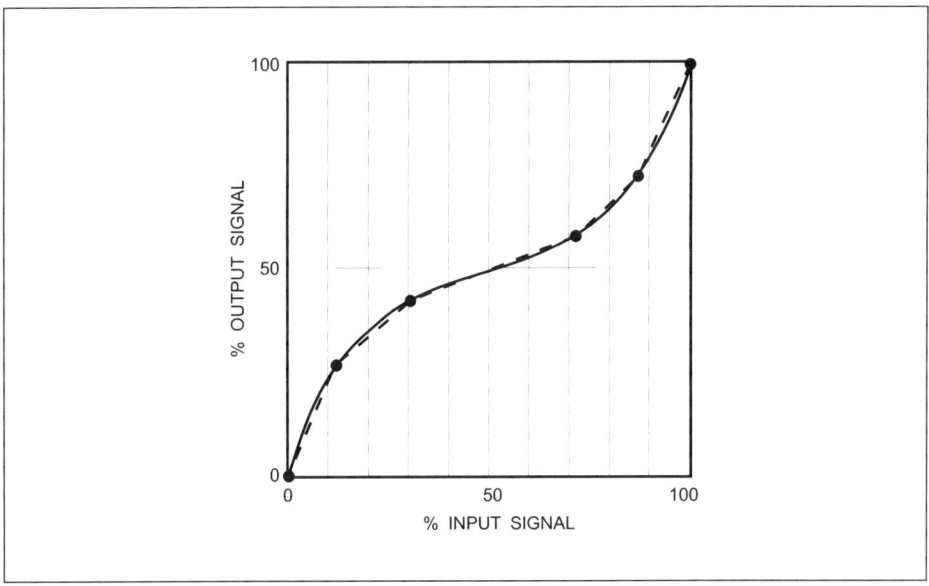

Figure 5-8. Function Generator with 5-Line Segment Approximation

Digital linearization of random curves offers more possibilities and can give better performance with much less calibration effort. Some of the linearization methods that can easily be programmed into digital receivers (indicators, recorders, controllers, computers, flow computers, etc.) are as follows:

(1) A group of straight line segments, similar to the method used in function generators, but with no manual calibration required.

(2) An exponential mathematical equation to match the flow characteristics of open channel meters.

$$\text{out} = (\text{in})^x \qquad (5\text{-}1)$$

where x might be 1.5, 1.53, 2.5, etc.

If the meter is to be used only over a certain portion of its flow range, be sure to match the desired curve very accurately in that section even at the expense of reduced accuracy at other flows, if such a compromise is necessary.

(3) A polynomial equation of sufficient order to match the required characteristic.

$$\text{out} = A + B\,(\text{in}) + C\,(\text{in})^2 + D\,(\text{in})^3 \qquad (5\text{-}2)$$

where A, B, C, and D are coefficients that can be determined by regression analysis to fit the curve closely.

(4) A hyperbolic equation.

$$\text{out} = \frac{A}{(\text{in} + B)} + C \qquad (5\text{-}3)$$

Most flow measurements require accuracy in terms of "% of rate," not "% of full scale." Therefore, the conformity of linearization must be much better at low flows than at high flows. Polynomial regression should be based on "% of rate" deviations.

> **Sometimes a very simple method can provide adequate linearization accuracy. Shift the usual straight line in/out relation of the receiving instrument so that the line does not go through the 0,0 point, but matches the nonlinear curve over the useful portion of the range. Figure 5-9 illustrates this situation.**

Don't be frightened by regression analysis. Many hand calculators can compute the coefficients of a polynomial from a group of X-Y coordinate points.

Compensation

Compensation of Flow Signals

Most flowmeter output signals represent the true flow rate only under specified fluid conditions. For liquids, variations in density or viscosity can degrade the accuracy of the signal. For gases, changes in pressure, temperature, molecular weight, and/or viscosity can ruin the meaning of the signal. Temperature not only affects fluid density and viscosity but also can cause the meter to expand or contract, causing an error.

"Compensation" means that the fluid conditions, which may vary, are measured and used along with the flowmeter signal (no longer a true flow signal) to compute the true flow rate (see Figure 5-10). An instrument to perform compensation is a "flow computer," which may be of greater or lesser complexity. To understand requirements of the flow computer, equations for mass flow rate must be derived as a function of the flowmeter signal and the various fluid properties, which can vary under operating conditions.

For this discussion, flowmeters are divided into three categories, according to their requirements for compensation systems:

(1) Linear flowmeters, which ideally measure volumetric flow at operating conditions (magnetic, turbine, vortex, positive displacement meters, etc.)

(2) Differential pressure meters, which measure pressure drop across a flow restriction (inferential flowmeters using orifices, flow tubes, Venturi tubes, etc.)

(3) Rotameters, which measure float position in a vertical tapered tube.

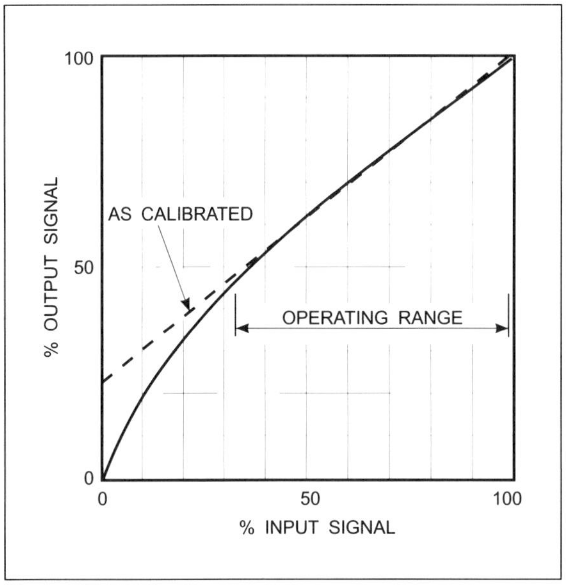

Figure 5-9. Calibration for Accuracy in a Limited Operating Range

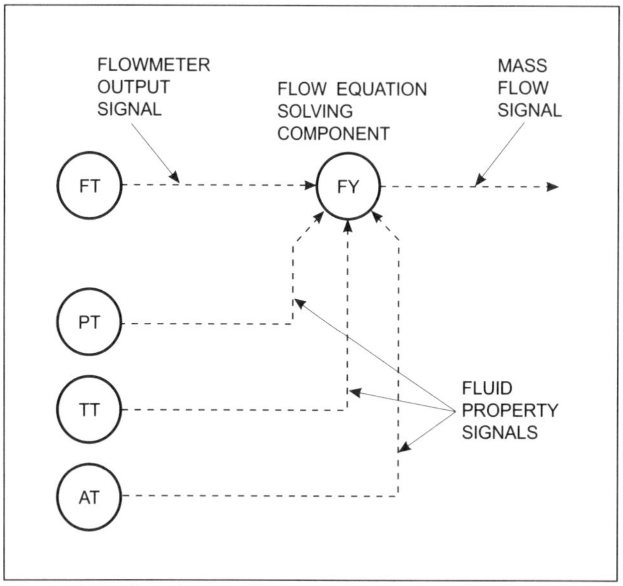

Figure 5-10. Flow Compensation System

Compensation of Linear Volumetric Meter Signals

Volumetric flow in these meters is calculated as:

$$Q = \frac{(\text{signal})}{KF_t} \tag{5-4}$$

where:

The signal is usually a pulse frequency but can be voltage or current.

Q = the volumetric flow rate.

F_t = the thermal expansion factor to account for the expansion or contraction of the flowmeter due to temperature.

K = the "meter factor," which scales the signal to the flow rate. Units of K are typically cycles/gallon, cycles/ACF, cycles/liter, etc.

Mass flow is:

$$W = \rho \times Q = \frac{(\text{signal})(\rho)}{K(F_t)} \tag{5-5}$$

where ρ is fluid density.

The signal represents true mass flow only when K, F_t, and density remain constant. Otherwise, additional measurements must be made to evaluate K, F_t, and density from which the mass flow can be computed.

K, the meter factor, is inherently constant in some flowmeters. In other types, it can change with operating conditions and is usually presented in a graph. Sometimes it is plotted against Reynolds number, but often it is plotted against the ratio of frequency divided by kinematic viscosity in units of hertz/centistoke. A typical calibration curve for a turbine meter is shown in Figure 5-11.

The K factor is often constant when measuring gas, because viscosity is low and hertz/cSt is very high.

It is possible to measure density and viscosity continuously, but, for a known fluid composition, a simple temperature measurement can be used to compute both density and viscosity indirectly.

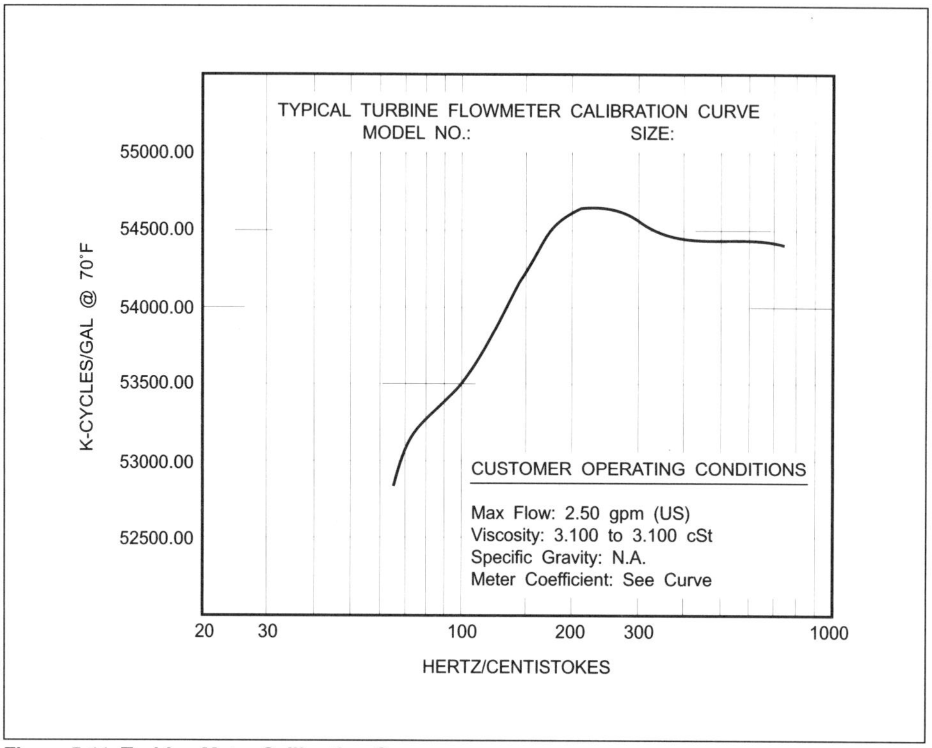

Figure 5-11. Turbine Meter Calibration Curve
(Couresty of Fischer & Porter Co.)

Compensation of Differential Pressure Meter Signals

These signals represent neither volume nor mass flow rate. They are merely the differential pressure, and they are highly nonlinear with flow. The working equation for mass flow is:

$$W = (\text{constant})\, K\, F_a\, Y \sqrt{(\text{signal})(\rho)} \qquad (5\text{-}6)$$

where:

K = a flow coefficient, which may vary with Reynolds number under certain conditions.
F_a = the thermal expansion factor.
ρ = fluid density, which can be computed from a temperature signal for liquids or from temperature and pressure signals for gases.
Y = the gas expansion factor, which applies only to gas flow and relates to the ratio of the measured differential pressure to the line pressure.

When the differential is greater than about 2% of the line pressure, the Y factor becomes a significant source of error and must be included as a variable in the compensation calculation. Y can usually be expressed as a linear function.

$$Y = 1 - (\text{constant})\left(\frac{\Delta P}{P}\right) \qquad (5\text{-}7)$$

where:

ΔP = the differential pressure
P = absolute pressure

If a gas acts as an ideal gas, if F_a is constant, and if Y is essentially equal to 1.0, then the flow equation is greatly simplified. It is:

$$W = (\text{constant}) \sqrt{\Delta P \left(\frac{P}{T}\right)} \qquad (5\text{-}8)$$

where T is absolute temperature.

Self-contained components are available to perform the mathematical functions under the square root sign in Equation (5-8). These "multiplier-dividers" receive three input signals—the flowmeter signal, an absolute pressure signal, and a temperature signal. The output of the multiplier-divider is then sent to a square root extractor for linearization, producing the linear mass flow signal (see Figure 5-12).

Digital recorders and controllers can perform these tasks and also include compensation for Y factor, F_a factor, and K if necessary.

If the pressure is high enough so that variations in the barometric pressure are insignificant, a gage pressure transmitter can be used instead of the absolute pressure transmitter. A constant barometric pressure is assumed. When above about 50 psig, it is usually safe to use this cost-saving approach.

Compensation of Rotameter Signals

The flow equation for a rotameter is somewhat like that for a differential pressure meter. It is, however, more complex because of the buoyancy of a liquid on the float. Since gases have a negligible density compared to that of the metal float, there is no buoyancy, and the effect of gas density is exactly the same as its effect in a differential pressure meter. Equations are:

$$\text{Liquids:} \quad W = (\text{constant})\, K\, (\text{signal}) \sqrt{(\rho_f - \rho)\left(\frac{\rho}{\rho_f}\right)} \qquad (5\text{-}9)$$

$$\text{Gases:} \quad W = (\text{constant})\, K\, (\text{signal}) \sqrt{\rho} \qquad (5\text{-}10)$$

where ρ is the fluid density and ρ_f is the float density.

Notice that the density is under the square root sign, but the signal is not. The signal is inherently linear, not squared, but the compensation for density must include the square root function. The thermal expansion factor is usually considered a constant for most rotameter applications and is not shown in the equation. The meter factor, K, is not necessarily constant. For very small meters or for high viscosity fluids, K can require compensation.

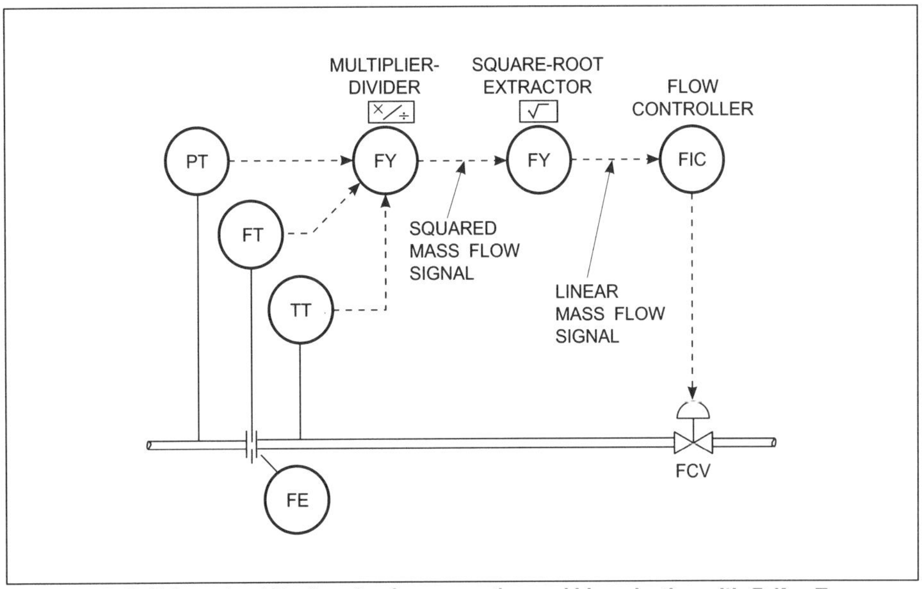

Figure 5-12. Self-Contained Devices for Compensation and Linearization with Orifce-Type Flow Measurement of Gas if Gas is Ideal, Y = 1.0, and F_a = 1.0

Measurement of the Fluid Properties Needed for Flow Compensation

LIQUID DENSITY MEASUREMENT

The density of liquids over a limited range of temperature very often has a linear decline with rising temperature. Only rarely does pressure affect a liquid density. Various liquid density transmitters are commercially available, but it is common practice to measure temperature and then calculate the density indirectly from the temperature signal. The equation becomes:

$$\rho = A - B(T) \tag{5-11}$$

where A and B are constants for a specific liquid in the range of operation, and T is temperature.

If the composition of the liquid can change, the density cannot be predicted from the temperature, and a measurement of density is required.

GAS DENSITY MEASUREMENT

Gas density can be measured directly, but it is almost always computed indirectly from pressure and temperature signals. If the composition of the gas is variable, it may be necessary to include a specific gravity transmitter. More commonly, the user can manually introduce a value for specific gravity (SG) or molecular weight (MW) whenever it is known that there has been a significant change.

If operating conditions remain in a range of pressure and temperature where the gas acts as an ideal gas, the equation is:

$$\text{Ideal Gas Density} = \frac{PMW}{RT} = (\text{constant})\left(\frac{P}{T}\right) \tag{5-12}$$

If the gas does not act as an ideal gas, the Z-factor must be included in the density equation. The Z-factor can be computed first from temperature and pressure and then introduced into the density equation. A simple approximating equation for gas density can be made to avoid two separate computations and give quite accurate results over moderate ranges of pressure and temperature. It is:

$$\text{Non–Ideal Gas Density} = (\text{constant})\left(\frac{P - P_0}{T + T_0}\right) + (\text{constant}) \tag{5-13}$$

where P_0 and T_0 are suitably chosen constants.

Minimum density will occur at lowest pressure and highest temperature. In order to measure the required maximum flow rate, the flowmeter must be sized large enough to handle maximum flow at minimum fluid density.

It is very common to encounter non-ideal gas situations. Many flowmetering systems measure incorrectly because this condition has been ignored. The common examples of non-ideal gases, steam and carbon dioxide, are nearly always non-ideal under typical operating conditions.

THERMAL EXPANSION FACTOR, F_t AND F_a

The physical dimensions of the flowmeter can change enough with temperature to significantly affect the measurement. Therefore, if measuring fluids whose temperature can change greatly, compensation for expansion must be made. Example equations, if the meter is calibrated at 60°F are:

$$\text{For linear volumetric meters} \quad F_t = 1 + \beta(60 - T) \tag{5-14}$$

where T is temperature in °F, and β is the cubical coefficient of expansion of the meter body material. (β is about 2.6×10^{-5} for 316 ss.)

$$\text{For differential pressure meters} \quad F_a = 1 + 2\alpha\,(T - 60) \qquad (5\text{-}15)$$

where T is temperature °F, and α is the linear coefficient of expansion of the flow restriction material. (2α is about 1.8×10^{-5} for 316 ss, but is not constant over very wide temperature changes.) For "impact" or "target" flowmeters, use alpha instead of 2α. If the pipe for an orifice flowmeter is of a material different from the orifice plate, a second thermal factor can be included.

Coefficients of cubical expansion and linear expansion are often shown in texts and handbooks. The coefficient of cubical expansion is approximately three times the linear coefficient.

FLUID VISCOSITY MEASUREMENT

To calculate Reynolds number, the viscosity must be known. In the case of gas flow, the Reynolds number is usually so high that its effect is negligible and viscosity is of no concern. Many types of liquid flowmeters are greatly affected by Reynolds number. Various viscosity transmitters are available, but when the fluid composition is constant, it is common to measure temperature and compute viscosity indirectly. A typical equation for viscosity is:

$$\text{Liquid Viscosity} = Ae^{B(T)} \qquad (5\text{-}16)$$

where A and B are appropriate constants, and T is temperature.

MEASURING FLOW OF DISSOLVED OR SUSPENDED SOLIDS

In some applications the desired flow signal should represent the flow rate of solids in a solution or suspension, regardless of the amount of liquid. Analyzers can be used to determine concentration of some materials. Measurements of density and temperature can be combined to determine concentration indirectly. Once the concentration has been established, it is multiplied by the flow rate of the entire solution or suspension to obtain mass flow of the solids in the stream.

Does a Flowmetering System Require Compensation?

How bad would it be if compensation were omitted? Table 5-1 shows the magnitude of flow error in an uncompensated flow signal caused by changes in fluid properties. The error is expressed in "% of rate" in each case. Values are approximate and can vary for different fluids.

Flow Computers

Due to the advent of the microprocessor, flow computers were developed to solve the various linearization and compensation equations described above. They are digital devices and, therefore, are very accurate and versatile. There are two basic types of flow computers. One incorporates most of the operations normally encountered and requires the user only to enter various numbers and select the desired functions before putting it into service. The other type is not preloaded with certain functions but allows the user to enter a program to perform any conceivable flow equation. Each has its advantages. The first type is easier to start up, provided the user is familiar with the available functions. The second type is much more versatile but requires that the user be more familiar with the programming language and flow equations.

At present the trend is toward digital receivers, recorders, controllers, and distributed control systems. Most of these can perform the mathematical functions required for linearization and compensation of flow signals at no extra cost. These digital devices can be configured by the user to add, subtract, multiply, divide, square root, drop out, linearize, totalize, solve for logarithms and exponents, etc. This provides great flexibility and interchangeability, and eliminates the requirement for intermediate instruments.

Table 5-1. Effect of Fluid Properties on Flowmeter Accuracy

Change	Volumetric Meters		Diff. Pressure Meters		Rotameters	
	Liquid	Gas	Liquid	Gas	Liquid	Gas
Density up 1% of value	-1	-1	-0.5	-0.5	-0.4	-0.5
Temperature change:						
up 10 °C at -100 °C	*	+6	*	+3	*	+3
up 10 °C at 20 °C	+0.2*	+3.4	+0.1*	1.7	+0.1*	+1.7
up 10 °C at 200 °C	+0.6*	+2	+0.6*	+1	+0.6*	+1
Pressure change:						
up 1 psi at -10 psig	0	-20	0	-10	0	-10
up 1 psi at 0 psig	0	-7	0	-3.5	0	-3.5
up 1 psi at 35 psig	0	-2	0	-1	0	-1
up 1 psi at 85 psig	0	-1	0	-0.5	0	-0.5
F_a due to temperature up 100 °C	-0.2	-0.2	-0.2	-0.2	-0.2	-0.2
Meter Factor Changes:						
up 1%	+1	+1	+1	+1	+1	+1

* Values shown are for water; may be much higher for other liquids.

The flow computer remains a very useful instrument for data logging of flow information and for applications where there are no other digital receivers needed.

Totalization of Flow Signals

Flow signals should be regarded as flow rate signals. This can be compared to the speedometer of an automobile. The pointer, or the digital speed indicator, goes up as the car goes faster and goes down as the car goes slower. When the car stops, the speedometer goes to zero. When the flow rate stops, the flow signal goes to zero.

On the other hand, the mileage counter (odometer) reads the total accumulated miles driven since the car was new. It offers no information about how fast the car has been going. When the car stops, the odometer does not go to zero. It merely stays at its last reading. When the car starts up again, the odometer will start counting upward again.

Often, the primary purpose of flow measurement is to display the accumulated amount of fluid that has passed through the meter. This permits daily or monthly readings for studying process usage of chemicals or raw materials as well as finished products created. Monthly billing for water and gas would be impossible without flow instruments to provide the accumulated totals. It is clear that devices are needed that can receive a flow signal as input and generate an output representing accumulated flow (see Figure 5-13). These instruments are called "totalizers" or, sometimes, "integrators."

The word "totalizer" is occasionally misused to mean a device that adds two signals together and retransmits their sum. Properly, such instruments should be called "summers" or "adders."

Many of the inherently volumetric flowmeters transmit electrical pulses at a frequency proportional to flow rate. The receiving instrument must convert the frequency into a meaningful display of flow rate. It would seem that totalization were only a matter of counting the pulses. Unfortunately, the signal is often slightly nonlinear, and the pulses are not scaled directly to gallons, liters, or other volumetric units. Compensation for density may be required. It is true, however, that without these conditions, totalization of a frequency signal can be quite straightforward. Figure 5-14 shows such a system.

The most common situation is to have a flow rate signal that must be totalized (or integrated) to generate the totalized value as shown on an odometer counter. The signal may be a continuously variable signal, or it may be a pulse frequency. The totalizer has a very accurate internal timing system, and the input signal is sampled at regular intervals. Depending on the magnitude of the flow signal, a proportionately sized increment is added to the accumulator output. The sampling periods are short, i.e., 5-10 per second, and the resulting accuracy can be excellent. The accuracy of most modern totalizers is often better than the accuracy of the flow signal itself.

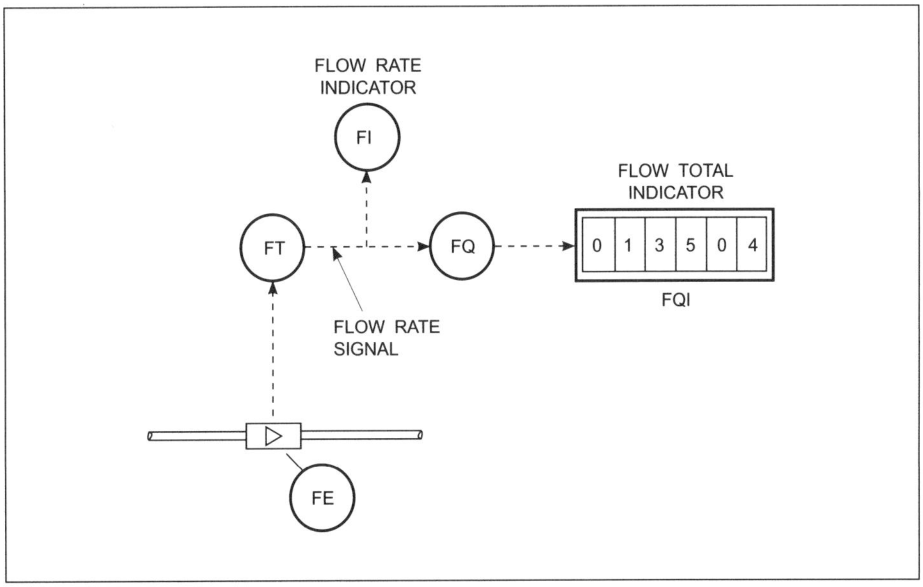

Figure 5-13. Flow Measurement System with Totalization

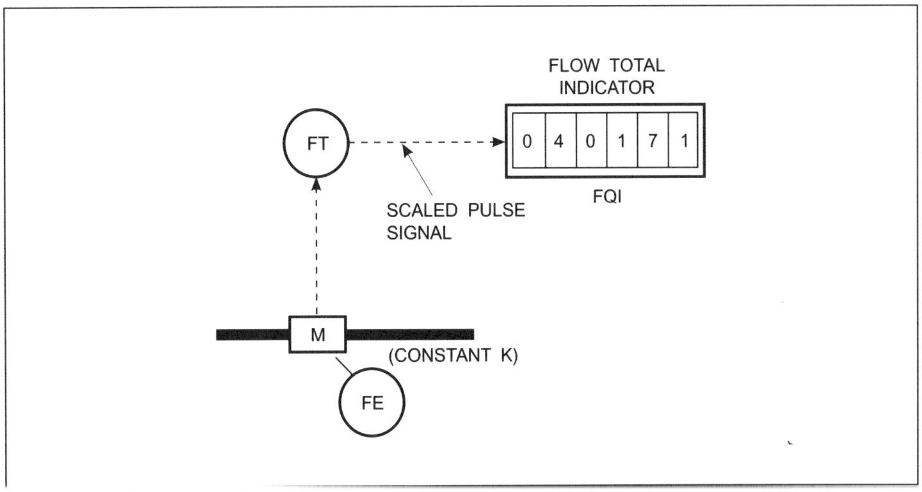

Figure 5-14. Simplified Totalization System if Meter Factor is Constant and Scaled Pulse Output is Provided

Totalizers are often built into recorders or controllers. A common requirement is to retransmit scaled pulses to remote mounted counters or to predetermining batch counters. Conversely, batch control capability is often included in the totalizer itself. Totalizers are not limited to counting upward. Some flowmeters can measure flow in either direction, and some totalizers must be able to count up or down.

Dropout Feature for Totalizers

When the flow is shut off completely, a volumetric flowmeter will stop sending pulses. The blades of a turbine meter and the rotor of a positive displacement meter stop turning, producing a perfect zero frequency signal. However, meters that transmit continuously varying signals can be slightly in error at zero flow. A totalizer could then continue to count very slowly. If a flowmeter is shut off for a week, it is possible that a large excess count could appear. Therefore, all totalizers should include a zero dropout feature, which assures a perfect zero when the input signal drops below some specified minimum value.

Accuracy of Flow Totalizing Systems

It is very difficult to evaluate the accuracy of a totalizing system. The main source of error is usually the flowmeter itself. In terms of "% of rate," most flowmeters are more accurate at high flow than at low flow. The accuracy of the accumulated total, therefore, depends on what portion of the total passed through the meter at high flow and what portion at low flow. A system could run at very low flow for ninety percent of the time. This would accumulate a small amount at low accuracy. Then, for ten percent of the time, the flow is very high, and a large amount is accumulated at high accuracy. The small amount at low accuracy becomes insignificant and the overall total is very accurate.

Scaling Factor for the Totalizer

It is important to choose units for the counter so that the reading is easily understood. It is best to choose factors such as 10, 100, 1000, etc. If, for example, one count equals 1000 pounds of material, the totalizer would bear a label stating " x 1000 = lbs." There is no particular advantage to counting at high speeds. The accumulated number of counts should provide adequate resolution between readings. If the counts are very frequent, it becomes difficult to read the counter, and the counter can fill up and "roll over" to zero too often.

Additional Applications for Totalization

Totalization can be used to derive the average flow rate over a specified time period. The totalizer is set to zero at the start of each period. The counter is not scaled to read total flow as is normally done; instead, it is set to read flow rate divided by the selected time period. At the end of each period the counter will, therefore, read the exact average flow rate for that period. This number is transferred to a display counter, and the totalizer is immediately reset to zero for the next time period.

A straightforward timer can be constructed from a totalizer. The input signal is made to be a constant rather than a flow rate signal. Then the totalizer is scaled so that the counter reads in seconds, minutes, hours, etc. (see Figure 5-15).

A "running time average" of flow rate can be provided by using two totalizers. One is a conventional flow totalizer; the other is a timer as described above. The output of the first is continuously divided by the output of the second totalizer. When plant operation is shut down, both totalizers are stopped but not reset. The result is a constant display of the average flow rate since the totalizers were last simultaneously set to zero. This can be a very valuable piece of information when a process flow is shut off at times, and the operator wants to know the average flow rate during operation regardless of downtime (see Figure 5-16).

Control of batch addition is commonly performed with a flowmeter and a totalizer operating an on/off valve. The totalizer is initially set to zero. The desired batch quantity is in another counter. When the "start" button is pushed, the valve opens and the totalizer starts to totalize the flow quantity. That total is compared to the target quantity, and, when the two are equal, the valve is automatically closed. Many refinements to the system can be made. Gradual opening of the valve is used to prevent a sudden surge in flow. Often, when near the end of the batch, the valve is partially closed to provide a dribble flow, which can be stopped very accurately as the batch is actually finished. This also prevents violent water hammer caused by sudden closing of the valve (see Figure 5-17).

Totalizers are used in some blending systems. A simple flow ratio control system is quite capable of mixing one flow stream with another at a constant ratio. If the controlled line happens to be too high for a little while, the mixture will be off ratio for that period but will return to the correct mix as the controller recovers. For a continuous process, this is the best action; only a small portion of the product was off specification during the abnormal period. If, however, the mixed

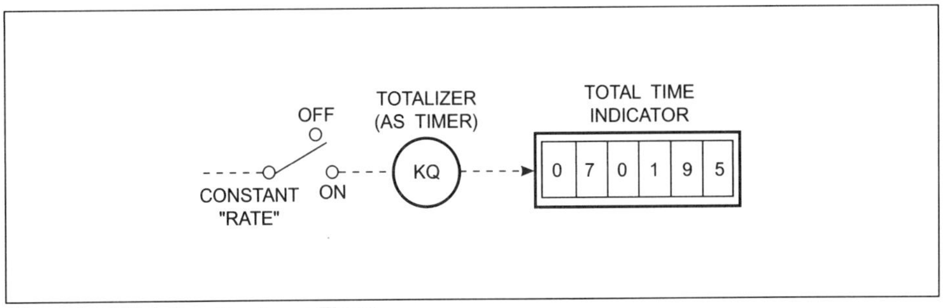

Figure 5-15. Totalizer Used as a Timer

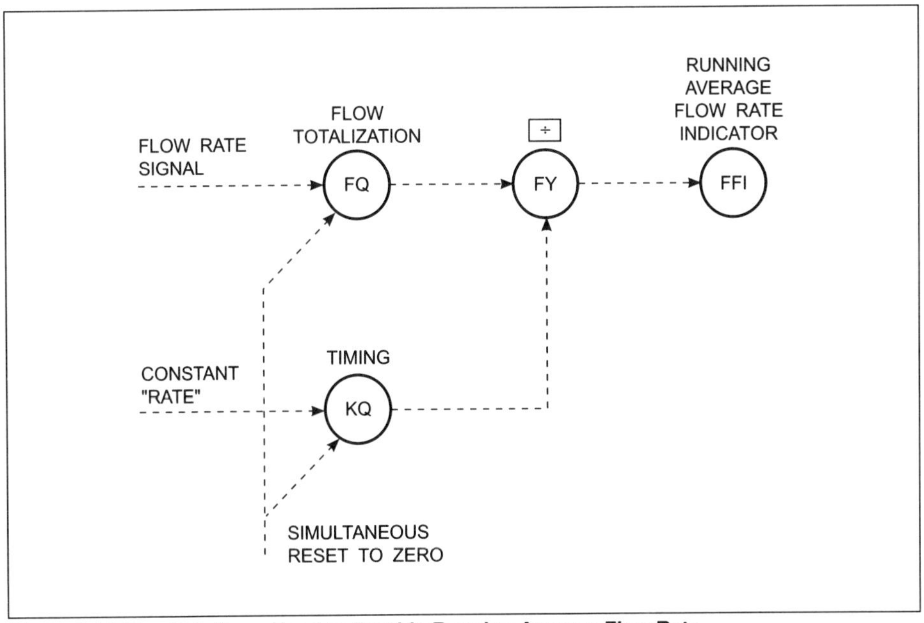

Figure 5-16. Two Totalizers Used to Provide Running Average Flow Rate

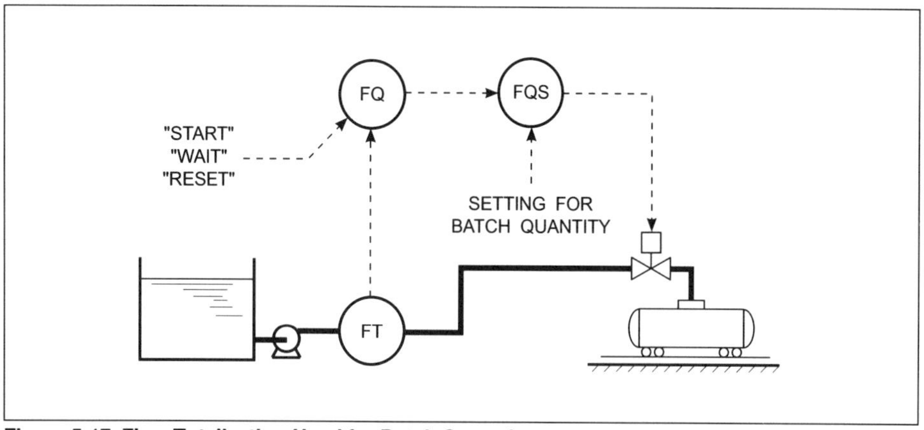

Figure 5-17. Flow Totalization Used for Batch Control

materials are being fed into a holding tank, and the aim is to produce a tankful of the proper mixture, the temporary off-spec material will destroy the accuracy of the mix in the entire tank. For these cases, flow rates are totalized, and the ratio of the totals is controlled. If a control upset occurs, the totalized values will reflect that fact, and the system will return the totalized ratio to its proper value. Figure 5-18 shows totalized ratio control for batch blending.

Energy flow is often totalized to compute the heat flow into a process stream as it passes through a heat exchanger. Heat flow is computed as:

$$Q = m \Delta T c_p \qquad (5\text{-}17)$$

where:

Q = heat flow
m = mass flow of fluid
ΔT = temperature difference (out - in)
c_p = heat capacity of the fluid

Often, the heat capacity of the fluid is known and constant, so the system requires only two temperature transmitters and one flowmeter (see Figure 5-19).

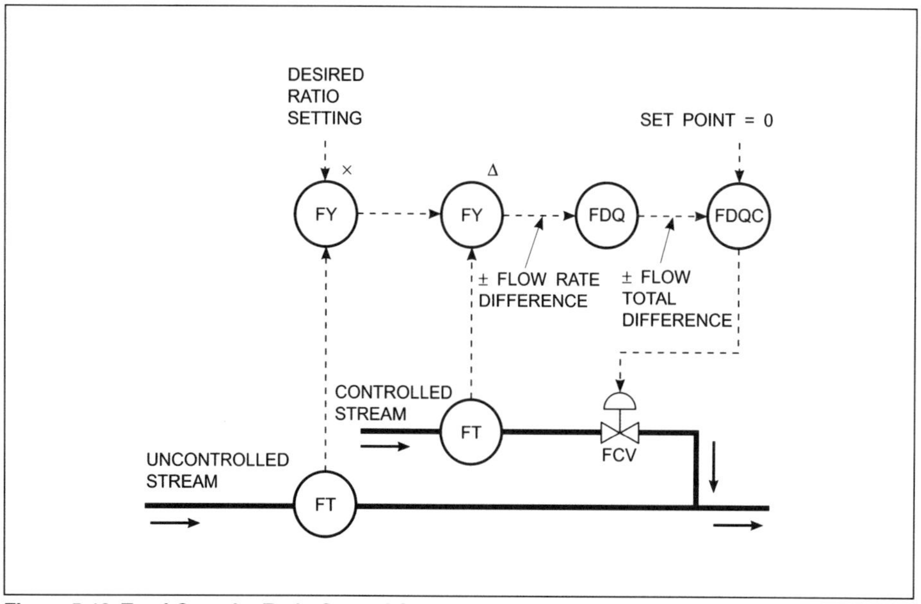

Figure 5-18. Total Quantity Ratio Control System

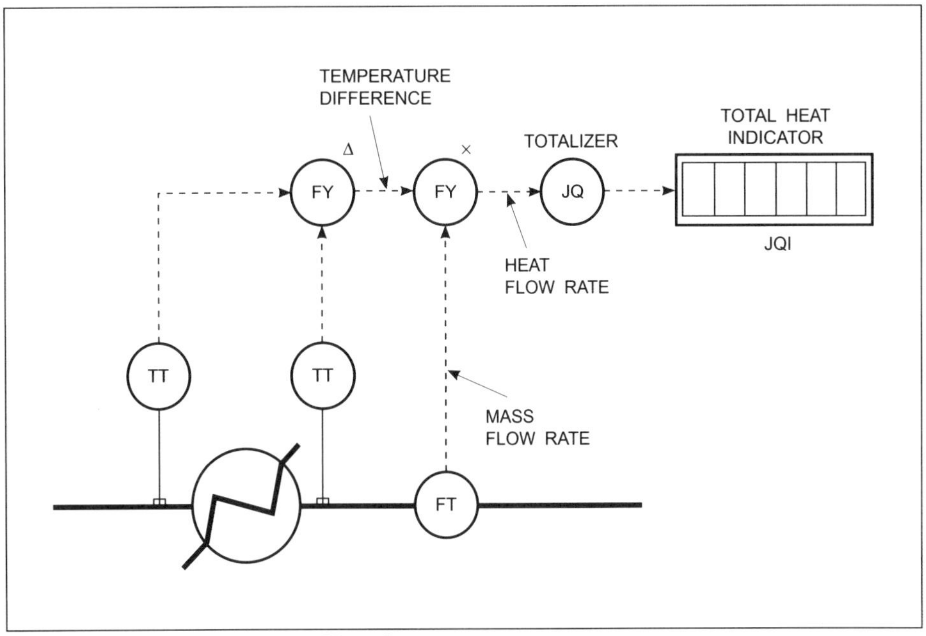

Figure 5-19. Totalization in a Btu Meter System

About the Author

Bill Buzzard served with the 42nd Infantry Division in Europe from 1944 to 1946. After earning his degree in Chemical Engineering from the Massachusetts Institute of Technology, he spent 5 years in pulp mill operation for Westvaco, Inc. He was with Fischer & Porter Co. for the next 37 years as Product Manager, Application Engineer, Project Manager, Chief Technologist, and Manager of the Application Department before his retirement in 1989.

A Charter Member of the Delaware Valley ISA Section, Mr. Buzzard has presented numerous technical papers and ISA training sessions. He is the author of the *Fischer & Porter Flowmetering Orifice Sizing Handbook.*

6

Field Calibration

Calibration can be defined as the comparison of a measuring instrument, with specified tolerances but an undetermined accuracy, to a measurement standard of known accuracy for the purpose of determining and/or eliminating by adjustment any out-of-tolerance condition.

The use of an uncalibrated instrument creates the potential for incorrect measurement and resultant erroneous conclusions and decisions. It is calibration that provides confidence in a measurement and assurance that an instrument has the accuracy required to maintain a product or process within specification.

Precise measurement depends on the use of properly calibrated instruments.

Calibrations can be performed in any location. In general, this chapter will discuss those calibrations performed by the user, rather than those performed by the equipment manufacturer or an independent laboratory. The proper overall approach towards calibration applies, however, no matter where the procedure is physically performed.

General Calibration Requirements

Scope

A specific calibration can range from a simple dimensional check to extensive measurements of multiple variables. Accordingly, before the details of a calibration can be planned, an initial step is to determine the appropriate scope. The first decisions that need to be made are (1) which variables should be measured and (2) what accuracy must be maintained. These decisions need to be based not only on the accuracy desired from an application standpoint, but also upon the capabilities of the device being calibrated.

The device to be calibrated must exhibit sufficient range, resolution, and freedom from drift to enable transfer of the desired level of accuracy from the calibration standard used. It is normally a futile effort to try to make a highly accurate instrument out of a low-precision device through calibration.

Types of Calibration Errors

Some element of error exists in all measurements, no matter how carefully they are conducted. The magnitude of the error can never be exactly determined, but the experimental uncertainty — the possible value of the error — can be estimated. The possible error in a measurement or calibration is referred to as the measurement uncertainty. In practice, however, it is often referred to simply as the error.

In order to determine the validity of a calibration, those factors that may cause uncertainty or error must be evaluated and the overall measurement uncertainty estimated. As an aid in this estimate, errors are commonly classified in two broad types. Systematic errors are fixed errors that cause repeated readings, with the same input, to be in error by roughly the same amount due to some unknown reason. Random errors are those in which repeated readings, with the same input, result in different responses. In general, large random errors may be due to the failure to control a variable that affects the measurement, for example, the supply voltage to the equipment. Calibration systems are often qualified by taking several repeat measurements before formally recording data. If the measurements repeat, the system is assumed accurate. This method runs the risk of overlooking a large systematic error.

From its definition, it can be seen that a good estimate of random error can be obtained simply by taking repeat runs. Systematic errors normally must be determined by more analytical methods. The systematic error of calibration equipment or standards is normally specified by the manufacturer.

Accuracy of Standards

Since calibration is a comparison, the standard used in any calibration should have a greater accuracy (lower measurement uncertainty) than the device being calibrated. The relationship between the two can be expressed as a ratio, commonly referred to as the accuracy ratio or, alternately, as the accuracy-to-uncertainty ratio. This ratio is calculated by dividing the accuracy tolerance of the device being calibrated by the specified uncertainty of the measurement standard system. The ratio can be based on absolute numbers or percentages, provided that any percentages that are expressed as a percent of full scale are referenced to the *same* full scale value. When multiple standards are used, one method of estimating the system uncertainty is to sum each possible error in the system. Realistically, the probability that this will occur is very low, as all errors would have to be at their maximum value and in the same direction. A more common method for estimating system uncertainty is to calculate the root mean square average. The formula for RMS system uncertainty is

$$\pm [\Sigma(E_i)^2]^{1/2}$$

where E_i is the error associated with each element in the system.

A 4:1 accuracy ratio should be maintained.

Many calibration procedures supply an estimate of the accuracy ratio. Traditionally, an accuracy ratio value of 10:1 has been considered desirable, but the tight tolerances of many common modern instruments generally preclude achieving this value. The National Conference of Standards Laboratories (NCSL) Recommended Practice for Calibration Procedures [Ref. 1] recommends that at least a 4:1 ratio be maintained and that, if it cannot, either the accuracy tolerance of the device under test be increased or improved statistical approaches toward generating and analyzing the data be undertaken. In a similar manner, Military Standard MIL-STD-45662A requires a 4:1 ratio be maintained on all calibrations performed for the U.S. Government unless otherwise specified in the contract requirements.

Assume an instrument is to be calibrated to a tolerance of 1% of reading. The accuracies of the three standards to be used in the calibration are:

Time Standard = 0.05% of Reading

Volume Standard = 0.20% of Reading

Temperature Standard = 0.10% of Reading

The accuracy ratio is: $$\frac{1}{[(0.05)^2 + (0.20)^2 + (0.10)^2]^{1/2}} = \frac{1}{0.23} = 4.4$$

Since the ratio is greater than four, the standards have sufficient accuracy to permit calibration of the test instrument to the desired ± 1% of reading tolerance.

A flowmeter is supplied to a calibration laboratory with a request that it be calibrated to a tolerance of ± 1% of reading over a range of 5 to 200 gpm.

The accuracies of the standards to be used in the calibration are:

Time Standard = 0.05% of reading

Master Flowmeter 1 = 0.2% of reading, 50–500 gpm

Master Flowmeter 2 = 0.05% of full scale, 5–50 gpm

From 50 to 200 gpm, the accuracy ratio is:

$$\frac{1}{[(0.05)^2 + (0.2)^2]^{1/2}} = \frac{1}{0.206} = 4.85$$

At 5 gpm, the accuracy of Master Flowmeter 2, as a percent of reading, is $\frac{0.05 \times 50}{5} = 0.5\%$ **and the accuracy ratio is:**

$$\frac{1}{[(0.05^2 + (0.5)^2]^{1/2}} = \frac{1}{0.503} = 1.99$$

The low accuracy ratio at the lowest flows is a result of the accuracy statement of Master Flowmeter 2. Good laboratory practice would dictate using a more accurate master for this portion of the calibration or opening the tolerance of the test meter over part of its range.

General Methods of Flow Calibration

In general, flow measurement devices are calibrated by three methods: (1) a "wet" calibration using actual fluid flow; (2) a dry calibration using flow simulation by electronic or mechanical means; or (3) a measurement check of physical dimensions and use of empirical tables relating flow rate to these dimensions.

The simplest, lowest cost method of calibration that should be selected is one that is capable of achieving the required accuracy tolerance for the particular application. Calculating the accuracy ratio is the key to proper selection. If the ratio is found to be low, either the tolerance should be opened or a more accurate method should be used.

Many measurements are made using a combination of several instruments acting together. It is often appropriate to calibrate different portions of such a system by different methods. In the case of a flowmeter with a transmitter and an indicator, a common practice is to wet calibrate the flowmeter and separately dry calibrate the transmitter and indicator. The flowmeter itself can often be further separated into a wetted primary (the portion generating the flow signal) and a flow secondary (the portion providing the signal processing).

Whenever portions of a system are separately calibrated, adequate accuracy ratios should be maintained for each calibration and the overall system accuracy calculated by the method previously described.

Wet Calibration

A wet calibration, using actual fluid flow, normally provides the highest calibration accuracy for a flowmeter and is used where accuracy is a prime concern or when the form of the meter does not lend itself to other methods. Precision flowmeters are usually wet calibrated at the time of manufacture. After some period of service, they are often removed by the user and recalibrated. The actual interval between these calibrations varies widely depending on the user's experience with the meter and the importance of the measurement. The recalibration can be performed at the flowmeter manufacturer's facility, in an independent laboratory, or at the user's site when adequate facilities are available. When the calibration is performed at an outside facility, the user should specify the flow range to be calibrated, the number of points to be taken, and the fluid viscosity (if other than water at room temperature is to be used). If the flowmeter is used for billing or other accounting purposes, it may be desirable to obtain both an "as received" and an "after adjustment" calibration. The "as received" results enable evaluating what magnitude of error existed in the meter prior to calibration. If this error is found to be large, the user should evaluate the impact of the error on prior measurements and determine whether additional action is required.

The procedures to be followed in calibrations by outside laboratories are normally left to the discretion of the laboratory but can be specified if desired. A system calibration, where all elements of the flow measuring system are calibrated together, can be specified when accuracy is critical.

Wet calibrations are usually performed with water, air, or hydrocarbon fuels using one or more of four basic standards: weight tanks, volumetric tanks, pipe provers, or master flowmeters.

WEIGHT AND VOLUME STANDARDS

Three methods of calibration are in general use when a weight or volume tank is used as the standard. In the static calibration method, the flow is quickly started to begin the test, held steady during the test, and then shut off at the end of the test. The totalized reading from the flowmeter is compared with the weight or volume collected and the meter performance is calculated. A static system operates best with flowmeters that have minimum sensitivity to low flow rates. It does not give optimum results with high-performance digital output meters, such as vortex flowmeters, because of errors obtained during the short periods of low flow at the beginning and end of the test.

A weight or volume standard can also be used in a completely dynamic mode. In this method, the flow is set at a steady rate before the start of the test. The flowmeter reading and initial weight or volume are read simultaneously to start the test and again, after the desired collection period, to end the test. Completely dynamic systems are limited by speed of response considerations and the general difficulties encountered with "on the fly" readings. Because of these problems, this method is not widely used.

HYBRID DYNAMIC START-AND-STOP, STATIC READING SYSTEMS

Hybrid dynamic start-and-stop, static reading weight and volume systems have been developed to provide more accurate liquid calibrations than purely static or dynamic systems. In such systems, the desired test flow rate is first obtained while the flow is diverted past the weight or volume flow standard. The test run is initiated by diverting the flow into the standard and completed by diverting it out of the system. The flow signal (which should be digital) is gated on and off by the diversion. The weight or volume is then read after an appropriate settling time and compared with the totalized flowmeter reading. Figure 6-1 illustrates the basic elements of a weighing standard of this type.

The key to the performance of a dynamic start-and-stop static reading system is the design of the flow-diverter valve that switches the flow in and out of the standard. In a well-designed system, the actual diversion time is much smaller than the collection time, and the flow pattern through the diverter is relatively independent of flow rate. Under these conditions, the limiting factor in system uncertainty is the basic accuracy and resolution of the weight or volume standard. With considerable care, errors can be reduced to less than 0.1% of reading. Relatively few calibration systems of this type are in field use because of their cost.

POSITIVE DISPLACEMENT METER PROVERS

Positive displacement meter provers utilize the principle of the displacement of a known liquid volume through the test meter. The displacer, either a spheroid or a piston, is moved by the flow or an external power source through an accurately calibrated section of pipe commonly referred to as the measuring section. Precisely located detectors at opposite ends of the calibrated section are actuated by the displacer so that a known volume is displaced between the trigger points while the output of the test meter is simultaneously recorded.

There are two general types of provers: unidirectional and bidirectional. In a unidirectional prover, the displacer travels in one direction only. It is returned to the initial position after each test run through a piping loop that is valved closed during the measurement cycle. Bidirectional provers are similar, except the displacer returns back through the same pipe by means of a four-way diverter valve. An advantage of bidirectional provers is some canceling of detector switch position errors between the forward and reverse cycles.

A variation on the bidirectional design is the small volume or compact prover. As the name implies, small volume provers use a smaller displacement volume than the other types, an advantage in portability. The basic outline of this design is shown in Figure 6-2. In operation, the poppet valve is first closed, launching the piston in the flow direction past the optical detection points which, together with the metering section bore, define the flow volume. After passing the last switch, the poppet valve opens allowing continuing flow through the prover. Hydraulic power is used to return the piston to the upstream (start) position.

Resolution is maintained, despite the small volume, by a system to resolve partial signal pulses from the meter under test and through the use of multiple prover runs at the same flow rate.

Provers are often skid or trailer mounted. They have historically been widely used in oil production to prove turbine meters but in recent years are coming into general use. They perform best with high-resolution digital flowmeters since time runs are relatively short [Ref. 2].

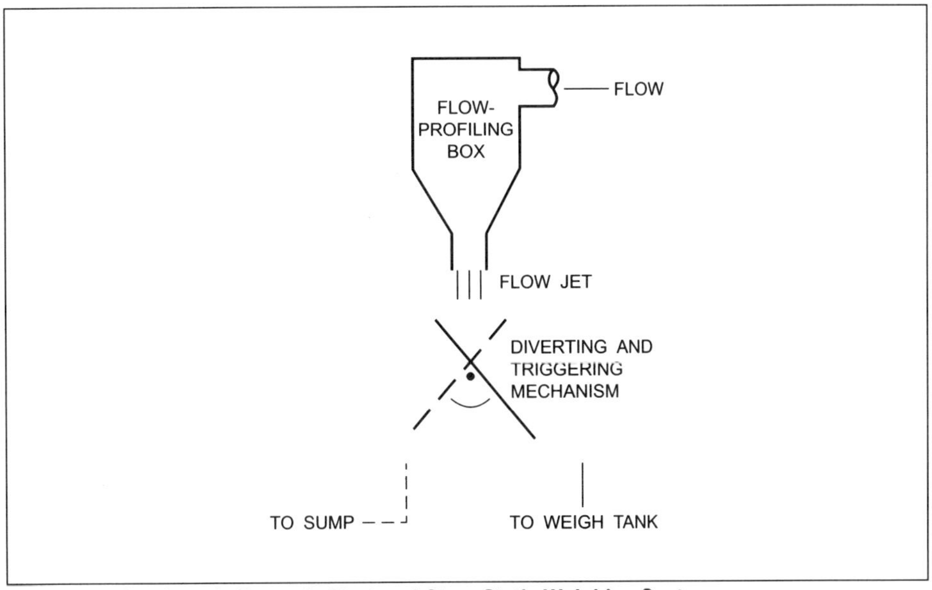

Figure 6-1. Gravimetric Dynamic Start-and-Stop, Static Weighing System

MASTER FLOWMETERS

Perhaps the most common method of wet calibration is to compare the output of the meter under test with one or more flowmeters that have been certified against high accuracy standards, commonly called master flowmeters. A small master meter comparison test stand can be constructed at less cost than other wet calibration methods. If properly designed and maintained, it is capable of calibrating flowmeters to a tolerance of ± 1% of reading or less.

Flowmeters to be used as masters should be selected by considering the desired range of the test stand, the required accuracy, the properties of the test fluid, and the allowable head loss. Typically, a master should have a digital output and maintain a constant percent of rate-type accuracy over at least a 10:1 turndown (or flow range). This permits testing meters one size smaller or one size larger than the master to be calibrated. Using multiple masters permits an even wider calibration capability. For example, a combination of 1-in. and 3-in. master flowmeters can provide a calibration capability for ½-in. to 4-in. sizes (see Figure 6-3). An important advantage of such a multiple master arrangement is the ability to check one master against the other to ensure no shifts have occurred. This is especially advantageous when masters such as turbine flowmeters are used. Turbines are very accurate, provide a high resolution digital output, and are relatively easy to use, but have the disadvantage that their calibration is somewhat dependent on the amount of bearing friction, particularly at low flows. Fluid contamination, for example, can cause a sudden shift in bearing friction and a resultant change in performance. The use of multiple turbine masters with overlapping ranges, piped so that they can be placed in series, enables checking their frequency ratio at several flow rates to make sure it has not changed. This provides good confidence in meter accuracy, except at lower flows on the smaller master where no overlap is possible [Ref. 3].

Regardless of the desired rig accuracy, the type of masters used, or the cross-checking capability, masters should be periodically calibrated against standards that are traceable to the National Institute of Standards and Technology (NIST). This certification will assist in user acceptance of test results when they do not

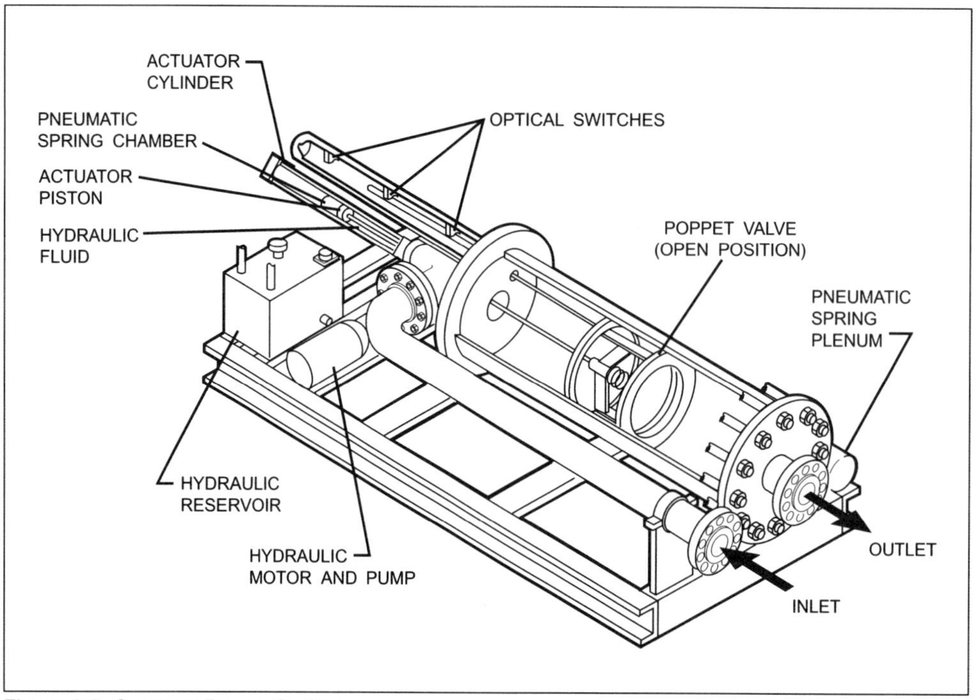

Figure 6-2. Compact Prover Outline

meet expectations or fail to resolve process problems. Without this type of documentation, a great deal of time can be spent on repeated runs and cross checks to satisfy users that the results are correct.

PIPING

Test stands should be constructed to provide sufficient straight piping upstream of both the test and master flowmeters to avoid significant velocity profile distortion and resultant errors. The calibration of different size meters will require changing the pipe sections immediately upstream and downstream of the test meter. These pipe sections are best constructed of seamless pipe using concentric reducers and can optionally contain flow straighteners to provide some smoothing of profile disturbance due to upstream elbows and valves.

The exact length of straight upstream pipe required depends on both the type of meter being calibrated and the design of the piping and valving upstream of the meter sections. Some recommendations are supplied by flowmeter manufacturers for their products. A general reference is ISO 5167 [Ref. 4], which specifies the length of straight pipe required for orifice plates after various disturbances, as a function of the orifice bore/pipe bore (β) ratio. Table 6-1 lists these requirements for a 0.7 β ratio plate as a general guide to the relative disturbance created by various fittings.

To minimize the size of the test stand, upstream bends in multiple planes should be avoided, reducers rather than expanders used where possible, and control valves located downstream. The general outline of a test stand incorporating these recommendations is shown in Figure 6-3.

TAKING WET CALIBRATION DATA

Flowmeters with digital outputs are normally calibrated using master flowmeters by holding a constant flow rate and simultaneously gating (on) both the master and test outputs. After sufficient counts have been accumulated to provide good resolution, the outputs are gated (off) and totals compared. The procedure is repeated over the desired flow range, and the results tabulated or graphically displayed. If required, adjustments can be made and the process repeated. Figure 6-4 shows the results of an actual calibration performed on a magnetic flowmeter using another magmeter as a master.

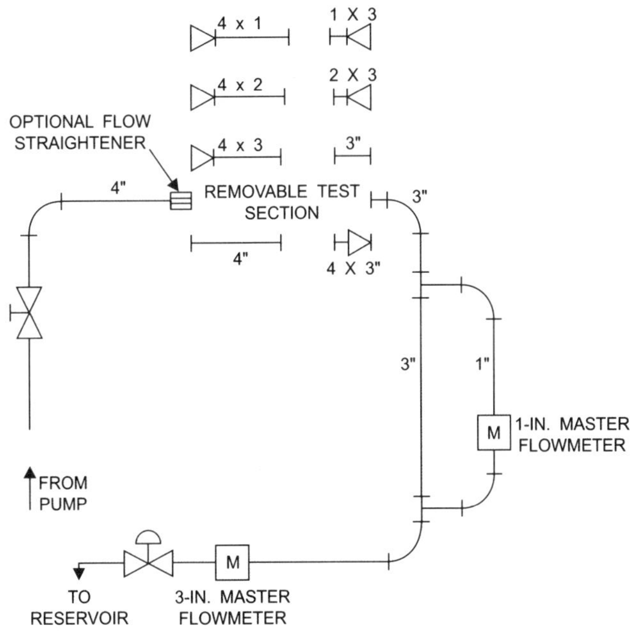

Figure 6-3. Flow Test Piping for 1-in. to 4-in. Meters

Table 6-1. Upstream Straight Pipe Requirements for a 0.7 β Orifice Plate per ISO 5167

One 90° Elbow	Two or More 90° Bends (Same Plane)	Two or More 90° Bends (Diff. Planes)	Reducer 2D to D Over 1.5 to 3D	Expander 0.5D to D Over 1 to 2D	Globe Valve (Open)	Gate Valve (Open)
14D	18D	31D	7D	15D	16D	10D

1. All straight lengths are expressed as a multiple of the pipe diameter. They shall be measured from the upstream face of the primary face of the primary device.
2. A straight downstream length of 3.5D is required for all fittings listed in this table.
3. These values are based on ± 0.5% "additional measurement uncertainty." For zero additional uncertainty, the lengths should be doubled.

The gated pulse technique averages the flow readings over a period of time, a distinct advantage when the signal of the test flowmeter fluctuates or has a significant noise content.

Test flowmeters with analog outputs can be calibrated using an analog-to-digital converter and the gated pulse technique or, alternately, by instantaneous comparison of the test output with that of the master and adjustment of the test meter until they coincide. This method is much faster than the gated pulse technique and provides sufficient accuracy for many applications.

Wet calibrations can be performed on the complete instrument, that is, the flow primary plus the secondary and any associated readout or, alternately, on the primary portion only. The remainder of the system can then be dry calibrated. Wet calibrating the primary only is a common practice with differential pressure-producing flowmeters, such as Venturi tubes.

A lab standard differential pressure readout, previously certified, is used to indicate the differential. Since there is normally no interaction between a differential-producing primary and its secondary, this method provides adequate overall system accuracy.

TEST REPORT

SERIAL NUMBER: __________ DATE: 18 JAN 1989

METER SIZE: 1.0 in. CALIBRATOR: __________

MODEL NUMBER: __________

	MASTER METER			TEST METER		
Run No.	Freq., Hz	K, cycles/gal	Actual Flow, gpm	Freq., Hz	Ind. Flow, gpm	Difference, % Rate
01	365.125	1468.59	35.345	876.90	35.311	-0.098
02	441.175	1469.42	18.014	446.70	17.988	-0.148
03	116.533	1475.60	04.738	117.48	04.731	-0.161

Meter Factor = 0.811

Full Scale gpm = 53.69

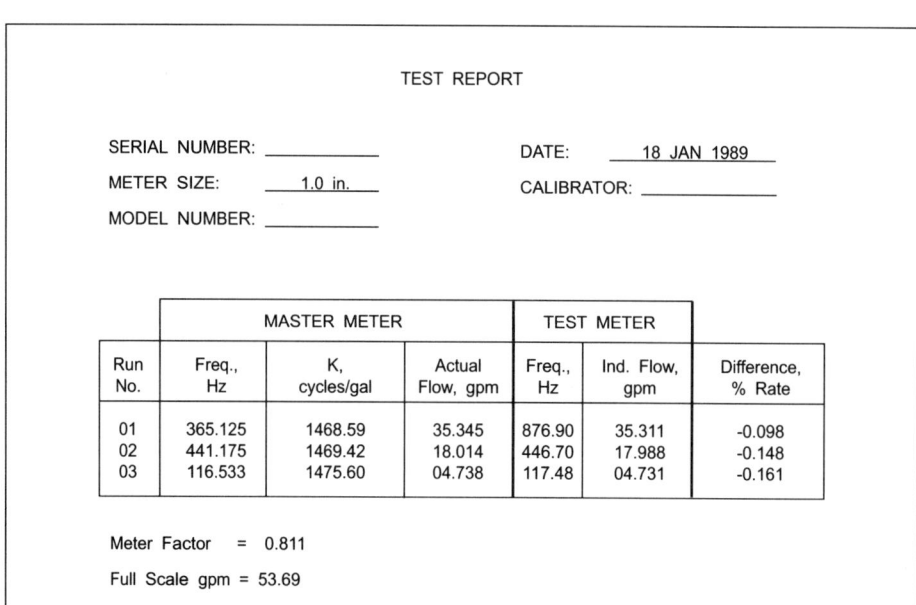

Figure 6-4. Flow Calibration Using a Master Meter

Certain meters, such as magnetic flowmeters, can exhibit some interaction between primary and secondary. When high accuracy is a requirement, they can either be calibrated together, or the primary wet can be calibrated using a master secondary of the same type.

Whether to calibrate the system or the primary only is a judgment that must be made each time a test is planned. A system calibration provides the highest accuracy and confidence that no unforeseen interaction has created a calibration error that would make performance in the actual application different from that on the test bench. The question is, however, whether for the particular application this added confidence is significant enough to offset the inconvenience of removing the secondary and/or readout. In many installations, this is very difficult as they are hard-piped and hard-wired in place.

Dry Calibration

A dry calibration is a calibration procedure performed on a flowmeter without subjecting it to a fluid medium. In a wet calibration, the flowing fluid produces the input signal; in a dry calibration, the input signal is introduced somewhere in the input portion as a frequency, millivoltage, differential pressure, or dimensional change. A portion of the fundamental flow transducer is, therefore, bypassed and not checked. Because of this, a dry calibration has a larger measurement uncertainty than a wet calibration, and the overall accuracy of the flow device must be inferred. The input signal for a dry calibration typically must be provided by a measurement standard or array of measurement standards. Measurement of the output signal requires the use of other measurement standards. To minimize errors, these standards should be connected in accordance with a defined calibration procedure for the specific meter type being tested.

Certain flowmeters, in particular those that are microprocessor-based, contain internal reference signals that can be connected by the user using software routines. Dry calibration of these meters may be entirely self-contained, making the procedure extremely quick and convenient.

Calibration Procedures

An accurate calibration is obtained through meticulous attention to the details of the test array and the recording of results. A vital step in achieving this result is obtaining and following a specific procedure for the calibration to be performed. A good procedure should provide a list of the necessary equipment for the calibration and the rated accuracy, or estimated uncertainty of each piece of equipment. It should also furnish a diagram or schematic showing the interconnection of the equipment and other pertinent information, such as the required warm-up time. The body of the procedure should provide a step-by-step method of performing the calibration and recording the results. Some procedures may also supply a form to facilitate taking data and an estimate of the uncertainty associated with the procedure.

Sources for Calibration Procedures

Writing a calibration procedure is a difficult, time-consuming effort. Although sometimes necessary, it can often be avoided or simplified by obtaining existing procedures from established sources. The easiest and most common source is the instruction book furnished by the flowmeter manufacturer. Most manufacturers expend considerable effort in presenting easy-to-follow instructions for the calibration of their equipment. Other sources are ISA Recommended Practices and the Government Industry Data Exchange Program (GIDEP). The exact equipment specified in these procedures often is not available, but the procedures can usually be adapted to use existing equipment without introducing significant systematic errors.

Making Calibration Adjustments

Most flowmeter secondaries, whether mechanical, pneumatic, or electronic, contain zero and span adjustments. Electronic secondaries often also have an adjustment to enable scaling the output to the desired flow units. Some secondaries have factory set zeros and typically do not require further zero adjustment. Other electronic secondaries, particularly those that are microprocessor-based, contain built-in reference signals that can be indexed either manually or by using software. These may allow checking/setting of zero, span, and scaling without connecting external equipment.

ZERO ADJUSTMENT

Adjustment of zero, when required, is performed by simulating an input to the test secondary that corresponds to a zero flow condition. The zero adjustment potentiometer or linkage, etc., is then adjusted in accordance with the calibration procedure. Proper zero adjustment is critical to accurate flowmeter performance since inaccuracies or drift in the zero adjustment will cause a shift in the flowmeter output at all flows. Since this results in constant error over the entire flow range, it can be expressed as a percent of full scale error.

SPAN ADJUSTMENT

Adjusting the span is performed by simulating an input to the test secondary that is equivalent to a full-scale flow at operating conditions. The span adjustment is then set in accordance with the calibration procedure. Inaccuracies or drift in the span adjustment cause errors in the flowmeter output that diminish linearly as flow rate decreases. They are, therefore, expressed as a percent of actual flow rate.

SCALING ADJUSTMENT

Scaling factor adjustments are often available on flowmeters that have frequency outputs. They are used to set the output of the flowmeter to read directly in gallons or other desired units. Scaling factors are normally set using thumbwheel switches or potentiometers in accordance with the manufacturer's specification. Care should be taken in performing the necessary calculations since any mistake in setting the scaling will appear as a span error.

Mechanical Measurements

When a wet calibration capability is not readily available, a check of mechanical dimensions is sometimes used to infer the performance of a flow primary. While not a true calibration, this method can sometimes provide adequate assurance of accuracy in noncritical applications. It is most commonly used with square edge concentric orifice plates where the diameter of the bore and the condition of the edge can both be checked. Of these two, the edge condition is by far the most critical as the edge can become nicked and rounded in service long before there is any measurable change in the bore. An acceptably sharp edge has commonly been defined as one with a radius equal or less than 0.0004d where "d" is the bore of the orifice. An orifice with rounding exceeding this will have a higher flow coefficient than that obtained from prediction tables and, thus, pass more flow for the same differential. The effect can be quite pronounced with errors of up to 8% cited for a radius of 0.05d [Ref.5].

A reasonably precise method of assessing edge sharpness exists where a lead foil impression is enlarged and projected. In practice, this is considered overly sophisticated. A visual inspection of the edge with a good light source is normally accurate enough to decide whether to keep or replace the orifice plate.

Floats in glass variable area flowmeters can also be checked for wear or nicking of the sharp edge normally used for metering. If there is no edge wear or other evidence of mechanical damage, it can be inferred that the calibration is unchanged.

In many respects, the shedding surface of a vortex flowmeter is analogous to the edge of an orifice plate or variable flowmeter float. If mechanical inspection of the shedding surface verifies that there has been no wear, it can be inferred that the shedding frequency is unchanged. If, however, wear is noted, there is little published information to enable estimating the shift in meter output. A wet calibration is required to precisely determine performance.

Another example of a mechanical check that can be performed is a turbine meter spin test. This check is used to determine if a major change in flow performance has taken place. First, the turbine rotor is accelerated in air to a reference rpm (overspeeding should be avoided). Then the driving force is removed and the time for rotation to cease measured. The time is compared with that obtained on a meter known to meet performance specifications. Times within perhaps 10%, combined with no evidence of mechanical damage, are used to infer performance is unchanged. Conversely, significantly slower times indicate probable excessive bearing friction normally associated with low-end calibration shifts.

References

1. NCSL RP-3-1988, National Conference of Standards Laboratories Information Manual Recommended Procedure 3, "Calibration Practices."
2. *Manual of Petroleum Measurement Standards,* Chapter 4, "Proving Systems," American Petroleum Institute, 1978.
3. Burgess, T.; Hernandez, A.; and Hopkins, R., "An Accurate Flow Calibration System Using Combined Primary and Secondary Standards," in *Flow, Its Measurement and Control in Science and Industry,* Research Triangle Park, N.C.: Instrument Society of America, 1981.
4. ISO 5167-1980(E), "Measurement of Fluid Flow by Means of Orifice Plates, Nozzles and Venturi Tubes Inserted in Circular Cross-Section Conduits Running Full."
5. Crockett, K., and Upp, E., "The Measurement and Effects of Edge Sharpness on the Floor Coefficients of Standard Orifices," in *Journal of Engineering,* New York, NY: American Society of Mechanical Engineers, 1973.

About the Author

Thomas H. Burgess is the Director of Quality Assurance for Fischer and Porter Company. Mr. Burgess has over 30 years experience in the development, test, and application of flow measuring equipment and holds ten patents in the field. He is a graduate of Stevens Institute of Technology and holds a Master's Degree from Drexel University.

7

Installation and Maintenance

Flowmeter installation and maintenance must be performed correctly to achieve the stated performance of a flowmeter. The effects of improper installation and maintenance vary from a reduction in accuracy to a flowmeter that does not operate at all. Attention to installation detail, piping requirements, and maintenance are critical to flowmeter performance.

Piping Considerations

Meter Operation

It is preferred to install flowmeters in vertical piping with flow upwards. However, this is not always practical. Liquid (gas) flowmeters require that the flowmeter be full of liquid (gas) to achieve accurate measurement. Often, liquid (gas) flowmeters must be located at the lowest (highest) point in the piping to achieve this end. Some flowmeters require special orientation, such as in the case of the magnetic flowmeter, which requires that the electrodes be mounted in the horizontal plane.

The proper physical installation of a flowmeter in the piping is critical to achieving accurate measurement.

Upstream and Downstream Piping

Most flowmeters require that the pipe installed upstream and downstream of the flowmeter be straight, be the same inside diameter as the flowmeter, have a smooth inside surface, not be out of round, and contain no pipe fittings (for example, tees, reducers, valves, etc.) to develop a uniform velocity profile with minimum distortion and/or swirl upstream of the flowmeter (see Figure 7-1). Flow conditioners may be used in upstream piping to minimize upstream straight run requirements. Upstream and downstream straight runs are usually expressed as multiples of inside pipe diameter, and these requirements vary with flowmeter technology, design, and manufacture. Downstream straight run is required for flowmeters whose performance is affected by the downstream piping. Control valves and abrupt enlargements must be installed an adequate distance downstream of the flowmeter to minimize downstream pipe effects.

Pressure Taps

Pressure taps should be located upstream of the flowmeter. The tap must be flush with the inside of the pipe and all burrs removed to ensure it will not disturb the velocity profile of the flow (as depicted in Figure 7-2). Liquid and gas pressure taps may be located off the side of the pipe. Gas pressure taps may also be located off the top of the pipe.

Steam pressure taps should be located off the side of the pipe, but may be located off the top when a pigtail is used. There should be a "dead leg" between the flowmeter and the pipe so condensate can form a liquid barrier to keep live steam away from the transmitter.

A pigtail is a pipe that has been bent in a complete circle, as shown in Figure 7-3, and is used for steam service where the pressure tap is located off the side or top of the pipe. The steam is allowed to condense in the tube and form a liquid barrier that isolates the pressure transmitter or gage from the process temperature and dampens pressure spikes, thereby protecting the transmitter from damage. Pigtails can also cause problems. In some applications, the pressure signal may be dampened or may become unstable (bouncing) as a result of installation of a pigtail. In other applications, the fluid can solidify at lowered temperatures and cause a loss of signal. Pigtails should be used only when required.

Temperature Taps

Temperature probes should be locate downstream of the flowmeter so the well can be in the flow stream for accuracy while not disturbing the velocity profile upstream of the flowmeter (as depicted in Figure 7-2).

Pressure and Differential Pressure Flowmeter Impulse Tubing

Liquid differential pressure transmitters should be located below the pipe with the impulse tubing constantly sloping upward towards the pipe so gas bubbles do not collect in the impulse tubing and cause a bouncy signal.

Gas differential pressure transmitters should be located above the pipe with the impulse tubing sloping constantly downwards toward the pipe so any condensate will drain into the pipe. Liquid collecting in the impulse tubing can cause differential pressure signal errors. The liquid can also solidify, plugging the impulse piping, and cause a loss of differential pressure signal.

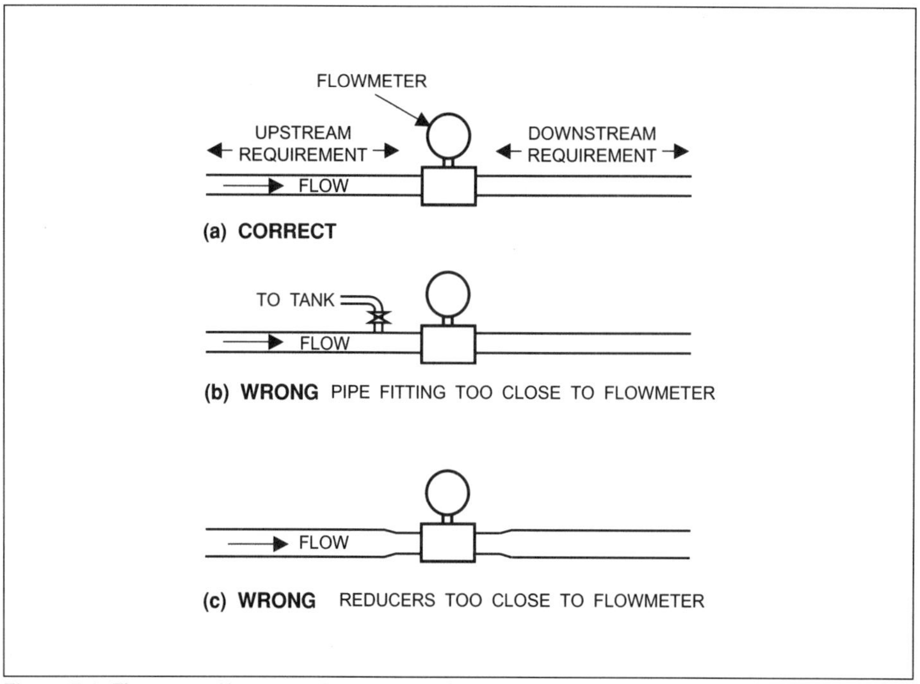

Figure 7-1. Flowmeter Upstream/Downstream Straight Run

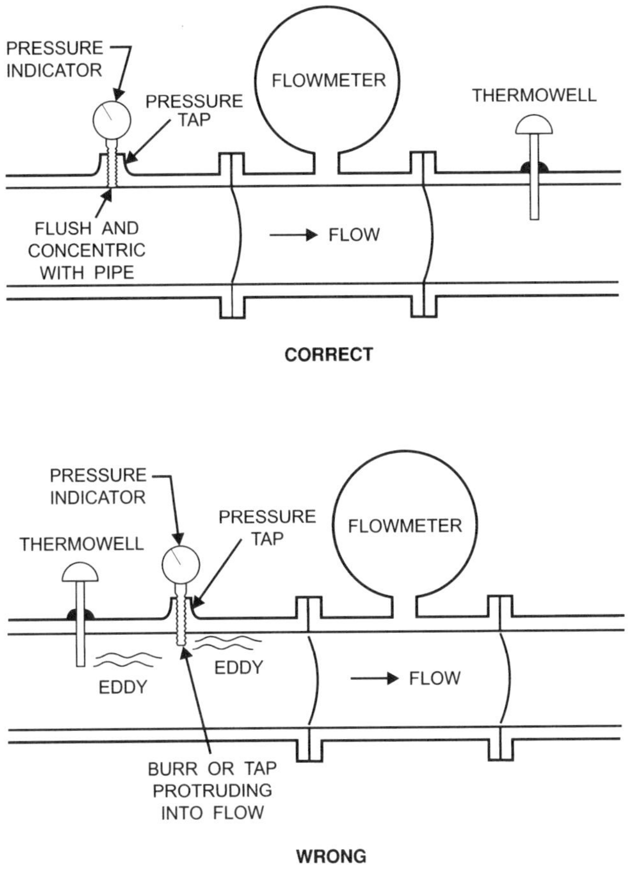

Figure 7-2. Flowmeter Pressure and Temperature Taps

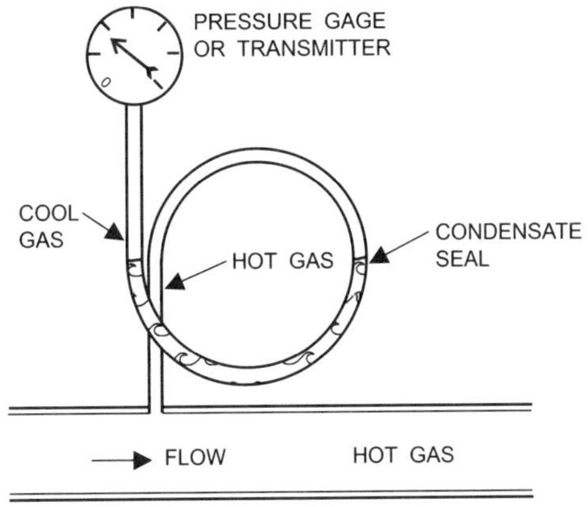

Figure 7-3. Pigtail—Principle of Operation

Steam differential pressure transmitter tubing should slope downwards toward the differential pressure flowmeter and include a "dead leg" or pigtail. If either the condensate levels or both sides of the differential pressure transmitter cannot be kept at the same height, the difference in elevations (levels or diaphragms) will cause a differential pressure offset that must be compensated for by offsetting the transmitter calibration.

Piping Supports

The pipe upstream and downstream of the flowmeter element should be properly supported so that no load or torque is transmitted into the meter, vibration is minimized, and the pipe is supported when the flowmeter is removed for service. Pipe supports should be designed to ensure that the flowmeter does not support the pipe (see Figure 7-4). Load of any type should not be transferred to the meter except in the case of very small meters designed to transfer pipe stresses.

When installing the flowmeter, the upstream and downstream piping should be aligned before the flowmeter is installed. The flowmeter should not be used to pull the piping into alignment. Torque from misaligned pipes can be transmitted into the flowmeter element and cause flow measurement error.

Welded Fittings and Flanges

Smooth straight upstream pipe should be installed even if the flanges are welded. The use of different types of fittings call for different operations to be performed to achieve a smooth inside pipe surface that minimized turbulence.

When welded fittings are installed, socket welded fittings are preferred. They have a socket machined into them to hold the pipe and fitting in close alignment and are welded externally to form a smooth connection on the inside pipe wall (if the pipe end is properly prepared). Socket weld fittings must be installed with the pipe fitted completely inside the socket so no void is created inside the pipe (see Figure 7-5).

When weld neck flanges are installed, the inside of the pipe should be reamed to remove any welding beads that may protrude into the flow steam and cause turbulence.

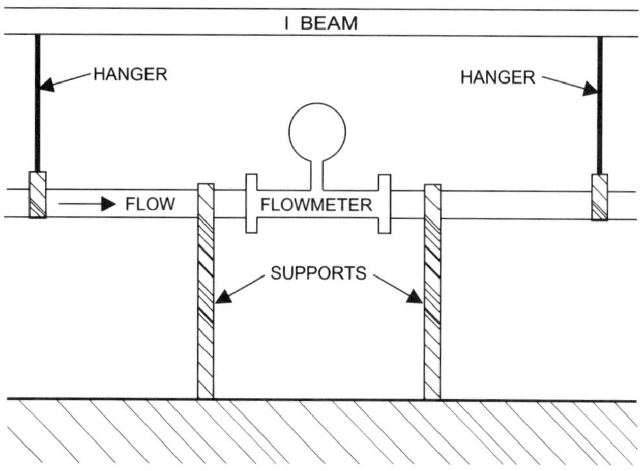

Figure 7-4. Flowmeter Piping Supports

Slip-on flanges should also be reamed and welded front and back. Care must be taken to ensure the flange and pipe end are flush to eliminate voids between opposing flanges. This is possible if the surface is machined after welding.

Gaskets

Care must be taken to ensure a smooth, straight inside pipe surface upstream and downstream of the flowmeter.

The center hole in the gasket should be the same size as the inside diameter of the pipe. Also it should be concentric and as thin as possible so there are no obstructions or voids created in the flow stream (see Figure 7-6). Changes of the inside diameter of the pipe upstream and downstream of the flowmeter element will cause turbulence and potential flowmeter measurement error.

Self-centering gaskets are preferred and full-faced gaskets make gasket alignment easier. The center hole is also the same size as the inside diameter of the pipe.

If the flowmeter was shipped with gaskets, bolts, and/or alignment tools, use them. Do not use plant substitutes.

Flowmeter Alignment Devices

To center the flowmeter in the pipe, some flowmeters utilize alignment devices such as rings that go on shoulders cut into the flow element (see Figure 7-7) or centering bushings that go over the flange bolts. Care must be taken to assure that the tools are used properly and that the correct bolts are used.

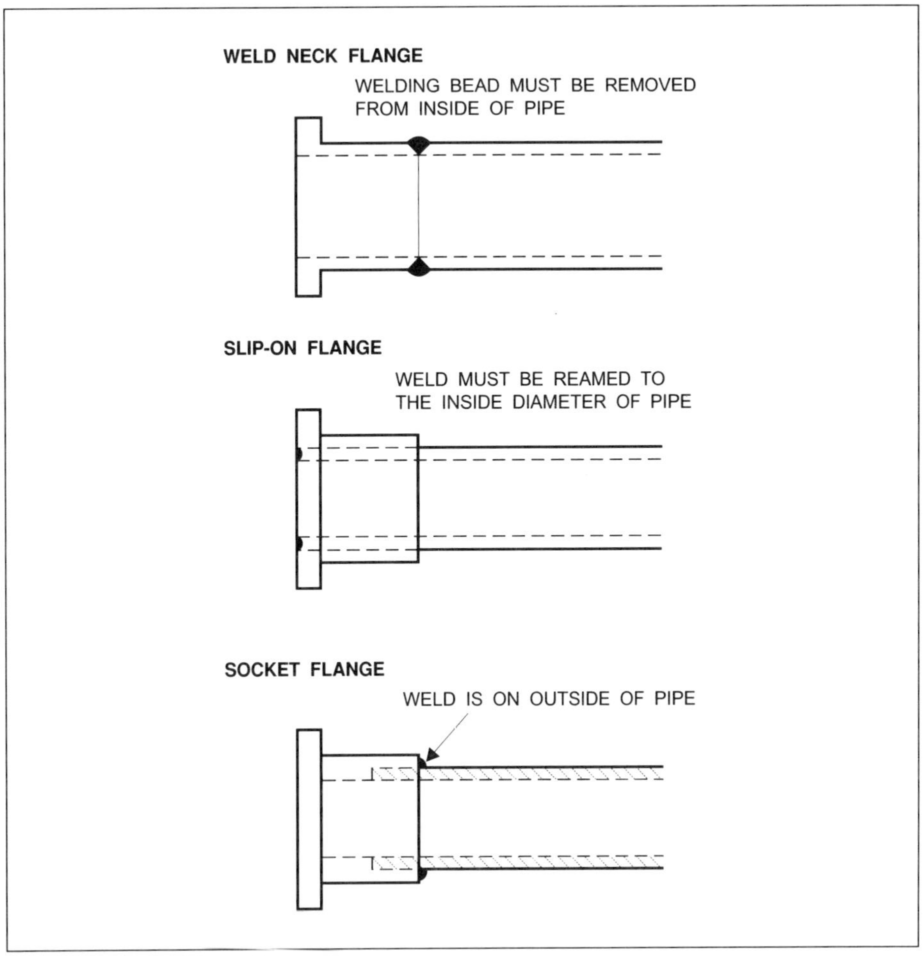

Figure 7-5. Piping Flange Connections

Do not insulate the transmitter electronics.

Insulation

Some flowmeters require insulation due to freezing and condensation, for personnel protection, and to keep radiant heat from damaging the transmitter electronics (see Figure 7-8). This insulation should be on the piping and the parts of the flow element recommended by the manufacturer.

In some applications, the transmitter electronics will fail because the flowmeter was not insulated or because it was insulated improperly.

The flange bolts will not provide dependable electrical continuity with the pipe.

Electrical Considerations

Grounding of piping and flowmeter body is often overlooked or performed improperly. Most flowmeters require the meter body to be grounded to the upstream and downstream piping to keep the flowmeter from being at different electrical potential than the pipe, and to prevent static buildup in the flowmeter from the flow fluid (see Figure 7-9). The preferred grounding method is to drill and tape the flanges to connect a ground strap from the flowmeter to each flange. Care should be taken to ground all electrically insulated *wetted* parts of the flowmeter.

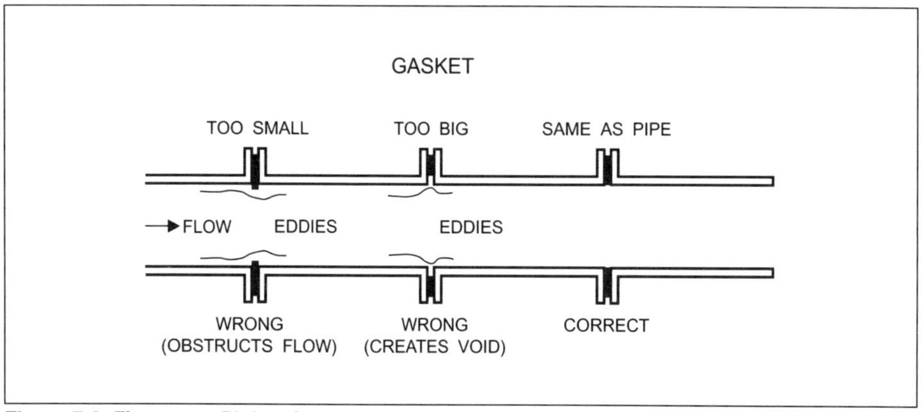

Figure 7-6. Flowmeter Piping Gasket Installation

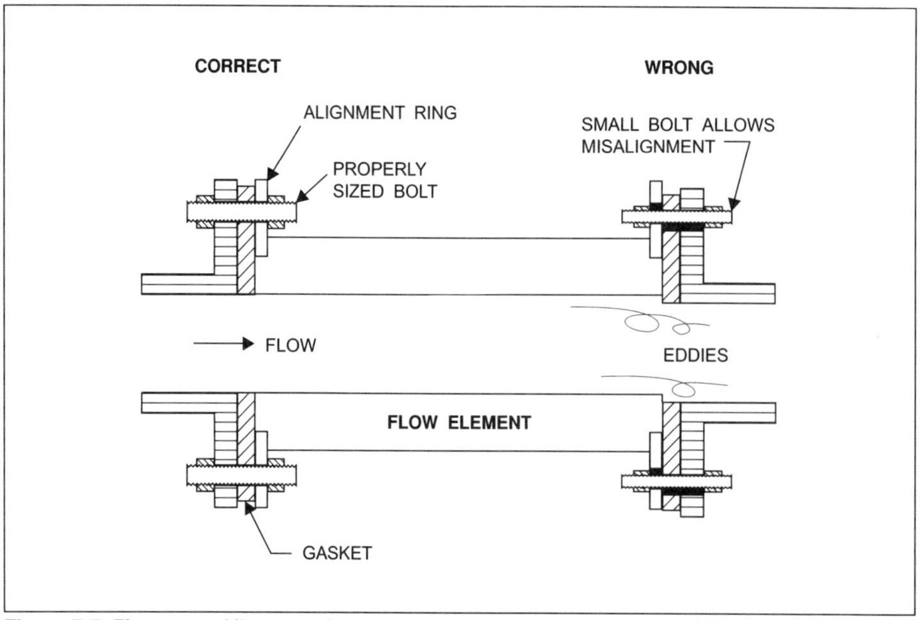

Figure 7-7. Flowmeter Alignment Devices

If the flowmeter is installed in a nonmetallic pipe, metal grounding rings upstream and downstream can be installed. The hole must be centered with the pipe and gasket and be of the same diameter as the inside diameter of the pipe.

Power

Most flowmeters are powered by 24 V DC, 110 V AC, or 220 V AC. Power (110 V AC and 220 V AC) should be routed to the flowmeter in a conduit system separate from the signal conduit system. For this reason, 24 V DC (2-wire) transmitters are usually preferred because they do not require a separate power con-

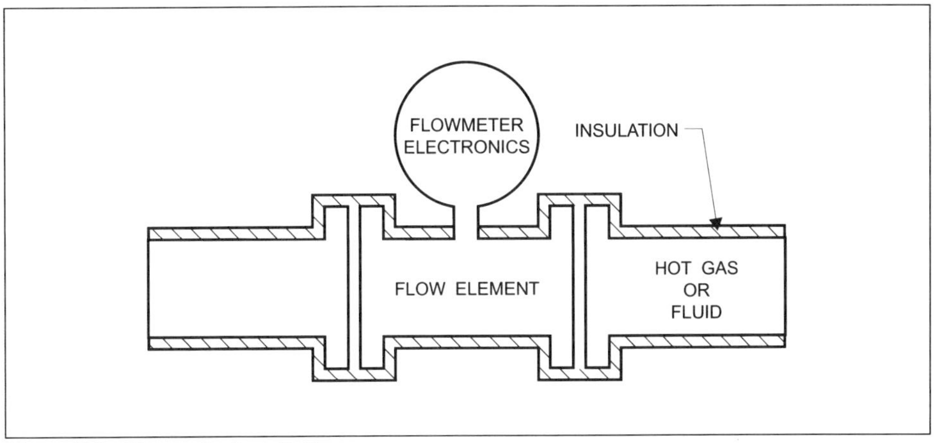

Figure 7-8. Flowmeter Installation

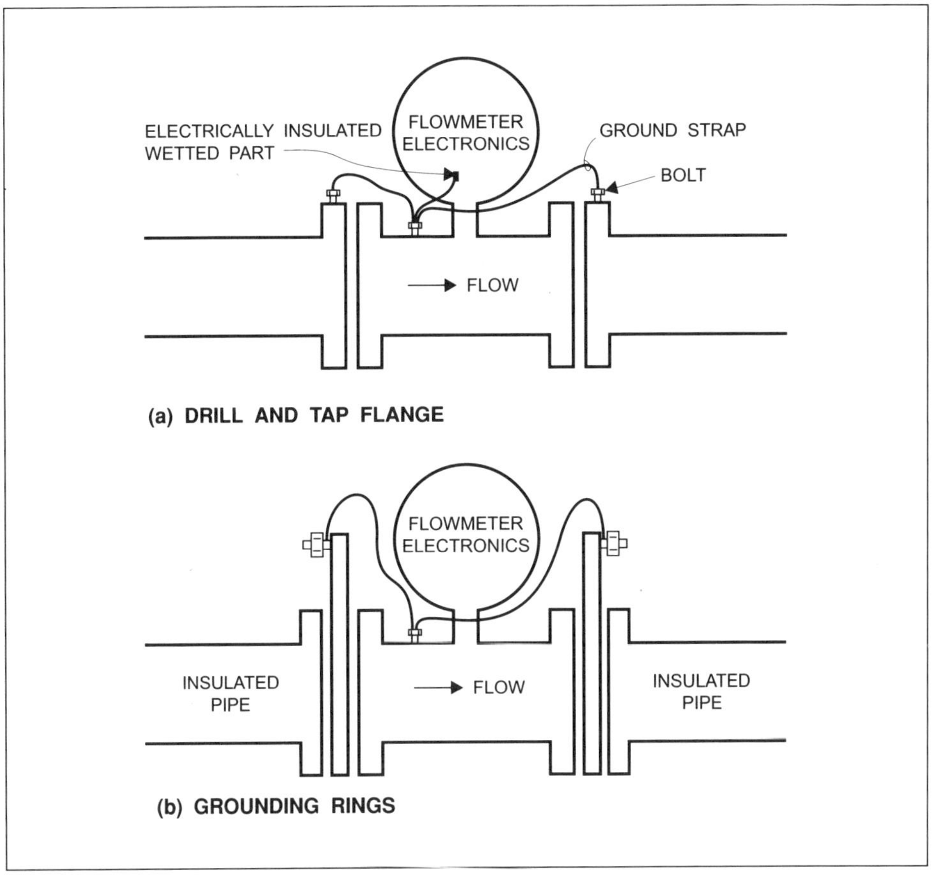

Figure 7-9. Flowmeter Grounding

duit. Regardless of power source, there should be a convenient method to turn the power off for service. This can be at the flow transmitter with a small plug or, in the case where there are several instruments using the same power source, a box with fuses and/or switches in a somewhat central location.

Power should be provided from a source dedicated to instruments. Utility outlets and lighting circuits should not be used because they can be tripped by overloading the circuit with other loads or inadvertently when performing maintenance on other equipment.

Signal Wiring

Signal wires should be shielded twisted pairs with separate shields for each signal. The shield should be continuous from the flowmeter to the sensing device and grounded at one end only, typically at the end closest to the sensing device (recorder, datalogger, controller, or input/output system). The end of the shield that is not connected should be insulated so it does not come in electrical contact with the other conductors or the flowmeter body (see Figure 7-10).

Location Considerations

Lighting

The flowmeter and flow transmitter should be located where the technician will have sufficient light while servicing the instrument, including at night, if applicable.

Clearance and Accessibility

Instruments located high in the middle of nowhere should be avoided.

The area around the flowmeter and flow transmitter should be accessible and clear of uninsulated hot or cold pipes that would impair the ability of the technician to work safely. A sufficiently large platform for personnel and/or equipment may be required.

Maintenance

Most properly applied and installed flowmeters require very little maintenance and can operate for extended periods of time with few problems. A few, however, may require some routine service. Maintenance problems and frequency of routine maintenance vary with the process fluid, type of flowmeter, and nature of the upset condition.

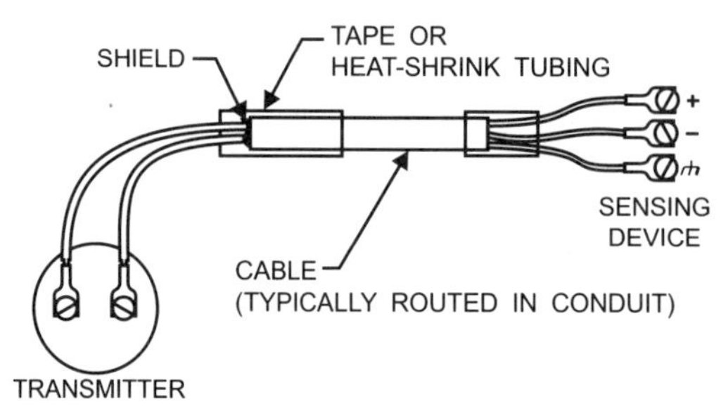

Figure 7-10. Signal Wiring Shielding

Routine Maintenance

Some flowmeters lose their accuracy as they get dirty from the process fluid. This change can occur slowly and can alter the dynamic response and/or the accuracy of the measurement.

ROTATING FLOWMETERS

Rotating meters such as turbine meters have bearings that may need to be lubricated periodically. Some rotating flowmeters may require periodic physical checks or wet calibration to detect bearing wear and/or damage.

MAGNETIC AND THERMAL FLOWMETERS

Magnetic and thermal flowmeters are required to maintain electrical and thermal conductivity, respectively, with the fluid. They may require periodic cleaning of their sensors, depending on the process fluid.

Loss of conductivity may manifest itself as a loss of output or full scale output depending on the type of meter.

VORTEX SHEDDERS

In some applications, coating can build up on wetted parts and require periodic checking and cleaning. This usually manifests itself as an unstable output with a known stable flow, or a complete loss of output.

Routine Calibration

Flowmeters can be checked for accuracy by putting a known flow through the flowmeter and checking the output (wet calibration) or by following the dry calibration procedure of checking the transmitter with a simulated flow signal.

A wet calibration is preferred but may not be practical. A dry calibration check can help to determine whether a problem is in the sensing element or the electronics.

Some flowmeters have internal standards and some self-checking features. Microprocessor-based transmitters may have more elaborate on-line diagnostics.

DRY CALIBRATION

Dry calibration is the use of simulated signal (voltage, resistance, differential pressure, current, etc.) to perform a calibration. This type of calibration check can help to locate problems in the flow element, transmitter, or receiving device and can be used to change the full-scale output of the flowmeter without removing the flow element from the process.

WET CALIBRATION

A wet calibration is performed when the flowmeter output is compared with the flow of a known quantity of fluid passed through the flowmeter. The flowmeter can be removed from the process and installed on a flow bench for calibration against a known standard or, alternatively, a prover or master meter may be installed in series with the flowmeter.

Wet calibration is the best way to resolve discrepancies and can help to detect problems that may involve piping or process errors.

MATERIAL BALANCE

This method uses the amounts of fluid in a process and process considerations to determine the flow independently of the flowmeter in question.

For example, the flow can be estimated by checking the time it takes at a given flow rate to empty or fill a known tank. If a company buys a product in large containers, such as tank trucks or rail cars, a monthly audit could be used to check the flow of raw material through a flowmeter.

Troubleshooting

Do not approach the meter as if it were broken!

Quite often the flowmeter is blamed for a process problem. Always ensure that the process is operating smoothly and all pumps and valves are working properly before checking the flowmeter.

After a flowmeter has been in service, it may generate a stable signal, but that signal may not be correct. Even if the flow signal appears to be correct, a number of reasons may make it desirable to check the flowmeter for accuracy.

An erratic flow signal can be caused by foreign particles inside the flow element or internal surfaces, or by sensor coating. This can also cause a sudden stoppage of the flow signal.

Various flowmeter troubleshooting techniques are available, including the following:

(1) Removal, examination, and replacement of the flow element (where possible)

(2) Testing or replacement of the flow transmitter (used especially where the process and piping are such that the flowmeter element cannot be removed from service)

(3) Analyze the installation for problems

Note that replacement parts should be *known to be good* before installation. The following is an abbreviated synopsis of typical flowmeter problems.

When observing the wetted parts of a flowmeter element, anything that alters its geometry (shape) potentially affects its accuracy.

Vortex shedders, especially those that are not "acceleration compensated" are susceptible to vibration. This may be caused by pumps or other structural movement. To minimize this problem, the flowmeter element should be supported with a rubber pad installed between the support and the piping. Sometimes, rotating the flowmeter out of the plane of vibration is required.

If the signal of a mass flowmeter suddenly jumps or falls but returns to normal flow signal, the problem most often is a loose connection between the flow element cable and the flow transmitter. If a thorough check shows no connection problems, the most likely problem is a bad drive coil inside the flow element.

If care has been taken to install differential pressure flowmeters correctly, the most common problem is plugging of the impulse tubing. In some extreme conditions, purges may be needed to keep the impulse tubing clear. Care must be taken to ensure the same purge flow through each impulse tube to minimize flow errors due to unbalanced pressures.

Turbine flowmeters will slow down as they wear and will eventually stop altogether. Bearing wear can sometimes be detected when a high flow is needed to get the turbine started.

AC magnetic flowmeters can drift and stop indicating flow when the sensors get coated. If magnetic flowmeters are subjected to extreme heat, they can become temperature sensitive.

In one application, a magnetic flowmeter was removed from service and checked on a calibration flow bench. The calibration was correct and the flowmeter was returned to service, only to fail one hour later. This "exercise" was performed several times until the flow bench fluid was heated and a calibration was performed on the hot fluid.

Unbeknown to the instrument technician, the pipe had been steamed out by a production operator and overheated. The flow signal at room temperature was accurate, but at the process temperature that was well within the operating range of the flowmeter, the flowmeter was inaccurate. Understanding the cause of the problem allowed the installation of another type of flowmeter that could be steamed without damage.

References

1. Spitzer, David W., *Industrial Flow Measurement*, Instrument Society of America, 1990.
2. Denny, Hugh W., *Grounding for the Control of EMI*, 1st edition, Don White Consultants, Inc., 1983.
3. Bean, Howard S., *Fluid Meters*, The America Society of Mechanical Engineers.
4. DeCarlo, Joseph P., *Fundamentals of Flow Measurement*, Instrument Society of America, 1984.
5. Spitzer, David W., and Carman, Theron A., "Practical Cost-Effective Onsite Flowmeter Proving," Paper Number 86-2828, Instrument Society of America, 1986.

About the Author

Theron A. Carman is presently working for Nepera Chemical, Inc., as a Senior Instrumentation Technician responsible for the preparation, installation, calibration, and repair on a plantwide basis. Before working in a similar capacity for Abex Corp. in the Engineering Test Center, Mr. Carman was in the U.S. Army Signal Corps working on and teaching the techniques for the use of microwave radio for telephone communications. He also worked for Solitron Devices in the manufacture of microwave diodes.

8

Differential Pressure Flowmeters

For many years differential pressure types of flowmeters have been the most widely applied flow rate measuring device for fluid flows in pipes that require accurate measurement at reasonable cost. Although a number of different types of flow rate-measuring devices are now available, the differential pressure type of flowmeter still makes up the largest segment of the total flow measurement market.

This type of device has a flow restriction in the line that causes a differential pressure or "head" to be developed between the two measurement locations. Differential pressure flowmeters are also known as head meters, and, of all the head meters, the orifice flowmeter is the most widely applied device. Italian physicist Giovanni B. Venturi (1746-1822) in 1797 performed the first recorded work that used orifices for the measurement of fluid flows.

Advantages:
Simple construction
Relatively inexpensive
No moving parts
Transmitting instruments are external
Low maintenance
Wide application of flowing fluid
Ease of instrument and range selection
Extensive product experience and performance data base
An abundance of application and selection guides
Readily available standards and codes of practice

Disadvantages:
Flow rate is a nonlinear function of the differential pressure
Low flow rate rangeability with normal instrumentation

Operating Principle

Differential pressure flowmeters have a change in flow cross section that can be described as a restriction placed in the flow line that causes the velocity of the flowing fluid to change. The difference in pressures between the two measurement locations of the flowmeter is the result of the change in the flow velocities. The volume flow rate through the cross-sectional area is given by,

$$Q = A \times \bar{v} \tag{8-1}$$

where:

Q = the volumetric flow rate
A = flow in the cross–sectional area
$\bar{v}$ = the average fluid velocity

Using Equation (8-1) and the theory of conservation of mass, the equation of continuity states the relationship between the velocity and fluid flows for incompressible fluid in a closed conduit as:

$$Q = A_1 \times \bar{v}_1 = A_2 \times \bar{v}_2 \tag{8-2}$$

where the subscripts refer to the cross sections 1 and 2 of Figure 8-1. For ideal fluid (no frictional losses), Bernouli's equation states that the sum of the static energy (pressure head), the kinetic energy (velocity head), and the potential energy (elevation head) of the fluid is conserved in the flow across a constriction in a pipe. Bernouli's equation at each flow cross section is given by:

$$\frac{P}{\rho} + \frac{\bar{v}^2}{2g} + z = \text{constant} \tag{8-3}$$

where:

P = static pressure
ρ = fluid density
g = acceleration due to gravity
z = elevation head of the fluid from a reference datum

Using Equation (8-3) for the flowing condition of Figure 8-1,

$$\frac{P_1}{\rho} + \frac{\bar{v}_1^2}{2g} + z_1 = \frac{P_2}{\rho} + \frac{\bar{v}_2^2}{2g} + z_2 \qquad (8\text{-}4)$$

At any flow cross section, a change in velocity causes a change in line pressure; i.e., with a decrease in velocity (kinetic energy), an increase in pressure (static energy) occurs. Certain physical laws allow this change of kinetic energy to static energy to be expressed in terms of the changes in pressure and velocity. Two equations from the laws of physics relating to the free fall of objects starting from rest and distance traveled are used to develop the differential pressure flow equation. The velocity of a free falling object starting from rest is given by:

$$\bar{v} = gt \qquad (8\text{-}5)$$

and the distance travelled by the object is:

$$h = \frac{1}{2} gt^2 \qquad (8\text{-}6)$$

where:

t = elapsed time
h = the change in height

Substituting the value of t from Equation (8-5) to Equation (8-6):

$$h = \frac{\bar{v}^2}{2g} \qquad (8\text{-}7)$$

or

$$\bar{v} = \sqrt{2gh} \qquad (8\text{-}8)$$

Equation (8-8) can be applied to determine the velocity of liquid through a hole near the bottom of a tank (Figure 8-2) with a liquid level of h from the center of the hole. The flow rate (velocity) through the hole is proportional to the

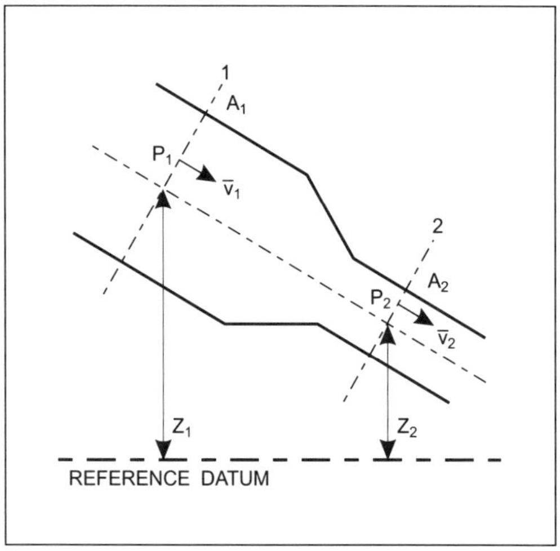

Figure 8-1. Flow through Closed Conduit

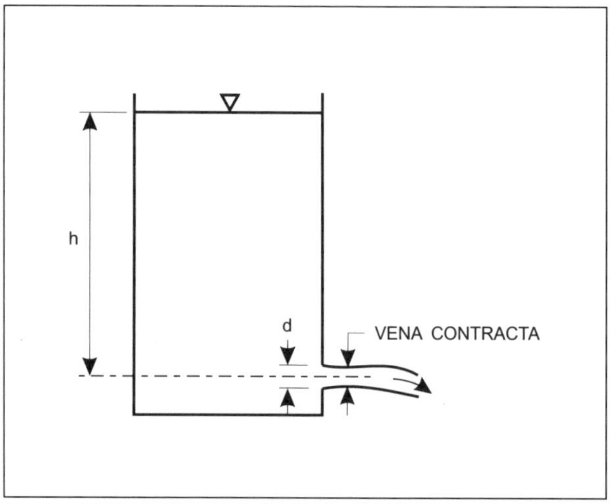

Figure 8-2. Flow through the Hole of a Tank

square root of the head pressure, h exerted by the liquid at the hole. The same relationship can be obtained by using Equation (8-4). This relationship of the pressure and velocity is the basis for the differential pressure meter flow calculations. Substituting velocity from Equation (8-8) in Equation (8-1), the flow rate through the hole is given by:

$$Q = A \times \sqrt{2gh} \tag{8-9}$$

where A is the cross-sectional area of the hole.

Consider a tank full of water with a round hole in the side near the bottom (Figure 8-2). The hole is d feet in diameter and is located h feet below the surface of the water. Under ideal conditions, the velocity of water flowing through the hole is given by the Equation (8-8). Due to certain physical phenomena, actual flow through the hole is less than the theoretical value and is about 98% of ideal flow condition. This factor of 0.98 is the ratio of the actual velocity to the ideal velocity and is called the *velocity coefficient*, C_v. The area of the hole is $\pi\, d^2/4$. Therefore, the volumetric flow rate from the tank calculated from Equation (8-9) is:

$$Q = \left(\pi \frac{d^2}{4}\right) \times C_v \times \sqrt{2gh} \tag{8-10}$$

The volumetric measurement indicated by Equation (8-10) has an error caused by the fact that the jet of water escaping from the restriction contracts in diameter as it leaves the restriction. At a distance of approximately one-half of the restriction diameter downstream from the restriction, the jet attains a minimum value of cross-sectional area. The point at which the jet has its smallest diameter is called the *vena contracta.* When the hole is a sharp-edged orifice, the cross-sectional area of the jet at the vena contracta is about 62% of that of the orifice. This factor of 0.62 caused by the contraction of the flow cross section is known as the *contraction coefficient*, C_c. Therefore, the actual flow rate though the hole in the tank is:

$$Q = C_c \times C_v \times \left(\pi \frac{d^2}{4}\right) \times \sqrt{2gh} \tag{8-11}$$

The product of the velocity coefficient and the contraction coefficient is called the *discharge coefficient*, C_d or C. Therefore, Equation (8-11) of actual flow rate through a restriction reduces to:

$$Q = C \times A \times \sqrt{2gh} \tag{8-12}$$

When using Equation (8-12) for differential pressure flow devices in pipes, h becomes the differential head that exists between two specific locations in the piping, relative to the restriction specified. The pressure head is commonly expressed as a height of liquid column, such as inches of water column. The locations at which the differential pressure measurements are made are called pressure taps. Pressure tap locations vary with type of device.

Determine the flow rate of water through the hole of the tank shown in Figure 8-2 where the hole diameter is 8 inches and the height of water from the center of the hole is 18 feet. Assume that the coefficient of velocity is 0.98, coefficient of contraction is 0.62, and acceleration due to gravity is 32.2 ft/sec^2.

(continued)

(continued)

Solution:
The cross-sectional area of the hole is:

$$A = \pi\left(\frac{4}{12}\right)^2 = 0.349 \text{ ft}^2$$

Using Equation (8-12):
The flow rate is given by:

$$Q = C_c \times C_v \times A \times \sqrt{2gh}$$
$$Q = 0.62 \times 0.98 \times 0.349 \times \sqrt{(2 \times 32.2 \times 18)}$$
$$= 7.22 \text{ ft}^3\text{/sec}$$

Using Bernouli's equation:
Let the top surface of water in the tank be flow section area 1 and the hole exit plane be flow section 2 from Equation (8-3):

$$\frac{P_1}{\rho} + \frac{\bar{v}_1^2}{2g} + z_1 = \frac{P_2}{\rho} + \frac{\bar{v}_2^2}{2g} + z_2$$

Let the reference datum be the center of the hole; thus, z_2 is zero and z_1 is 18 feet. Pressures at the surface and at the discharge plane are atmospheric; hence, the pressure terms on both sides of the equation are equal. If the tank diameter is orders of magnitude larger than the hole diameter, the change in the surface level (i.e., the velocity at cross section 1) is negligible or zero. Therefore, Bernouli's equation reduces to the form:

$$\frac{\bar{v}_2^2}{2g} = z_1$$

or

$$\bar{v}_2 = \sqrt{2gz_1}$$

Using Equation (8-1) and the correction factors for coefficients of velocity and contraction:

$$Q = C_c \times C_v \times A \times \sqrt{2gz_1}$$
$$= 0.62 \times 0.98 \times 0.349 \times \sqrt{(2 \times 32.2 \times 18)}$$
$$= 7.22 \text{ ft}^3\text{/sec}$$

Reynolds Number

For fluids with no viscosity, the velocity profile of the fluid in a pipe would be uniform across the pipe section. In "real" fluids, the fluid viscosity slows down the flow near the pipe wall, and the velocity at the wall approaches zero. Thus, the flow profile is nonuniform across the pipe cross section. A nondimensional flow parameter, Reynolds number, can describe the nature of the fluid flow in a pipe. This parameter is named after the scientist Osborne Reynolds who discovered the correlation between the flowing condition in the pipe and this nondimensional parameter. The pipe Reynolds number is defined as:

$$\text{Re}_\text{D} = \bar{v}\rho\frac{D}{\mu} = \frac{3160 Q_{\text{gpm}} \text{SG}}{\mu_{\text{cP}} D_{\text{in.}}} \qquad (8\text{-}13)$$

where:

D = the pipe diameter
μ = the absolute viscosity
SG = the fluid specific gravity

For Reynolds numbers less than approximately 2000 the flow is in the laminar flow regime; i.e., the fluid direction and velocity at any point in the flow section does not vary with time. For Reynolds numbers above 4000 the flow is in the turbulent flow regime; i.e., the instantaneous fluid direction and velocity at any point in the flow cross section vary with time; but for a steady flow rate the average fluid direction and velocity through the cross section remain unchanged and satisfy the continuity equation. For Reynolds numbers between approximately 2000 and 4000 flow is in the transitional flow regime; i.e., the flow may intermittently change from the laminar to turbulent regime and vice versa. Most flowmeter applications are in the turbulent flow regime; however, laminar flows occur at low flow rates, medium or high fluid viscosities, or if the piping is oversized for the given flow.

Determine the Reynolds number for water at 60°F flowing through a 3.067 inch pipe at 25 gallons per minute. Is the flow laminar or turbulent?

Solution:

The specific gravity of water at 60°F is normally 1.0, and the viscosity is normally 1.0 cP. The pipe Reynolds number is:

$$\mathrm{Re} = \frac{(3160)(25)(1.0)}{(1.0)(3.067)} = 25{,}575$$

The pipe Reynolds number is greater than 4,000; hence, the flow is turbulent.

Types of Differential Pressure Flowmeters

Differential pressure flowmeters operate on the principle of developing a differential pressure across a restriction that can be related to the fluid flow rate. Measurement accuracy is reduced when an uncalibrated meter is installed in the pipe and/or when monitoring instruments are not calibrated periodically. This can occur when efforts are made to lower the flowmeter cost or to reduce operating expenses.

Optimum measurement accuracy is maintained when the flowmeter is calibrated, the flowmeter installed in accordance with standards and codes of practice, and the transmitting instruments are periodically calibrated.

The most commonly used differential pressure flowmeter types are:

(1) orifice,

(2) Venturi,

(3) nozzle,

(4) Pitot-static tube,

(5) elbow, and

(6) wedge.

Other special designs of differential pressure devices include:

(1) "V-cone" flowmeter,

(2) spring-loaded variable aperture,

(3) laminar flow element,

(4) Dall tube,

(5) Elliot-Nathan flow tube,

(6) Dall orifice, and

(7) Epiflo.

Orifice

The most commonly applied orifice is a thin, *concentric*, and flat metal plate with an opening in the plate (Figure 8-3), installed perpendicular to the flowing stream in a circular conduit or pipe. In most applications, a sharp edged hole is bored in the center of the orifice plate. As the flowing fluid passes through the orifice, the restriction causes an increase in velocity. A concurrent decrease in pressure occurs as potential energy (static pressure) is converted into kinetic energy (velocity). As the fluid leaves the orifice, its velocity decreases and its pressure increases as kinetic energy is converted back into potential energy according to the laws of conservation of energy. However, there is always some permanent pressure loss due to friction, and the loss is a function of the ratio of the diameter of the orifice bore (*d*) to the pipe diameter (*D*). This ratio (*d*/*D*) is known as beta (β) ratio.

For dirty fluid applications, a concentric orifice plate will eventually have impaired performance due to dirt buildup at the plate. Instead, *eccentric* or *segmental* orifice plates (Figure 8-4) are often used. Measurements are typically less accurate than those obtained from the concentric orifice plate. Eccentric or segmental orifices are rarely applied in current practice.

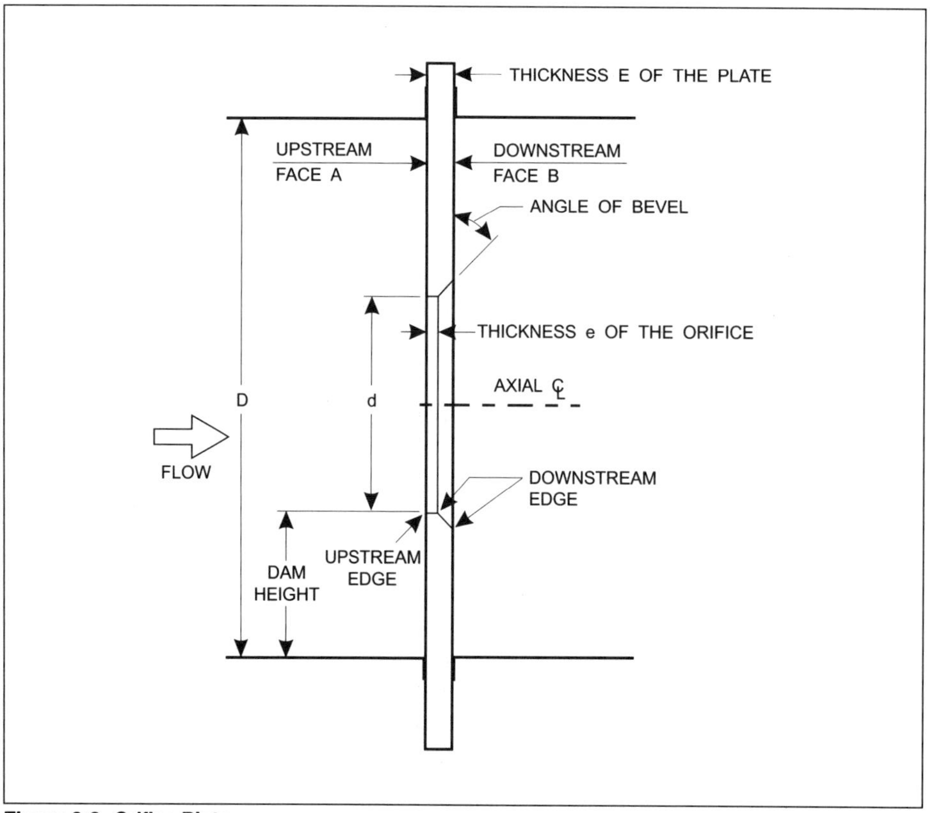

Figure 8-3. Orifice Plate

When the pipe Reynolds number is less than 10,000, the upstream edge of the orifice can be contoured to achieve a more constant and predictable discharge coefficient at lower Reynolds numbers. The change in the value of the coefficient of discharge for the square-edged orifice over the Reynolds number range of 5,000 to 10,000 can be as high as 30%, while the changes for the contoured orifice over the same Reynolds number range could be of the order of only 1 to 2%. The contoured orifice is either rounded or conical (Figure 8-5). The rounded orifice is known as quarter round or quadrant orifice and is widely used in North America; the conical orifice is prevalent in Europe.

The radius of the quadrant orifice is a function of β . This bore is specifically designed for viscous flows, such as syrups and slurries, having pipe Reynolds numbers as low as 250 and possibly as high as 100,000.

An integral orifice plate is a machined concentric orifice assembly that is either directly attached to, or mounted inside, the transmitter. Integral orifice flowmeters are used for low flow rates in small line sizes, typically in ½ to 1½ inch line sizes.

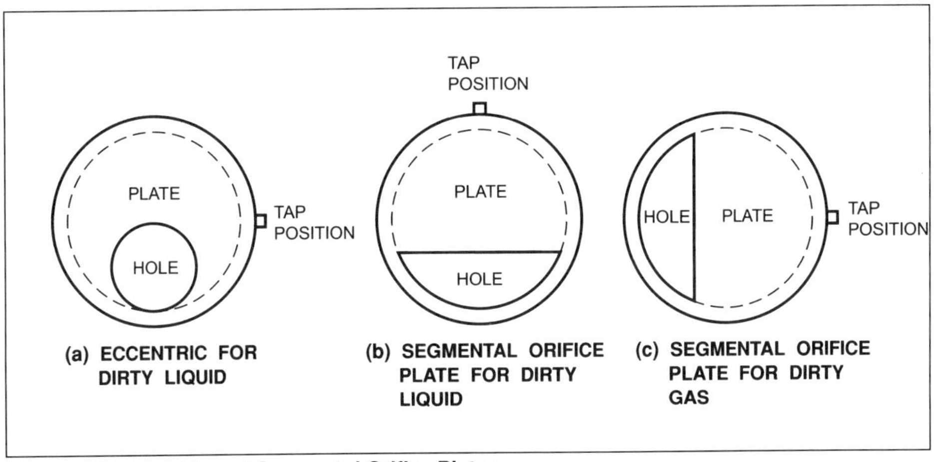

Figure 8-4. Eccentric and Segmental Orifice Plates

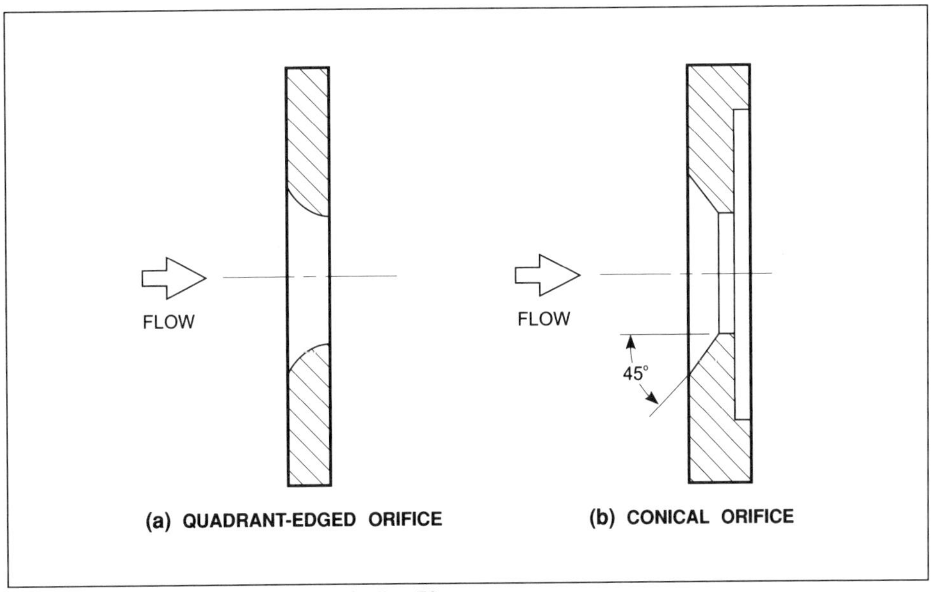

Figure 8-5. Low Reynolds Number Orifice Plates

Venturi

A Venturi is a restriction with a relatively long passage with smooth entry and exit (Figure 8-6). It produces less permanent pressure loss than a similar sized orifice but is more expensive. It is often used in dirty flow streams since the smooth entry allows foreign material to be swept through instead of building up as it would in front of an orifice.

Nozzle

Flow nozzles have a smooth entry and sharp exit (Figure 8-7). For the same differential pressure, the permanent pressure loss of a nozzle is of the same order as that of an orifice, but it can handle dirty and abrasive fluids better than an orifice can. Note that for the same line size, flow rate, and β , the differential pressure at the nozzle is lower than the differential pressure for an orifice; hence, the total pressure loss is lower than that of an orifice. Nozzles are primarily used in steam service because of their rigidity, which makes them dimensionally more stable at high temperatures and velocities than orifices.

A useful characteristic of nozzles is that they reach a critical flow condition, that is, a point at which a further reduction in downstream pressure does not produce a greater velocity through the nozzle. When operated in this mode, nozzles are very predictable and repeatable. For this reason nozzles have become

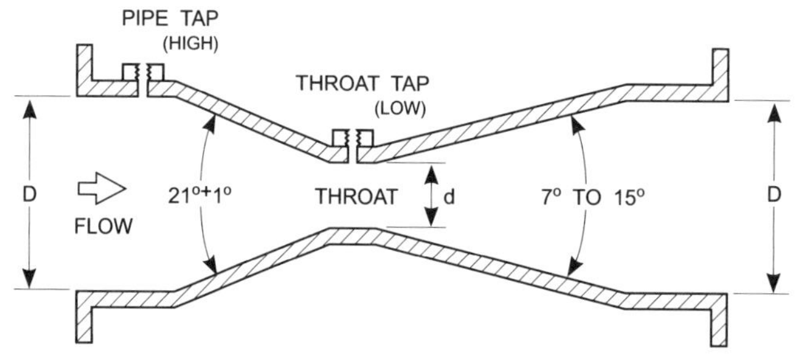

Figure 8-6. Venturi Tube

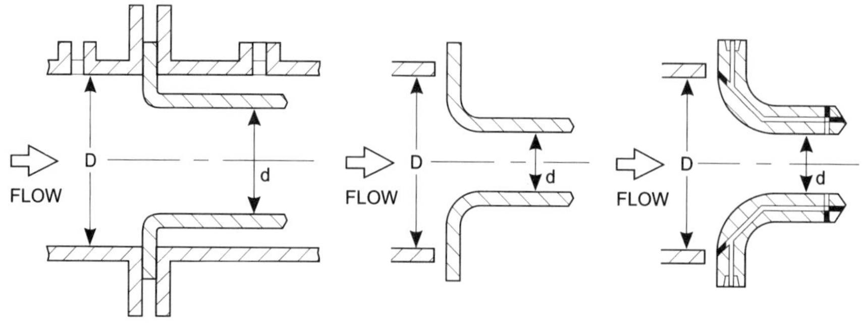

Figure 8-7. ASME Long Radius Flow Nozzle

popular for testing other gas meters. Standard flow nozzles reach critical flow when the pressure drop across the nozzle is approximately fifty percent of the upstream absolute pressure. When a divergent section is added to the flow nozzle, the choking pressure ratio is only about five to ten percent of the upstream absolute pressure. Therefore, critical flow can occur with a downstream absolute pressure of 90 to 95 percent of the upstream absolute pressure.

Pitot Tube and Pitot-Static Tube

A Pitot tube is a point velocity measuring device (Figure 8-8). It has an impact port; as fluid hits the port, its velocity is reduced to zero and kinetic energy (velocity) is converted to potential energy (pressure head). The pressure at the impact port is the sum of the static pressure and the velocity head. The pressure at the impact port is also known as stagnation pressure or total pressure. The pressure difference between the impact pressure and the static pressure measured at the same point is the velocity head. The flow rate is the product of the measured velocity and the cross-sectional area at the point of measurement. The Pitot tube has negligible permanent pressure drop in the line, but the impact port must be located in the pipe where the measured velocity is equal to the average velocity of the flowing fluid through the cross section.

The Pitot-static tube (Figure 8-9) has an integral construction of a Pitot tube and an annular tube with static pressure ports. The differential pressure between the two ports is the velocity head. This indication of velocity combined with the cross-sectional area of the pipe provides an indication of flow rate. The impact port must be located to measure the average fluid velocity.

The averaging Pitot has multiple impact pressure ports normal to the flow in the conduit (Figure 8-10). The impact pressures from all the ports are hydraulically averaged to provide a more accurate average flow rate indication, regardless of changes in the velocity profile of the fluid in the cross section of the conduit. Other designs of averaging Pitot flowmeters include those for non-circular ducts using a grid-mesh type.

Elbow Meter

This flowmeter is a pipe elbow in which a pressure differential is created by centrifugal force between the inside diameter and the outside walls of the pipe elbow. This flowmeter does not introduce any additional pressure loss in the sys-

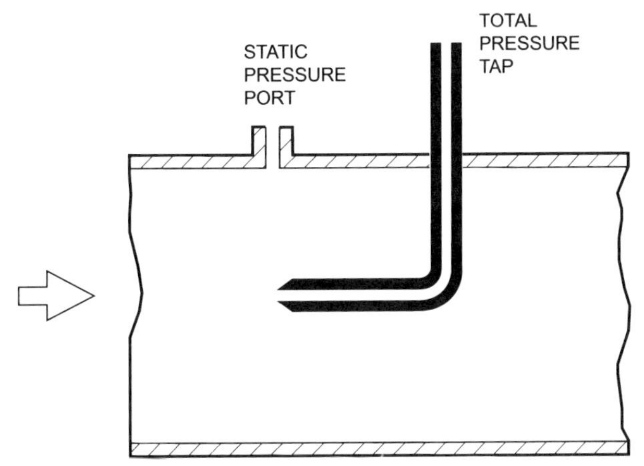

Figure 8-8. Pitot Tube

tem other than that already caused by the elbow, but its pressure differential is very low. Elbow flowmeters are used primarily for balancing loads among compressors and pumps in multiple-unit stations, performing efficiency tests, and providing flow signals using existing elbows in the piping.

Wedge Meter

This flowmeter consists of a wedge-shaped element placed perpendicular to the flow at the top of the conduit (Figure 8-11). The bottom of the conduit is un-

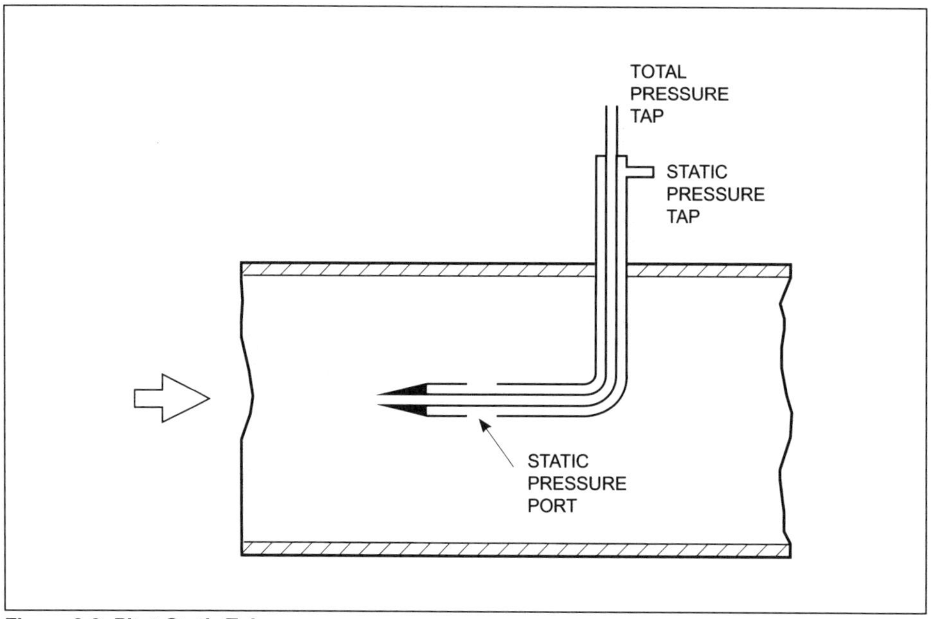

Figure 8-9. Pitot-Static Tube

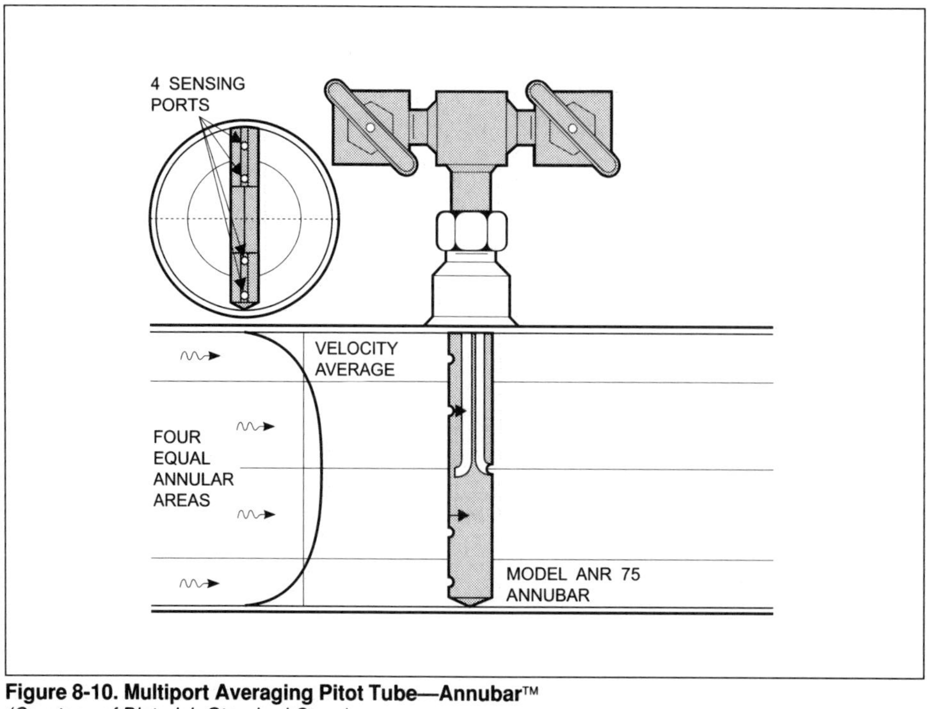

Figure 8-10. Multiport Averaging Pitot Tube—Annubar™
(Courtesy of Dieterich Standard Corp.)

restricted, making the wedge meter particularly useful in slurry measurement. Another type of wedge meter is available with two vertical wedges in the line.

V-Cone Flowmeter

This flowmeter consists of a V-shaped cone element placed at the center of the pipe, leaving an annular space for the passage of fluid (Figure 8-12). The beta ratio of this device is the ratio of the diameter of an axisymmetric hole that is equal in area to the annular opening of the V-cone flowmeter to the pipe diameter. The differential pressure generated by this device is lower than that of an orifice flowmeter with the same line size, beta ratio, and flow rate; hence, the permanent pressure loss is also lower.

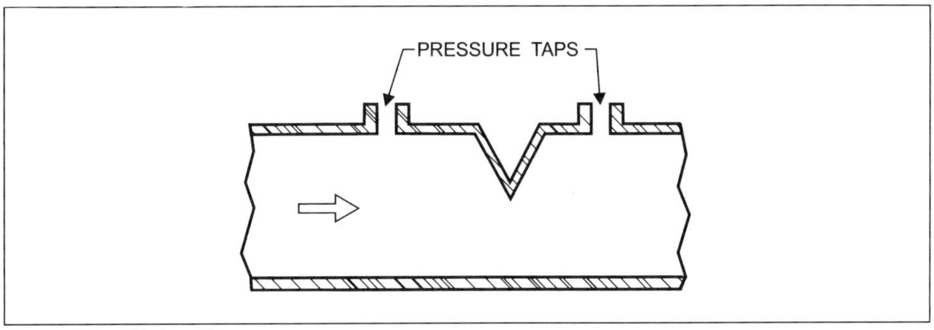

Figure 8-11. Wedge Meter

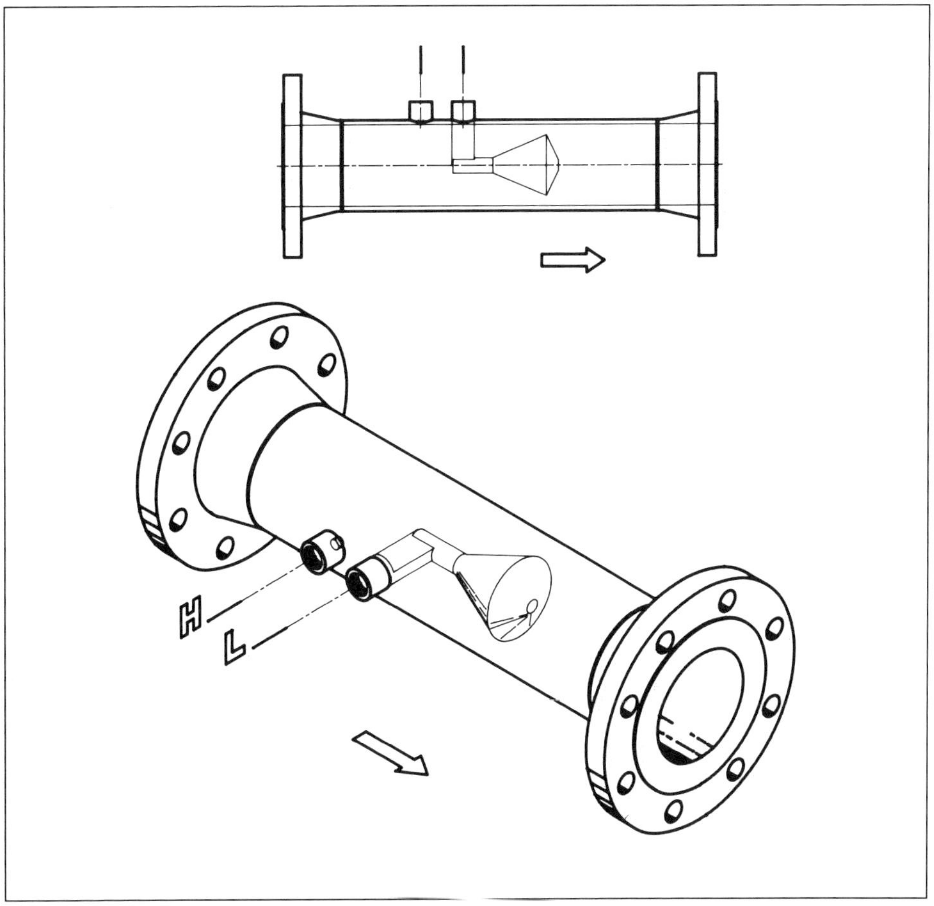

Figure 8-12. V-Cone Flowmeter
(Courtesy of McCrometer Division, Ketema, Inc.)

Spring-Loaded Variable Aperture

In an effort to extend the flow range of orifice flowmeters, a few proprietary designs that use a fixed orifice with a spring-loaded cone are available (Figure 8-13). With the change in flow rate, the differential pressure across the spring-loaded cone varies. To balance the force on the cone-spring assembly, the cone repositions with respect to the fixed orifice. Repositioning of the cone changes the aperture for the flow through the meter. Flow rate has an unique relationship with the differential pressure of the flowmeter and the position of the spring-loaded cone.

All designs are proprietary, and information for these devices must be obtained from the manufacturer. Performance data for these types of flowmeters is limited.

Laminar Flow Element

A laminar flow element is a flowmeter in which the differential pressure is linearly proportional to the flow. This phenomenon occurs when fluid flow through the flowmeter is in the laminar flow regime. The relationship between the differential pressure and the flow rate of fully developed laminar flow is defined by the Hagan-Poiseuille equation:

$$Q = \frac{\text{constant} \times \Delta P}{\mu_{cP} \times L} \tag{8-14}$$

where the constant is a proprietary multiplying factor for the flowmeter (its value is a function of the flowing fluid at a reference operating condition) and ΔP is the differential pressure across the flowmeter.

Equation (8-14) states that the flow rate is linearly proportional to the differential pressure and inversely proportional to the viscosity of the flowing fluid. Fluid viscosity is a function of the fluid temperature; thus, correction for changes in viscosity due to variation of fluid temperature is essential for accurate measurement. The effects of temperature on fluid viscosity can be minimized if the temperature of the flowing fluid in the flowmeter is controlled. The correction factors for temperature and pressure as a function of the difference from the reference temperature and pressure are provided by the manufacturer.

A number of different proprietary configurations of the laminar flow elements are available (Figure 8-14). Most designs utilize a capillary or tube bundle in the flow section. The value of the pipe Reynolds number (Equation (8-13)) is proportional to the diameter of the pipe for a given viscosity of the fluid. Therefore, the

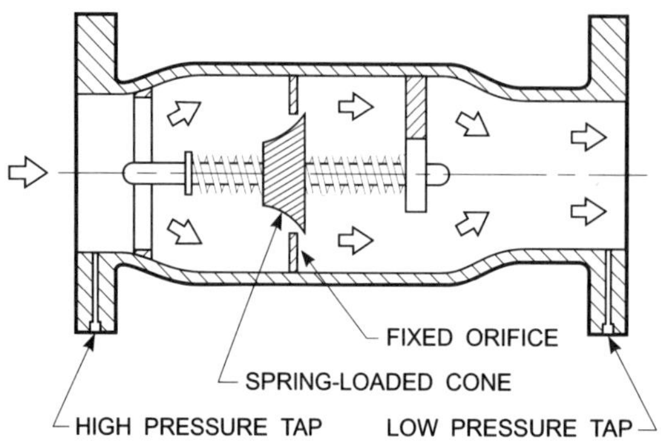

Figure 8-13. Spring-Loaded Variable Aperture

flow through a properly designed small diameter pipe can be made to be laminar; i.e., even when the pipe Reynolds number through a line is turbulent, by forcing the fluid to flow through a number of small diameter tube bundles and the small cross-sectional void areas between adjacent tubes of the bundle, flow through each of the flow path can be at a laminar flow condition. Laminar flow elements are often used in the automotive industry to measure the intake air flow rates during engine testing.

Special Types of Differential Flowmeters

A number of specially designed differential pressure devices have been developed over the years. A few of those designs, shown in Figure 8-15, include the Dall tube, the Elliot-Nathan flow tube, Epiflo, and so on. A Dall tube installed between two flanges similar to an orifice plate is called a Dall orifice flowmeter. These designs were developed in Europe and are rarely used in North America. Test data for these flowmeter is very limited. Even in Europe, recent installations of these specialty flowmeters has dropped significantly.

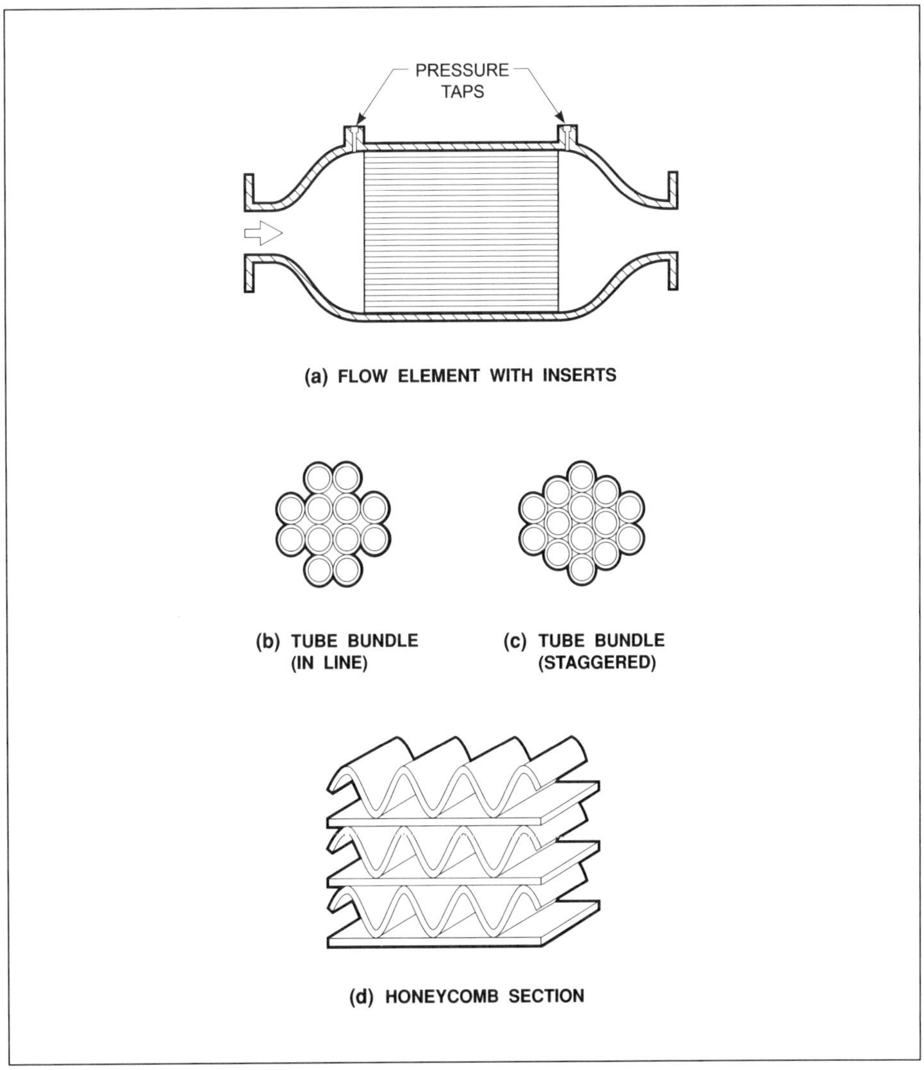

Figure 8-14. Laminar Flow Element

A special design of a nozzle with a diffuser cone on the downstream side can significantly reduce the permanent pressure loss across the nozzle. Such a device is generically known as a lo-loss tube.

Orifice Flowmeters

As the flow approaches the orifice plate (Figure 8-16), there is a slight increase in pressure on the upstream side of the orifice plate. After passing through the orifice bore restriction, the flow velocity increases and the pressure drops. This is a conversion of pressure head energy to velocity head energy. As the flow leaves the orifice, the velocity decreases and the pressure increases. Although the flow velocity downstream of the orifice recovers to the velocity upstream of the orifice plate, there is a permanent pressure loss across the flowmeter.

There are five common locations for the differential pressure taps (Figure 8-17). Flange taps are located 1 inch upstream and 1 inch downstream from the respective faces of the orifice plate. Pipe taps or full flow taps are located 2.5 pipe diameters upstream and 8 pipe diameters downstream from the respective faces of the orifice plate.

Vena contracta taps are located at one diameter upstream of the inlet face of the orifice plate, while the location of the downstream tap depends on the location of the vena contracta (which is a function of the beta ratio). With increasing beta ratio, the distance of the downstream tap decreases. For a beta ratio range of 0.8 to 0.15, the downstream tap location would change from 0.37 to 1.03 pipe diameters downstream of the upstream face of the orifice plate. The tolerance of the tap location is also a function of the beta ratio. Since the location of the

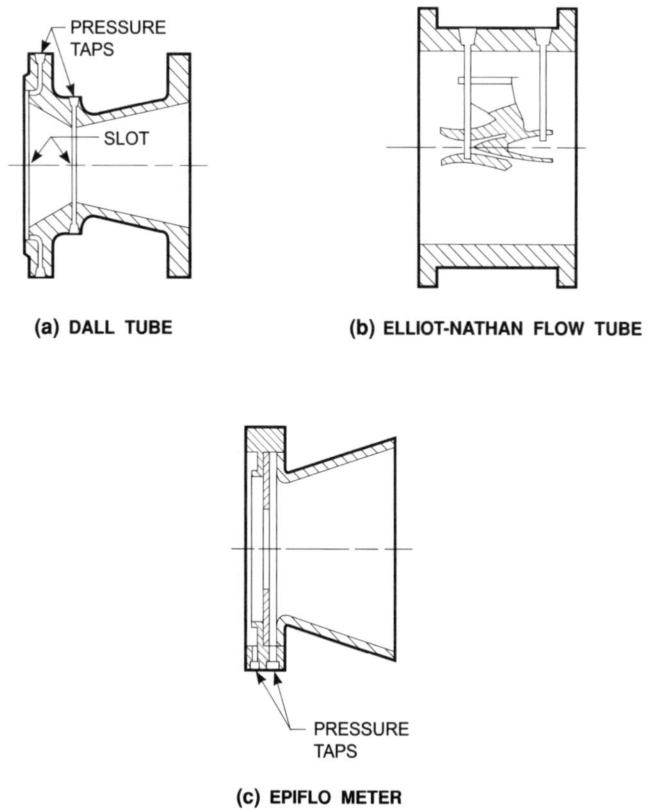

Figure 8-15. Schematics of Some Differential Pressure Flowmeter Designs

downstream tap is unique to the beta ratio of the meter tube this type of tap is not widely used.

Radius taps or D and D/2 taps are located 1.0 pipe diameter upstream and 0.5 pipe diameter downstream from the upstream face of the orifice plate. Corner taps are located at each face of the orifice plate. Note that the locations of the radius or D and D/2 taps and vena contracta taps are measured from the upstream face of the orifice plate, while all other taps are located with reference to both faces of the orifice plate.

The differential pressure measured across full flow taps is the permanent pressure drop across the orifice. The differential pressure measured across flange taps and radius taps is close to the maximum pressure drop through the orifice and will be greater at a given flow rate than the pressure drop measured across pipe taps. Flange taps are the most commonly applied taps in North America for pipe sizes of 2 inches and larger. Corner taps are typically used in pipe sizes of 1.5 inch and smaller. Radius taps are typically used in large pipe sizes. Corner taps are applied in Europe, and rarely in North America. Vena contracta taps are rarely applied, since the location of the vena contracta moves with orifice size and flow velocity.

The permanent pressure loss across an orifice or a nozzle is a function of the differential pressure across the primary element and β. An approximate value of the permanent pressure loss of an orifice or a nozzle is:

$$h_{\text{loss}} = h \times (1 - \beta^2) \qquad (8\text{-}15)$$

where h is the differential pressure measured across the primary element.

The line pressure upstream of a 4-inch orifice flowmeter is 40 psig. The orifice bore is 2.5 inches, and the differential pressure at the flange tap is 87 inches of water column. What is the line pressure downstream of the flowmeter? One inch of water is 0.03612628 psi.

Solution:

$$\begin{aligned}\text{Beta ratio} &= \frac{2.5}{4} \\ &= 0.625\end{aligned}$$

Using Equation (8-15), the estimated permanent head loss is:

$$\begin{aligned}h_{\text{loss}} &= 87 \times (1 - 0.625^2) \\ &= 53.02 \text{ inches of water} \\ &= 0.03612628 \times 53.02 \\ &= 1.915 \text{ psi}\end{aligned}$$

Therefore, the downstream line pressure is approximately 38 psig.

Flange taps are most commonly used for several reasons:

(1) More experimental data are available for flange tap coefficients than for any other tap locations.

(2) Flange tap meter tubes are shorter and less expensive than pipe tap meter tubes in smaller sizes.

(3) Differential pressure transmitters can be directly connected (close-coupled) to flange taps, allowing shorter impulse lines.

(4) The measured differential pressure across flange taps is higher than the permanent pressure loss across the orifice, allowing increased instrument resolution.

The static pressure can be measured at the upstream or the downstream differential pressure tap. At high static pressures, the difference between the upstream and downstream static pressures becomes relatively insignificant. At low pressures, however, this difference can be a large percentage of the total static pressure, and proper corrections must be made. The downstream connection is the most commonly used in gas measurement because the magnitude of the correction is only half of that with an upstream pressure connection.

Orifice Plate

The orifice plate is the heart of the orifice meter. It is the device that restricts the flow and thus develops the differential pressure that is proportional to the square of the flow rate. The quality, installation, and maintenance of the orifice plate are among the greatest influence factors on the accuracy of measurement achieved.

Orifice plates have two designs: the paddle plate and the universal plate (Figure 8-18). The paddle plate is used in orifice flanges and has a handle to facilitate installation between flanges. Generally, the nominal line size, the flange rating, the orifice bore diameter, and the word "INLET" are stamped on the upstream side of the handle. To assure that the orifice plate is centered in the line, the outside diameter of a paddle plate varies with the ANSI pressure rating of the flanges in which it is to be installed. When the outside diameter of the orifice plate is equal to the diameter of the bolt circle plus one bolt hole diameter minus two bolt diameters, all of the flange bolts will be forced to the outside of the bolt holes when the orifice plate is properly installed. Other methods to ensure centering of a paddle-type plate include (a) making the outside diameter of the orifice plate equal to the inside diameter of the bolt holes along with placing sleeves over the flange bolts and (b) installing dowel pins in the flanges.

The universal orifice plate is a circular plate designed for use in orifice fittings or in the plate holders of ring-type joint (RTJ) orifice flanges (Figure 8-19). Since the universal orifice plate is placed in a holder, the outside diameter is uniform for all pressure ratings in any given size. However, when using an orifice fitting, the internal diameter of the meter tube must be specified because the orifice plate is held in an orifice plate sealing unit. Seals are made of rubber, Viton,™ Teflon,™ or metal that is placed around the outside diameter of the orifice plate to provide

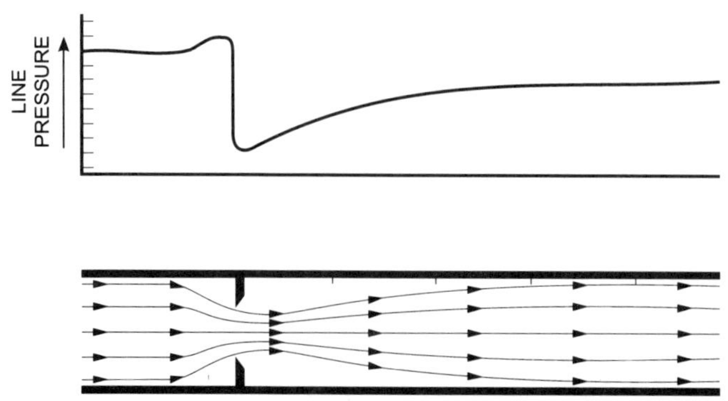

Figure 8-16. Flow Velocity and Pressure Distribution of an Orifice Flowmeter

a pressure seal on the upstream and downstream sides of the plate. The internal diameter of this orifice plate sealing unit must be equal to the internal diameter of the meter tube and orifice fitting.

The most common orifice bore is the concentric bore, where the orifice is a circle bored in the center of the orifice plate. This is the only type generally accepted for use in custody transfer measurement, since adequate data is not available for other bores. When measuring a wet gas such as saturated steam, a weep

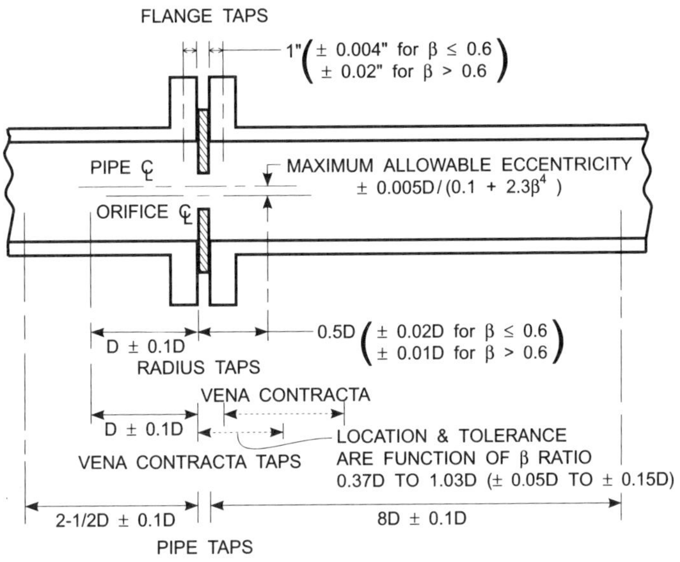

(a) FLANGE TAPS, RADIUS OR D-D/2 TAPS, VENA CONTRACTA TAPS, AND PIPE TAPS

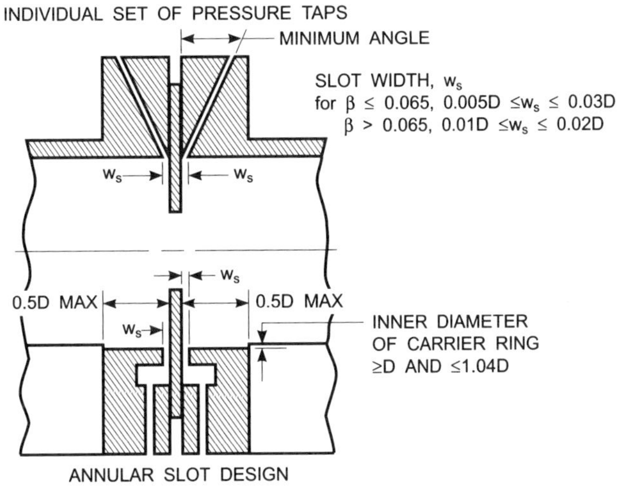

(b) CORNER TAP - INDIVIDUAL PRESSURE TAP AND ANNULAR SLOT DESIGN

Figure 8-17. Pressure Tap Spacing

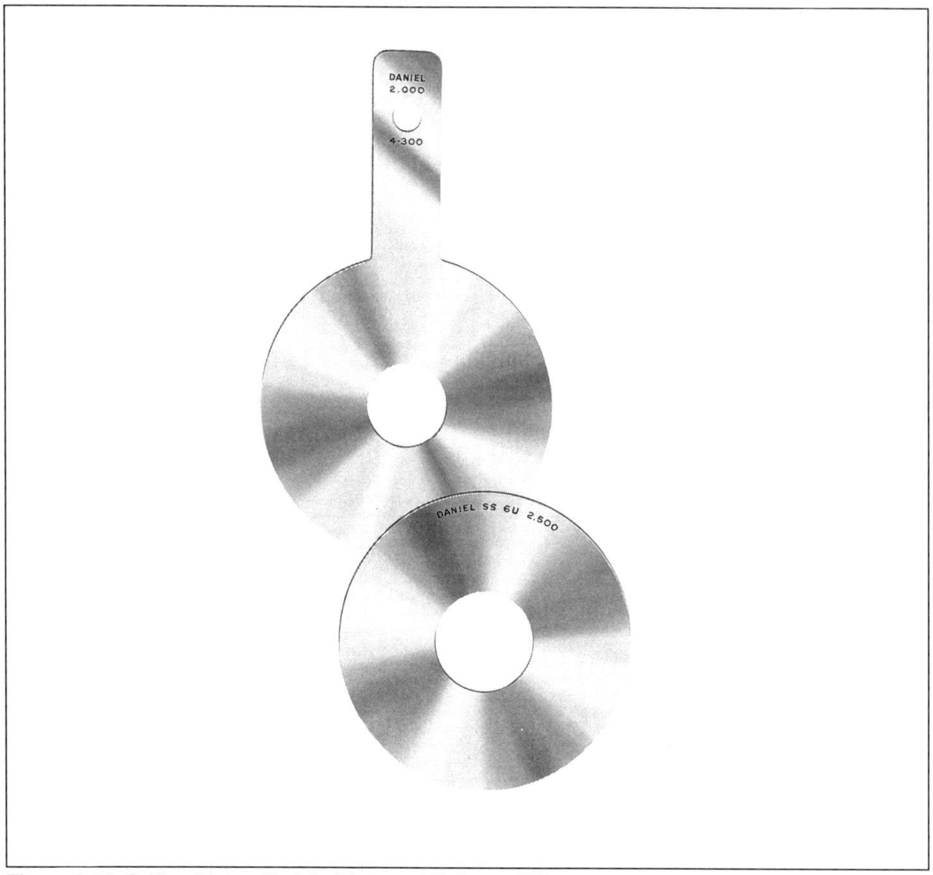

Figure 8-18. Orifice Plate—Paddle Plate and Universal Plate
(Courtesy of Daniel Industries, Inc.)

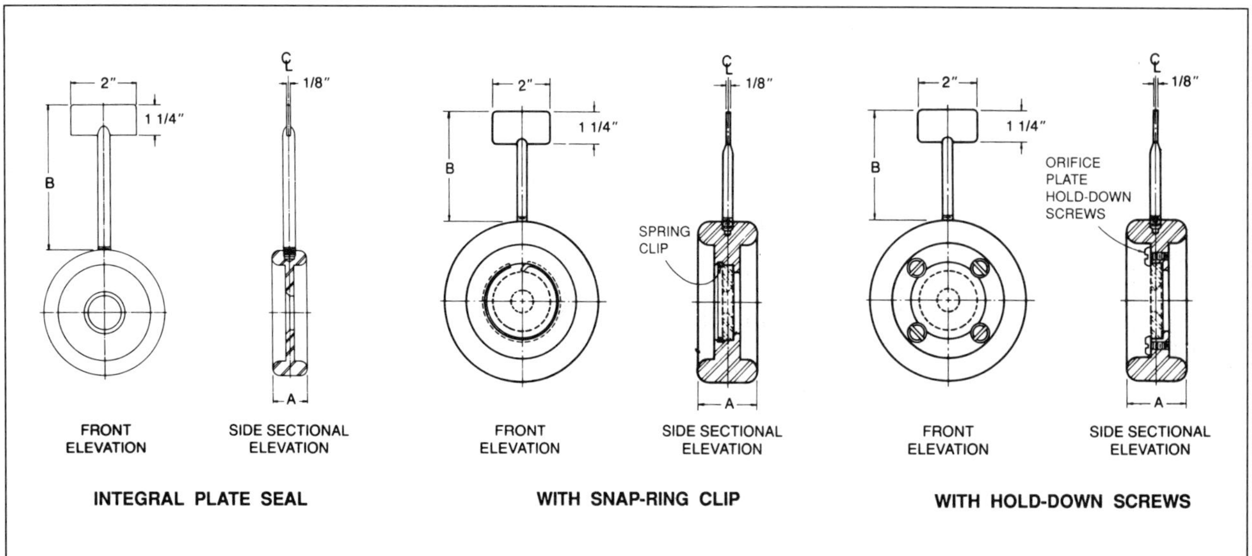

Figure 8-19. Ring-Type Joint Orifice Plate Holder
(Courtesy of Daniel Industries, Inc.)

hole is sometimes placed in a concentrically bored orifice plate. This is a small hole in the plate, and the plate is installed with the hole near the bottom pipe wall to allow the passage of liquids. The area of the weep hole must be considered when calculating a flow coefficient. An orifice plate with weep hole should not be used for accurate measurement since they plug easily, and its condition cannot be verified without removing the orifice plate.

Though the orifice plate seems like a relatively simple device, it is a precision measurement instrument. Critical items include:

(1) the flatness of the orifice plate,

(2) the smoothness of the orifice plate surface,

(3) the cleanliness of the orifice plate surface,

(4) the sharpness of the upstream orifice edge,

(5) the diameter of the orifice bore, and

(6) the thickness of the orifice edge.

An orifice plate is acceptable if it does not depart from flatness along any diameter by more than 0.010 inch per inch of dam height $(D - d)/2$ (Figure 8-3). Flatness can be checked by placing a machinist's straight edge across the face of the orifice plate and using a feeler gage to measure the departure from flatness.

The orifice plate shall have a surface smoothness not exceeding 50 microinches.

The upstream edge of the orifice shall be square and sharp so that it will not reflect a beam of light when viewed without magnification. A simple way to check the edge sharpness is to hold the orifice plate in a bright beam of light, such as bright sunlight. A sharp orifice should not reflect any light when viewed with the naked eye.

The diameter of the orifice bore should be measured on three equally spaced diameters. Using these three dimensions, calculate the average bore diameter. No measured diameter shall differ from any other measured diameter by an amount greater than the tolerances given in Table 8-1. To assure that the orifice bore diameter is within the required tolerances, it is necessary to check with an internal micrometer or vernier caliper. The vernier caliper should be of the non-knife edge type. The orifice bore diameter should be verified during each visual inspection of the orifice plate. A scale, ruler, or tape measure can be used for verification; the measured bore diameter should be compared to the bore diameter stamped on the plate and recorded on the meter record.

The orifice diameter should be determined when the orifice plate is at 68°F (20°C). If the orifice plate is operating at a temperature other than 68°F (20°C), average bore diameter should be adjusted using the following formula:

Table 8-1. Practical Tolerance for Orifice Bore (in Inches)

Orifice Bore(d)	Tolerance
0.250	0.0003
0.375	0.0004
0.500	0.0005
0.625	0.0005
0.750	0.0005
0.875	0.0005
Over 1.000	0.0005 per inch of diameter

For 304 and 316 stainless steel,

$$d = d_{meas} \times [1 + 0.0000185 \times (T_{°F} - 68)] \tag{8-16}$$

For Monel,

$$d = d_{meas} \times [1 + 0.0000159 \times (T_{°F} - 68)] \tag{8-17}$$

where d_{meas} is the actual measurement at temperature $T_{°F}$.

For a stainless steel orifice plate, a change of 27°F (15°C) will cause a change of 0.0005 inch per inch of bore diameter. For a Monel plate, a change of about 31°F (17°C) will cause the same change. It is, therefore, important to consider the temperature of the orifice plate when measuring the bore diameter or making precise flow calculations.

Limits of the thickness of the orifice edge are tabulated in Table 8-2. The thickness of the orifice plate shall not exceed 1⁄50 of the line internal diameter (D) or 1⁄8 of the orifice bore diameter (d), with the smaller of the two governing. This is most commonly achieved by beveling the downstream edge of the orifice. The bevel is machined on a 30 to 45 degree angle to the desired edge thickness, with 45 degrees being most common. An alternative is the bore and counterbore method in which, instead of beveling at the normal 45 degrees, the plate is counterbored to the desired edge thickness. When using a beveled plate, it is essential to insert the plate with the sharp edge upstream and the bevel downstream. A bore and counterbore plate must be installed with the flat side upstream and the counterbore downstream.

The orifice plate thickness and orifice edge thickness should be checked with a micrometer or caliper to achieve measurement within the required tolerances. If the orifice plate is being used in an orifice fitting, the orifice plate thickness should be compared to the distance between the orifice tap holes in the fitting. The orifice tap hole separation is 2 inches plus the orifice plate thickness. For a 1⁄8-inch thick orifice plate, the orifice tap hole separation should be 2 1⁄8 inches. For proper location of the tap with respect to the face of the orifice plate, the gasket thickness is taken into account in the fabrication of the flowmeter.

When sizing an orifice plate, selection of proper β (d/D) is important. The ANSI/API 2530 Standard recommends that beta ratios be limited as follows:

(1) With meters using flange taps, β shall be between 0.15 and 0.70.

(2) With meters using pipe taps, β shall be between 0.20 and 0.67.

Accepted empirical equations predict the discharge coefficients of orifice flowmeters with an uncertainty of ± 0.5 percent for flange taps and about ± 0.75 percent for pipe taps when beta ratios are within the above limits. These limits may be exceeded when additional flow uncertainty is acceptable. When the above limits are exceeded, uncertainty may be greater than ± 1 percent for meters with flange taps and may be as high as ± 1.5 percent for meters with pipe taps. A practical consideration in sizing orifice plates is to use "even" bore sizes. Even bores increase in increments of 0.125 inch through 4.000-inch bores, in increments of 0.250 inch through 12.000-inch bores, and in increments of 0.500 inch through 21.500-inch bores. Plates with these bores are generally stocked in the popular pipe sizes by major suppliers of orifice measurement equipment, reducing delivery times and (often) cost. With fewer orifice bore sizes in use, and all sizes in even fractional increments, the chance of mistake in flow calculations is greatly reduced.

Maintenance of an orifice plate consists of cleaning any foreign material from the surfaces and visually inspecting the orifice plate to see that it has not been physically damaged. The plate flatness and tolerances are shown in Table 8-3. On

Table 8-2. Orifice Plate Dimensions *(Courtesy of American Petroleum Institute, API 2530)*

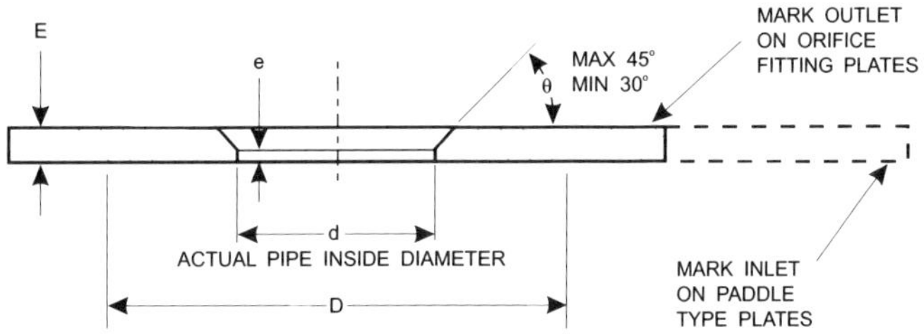

	Nominal Inside Diameter, in Inches																		
Published Inside Diameter	**2**	**3**		**4**		**6**		**8**		**10**	**12**		**16**		**20**	**24**		**30**	
	1.687		2.624	3.152						9.562					18.812				
	1.939		2.900	3.438		4.897	5.761		7.981	10.020		11.938		15.000	19.000		23.000		29.000
	2.067	2.300	3.068	3.826	4.026	5.187	6.065	7.625	8.071	10.136	11.374	12.090	14.688	15.250	19.250	22.624	23.250	28.750	29.250
	Orifice Plate Thickness, *E*, in Inches																		
Minimum	0.115	0.115	0.115	0.115	0.115	0.115	0.115	0.115	0.115	0.115	0.175	0.175	0.175	0.175	0.240	0.240	0.240	0.370	0.370
Maximum	0.130	0.130	0.130	0.130	0.130	0.163	0.192	0.254	0.269	0.319	0.379	0.398	0.490	0.500	0.505	0.505	0.505	0.562	0.578
Recommended	0.125	0.125	0.125	0.125	0.125	0.125	0.125	0.125	0.125	0.250	0.250	0.250	0.375	0.375	0.375	0.375	0.375	0.500	0.500

Orifice Diameter *d*	**$e \leq d/8$**	**Maximum Orifice Edge Thickness, e, in Inches**																		
0.250	1/32	× 1/32	× 1/32	1/32	1/32	1/32														
0.375	3/64			× 3/64	3/64	3/64														
0.500	1/16				× 1/16	1/16	1/16	1/16												
0.625	5/64					× 5/64	5/64	5/64												
0.750	3/32						× 3/32	3/32												
0.875	7/64							× 7/64	7/64	7/64										
1.000	1/8								1/8	1/8	1/8									
1.125	9/64								× 9/64	9/64	9/64									
1.250	5/32									× 5/32	5/32	5/32	5/32							
1.375	11/64										11/64	11/64	11/64							
1.500	3/16										× 3/16	3/16	3/16	3/16	3/16					
1.625	13/64											13/64	13/64	13/64	13/64					
1.750	7/32											× 7/32	7/32	7/32	7/32					
1.875	15/64												× 15/64	15/64	15/64					
2.000	1/4													1/4	1/4	1/4				
2.250	9/32													× 9/32	9/32	9/32				
2.375	19/64														× 19/64	19/64	19/64	19/64		
2.500	5/16															5/16	5/16	5/16		
2.750	11/32															11/32	11/32	11/32		
2.875	23/64															23/64	23/64	23/64	23/64	23/64
3.000	3/8															× 3/8	3/8	3/8	3/8	3/8
3.250	13/32																13/32	13/32	13/32	13/32
3.500	7/16																× 7/16	7/16	7/16	7/16
3.625	29/64																	× 29/64	29/64	29/64
3.750	15/32																		15/32	15/32
4.000	1/2																		1/2	1/2
4.250	17/32																		17/32	17/32
4.500	9/16																		× 9/16	9/16
4.625	37/64																			× 37/64
4.750	19/32																			
5.000	5/8																			

Notes:

1. The maximum edge thickness is defined by $e \leq D/50$ or $e \leq d/8$, whichever is smaller.
2. Orifice edge thickness marked with × in this table is the maximum for that particular meter tube diameter and is applicable to all larger orifice diameters for that meter tube diameter.
3. Orifice diameters smaller than those marked × are defined by $e \leq d/8$.
4. Orifice plates of which the edge thickness meets the value $e \leq D/30$ need not be rebeveled unless reconditioning is required for other reasons.
5. All dimensions are in inches. For ease in machining, the next smaller value of *e* in even multiples of 1/16 or 1/32 inch may be used where *e* is given in 1/64ths.
6. Orifices used to measure dual directional flows must not be beveled. Where *e* exceeds the above limits, the flow constant F_b may be subject to higher uncertainty.

Table 8-3. Orifice Plate Flatness Tolerance *(Courtesy of American Petroleum Institute, API 2530)*

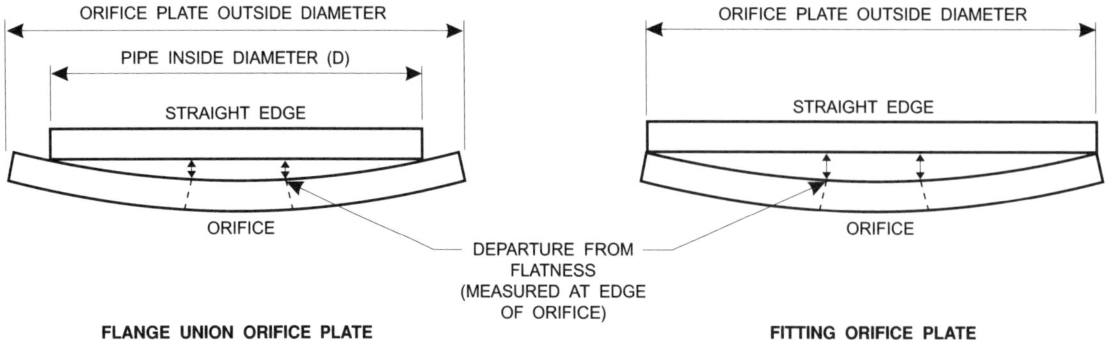

Orifice Diameter, Inches	Maximum Departure from Flatness, in Inches										
	Nominal Meter Tube Size, Inches										
	2	3	4	6	8	10	12	16	20	24	30
0.125											
0.250	0.009										
0.375	0.008										
0.500	0.008	0.013									
0.625	0.007	0.012	0.017								
0.750	0.007	0.012	0.016	0.027							
0.875	0.006	0.011	0.016	0.026	0.036						
1.000	0.005	0.010	0.015	0.025	0.035	0.046					
1.250	0.004	0.009	0.014	0.024	0.034	0.044	0.054				
1.500	0.003	0.008	0.013	0.023	0.033	0.043	0.053	0.067			
1.750		0.007	0.011	0.022	0.032	0.042	0.052	0.066			
2.000		0.005	0.010	0.020	0.030	0.041	0.050	0.065	0.085		
2.250		0.004	0.009	0.019	0.029	0.039	0.049	0.063	0.083		
2.500			0.008	0.018	0.028	0.038	0.048	0.062	0.082	0.102	
2.750			0.006	0.017	0.027	0.037	0.047	0.061	0.081	0.101	
3.000			0.005	0.015	0.025	0.036	0.045	0.060	0.080	0.100	0.130
3.250				0.014	0.024	0.034	0.044	0.058	0.078	0.098	0.128
3.500				0.013	0.023	0.033	0.043	0.057	0.077	0.097	0.127
3.750				0.012	0.022	0.032	0.042	0.056	0.076	0.096	0.126
4.000				0.010	0.020	0.031	0.040	0.055	0.075	0.096	0.125
4.500				0.008	0.018	0.028	0.038	0.052	0.072	0.092	0.122
5.000					0.015	0.026	0.035	0.050	0.070	0.090	0.120
5.500					0.013	0.023	0.033	0.047	0.067	0.087	0.117
6.000					0.010	0.021	0.030	0.045	0.065	0.085	0.115
6.500						0.018	0.028	0.042	0.062	0.082	0.112
7.000						0.016	0.025	0.040	0.060	0.080	0.110
7.500						0.013	0.023	0.037	0.057	0.077	0.107
8.000							0.020	0.035	0.055	0075	0.105
8.500							0.018	0.032	0.052	0.072	0.102
9.000							0.015	0.030	0.050	0.070	0.100
9 (see note)											

Note: For larger sizes, calculate maximum departure from flatness equals 0.005 $(D - d)$.

plates that are new to the installation, the actual dimensions of the orifice bore should be verified with a micrometer. Visual inspection should check items such as:

(1) material buildup on the surface of the plate, which includes such materials as paraffin and salt encrustation;

(2) surface defects, such as deep scratches or surface etching;

(3) liquid film on the plate surface;

(4) liquid level line on orifice plates in gas applications indicating liquid build-up in the meter tube; and

(5) orifice plate installed backwards.

If the orifice plate is dirty, it should be cleaned and reinstalled in the line. If it is in backwards, it should be reinstalled correctly. If it has defects, it should be replaced. In all cases, the differential pressure after replacement should be compared with the differential pressure before inspection to determine the extent of error that might have existed.

Orifice Plate Holder

The orifice plate is held in the line in an orifice plate holder, which may be a pair of orifice flanges or an orifice fitting. The original method of holding the orifice plate was to place it between two flanges. This is still the least expensive method and is used extensively where frequent orifice plate changes or inspections are not required. Orifice fittings simplify removal of the orifice plate for inspection or changing and assure that the orifice plate is properly centered in the line.

Orifice Flange

The orifice flange is a special flange with tap holes drilled in a location such that they will be located 1 inch from each face of the orifice plate when $\frac{1}{16}$-inch gaskets are used on either side of the orifice plate. Orifice flanges are equipped with jack screws to permit spreading them apart when it is necessary to remove the orifice plate. Flanges without jack screws may require flange spreaders.

The most common type of orifice flange is the raised face (RF) flange. The face of this flange has a flat surface, and the orifice plate is held between gaskets between these flat surfaces. For high pressure service (900 lb ANSI ratings and above) the ring-type joint (RTJ) flange is often used. The pressure seal with this type of flange is a ring that fits in grooves in the face of each of the flanges, and a special orifice plate holder combines the ring and a universal orifice plate in a single assembly.

In orifice flange unions, gaskets provide a seal to contain line pressure as well as prevent flow from bypassing the orifice. The gasket material must be compatible with the flowing fluid, and the gasket should be $\frac{1}{16}$-inch thick so the orifice tap holes will be the proper distance from the face of the orifice plate. The outside diameter of the gasket should be the same as the outside diameter of the orifice plate to assure centering of the gasket in the line. The internal diameter of the gasket should be the same as the internal diameter of the meter tube to prevent an abrupt change in meter tube diameter at the face of the plate.

Where ease of orifice plate removal is desired, two types of orifice fittings are available. The less expensive type is the single chamber fitting. This fitting has a slot in which the orifice plate is held and a simple closure on the top that, when opened, permits easy removal of the orifice plate and its holder. The pressure must be removed from the line before the fitting can be opened, but there is no need to spread flanges or remove gaskets when changing the orifice plate. The orifice fitting also assures that the tap holes remain exactly 1 inch from the face of the plate and that the plate is centered in the line when the plate holder is properly installed.

The other design is a double-chamber orifice fitting that allows removal of the orifice plate without depressurizing the line or interrupting the flow. This orifice fitting has two chambers: the line or lower chamber and an upper chamber that can be isolated from the line. The orifice plate carrier is mechanically moved from the line chamber into the upper chamber, and a valve is closed to isolate the upper chamber from the lower chamber. After depressurizing the upper chamber, the cover can be removed to extract the orifice plate from the fitting without interrupting flow. The orifice plate can be installed by reversing the procedure.

Seals and Gaskets

In orifice fittings, it is necessary to provide a sealing unit around the orifice plate to prevent flow from bypassing the orifice. Several orifice fitting manufacturers use a sealing unit that mounts directly on the universal type of orifice plate (Figure 8-20). Others mount the orifice plate in a metal plate holder with O-ring grooves and use O-rings to seal the plate to the carrier and the carrier to the fitting.

The dual seal is the simplest and most practical orifice plate sealing device for typical flowing streams. Seals are precision molded nitrile synthetic rubber or Viton if swelling is a problem. It provides four rubber-to-metal sealing surfaces 360° around the plate. The unit positively seals against both outer faces of the plate and against both seats of the orifice fitting to prevent leakage. The dual seal is used in pressure ratings to 600 lbs ANSI and in temperatures from +20°F to +275°F (-7°C to +135°C).

The snap seal ring is a removable orifice plate holder designed for use in services where elastomer seal swelling is a problem. The snap seal ring unit consists of two symmetrical metal rings, each one having an O-ring on both sides for a positive seal on the plate side and the fitting side of the ring. The assembled unit provides a full 360° rubber-to-metal seal around both sides of the orifice plate, even in the absence of any pressure differential. The snap seal ring is designed for service temperatures ranging from -67°F to +437°F (-55°C to +225°C).

The Teflon seal is designed for difficult, corrosive flows and for higher temperatures than the previous units can handle but is not recommended for natural gas service. Teflon is not flexible and is likely to leak in gas service. The Teflon seal has proven effective in such flows as dilute sulfuric acid, fuming nitric acid, hydrazine, liquid oxygen, and other fluids from -65°F to +500°F (-51°C to +260°C).

The metal seal is a stainless or cadmium-plated carbon steel clip-ring assembly recommended for high pressures and for temperatures to +1200°F (650°C). Metal seal units of cadmium-plated carbon steel are recommended for services to +600°F (+315°C), standard 316 stainless steel units to +1000°F (+520°C), and 316 stainless steel units with an Inconel™ spring to +1200°F (+650°C).

Orifice Fitting Design

A common configuration for an orifice fitting is a welded connection on the upstream side and a flanged connection on the downstream side. When using this type of fitting, the upstream tube is welded to the fitting and the inside of the weld is reached through the fitting for grinding. The downstream tube is welded to a flange that is mated to the flange on the fitting. To assure perfect alignment, dowel pins may be inserted in the fitting with matching holes in the meter tube flange.

The choice of an orifice plate holder varies with the installation. The following are some general considerations.

ORIFICE FLANGE UNION

This device is used when the plate will not be changed often, such as a flow control application with a steady flow rate.

SINGLE-CHAMBER ORIFICE FITTING

This is used when the plate must be removed for inspection or size changing at frequent intervals and it is possible to interrupt the flow or bypass the meter run for a short period of time to depressurize the line. Most applications for this type of fitting are in line sizes of 6 inch and smaller.

Figure 8-20. Orifice Plate Seals
(Courtesy of Daniel Industries, Inc.)

DOUBLE-CHAMBER ORIFICE FITTING

This is the most expensive but also the most flexible of orifice fittings. The orifice plate can be removed from the line without interrupting the flow. When flow cannot be interrupted, and cost and delivery are not critical, the double-chamber fitting is the fitting of choice.

Orifice Meter Tube

The complete meter tube assembly consists of an orifice plate, an orifice plate holder, straight pipe upstream and downstream of the orifice plate, and possibly straightening vanes. A meter tube is the straight upstream pipe of the same size between the orifice and the nearest pipe fitting and the similar downstream pipe between the orifice and the nearest pipe fitting. There shall be no pipe connections within these distances other than the orifice pressure taps. A thermometer well may be installed in the meter tube upstream of the straightening vanes, if used.

The selection of the proper size of meter tube is determined by the volume of fluid to be measured. The first decision in sizing the meter tube is to determine the range of the differential pressure instrument that will be used. After this decision has been made, a calculation can be made of the orifice diameter necessary to produce the proper differential pressure at the maximum flow rate for a given line size. The following are factors that must be considered in selecting the meter tube diameter:

(1) The β ratio should be within the limits of the accepted standard. The following are allowable limits of ANSI/API 2530:

  (a) When using flange taps, β shall be between 0.15 and 0.70.

  (b) When using pipe taps, β shall be between 0.20 and 0.67. Many companies, especially for custody transfer, require a minimum β of 0.20 in all cases and a maximum β of 0.60.

(2) Standard nominal pipe sizes should be selected; e.g., 2, 3, 4, 6, 8, 10, 12, 16, 20, 24, 30 inch. The wall thickness, or schedule, of the pipe depends upon the maximum working pressure of the line. If a choice of more than one size is possible, consideration should be given to the following:

  (a) Will future flow rates and/or operating pressures be higher or lower than the initial design conditions?

  (b) What is the relative cost of the larger versus the smaller line size?

  (c) Will the selection provide uniformity of size in a large installation with many meter tubes?

The length of the meter tube, both upstream and downstream of the orifice, is determined by the diameter of the pipe and by the upstream and downstream piping configuration. Any major distortion of the flow profile could generate flow measurement errors. There are many piping configurations for which an orifice flowmeter will not produce results within the expected uncertainty of the flowmeter. For accurate measurement, it is important that at least the recommended minimum lengths be used and that no obstructions be placed in the meter tube within these distances from the orifice plate.

A few of the more common types of piping installations (Figures 8-21 through 8-25) have been studied for their effect on the metering accuracy. Since many upstream piping and fitting configurations will result in a swirl and may take many diameters of straight lengths to dissipate, it is recommended that a flow conditioner be used in flow measurement installations where accuracy is important. Flow conditioners are used to enable the use of shorter upstream lengths. A number of flow conditioning devices and some proprietary designs are available.

For these configurations, recommended minimum meter tube lengths for accurate measurement are given in a number of different standards; e.g., ANSI/API 2530, AGA 3, ASME MFC-3M, and ISO 5167. As β increases, the required length of the meter tube increases. Some of these dimensions vary between different standards.

The recommended minimum lengths from API 2530 is given in Table 8-4 for the piping installations of Figures 8-21 through 8-25. The flow conditioners specified by API 2530 are of the tube bundle type. When the piping installation is not explicitly covered in installation sketches in Figures 8-21 through 8-25, it is recommended that the meter tube lengths match the lengths recommended by the partially closed valve upstream of the meter tube (Figure 8-21 and Table 8-4). For installations of Figure 8-22, if two ells are closely (less than $3D$) preceded by

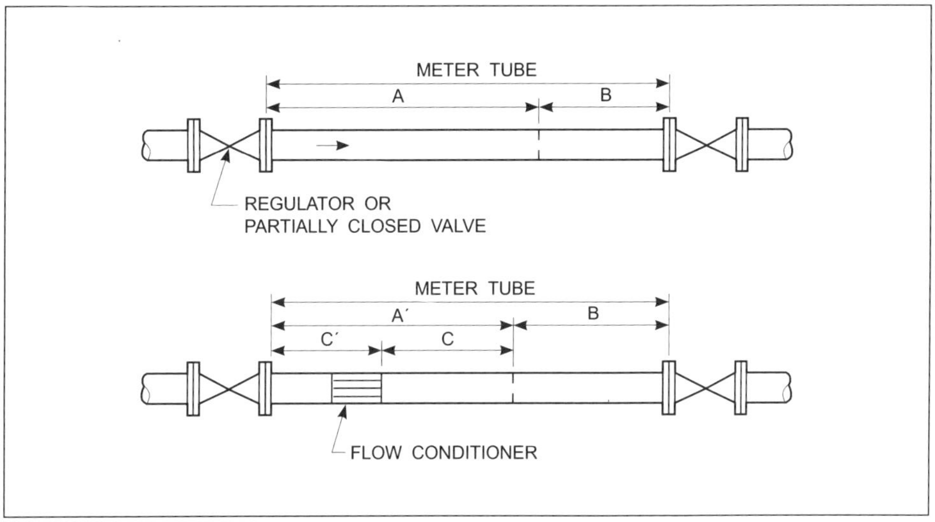

Figure 8-21. Partially Closed Valve Upstream of Meter Tube

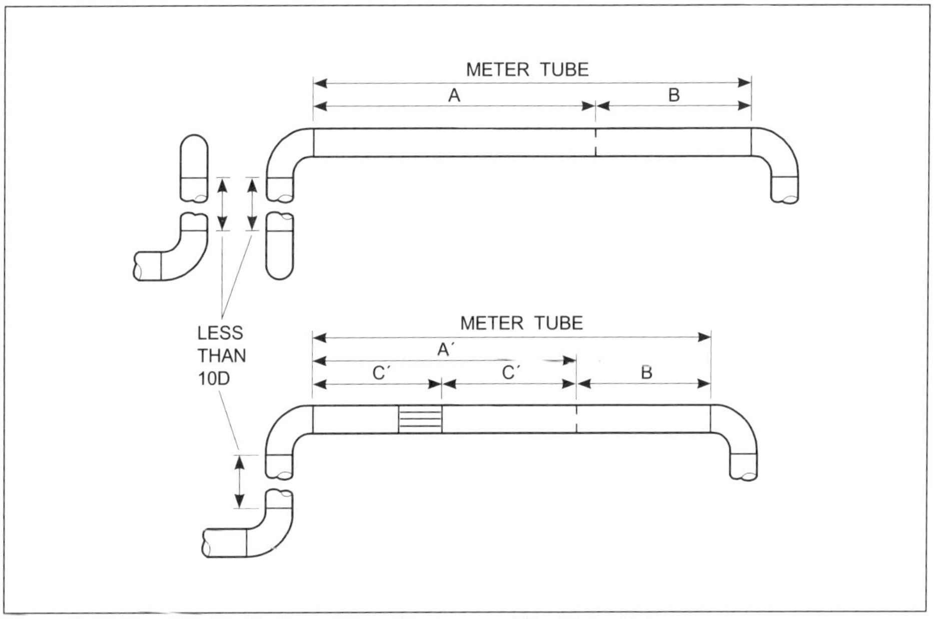

Figure 8-22. Two Ells Not in Same Plane Upstream of the Meter Tube

Table 8-4. Minimum Meter Tube Lengths in Terms of Pipe Diameters and Beta Ratio
(From ANSI/API 2530)

Installation	Dimension	Beta Ratio							
		0.1	0.3	0.4	0.5	0.6	0.67	0.7	0.75
Figure 8-21	A	14.0	16.0	18.0	25.2	30.2	36.0	38.8	44.5
	B	2.5	3.0	3.5	3.7	4.0	4.2	4.3	4.5
	A′	9.5	9.5	9.5	10.3	12.2	14.2	15.3	17.5
	C′	4.5	4.5	4.5	5.3	6.7	8.0	9.0	10.5
	C	5.0	5.0	5.0	5.0	5.5	6.2	6.3	7.0
Figure 8-22	A	14.0	16.0	18.0	21.0	25.0	28.8	31.3	35.3
	B	2.5	3.0	3.4	3.7	4.0	4.2	4.3	4.5
	A′	9.5	9.5	9.6	10.0	11.3	12.8	13.5	15.0
	C′	4.5	4.5	4.6	5.0	5.8	6.6	7.0	7.8
	C	5.0	5.0	5.0	5.0	5.5	6.2	6.5	7.2
Figure 8-23	A	8.5	8.5	8.5	10.2	13.7	17.4	18.9	21.5
	B	2.5	3.0	4.0	3.7	4.0	4.2	4.3	4.5
	A′	8.5	8.5	8.5	9.2	10.3	11.7	12.0	13.5
	C′	3.5	3.8	4.0	4.2	4.7	5.5	5.6	6.4
	C	5.0	5.0	5.0	5.0	5.6	6.2	6.4	7.1
Figure 8-24	A	6.0	6.0	6.0	6.9	9.3	12.3	13.9	16.6
	B	2.3	3.0	3.2	3.7	4.0	4.2	4.3	4.5
Figure 8-25	A	6.0	6.0	6.2	7.2	9.8	11.6	12.1	13.4
	B	2.5	3.0	3.3	3.7	4.0	4.2	4.3	4.5

Notes:

1. When pipe taps are used, lengths A, A′, C shall be increased by two pipe diameters and B by eight pipe diameters.
2. When diameter of orifice may require changing to meet different conditions, the lengths of straight pipe should be those required for the maximum orifice-to-pipe diameter ratio that may be used.

a third that is not in the same plane as the middle or second ell, the piping requirements of A should be doubled.

For installations with reducers or expanders, straightening vanes will not reduce required upstream length of straight pipe (length *A* of Table 8-4). Straightening vanes are not required for the reducers, but they may be required because of other fittings that precede the reducer. Length *A* is to be increased by an amount equal to the length of the straightening vanes whenever they are used. If it is suspected that the flowing fluid may be partially condensed, the expander type of installation, as well as any other configuration that might create two-phase flow in the meter tube, should be avoided.

API 2530 is the most widely accepted standard for orifice metering of natural gas. The straight run requirements given by the ASME-MFC-3M and ISO 5167 (for fluids other than natural gas) usually have more stringent requirements than those of API 2530.

The concentricity of the pipe (sometimes referred to as ovality) must meet certain requirements. According to ANSI/API 2530, the pipe concentricity must be maintained within 3 percent of the inside diameter of both the upstream and downstream sections of the orifice meter tube.

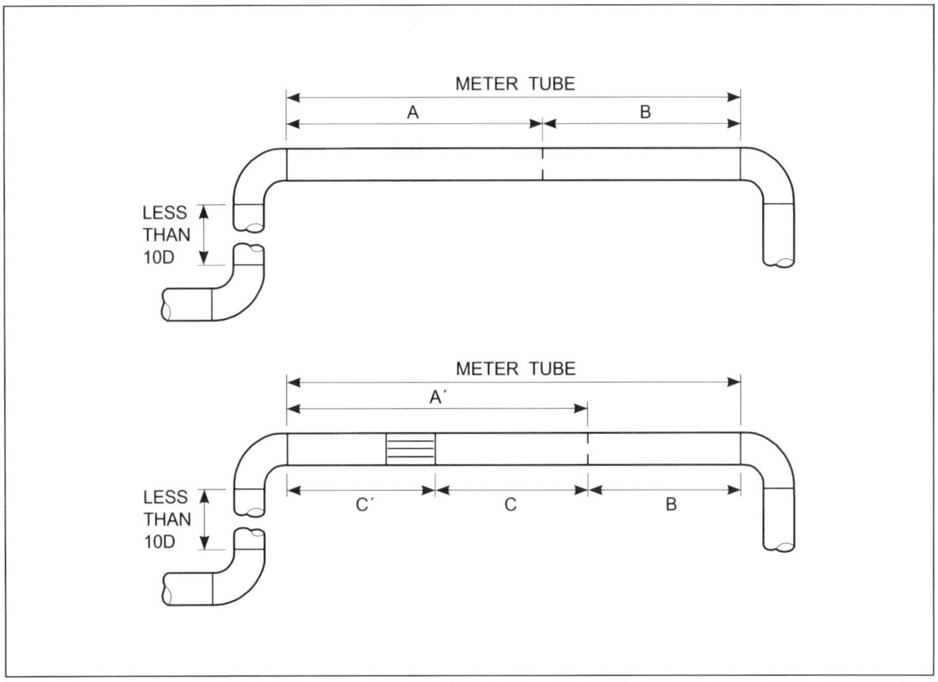

Figure 8-23. Less Than Ten Pipe Diameters (*D*) between Two Ells in Same Plane Upstream of Meter Tube

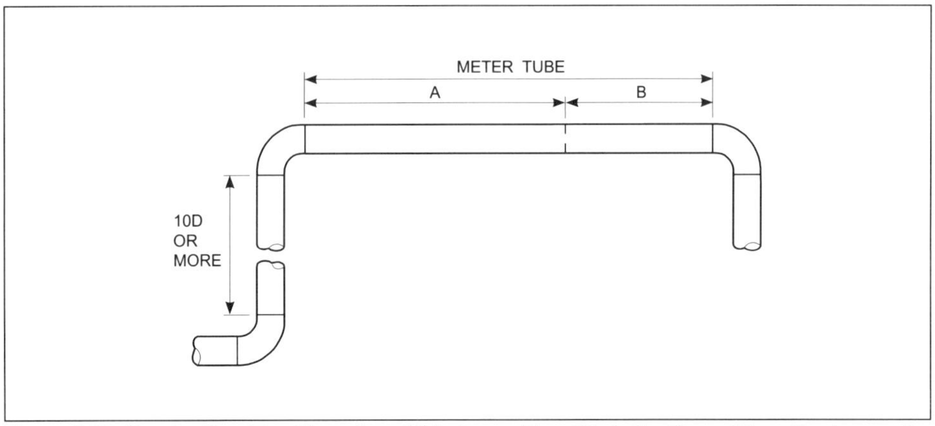

Figure 8-24. Greater than Ten Diameters (*D*) between Two Ells in the Same Plane Upstream of Meter Tube

The inside pipe walls of the meter tube shall be smooth, with a roughness not to exceed 300 microinches. If necessary, the inside pipe walls may be machine ground, coated, or polished to meet this requirement. Grooves, scoring, pits, ridges resulting from seams, distortion caused by welding, offsets, and so on, that affect the inside diameter by more than 0.5% shall not be permitted. This becomes extremely important at the point where the meter tube is welded to the orifice plate holder. This area should be ground smooth so that it is of the same internal diameter as the pipe and is concentric within tolerances. Most defects of this type can be detected by visual inspection, and they may be corrected by filling in grooves or pits and/or grinding or filing the ridges. As a comparison or reference, an orifice plate has a roughness of 50 microinches or less, and good quality pipe or tubing has a roughness of about 150 microinches. Roughness scales are available that have areas with various roughness for comparison to the meter tube visually and by feel.

When a corrosive condition is recognized where the meter tube inner surfaces could become severely pitted within a reasonably short period of time, plastic

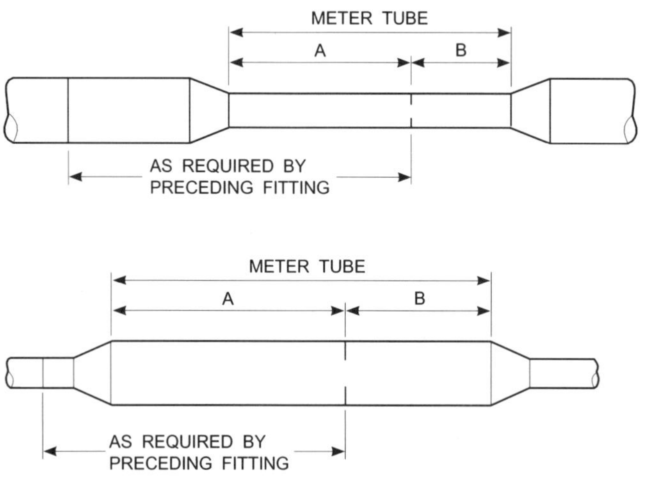

Figure 8-25. Reducer or Expander Upstream of MeterTube

coating may be applied to the internal surfaces of the meter tube. The meter tube diameter and internal wall smoothness should be checked after the coating has been applied.

The amount of pipe that can be saved by use of straightening vanes varies with the upstream piping configuration.

In some cases the required length of the upstream meter tube can be reduced by installing a flow conditioner upstream of the pipe section of the orifice plate. A commonly used flow conditioner or straightening vane is the tube bundle. The purpose of the straightening vanes is to remove swirl from the flow in the line as it approaches the orifice plate, resulting in a shorter upstream meter tube length requirement for many applications. Other flow conditioners remove abnormalities in the flow profile.

Note that in the case of a reducer or expander upstream of the meter tube (Figure 8-25), the straightening vane does not provide any length reduction but actually increases upstream meter tube length by the length of the vane. In small diameter meter tubes, the cost of the additional pipe is often less than the cost of straightening vanes. Regardless of cost, the use of straightening vanes is often necessary where space is at a premium.

An orifice flowmeter is to be installed in a 4-inch line with a piping configuration of two upstream 90° elbows and one 90° elbow downstream. The two upstream elbows are separated by 24 inches of straight pipe. Available space between the upstream and downstream elbows is 72 inches. For the operating condition the beta ratio is 0.65. Determine the meter installation lengths using data from an accepted standard for reliable measurement.

Solution:

Two upstream elbows in plane are separated by 6 pipe diameters (24 inches in 4-inch line). The meter installation is similar to Figure 8-23. Minimum straight pipe lengths upstream and downstream of the meter for $\beta = 0.65$ is approximated from Table 8-4. Linear interpolation between $\beta = 0.60$ and beta = 0.67 gives the following:

(continued)

(continued)

$$A = 16.34 \times D = 65.4 \text{ inches}$$

$$B = 4.91 \times D = 16.6 \text{ inches}$$

$$A' = 11.3 \times D = 45.2 \text{ inches}$$

$$C' = 5.27 \times D = 21.1 \text{ inches}$$

$$C = 6.03 \times D = 24.1 \text{ inches}$$

Minimum upstream straight pipe length required for the meter without a flow conditioner is 82.0 inches ($A + B$) and is more than the available space. With a flow conditioner installed upstream of the orifice plate a minimum length of 61.8 inches ($A' + B$) is required. Available space is 72 inches. The meter tube could be installed with B = 18 inches, C' = 24 inches, and C = 30 inches. Note that available additional length is distributed to all dimensions and proportionally more to the upstream lengths than to the downstream length. Increased upstream lengths have a favorable effect on the flowmeter performance.

The effect of the orifice plate concentricity, pipe roughness, and pressure tap geometry have a greater influence on the measurement accuracy of small flowmeters, especially below 2 inch. To improve measurement accuracy, small line size orifices are installed in honed meter runs with a corner-tap design and annular grooves on each side of the plate. The measurement accuracy of an uncalibrated honed flow section is approximately ± 0.75 percent. The line size of honed flow sections is between ½ and 1½ inch. The beta ratio range is 0.1 to 0.8, and the pipe Reynolds number should be greater than 1,000. Occasionally, flange taps are used in line sizes between 1 and 1½ inch where beta ratio range is limited to 0.15 to 0.7. Uncalibrated flange tap honed meter sections have reduced accuracy compared to the corner tap honed orifice meter run.

Connections on meter runs for thermometers, gages, samplers, gravitometers, and so on, should be located downstream from the orifice (Figure 8-26). For

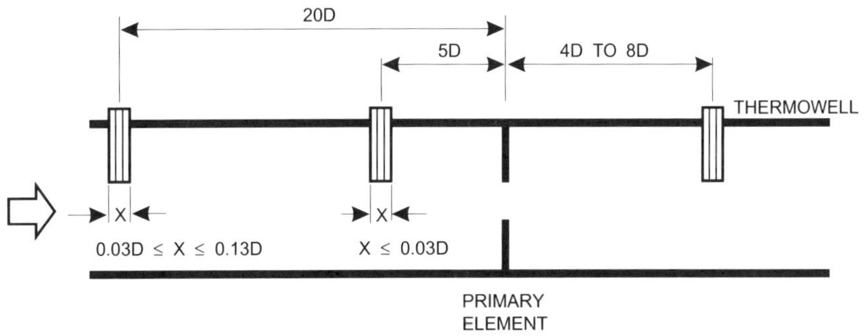

NOTES:

1. ANSI 2530 recommends that the upstream thermowell be located 12 to 36 inches upstream of a flow conditioner.

2. No recommended location for downstream measurement by ISO.

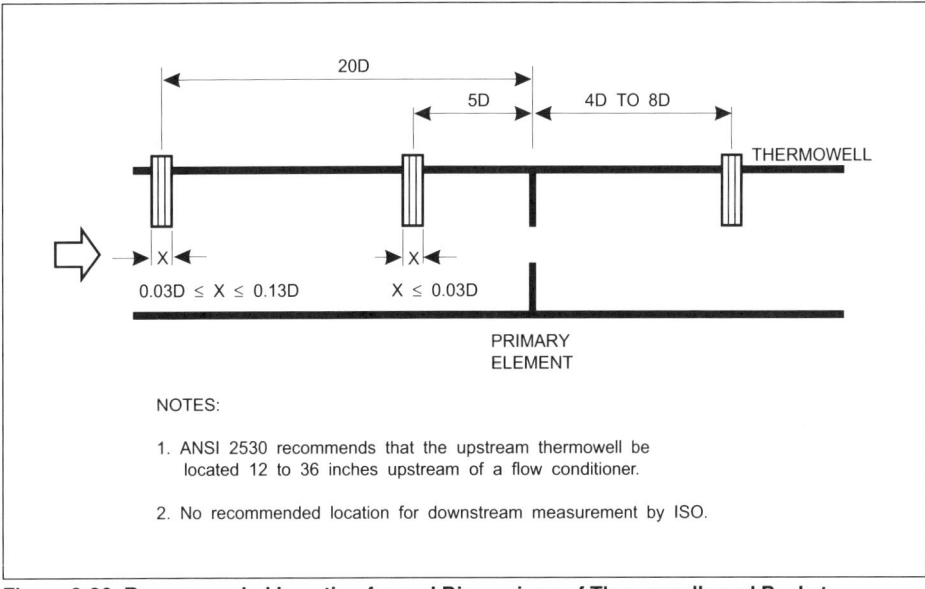

Figure 8-26. Recommended Location for and Dimensions of Thermowells and Pockets

flange taps, connections shall not be nearer than 4.5 nominal pipe diameters to the orifice. For pipe (full flow) taps, connections shall not be nearer than 12 nominal pipe diameters to the orifice. Certain dimensional constraints are recommended for upstream locations. If a flow conditioner is installed, thermowells upstream of the orifice plate must be 12 to 36 inches upstream of the straightening vane.

The mean inside diameter of the upstream meter tube shall be determined by measuring the internal diameter with a micrometer on four equally spaced axes at a point located one inch from the upstream face of the orifice plate. The average of these four measurements will be used as the internal diameter (*D*) of the upstream meter tube for flow calculations. Check measurements shall be made at two or more additional cross sections covering at least two pipe diameters from the face of the orifice plate or past the flange or fitting weld, whichever is greater. Measurement of the inside diameter of the downstream meter tube shall be made in a plane one inch downstream from the downstream face of the orifice plate with check measurements at two other cross sections similar to the upstream tube. The maximum allowable tolerance between the maximum and the minimum of all measurements (pipe concentricity) shall not exceed the tolerances shown in Figure 8-27.

Meter tubes using flange taps shall have the center of the upstream pressure tap hole placed 1 inch from the upstream face of the orifice plate. The center of the downstream tap hole shall be 1 inch from the downstream face of the orifice plate. If the pressure tap holes are located by measuring from the bearing face of an orifice flange, allowance must be made for the thickness of the gasket, which is typically 1⁄16 inch. The tolerance on the tap hole location is ± 0.030 inch for 4 inch and larger pipe with flange taps and a beta ratio of 0.75 (0.015 inch for under 4-inch pipe with flange taps) as shown in Figure 8-28.

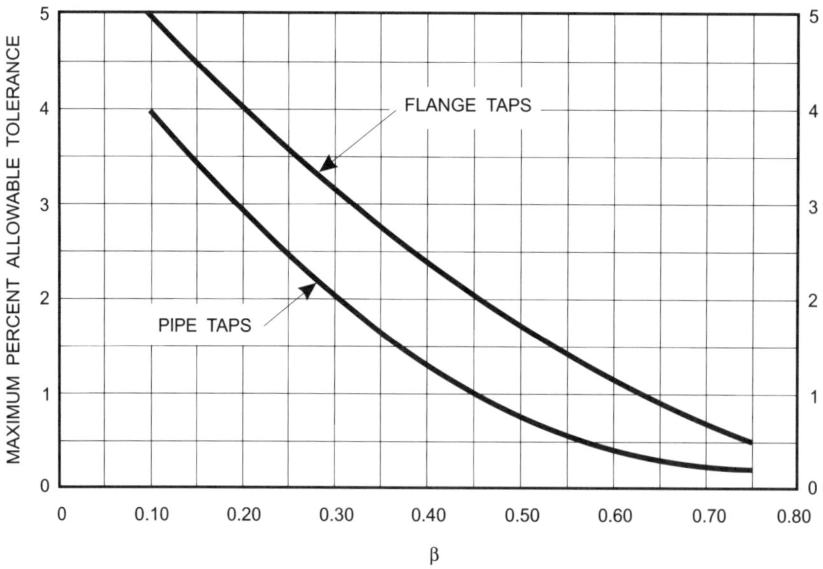

NOTE: It is recommended that for new installations in which the orifice-to-pipe diameter ratio is likely to be changed, the tolerance permitted for variations in pipe size, as shown in Figure 8-27, be the same as that given for the maximum orifice plate diameter ratio β = 0.75.

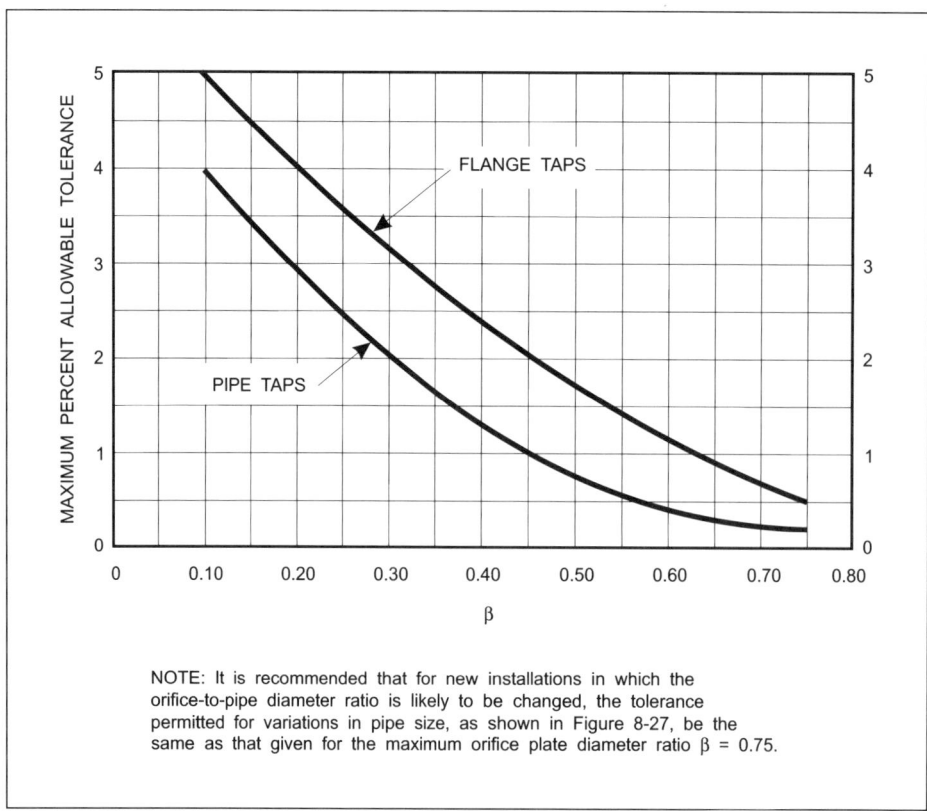

Figure 8-27. Maximum Percent Allowable Meter Tube Tolerance
(From API 2530)

Tap locations with respect to the orifice plate for different types of pressure tap configurations are shown in Figure 8-17. Pressure tap holes shall be drilled radial to the meter tube; that is, the centerline of the tap hole shall approximately intersect and form a right angle with the axis of the meter tube. The diameter of the pressure tap holes at the inner surface of the pipe shall be in accordance with Table 8-5. The edges of the pressure tap holes on the inner surface of the meter tube shall be free from burrs and may be slightly rounded. The finished tap hole shall be ± 1⁄16 inch from the selected nominal tap hole diameter along the drilled length of the hole.

Table 8-5. Recommended Meter Tube Pressure Tap Holes (dimensions in inches)

Nominal Diameter (pipe inside)	Nominal Tap Hole Diameter		
	Recommended Nominal	Maximum	Minimum
Under 2	1⁄4	1⁄4	1⁄4
2 through 3	3⁄8	3⁄8	1⁄4
4 and larger	1⁄2	1⁄2	1⁄4

Meter tubes using flange taps may have the outer ends of the pressure tap holes in the flanges or fittings drilled out and threaded to receive the desired size of pressure piping. Meter tubes using pipe taps shall have a hole of the proper size drilled through the pipe wall. This hole shall not be threaded. A fitting should be fastened to the pipe at this point, exercising great care to make sure that the inside of the pipe is not distorted in any way. If the fitting is welded to the pipe, the tap hole should not be drilled until after the welding is done.

If a meter tube has both flange and pipe or radius taps, it is recommended that the unused set of taps be plugged and ground smooth on the inside pipe wall.

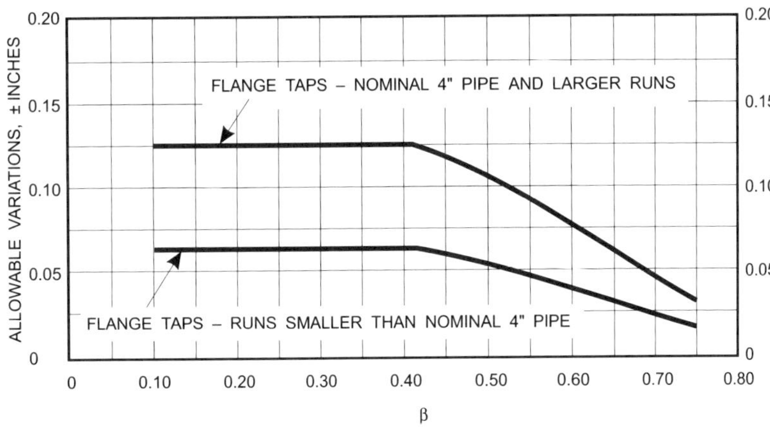

NOTES:

1. It is recommended that a maximum β = 0.75 ratio be used in the design of new installations.
2. For pipe taps, use plus or minus ten times the tolerance given.

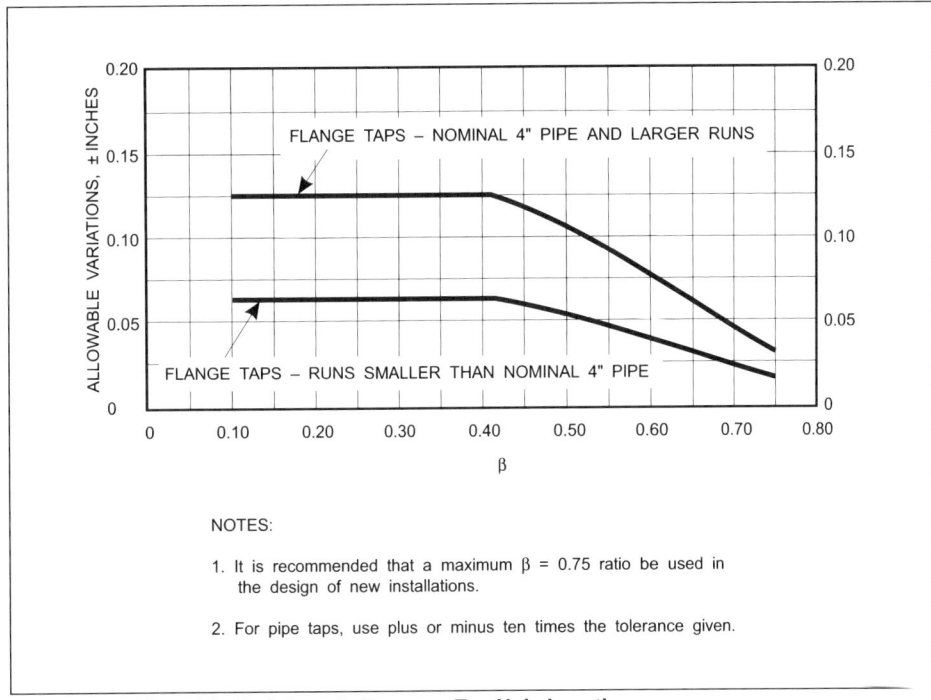

Figure 8-28. Allowable Variations in Pressure Tap Hole Location
(From API 2530)

Pulsation in the flow line contributed by a reciprocating device and/or a rotary device, valve action, line resonance, and any other repetitive piping disturbances can result in sudden changes in both the velocity and pressure of the flowing fluid. For reliable measurement, flow pulsation at the meter should be minimized. The following are some recommendations for reducing pulsation effects on flowmeter performance:

(1) Locate the meter tube in a favorable location with respect to the source of pulsation; e.g., upstream of the regulator or by increasing the distance of the meter from the source of pulsation.

(2) Install devices between the source and the meter tube to reduce amplitude of fluctuation.

(3) Operate the flowmeter at the highest possible differential pressure, i.e., relative change due to fluctuation is negligible compared to the actual differential pressure.

(4) Use smaller sized meter tubes and keep essentially the same orifice diameter while still maintaining the highest practical limits of differential.

(5) Shorten tube lengths connecting the pressure tap to the pressure transmitter to eliminate any low frequency resonance in the pressure tap line.

Operating Constraints

Rangeability of differential pressure devices is limited by the accuracy of the differential pressure measuring devices, especially because the developed differential pressure is proportional to the square of the flow velocity. As a result, the measured differential pressure of 12:1 covers a flow turndown or rangeability of about 3.5:1. Differential pressure transducers with different calibrated ranges can be "stacked" or installed in parallel across the orifice plate to achieve flow turndown of 10:1 or better. With the availability of "smart transmitters" with wide rangeability and improved accuracy, measurement precision similar to that of the "stacked" transmitters may be achieved by one "smart transmitter." The operating range and many other parameters of smart transmitters can also be changed during operation through a digital communicator. However, accuracy of the transmitter over the low differential pressure measurement must be carefully evaluated to determine the overall precision of the flow measurement. Operational difficulties, on-site calibration, long-term repeatability and accuracy, and effects of ambient temperature at the smart transmitter should also be evaluated for specific applications.

Approximate Reynolds number constraints for different orifice meters are given in Table 8-6. These constraints must be met at all operating conditions, including low flow and temperature.

Flow Equation

From the equation of continuity (Equation (8-2)) and Bernoulli's equation (Equation (8-3)), the theoretical mass flow rate of an incompressible fluid through the pipe in Figure 8-1 can be expressed as:

$$Q_{\text{theo}} = \frac{\pi}{4} \cdot D_2^2 \, \frac{\sqrt{2g \cdot (P_1 \;\; P_2)\, \rho}}{\sqrt{1-(D_2/D_1)^4}} \qquad (8\text{-}18)$$

Table 8-6. Reynolds Number Constraints for Different Orifice Flowmeters

	Pipe Reynolds Number (Re_D)
Concentric (under 2 inches)	1000+
2 inches and over	5000*d*+
Conical	$250\beta < Re_D < 200{,}000\beta$
Eccentric	10,000 - 1,000,000
Integral	1000 *d/D*+
Quadrant	250 - 3200 < Re_D < 60,000 - 280,000
Segmental	10,000 - 1,000,000

If the flow section 2 of Figure 8-1 is the bore of the orifice or a nozzle and differential pressure is the difference of the pressures at the two flow sections, Equation (8-18) reduces to the form:

$$Q_{\text{theo}} = \frac{\pi}{4} \cdot d^2 \cdot \frac{\sqrt{2g \cdot (P_1 - P_2) \cdot \rho}}{\sqrt{1 - \beta_4}} \qquad (8\text{-}19)$$

For the differential pressure expressed in inches of water, the bore diameter in inches, the acceleration due to gravity as 32.17405 ($lb_m ft/lb_f sec^2$), and density of flowing fluid in lb/ft^3, Equation (8-19) reduces to the form:

$$(Q_{\text{pps}})_{\text{theo}} = 0.00997019 \cdot d^2 \frac{\sqrt{h_w \cdot \rho}}{\sqrt{(1 - \beta^4)}} \qquad (8\text{-}20)$$

where Q_{pps} is the mass flow rate in pounds per second, h_w is the differential pressure in inches of water, ρ is the density of the flowing fluid in lb/ft^3, and d is the bore diameter in inches.

From Equation (8-20) the volumetric flow rate through the meter tube is

$$(Q_{\text{cfs}})_{\text{theo}} = 0.0997019 \cdot d^2 \frac{\sqrt{h_w}}{\sqrt{\rho \cdot (1 - \beta^4)}} \qquad (8\text{-}21)$$

where Q_{cfs} is the volumetric flow rate in cubic feet per second.

For compressible fluid flow, the lower pressure at the restriction results in an expansion of the fluid at the bore. Therefore, the differential pressure between the throat and the line is different than for incompressible fluid flows. For gas (vapor) flows, the fluid density is not constant between the line and the bore. The adiabatic gas expansion factor Y is used to compensate for the expansion of the fluid at the bore. The subscript of gas expansion factor Y is 1 when the reference pressure is the upstream line pressure and the flowing fluid density is at the reference pressure. The subscript is 2 when the downstream pressure is the reference pressure and the flowing fluid density is derived at the reference pressure. From certain physical laws, the gas expansion factor is given as:

$$Y_1 = \frac{\sqrt{(1 - \beta^4 \cdot k/(k-1) \cdot P_r^{(2/k)} \cdot (1 - P_r^{(k-1)/k})}}{\sqrt{(1 - \beta^4 \cdot P_r^{(2/k)}) \cdot (1 - P_r)}} \qquad (8\text{-}22)$$

where P_r is the ratio of the absolute downstream line pressure to the absolute upstream line pressure, and k is the ratio of the specific heats of the flowing fluid. To obtain the flow rates of compressible fluids, Equation (8-20) and Equation (8-21) are multiplied by the gas expansion factor. The mass flow rate equation for compressible fluid is:

$$Q_{pps} = 0.0997019 \cdot Y_1 d^2 \frac{\sqrt{h_w \rho_1}}{\sqrt{1-\beta^4}} \tag{8-23}$$

When the line pressure is measured at the upstream tap location, the gas expansion factor is defined by Equation (8-22). For reference pressure measurement at the downstream tap location, the adiabatic gas expansion factor is expressed as:

$$Y_2 = Y_1 \frac{\sqrt{1}}{\sqrt{1-x_1}} = Y_1 \sqrt{1+x_2} \tag{8-24}$$

where x_1 is the ratio of the differential pressure to the absolute upstream line pressure, $(P_1 - P_2)/P_1$, and x_2 is the ratio of the differential pressure to the absolute downstream reference pressure P_2. Equation (8-23) becomes:

$$(Q_{pps})_{theo} = 0.0997019 \cdot Y_2 \cdot d^2 \frac{\sqrt{h_w \cdot \rho_2}}{\sqrt{1-\beta^4}} \tag{8-25}$$

The theoretical flow equation is developed with the ideal conditions and no viscous or frictional losses. The true flow rate for a given differential is always less than the theoretically calculated value. Two specific effects were previously discussed: the coefficient of velocity, C_v, and coefficient of contraction, C_c. For many devices the actual low pressure reading is not obtained at the throat but is measured at the downstream tap location; e.g., pipe tap is 8 pipe diameters downstream and radius tap is ½ pipe diameter downstream of the upstream face of the plate. Hence, the flow rate equation must be modified to account for the changes in the differential pressure readings at the actual tap locations.

How closely the true flow can be calculated from the theoretical equation depends almost entirely on the shape of the constriction, especially the approach and recovery sections of the meter. For both abrupt and contoured reduction in flow cross section, the assumption of the uniform velocity profile is valid at high Reynolds numbers. For the same flowing conditions, an increase in fluid viscosity or a decrease in flow velocity results in a decrease in the pipe Reynolds number, thereby affecting the velocity profile. The velocity profile change for a contoured contraction in the pipe is different from that of an abrupt change in flow cross section. Hence, the theoretical equation is adjusted for these effects with empirically determined correction factors. One factor is the discharge coefficient, C, which corrects for the variations in velocity profile (Reynolds number), tap location, form of the contraction, and other effects.

DISCHARGE COEFFICIENT

For any given primary element, the discharge coefficient is derived from laboratory test data of the actual flow rates and ideal theoretical flow rate value. The discharge coefficient is defined as:

$$C = \text{(true flow rate)/(theoretical flow rate)} \tag{8-26}$$

In some equations, the discharge coefficient is combined with the velocity of approach, E, and redefined as flow coefficient K. The flow coefficient is defined as:

$$K = C / \sqrt{(1-\beta^4)} = E \cdot C \tag{8-27}$$

where $\beta = \frac{d}{D}$ and $E = \frac{1}{\sqrt{(1-\beta^4)}}$ (the velocity approach factor).

DISCHARGE COEFFICIENT FOR DIFFERENT DEVICES

The discharge coefficient for a given device can be presented in different forms; the following are various methods widely used.

(1) The discharge coefficient may be plotted as a function of Reynolds number based on actual laboratory calibration data. Since actual data accounts for all physical constraints, this is the most accurate method of achieving precise measurement. Extrapolation of the value of the discharge coefficient for Reynolds numbers outside the actual calibration range should be avoided if possible. Certain applications may generate Reynolds numbers for which no laboratory calibration facility may be able to match the actual Reynolds number range; e.g., high temperature steam flows or certain cryogenic applications. For such special cases discharge coefficient data extrapolation over high Reynolds number ranges may be the only viable option.

(2) The discharge coefficient for a given device is often obtained from a graphic data base of different meters and line sizes. These graphs are usually developed from actual data obtained from previous calibrations of similar devices and the discharge coefficient data are plotted as a function of Reynolds number.

(3) The flow coefficient is often presented as a constant number for a given device with fixed geometry. The value of the constant usually corresponds to the flow coefficient of the device at very high Reynolds number. The correction factor as a function of the pipe or bore Reynolds number is either plotted or tabulated for flow computation necessary to determine the actual flow rate for a differential pressure generated by the flowing fluid.

(4) Based on numerous points of many carefully conducted tests on standardized primary elements, empirical equations have been developed that predict the discharge coefficient from the bore and pipe diameter dimensions. There are accepted equations in a number of standards; therefore, for most flowmeter applications actual flow calibration is seldom performed.

If measurement is critical and improved accuracy is desired, the primary element should be wet flow calibrated. For Reynolds number, pipe size, applications outside the specified range of the equation, nonconforming upstream or downstream straight pipe lengths, etc., the meter tube should be wet flow calibrated, preferably over the actual operating Reynolds number range.

Most of the differential producing devices have a generalized equation for the discharge coefficient at infinite Reynolds number and a correction term for the change in discharge coefficient at other Reynolds numbers. In some orifice meter equations, the correction factor also includes the location of the upstream/downstream tap.

DISCHARGE COEFFICIENT EQUATION

A generalized correction term for the discharge coefficient for many differential pressure primary elements in the turbulent flow regime ($Re_D > 4000$) is given as:

$$C = C_\infty + b/Re_D^n \qquad (8\text{-}28)$$

where C_∞ is the discharge coefficient at infinite Reynolds number, b is a constant, and n is the exponent for the pipe Reynolds number.

From Equation (8-28) the flow coefficient is defined as:

$$K = E\,(C_\infty + b/Re_D^n) = [1/\sqrt{(1-\beta^4)}\,]*[C_\infty + b/Re_D^n\,] \qquad (8\text{-}29)$$

The infinite Reynolds number discharge coefficient may be a constant or a function of the dimensions of the primary element or tap location. For Reynolds number correction, the value of b could be a function of the flowmeter dimensions while the Reynolds number exponent, n, is a constant and depends on the primary element.

For orifice flowmeters, location of the taps influences the value of the discharge coefficient. Based on experimental data for different flowmeter sizes and

Reynolds number ranges, empirical equations for the discharge coefficient have been developed and are used to determine the flow rate. There is more than one equation for orifice flowmeters. The values of the discharge coefficients predicted by each equation for the same orifice flowmeter and flowing conditions differ, but for the typical applications of the orifice flowmeter, the percentage change in the discharge coefficient between two equations is usually less than the measurement uncertainty.

ISO AND ASME EQUATION

A generalized empirical equation for the value of the discharge coefficient of an orifice flowmeter was developed by a French scientist, J. Stolz, from experimental data base of different line sizes, beta ratios, and wide ranges of Reynolds number. The Stolz orifice equation, accepted by the International Organization for Standardization (ISO) and ASME for line sizes of 2 inch and larger, is:

$$C = 0.5959 + 0.0312*\beta^{2.1} - 0.1840*\beta^{8} + 0.0900 L_1 * \beta^4 / (1 - \beta^4) - 0.0337 L_2' * \beta^3 + 91.71 * \beta^{2.5} / Re_D^{0.75} \quad (8\text{-}30)$$

where:

L_1 = dimensional correction for upstream tap location
= l_1 / D, l_1 measured from upstream face of the plate

L_2' = dimensional correction for downstream tap location
= l'_2 / D, l'_2 measured from downstream face of the plate

Note that the distances of the upstream and downstream taps are measured from the respective face of the orifice plate, and the last term in Equation (8-30) is the Reynolds number correction term. The upstream tap location correction factor has a limit on the pipe diameter. The value of correction factor L_1 in Equation (8-30) cannot be greater than 0.4333; i.e., the coefficient $0.0900L_1$ is 0.0390 or less, and the value of 0.0390 corresponds to a line size of 2.3 inches.

For Corner Taps:

Tap location terms l_1, and l_2 for corner taps are zero and Equation (8-30) reduces to the form:

$$C = 0.5959 + 0.0312*\beta^{2.1} - 0.184*\beta^{8} + 91.71*\beta^{2.5} / Re_D^{0.75} \quad (8\text{-}31)$$

For Flange Taps:

The taps are located one inch upstream and downstream of the respective face of the plate. Equation (8-30) for the flange tap with diameter limits is as follows:

For $D \geq 2.3$ inch, $L_1 = L'_2 = 1/D$,

$$C = 0.5959 + 0.0312*\beta^{2.1} - 0.1840*\beta^{8} + 0.0900*\beta^4 / [\, D(1 - \beta^4)\,] - 0.0337*\beta^3 / D + 91.71*\beta^{2.5} / Re_D^{0.75} \quad (8\text{-}32)$$

For $2 \leq D < 2.3$ inch, $L_1 = 0.4333$, $L'_2 = 1/D$,

$$C = 0.5959 + 0.0312\,\beta^{2.1} - 0.184\,\beta^{8} + 0.0390\,\beta^4 / (1 - \beta^4) - 0.0337\,\beta^3 / D + 91.71\,\beta^{2.5} / Re_D^{0.75} \quad (8\text{-}33)$$

For D and $D/2$ Taps:

The upstream tap location correction factor, $L_1 = 0.4333$. The downstream tap location of the radius tap is measured from the upstream face of the orifice plate.

So the L'_2 term is a function of plate thickness. For standard plate thickness, $L'_2 = 0.47$ is acceptable and Equation (8-30) takes the form:

$$C = 0.5959 + 0.0312*\beta^{2.1} - 0.1840*\beta^8 + 0.0390*\beta^4/(1-\beta^4) - 0.01584*\beta^3 + 91.71*\beta^{2.5}/\mathrm{Re}_D^{0.75} \quad (8\text{-}34)$$

ANSI/API AND AGA EQUATION

The empirical equation for the discharge coefficient value defined by ANSI/API 2530, AGA Report-3, or GPA 8185-85 is very widely used and is commonly referred to as the API or AGA Equation. This equation is also known as the Buckingham Equation and is in the ASME *Fluid Meters* publication [Ref. 4]. The empirical equation of the flow coefficient for flange taps and infinitely large Reynolds number is:

$$K_{f_\infty} = \frac{0.5993 + 0.007/D + (0.364 + 0.076/\sqrt{D})*\beta^4 + 0.4*(1.6 - 1/D)^5*[(0.07 + 0.5/D) - \beta]^{5/2} - (0.009 + 0.034/D)*(0.5-\beta)^{3/2} + (65/D^2 + 3)*(\beta - 0.7)^{5/2}}{1.01245 - 0.075*\beta + 0.135*\beta^2 - 0.063*\beta^3 + 0.00795/\sqrt{D}} \quad (8\text{-}35)$$

For orifice flowmeters with pipe taps and infinitely large Reynolds number, the flow coefficient is:

$$K_{p_\infty} = \frac{0.5925 + 0.0182/D + (0.440 - 0.06/D)*\beta^2 + (0.935 + 0.225/D)*\beta^5 + 1.35*\beta^{14} + (1.43/\sqrt{D})*(0.250-\beta)^{5/2}}{1.01358 - 0.075*\beta + 0.135*\beta^2 - 0.063*\beta^3 + 0.0131/D} \quad (8\text{-}36)$$

In the numerator of Equations (8-35) and (8-36), the signs of the terms with fractional exponents (last three terms of Equation (8-35) and the last term of Equation (8-36) become negative for some values of β. In such cases, these terms are to be neglected and their values treated as zero.

For any value of bore Reynolds number, $\mathrm{Re_d}$, the discharge coefficient of an orifice flowmeter for flange taps is:

$$C = \sqrt{(1-\beta^4)}*Y*K_{f_\infty}\,[1 + d(830 - 5000*\beta + 9000*\beta^2 - 4200*\beta^3 + 530/\sqrt{D})\,/\,\mathrm{Re_d}\,] \quad (8\text{-}37)$$

For pipe taps:

$$C = \sqrt{(1-\beta^4)}*Y*K_{p_\infty}\,[1 + d(905 - 5000*\beta + 9000*\beta^2 - 4200*\beta^3 + 875/D)\,/\,\mathrm{Re_d}\,] \quad (8\text{-}38)$$

where the $K_{f\infty}$ and $K_{p\infty}$ values are defined by Equations (8-35) and (8-36), and Y is the expansion factor. For incompressible fluids the value of $Y = 1$ (exactly).

For compressible fluids the value of the expansion factor depends on the reference pressure. If the absolute static pressure is taken at the upstream differential pressure tap, then the value of the expansion factor for flange taps is:

$$Y_1 = 1 - (0.41 + 0.35*\beta^4)(x_1/k) \quad (8\text{-}39)$$

and for pipe taps is:

$$Y_1 = 1 - [0.333 + 1.145*(\beta^2 + 0.7*\beta^5 + 12*\beta^{13})]*(x_1/k) \quad (8\text{-}40)$$

where:

$$x_1 = (P_1 - P_2)/P_1 = h_W/(27.707*P_1) \quad (8\text{-}41)$$

h_w = differential pressure in inches of water at 60°F, and
k = the ratio of the specific heat of the gas at constant pressure to the specific heat of the gas at constant volume

For Equations (8-39) and (8-40), the differential pressure ratio, x_1, should not exceed the value of 0.2. For flange taps, Equation (8-39) is applicable over the β ratio range of 0.10 to 0.80, and Equation (8-40) for pipe taps may be used over a range from 0.10 to 0.70. If the absolute static pressure is taken at the downstream tap, then the value of expansion factor Y_2 can be calculated by modifying Equations (8-39) and (8-40):

For flange taps:

$$Y_2 = \sqrt{(1 + x_2)} - (0.41 + 0.35\beta^4)\{x_2/[k\sqrt{(1 + x_2)}]\} \quad (8\text{-}42)$$

For pipe taps:

$$Y_2 = \sqrt{(1 + x_2)} - [0.333 + 1.145*(\beta^2 + 0.7\beta^5 + 12\beta^{13})] *\{x_2/[k\sqrt{(1 + x_2)}]\} \quad (8\text{-}43)$$

where $x_2 = (P_1 - P_2)/P_2 = h_w/(27.707*P_2)$ (8-44)

A 12-inch orifice flowmeter has a beta ratio of 0.54. The line pressure at the upstream flange tap is 60 psig. The differential pressure is 82 inches of water. The flowing fluid is natural gas with specific heat ratio k = 1.3. Determine the expansion factor.

Solution:

Using Equation (8-41) the differential pressure ratio,

$$x_1 = 82/[27.707*(60 + 14.696)] = 0.03962$$

The differential pressure ratio x_1 is less than 0.2; hence, Equation (8-39) is applicable.

For a flange tap orifice flowmeter with a beta ratio of 0.54 and flowing fluid specific heat ratio k = 1.3 and using Equation (8-40), the expansion factor is:

$$Y_1 = 1 - [0.41 + 0.35*(0.54)^4]*[0.03962/1.3] = 0.98650$$

NEW AGA/API EQUATION

Based on a large number of test data over a wide Reynolds number range and different pipe and bore sizes, a new AGA/API discharge coefficient equation, the

Reader-Harris & Gallagher (R-G) equation, was developed in 1989. The discharge coefficient by the new equation is:

For corner taps:
For $D \geq 2.8$ inch:

$$C = 0.5961 + 0.0291*\beta^2 - 0.2290*\beta^8 + 0.000511(10^6/\,Re_D)^{0.7} + [0.0210 + 0.0049(19000*\beta/\,Re_D)^{0.8}]*\beta^4*(10^6/\,Re_D)^{0.35} \quad (8\text{-}45)$$

For $2 \leq D < 2.8$inch:

$$C = 0.5961 + 0.291*\beta^2 - 0.2290*\beta^8 + 0.003(1-\beta)*(2.8-D) + 0.000511*(10^6*\beta/\,Re_D)^{0.7} + [0.0210 + 0.0049 *(19000*\beta/\,Re_D)^{0.8}]*\beta^4*(10^6/\,Re_D)^{0.35} \quad (8\text{-}46)$$

For flange taps:
For $D \geq 2.8$ inch:

$$C = 0.5961 + 0.0291\beta^2 - 0.229\beta^8 + 0.000511*(10^6*\beta/\,Re_D)^{0.7} + [0.0433 + 0.0712*e^{(-8.5*L1)} - 0.1145*e^{(-6*L1)}]*[(\beta^4/(1-\beta^4)] *[1 - 0.23*(19000*\beta/\,Re_D)^{0.8}] - 0.0116*[(2*L_2'/(1-\beta)) - 0.52 *[2*L_2'/(1-\beta)]^{1.3}]*\beta^{1.1}*[1 - 0.14(19000*\beta/Re_D)^{0.8}] + [0.0210 + 0.0049(19000*\beta/Re_D)^{0.8}]*\beta^4*(10^6/Re_D)^{0.35} \quad (8\text{-}47)$$

For $2 \leq D < 2.8$ inch:

$$C = 0.5961 + 0.0291*\beta^2 - 0.229*\beta^8 + 0.003(1-\beta)*(2.8-D) + [0.0433 + 0.0712e^{(-8.5*L1)} - 0.1145e^{(-6*L1)}]*[(\beta^4/(1-\beta^4)] *[1 - 0.23*(19000*\beta/Re_D)^{0.8}] + 0.000511(10^6*\beta/Re_D)^{0.7} - 0.0116*[(2*L_2'/(1-\beta)] - 0.52*[2*L_2'/(1-\beta)]^{1.3}]*\beta^{1.1} *[1 - 0.14*(19000*\beta/Re_D)^{0.8}] + [0.0210 + 0.0049*(19000*\beta/Re_D)^{0.8}] *\beta^4*(10^6/Re_D)^{0.35} \quad (8\text{-}48)$$

where L_1 and L'_2 are the ratios of the distances of the upstream and downstream taps from the respective face of the orifice plate to the inside pipe diameter, i.e., $L_1 = L'_2 = 1/D$ for flange taps.

The discharge coefficient equation for other primary devices is available in standards and literature. A comprehensive list of the discharge coefficients for a number of different primary elements are listed in Table 8-7.

Determine the discharge coefficient of a 4-inch orifice flowmeter with a 2-inch bore plate and a pipe Reynolds number of 200,000. Flowing fluid is incompressible liquid and the meter tube has flange taps.

Solution:

(a) Using the ISO/ASME equation:

$$\beta = 0.5, L_1 = L_2' = 1/D = 0.25, \text{and } D > 2.3 \text{ inch}$$

(continued)

(continued)

Using Equation (8-31):

$$\begin{aligned} C &= 0.5959 + 0.0312(0.5)^{2.1} - 0.184(0.5)^{8} + (0.0900*(0.5)^{4} / [4 \\ &\quad *(1-(0.5)^{4}] - 0.0337*(0.5)^{3} / 4 + 91.71*(0.5)^{2.5} / (200000)^{0.75} \\ &= 0.60462 \end{aligned}$$

(b) Using the ANSI/API 2530 equation:

From Equation (8-35),

$$K_{f_\infty} = \frac{\begin{aligned} &0.5993 + 0.007/4 + (0.364 + 0.076/\sqrt{4})(0.5)^{4} + 0.4(1.6 - 1/4) \\ &*[(0.07 + 0.5/4) - 0.5\,]^{5/2} - (0.009 + 0.034/4)*[0.5 - 0.5]^{3/2} \\ &+ (65/4^{2} + 3)*[0.5 - 0.7]^{5/2} \end{aligned}}{1.01245 - 0.075(0.5) + 0.135(0.5)^{2} - 0.063(0.5)^{3} + 0.00795/\sqrt{4}}$$

For the given flowmeter dimensions, the last three terms in the numerator with fractional exponents are either zero or negative and from the limits of Equation (8-35) those terms are equal to zero. Therefore, the flow coefficient $K_{f\infty} = 0.62318$. The bore Reynolds number, $Re_d = Re_D/\beta$; i.e., $Re_d = 400{,}000$. For an incompressible fluid $Y = 1$ and using Equation (8-37):

$$\begin{aligned} C &= \sqrt{1-(0.5)^{4}}*(1)*(0.62318)*[1 + 2*(830 - 5000*(0.5) \\ &\quad + 9000*(0.5)^{2} - 4200*(0.5)^{3} + 530/\sqrt{4})/400000] \\ &= 0.60436 \end{aligned}$$

(c) Using new AGA/API equation:

$L_1 = L'_2 = 1/4 = 0.25$ and for $D > 2.8$ the value of M_1 in Equation (8-44) is zero. From Equation (8-47) the discharge coefficient:

$$\begin{aligned} C &= 0.5961 + 0.0291*0.5^{2} - 0.2290*0.5^{8} + [0.0433 + 0.0712*e^{(-8.5*0.25)} \\ &\quad - 0.1145*e^{(-6*0.25)}]*[0.5^{4}/(1-0.5^{4})]*[1 - 0.23 \\ &\quad *(19000*0.5/200000)^{0.8}] - 0.0116[2*0.25/(1-0.5) \\ &\quad - 0.52\,[2*0.25/(1-0.5)]^{1.3}]*0.5^{1.1}*[1 - 0.14 \\ &\quad *(19000*0.5/200000)^{0.8}] + 0.000511[10^{6}*0.5/200000]^{0.7} \\ &\quad + [0.0210 + 0.0049(19000*0.5/200000)^{0.8}]*0.5^{4}*[10^{6}/200000]^{0.35} \\ &= 0.60495 \end{aligned}$$

Using the same calculation steps, values of discharge coefficient at Reynolds numbers of 20,000 and 2,000,000 are computed for the same flowmeter.

Reynolds Number	ISO/ASME	Old ANSI/API	New AGA/API
20,000	0.61255	0.61305	0.61230
200,000	0.60462	0.60436	0.60495
2,000,000	0.60321	0.60349	0.60286

Note that the maximum deviation between two values of the discharge coefficient at any of the three Reynolds numbers is 0.12%, which is of the order of the measurement uncertainty of an orifice flowmeter.

Table 8-7. Value of the Discharge Coefficient for Different Primary Devices *(Adapted from R. W. Miller, Flow Measurement Engineering Handbook, © 1989, McGraw-Hill Publishing Company, with permission)*

Primary Device	Discharge Coefficient, C
Venturi:	
Machined inlet	0.995
Rough-cast inlet	0.984
Rough-welded sheet-iron inlet	0.985
Universal Venturi tube*	0.9797
Lo-loss tube*	$1.005 - 0.471\beta + 0.564\beta^2 - 0.514\beta^3$
Nozzle:	
ASME long radius	$0.9975 - 6.53 / \sqrt{Re_d}$
ISA	$0.9900 - 0.2262\beta^{4.1} - [13900\beta^2 - 26213\beta^{4.15}] / Re_D^{1.15}$
Venturi nozzle (ISA inlet)	$0.9558 - 0.196\beta^{4.5}$
Orifice:	
ASME or ISO Equation:	
Corner taps	Equation (8-31)
Flange taps	
$D >= 2.3$ inch	Equation (8-32)
$2 <= D < 2.3$ inch	Equation (8-33)
D and $D/2$ taps	Equation (8-34)
ANSI/API 2530 Equation:	
Flange taps	Equations (8-37) and (8-35)
Pipe taps	Equations (8-38) and (8-36)
New API/AGA Equation:	
Corner taps	
$2 <= D < 2.8$ inch	Equation (8-46)
$D >= 2.8$ inch	Equation (8-45)
Flange taps	
$2 <= D < 2.8$ inch	Equation (8-48)
$D >= 2.8$ inch	Equation (8-47)
Eccentric orifice (flange taps)	
180° tap	
$D <= 4$ inch	
$0.5875 + 0.3813*\beta^{2.1} + 0.6898\beta^8 - 0.1963\beta^4/(1 - \beta^4) - 0.3366\beta^3 + [7.3 - 15.7\beta + 170.8\beta^2 - 399.7\beta^3 + 332.2\beta^4] / Re_D^{0.75}$	
$D > 4$ inch	
$0.5949 + 0.4078\beta^{2.1} + 0.0547\beta^8 + 0.0955\beta^4/(1 - \beta^4) - 0.5608\beta^3 - [139.7 - 1328.8\beta + 4228.2\beta^2 - 5691.9\beta^3 + 2710.4\beta^4] / Re_D^{0.75}$	

* Manufacturers should be consulted for the exact coefficient.

Table 8-7. Value of the Discharge Coefficient for Different Primary Devices (continued)

Primary Device	Discharge Coefficient, C
90° tap	
$D >= 4$ inch	
	$0.6284 + 0.1462\beta^{2.1} - 0.8464\beta^8 + 0.2603\beta^4/(1-\beta^4) - 0.2886\beta^3 + [69.1 - 469.4\beta + 1245.6\beta^2 - 1287.5\beta^3 + 486.2\beta^4]/Re_D^{0.75}$
90° tap	
$D < 4$ inch	
	$0.6276 + 0.0828\beta^{2.1} + 0.2739\beta^8 - 0.0934\beta^4/(1-\beta^4) - 0.1132\beta^3 + [103.2 - 898.3\beta + 2557.3\beta^2 - 2977\beta^3 + 1131.3\beta^4]/Re_D^{0.75}$
Segmental flange tap	
$D <= 4$ inch	
	$0.5866 + 0.3917\beta^{2.1} + 0.7586\beta^8 - 0.2273*\beta^4/(1-\beta^4) - 0.3343\beta^3$
$D > 4$ inch	
	$0.6037 + 0.1598\beta^{2.1} - 0.2918\beta^8 + 0.0244*\beta^4/(1-\beta^4) - 0.0790\beta^3$
Quadrant (corner and flange taps)	
$D > 1.5$ inch	
	$0.7746 - 0.1334\beta^{2.1} + 1.4098\beta^8 + 0.675*\beta^4/(1-\beta^4) + 0.3865\beta^3$
Honed orifice meter run	
½ to 1–½ inch	
Flange taps	$[0.5980 - 0.468(\beta^4 + 10*\beta^{12})]\sqrt{(1-\beta^4)} + (0.87 + 8.1*\beta^4)\sqrt{[(1-\beta^4)/Re_D]}$
Corner taps	
	$[0.5991 + 0.0044/D + (0.3155 + 0.0175/D)*(\beta^4 + 2*\beta^{16})]\sqrt{(1-\beta^4)} + [0.52/D - 0.192 + (16.48 - 1.16/D)*(\beta^4 + 4\beta^{16})]*\sqrt{[(1-\beta^4)/Re_D]}$
Conic orifice $D >= 1$ inch	
$250 <= Re_D <= 5000\beta$	0.734
$5000\beta < Re_D < 300{,}000\beta$	0.730
Integral flow assembly:	
Quadrant edged	
$D =$ ½ inch	$1.1126 - 99.13\beta^2 + 8006\beta^4 - 26900\beta^8 - (10.72 - 3823\beta^2 + 309300\beta^4)/\sqrt{Re_d}$
Square edged	
$D =$ ½ inch	$0.6479 - 0.3505\beta^2 + 0.3853\beta^4 + 4.645\beta^8 - (0.4356 - 33.49\beta^2 + 88.33\beta^4)/\sqrt{Re_d}$
$D = 1$ inch	$0.6050 - 0.1837\beta^2 + 0.6615\beta^4 - 1.094\beta^8 + (1.646 + 2.394\beta^2 - 4.899\beta^4)/\sqrt{Re_d}$
$D =$ 1–½ inch	$0.6122 - 0.1076\beta^2 + 0.3416\beta^4 - 0.684\beta^8 + (0.2368 + 14.3\beta^2 - 12.86\beta^4)/\sqrt{Re_d}$

THERMAL EXPANSION FACTOR

The deviation of the temperature at the primary element from a base temperature may change the dimensions of the pipe and bore and influence the flow calculation. Base temperature for the thermal expansion factor is 68°F. If the coefficient of thermal expansion for the primary element and the pipe are α_e and α_p, then the thermal expansion factor correction (F_a) is:

$$F_a = 1 + 2*(\alpha_e - \beta^4\alpha_p)*(T_{°F} - 68)/(1 - \beta^4) \qquad (8\text{-}49)$$

where $T_{°F}$ is the temperature in Fahrenheit.

If $\alpha_e = \alpha_p = \alpha$, Equation (8-49) reduces to:

$$F_a = 1 + 2\alpha\,(T_{°F} - 68) \qquad (8\text{-}50)$$

BASIC ORIFICE FACTOR

The flow coefficient values of flange and pipe taps at infinite Reynolds number are defined for the ANSI/API and AGA equations by Equations (8-35) and (8-36). The value of $K_{f\infty}$ or $K_{p\infty}$ is a constant for known line and bore dimensions. The value of basic orifice factor F_b is tabulated for standard line and bore sizes as:

$$F_b = 338.178 * d^2 * K_\infty \qquad (8\text{-}51)$$

For a standard line and bore size, the value of K_∞ can be determined from Equation (8-51) by taking the value of F_b from the flange or pipe tap tables in the ANSI/API 2530 standard.

In ANSI/API 2530, the value of the basic orifice factor for a 6-inch Schedule 40 orifice flowmeter with flange taps and a 3.5-inch bore is given as 2655. Determine the discharge coefficient for a bore Reynolds number of 500,000.

Solution:

From Equation (8-51), the flow coefficient at the infinite Reynolds number is:

$$\begin{aligned} K_{f_\infty} &= F_b / (338.178 * d^2) \\ &= 2655 / [338.178 * (3.5)^2] \\ &= 0.64089 \end{aligned}$$

Using Equation (8-37) for $D = 6.065$, $d = 3.5$; beta = 0.577, $K_\infty = 0.64089$, the discharge coefficient is:

$$\begin{aligned} C &= \sqrt{[1 - (0.577)^4]} * (0.64089) * [1 + 3.5 * (905 - 5000 * (0.577) \\ &\quad + 9000 * (0.577)^2 - 4200 * (0.577)^3 + 875/\sqrt{6.065}) / 500000] \\ &= 0.60671 \end{aligned}$$

PERFORMANCE

The accuracy of a properly installed uncalibrated square-edged orifice plate is ± 0.6 percent of the measured flow rate. Excluding the uncertainty of the secondary instruments, meter performance is typically worse due to other influence factors and bias error. Inherent uncertainties and limitations of commonly used differential flowmeters are given in Table 8-8.

For a concentric orifice flowmeter with flange taps, available data are:

$$\begin{aligned} Q_{gpm} &= 100 \text{ gpm (normal)} \\ SG &= 1.2 \\ \mu_{cP} &= 10 \text{ cP} \\ d &= 0.75 \text{ inch} \\ D &= 2.067 \text{ inch} \end{aligned}$$

Are the application constraints satisfied?
Solution:

$$\begin{aligned} R_D &= (3160 * Q_{gpm} * SG) / (\mu_{cP} * D) \\ &= (3160 * 100 * 1.2) / (10 * 2.067) \\ &= 18345 \end{aligned}$$

The beta ratio, 0.75/2.067 = 0.3628, Re_D = 18345 and $Re_d = Re_D/\beta$ = 50565. From Table 8-8 for a data uncertainty of ± 0.6, the minimum pipe Reynolds number, Re_D, should be greater than 10,000. For Re_D = 10,000, the minimum accurate flow rate is,

$$\begin{aligned} Q &= [10{,}000 * D * \mu_{cP}] / (3160 * SG) \\ &= (10{,}000 * 2.067 * 10) / (160 * 1.2) \\ &= 54.5 \text{ gpm} \end{aligned}$$

Therefore, the flow rate should not be lower than 54.5 gpm for the flowmeter measurement accuracy to be within an allowable tolerance of ± 0.6.

Flow Rate

The generalized volumetric flow rate equation for a differential pressure device in actual cubic feet per second is

$$Q_{acfs} = 0.0997019 * C * Y_1 * d^2 * F_a * \sqrt{h_w / [\rho_1 * (1 - \beta^4)]} \qquad (8\text{-}52)$$

From Equation (8-52) the mass flow rate in pounds per minute

$$Q_{ppm} = 5.982114 * C * Y_1 * d^2 * F_a * \sqrt{(h_w * \rho_1) / (1 - \beta^4)} \qquad (8\text{-}53)$$

The flow rate can be computed in mass or volume units. In the generalized equation the numerical constant of Equation (8-53) will change for different flow rate units. The values of the constants for different flow rate units are listed in Table 8-9.

An ASME long radius wall-tap nozzle has pipe I.D. of 8.071 inches and a 3-inch bore diameter. Specific gravity of the flowing fluid is 0.95 and viscosity is 1.5 cP. The thermal expansion factor for the nozzle and the pipe is 0.0000185. The fluid temperature is 90°F and the differential pressure is 90 inches of water column. Determine the volumetric flow rate in gallons per minute. Assume that the density of water is 62.3 lb/ft^3.

(continued)

(continued)

Solution:

The beta ratio of the nozzle is 0.3717 and is within the allowable limits of the nozzle (Table 8-8). From Equation (8-50), the thermal expansion factor

$$F_a = 1 + 2 * (0.0000185) * (90 - 68)$$
$$= 1.000814$$

For liquids, the expansion factor $Y_1 = 1$. The density of the fluid is the product of the specific gravity and the density of water, i.e., $0.95 \times 62.3 = 59.185$ lb/ft^3. From Table 8-7, for ASME long radius wall-tap nozzle, $C = 0.9975 - 6.53/\sqrt{Re_d}$. From Table 8-8, the pipe Reynolds number should be between 10,000 and 2,500,000. Since the flow rate is to be determined, the Reynolds number is initially unknown. The solution is an iterative process, but for most applications one stage of iteration generates precise results; i.e., within +0.01% of the actual result.

First Iteration

The Reynolds number is assumed to be infinitely large; i.e., the Reynolds number correction factor is neglected and the discharge coefficient is 0.9975. From Equation (8-52) and Table 8-9, the flow rate is:

$$Q_{gpm} = 44.7428(0.9975)(1)(3^2)(1.000815)\sqrt{90 / [59.185 *(1 - 0.3717^4)]}$$
$$= 489.243 \text{ gpm}$$

From Equation (8-13) the bore Reynolds number is:

$$Re_d = 3160 \times 489.243 \times 0.95 / (1.2 \times 3)$$
$$= 407{,}974$$
$$\text{and} \quad Re_D = 151{,}644$$

Second Iteration

For bore a Reynolds number, $Re_d = 407{,}974$, the discharge coefficient:

$$C = 0.9975 - 6.53 / \sqrt{407{,}974}$$
$$= 0.98728$$

The flow rate using the new value of the discharge coefficient

$$Q_{gpm} = 44.7428(0.98728)3^2 (1.000815)\sqrt{90 / [59.185 *(1 - 0.3717^4)]}$$
$$= 495.403 \text{ gpm}$$

For the new flow rate, the bore Reynolds number, $Re_d = 413{,}111$. For the new Reynolds number, the discharge coefficient, $C = 0.98734$. The change in the value of the discharge coefficient after the first and second iterations is about 0.007%, which is essentially unchanged. All flow and meter parameters are within the acceptable limits. In most applications, the discharge coefficient and flow rates computed after the second iteration is within 0.1% of the actual value.

Table 8-8. Recommended Restrictions and Tolerance for Discharge Coefficients *(Adapted from R. W. Miller, Flow Measurement Engineering Handbook, © 1989, McGraw-Hill Publishing Company, with permission)*

Primary Device	Pipe Size (inches)	Reynolds Number	Beta Ratio β	Coefficient Tolerance
Venturi:				
Rough Cast	4 to 32	$2 \times 10^5 < Re_D < 10^6$	0.3 to 0.75	±0.7%
Machined Inlet	2 to 10	$2 \times 10^5 < Re_D < 10^6$	0.4 to 0.75	±1.0%
Welded Sheet-Iron Inlet	8 to 48	$2 \times 10^5 < Re_D < 10^6$	0.4 to 0.7	±1.5%
Universal Venturi Tube	> = 3	$Re_D > 7.5 \times 10^4$	0.2 to 0.75	±0.5%
Lo-Loss Tube	3 to 120	$1.25 \times 10^5 <= Re_D < 3.5 \times 10^6$	0.35 to 0.85	±1.0%
Nozzle:				
ASME Long Radius Pipe-Wall Tap	2 to 16	$10^4 <= Re_D < 2.5 \times 10^6$	0.2 to 0.8	±2.0%
ASME Long Radius Throat Tap	2 to 16	$Re_d <= 10^5$	0.2 to 0.5	±0.8%
	2 to 16	$Re_d > 10^5$	0.2 to 0.5	±2.0%
1932 ISA Flow Nozzle	2 to 40	$2 \times 10^4 < Re_D < 10^6$	0.32 to 0.8	±1.0%
Venturi Nozzle	3 to 20	$2 \times 10^5 < Re_D < 2 \times 10^6$	0.3 to 0.75	$\pm 1.2 \pm 1.5\beta^4$%
Orifice:				
Corner, Flange, *D*,and *D*/2 Taps	2 to 36	$10^4 < Re_D < 10^7$	0.2 to 0.6	±0.6%
		$Re_d > 5000 * D$	0.6 to 0.75	$\pm\beta$%
		$2 \times 10^3 < Re_D < 10^4$	0.2 to 0.75	$\pm 0.6 \pm \beta$%
Pipe Tap	2 to 36	$10^4 < Re_D < 10^7$	0.2 to .5	±0.8%
			0.51 to 0.7	±0.6%
Vena Contracta	$D >= 2$	$Re_d >= 5000 * D$	0.1 to 0.8	±1.0 to ±2.5%
Honed Flow Section:				
Corner Tap	0.5 to 1.5	$Re_D > 1000$	0.1 to 0.8	±0.75%
Flange Tap	1 to 1.5	$Re_D > 1000$	0.15 to 0.7	±0.75%
Eccentric Orifice:				
Flange and Vena Contracta Taps	4	$10^4 < Re_D < 10^6$	0.35 to 0.75	±2.0%
	6 to 14	$10^4 < Re_D < 10^6$	0.35 to 0.75	±1.5%
Segmental Orifice:				
Flange and Vena Contracta Taps	4 to 14	$10^4 < Re_D < 10^6$	0.35 to 0.85	±2.0%
Quadrant Edge Orifice:				
Flange and Corner	1 to 4	$250 < Re_D < 6 \times 10^4$	0.25 to 0.6	± 2 to ± 2.5%

Table 8-9. Generalized Flow Rate Equation for Different Units

Flow Rate Unit	Constant	Equation
Gallons per minute	44.7428	(8-52)
Barrels per day	1534.26	(8-52)

(Differential pressure in inches of water column)

FLOW RATE FOR COMPRESSIBLE FLUIDS

A change in temperature and pressure results in a change in volume of the fluid, and the effect is more pronounced in gases than in liquids. Hence, flow rate units are expressed in actual volume or standard volumetric flow rate. The standard conditions are also known as base conditions. In the United States, the standard cubic foot is the most commonly used unit for gas volume. In ISO 5024 for natural gas or petroleum gas, standard pressure and temperature are 14.696 psia and 59°F (15°C), respectively; while for ANSI/API 2530 the base pressure is 14.73 psia and 60°F is the base temperature. The base pressure and temperature values depend on the industry, the country, standard practices, and mutually agreed terms of contractual requirements.

When gas densities at the flowing condition and the base condition are known, the flow rates in actual and base conditions are:

$$(Q_{\text{scfs}})_b = (\rho_f / \rho_b) Q_{\text{acfs}} \tag{8-54}$$

where $(Q_{\text{scfs}})_b$ is the flow rate in standard cubic feet per second at the selected base condition, Q_{acfs} is the volumetric flow rate in actual cubic feet per second, ρ_f is the density of the fluid at the flowing condition, and ρ_b is density of the fluid at the base condition.

If the density is calculated from the Non-Ideal Gas Law, Equation (8-54) can be written as

$$(Q_{\text{scfs}})_b = [(Z_b T_b P_f) / (Z_f T_f P_b)] Q_{\text{acfs}} \tag{8-55}$$

For perfect gases, the gas compressibility factors are $Z_b = Z_f = 1$. For ASME standard base pressure, $P_b = 14.696$ psia and base temperature, $T_b = 68$°F (20°C), Equation (8-55) becomes:

$$(Q_{\text{scfs}})_b = 35.9057 [(Z_b P_f) / (Z_f T_f)] Q_{\text{acfs}} \tag{8-56}$$

A 6-inch schedule 40 machined inlet Venturi is installed in a natural gas flow pipe line. The throat diameter is 2.5 inches. Line pressure measured at the upstream tap is 130 psig and the temperature is -10°F. The differential pressure is 54 inches of water. The base temperature and pressure are 68°F and 14.696 psia. The compressibility factors of the gas are 0.9678 and 0.9982 at the flowing and base conditions. The specific gravity of the gas is 0.6. Determine the flow rates for the actual and base conditions. Viscosity of the gas at the flowing condition is 0.0125 cP and the specific heat ratio is 1.3. The expansion coefficient of the Venturi is 0.0000067.

Solution:

The inside diameter of the pipe is 6.067 inches and the beta ratio is 0.41207. The flowing pressure, $P_1 = 144.696$ psia and temperature, $T = [459.67 + (-10)] = 449.67$°R. The pressure ratio $P_r = 1 - x_1$, and from Equation (8-41), $P_r = 1 - 54/(27.707*144.696) = 0.98653$.

(continued)

(continued)

From Equation (8-22), the expansion coefficient is:

$$Y_1 = \frac{\sqrt{[1-(0.41207)^4]*[(1.3)/(1.3-1)]*(0.98653)^{2/1.3}} *\sqrt{[1-(0.98653)^{(1.3-1)/(1.3)}]}}{\sqrt{[1-(0.41207)^4*(0.98653)^{2/1.3}]*(1-0.98653)}}$$
$$= 0.991899$$

From Equation (8-50), the correction factor for thermal expansion is:

$$F_a = 1 + 2 * 0.0000667(-10 - 68)$$
$$= 0.98959$$

The flowing gas density from the Non-Ideal Gas Law is:

$$\rho_f = 2.698825(SG)\, P_f / z_f T_f$$
$$= 2.698825(0.6)(144.696) / (0.9678*449.67)$$
$$= 0.53840 \text{ lb}_m/\text{ft}^3$$

From Table 8-7 and Equation (8-52), the discharge coefficient is 0.995 and the volumetric gas flow rate is:

$$Q_{acfs} = 0.0997019 \times 0.995 \times 0.991899 \times (2.5)^2 \times 0.98959$$
$$*\sqrt{54 / [0.53840*(1-0.41207^4)]}$$
$$= 6.184856 \text{ acfs}$$
$$= 371.09 \text{ acfm}$$

The pipe Reynolds number for the gas is:

$$Re_D = [379 \times Q_{acfm} \times \rho] / [\mu_{cP} \times D]$$
$$= (379 \times 371.09 \times 0.53840) / (0.0125 \times 6.067)$$
$$= 998{,}479$$

From Equation (8-56), the flow rate at the base conditions is

$$(Q_{scfs})_b = 35.9057*[(0.9982*144.696) / (0.9678*14.696)]*6.184856$$
$$= 2255.186 \text{ scfs}$$
$$= 135{,}311 \text{ scfm}$$

Sizing of Flowmeter

Many factors must be considered in sizing of a differential pressure flowmeter. A number of interacting parameters affect the dimensions of a meter. A method of sizing orifice plates or other differential pressure producing devices based on the ASME/ISO equation is presented in *Flow Measurement Engineering Handbook* by R. W. Miller [Ref. 2] and is similar to the method presented in *Principles and Practice of Flow Meter Engineering,* Ninth Edition, by L. K. Spink.

Precise sizing calculations are usually performed on a computer. A number of computer programs for sizing differential flowmeters are commercially available. The computation can be based on any selected equation. Orifice sizing slide rules are also available to estimate the approximate value of the β ratio. For a given flowmeter, the slide rule can be used to estimate flow rate of a known fluid for a specified differential.

A meter sizing factor based on flowmeter dimensions and constant parameters for the flowing condition is developed from the flow rate equation. For expansion factor $Y_1 = 1$ and replacing d by βD, Equation (8-53) can be rewritten as:

$$C \sqrt{\beta^4 / (1 - \beta^4)} = Q_{\text{ppm}} / [5.982114 * F_a * D^2 * \sqrt{(\rho_1 * h_w)}] \qquad (8\text{-}57)$$

For estimating beta ratio, the right-hand side of Equation (8-57) is modified as the meter sizing factor:

$$S_m = Q_{\text{ppm}} / [5.982114 * F_a * D^2 * \sqrt{(\rho_1 * h_w / F_p)}] \qquad (8\text{-}58)$$

where F_a is the thermal expansion factor (Equation (8-50)) and F_p is the liquid compressibility factor correction.

For gases: $F_p = 1.0$

For most liquids:

$$F_p = 1.0 + Z_L * P_1 / 10{,}000 \qquad (8\text{-}59)$$

and

$$Z_L = -0.461 + 2.69 * T_r - 5.163 * T_r^2 + 3.521 * T_r^3 \qquad (8\text{-}60)$$

where T_r is the ratio of the absolute fluid temperature to the absolute critical temperature of the liquid.

Determine the compressibility factor correction for water at 100 psig and 80°F. The critical temperature and pressure for water are 705.455°F and 3203.8 psia, respectively.

Solution:

The reduced temperature is:

$$T_r = (459.67 + 80) / (459.67 + 705.455)$$
$$= 0.46319$$

From Equation (8-60), the liquid compressibility factor is:

$$Z_L = -0.461 + 2.69*(0.46319) - 5.165*(0.46319)^2 + 3.521*(0.46319)^3$$
$$= 0.02676$$

From Equation (8-59),

$$F_p = 1 + 0.02676*(100 + 14.696) / 10000$$
$$= 1.000307$$

Note: For many operating conditions the F_p value is almost unity, but for many liquified hydrocarbons the deviation of F_p from unity can be significant.

To determine an approximate beta ratio value, the right-hand side of Equation (8-57) is replaced by Equation (8-58) and for first iteration value of beta ratio, β_0, determined from Equation (8-57) is:

$$\beta_0 = [1 + (C / S_M)^2]^{-1/4} \qquad (8\text{-}61)$$

Assuming a high Reynolds number for the flowing condition, the Reynolds number correction factors in the discharge coefficient term of Table 8-7 can be neglected for approximation of β_0. The value of S_M can be modified for certain factors; i.e., steam quality, drain or vent holes, etc. From Table 8-7 the approximate sizing equation for β_0 is listed in Table 8-10.

Table 8-10. Approximate Sizing Equation for β_0 *(Adapted from R. W. Miller, Flow Measurement Engineering Handbook, © 1989, McGraw-Hill Publishing Company, with permission)*

Type of Flowmeter	$1/\beta_0^4$ Equation
Venturi:	
Machined Inlet	$1 + (0.995/S_M)^2$
Rough-Cast Inlet	$1 + (0.984/S_M)^2$
Rough-Welded Sheet-Iron	$1 + (0.985/S_M)^2$
Universal Venturi Tube	$1 + (0.9797/S_M)^2$
Lo-Loss Tube	$1 + (0.92/S_M - 0.31)^2$
Nozzle:	
ASME Long Radius	$1 + (0.9975/S_M)^2$
ISA	$1 + (0.9944/S_M - 0.118)^2$
Venturi Nozzle (ISA Inlet)	$1 + (0.989/S_M - 0.09)^2$
Orifice:	
Corner, Flange *D* and *D*/2 Taps	$1 + (0.605/S_M)^2$
2½*D* and 8*D* Taps	$1 + (0.6/S_M + 0.55)^2$
Eccentric	$1 + (0.607/S_M 0.088)^2$
Segmental	$1 + (0.634/S_M - 0.062)^2$
Quadrant ($\beta \leq 0.6$)	$1 + (0.76/S_M + 0.26)^2$
Conic, Corner (β <+ 0.3)	$1 + (0.734/S_M)^2$
Honed Flow Section:	
Corner Tap	$1 + (0.61/S_M)^2$
Flange Tap	$1 + (0.605/S_M)^2$

Procedure for Selection of Bore Size

These steps can be followed to select the approximate size of the bore diameter for a number of differential pressure flowmeters.

Step 1: Define the normal flow rate. If not known, the design flow rate is estimated at 80% of the maximum allowable flow rate.

Step 2: Select the differential pressure for the flowmeter and the maximum range of the differential pressure device. The maximum differential pressure would be generated at the maximum flow rate. For compressible fluids, the limits of the differential pressure ratio from Equation (8-41) is 0.2. For the selection of beta ratio, the differential pressure ratio, x_1, should be limited to a maximum of 0.04; i.e., the ratio of the maximum differential pressure in inches of water column to the line pressure in psia should not exceed the value of 1.

The design flow rate or normal flow rate and the differential pressure at the design flow are used for subsequent calculations.

Step 3: Calculate the Reynolds number at the design flow rate and operating condition. The line size and the Reynolds number limits given in Table 8-8 should be checked for proper selection of flowmeter type.

Note that the discharge coefficient equation of some types of differential devices do not have any Reynolds number correction factors (Table 8-7). The Reynolds number range for the expected operating flow rates should be evaluated to define the acceptable operating flow range of the flowmeter; e.g., universal Venturi tube should have a pipe Reynolds number greater than 75,000.

Step 4: Calculate the meter sizing factor, S_M, at the design flow rate and operating temperature and pressure.

Step 5: From Table 8-10 select the appropriate equation and calculate the value of β_0. The value of β_0 should be between the limits specified for the type of the flowmeter (Table 8-8); i.e., the beta ratio limits are 0.2 to 0.5 for ASME long radius throat tap nozzle.

When β_0 is outside the allowable limits or if the desired beta ratio is different from the calculated beta, return to Step 2 and select a higher or lower differential pressure or a larger or smaller pipe size to decrease or increase beta, respectively.

For an iterative solution for improved accuracy, proceed to Step 6. If the desired accuracy is of the order of ± 2.0%, then skip to Step 10.

Step 6: If the discharge coefficient term has a Reynolds number correction factor or is a function of the beta ratio, recalculate the discharge coefficient for the new beta ratio value (Table 8-7).

Step 7: For liquids $Y_1 = 1$. For gases and vapors, calculate the gas expansion factor (Equation (8-22)) for the line pressure, differential pressure, and the value of the beta ratio.

Step 8: Calculate the next estimate of β ratio:

$$\beta_1 = [1 + (C*Y_1/S_M)^2]^{-1/4}$$

Step 9: Repeat Steps 6, 7, and 8 until two consecutive iterations of β differ by less than 0.0001.

Step 10: Calculate the bore of the flowmeter using $d = \beta D$.

Size the bore of a flange tapped concentric orifice plate to measure a maximum flow rate of 200 gpm of liquid in a 3-inch schedule 40 pipe at 100 psig line pressure and 90°F. The specific gravity of the fluid at the operating condition is 0.82, and the viscosity is 0.5 cP. The critical temperature of the liquid is 800°F. The design flow rate is 180 gpm at 80 inches of water column. The thermal expansion factor of the meter is 0.00001.

Solution:

Step 1: **Full scale flow 200 gpm**

Design flow 180 gpm

Step 2: **Let the full scale range of the differential pressure device be approximately 100 inch of w.c.**

(continued)

(continued)

Differential Pressure:

	@ design flow	80 inches
	@ maximum flow rate	
	$80 \times (200/180)^2$	98.77 inches

Step 3: For the design condition

Q_{gpm} = 180 gpm
SG = 0.82
μ_{cP} = 0.5 cP
D = 3.068 inches

From Equation (8-13)

$$Re_D = (3160 \times 180 \text{ gpm} \times 0.82) / (0.5 \text{ cP} \times 3.068 \text{ in.}) = 304{,}052$$

From Table 8-8, the Reynolds number is within the acceptable range.

Step 4: Flow rate in pounds per minute:

$$Q_{ppm} = 180 \text{ gpm} \times 8.334 \text{ lb/gal} \times 0.82 = 1230.1 \text{ lbs/min}$$

From Equation (8-50):

$$F_a = 1 + 2*(0.00001)*(90 - 68) = 1.00044$$

$$T_R = (459.67 + 80) / (800 + 459.67) = 0.43636$$

From Equation (8-60):

$$Z_L = -0.461 + 2.69*T_r - 5.165*T_r^2 + 3.521\, T_r^3 = 0.022272$$

$$\rho_1 = 0.82 \times 62.3362 = 51.1157$$

h_w = 80 inches of w.c.

From Equation (8-59):

$$F_p = 1 + 0.022272*(100 + 14.696) / 10000 = 1.000255$$

From Equation (8-58):

$$S_M = 1230.1/ [5.982114*(1.00044)*(3.068)^2* \sqrt{(80*51.1157/1.000355)}\,] = 0.34152$$

Step 5: From Table 8-10, for $Re_D = 304052$

$$1 / \beta_0^4 = 1 + (0.60 / S_M)^2 \quad \text{and}$$

$$\beta_0 = [1 + (0.60 / 0.34152)^2]^{-1/4} = 0.70333$$

Step 6: Using Equation (8-32):

$$C = 0.5959 + 0.0312*\beta^{2.1} - 0.1840*\beta^8 + \frac{0.0900*\beta^4}{[D\,(1 - \beta^4)]} - 0.0337*\beta^3 / D + 91.71*\beta^{2.5} / Re_D^{0.75}$$

For $\beta = \beta_0 = 0.70333$ and $Re_D = 304{,}052$, $C = 0.608403$.

Step 7: For liquid, expansion factor $Y_1 = 1.0$

(continued)

(continued)

Step 8: Calculate the next estimate of beta ratio.

$$\beta = [1 + (C*Y_1 / S_M)^2]^{-1/4}$$
$$= [1 + (0.608403*1 / 0.34152)^2]^{-1/4}$$
$$= 0.69964$$

Step 9: The change in beta ratio value is 0.0015, which is greater than 0.0001; therefore, repeat Steps 6, 7, and 8.

Iteration: 1

Step 6: For $\beta = 0.69964$ and $Re_D = 304{,}052$, from Equation (8-32):

$C = 0.608455$

Step 7: $Y_1 = 1.0$.

Step 8: Calculate the next estimate of β for discharge coefficient, $C = 0.608455$:

$\beta = 0.699615$.

Step 9: The change in beta ratio value of 0.0000088 is less than 0.0001. *No further iteration is required.*

Step 10: Required orifice bore diameter is 2.1464 inches. The beta ratio is 0.699615, which is less than 0.75 limit indicated in Table 8-8.

Size the bore of a universal Venturi tube required to measure a flow rate of 800 scfm of an ideal gas through a 4-inch schedule 40 pipe at 50 psig and 800°F with a viscosity of 0.018 cP at flowing conditions and a specific gravity of 0.65 at base conditions of 68°F and 14.7 psia. The specific heat ratio of the gas is 1.3. The thermal expansion coefficient of the meter is 0.0000098.

Solution:

Step 1: Design flow 800 scfm

Maximum design flow 1,000 scfm

Step 2: Maximum differential pressure = 30 inches of w.c.

Differential pressure

@ design flow rate $(0.8^2 \times 30) = 19.2$ inches of w.c.

$$\text{Differential pressure ratio} = \frac{19.2}{[27.707 \times (50 + 14.7)]} =$$

0.0107

Step 3: Calculate the Reynolds number and $\Delta P/P_1$ at the design condition.

$$Q_{acfm} = 800*[(460 + 80) / (460 + 68)]*[14.7 / (50 + 14.7)]$$
$$= 185.89 \text{ acfm}$$

For base condition, density of air:

$$\rho_{air} = (14.7 \times 144) / [53.34*(68 + 460)]$$
$$= 0.07516$$

(continued)

(continued)

Density of the gas at flowing condition is:

$$\rho_f = 0.65 \times 0.07516 \times \frac{(460+68)}{(460+80)} \times \frac{(14.7+50)}{14.7}$$
$$= 0.210246$$
$$D = 4.026$$
$$\text{Re}_D = 379 \times 185.89 \times \rho_f / (\mu_{cP} \times 4.026)$$
$$= 204{,}512$$

Step 4: Calculate the meter sizing factor.

$$Q_{ppm} = 185.89 \times 0.21036$$
$$= 39.083 \text{ lb/min}$$
$$F_a = 1 + 2 \times (0.0000098) \times (80 - 68)$$
$$= 1.000235$$
$$F_p = 1 \text{ for gas}$$

From Equation (8-58)

$$S_M = 39.083 / [5.982114 * F_a * 4.026\sqrt{(19.2 * 0.210246 / 1)}]$$
$$= 0.200572$$

From Table 8-7 for $C = 0.9797$,

$$1 / \beta_0^4 = 1 + (0.9797 / 0.200573)^2$$
$$= 24.8584$$
$$\beta_0 = 0.44785$$

Step 6: The discharge coefficient does not depend on Reynolds number or β ratio, so no iterations are necessary.

Step 7: $x_1 = 19.2 / [27.707 \times (50 + 14.7)] = 0.01071$

$$P_r = 1 - x_1 = 0.98929$$
$$k = 1.3$$

From Equation (8-22), $Y_1 = 0.993463$

Step 8: Calculate the next estimate of β ratio.

$$\beta = [1 + (C * Y_1 / S_M)^2]^{-1/4}$$
$$= [1 + (0.9797 \times 0.993463 / 0.200572)^2]^{-1/4}$$
$$= 0.4492588$$

Step 9: The change in β ratio value is 0.00147, i.e., greater than 0.0001. Repeat Steps 6, 7, and 8.

Iteration: 1

Step 6: Discharge coefficient is independent of Re_D and β ratio.

Step 7: For new β ratio 0.4492588, compute gas expansion factor, $Y_1 = 0.9934583$

Step 8: New β value.

$$\beta = [1 + (0.9797 \times 0.993458 / 0.200572)^2]^{-1/4}$$
$$= 0.4492598$$

Step 9: Change in β value, 0.000001

No further iteration necessary

Step 10: The bore diameter, $d = 0.4492598 \times 4.026 = 1.8087$ inches.

Critical Flow Venturi Nozzle

When a gas accelerates through a nozzle, its density decreases. The mass flow can be progressively increased until the velocity at the throat is sonic. A further decrease in the downstream pressure does not increase the mass flow rate. This phenomenon is referred to as choked or critical flow.

For liquid flows, if the pressure is reduced below the vapor pressure in the flowmeter, cavitation restricts the flow, and decreasing downstream pressure does not increase the flow rate through the nozzle. For gases at the critical pressure ratio of the absolute downstream pressure to the absolute upstream pressure, the velocity of the fluid reaches the sonic velocity at the throat. For pressure ratios above the critical pressure ratio, the velocity at the throat is subsonic. For pressure ratios below the critical pressure ratio, the flow rate is constant, and the velocity is sonic at the throat. In critical flow conditions, the flow rate can be increased only by increasing upstream pressure.

The flow rate through the nozzle at choked condition is:

$$Q_{\text{ppm}} = 36.593178 * C * d^2 * Y_{\text{CR}} * F_{\text{TP}} * P_{f1} * \sqrt{\text{SG} / T_1} \qquad (8\text{-}62)$$

where:

C = the discharge coefficient
d = the throat diameter in inches
T_1 = the absolute pressure upstream of the nozzle inlet
SG = specific gravity
F_{TP} = factor correcting static pressure total pressure
Y_{CR} = the critical flow function

Discharge coefficient for critical flow Venturi nozzles:
Toroidal throat

$$10^5 < \text{Re}_\text{d} < 2.5 \times 10^7 \qquad C = 0.9935 - \frac{1.525}{\sqrt{\text{Re}_\text{d}}}$$

Cylindrical

$$3.5 \times 10^5 < \text{Re}_\text{d} \le 2.5 \times 10^6 \qquad C = 0.9887$$

$$2.5 \times 10^6 < \text{Re}_\text{d} \le 2 \times 10^7 \qquad C = \frac{1 - 0.2165}{\text{Re}_\text{d}^{0.2}}$$

Critical flow function:

$$Y_{\text{CR}} = \sqrt{(k / Z_f) * [2 / (k+1)]^{(k+1)/(k-1)}} \qquad (8\text{-}63)$$

where Z_f is the gas (vapor) compression factor at the flowing condition.

$$F_{\text{TP}} = 1 / [1 - (k/2) * [2 / (k+1)]^{(k+1)/(k-1)} * \beta^4] \qquad (8\text{-}64)$$

For critical flow Venturi nozzles, the only measurements required are the upstream pressure and temperature or the density of the fluid upstream of the critical Venturi nozzle. Flow rate is calculated from thermodynamic considerations. Most common applications of critical flow Venturi nozzles have been for testing, calibration of other meters, and flow control applications.

References

1. Spitzer, D. W., *Industrial Flow Measurement*, Research Triangle Park, NC: Instrument Society of America, 1990.

2. Miller, R. W., *Flow Measurement Engineering Handbook*, Second Edition, NY: McGraw-Hill, 1989.

3. Howard, A. T. J., *Flowmeters: A Basic Guide and Source Book for Users*, London: MacMillan Press Limited, 1979.

4. "Fluid Meters," Report of ASME Research Committee on Fluid Meters, 6th Edition, NY: American Society of Mechanical Engineers, 1971.

5. ANSI/API 2530, AGA Report No. 3, GPA 8185-85, "Orifice Metering of Natural Gas and Other Related Hydrocarbons," Arlington, VA: American Gas Association.

6. ASME MFC-3M-1989, "Measurement of Fluid Flow in Pipes Using Orifice, Nozzle, and Venturi," NY: American Society of Mechanical Engineers.

7. ISO 5167, "Measurement of Fluid Flow by Means of Orifice Plates, Nozzles, and Venturi Tubes in Circular Cross-Section Conduits Running Full," International Organization for Standardization (ISO).

8. ANSI/ISA-S51.1-1979, "Process Instrumentation Terminology," Research Triangle Park, NC: Instrument Society of America.

9. ANSI/ISA-S5.1-1984, "Instrumentation Symbols and Identification," Research Triangle Park, NC: Instrument Society of America.

10. ASME/ANSI MFC-7M-1987, "Measurement of Gas Flow by Means of Critical Flow Venturi Nozzles," NY: American Society of Mechanical Engineers.

About the Authors

Dr. Zaki D. Husain, Ph.D., is Director of Flow Measurement and Technical Services at Daniel Flow Products, Inc., Houston, Texas, where he is responsible for operations of flow lab and flowmeter calibration facilities, flow product development, fluid dynamic research, and technical support for the plant and Daniel customers. Dr. Husain is a Mechanical Engineer and has been active in flow measurement and instrumentation over the last fifteen years. He received his bachelor's and master's degrees in Mechanical Engineering in 1968 and 1974 from Bangladesh University of Engineering. In 1982, he received his doctoral degree from the University of Houston in Fluid Dynamics—Experimental Research in Turbulence.

Dr. Husain has five journal publications and over twenty publications and presentations in conferences and symposia in the United States, Canada, Europe, and Asia. He taught as a full-time member of the faculty at the Bangladesh University of Engineering for seven years and is now a part-time lecturer at the Department of Chemical Engineering of the University of Houston, Texas.

Dr. Husain serves on the American Society of Mechanical Engineers Codes & Standards Committee on Measurement of Fluid Flow in Closed Conduits and is Chairman of a Subcommittee. He is active with the Flow Measurement Committee of the Instrument Society of America. He has also served on the Codes and Standards Committee of Fluid Meters of the American Water Works Association.

M. J. "Joe" Sergesketter (1931-1990) worked in instrumentation and measurement for over thirty years and was involved in training activities for over twenty years. He was active in developing technical training programs for the University of Texas Petroleum Extension Service, for the Instrument Society of America, and for several major corporations. Mr. Sergesketter was a graduate of Purdue University and the U.S. Army Command and General Staff College. A Registered Professional Engineer in the States of Texas and Louisiana, he worked as a consultant and for Shell Oil Company, ITT Barton, and Daniel Industries. While with Shell, Mr. Sergesketter was involved in the initial field testing of LACT units and displacement meter provers. He had extensive experience in field measurement applications, especially the area of light liquid hydrocarbon measurement, and conducted measurement audits for several production and pipeline companies.

9

Magnetic Flowmeters

Magnetic flowmeters (magmeters) are designed to measure the flow of electrically conductive liquids in a closed pipe. They are volumetric flow measuring devices that have been commercially available since the mid 1950s. Sizes range from $1/10$ inch to about 96 inches from most manufacturers. This covers a flow range of about 0.01 gallon per minute to about 500,000 gallons per minute. Early units were flanged devices that bolted to adjacent pipe flanges much the same as an ordinary section of flanged pipe. They were large, heavy, and expensive, but they offered several advantages over other flowmeters available at that time.

Advantages:
Obstructionless design
Linear output
Corrosion-resistant wetted parts
High accuracy

With the obstructionless design, there are no moving parts to wear and no pressure drop other than that offered by a section of pipe with equal length and inside diameter. The only wetted parts are the electrodes and an insulating liner. These can be selected for compatibility with the most corrosive of chemicals as well as to meet sanitary requirements for food applications such as milk and other liquid dairy products. The output signal is linear and directly proportional to the flow velocity. Accuracy over a wide range (typically 10 to 1) has evolved from 1% of full scale reading to 1% of rate as standard. Higher accuracies are available for special applications. The basic fundamentals of operation, selection, installation, and maintenance will be discussed in this chapter. The discussion will be basic and generic. Specifics on availability of features and options can be obtained from the various magmeter manufacturers.

Operating Principle

Magnetic flowmeters operate on the principle of Michael Faraday's Law of Electromagnetic Induction. Without getting involved in the mathematics of this theory, it can be simply stated as:

$$E = \text{constant} \times B \times L \times \bar{v} \qquad (9\text{-}1)$$

where:

E = magnitude of the voltage
B = magnetic field density
L = path length
$\bar{v}$ = average velocity of the medium

Equation (9-1) implies that a voltage is developed when a conductor is passed through a magnetic field. It further states that the voltage developed is proportional to the density of the magnetic field, the length of the conductor, and the velocity of the conductor moving through the field. There is nothing in the equation about temperature, pressure, density, or viscosity because the magmeter develops its signal independent of these parameters.

In the conventional construction of a magnetic flowmeter, coils are mounted on the outside of a nonmagnetic pipe section. Voltage, typically 120 or 240 volts at 50 or 60 hertz, is applied to the coils. As current passes through the coils, a magnetic field is generated inside the pipe section. Liquid passes through the pipe section perpendicular to the plane of the magnetic field. This is schematically illustrated in Figure 9-1.

As conductive liquid flows through this pipe section, a voltage is generated. The voltage is extracted through a pair of electrodes that are installed on opposite sides of the pipe. When the pipe section is constructed of a conductive material such as stainless steel, it is lined with a nonconductive material to insulate the pipe from the electrodes and prevent the flow voltage from being dissipated into the pipe section.

The magnetic field density is fixed for each magmeter size. The length of the conductor is essentially the distance between the electrodes and is fixed by meter size. This leaves velocity as the only variable in Equation (9-1). Consequently, it can be said that magmeters are velocity-measuring devices. Note that the conductor length is not simply the straight line distance between the electrodes but the sum of an infinite number of conductors that make up the cross-sectional area of pipe at the electrodes. All the velocities in this slice are summed to get an accurate flow measurement with minimal effects from flow profile. In addition, a coil design feature introduced in 1967 characterized the magnetic field to make the magmeter even less sensitive to flow profile problems. Prior to this, a magnetic field of uniform density was generated across the pipe and extended for considerable distance up and downstream from the electrode plane. The uniform magnetic field design is still used by some manufacturers.

Liquid particles flowing through a magmeter do not all move at the same velocity and do not generate a voltage of the same magnitude in a uniform field. A signal-generating coefficient, which varies with the radial and axial displacement of the individual liquid particles from the electrodes, must be applied. This is a geometric phenomenon that depends only upon the dimensional characteristics of the flowmeter and the location of the liquid particles. Compensation for these variations can be achieved by shaping the coils so that the magnetic flux is greatest where the signal-generating coefficient is lowest, and vice versa. A graphic representation of the coefficients is shown in Figure 9-2. Shaping the field reduces errors that result from nonsymmetrical flow patterns. Because of this, the magmeter rates very high among the flowmetering devices that are least affected by piping configurations. More about piping effects will be discussed in the section on installation.

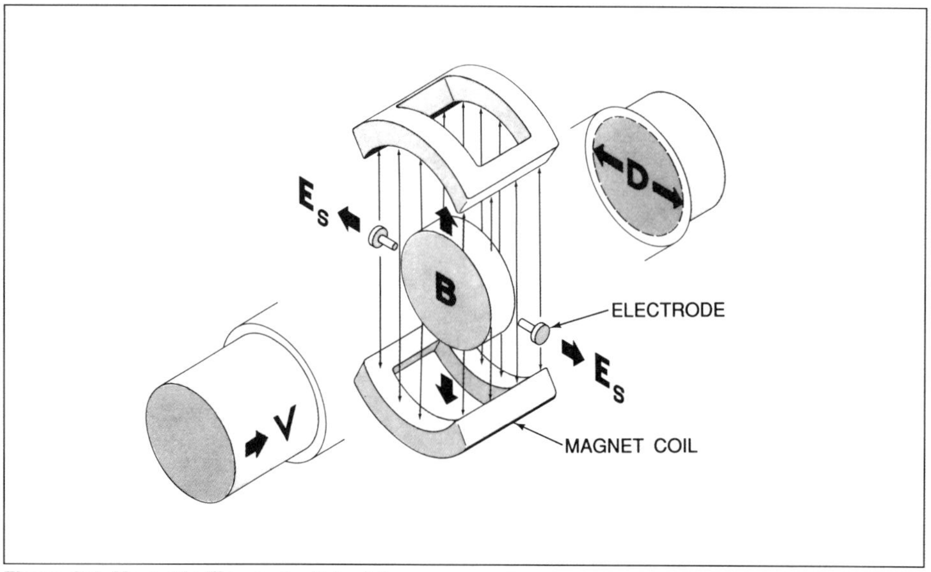

Figure 9-1. Magnetic Flowmeter Principle
(Courtesy of Fischer & Porter Company)

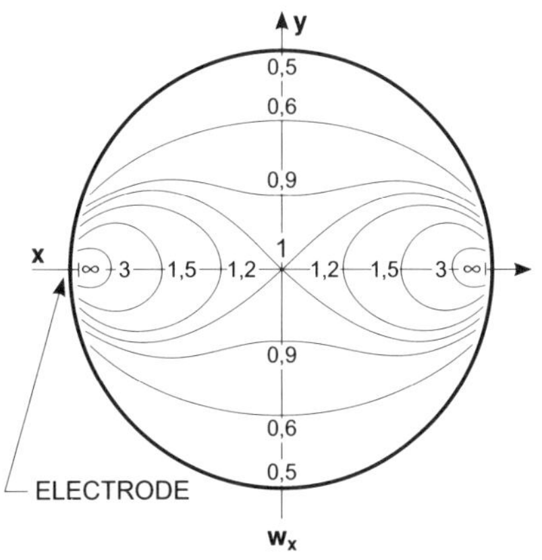

Figure 9-2. Weighted Compensating Coefficients

The System

The voltage developed at the electrodes is an extremely low-level signal. The actual voltage varies by model and type and from manufacturer to manufacturer. For example, the flow signal could be on the order of 250 microvolts per foot per second (μV/ft/sec). Most manufacturers have established an arbitrary maximum flow velocity for each size magmeter of about 30 feet per second (ft/sec). At 250 μV/ft/sec, the maximum output would be 7.5 millivolts (mV). By comparison, a 1.5-volt flashlight battery has 200 times the voltage that is developed in a magnetic flowmeter at its maximum output. In order to use this low-level signal, a transmitter or signal converter is provided to amplify, condition, and present a more usable signal, usually in the form of 4 to 20 milliamp direct current (mA DC). A frequency output such as 0-1000 Hz might also be used. In addition, some systems offer an additional scaled frequency output for totalization.

Remember that the output signal from the magmeter is strictly linear. Therefore, the transmitter output signal is 4 mA DC at zero flow and 20 mA DC at 100% flow, with all points in between being directly proportional to velocity.

The combination of the magmeter and the transmitter is considered as a system. A typical system, schematically illustrated in Figure 9-3, shows a transmitter mounted remote from the magmeter. Some systems are available with transmitters mounted integral to the magmeter. Each device is individually calibrated

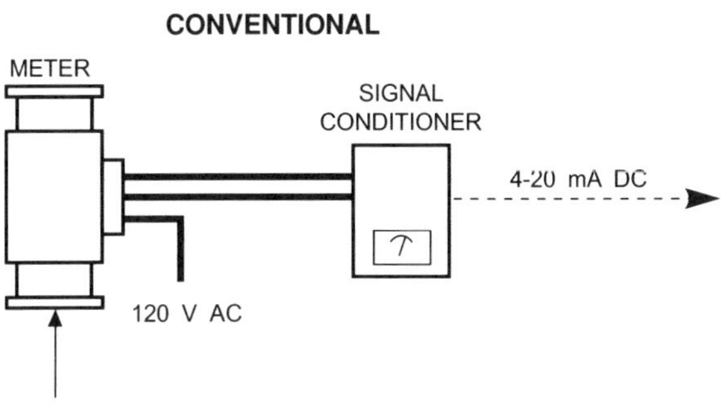

Figure 9-3. Magnetic Flowmeter System

during the manufacturing process, and the accuracy statement of the magnetic flowmeter includes both pieces of equipment. One is not sold or used without the other.

Rangeability

Most manufacturers provide transmitters that are interchangeable with the same series magmeter without the need to go through an further calibration other than to change or set the range, which is done in the transmitter and is not a calibration procedure, although it is commonly referred to as calibration.

Rangeability is herein defined as the ability of the magmeter system to provide an output that is within the specified accuracy from some upper limit to some lower limit. The upper limit can be the capacity of the magmeter or some lower range value. Although magmeters are designed to measure maximum flows of 30 ft/sec and higher, most flows do not reach such high velocities. Typical maximum flows or upper range values are more on the order of 6 to 10 ft/sec. The user wants an output signal of 4 mA DC at zero flow and 20 mA DC at the true maximum flow and not at the 30 ft/sec capacity of the meter. This is accomplished by setting the range in the transmitter. There are many different methods to accomplish this, but only one will be discussed here. In order to maintain a train of thought relating to flow velocities, a range-setting method based upon velocity settings will be examined. Consider the following equation:

$$v_{max} = \frac{0.4085\,(Q_{max})}{(D)^2} \qquad (9\text{-}2)$$

where:

v_{max} = maximum liquid velocity, ft/sec
Q_{max} = maximum desired volumetric flow rate, gpm
D = effective diameter, in., shown on the magmeter nameplate
0.4085 = conversion constant

Assume a size two-inch magmeter with an effective diameter of 2.078 inches. The capacity of a magmeter of this size is nominally 317 gpm at a flow velocity of 30 ft/sec. The real flow in this example will be 50 gpm, and, for a two-inch magmeter, the maximum flow velocity should be approximately 5 ft/sec. Putting these values into Equation (9-2) yields:

$$v_{max} = \frac{0.4085(50)}{(2.078)^2}$$

$$= 4.73 \text{ ft/sec}$$

In the transmitter, there is a range device calibrated in ft/sec, which is set at 4.73. This establishes the full scale output of 20 mA DC at 50 gpm. If the maximum flow requirement should change, the new v_{max} can be calculated and set into the range device of the transmitter; no other changes are required. There are some rather complicated methods for setting ranges, but most manufacturers offer a simple range-setting procedure similar to the one used in this example.

Now that a maximum flow rate can be established, what accuracy can be expected from a typical magmeter rated at 1% of actual flow (1% rate), assuming a typical analog-type transmitter with a rangeability of 10 to 1? Although it is not specifically stated in each accuracy statement, accuracies are based upon frequency output signals, because current output accuracy is typically ±0.1% of the full scale value worse than that of the frequency output. Accuracies are commonly stated as percent of rate for flows above a given value and in percent of full scale for flows from zero up to the given value. At this point, it might be worthwhile to look at the difference between percent of rate and percent of full scale accuracies. The difference is illustrated in Table 9-1, which shows that there is a significant error at low flow rates for percent of full scale accuracies, while the 1% accuracy remains constant over the complete 10 to 1 rangeability for the percent of rate accuracy. Consequently, flows from 100% of the maximum setting (50 gpm) to 10% of maximum setting (5 gpm) will be measured with an accuracy of 1% rate. The error at 100% will be ±0.5 gpm, and the error at 10% will be 0.05 gpm. For current output, it is necessary to add ±0.1% of the full scale value. This means that at 100% of flow the accuracy is ±1.1% of rate. At 50% of the maximum setting (25 gpm) the accuracy is ±1.2% of rate, and at 10% of the maximum range setting, the accuracy is ±2.0% of rate. If this were a percent of full scale device, the frequency output error at 10% of max flow would be ±5 gpm or ±10% of rate.

What happens at flows that are below 10% of the range setting? The accuracy gradually deteriorates at a predictable rate. For example, one statement says that for the 1% of rate used here, the accuracy becomes 0.1% of full scale for flows below 10% of the range setting. This appears graphically in Figure 9-4. In order to be consistent, the curve is plotted to show the accuracy in terms of percent of rate. The lower curve applies to the frequency output and the upper curve to the analog output. Using 50 gpm as the range setting, the error at 0.1% of full scale is a constant ±0.05 gpm for all flows at or below 5 gpm. If this is converted to percent of rate, the error at 5% of range setting (2.5 gpm) becomes ±2% of rate; at 2-1/2% of range setting (2.5 gpm) it is ±4% of rate. Check the graph in Figure 9-4 to confirm this. Also observe from the curve and the example that, for flows below 10% of range setting, the accuracy deteriorates by one half every time the flow rate is reduced by one half. That is, accuracy at 10% of range setting is ±1%

Although it is not specifically stated in each accuracy statement, accuracies are based upon frequency output signals, because current outputs are typically ±0.1% of the fulll scale value worse than the frequency output.

Table 9-1. Accuracy Comparison - % of Actual Flow

% of Flow	1% Full Scale Accuracy	1% Rate Accuracy
100%	1%	1%
75%	1.33%	1%
50%	2%	1%
25%	4%	1%
10%	10%	1%

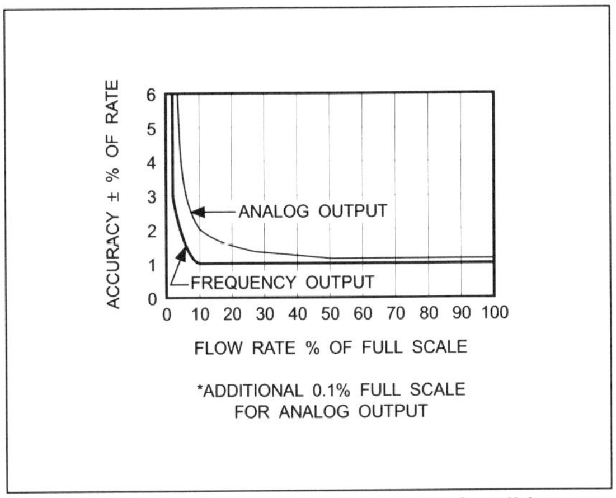

Figure 9-4(a). Accuracy under Reference Conditions

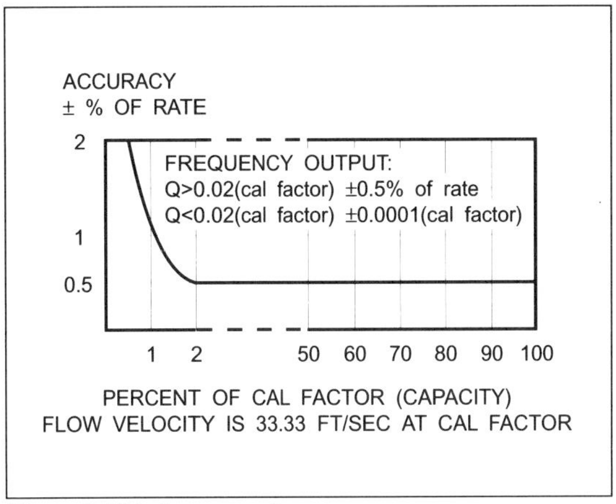

Figure 9-4(b). Accuracy Curve

of rate; at 5% of range setting it is ±2% of rate, and 2-2½% of range setting it is ±4% of rate.

Range Limits

Magnetic flowmeters are said to have at least 100 to 1 rangeability. This subject will be treated separately for analog-type transmitters and microprocessor-type transmitters. For analog-type transmitters this statement takes into consideration a rangeability of 10 to 1 and the ease of range setting. If the range is set at the capacity of the magmeter (30 ft/sec), the standard accuracy from 30 ft/sec down to 3 ft/sec is achieved. If the range is changed to 10% of the capacity (3 ft/sec), the standard accuracy applies to 0.3 ft/sec. That's 100 to 1 (30 ft/sec to 0.3 ft/sec), but it requires a range change to get it. Why not set the range at 1 ft/sec and use the 10 to 1 rangeability to go all the way down to 0.1 ft/sec? This can be done but not with the same rangeability. There is a practical limit to how low the range can be set and still achieve a rangeability of 10 to 1 holding the standard or optional accuracy. That limit is generally 3 ft/sec. Ranges can be set as low as one ft/sec, but the standard accuracy is not attained below 0.3 ft/sec. This reduces the rangeability from 10 to 1 to about 3.3 to 1. Some accuracy statements limit the flow velocity to a minimum of 3 ft/sec in order to maintain the stated accuracy. For these, the rangeability is normally less than 10 to 1. For example, a range setting of 9 ft/sec is a rangeability of only 3 to 1 (9 ft/sec to 3 ft/sec).

Most range limits are relatively straightforward, and rangeability is not much of a problem. But what if the maximum flow never reaches 3 ft/sec? One solution is to reduce the meter size to get the required maximum velocity, unless piping restrictions will not allow such a reduction.

For example, the velocity for a max flow of 5 gpm in a size one-inch magmeter is about 2 ft/sec. The velocity for this same flow in a size ½-inch magmeter is about 8 ft/sec. Provided there are no piping restrictions to prevent a reduction in size, using reducers to go from a one-inch line to a ½ inch magmeter is feasible for this application. If piping restrictions will not allow a reduction in size, the rangeability will be limited to about 6.7 to 1 (5 gpm to 0.75 gpm).

Rangeability is sometimes expressed in terms of the magmeter capacity (CAL factor), as is the case for some microprocessor-type transmitters. The standard accuracy for many of these devices is ±0.5% of rate for all flows from the capacity of the magmeter down to some small percentage of the capacity. For example, one statement holds ±0.5% of rate accuracy from 100% of the magmeter capacity down to 2% of the capacity (see Figure 9-4). This is a rangeability of 50 to 1. As in the analog-type transmitter, flows below 2% of the capacity are half as good every time the flow is reduced by one half. Consequently, at 1% of the magmeter capacity, the accuracy is 1% of rate. The end result is a 50 to 1 rangeability at ±0.5% of rate and 2 to 1 at ±1% of rate, for a total rangeability of 100 to 1 at no worse than ±1% of rate. Changing the range does not extend the accuracy as in the analog-type transmitter example; no matter where the range is set, the accuracy does not extend below 2% of the magmeter capacity. The lowest range setting permitted in this system may also be expressed in terms of the capacity, such as 5% of capacity. For example, a size 4-inch magmeter with a capacity of 1200 gpm would have an accuracy of ±0.5% for all flows from 1200 gpm down to 24 gpm no matter where the range was set. The range could be set from 1200 gpm down to 60 gpm. At a range setting of 1200 gpm, the rangeability is 50 to 1, and, at 60 gpm, it is 2½ to 1.

Most flow applications do not require rangeabilities greater than 10 to 1. However, there are instances where flows do exceed 10 to 1, and the wide rangeability of this type of transmitter is useful. For example, plant intake of water could be very much higher during the day (10,000 gpm) than at night (600 gpm), resulting in a rangeability requirement of nearly 17 to 1. A size 20-inch magmeter with a capacity of 29,375 gpm would yield an accuracy of ±0.5% of rate from 29,735 gpm down to 588 gpm. At 294 gpm, the accuracy would be ±1% of rate. If a chart with a range of 0 to 10,000 were used to record flow, setting the range at 10,000 gpm would set the current output to be 4 mA DC at zero flow and 20 mA DC at 10,000 gpm. The velocity at 10,000 gpm for this example would be about 10 ft/sec, and at 600 gpm, about 0.6 ft/sec. Both velocities are acceptable for this application.

Low-Flow Cutoff

Below 1 or 2% of range setting, most magmeter systems have a cutoff built into the signal converter that drops the output to zero (4 mA DC) when the flow drops to that level. One of the reasons for this is that, at what is supposed to be zero flow, there may be sufficient movement in the pipe such as sloshing or enough electrical noise in the system to produce an output signal even when there is no real flow in the system. The dropout feature can be removed completely for special applications that require flow to be measured to these low levels. It can also be elevated to higher values (as much as 10% in some systems). Except for some microprocessor-based transmitters, a change in the dropout feature requires modification of the transmitter circuitry.

Empty Pipe

Magnetic flowmeters are designed to measure conductive liquids only. If air or gas is mixed with the liquid, the output becomes unpredictable. The following is an examination of one of the problems that can occur at conditions of no flow.

Some piping configurations (by design or because of faulty equipment) allow the magmeter to drain empty during conditions of no flow. As liquid drains below the level of the electrodes, the transmitter sees a condition for which it was not designed to respond. It was designed to respond predictably to flow changes of zero to 100% in a pipe full of a conductive liquid. In an empty pipe condition, it will still respond but with an unpredictable, sporadic output. This can be an intolerable situation if these outputs end up as counts on a totalizer and/or pen movements on a chart.

The zero return feature was designed to correct this problem. It is included as a standard part of the transmitter circuitry or is available as an option. It is normally activated by an external dry contact such as the contacts on a pump. Operation of the circuit requires running two wires from the transmitter to a pair of normally open contacts on the pump. When the pump stops, the contact closes, and the zero return is activated. When the zero return is activated, the transmitter output goes to its normal zero output. This is typically 4 mA DC or 0 Hz if a frequency output is used. If a totalizer output is also available, it will be zero as well. If the pump is not located near the transmitter or if the system operates on gravity feed, the external contact may be located on a valve or a flow switch. In some installations, an external contact may be difficult or impractical to use. This problem has been addressed in the design of some microprocessor-based transmitters. Some of these units have an empty pipe detection circuitry that senses when the pipe drains empty. When this occurs, an internal zero return circuit is automatically activated, and the outputs go to zero. There is no requirement for external wiring to activate this system. It is all integral to the transmitter. Some initial on-site calibration of this circuit may be required, but its operation is completely automatic.

Construction

The two major components of a magnetic flowmeter system, as was mentioned earlier, are the magmeter and the signal converter or transmitter.

The basic components of the magmeter are the body (pipe section), the electrodes, the liner, magnet coils, and electronics. In the traditional design, the body is made up of a pipe section with a flange welded to each end.

In 1981, a flangeless design (wafer style) was introduced for sizes up to and including four inches. This design has been extended to 6- and 8-inch sizes by some manufacturers. The major advantage of the flangeless design is a reduction in size and weight without sacrificing performance.

Reduction in the size of the magmeter has been an ongoing project, and significant progress was made as far back as 1967. At that time, the face-to-face dimension of the body was reduced to 1.5 times the meter size. For example, the face-to-face (overall) dimension of a size 24-inch meter was reduced from a typical 64 inches to 36 inches or 1.5 times 24. Research showed that the magmeter could not be made shorter than 1.5 times the meter size without causing a span shift. The error could be as great as 6% of the range setting. A number of papers have been published on this subject, at least two of which have appeared in recent years as ISA reprints [Refs. 1, 2]. According to Reference 2, 1.3 diameters is acceptable if an error of 0.2% of rate is acceptable. The paper further states that even with the non-uniform magnetic field, attempts to make the face-to-face dimension shorter than 1.5 D resulted in a problem, because the signal output depends on the conductivity of the inner wall of adjacent pipes.

In the conventional magmeter design, the magnet coils are mounted on the outside of the meter body (see Figure 9-5). This means that the magnetic field must pass through the body into the pipe area where the process will flow and develop the flow signal. Consequently, the meter body must be constructed of a magnetically nonpermeable material. If the body is metal, it is generally made of stainless steel with strict requirements regarding the magnetic qualities of the material. If the body is of nonmetal insulator material, magnetic permeability is not a concern nor is insulation of the body from the electrode. One nonconventional design has a carbon steel (magnetically permeable) body. In this design, the coils are encapsulated in epoxy inside the body, which is part of the magnetic field circuit (see Figure 9-5). It was this concept that facilitated the initial reduction in magmeter size and weight. It also made possible the continuous submergence design, which allows the magmeter to operate completely submerged in water to a depth of 30 feet.

Electrodes are almost always made of some type of metal. They are installed through the meter body and are in contact with the process liquid. If the meter body is also made of a metal, it must be insulated from the electrode. This is done by installing a liner that goes between the body and the electrode to prevent the flow signal from being shorted to the body. A few electronic components are included to aid in calibration.

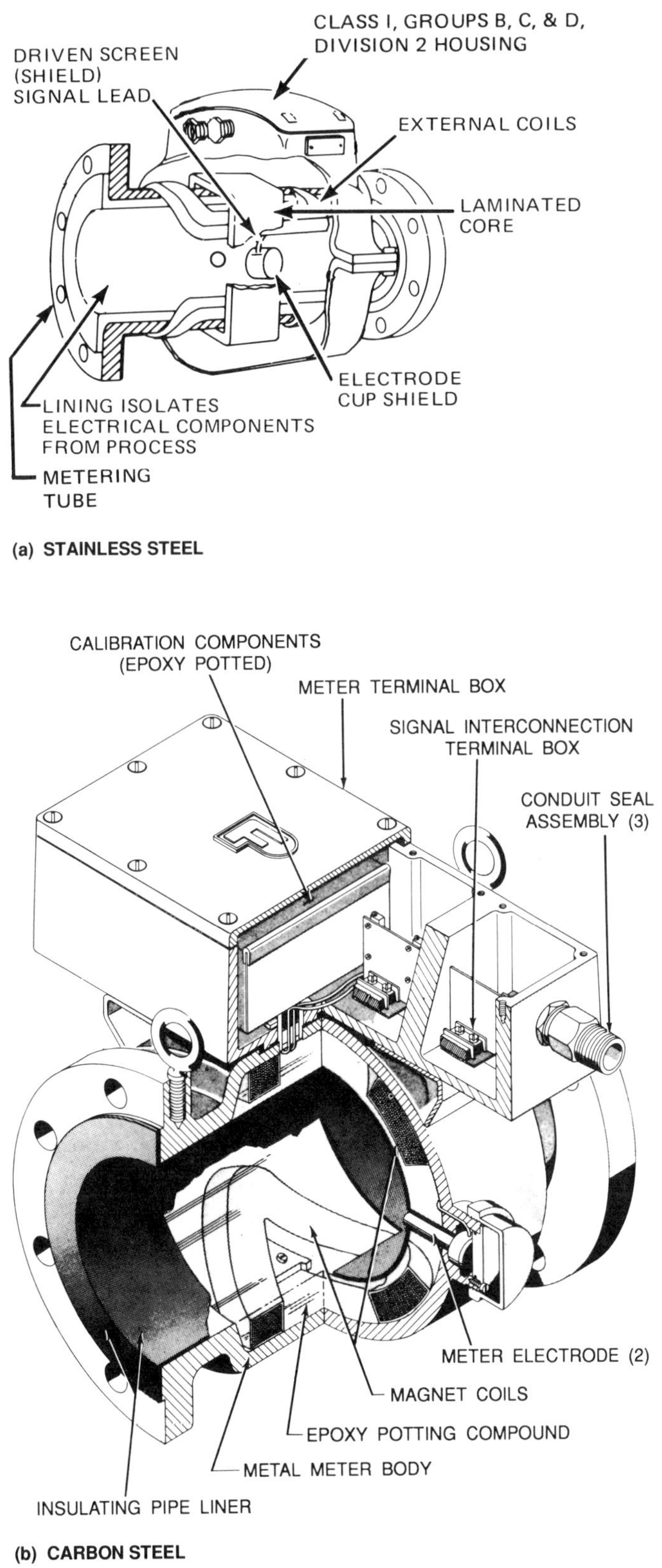

Figure 9-5. Magmeters with Steel Bodies

Table 9-2. Material Selection Guide

Process Liquid at Maximum Operating Temperature of 212°F (100°C)	Liner			Electrode Material and Average Penetration per Year										Notes
	TEFLON™	POLYURETHANE	NEOPRENE	316 SS	HASTELLOY C™	MONEL™	TANTALUM	TITANIUM	ALLOY 20	PLATINUM	HASTELLOY B™	NICKEL	ZIRCONIUM	
Acetic Acid (70% max)	A	N	N	A	A	A	A	A	A*	A	A	N	A	*40% max
Beer	A	N	N	A	A	A	A	A	A	A	A	A	A	
Black Liquor	A	N	N	B	B	U	N	U	U	A	U	U	A	
Brine	A	B	A	A	A	A	U	A	A	A	A	A	A	
Caustic Soda (50% max)	A	N	N	B*	U	A	N	B	B	A*	A	A	A	*75°C max
Chlorine Water	A	N	N	N	A	N	A	A	N	A	N	N	N	
Ferric Chloride (50% max)	A	N	A	N	N	N	A	A	N	A*	N	N	N	*75°C max
Hydrochloric Acid (40% max)	A	N	A	N	N	N	A	N	N	A	N	N	A	
Milk	A	N	N	A	U	B	A	U	B	A	U	N	U	
Sodium Hydroxide (50% max)	A	N	N	B*	U	A	N	B	B	A	U	N	U	
Sodium Hypochlorite (20% max)	A	N	N	N	A	N	A	A	N	A	A	N	A	
Sulfuric Acid (1-20%)	A	N	A	N	N	B	A	N	B	A	A	N	B	
Sulfuric Acid (21-40%)	A	N	B	N	N	B	A	N	B	A	B	N	B	
Sulfuric Acid (41-70%)	A	N	N	N	N	N	A	N	B	A	B	N	B	
Sulfuric Acid (71-100%)	A	N	N	N	N	N	A	N	N	A	A*	N	N	*90% min

A - Less than 0.002 inch (2 mils)
B - Less than 0.020 inch (20 mils)
N - Not recommended
U - No data available

Materials of Construction

The survival of a magnetic flowmeter in a given application depends primarily on the proper selection of liner and electrode materials. These are the only parts of the magmeter that are in contact with the process (see Figure 9-6). Excessive wear on these parts could cause the meter to cease functioning and could result in damage to other parts as well. See Table 9-2 for an abbreviated guide to the selection of materials of construction.

The principal factors to consider when making liner and electrode material selections are the chemical makeup, operation temperature, and abrasive characteristics of the process. In the majority of applications, the concern is for the effect of these three process parameters on the liner and electrodes. In some applications, however, the process itself could be affected by contact with the liner or electrode materials, which is especially true for sanitary food processes. The selected materials of construction could be acceptable based upon chemical, temperature, and abrasive characteristics of the process but might be unacceptable because they could contaminate the process. In addition, some materials act as a catalyst to the process. For example, platinum, which is perhaps the most inert of all electrode materials, accelerates decomposition of hydrogen peroxide.

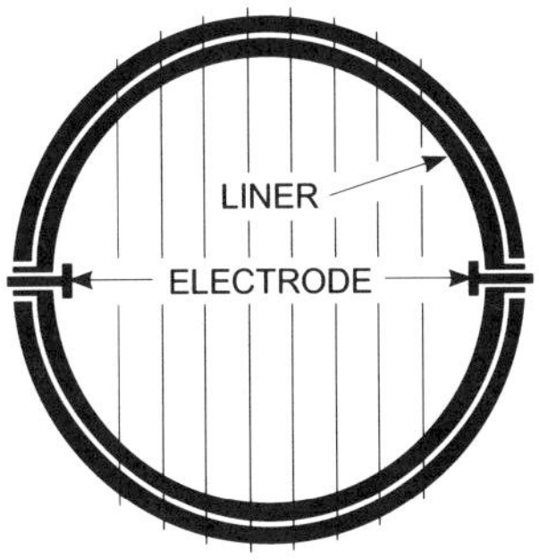

Figure 9-6. Liner and Electrode

Cleaning

The manner in which the process lines are cleaned must not be overlooked. Liner and electrode materials must be compatible with the cleaning materials as well as the process materials. Steam cleaning should be limited to about 300°F (149°C) and to high temperature liners such as Teflon™. Rapid cooling of the process lines after steam cleaning could result in the creation of a partial vacuum in the line. This could cause the Teflon liner to collapse. Installation of a vacuum breaker in the process line is recommended when this condition exists.

Tantalum is an excellent electrode material for use in a ferric chloride process. However, if sodium hydroxide (caustic) were used to clean this same line, the electrode would be destroyed. How long it would take for this to happen depends upon the concentration of the sodium hydroxide and the length of time the cleaning process requires.

Liner Materials

A wide variety of materials are used to line magmeters, but most are lined with one of four basic materials: Teflon, Tefzel™, polyurethane, and rubber, to which this discussion will be limited. Polyurethane and neoprene are both synthetic rubbers that are used extensively in abrasive processes. Other types of rubber, both natural and synthetic, are available, and a few are acceptable for sanitary applications. However, in sizes up to about 12 inches, Teflon is preferred by most users for sanitary processes. Above 12 inches, cost can be a factor. Food grade rubbers that meet the sanitary requirement are available and are much less expensive than Teflon in sizes larger than 12 inches. All liner materials are not available in every size magmeter. Generally, Teflon is available in sizes ¼ inch through 24 inches, while polyurethane and rubber are available in 2-inch and larger sizes.

TEFLON

The Teflon lining of a magmeter is a sleeve that is inserted into the magmeter body and flared over the face of the body flanges. It is not normally bonded to the body, but is held in place by the electrodes that are inserted from the inside of the lined body and pulled up against the liner to make a seal between the electrode and the liner (see Figure 9-7). The flared liner covers part of the flange face, stopping just short of the flange bolt circle. This results in a raised face

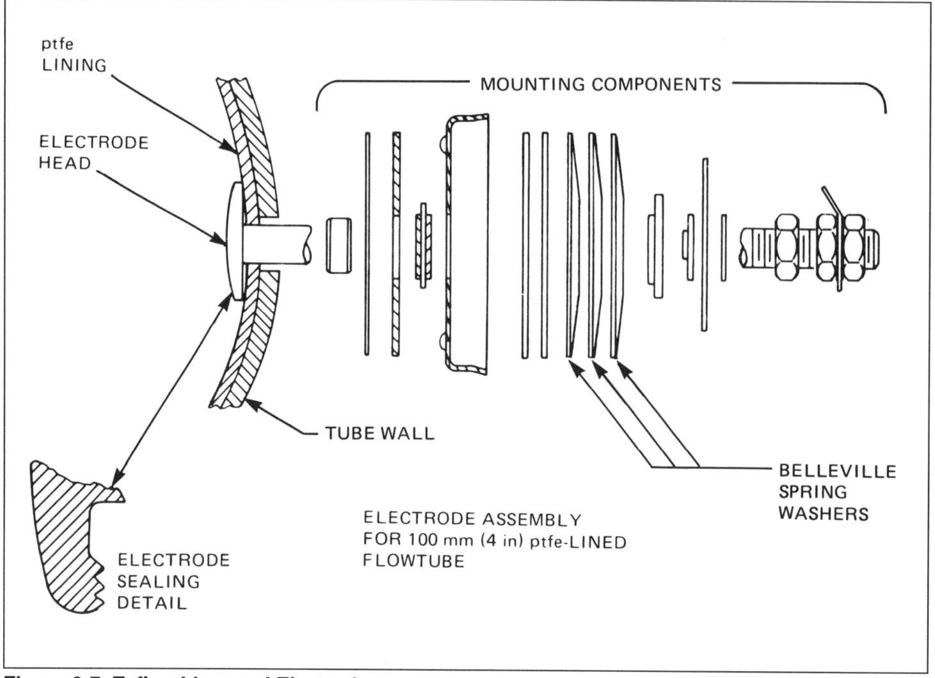

Figure 9-7. Teflon Liner and Electrode

flange configuration (see Figure 9-8). Flaring does not lock the liner in place over the face of the flange; left unattended, it will move away from the flange face. This could result in damage to the liner during installation because the liner would extend beyond the mounting space between the mating pipe flanges. Shipping covers are usually provided to hold the liner tight against the flange face. These should not be removed until just prior to installation in the pipeline.

Devices called Teflon liner protectors have been developed to hold the liner against the flange face during shipping and installation. They are attached to the flange and are intended to remain on the magmeter while it is in service. One such device is shown in Figure 9-9. They protect the liner during shipping, installation, and removal. In addition, they protect the leading edge of the liner, where it flares over the flange, from abrasion by process particles. Where the liner diameter is slightly smaller than the pipe diameter, a step is formed, and head-on collisions with suspended particles could occur. The liner protector has a diameter slightly smaller than the liner, which prevents these head-on collisions and extends the life of the liner. The protector also serves as a grounding ring if it is required.

Figure 9-8. Teflon-Lined Magmeter

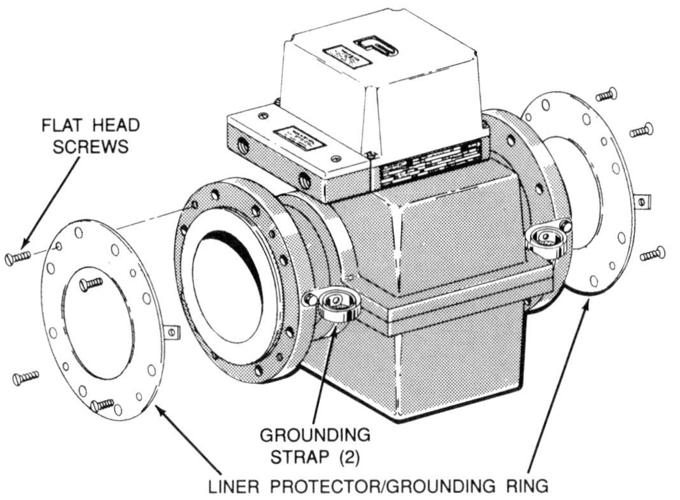

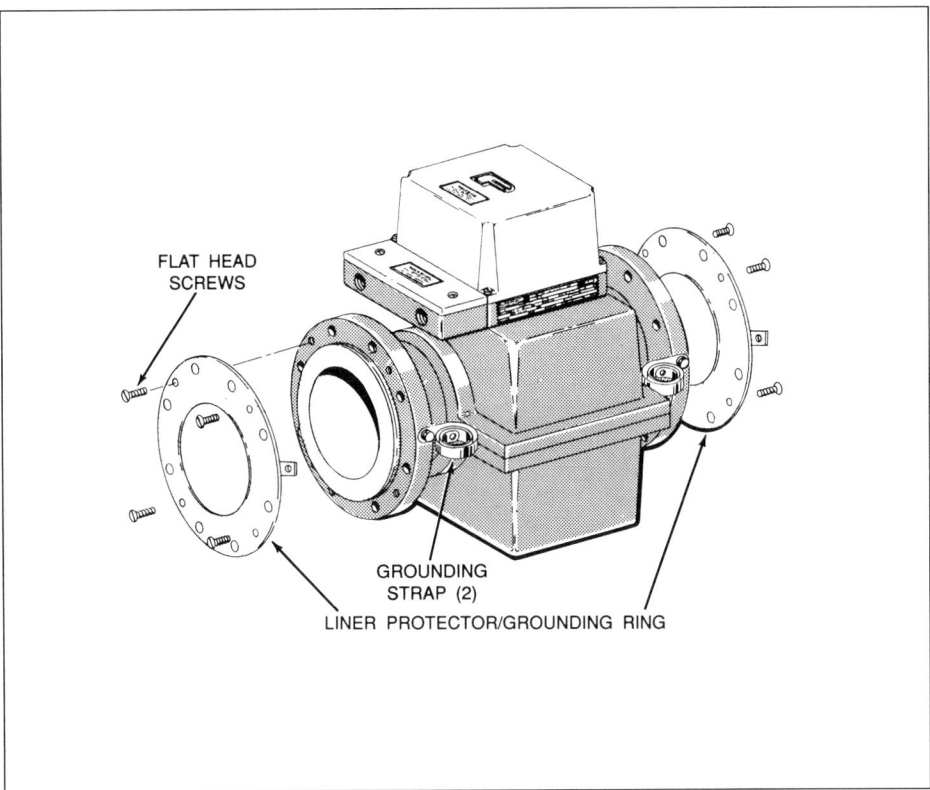

Figure 9-9. Teflon Liner Protector

Teflon is an exceptionally good material for highly corrosive chemicals and high temperature service. However, it breaks down in the presence of nuclear radiation and is not recommended for that service. The upper temperature limit for Teflon-lined magmeters is determined more by other components than the liner itself. Coil insulation and electronic components located in the magmeter could fail before the temperature gets high enough to damage the Teflon. Process operating temperatures are normally limited to about 360°F (182°C), but consult the manufacturer's specification to be certain. Teflon does not hold up as well as polyurethane or rubber in abrasive service, although it does quite well if the suspended solids are fine particles and the maximum flow velocity does not exceed about 10 ft/sec. In applications where there are suspended particles and the temperature is above 190°F (88°C) or there are corrosive chemicals in the process, Teflon may be the compromise liner of choice.

POLYURETHANE

Polyurethane is normally applied as a liquid in a spinning operation similar to rotomolding. It is spun into the meter body over a bonding material and allowed to flow out over the face of the flange and stop just short of the flange bolt circle. This forms a raised-face flange configuration similar to that of the Teflon liner (see Figure 9-10). Polyurethane is an extremely rugged liner material and is widely used to line pipes and magmeters that are used in abrasive service. It is much less corrosion-resistant than Teflon, and its upper temperature limit is only 190°F (88°C). Consequently, the chemical and temperature characteristics of both the process and the cleaning materials must be relatively mild if polyurethane is to be the selected liner material.

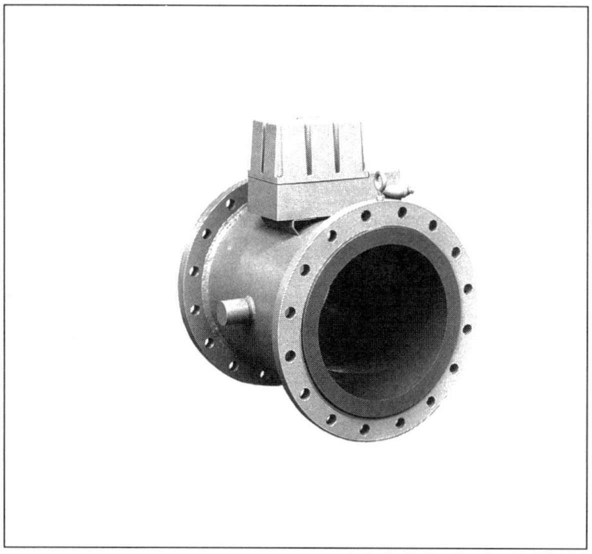

Figure 9-10. Polyurethane-Lined Magmeter

Figure 9-11. Rubber-Lined Magmeter

RUBBER

Rubber liners are placed in the magmeter body over a bonding material and are brought out over the full face of the body flange. This results in a flat-face flange configuration (see Figure 9-11). Rubber has excellent abrasion resistance, and like polyurethane, has received wide acceptance as a liner material for abrasive processes. Its temperature limit varies from about 160°F (71°C) to about 200°F (93°C) depending upon the type of rubber selected. It is considerably more corrosion resistant than polyurethane but much less resistant to chemical attack than Teflon.

TEFZEL

Tefzel is a fluoropolymer with temperature and corrosion resistant characteristics similar to Teflon. Temperature limits range from 250°F (120°C) to 300°F (149°C) depending on the magmeter manufacturer. Tefzel liners are generally available in two forms. One is a glass-filled injection molded process that provides a rough, durable liner (see Figure 9-12). The other is an unfilled form that is rotomolded into a liner much the same as polyurethane except at a much higher temperature. Unlike Teflon, Tefzel holds up very well in processes where nuclear radiation is present. In fact, it is approximately 300 times better than Teflon for this service.

Magmeter Bodies

In addition to stainless steel and carbon steel, magmeter bodies have been made of fiberglass, plastic, and ceramic. All of these non-metal materials are insulators, and as such, do not require liners to insulate them from the electrodes. Fiberglass and plastics offer economy in manufacturing costs but are limited in applications to relatively low process pressures (ANSI Class 150) and temperatures (about 250°F) and mild chemicals. Ceramic used in magmeter construction is normally high-purity aluminum oxide on the order of 99.5%. It is the most abrasion-resistant material used in the construction of magmeters. It can be used in high temperature processes (360 to 400°F) and with highly corrosive chemicals. It can withstand pressures up to ANSI Class 300 in sizes up to 3 inches. As the size increases, the pressure rating is drastically reduced. Careful attention

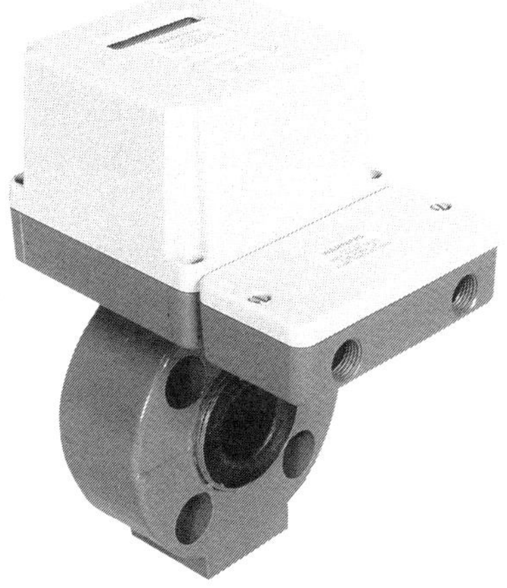

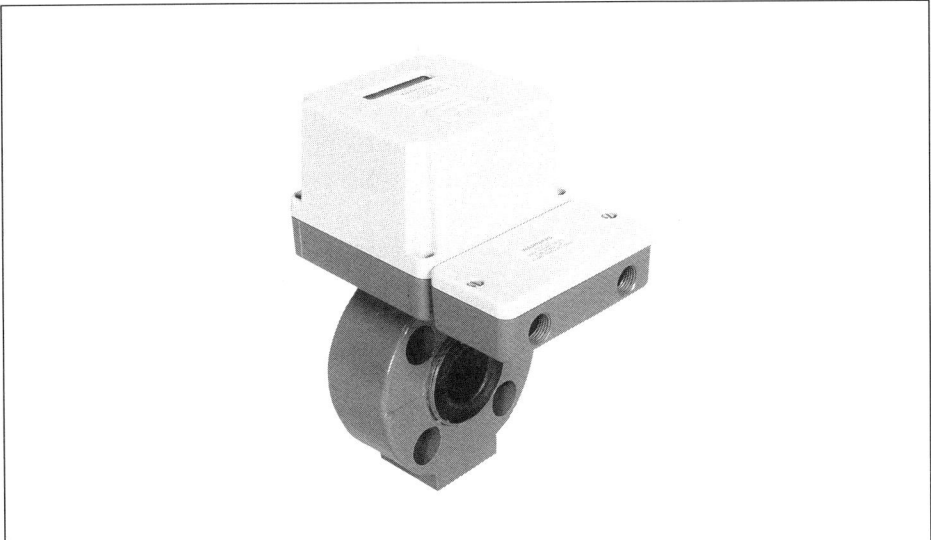

Figure 9-12. Tefzel-Lined Magmeter

should be paid to installation and maintenance instructions provided by the manufacturer. Fiberglass bodies are generally available in sizes 14 through 36 inches and plastic bodies in sizes ½ through 4 inches. Ceramic bodies are available from most manufacturers in the flangeless (wafer) design in sizes 1⁄10 through 4 inches but are also available in sizes 6 through 12 inches.

Electrodes

Electrodes are available in a wide assortment of materials. Included in that assortment are 316 stainless steel, Alloy 20, Hastelloy B, Hastelloy C, titanium, tantalum, Monel, nickel, zirconium and platinum. Tantalum and platinum are the most chemically compatible and the most expensive.

A rule of thumb for selecting electrodes is to pick the lowest cost material that does the job.

Selection

Many publications list the compatibility of the various liner and electrode materials with chemical processes. Two references used frequently by this author are *Corrosion Data Survey* by the National Association of Corrosion Engineers and *Handbook of Corrosion Resistant Piping* by Philip A. Schweitzer, P.E. Using these and similar references, some manufacturers have prepared liner and electrode selection guides that list their liner and electrode materials and their compatibility with various processes. These lists are generally available upon request. They are for reference only and do not constitute a guarantee regarding compatibility of material and process. Selections should be made on the basis of compatibility of specific materials with specific chemicals over the pH and concentration range expected in the process.

The liner and electrode must be compatible with the process fluid under all process conditions, pH, and composition.

Materials selected on the basis of pH only might not be compatible with the process, resulting in attack on the magmeter materials. For example, heavily chlorinated water can have a low pH (acidic) or high pH (alkaline) depending upon whether the source of chlorine is chlorine gas dissolved in water or sodium hypochlorite in water.

Not all material combinations are compatible with the full range of process conditions. At low pH, chlorine gas can permeate a Teflon liner and damage the meter housing; at high pH, several electrode materials are subject to attack.

The Magnetic Field

As was pointed out earlier, the strength of the magnetic field is fixed by size. A quick look at Equation (9-1) shows that as the meter size increases, the magnetic field strength must decrease proportionately in order to generate the same voltage for a given change in flow velocity regardless of the size of the magmeter. This simplifies the job of the transmitter to recognize the flow change and to output the same value for that change regardless of the size magmeter to which it is connected. At the present time, two basic methods being used to excite the magnetic field; they are commonly referred to as AC and pulsed DC excitation. These are graphically illustrated in Figure 9-13. AC excitation was used in the earliest designs and is still in use on a limited basis. Pulsed DC excitation was introduced in 1974 and is currently the most used method for coil excitation in magmeters. Some manufacturers do not even offer an AC-type magmeter system.

AC Excitation

In AC excitation, line voltage (typically 120 or 240 volts at 50 or 60 hertz) is applied directly to the magnet coils. This generates a magnetic field in the meter body that varies with the frequency of the applied voltage. The variation follows the pattern of a sine wave as shown in the left drawing of Figure 9-13, which means that the flow signal at a constant flow velocity will also look like a sine wave. The amplitude of the sine wave will be proportional to the flow velocity. This system produces an accurate, reliable, fast responding magmeter, but it has its drawbacks in some applications.

Faraday's Law states that a conductor passing through a magnetic field will generate a voltage. This law also states that, if a stationary conductor is located in a changing magnetic field, a voltage will be developed, which is what happens in a magnetic flowmeter. The flow signal is picked up by electrodes and passed through wires connected between the electrodes and the signal converter (transmitter). These wires are located within the magnetic field that was generated to develop a flow signal. The varying or changing magnetic field induces a voltage in the electrode loop. Consequently, the transmitter sees two signals. One of these is the flow signal, and the other can be called noise. For practical purposes, anything that is not a flow signal is noise, and a way must be found to separate the two.

AC voltages can be in-phase or out-of-phase with each other. If the signals are out of phase, the two can be separated with circuitry in the electronics. Phase detection circuits can be used to identify the flow signal and pass it along while rejecting the noise signal. This method of eliminating noise has been referred to

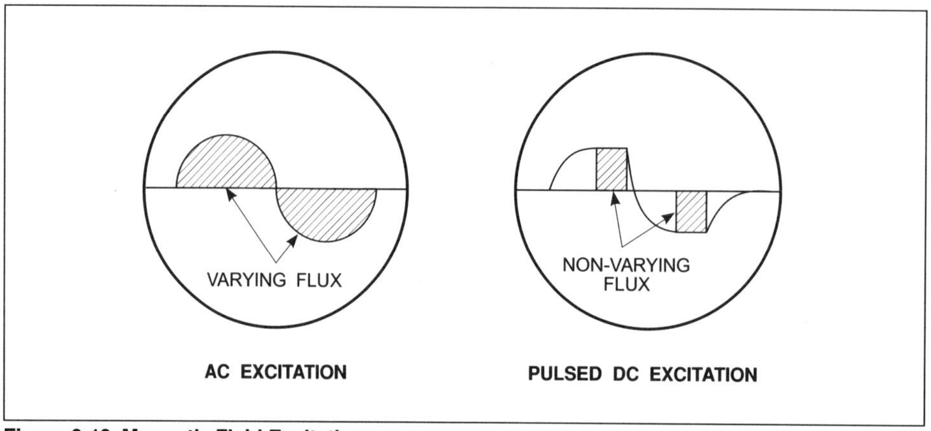

Figure 9-13. Magnetic Field Excitation

as quadrature rejection, and it does the job well for those signals that are out of phase with each other. However, those voltages that are in phase with each other or are of the exact opposite phase must be dealt with. This noise can be adjusted out if a condition can be created that will show that the noise is present and how much there is.

A full pipe, no flow condition could provide the information necessary to eliminate the in-phase noise. This requires setting the magmeter up so that it is full of the process liquid; there is no liquid movement through the meter, and power is applied to the system. Under these conditions, the output of the transmitter should be zero (4 mA DC in this example). If the output is greater than or less than 4 mA DC, noise is present. The transmitter can be adjusted so that the output is 4 mA DC or whatever the normal zero output signal is. This procedure constitutes the zero adjust, and it is required during initial start-up of any AC-type magnetic flowmeter. The zero adjustment can be made by several methods. In some systems, a current meter is required to be connected to the output of the transmitter so that the zero can be adjusted until the meter reads the proper value. A simpler system used by some manufacturers is an LED blink light located on the transmitter near the zero adjustment. If the light is on during zero flow conditions, the noise is in-phase and the output is too high. If the light is out during these conditions, the noise is of opposite phase and the signal is too low. If the light blinks, there is minimum noise signal present. A screwdriver is used to adjust the zero until the light blinks slowly. In many applications, no further adjustments are necessary.

Sometimes external inductive signals are injected into the electrode loop. This can come from inductive motors located near the magmeter. As the motor is turned on and off, the influences on the electrode loop change and so does the zero. Relocating the magmeter or shielding it could be a solution.

A common and serious problem is electrode coating. Process particles deposited on the electrodes will cause the zero to shift—the major cause of zero shift in AC-type magmeters. This will be covered in detail later in this chapter.

Pulsed DC Excitation

In the pulsed DC magmeter, line voltage is still the basic source of power, but, instead of being applied directly to the coils, it is first applied to a magnet driver circuit. This circuit sends low-frequency pulses to the coils to generate the magnetic field. The frequency of these pulse in the first pulsed DC systems was 3$\frac{3}{4}$ Hz for 60-Hz line voltages. Today, frequencies of 7.5, 15, and 30 Hz are also used.

The main reason for developing the pulsed DC system was to eliminate the zero shift problem; that goal was achieved. To understand how this was done, start by looking at the drawing on the right in Figure 9-13. This drawing shows that the magnetic field goes through a period of time when the strength is varying, and then it reaches a period of time when the field strength remains constant. During the time that the field is changing (varying), noise is induced into the electrode loop. During the time that the field strength remains constant, noise is not injected into the loop. Remember that it takes a changing magnetic field to induce a voltage in a stationary conductor such as the electrode lead. While the magnetic field is not changing, only the flow signal is present. This is because the major source of the zero shift, a changing magnetic field, has been isolated from the flow signal. Zero shift is, therefore, essentially a problem associated with the AC magmeter, in which the magnetic field is constantly changing during measurement of the flow signal. Consequently, most pulsed DC systems do not even have a manual zero adjustment on the transmitter.

The key to eliminating zero shift is to take the flow measurement during the time that the magnetic field is not changing.

The virtues of the methods used to create a pulsed DC magnetic field may be extolled in literature of the various manufacturers. They imply that one system may establish a more stable zero than another. The fact of the matter is that they all work. If there is more stability in one system than another, it is in the system that operates the coils at a lower frequency than the other. It has been established, for example, that a pulsed DC system operating at a coil frequency of 7.5 Hz is more stable than a system operated at 30 Hz. The advantages of higher coil operating frequencies are discussed later.

An additional feature of the pulsed DC system is low power consumption. AC type magmeters consume considerably more power than do the pulsed DC type. In the AC type, power consumption is more or less a function of size, and sizes larger than 4 inches could require from 250 to 1000 volt amps. Today's pulsed DC systems require approximately 15 to 20 volt amps regardless of size. Because of the reduced power consumption, the pulsed DC system can be powered by 24-volts DC. This is not a two-wire system but simply one that can be powered by a 24 volt DC power supply.

There is no real indication that either of these systems is more accurate than the other. If zero shift is a problem, then the long term accuracy would be better in the pulsed DC system simply because the zero would not move around. Most of the new transmitter development that results from changing technology is going into the pulsed DC units. This may provide some small overall improvement in the system accuracy. The standard accuracy appears to be moving from 1% of rate to ½% of rate on frequency outputs of these units. Optional ¼% of rate accuracy is possible in some applications over a limited rangeability.

Although the pulsed DC method of magnetic field generation has become the predominant method, it has not replaced the AC method entirely. Each system has advantages over the other, but the pulsed DC system is being improved and is gradually acquiring the advantages of both systems. The following section will discuss some of the major problems in each system and the solutions to these problems.

Electrode Coating

One of the most common process problems encountered by the magmeter is electrode coating. Coatings are particles that separate from the process and attach themselves to the pipe wall, grounding devices, the magmeter liner, and the magmeter electrodes. The coating material can have a higher conductivity, a lower conductivity, or the same conductivity as the process as a whole. Very heavy coatings that significantly change the diameter of the magmeter and the pipe cause errors due to increased velocity and reduced area at the electrodes. This type coating is not the subject of this discussion because it is more of a piping problem than it is a magmeter problem. The concern here is with the thin coatings that result in span and zero shifts. Thin coatings that are of the same conductivity as the overall process conductivity have no noticeable affect on the output signal. The two remaining types of coating will be the basis for this discussion.

Conductive Coatings

A coating that is more conductive than the process is more serious and more difficult to deal with than a coating that is less conductive than the process. Higher conductivity coatings may be present in industrial waste, where metals could plate out, or in slurries, where fine metal particles might be precipitated. If coatings were limited to the electrode, the conductive type of coating would not be a real problem. It would act like a small extension of the electrode, and the effects would be minimal. However, coatings may cover the liner of the magmeter as well as the electrodes. With insulating (lower conductivity) coatings, this does not present a problem because the liner is also an insulator. However, a conductive coating on the liner constitutes a low resistance path or short circuit between the electrodes. The result is a reduced flow signal. The signal could actually drop to a zero output over a period time if the coating is highly conductive.

Not much can be done to remove conductive coatings from the magmeter liner. The ideal solution, from a flowmeter standpoint, is to remove the problem particles from the process before they reach the point of flow measurement. If this is not possible or not practical, periodic cleaning of the line and the magmeter may be an acceptable solution. A dense, slick liner material such as Teflon or ceramic, for sizes in which these materials are available, may reduce coating. If these suggestions cannot be implemented, a magnetic flowmeter may not be the proper choice of measuring device for this process. Fortunately, conductive coatings are encountered in only a small percentage of applications.

Coating takes place more often at low flow velocities than at high velocities, so sizing the magmeter for flows of 5 ft/sec or better will reduce the probability of coating.

Insulating Coatings

The most common type of coating is one that is less conductive than the process as a whole. These coatings contribute significantly to shifts in the magmeter zero and span.

ZERO SHIFT

As was pointed out earlier, zero shift is a problem inherent to the AC-type magmeter. The zero shift problems of AC-type magmeters have been most prevalent in wastewater treatment facilities. Processes such as raw sewage, settled sewage, primary sludge, mixed liquor, return and waste-activated sludge, thickened sludge, digester sludge, and digester supernatant can leave an insulating type of deposit on the electrodes, which can change the characteristics of the electrode circuit. Coupled with the varying flux of the AC-generated magnetic field, the coating results in a zero shift. Shifts as large as ±20% of the full scale flow value may occur.

SPAN SHIFTS

Insulating coatings add resistance to the electrode circuit, which causes some of the flow signal voltage to be dropped across the electrodes. Ideally, all of the flow signal should be dropped across the input to the transmitter. If this were possible, a process flowing at 100% of its span setting would produce a 100% output signal at the transmitter. In actual service, this does not happen. Some of the process-generated signal is lost before it reaches the signal converter. Some of it is dropped across the impedance created by the coating, and the remainder is dropped across the input impedance of the transmitter. It is the ratio of the coating impedance to converter impedance that determines how much of the flow signal is lost.

Initially, the electrode signal will increase slightly as the coating builds up until such time that the added impedance between the liquid and the electrode becomes about 2% of the transmitter input. The signal will then decrease as the coating impedance increases, until it is entirely lost. During the increase in impedance, the electrode becomes more susceptible to electrostatic pickup from nearby line frequency devices. This could result in a noisy output signal as well as a reduction in span.

State-of-the-art electronic technology is used to design high input impedances into modern transmitters. These impedances are typically on the order of 10^9 to 10^{12} ohms. The coating impedance varies widely but is considerably less than that of the transmitter in virtually every application. This is particularly true for wastewater processes. The high-input impedance transmitters available today provide adequate protection against span shifts caused by typical coatings found in wastewater and most other processes. Consequently, span shifts are generally not a problem.

A realistic worst condition used by at least two magmeter manufacturers is 10^7 ohms (10,000,000 ohms) impedance for most coatings. At this impedance, serious span shifts could result for transmitters with low input impedances of 10^6 to 10^8 ohms. Using these values, errors of 90% for a transmitter impedance of 10^6 and 9% for a transmitter impedance of 10^8 ohms can be calculated. A transmitter with 10^{11} ohms input impedance yields a span error of only 0.01%. Most transmitters manufactured today have input impedances between 10^7 and 10^{11} ohms. For purposes of illustration, transmitters with input impedances 10^6 to 10^{12} ohms will be used here.

The percent span reduction for coating processes can be determined by applying the ratio of the coating impedance and the transmitter input impedance to the flow signal voltage. In an electrical series circuit, the voltage drops across the impedances in the circuit must equal the applied voltage. An equivalent magmeter circuit is shown in Figure 9-14 where the applied voltage is the flow signal picked up at the electrodes, and the impedances are the electrode coating and the input to the transmitter.

$$V_{total} = V(R_t) + V(R_c) \qquad \textbf{(9-3)}$$

$$\%\ \text{span reduction} = \frac{R_c}{(R_c + R_t)} \times 100$$

V = Flow voltage generated by the process
R_c = impedance of the coating
R_t = input impedance of the transmitter

If a typical value of 10^9 ohms input impedance for the transmitter and the extreme value of 10^7 ohms for the electrode coating, the calculated reduction is about 1%.

$$\%\ \text{Reduction} = \frac{10^7}{(10^9 + 10^7)} \times 100 = 1\%$$

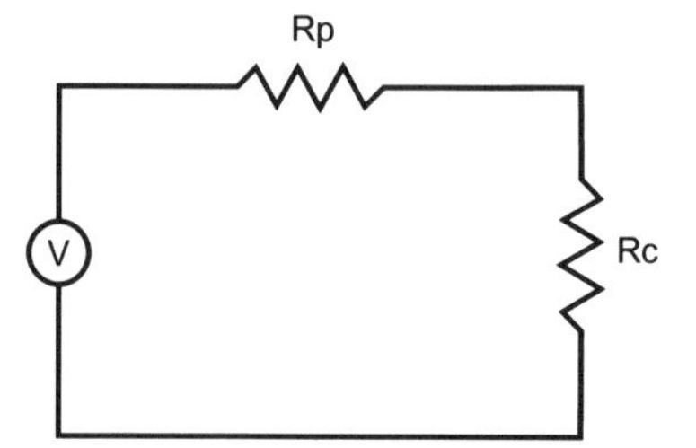

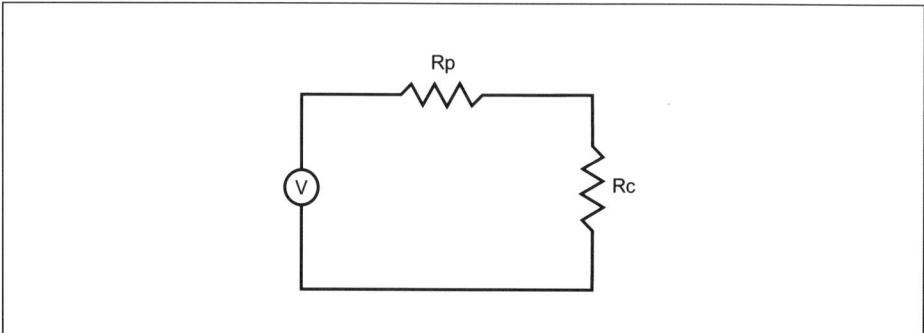

Figure 9-14. Equivalent Impedance Circuit

Table 9-3 was prepared by running through this calculation for various converter input impedances and using two different high coating impedance values (10^7 and 10^6 ohms). Values have been rounded off to the nearest one hundredth.

Table 9-3. % Span Reduction at Various Transmitter and Coating Impedances

Transmitter Impedance	Coating Impedance	
	10^7 ohms	10^6 ohms
10^6	91%	50%
10^7	50%	9%
10^8	9%	1%
10_9	1%	0.1%
10^{10}	0.1%	0.01%
10^{11}	0.01%	0.001%
10^{12}	0.001%	0.0001%

Table 9-3 shows that for worst case conditions, significant span reductions are possible for transmitters with input impedances of 10^8 ohms or less. The table also shows considerable improvement in span stability where the coatings are 10^6 ohms or less. Therefore, it is important to ensure that transmitters with at least 10^9 ohms input impedance are selected for services where electrode coating is possible. Such transmitters are readily available. In fact, it should not be difficult to find transmitters with an input impedance of 10^{11} ohms. They are available, and they provide added insurance of successful flow measurement even for those coatings that exceed 10^7 ohms.

A few coatings are such good insulators (greater than 10^7 ohms) that they produce span shifts even when a very high-input impedance transmitter is used. It would be a Herculean task to identify every process that contains coatings of this type. However, three known coatings could produce serious problems even for a transmitter with an input impedance of 10^{12} ohms. They are iron oxide, calcium and magnesium carbonates, and certain types of resin. Iron oxide is a precipitate that has been known to coat the electrodes of magmeters used in some process lines of an aluminum manufacturing plant. The paper industry uses various resins for coating paper. Calcium/magnesium deposits can be precipitated from hard water. Except for the calcium/magnesium deposits, these coatings are relatively limited to select industries. Hard water can be present in applications where water is part of the process makeup, and some form of water softening is not utilized. There is a need, in these applications, to prevent electrode coating or to provide a means to clean the electrodes periodically.

Solutions to Electrode Coating

The insulating type electrode coating and the resultant zero shift in the early AC-type magmeters was recognized as a problem soon after magmeters were introduced, and methods to prevent or reduce electrode coating began to appear.

BRUSH CLEANOUT

In 1958, brush cleaning of the electrode was introduced. In this special electrode design, a small brush was inserted into a hollowed out electrode, and cleaning of the electrode was done manually. The flow signal was interrupted during the cleaning process.

BURN-OFF

Shortly after brush cleanout was introduced, the burn-off method was developed. In this method, line voltage (120 V AC) is applied across the electrodes. Current flows through the electrode circuit; as it does, bubbles form on the electrode and the coating is "boiled" off. This approach can be satisfactory at times, and it is still used on occasion. It has some drawbacks, however, including the loss of flow signal during the burn-off period. Some stubborn coatings can take as much as ten to fifteen minutes to burn off, which is all lost signal time. A second drawback is the possible discoloration or pitting of the electrode.

THE SLUDGEMETER

In 1963, the heated tube sludge-type magmeter was introduced. The original offering featured a heated tube design that utilized an aluminum tube to generate heat through eddy currents induced in the tube by the changing flux of the AC magnetic field. Two manufacturers followed this design with one that utilized heater blankets wrapped around the metering tube and one that used parallel wiring of the magnet coils to generate heat through high power consumption. The theory behind this design was that by heating the tube, the process liquid would also be heated. The intent was to keep grease in the sludge from forming a hard coating on the electrodes and liner. This method was marginally effective. In fact, it compounded the problem in some instances where the process pipe drained empty. This caused light coatings left on the liner and electrodes to be baked and form hard coatings that remained when the process was turned on. The aluminum tube design is no longer available. It may still be possible to get one of the two remaining designs, but far better systems than these are now available.

SELF-CLEANING ELECTRODES

In 1967, the bullet-nose or self-cleaning electrode was introduced. Prior to this design, electrodes were designed so that the tip of the electrode was flat and flush with the inner surface of the liner. The self-cleaning electrode is rounded and extends beyond the inner wall of the liner into the flow stream about ¼ inch. This protrusion into the flow stream limits the design to sizes one inch and larger. As the velocity of the process approaches 5 ft/sec, the protruding electrode creates turbulence. This turbulence scrubs the electrode and keeps at least part of it clean. It is not necessary to keep the entire electrode tip clean. This has been an extremely effective method for electrode cleaning for most coating processes, and it does not affect the flow signal. Depending upon the severity of the coating, the process must reach a velocity of 5 ft/sec occasionally or all the time. A particularly troublesome coating process such as latex could require flow velocities of 12 to 15 ft/sec all the time. However, this is an exception. Normal flow velocities of 5 to 8 ft/sec are adequate for most applications.

REMOVABLE ELECTRODES

Removable electrodes are available from some manufacturers in some sizes and in some liner materials. Whether or not it is practical to remove electrodes for cleaning is highly debatable. Electrode wiring could be damaged, and the seal between electrode and liner could be broken. As a rule, magmeters with polyurethane and rubber liners have electrodes installed by inserting them through the meter body from the outside the body. This is schematically illustrated in Figure 9-7. Electrodes in Teflon-lined magmeters are installed from inside the body (see Figure 9-7). Electrodes installed from outside the body can sometimes be removed without removing the magmeter from the process line, but the process line must be drained of liquid to a point below the level of the electrode. For electrodes installed from inside the pipe, the magmeter must be removed from the process line. A few manufacturers offer a special electrode design that allows the electrode to be removed without stopping the flow of

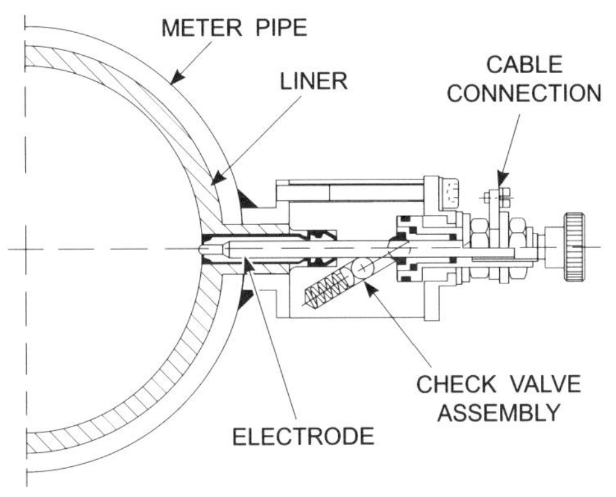

Figure 9-15. Removable Electrode

process in the line. One such design is shown in Figure 9-15. Regardless of how it is done, the flow signal is not available during electrode removal and replacement.

ULTRASONIC ELECTRODE CLEANING

In 1967, another electrode cleaning method called ultrasonic electrode cleaning was introduced for magmeters in sizes one inch and larger. In this method, a crystal is attached to the back of each electrode, and an ultrasonic generator applies a high-voltage, high-frequency signal to the crystal. The crystal then vibrates, causing the electrode to vibrate with it. If the coating on the electrode is relatively hard, the vibrations will cause the coating to break up and be washed downstream in the process. It also causes cavitation at the face of the electrode to aid in cleaning. The ultrasonic generator supplies the excitation voltage to drive the crystals. It can be wall mounted and dedicated to one magmeter to supply continuous cleaning, or it can be a portable unit to supply periodic cleaning to a number of units. An advantage of ultrasonic electrode cleaning is that it does not interrupt the flow signal. Although this method works well against hard coating materials, it is not effective against soft, sticky processes such as latex. Ultrasonic electrode cleaning is still available from some manufacturers, but for most applications, it is not needed.

THE MODERN APPROACH

Although there were generally effective methods to handle coating processes, research continued for an even more effective way to prevent zero shift. That way was developed and introduced in 1974 as the pulsed DC system. This design takes a different approach to solving the electrode coating problem. In essence, it ignores the coating insofar as the zero shift is concerned. How the zero shift is eliminated in the pulsed DC system was covered earlier. It makes no difference whether the electrode is coated or not coated. The result is the same—no zero shift! Fifteen years of field experience confirms this. A good example is ferric chloride. This is a highly corrosive process that leaves an insulating type of coating on the electrodes and the liner. Serious zero shifts occur in AC-type magmeters used to measure this process flow, but they do not occur in the pulsed DC type.

Span shifts resulting from electrode coating were also discussed earlier. That discussion pointed out that if the magmeter transmitter has a high-input impedance, span shifts will be rare if they occur at all. The selection of such a transmitter, whether it be of the AC type or the pulsed DC type, ensures that any possibility of span shift will be minimal.

It is incumbent upon the user and consulting firms to take full advantage of modern day technology in selecting an appropriate magmeter for applications involving coating processes. With few exceptions, the combination of a pulsed DC system and a high-input impedance transmitter solves the problem of coating applications and does so at minimum cost. Optional cleaning devices such as ultrasonic cleaning and removable electrodes are expensive and normally are not required. Some manufacturers do not offer either of these options but rely on pulsed DC and high-input impedance transmitters to obtain the desired results.

EXCEPTIONS

There are, however, those few insulating-type coatings with very high impedances that require special attention. In the case of iron oxide and calcium coatings, ultrasonic cleaning should be an effective method to prevent the coating from building up to the point that it will cause a span shift. Both these materials are hard and brittle. They respond extremely well to the vibrations of ultrasonic electrode cleaning. And remember that this is done without interrupting the flow signal. In addition, the ultrasonic electrode cleaning option is available with the self-cleaning (bullet nose) type of electrode. If the flow velocity can be held to a minimum of 5 ft/sec, ultrasonic cleaning might not be needed.

When in doubt, consider selecting a magmeter with the ultrasonic cleaning crystals installed in the magmeter, but hold off on purchasing the ultrasonic generator. If it is determined during service that ultrasonic electrode cleaning is necessary, the generator can be purchased and installed at that time. Waiting would also make it easier to determine whether a dedicated or a portable generator should be used.

The final coating to be considered here is resin. This may or may not be responsive to ultrasonic cleaning. Resins were high on the list of coatings that prompted the development of ultrasonic electrode cleaning. They do not present as great a problem today as they did in 1967 because pulsed DC eliminates the zero shift. With good flow velocities, the self-cleaning electrode reduces the span problem and remains an excellent electrode cleaning method. For resins that form a hard coating on the electrode, ultrasonic cleaning is still a viable option for meters in sizes larger than ½ inch.

THE ELECTRODELESS MAGMETER

In all the discussion about insulating-type coatings, the culprit has been the electrode. If there were no electrode to coat, there would be no problem. If a magmeter could be designed without an electrode, there would be no concern for the thin insulating-type coatings. This has not happened. There is no such thing as an electrodeless magmeter. However, there is a design that is the next best thing. The electrodes have been redesigned and have been removed from contact with the process. In this design, the electrodes are large plates that are mounted between the liner and the stainless steel meter body or on the outside of a non-conductive meter body. The flow signal is capacitively coupled to the transmitter through a high-impedance amplifier. The concept of a capacitively coupled signal has been around for quite a long time. Evaluation units were installed in the field as early as 1974, and the product has been for sale since shortly thereafter.

The early designs had the electrode plates located between the liner and the meter body, as shown schematically in Figure 9-16. The product was offered in sizes 2 through 20 inches and can still be purchased. The electrode is located inside the magmeter pressure vessel. Changing process pressures cause small movements of the flexible liner, which are translated to the electrode package and result in distortion of the flow signal. Attempts to correct this problem have met with only partial success. Consequently, this product has not been widely accepted as a solution to the electrode coating problem.

In 1987, a promising new capacitively coupled magmeter became available. The meter body in this new offering is ceramic (99.5% aluminum oxide), and the electrode package is attached to the outside of the meter body. The electrode is not affected by changes in the process pressure because it is located outside the pressure vessel of the magmeter. Initial reports from the field indicate that this design does exceptionally well in insulating-type coating processes, such as a process called rubber fines. This is not a very corrosive process, but it does leave a highly insulating type of coating on the electrodes. Large span shifts require removing the magmeter from the process line and cleaning the electrodes manually about every six to eight weeks. An electrodeless magmeter evaluated in this service was performing well without span shift for over a year at last report. Unfortunately, it is only available in sizes one inch through four inches as of this writing. However, most of all magmeter applications fall within these sizes.

Magnetic flowmeters have come a long way in their ability to resist the effects of electrode coating. The key to success is knowing enough about the process to recognize that coating can occur, knowing what types of magmeters are available for the service, and selecting the right one for the job. When in doubt, contact a magmeter manufacturer for application assistance. Any magmeter manufacturer should have competent applications personnel who can make recommendations on which magmeter system is best for a given process.

Speed of Response and Recovery

Speed of response is the time it takes for the transmitter output to show that a change in flow has occurred. It is normally expressed as the time it takes show a 100% change in flow. For the AC system, this is about one second. Until 1983, the response time for the pulsed DC system was about 4 seconds. In 1983, it became about one second for some pulsed DC systems. Today, pulsed DC response time can be less than one second, which approaches the speed of a turbine meter. This is important for batching operations where the batch time is extremely short. In some applications, the batch time can be as short as several seconds.

Speed of recovery is the time it takes for the magmeter to recover from an empty pipe condition and produce a normal flow signal. Unpredictable outputs that occur during empty pipe conditions were discussed earlier. When the magmeter is filled with liquid, the unpredictable output continues until the transmitter has time to recover. This recovery time is one second for the AC system but about 20 seconds for some pulsed DC systems. Design changes in the pulsed DC system have reduced its recovery time to that of the AC system.

Process-Generated Noise

A phenomenon associated with the pulsed DC system appears in some slurry processes and is the result of particles in the process impinging upon the

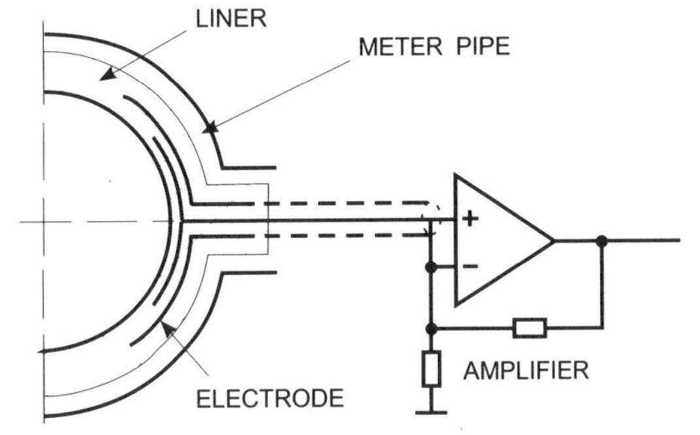

Figure 9-16. Capacitive Electrode Flexible Liner

electrodes. It is manifested as noise in the output signal. In severe cases, it can cause the output to fluctuate as much as 60% above and below the true flow signal. In hard slurries such as coal and phosphate, the noise developed is proportional to the size of the solids in the slurry, the hardness of the solids, the amount of solids, and the flow velocity. Paper stocks are more complicated than the hard slurry, and there is less understood about them than about the less complex hard slurries. The chemical additives found in various paper stocks make it difficult to predict accurately whether or not noise will be a problem in the application. Generally speaking, paper stocks with less than 4% consistency should not produce sufficient noise to adversely affect the output signal. In addition, chemicals added to any process result in nonhomogeneous conductivities. This means that pockets of different conductivity are present in the process. If the chemical is added upstream of the magmeter and the pockets do not combine to form a relatively constant conductivity, a noisy output can be expected.

Over the years, three solutions have been developed to solve the noise problem. Each has been partially to fully successful depending upon the severity of the problem. An early solution was a hard electrode tip made from a material such as tungsten carbide. This reduced the level of process-generated noise significantly. It was a satisfactory solution to many of the slurry processes. Another solution increased the magnet coil operating frequency: It was discovered that doubling the coil frequency reduced the noise level by about half. This is the main reason that coil frequencies have gone from 3 ¾ Hz to 7.5,15, and 30 Hz. This, coupled with the hardened electrode tip, was the solution for some of the processes that were noisy with the hardened electrode alone. Finally, a noise reduction function was developed for the microprocessor-based transmitter. This, coupled with the first two solutions, provides what appears to be the complete solution for process-generated noise. The noise reduction feature is covered in more detail later.

Conductivity

The first parameter to consider in deciding whether or not to use a magnetic flowmeter in a given application is conductivity. Does the process liquid have sufficient conductivity to meet the minimum requirement as a conductor? What is the minimum requirement? This requirement varies from manufacturer to manufacturer and from magmeter type to magmeter type. The conductivity unit of measure is microsiemens/cm or (μS/cm) micromhos/cm (μmhos/cm), which are the same in terms of actual value. Until recently the term micromho was used exclusively. Today, the word siemen is used particularly in the SI system, in place of mho, which is ohm spelled backwards. The ohm and the mho are reciprocals, hence the use of these terms to indicate resistance or conductivity. Either mhos or siemens are used to denote conductivity.

The standard minimum conductivity level has been 5 μS/cm, but some standard magmeters are available with minimum conductivity limits of 1 or 2 μS/cm. There are also some magmeters that require a minimum of 20 μS/cm. Consequently, it is important to know both the minimum conductivity of the process and the minimum conductivity level of the magmeter under consideration for the application. But where is all this information available? The conductivity limit of the magmeter is contained in the specifications from the manufacturer. The actual conductivity of many aqueous solutions and pure liquids can be found in Lange's *Handbook of Chemistry* [Ref. 3]. In addition, the magmeter manufacturer usually has a supplemental list of liquid conductivities that has been compiled by experience or actual test (see Tables 9-4, 9-5, and 9-6). If the information is not available from these sources, a conductivity test can be conducted on site or at a

Table 9-4. Electrical Conductivity of Aqueous Solutions

Chemical Name	Formula	% by Weight	Temp., °C	Conductivity in Microsiemens/cm
Acetic Acid	CH_3CO_2H	40 99.7	18	1.08×10^{3} 4.00×10^{-2}*
Ammonia	NH_3	1.60 30.50	15	8.67×10^{2} 1.93×10^{2}
Calcium Chloride	$CaCl_2$	5 35	18	6.43×10^{4} 1.37×10^{5}
Hydrochloric Acid	HCl	5 30	15	3.95×10^{5} 6.62×10^{5}
Hydrofluoric Acid	Hf	1.50 29.80	18	1.98×10^{4} 3.41×10^{5}
Nitric Acid	HNO_3	6.2 62.0	18	3.12×10^{5} 4.90×10^{5}
Phosphoric Acid	H_3PO_4	10 80	15	5.66×10^{4} 9.79×10^{4}
Sodium Carbonate	Na_2CO_3	5 15	18	4.51×10^{4} 8.36×10^{4}
Sodium Hydroxide	NaOH	1 50	18	4.65×10^{4} 8.20×10^{4}
Sulfuric Acid	H_2SO_4	5 94	18	2.09×10^{5} 1.07×10^{5}

*Conductivity too low for magnetic flowmeter

laboratory off site. The test must be conducted at the normal process temperature because temperature has an effect on the conductivity. When the process temperature dictates that the test sample be heated or if the liquid is hazardous, testing should be conducted on site. Shipping and disposing of any sample must been done according to OSHA standards.

A magmeter used on a relatively low-conductivity (1μS/cm) process one day can be used on a high-conductivity (1000 μS/cm) the next day with no change in calibration. How is that possible? Conductivity is the reciprocal of resistance; therefore, as the conductivity of the process drops, the resistance or impedance of the process increases. This creates a condition similar to that of electrode coating,

Table 9-5. Electrical Conductivity of Pure Liquids

Chemical Name	Temp., °C	Conductivity in Microsiemens/cm
Carbon Tetrachloride	18	4.0×10^{-2}*
Ethyl Alcohol	25	0.0013*
Furfural	25	1.5**
Glycol	25	0.3**
Methyl Alcohol	18	0.44**

*Conductivity too low for magnetic flowmeter
**Low conductivity application

Table 9-6. Conductivities of Miscellaneous Liquids

Name	Temp., °C	Conductivity in Microsiemens/cm
Black Liquor	93	5000
Fuel Oil	—	< 10^{-7} *
Water, New York City	25	72

*Conductivity too low for magnetic flowmeter

and the impedance of the process can be substituted for the impedance of the electrode coating in the example of insulating coating (above). It is necessary only to convert conductivity to impedance. First, convert microsiemens to siemens. A siemen (mho) is one million microsiemens (micromhos); therefore, a microsiemen is one millionth of a siemen (0.000001 siemen) and 1000 microsiemens is 0.0010000 (0.001) siemen. Divide the number by each of these values and substitute that number for R_C in the insulating coating example to get the percent span shift for each of these conductivities. Using 10^{10} as the transmitter input impedance, the following can be calculated:

$$\text{For 1 microsiemen/cm, \% error} = \frac{10^6}{(10^6 + 10^{10})} \times 100 = 0.01\%$$

$$\text{For 1000 microsiemens/cm, \% error} = \frac{10^3}{(10^3 + 10^{10})} \times 100 = 0.00001\%$$

It is clear from these two calculations that at conductivities as low as 1 µS/cm the span shift is insignificant and at higher conductivities the shift is practically non-existent.

Low Conductivity

Special designs allow magmeters to measure process flow where conductivities are well below 5 µS/cm. Even the standard meters that measure process liquids in the 1 - 5 µS/cm range require special circuitry. This is usually done by what are called driven electrode shields. A voltage from the transmitter is applied to a wire shield that surrounds the electrode wire. The driven shields provide protection against capacitance losses and process-generated noise. As the conductivity level drops below 5 µS/cm, electrical noise becomes a problem. This is manifested as an oscillating output signal.

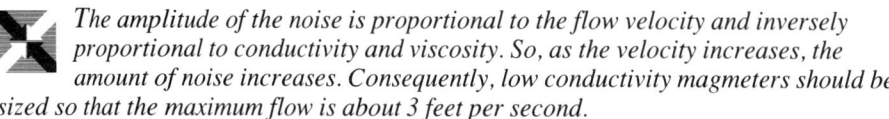

The amplitude of the noise is proportional to the flow velocity and inversely proportional to conductivity and viscosity. So, as the velocity increases, the amount of noise increases. Consequently, low conductivity magmeters should be sized so that the maximum flow is about 3 feet per second.

On the other hand, as either the conductivity or the viscosity drops, the amount of noise will increase. Normally the viscosity of the process is not a concern with magmeters. In fact, it still is not a factor as far as flow signal development is concerned. But it does affect the amount of noise that is present in the process, and this could make the flow measurement extremely difficult or even impossible.

Perhaps the most difficult of the low conductivity processes is deionized water. Water conductivities vary all over the place. Well water can be in the neighborhood of 40 µS/cm while deionized water can be as low as 0.04 µS/cm. Depending upon the amount of noise, it could be overcome by adding smoothing (damping) to the output signal. This slows down the response of the transmitter to flow changes, but it can reduce the noise in the output signal to a useable level.

Special Low-Conductivity Designs

Some low-conductivity designs go beyond the driven shield. They add preamplifiers and impedance matching devices between the magmeter and the transmitter in order to reach conductivities as low as 0.1 µS/cm. This type of magmeter is more expensive and requires longer delivery times than the standard magmeter. It is available in sizes ¼ through 4 inches and will operate in process temperatures as high as 360°F (182°C).

The electrodeless magmeter was discussed earlier as a solution to electrode coating. This is also a low-conductivity design and will operate with process conductivities as low as 0.05 µS/cm. It is also more expensive than the standard magmeter, but the ceramic body version is more readily available than the special design mentioned in the preceding paragraph. The noise problem discussed above in the section on low conductivity applies to the electrodeless design as well as to the standard design.

A good example here is an expansion where eight standard magmeters were installed in a food process. All meters started up well except one. It was determined that the conductivity in that line was lower than the minimum required for standard magmeters. An electrodeless magmeter was shipped from stores, and in several days the system was up and running.

The Transmitter

Up to this point, the discussion has been devoted almost exclusively to the primary sensing element of the magmeter system. It was mentioned that the low-level signal developed by this primary element had to be conditioned by a transmitter and presented in useable form such as 4-20 mA DC or perhaps an unscaled frequency of 0-1000 Hz, 0-2000 Hz, or some other value. The early transmitters did just that and very little more. The transmitter of today is a highly sophisticated device with a wide variety of features that simplify operator checks and adjustments and facilitate interface with computerized control systems. The degree of simplification or sophistication varies greatly, and one manufacturer may offer both a simplified version and one or more sophisticated types. There is, of course, a difference in price but probably not as much as one might think for the added features.

Basic Transmitter Outputs

The basic transmitter receives the signal from the magmeter and produces a typical 4-20 mA DC output signal that will be compatible with the input of one or more receiving devices such as an indicator, recorder, controller, or computer. These devices normally have input impedances of 250 ohms. The transmitter current output can operate into 0 to 1000 ohms at best and in some designs as low as 0 to 600 ohms. This means that the number of devices that can be driven by the transmitter is limited to the number of receivers whose total impedance does not exceed the maximum specified for the transmitter. This is normally not a problem. However, an application could require that one transmitter operate four receivers having 250 ohms input impedance each for a total of 1000 ohms. The transmitter with 0 to 1000 ohms output driving current will drive these four receivers. The transmitter with only 0 to 750 ohms driving current will not. An additional device such as a current-to-current converter might be required to make the system functional. If the output is a frequency rather than current, it is used as input to a digital device such as a computer or totalizer. The input requirements of the receiving device must be compared with the output specifications of the transmitter to ensure compatibility between the two. This is not as straightforward as the analog requirement.

Output Options

In addition to the standard analog output, some transmitters have an optional frequency output that is unscalable or one that can be scaled into engineering units. This output can be an active pulse such as 24 V DC to drive an electromechanical counter directly; it can also be a passive output such as an optically coupled transistor that does not supply a voltage to drive the receiving device. Counters and other receivers that work with passive input devices have an internal power source or use an external power supply to furnish the necessary drive voltage. Both types of frequency output can be scaled to engineering units. Some can be scaled so that the pulse that reaches the receiver represents the engineering value directly based upon the maximum range setting; for example, one pulse could represent one gallon, and the range setting could be 200 gpm, 300 gpm, or some other maximum value. Some scaled frequencies can provide only a direct engineering unit based upon the capacity of the magmeter; for example, in a size two inch magmeter with a capacity of 275 gpm, one pulse per gallon is possible for a maximum range of 275 gpm, and for a maximum range of 50 gpm, one pulse would represent about 0.18 gallon.

Knowing what type of scaling is being used could save a lot of time and confusion.

Frequency Scaling

There are several ways to scale frequency outputs so that each pulse represents a specific engineering unit. Basically, it involves a scaling device that divides the total incoming pulses in a given period of time by the total flow in the same period of time. The input to the scaler is expressed in terms of pulses per second at some maximum (such as 2,000 or 10,000 pulses per second) at the maximum range setting or the capacity of the magmeter. The flow units could be expressed in gallons per minute. It is necessary to change the pulses per second to pulses per minute or to change the flow to gallons per second.

Assume a scalable frequency of 10,000 pulses per second and a flow of 300 gpm. Change pulses per second to pulses per minute and divide by the maximum flow of 300 gpm.

10,000 pulses per second × 60 seconds per minute = 600,000 pulses per minute

600,000 pulses per min/300 gallons per min = 2,000 pulses per gallon

The scaler has switches with numerical values so that by selecting the proper switch settings the incoming pulses will be divided by 2,000 and each pulse leaving the scaler will represent one gallon. Scaling can be modified so that one pulse can represent 0.1 gallon or 10 gallons or some other value. Scalers do have limitations as to what scaling units are possible. Consult the manufacturer for this information.

Enclosures

Nearly all transmitters are enclosed in some type of field mount housing that carries an environmental rating of NEMA 4 and can be mounted on a wall or pipe near the magmeter or some distance away. The limit on this distance varies, and the manufacturer should be consulted in this regard. In general, the distance is limited by two factors:

(1) Where the magnetic field generating device (magnet driver) is located at the magmeter, the conductivity of the process liquid is the limiting factor. A rule of thumb is ten feet for every microsiemen/cm of conductivity. If the conductivity is 5 μS/cm the converter can be mounted fifty feet from the magmeter.

(2) Where the magnet driver is an integral part of the transmitter, the limiting factor is the coil voltage. In this configuration, which is becoming the predominant one, the coil energizing voltage is sent from the transmitter to the magmeter. If the distance is great enough, too much voltage will be lost in the cable between the transmitter and the magmeter. There will not be enough left to energize a proper magnetic field.

With either of these limitations, a distance of 200 feet between the magmeter and transmitter is not unreasonable.

Some transmitters can be mounted integral to the magmeter or mounted on top the magmeter. This location of the transmitter reduces installation costs and is quite popular where flow parameters do not change over long periods of time. However, all adjustments are located in the transmitter. If flow ranges or scaling requirements change frequently, it might be wise to locate the transmitter in a location where these adjustments can easily be reached. This is especially important if the magmeter is located in a pit, high above ground level, or some other area that is difficult to access. In addition, process conditions can dictate that the transmitter be mounted remote from the magmeter. Excessive pipe vibration and high process temperature could demand that the transmitter be located in a less hostile environment. If the magmeter is located in a pit that could become flooded, it is wise to remote mount the transmitter, especially if it is equipped with a readout and window that cannot keep water out for extended periods of time.

Readouts

A multitude of analog and digital readouts are available for mounting integral to the transmitter or remote from it. The physical size of the transmitter and the mounting requirement of the readout places limits on the size and style of the integral readout, while the variety of remote readouts is almost limitless. In recent years the liquid crystal display has been used increasingly as both an indicator and a totalizer. It has added flexibility and ruggedness to direct-reading indicators that is beyond the capability of scales required in the analog-type display. See Figure 9-17 for an example of a transmitter with an integral analog rate display. A liquid crystal totalizer and indicator integral to a transmitter is shown in Figure 9-18.

Liquid crystal indicators are available with small printed circuit boards that provide a means to calibrate an analog input to a direct-reading digital readout. This results in a nearly unlimited range of values for the full scale setting. Three or four digits with an adjustable decimal point makes this type readout very flexible and easy to use.

Figure 9-17. Transmitter with Analog Indicator

Figure 9-18. Transmitter with Liquid Crystal Indicator and Totalizer

Liquid crystal totalizers are available with battery backup, which ensures that totalization is not lost during power failure. The life of the battery can be as long as seven years. The battery also provides the necessary voltage to advance the count in the totalizer. This makes it possible for some of these devices to operate with scalers that do not have a live pulse output but a contact closure or optically coupled transistor output.

There are two drawbacks to the liquid crystal displays. One is the temperature limits beyond which the display will fade out or freeze up. These limits vary, but 32 to 120°F (0 to 50°C) is average. The display will not be damaged if the temperature range of −13 to +149°F (−25 to +65°C) is not exceeded. Again, this is an average temperature range. The operating range of the transmitter electronics, −40 to +149°F (−40 to 65°C) is greater than that of the display, and data will not be lost for operating temperatures within these limits. Displays that have faded or frozen up will show updated information when temperatures return to the operating limits of the display. The second drawback is readability. Most liquid crystal displays require some light to make them readable. However, some are available with backlighting. These displays are readable even in a darkened room.

The size limitation of the transmitter precludes the integral mounting of batching-type totalizers. Such totalizers are designed to provide a contact closure output after a settable number of counts. The contact closure is then used to perform some function in connection with the process, such as stopping a pump or closing a valve either fully or partially. If the application requires moving a specific quantity from one point to another (batching), the presettable totalizer provides a simple way to do the job inexpensively. These totalizers are available as electromechanical (requires live voltage to operate) as well as light-emitting diode or liquid crystal types that can operate with a contact closure input. They are larger than the standard totalizer and are normally available as panel- or desk-mounted devices.

The Microprocessor-Based Transmitter

The microprocessor-based transmitter first became available in 1983. Nearly every magmeter manufacturer now offers this type either as the standard or as an optional transmitter. Microelectronics make it possible for this type of transmitter to offer a lot of features, but it has the potential for some truly operational features that make it very operator friendly and very functional. One basic function of a true microprocessor-based transmitter is the ability to automatically measure flow bi-directionally (flow in either the forward or reverse direction).

Unfortunately, there is a very wide interpretation of what constitutes a microprocessor-based transmitter. It is wise to evaluate these offerings very carefully to determine what real features are offered. Do not assume that the word "microprocessor" in the transmitter specification means that it is equal to other microprocessor-based transmitters. It might even be wise to obtain a copy of the instruction bulletin or technical bulletin for the instrument. These publications cover operation and maintenance details that are not included in the instrument specifications.

This section will be devoted to defining the features that are available and desirable as minimum standard requirements for a microprocessor-based transmitter.

Memory

Data stored in memory of the transmitter is not lost when power is removed from the system. Some transmitters use a battery backup, while others use a type of EPROM that does not require battery backup to retain memory for as long as ten years.

Configuration

The microprocessor should not require any programming knowledge on the part of the operator. What the operator does is properly called configuring. This simply requires that certain known values be entered into memory through use of push buttons or a numerical key pad. Calculations should not be required of the operator for such parameters as range settings or totalizer scaling factors. This should all be done by the computer in the transmitter.

The microprocessor-based transmitter has what can be classified as two operating modes. One is the "on-line" mode where data is monitored, stored in memory, and displayed on a readout device. The other is the "configuration" mode where the various parameters (such as range, damping, scaling, and engineering units) can be changed. When the configuration mode is entered, the on-line functions should continue without interruption. There are microprocessor based transmitters that "freeze" the data at the last known value when the configuration mode is entered and some that drop the output signal to zero. There are those that continually update information and transmit an output that accurately reflects changes in flow when the transmitter is in the configuration mode. Specifications for the particular transmitter should clearly state what happens to the output signal and new data during configuration. If this information is not included in the specifications, the manufacturer should be consulted if it is important to maintain a flow signal for control, billing or any other reason.

Display

The display in a microprocessor-based transmitter is used to display instantaneous flow data, to display configuration data, and to display alarm conditions. Examples of all three types of data will be shown here, but not all microprocessor transmitters make all three functions available. Some have one-line displays, and some have two-line displays. Some have as few as four characters in the display, and some have as many as sixteen characters. Some display only numbers, while others display both numbers and letters (alphanumeric). Displays use primarily light-emitting diodes (LED) or liquid crystals (LCD).

The LED type of display is larger and brighter than the average LCD. It is, however, difficult to read in direct sunlight. Many are not available with alphanumeric characters. Consequently, all configuring must be done by using code numbers to represent the various parameters.

Most LCDs require some light in order to be easily read, but some are backlighted and are readable even in a dark room. LCDs are available with alphanumeric characters, which can make the microprocessor-based transmitter extremely user friendly. For example, selection of parameters from a menu is possible, while other operator functions appear in the display in clear language.

A display with push buttons is shown in Figure 9-19. The top line of the display shows instantaneous flow rate in percent of the full scale range setting. This can be changed to direct-reading flow rate by entering the configuration mode and selecting it from a menu. In this unit, the configuration mode is entered by pressing the third push button from the left marked C/CE. The other two push buttons are used to step through the menus and to enter or change parameters. Escape from the configuration mode is made by pressing CE again. The unit will return to the monitoring mode automatically if no changes to the data base are made for a period of about 20 seconds. The bottom line of the display shows the totalized flow. Some units always show both the instantaneous flow and the total flow. Some do not. The "F" in the display indicates data for the forward flow.

Most microprocessor-based transmitters are inherently bidirectional. They display and store data in both the forward and reverse directions automatically.

Figure 9-20 shows a display in the configuration mode. The top line shows the parameter to be changed. In this case, it is the range in the forward direction. A separate range can be set for each direction, and they need not be the same value. The second line of the display shows the value to be changed. In this case, the selected engineering unit is gpm. The numerical value of the range is set by entering the appropriate number using the push buttons on the left to call up the numbers one at a time. Any value within the range limits of the magmeter can be set in. Some units will even display an alarm message if an attempt is made to set the range outside these limits.

Figure 9-21 shows one of the alarm messages that may appear in some units. The error message appears on the top line of the display. This one states that flow has exceeded 130% of the range setting. This could happen if a valve were improperly positioned or if the range were inadvertently set below the actual flow.

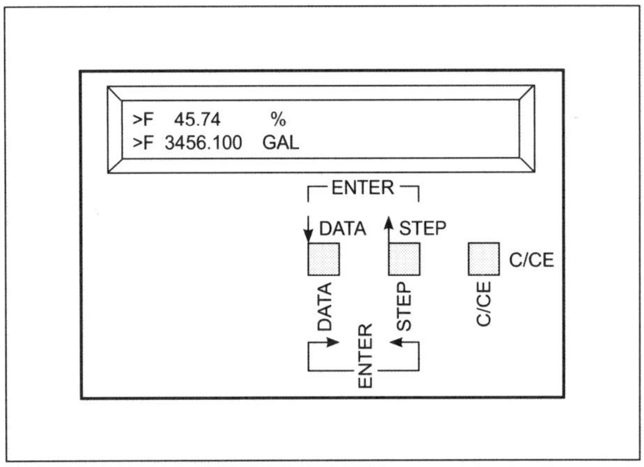

Figure 9-19. LCD with Push Buttons

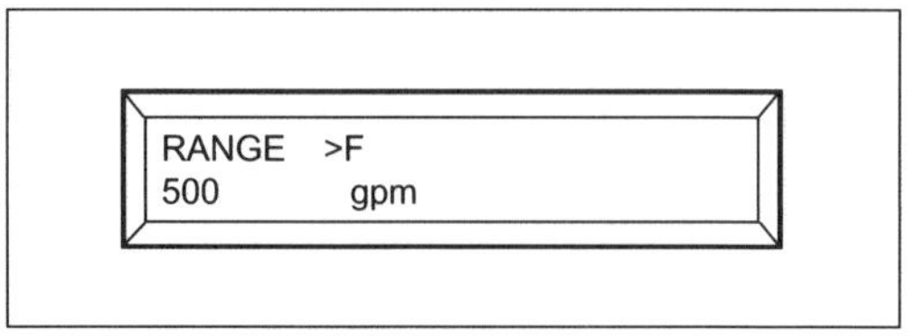

Figure 9-20. Display Showing Range Change

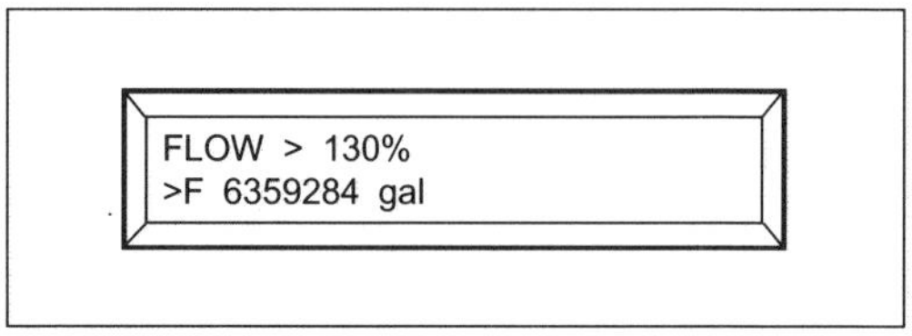

Figure 9-21. Display Showing Range Conditions

The error message may be accompanied by an alarm contact in the transmitter when an alarm condition occurs. The contact closes automatically upon alarm and can be used to operate a light or buzzer to announce the alarm condition in a remote location.

Self-Check

Most microprocessor-based transmitters have a self-check feature. This feature is used when it is suspected that the unit is not functioning properly and is used as a troubleshooting tool. When the self-check feature is selected, the transmitter is taken off line, and data is not updated. This is logical because the data is not correct if the unit is malfunctioning. The check can include one or more menus from which various software and hardware items can be selected for test. The EPROM, the data memory, the display, alarm contacts, calibration of span and outputs, and zero return are some of the items that can be tested.

It should be possible to perform the actual test by selecting the appropriate parameter from a self-check menu and reading the results in the transmitter display. It should also be possible to make calibration-type adjustments through software.

Communications

A full-blown microprocessor-based transmitter should be able to pass data over a data link to a computer terminal or some other remote control device located in a control room or some other convenient location. It should be possible through these devices to access all the information in the data base of the transmitter and to change it if necessary. Whatever functions the operator can perform at the transmitter should also be possible from the remote terminal. In addition, flow data as well as the parameters and performance of the flowmeter can be collected and stored at the remote computer for future evaluation. The transmitters that have this capability use one of two systems to communicate. One of these is a four-wire data link using an RS-232C or an RS-485 interface. The other system superimposes a high frequency signal on the 4-20 mA DC output signal, the most popular of which is the Hart Protocol.

The four-wire system is the faster of the two. It is capable of operating at speeds as high as 28,800 baud. The two-wire system saves costs on wire installation but is limited to speeds up to about 1200 baud. This low baud rate is fine for changing or collecting data from a small number of transmitters, but time becomes a critical factor when a large number of units require updating. The ideal data scan time is one that provides the shortest time between live data updates. This would be about one second, and at this scan rate, 100 updates are possible at 28,800 baud while only 3 updates are possible at 1200 baud.

A system that passes data over the 4-20 mA DC output normally includes a hand-held communications device that can be plugged into the transmitter analog output at strategic points. This allows an operator to scan the data base, perform diagnostic checks, and change data if necessary. This is all done without removing the cover from the transmitter, which is important if the unit is installed in a hostile environment. The four-wire system can be set up in a similar manner by using a remote keypad control unit to which the transmitters would be connected. When operated in this manner, both systems provide a method of communicating without requiring hookup to a computer or distributed control system.

Noise Reduction

Process-generated noise and its solutions, except the noise reduction feature, were discussed earlier. Noise reduction acts a little like a damping function in that it reduces or eliminates process-generated noise. However, it does not slow down the response to flow changes severely as does damping. Damping is used to smooth out pulsations in flow that are generally caused by pulsating-type pumps. It slows the response to every change in flow and to any noise that may

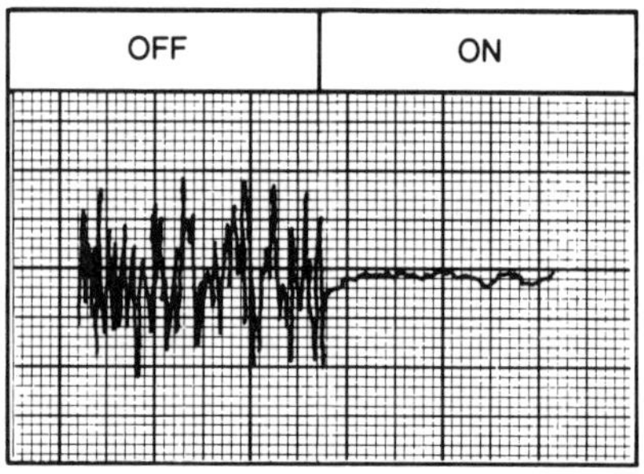

Figure 9-22. Noise Reduction

appear in the process. Noise reduction, on the other hand, selectively eliminates large, short-duration signals that are not flow changes but noise that has appeared at the electrodes. Figure 9-22, illustrating how effective the noise reducing software can be, shows a noise level in excess of 40% above and below the true flow with the function turned off. With the function turned on, the flow signal shows variations of only about 2%. While the 40% oscillations are unacceptable for control purposes, the 2% fluctuations are quite manageable.

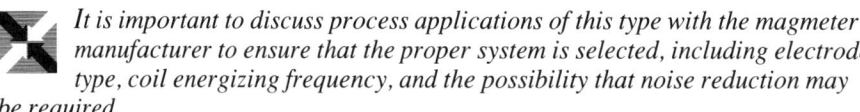

It is important to discuss process applications of this type with the magmeter manufacturer to ensure that the proper system is selected, including electrode type, coil energizing frequency, and the possibility that noise reduction may be required.

Batching

Some microprocessor-based transmitters can perform the batching function discussed earlier, as an optional feature. The batch size is entered into the internal totalizer by normal configuration. At the end of batch, the transmitter provides a dry contact closure or an optically coupled transistor output to end the batch. A second contact can be made available before the end of the batch for "topping off."

Summary

Microprocessor-based transmitters are rapidly becoming commonplace. They have the potential to be extremely easy to use while providing functional features that would have been expensive options in the conventional design, if they were available at all. The user need only determine what functional features are required for the job at hand and shop around. Shopping around is necessary because, as was pointed out earlier, not all microprocessor-based transmitters are created equal.

Calibration

Two calibrations are involved in a magmeter system: one is an electronic calibration that is performed on the transmitter, and the other is a hydraulic calibration that is performed on the magmeter. They are conducted independently of each other in the manufacturing process for systems that require standard accuracies. For systems that require better than the standard accuracies, some manufacturers will calibrate the magmeter and transmitter as a system. For example, if the standard system accuracy is specified as 1% of rate by the manufacturer, the independent calibrations will suffice. If this same system has an optional accuracy of ½% of rate, it is likely that the transmitter and magmeter will be calibrated together as a system to ensure, by reducing interaction tolerances, that they produce the optional accuracy.

Transmitter Calibration

A sophisticated electronic test fixture that simulates the output of a magnetic flowmeter is used to calibrate the transmitter. Inputs from the test fixture are compared to the outputs of the transmitter, which must agree within the manufacturer's tolerance. The test fixture is periodically checked against a laboratory standard that may be traceable to the National Institute of Standards and Technology (NIST), formerly the National Bureau of Standards (NBS).

A field calibrator is available from most magmeter manufacturers. Like the factory text fixture, it simulates the output of the magmeter. It has already been pointed out that range setting is not a calibration procedure, and a calibrator is not required for this purpose. This was a requirement in the early days of magmeter development, but technology has brought magmeters a long way since those days. Today, the calibrator is used primarily as a troubleshooting device. If the system appears to be malfunctioning, a check or the transmitter with a calibrator could isolate the problem to either the magmeter of the transmitter. If the transmitter is at fault, the calibrator can be used to further isolate the problem to a section of the transmitter.

Magmeter Calibration

A true calibration of the magmeter cannot be made with an electronic simulator. This must be done by passing a known volume of liquid through the magmeter and checking the output against the known value. This can be done in several ways. The liquid that passes through the magmeter can be diverted into a tank, and the tank can be weighed. This is a highly accurate means of calibrating a magmeter. However, it has some practical limitations. Each calibration should cover at least three flow points, and for a weigh tank calibration, this means emptying the tank after each run. This is time consuming, which slows down the production process and increases production costs. In addition, there is a physical limitation. Magmeters larger than about 20 inches can fill a typical weigh tank in a matter of seconds. Again, building larger weigh tanks with larger scales is expensive and only adds to the costs of manufacturing. However, master meters can do the job as well at a fraction of the cost.

A master meter frequently is a turbine meter or a magnetic flowmeter that is installed in the hydraulic flow loop with the meter to be calibrated. The master itself is calibrated by the weigh tank method and is normally accurate to better than 0.15%. This is the basis for calibrating the magmeter to 1%, ½%, and even ¼% in special cases. Using the master meters for calibration makes it possible to maintain flow through the magmeter for longer periods of time for a more accurate measurement without the need to stop the process and empty the weigh tank for another run.

A system such as the one described here maintains traceability to the National Institute of Standards and Technology through the weigh tanks. The master meter is certified through the weigh tanks. The weigh tanks are certified by weights maintained by the facility, and the weights themselves are certified at a facility that is traceable to the NIST, which could be a state or other government agency.

A copy of the data showing the results of the hydraulic calibration should be available for every magmeter purchased. The data should include information that is easily understood by the user. It could include the measured flow through the magmeter, the measured flow through the master meter, and the percent difference between the two. It should also include a statement that the system is traceable to the NIST and is maintained in accordance with the MIL Specification governing traceability, if the system is truly traceable.

One final point should be made about magmeter calibration: magmeters are not normally calibrated for specific user range. The magmeter's well-established linearity (proportionality between output and flow rate) makes it possible to calibrate a magmeter over a standard range. For example, a magmeter calibrated over a range of 1 to 10 feet per second will produce the same calibration factor for a user range of 3 to 30 feet per second. This makes it possible to accurately calibrate a meter for 120,000 gallons per minute at 30 feet per second using a calibration flow rate of 40,000 gallons per minute at a velocity of 10 feet per second. Standardized flow ranges per meter size make it possible to computerize the control of the calibration facility and minimize setup time.

Installation

No matter how accurate and reliable the magmeter system is, it will not perform satisfactorily if it is not installed properly.

The importance of proper magmeter installation cannot be overemphasized. For example, magmeters can measure the flow of conductive liquids in both the forward and reverse directions. However, flow in the forward direction develops a voltage that is opposite in phase or polarity from the flow in the reverse direction. Unless the system is designed for bidirectional flow, it cannot detect this difference. It is considered to be unidirectional. Consequently, many magmeters are built with a flow direction arrow to indicate the proper direction for flow through the meter. If a unidirectional magmeter is installed so that flow goes through it opposite to the direction of the flow arrow, the output from the transmitter will remain at zero flow indication even though there is known flow going through the meter. Removing the magmeter from the line, turning it around, and reinstalling it will correct the problem, but this could be very expensive and time consuming.

A simpler solution is to reverse the flow signal by reversing the connections of the flow signal leads. This is normally a simple procedure that can be accomplished at the transmitter terminals. If the flow signal leads are marked 1 and 2, connect the number 2 lead to the number 1 terminal and the number 1 lead to the number 2 terminal. The output signal should now be normal, assuming that there are no other problems with the system.

One of the microprocessor converters has a unique way to determine that the magmeter is installed in reverse of the flow. The flow direction indicator and the letter "R" blink to show that the meter has been installed opposite to the direction of flow. This is corrected simply by selecting the menu for flow indication and choosing reverse instead of normal. This restores the system to normal operation.

Orientation

The preferred orientation for a magmeter is vertical with flow upward through the meter. Also preferred is a sloping installation with flow again moving upward through the meter. Acceptable, and perhaps most common, is horizontal. In every case, precautions must be taken to ensure that the magmeter is filled with process liquid at all times during flow measurement. A vertical installation, with the pipeline carrying liquids upward, assures a filled magmeter under low flow conditions and minimizes wear on the meter lining by abrasive particles that may be in the process. Electrodes should be oriented so that they are not at the top of the pipe in a horizontal or sloping installation. This prevents entrained air from coming in contact with the electrodes and causing errors in the flow signal. Figure 9-23 illustrates these three piping orientations.

Minimum Piping Straight Run Requirements

Magnetic flowmeters are quite forgiving of piping configurations, but they are not immune. Some complicated configurations cannot be analyzed without actually duplicating the system and measuring the results. However, some basic rule of thumb guidelines, based upon studies done at manufacturing facilities and at independent laboratories, can be recommended. These guidelines cover elbows, reducers, valves, and pumps. Distances are measured from the centerline of the magmeter to the mating flange of these devices. The downstream side of the magmeter is much less critical than the upstream side. Essentially, all that is required of the downstream side is that sufficient backpressure is provided to keep the magmeter full of liquid during flow measurement. Two diameters downstream should be acceptable for any of the devices mentioned above.

Straight run upstream of a well-designed magmeter measured from center of meter (electrode) for standard accuracy (1% rate):

Elbow, 3 D
Pump, 10 D
Control Valve, 10 D*

For higher accuracy requirements, use twice the distances shown.

***Should be located downstream of the magmeter.**

Elbows should be located a minimum of three pipe diameters upstream from a well-designed magmeter. This applies to a single elbow or a double elbow in the same or a different plane. Control valves should always be located on the downstream side of the magmeter. Valves can create turbulence that may result in air pockets that will affect meter accuracy. If for some reason the control valve cannot be located downstream, then at least ten diameters are required between the meter and the valve. This requirement also applies to pumps.

Blocking valves should be operated fully open or fully closed. When open, the throat of the valve should be equal to or larger than the opening of the magmeter. Upstream diameters smaller than that of the magmeter could result in error signals caused by cavitation.

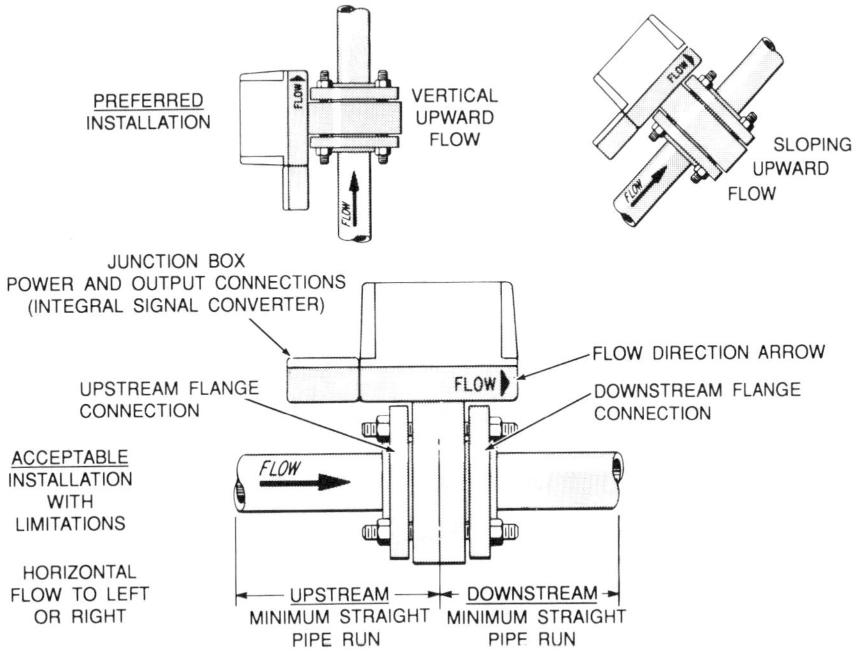

NOTES:
1. Meter pipe must be full. If meter is in gravity fed vertical pipe run, flow should be upward.
2. Meter must be oriented in accordance with flow direction.
3. When required, flow control valve must be located downstream of magnetic flowmeter.

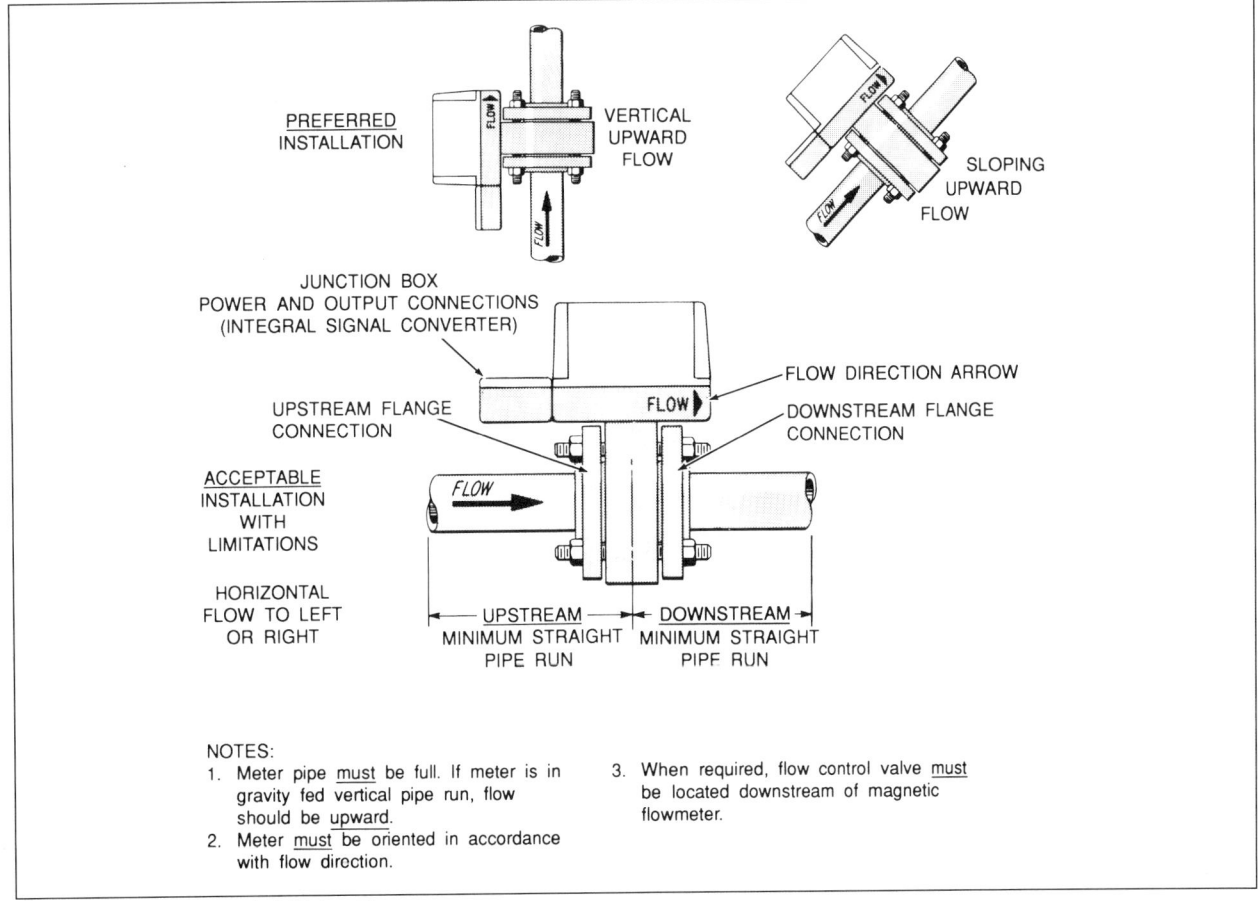

Figure 9-23. Recommended Piping Diagram

Reducing a line to a smaller diameter magmeter has little effect on accuracy. Reducers have often been installed immediately upstream of the magmeter with little or no adverse effect. This applies to either an eccentric or a concentric reducer. However, a conservative approach with a well-designed magmeter would dictate two or three diameters between the magmeter and the reducer.

These piping recommendations are valid for standard accuracies. For optional high accuracies, the upstream requirements of three and ten diameters of straight run should be increased to six and twenty diameters, respectively.

Grounding

Magmeter grounding is really a combination of standard grounding procedures and of bonding the meter body to the process liquid. The most important of these by far is the bonding, which is nothing more than ensuring that the meter body is in contact with the process liquid at each of its ends. If this is not done properly, the meter will function poorly, and in the case where there is no bonding, those flow signal circuits that are completed through the process liquid will not function at all.

Magnetic flowmeter instruction bulletins go into detail on how to properly bond and ground a magmeter. Separate installation instructions, including grounding, may be shipped with the magmeter in addition to the standard instruction bulletin.

Stray electrical currents are common in magmeter installations. These currents can develop as leakage from deteriorated insulation in motors and from capacitive or inductive coupling from motor windings and other conductors. Pipelines form excellent conductors for these stray currents. When a magmeter is installed in a pipeline, it becomes a part of the path for any stray current traveling down the pipeline or in the process itself. If these currents are allowed to pass through the magmeter, a zero shift may occur. The amount of error that this causes depends upon the magnitude of the stray current and the conductivity of the process. Bonding provides a short circuit by which the stray currents can be routed around the magmeter instead of through it.

Figure 9-24 shows a grounding procedure for a conductive pipe and a non-conductive or insulated pipe. A conductive pipe requires only that the grounding straps be attached between the flanges of the magmeter and then to a good earth ground. If the pipeline is made of a non-conductive material or is lined with a non-conductive material, then grounding rings or similar conductive devices are required to create a conductive path between the magmeter body and the process liquid. This allows stray currents that may be traveling along the pipeline in the process to pass from the liquid to the grounding ring, to the body of the magmeter, and back to the process on the other side of the magmeter. These currents follow the path of least resistance, and the metal grounding, strap, and body present less resistance than the process liquid.

Remember that anything that is in contact with the process must be compatible with the chemicals in it or be consumed by it. This includes the meter liner, electrodes, and grounding rings. Consequently, electrodes and grounding rings are often made of the same material. At times, however, the grounding ring may be made of a slightly less corrosion-resistant material than the electrode in order to reduce costs or delivery time. This is an acceptable practice because the grounding ring has much more material than the electrode and will take a much longer time to wear away. For example, tantalum or platinum are the preferred electrode material choices for sodium sulfate, but they are also the most expensive of the materials of choice. On the other hand, most of the less costly materials for grounding rings are rated only slightly lower on corrosion for this service than are tantalum and platinum.

The typical grounding ring is basically a paddle-type orifice plate with a bore equal to the nominal magmeter size, as shown in Figure 9-25. Its function is not to develop a pressure drop but simply to make contact with the process liquid.

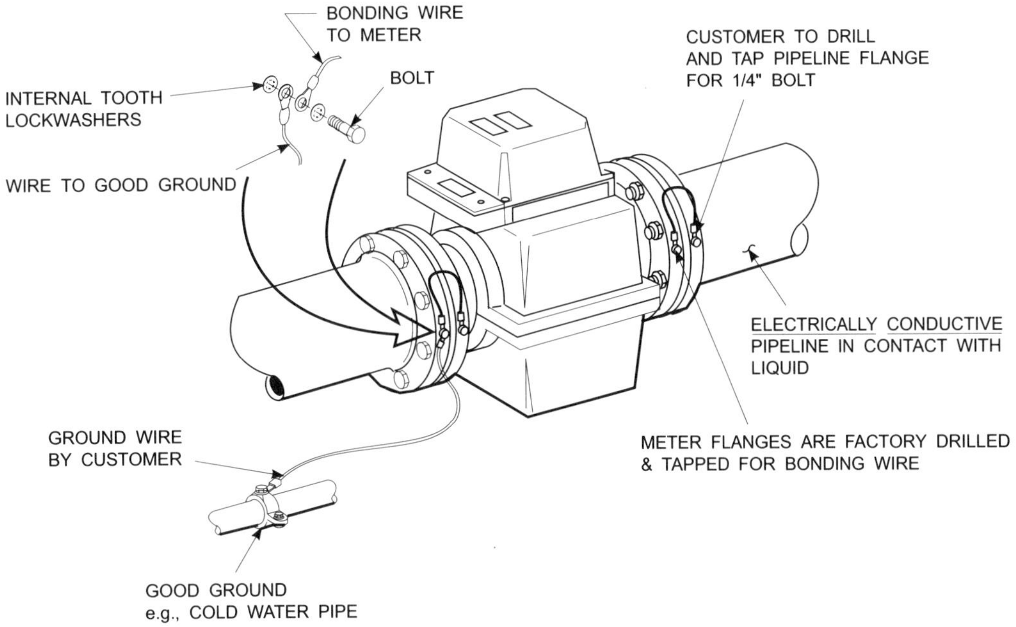

(a) CONDUCTIVE PIPE

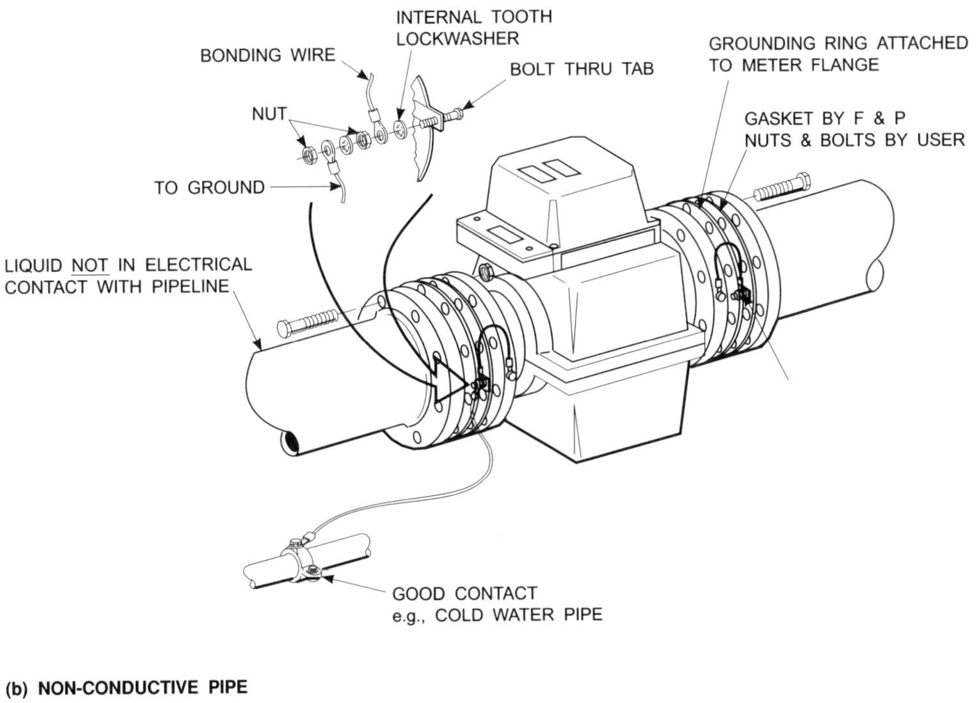

(b) NON-CONDUCTIVE PIPE

Figure 9-24. Grounding Procedure

Consequently, the thickness of the plate can be much less than a standard orifice plate in order to reduce costs.

Torquing

Not following installation instructions could void the warranty.

Magmeters are too often treated like another piece of pipe by the people who install them. Flange bolts are tightened well beyond what is required to provide a good seal between magmeter flanges and mating pipe flanges. As a result, some magmeters are damaged to the extent that they must be returned to the factory for rework. The installation instructions supplied by the manufacturer are often not even read until after the damage has been done. Teflon-lined magmeters are the most susceptible to installation damage, and even the Teflon liner protector cannot prevent damage from overtightening of flange bolts. Some magmeters are supplied with specific torque values while others are not, but all should be supplied with adequate installation instructions.

Hazardous Locations

Magmeters must sometimes be installed in hazardous locations. It is a misconception to think of these meters as explosion-proof, a term commonly used to describe requirements for this service. Hazardous locations are defined in Article

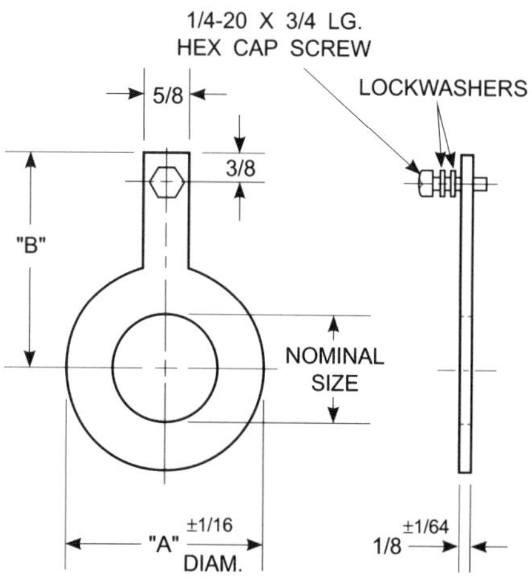

NORMAL SIZE		ANSI CLASS ISO	
mm	(inch)	"A"	"B"
15	(1/2)	1-7/8	2-7/8
25	(1)	2-5/8	3-1/4
40	(1-1/2)	3-3/8	3-5/8
50	2	4-1/8	4
60	(3)	5-3/8	4-5/8
100	(4)	6-7/8	5-3/8

NOTES:
1. All dimensions in inches, except where noted.
2. General tolerence = ± 1/8.
3. Grounding ring centered by flange bolts.

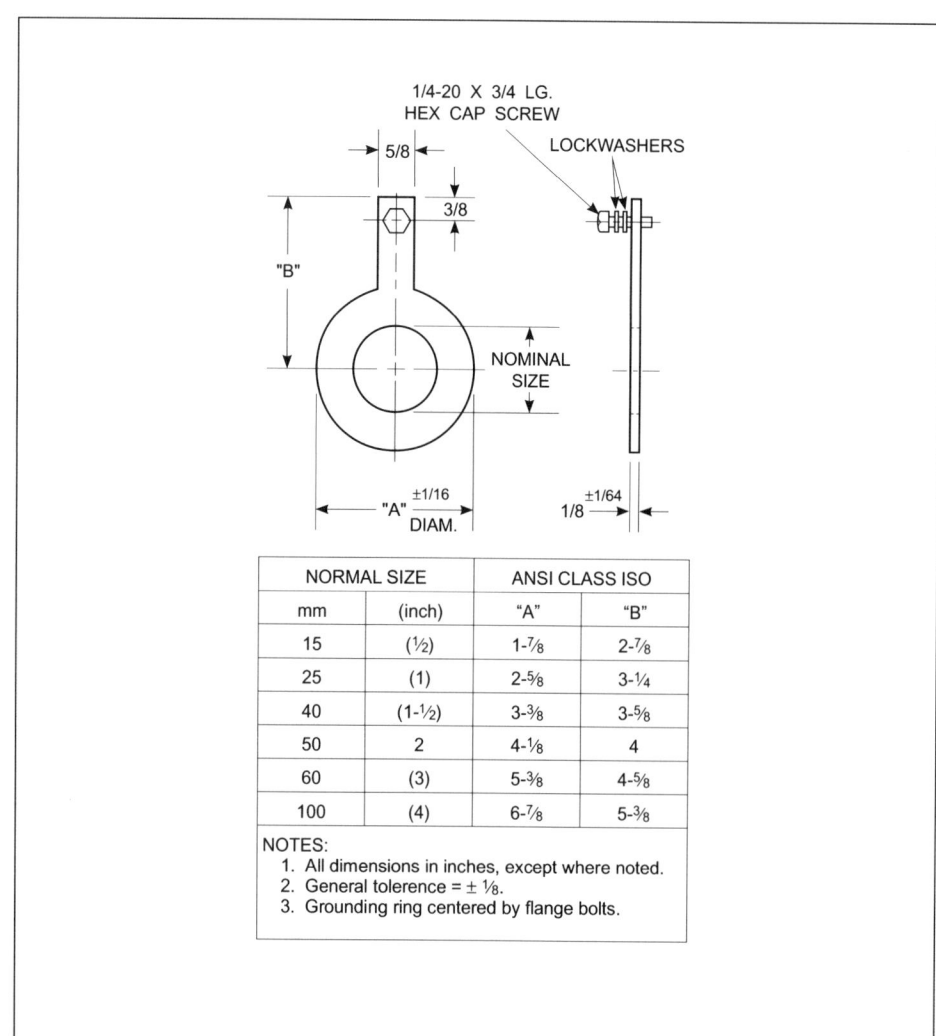

Figure 9-25. Grounding Ring

500 of The National Electric Code, ANSI/NFPA 70-1987. The conditions most commonly encountered for magmeter service are: General Purpose (Non-hazardous), Class I, II, and III, Division 2, and Division 1. Class I locations include flammable gases or vapors and flammable liquids. Class II locations contain combustible dusts, and Class III locations contain ignitable fibers or flyings. These conditions must be clearly defined in order to select the proper magmeter for the service.

Most standard magmeters manufactured in the United States are designed for service in Class I, II, and III, Division 2 hazardous locations. In general terms, this means that flammable gases, vapors and liquids, or combustible dust, or ignitable fibers or flyings may be present in the area but only if there is failure of containment. These materials will normally be confined within closed containers or closed systems. The failed containment need not be the magmeter. The meter designed for Class I, II, or III, Division 2 service must not have arcing or sparking contacts and must not have any surface temperatures exceeding 80% of the ignition temperature of the specified hazardous material.

The basic difference between a Division 1 location and a Division 2 location is that in Division 1 locations the hazardous material may be present even if there is no containment failure. Design requirements for this service are much more stringent than for Division 2.

It is assumed that the interior of the pipeline and the magmeter can be classified as a Division 1 location where the hazardous material is periodically or continuously present. For example, the pipeline could drain empty of liquid during no flow conditions. In the absence of the liquid, flammable vapors could collect in the pipeline and in the magmeter. Factory Mutual (an approval agency) requires that magmeters approved for Class I, Division 2, Groups A, B, C, and D hazardous locations must have electrodes that are intrinsically safe for the Class I, Division, Group A, B, C, and D hazardous location inside the pipe. However, this does not mean that the magmeter is "intrinsically safe."

Division 1 magmeters are often referred to as "explosion-proof." An explosion-proof enclosure is designed to contain an explosion of the most easily ignitable concentration of a specified mixture of gas or vapor and air. The enclosure is not gas-tight but is designed with specific gap or flame path dimensions. These flame paths will allow the explosion pressure to escape but will quench or cool any flame front, preventing ignition of the surrounding hazardous atmosphere. Use of this protection technique could result in a very expensive design, especially for large magmeters.

An alternative to "explosion-proof" is an "air-purged" design. This protection technique, described in NFPA 496-1986, will permit the use of Class I Division 2 magmeter, with air purging, in a Class I, Division I hazardous location. In principle, air-purged equipment is pressurized by a supply of nonhazardous air or gas that prevents any surrounding hazardous atmosphere from entering the enclosure.

The three most common agencies used for magmeter approvals in North America are Factory Mutual (FM), Canadian Standards Association (CSA), and Underwriter Laboratories (UL), in that order.

Hazardous Location Certifications in Europe

Within the European Economic Community (EEC), electrical apparatus designed for use in potentially flammable atmospheres must be certified by a national test facility. The EEC member countries have agreed to abide by the standards established by the European Committee for Electrotechnical Standardization (CENELEC). CENELEC members are the national electrotechnical committees of Austria, Belgium, Denmark, Finland, France, Germany, Ireland, Italy, Norway, Portugal, Spain, Sweden, Switzerland, and the United Kingdom.

A product certified to the CENELEC standards by a national test facility is automatically accepted by all member countries. Unfortunately, there is no reciprocal agreement between North America and Europe. North American manufacturers must submit their products to a European test facility such as BASEEFA in the U.K., LCIE in France, CESI in Italy, or PTB in Germany for certification to CENELEC standards. European manufacturers must submit their products to North American agencies such as CSA in Canada and UL or FM in the US for approval to those standards.

Work is proceeding in the standards-writing committees of organizations such as ISA, UL, CSA, and others in North America and the International Electrotechnical Commission (IEC) in Europe to harmonize the various standards related to hazardous locations. The eventual goal is universal acceptance of a product's safety certification anywhere in the world. In the meantime, agreements have been reached between some test facilities in Europe and Australia with agencies in North America to accept testing to each country's standards.

Maintenance

This section is not intended to be a comprehensive troubleshooting dissertation but a common sense guide for simple failures. Maintenance manuals should contain step-by-step procedures for the particular magmeter design.

Very little maintenance is involved in magnetic flowmeters. The calibrator discussed earlier can isolate problems in the transmitter. A few simple checks with an ohmmeter can identify component failures in the magmeter. There are only the magnet coils, the electrode circuit, and perhaps an interface board in the magmeter.

Some magmeters generate a reference voltage in the magmeter while others generate the reference in the transmitter. For the former, a good reference signal means that the coils are good. For others, a resistance reading of the coils should show that they are open, short, or acceptable. Resistance values for the coils should be available in the manufacturer's instructions.

Electrodes are checked for continuity or no continuity, but they are difficult to check in a magmeter that is installed and full of liquid.

Electrodes are insulated from the body of the magmeter and should show infinite resistance between the electrode and the meter body. However, with liquid in the line, an electrical path exists from the electrode through the liquid to the body, and a resistance value that includes the resistance of the liquid in the magmeter can be read. Consequently, a suspected short in the electrode circuit, such as loss of one half the flow signal, should be checked with the magmeter removed from the line and dry. On the other hand, an open electrode could be detected with the magmeter in or out of the pipeline.

Electrode coating in an AC-type magmeter could be manifested by difficulty in adjusting the zero or by sluggish response to changes in flow. This can be detected in both pulsed DC and AC types by a resistance check of the electrode circuit. If the resistance between the electrodes is measured with the magmeter in the line, full of process, and with clean electrodes, that resistance value can be a reference point. If, after a period of service, coating is suspected because of a span or zero shift, a second resistance reading can be taken. If the second reading is significantly higher than the reference reading, it would indicate that the electrodes may have an insulating-type coating. Appropriate action to clean the electrodes should return the meter to normal service.

Some transmitters can be field repaired by component or circuit board replacement. This normally requires recalibration after the repair to return the transmitter to its original calibration. Magmeters have few components that should be replaced in the field. Some designs permit replacement of the magnet coils, the spool body complete with liner, electrodes, wiring, and reference generating devices such as the magnet driver or the calibration components block. Many of the small magmeters, up to size four inch, are being manufactured as throwaway devices. In some cases, the manufacturer will supply another body on a repair/exchange basis, even though the meter body is not repairable.

References

1. Kuromori, K., and Mannherz, E., "Method for Calculating the Effects of Pipe Wall Contamination of the Calibration of Magnetic Flowmeters with Various Electrode Configurations," *ISA/81 Conference Proceedings,* Research Triangle Park, NC: Instrument Society of America, 1981.
2. Kayama, N.; Suzuki, K.; and Witlin, W., "Effects of the Inner Wall Conducting of Adjacent Connecting Pipes on the Signal of Magnetic Flowmeters," *ISA/84 Conference Proceedings*, p. 1181, Research Triangle Park, NC: Instrument Society of America, 1984.
3. Lange, N.A., *Handbook of Chemistry*, NY: McGraw-Hill, 1985.

About the Author

Raymond C. Mills, Jr., graduated in physics from Temple University and served in the U.S. Army from 1945 to 1966, where he was guidance and test officer with a Nike Hercules missile unit. Mr. Mills joined Fisher & Porter Co. after retirement from active duty and for the past 16 years has performed as Product Manager and Applications Engineer for magnetic flowmeters.

The author would like to thank Fred Kent, Elmer Mannherz, and Herb Shauger of Fischer & Porter Co. for their technical assistance.

10

Mass Flowmeters

Flowmeters that measure mass directly using the properties of mass, as opposed to those that measure volume or velocity, were developed and commercialized in the 1980s. Meters of this type have found wide application because the fluid measurement is virtually independent of changing fluid parameters. Many of the other flowmeter technologies are affected by changes in fluid density, viscosity, pressure, and/or temperature. Meters that measure mass directly, in effect, weigh the fluid as it passes through the meter, yielding a highly accurate measurement that is virtually independent of varying process conditions that often occur. Because of this unique ability, it is possible to use a mass flowmeter on a wide variety of process fluids without need for recalibration or compensation to specific fluid parameters.

Mass flowmeters can be used for the following applications:
Clean liquids
Dirty liquids
Slurries
Gases
Steam
Liquids with entrained gases

The Coriolis principle flowmeter is a true mass flowmeter because it uses the properties of mass to measure mass. It is relatively easy to apply and size. Because it possesses no moving parts, it exhibits low maintenance requirements and does not require frequent calibration. Wetted parts that can be constructed of a variety of materials make it adaptable to many corrosive fluids as well as fluids containing solid or fibrous particles. Coriolis flowmeters can also measure the density of the process fluid, making it possible to infer the flow of one component in a two-component flow stream (e.g., dry solids flow rate of a slurry). Coriolis flow sensors can be designed for intrinsically safe operation in hazardous locations.

Principles of Operation

Mass Flow

Coriolis principle mass flowmeters are manufactured in a variety of shapes, sizes, and materials of construction. All of these factors influence the sensitivity of the meter to flow rate; however, the basic principle of operation remains the same. Simply stated, Coriolis meters operate on the basic principles of motion mechanics. The fluid in motion through a vibrating flow tube is forced to take on an acceleration as it moves toward the point of peak amplitude of vibration. Conversely, the fluid decelerates as it moves away from the point of peak amplitude as it exits the tube. The moving fluid exerts a force on the inlet side of the tube in resistance to this acceleration, causing this side of the tube to lag behind its no-flow position. On the outlet side, the force exerted by the flowing fluid is in the opposite direction as the fluid resists the deceleration. This force causes the outlet side of the tube to lead ahead of its no-flow position. The result of these forces is a twisting reaction of the flow tube during flow conditions as it traverses each vibrational cycle. This is demonstrated with a U-shaped tube in Figure 10-1. Fluid at point B moves up and down faster than fluid at A or C. Fluid passing through the curved section must, therefore, change its overall velocity in space even though its speed relative to the tube may be constant. When the fluid moves away from the mounting axis, the tube motion lags behind point B. When the fluid moves toward the mounting axis, the tube motion leads ahead of point B.

The flow tube is vibrated at its natural frequency through the use of electromagnetic devices. The motion at any point on the tube represents a sine wave. With no flow through the tube, all points move in sequence or in phase with the driver (Figure 10-2(a)). As mass flow through the tube occurs, the inlet side motion of the tube lags the driver phase, and the outlet side motion leads the driver phase (Figure 10-2(b)). The time delay between S_1 and S_2 is directly proportional to the mass flow rate through the sensor.

Looking again at a U-shaped flow tube (Figure 10-3) in a simplified fashion, the basic equations that apply to this structure can be developed.

The mathematical derivation in Equation (10-1) describes the relationship between the mass flow rate and the behavior of the measuring element for a sensor

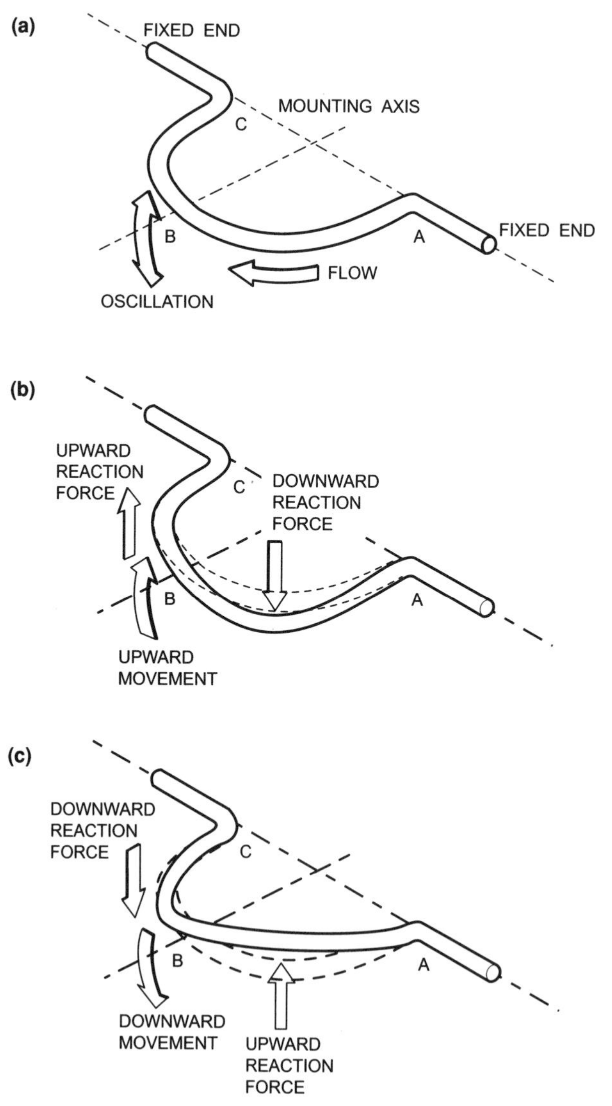

Figure 10-1. Coriolis Principle
(Courtesy of K-Flow)

with only one U-shaped tube. This equation applies as well to the implementation with two opposite measuring tubes.

Figure 10-3 shows a fluid with mass m flowing with a velocity v through a U-shaped tube oscillating with an angular velocity around axis 0-0.

The Coriolis force induced by the flow is described by the following equation:

$$F = 2m\omega\bar{v} \tag{10-1}$$

in which F (force), ω (its angular motion) and $\bar{v}$ (velocity) are vectors and m is the mass to be applied to a known point at a distance L from the axis 0-0. This equation is equivalent to $F = ma$ (Newton's Second Law for Rotational Motion).

The vectors for the input and output velocities of the fluid are in opposite directions. Looking at the measuring tube along the axis R-R, the forces F_1 and F_2 applied by the fluid to the input and output legs are in opposite directions but have the same amplitude.

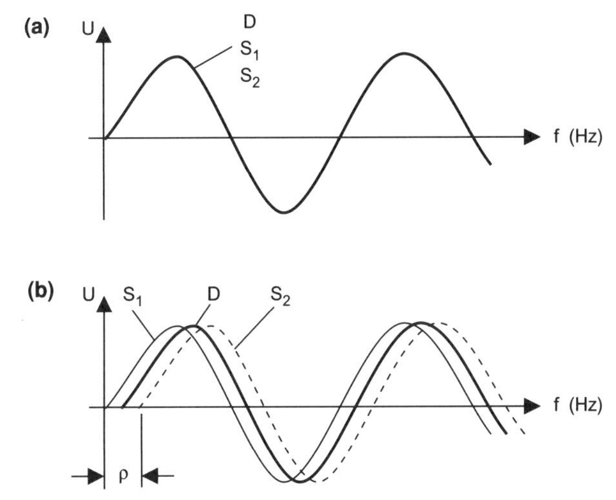

Figure 10-2. Sine Waves, (a) No Flow, (b) Mass Flow
(Courtesy of DanFoss)

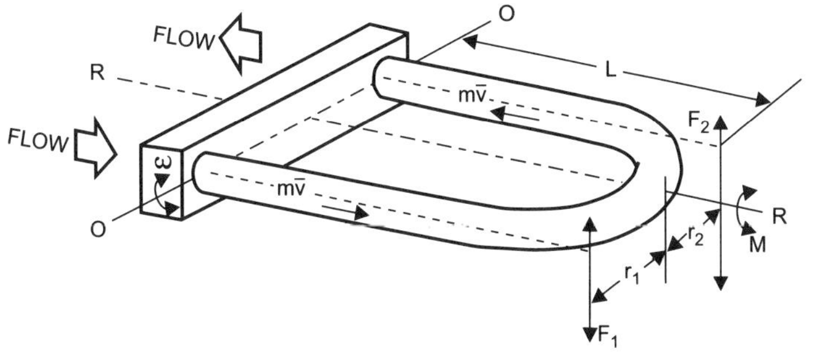

Figure 10-3. Oscillating U-Tube
(Courtesy of Micro Motion, Inc.)

Since the tube oscillates around the axis 0-0, the forces developed create an oscillating moment M around the axis R-R, with a radius r. It is expressed by:

$$M = F_1 r_1 + F_2 r_2 \qquad (10\text{-}2)$$

Since $F_1 = F_2$ and $r_1 = r_2$, it follows from Equations (10-1) and (10-2):

$$M = 2Fr = 4mv\omega r \qquad (10\text{-}3)$$

The mass m is defined as the product of the density ρ , the cross-sectional area A and length L of the measuring tube. The velocity is defined as unit of length L per unit of time t. The mass flow rate W is defined as the mass m that flows by a given point per unit of time. Therefore:

$$m = \rho AL \qquad (10\text{-}4)$$

$$\bar{v} = \frac{L}{t}$$

$$W = \frac{m}{t}$$

Thus, by substituting $W = \frac{m\bar{v}}{L}$, in which L is the length of the tube, Equation (10-3) becomes:

$$M = 4\omega rWL \qquad (10\text{-}5)$$

The moment M induces an angular deflection, or twist θ, of the measuring tube around the axis R-R. The twist is maximum at half the travel of the vibrating tube (Figure 10-3).

Nevertheless, the deflection caused by the moment is opposed by the force K corresponding to the elastic modulus of the tube. In general, for a spring subjected to a twisting moment, the torque T is defined by:

$T = K\theta$

Since $T = M$, the mass flow rate W can now be related to the deflection angle θ using Equation (10-5):

$$W = \frac{K\theta}{4\omega rL} \qquad (10\text{-}6)$$

The mass flow rate can be derived by measuring the deflection angle with two magnetic position sensors (Figure 10-4).

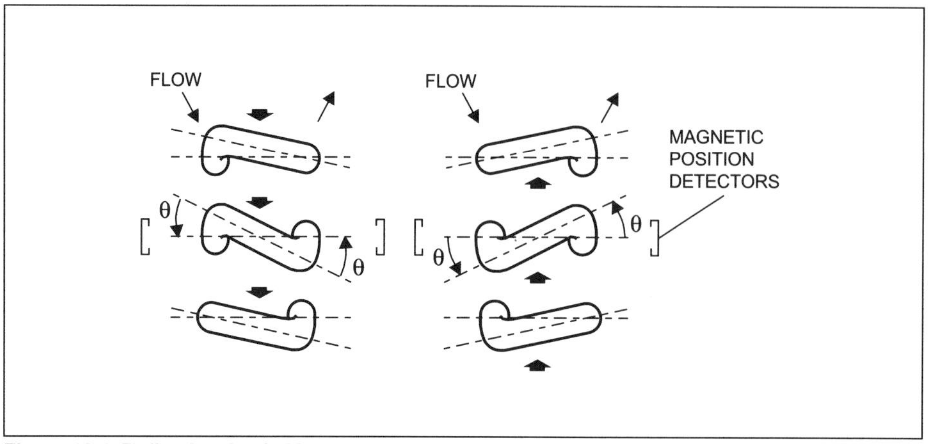

Figure 10-4. Deflection Angle Measurement
(Courtesy of Micro Motion, Inc.)

One signal processing method uses each sensor to measure θ as a function of the time it takes for each leg of the tube to cross the middle point of the travel corresponding to the total deflection. The time differential between the two legs is zero in no-flow conditions. When a flow is established (increasing the angle θ), the time differential between the signals corresponding to the high and low travels of the tube legs also increases.

These time differentials are interpreted as pulses of different lengths by the digital logic circuit.

The velocity v_t of the tube in the middle of its travel, multiplied by the time interval Δt, is related to θ by the following equation:

$$\sin \theta = \frac{v_t \, \Delta t}{2r} \tag{10-7}$$

If θ is small, it is almost equal to its sine. And, for a small angular rotation, v_t is equal to the product of ω by the length L of the tube.

Thus, if $\sin \theta = \theta$ and $v_t = \omega L$, Equation (10-7) becomes:

$$\theta = \frac{\omega L \, \Delta t}{2r} \tag{10-8}$$

Combining Equations (10-6) and (10-8), it follows that:

$$W = \frac{K \omega L \Delta t}{8 r^2 \omega L} = K \Delta t \tag{10-9}$$

Therefore, the mass flow rate is proportional only to the time interval Δt and some geometric constants.

W is independent of ω and, thus, independent of vibrating frequency of the measuring tube.

The term K in the equations above represents the elastic modulus of the tube. The elastic modulus of all metals varies as a function of temperature. Since K is a constant of proportionality, it must be compensated for temperature changes that occur in the process. Typically, the surface temperature of the flow tube is monitored using a platinum RTD. As temperature changes, the signal processing electronics will continuously adjust the constant of proportionality, which scales Δt to obtain the corrected mass flow rate measurement.

In practical implementations of the Coriolis principles in other than small line sizes, two flow tubes in parallel or in series are vibrated in opposition to each other to neutralize the vibration induced into the piping and mounting system. This allows one tube's twist to be measured with respect to the other tube's twist while rejecting piping vibration. The amplitude of vibration of the tube is controlled to a level that keeps the bending stress in the tubing to well below the elastic limits of the tube material. It should be emphasized that the amount of Coriolis twist induced by the mass flow is very small. One manufacturer claims that for a 0.040 inch amplitude of vibration, the tube only twists a maximum of 0.0014 inch at rated full scale flow. Figure 10-5 shows three popular configurations of Coriolis meter flow tubes.

Density

Since the mass flow measurement is not affected by the frequency of vibration of the tube, it is possible to allow the tube to be vibrated at its fundamental natural frequency. This has two advantages. Vibrating the tube at its natural frequency, much like a tuning fork, requires the least amount of energy delivered to the drive coil to keep the tube vibrating. Secondly, the natural frequency of the vibrating tube depends on the mass of the material contained within the tube. It is possible to measure the density of the process fluid with the same sensor used for mass flow. The following equations govern the density measurement implement with a Coriolis flow sensor. For a vibrating spring mass system,

$$\omega = 2\pi f = \sqrt{\frac{K}{m}} \tag{10-10}$$

where:

ω = angular frequency of oscillation
f = frequency of oscillation
K = spring constant
m = mass

The mass is comprised of the mass of the vibrating structure plus the mass of its contained liquid.

$$m = m_{\text{tube}} + m_{\text{liquid}} \tag{10-11}$$

The mass of the liquid is equal to density multiplied by volume.

$$m_{\text{liquid}} = \rho V \tag{10-12}$$

Substituting,

$$\omega = \sqrt{\frac{K}{m_{\text{tube}} + \rho V}} \tag{10-13}$$

Where period $T = \frac{1}{f}$, solving for ρ :

$$\rho = \frac{\frac{KT^2}{(2\pi)^2} - m_{\text{tube}}}{V} \tag{10-14}$$

By measuring the natural period of vibration and compensating for the change in spring constant with temperature, the process fluid density can be calculated.

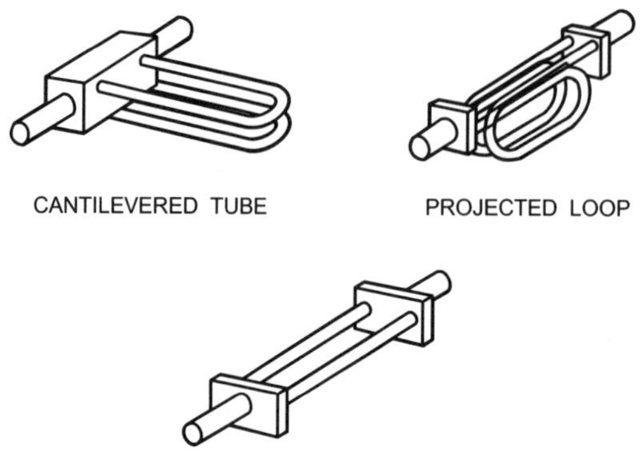

Figure 10-5. Coriolis Flowmeter Flow Tube Configurations
(Courtesy of K-Flow)

Construction

Construction techniques vary widely, but the following general principles are observed. The primary measurement section is made of highly corrosion-resistant metal tubing. The most common material used is seamless 316L stainless steel. Coriolis meters are also available in tubing made of titanium, Hastelloy®, and tantalum for chemical compatibility with process fluids that are corrosive to stainless steel. In parallel tube sensors, the ends of the tubing are rigidly attached to a flow splitter via tube to casting weldments. This flow splitter forms the transition between the process connection and the vibrating tube measurement section. The splitter allows approximately half the flow to pass through each tube. In most cases a brace bar is attached to the tubing a short distance away from the tube attachment joint to form a rigid beam against which the flow tubes are vibrated. This effectively decouples the vibrating tubes from the process piping system in which the meter is mounted. In some designs, the maximum stress from bending the tubes during vibration occurs at this attachment.

A drive coil on one tube and an opposing magnet on the other tube are mounted in the center, an equal distance from each end, to vibrate the tubes. Motion sensors are mounted on the inlet and outlet legs of the tube to sense the motion of each side leg. The most common type of motion sensors are electromagnetic, although some manufacturers have successfully applied optical sensors. Their signals are compared in the electronics to determine the amount of Coriolis twist. Finally, a miniature temperature sensor is attached to the exterior surface of one of the flow tubes to compensate for changes in spring constant due to temperature. Figure 10-6 shows a typical construction.

The entire tube assembly is usually encased in a welded sealed stainless steel enclosure to protect the tubes and sensors from their operating environment. Wires from the drive, motion sensors, and the temperature sensor are passed through a sealed feedthrough to a junction box for wiring the sensor to the electronic transmitter.

The electronic transmitter may be mounted close to the sensor but is capable of operating hundreds of feet away. Transmitters are available for field mounting as well as rack mounting from most manufacturers.

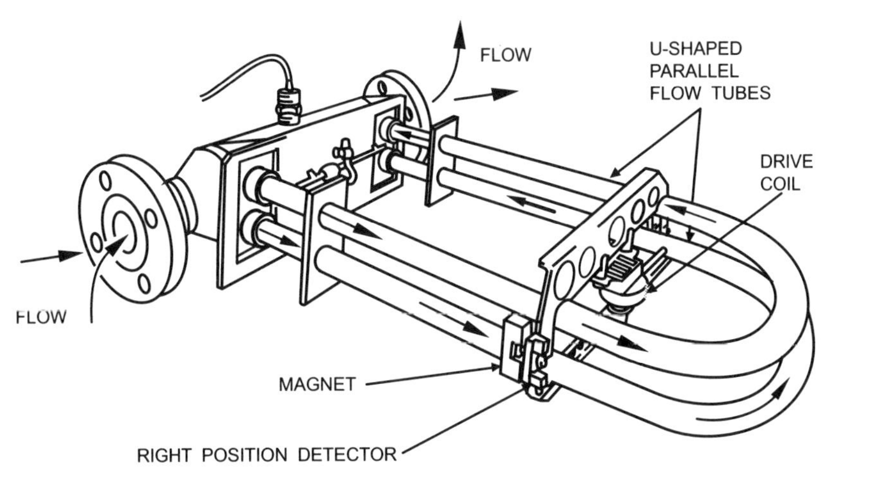

Figure 10-6. Typical Construction
(Courtesy of Micro Motion, Inc.)

Due to the power required to vibrate the flow tubes, Coriolis transmitters are not of the 2-wire, 4-20 mA type and most require separate wiring to power the device. Both AC- and DC-powered versions are available. A typical meter requires approximately 10 watts of power to operate. Output signals consist of various combinations of 4-20 mA analog and pulsed frequency outputs. Microprocessor-based computational electronics are common in Coriolis transmitters and many have the capability of direct digital communication with host control systems.

For use of the sensor in hazardous locations, most transmitters have built-in intrinsic safety barriers to limit the currents and voltages at the sensor. Sensors designed for intrinsically safe service can be safely installed in an appropriate Division 1 area, while the transmitter can be located in either an appropriate Division 2 or safe area. Consult manufacturer's literature and wiring techniques for specific requirements for hazardous locations.

Flowmeter Design

The effect of the Coriolis force on a semi-rigid vibrating tube system is, by any standard, small. Full scale flow will cause deflections of less than one-one thousandth of an inch (0.001 inch) in most designs. For a meter to be usable over a 100:1 range, deflections on the order of ten-one millionths of an inch (0.000001 inch) need to be measured accurately. To achieve high performance in industrial environments, sensor designs are meticulously developed, analyzed, and tested against field operating conditions (density, temperature, vibration, etc.). Fortunately, over the past few years, sophisticated computer modeling systems have become available that can thoroughly analyze complex vibrating structures to determine their behavior under various conditions. These tools have become essential in the development of state-of-the-art Coriolis flowmeters.

Primary design parameters for Coriolis flowmeters include chemical compatibility, pressure rating, flow range, pressure drop, signal amplitude, and signal-to-noise ratio. The design process used to effectively address each of these requirements is exceedingly complex because of many interdependencies. For example, the tube material, geometric shape, and wall thickness have major effects on the signal developed. Metals have different spring constants; long, bent tubes twist more easily than short, straight ones; thick walled tubes deform less due to Coriolis forces than do thin walled tubes, etc. The tube material, diameter, and wall thickness affect the pressure rating. The tube diameter, length, and radius of bends have major effects on the pressure drop. Numerous tradeoffs, therefore, occur in the design of each sensor.

Once the primary design goals are met, keeping the small Coriolis signals from being contaminated by external disturbances is a major design task. A computer-aided engineering tool known as finite element analysis is used to design against undesirable interactions of the flowmeter with its operating environment. Vibrating systems can react in a variety of different vibrational modes other than the one produced by the drive system. The finite element analysis technique allows the designer to adjust the tube geometry and placement of components to position undesirable vibration modes at frequencies that will not interfere with the fundamental measurement mode. This makes it possible to operate Coriolis mass flowmeters over a wide range of process conditions and in the presence of noisy pumps, compressors, and other equipment.

The design of the electronics that detect these small signals is equally critical. The Δt signal, which is the basis of the mass flow measurement, needs to be resolved to the nanosecond range. (One nanosecond is the time it takes for an electrical signal to travel one foot in a conductor.) Close attention is paid to timing delays through wiring, printed circuit board conductors, and logic gates in order to measure these short times accurately. Sources of electrical noise need to be eliminated or shielded against to preserve signal integrity. Finally, averaging of the individual measurements made each tube cycle is essential to achieve a smooth flow signal. Both analog and digital signal processing and filtering techniques are used in state-of-the-art Coriolis flowmeters.

Performance/Limitations

Coriolis mass flowmeters work effectively on a wide variety of liquid processes and slurries and have limited application on gases.

Meters are available covering line sizes from approximately $\frac{1}{16}$ inch up to 6 inches. Flow rates from a few grams per hour up to 10 tons/min can be measured. Because of the linear nature of the signal developed, Coriolis meters have a wide dynamic range. Typical turndown ratios are on the order of 25:1, and good accuracy over a 100:1 dynamic flow range can be obtained in some applications.

Typical measurement accuracy statements are between ± 0.15% and ± 0.25% of rate, plus a zero shift error, which one manufacturer terms the "zero stability." The zero shift error is due to small varying offsets in both the sensor and electronics. Most manufacturers specify an uncertainty in the meter at zero flow as a percentage of the meter capacity of the sensor. This uncertainty adds to the basic accuracy and can become the dominant portion of the overall accuracy at the lower end of the flow range. For example, a zero shift of ± 0.01% of meter capacity will contribute an additional ± 0.25% of rate to the accuracy at 4% of meter capacity. At 1% of capacity, the zero shift error accounts for an additional 1.0% of rate error. If accuracy limits are plotted as a percentage of flow rate, the graph will be as shown in Figure 10-7. Due to the zero shift error, accuracy is degraded as the flow rate approaches zero. More than one size flowmeter may be applicable for a given application; however, the larger flowmeter will exhibit less pressure drop but operate at a lower percentage of meter capacity. Due to the degradation in accuracy at the low end, mass flowmeters should be sized to obtain the best accuracy versus pressure drop trade-off.

To obtain the best possible low end performance, Coriolis flowmeters offer the ability to adjust zero at actual line operating conditions. After the flowmeter is installed and filled with the process fluid at process temperature, the zero can be optimized by blocking in the flow and making a zero adjustment. Since the zero offset may be slightly affected by temperature, most Coriolis meter manufacturers recommend that the zero be set at a temperature within 10°C of the actual operating temperature.

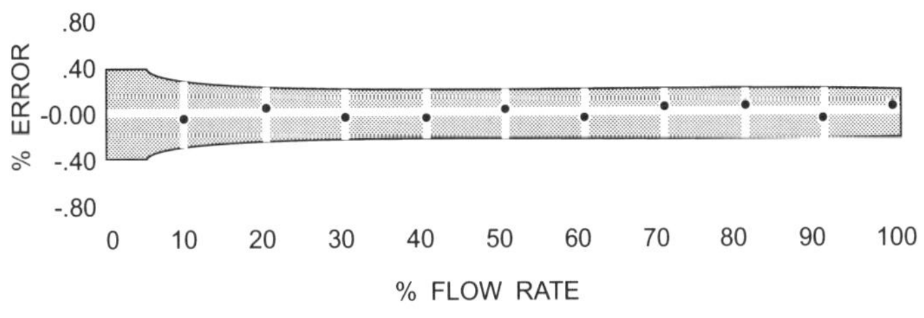

Figure 10-7. Performance Curves (Meter Calibration at Maximum Full Scale)

As is true with many other flowmetering technologies, the Coriolis meter's accuracy can be affected if air or gas pockets are present in the flowing fluid. Gas that is well mixed or entrained in the liquid poses little problem other than requiring slightly more drive power to keep the tubes vibrating. If gas breaks out, forming voids in the liquid, meter performance will degrade due to viscous damping of the vibrating tube. Small void fractions, up to 5% by volume, will cause noisy flow signals due to "sloshing" of the liquid inside the vibrating tube, which causes noise to be superimposed on the Coriolis twist signal being measured. Larger void fractions can cause the tube to stop vibrating altogether, because the energy absorbed by the sloshing liquid exceeds that available from the voltage- and current-limited drive system. Coriolis meters must vibrate in order to produce their measurement.

Air eliminators are recommended in such instances to prevent "slug flow" conditions detrimental to the Coriolis measurement. It should be noted that slug flow will not physically damage a Coriolis meter even though the signals may be corrupted.

Slug flow conditions can be detected by the flowmeter electronics in many cases. They manifest themselves as excessive drive power, a drop in measured density, or a reduction in amplitude of the motion sensor signals. Many manufacturers offer features that inhibit the output signals or produce an alarm when the meter is incapable of making an accurate measurement during "slugging" conditions. This is particularly useful in batching applications where the flow line is purged between runs, leading to slug flow during filling and emptying of the line. Flashing can also be detrimental to flowmeter performance. Tube imbalance due to varying amounts of gaseous fluid can cause a "bouncy" signal and degrade accuracy.

Sizing

The selection of a mass flowmeter is based on considerations such as flow rate, line size, pressure and temperature ratings, wetted parts, pressure drop (ΔP), fluid velocity, and accuracy. Of these considerations, the estimated pressure drop is often the most important aspect in determining meter selection.

Flow Rate and Line Size

The following "rules of thumb" should be applied when sizing meters: (1) The nominal bore size of the sensor should usually be equal to or less than the line size. (2) For optimal accuracy performance, the normal full scale flow rate should be as high as possible in the range of the meter, pressure drop requirements permitting.

Pressure and Temperature Ratings

The pressure and temperature ratings of the sensors, flow tube, and process connections should be considered when selecting a meter.

The sensor tube pressure rating should be rated higher than the maximum rated pressure of the line in which the sensor is to be installed. Care should be taken that the rating obtained from the manufacturer has been properly derated for the temperature range specified for the given sensor and that the rating is suitable for the application.

The process connections should also be selected for the maximum expected operating pressure. Table 10-1 shows the relationship between operating temperature and maximum rated pressure for some common industry standard process connections. This table is intended as an example and does not cover all potential fittings that are found in various manufacturers' catalogs. Various configurations of flanges for Hastelloy®, tantalum, and other material would have similar, but different, specifications, and the manufacturer should be consulted if such materials are utilized.

Table 10-1. Working Pressure Ratings for Process Connections (Pressure in psig)

Flange	Class	Mat'l	-20 to 100°F	200°F	300°F	400°F
ANSI	150 lb	316L	230	195	175	160
(per B 16.5)	150 lb	304	275	235	205	180
	300 lb	316L	600	505	455	415
	300 lb	304	720	600	530	470
	600 lb	316L	1200	1015	910	825
	600 lb	304	1440	1200	1055	940
	900 lb	316L	1800	1520	1360	1280
	900 lb	304	2100	1800	1580	1410
			15 to 250°F			
DIN (per DIN 2401, Part 2)	PN40	316L	580			
JIS (per JIS B 2212)	10K	316L	140			

Wetted Parts

Compatibility of a particular process fluid with a given material of construction is best determined by the user. Depending upon the design of the vibrating tube, stress corrosion can be a difficulty in some applications. The materials of construction should be selected such that they are not affected by pitting agents present in the fluid stream. The materials of construction should be "noble" to the fluid, and no allowance for corrosion should be made.

Although the standard material of construction for most manufacturers is 316L stainless steel, sensors are manufactured of other materials such as Hastelloy C®, Monel®, titanium, tantalum, and so on.

Pressure Drop for Liquids

The pressure drop curves available from the manufacturer are typically based on Darcy's formula, modified to account for sensor geometry. In the laminar flow regime, pressure drop is proportional to the flow velocity and the viscosity. In the turbulent flow regime, pressure drop is proportional to the square of the velocity and is approximately ¼ power of the viscosity (see Figure 10-8).

Laminar and turbulent flows in pipes are described by dimensionless value called the Reynolds number. Typically, a pipe Reynolds number of 2,000 or less denotes laminar flow, and a Reynolds number of 4,000 or greater denotes turbulent flow. Between these two approximate values lies the region of transition that is not mathematically defined. Typically, a Reynolds number of approximately 2,000 is used for the break point between laminar and turbulent regimes.

When utilizing a manufacturer's graph (such as in Figure 10-9) for sizing, the plotted pressure should be divided by the specific gravity of the process fluid relative to water, as these charts typically represent mass flow units for water.

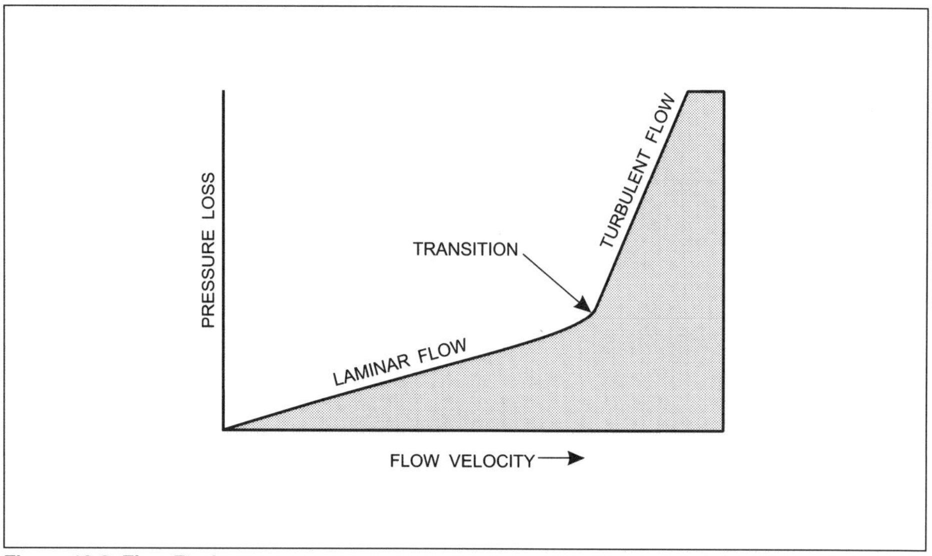

Figure 10-8. Flow Regimes

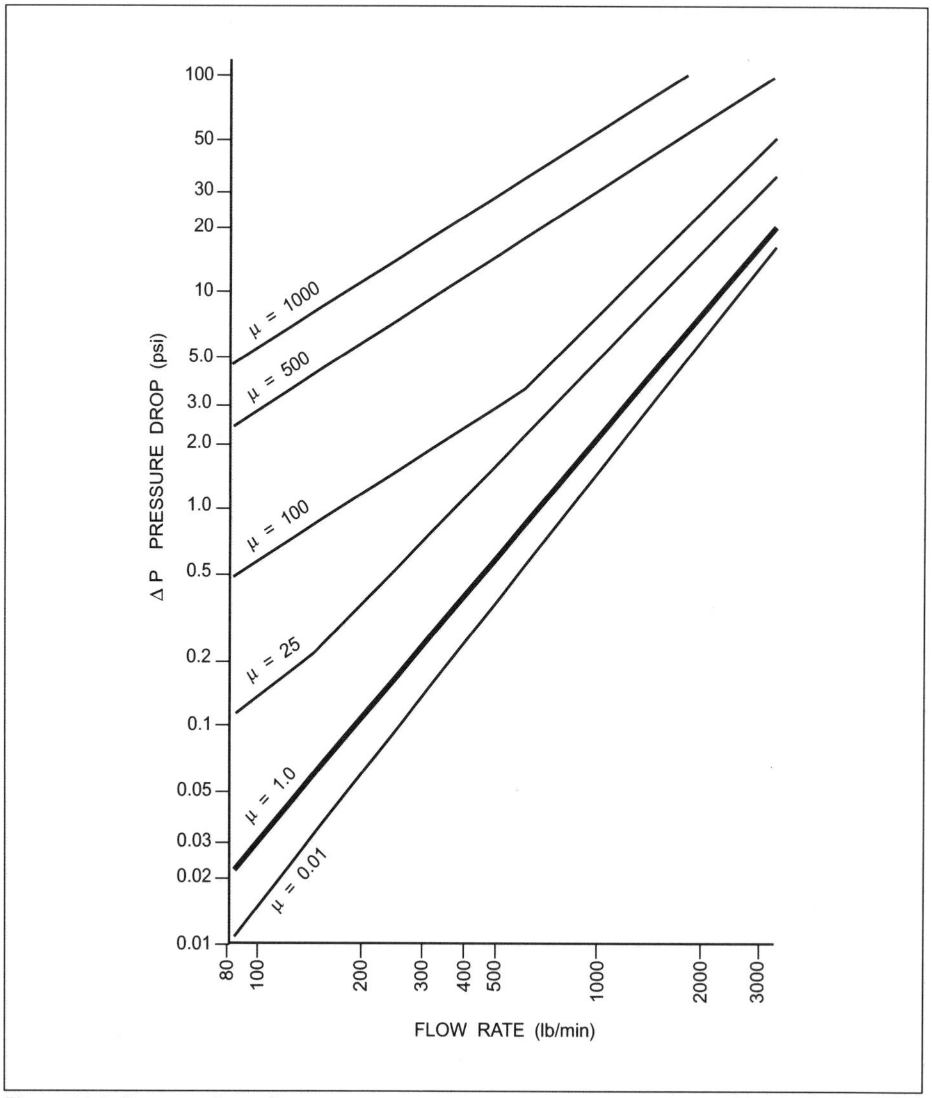

Figure 10-9. Pressure Drop Curves

For higher viscosity applications, contact the manufacturer.

Assuming a linear relationship of pressure drop versus viscosity can result in underestimation of pressure drops for high viscosity fluids in the laminar flow regime.

With regard to pressure drop, the selection of a meter for liquid applications requires the following information:

(1) Full scale flow rate

(2) Available pressure drop

(3) Process fluid viscosity at operating conditions

(4) Process fluid density

Pressure Drop For Gases

Due to the potential for a relatively low density condition in the gas phase, accurate measurement may not be possible at line pressures below 5 bar (72.5 psig). Since gas viscosity is typically less than 0.02 centipoise, turbulent flow is virtually assured. The manufacturers' charts for water flows can be used to estimate the pressure drop for the gas application by taking the ratio of gas density at operating conditions to the density of water at standard conditions. This ratio is then divided into the pressure drop for water to achieve the pressure drop value that would be seen in a gas application.

The gas pressure drop is also dependent on the total line pressure. If the calculated pressure drop is more than 10% of the absolute line pressure, the density of the gas will change significantly through the meter and will cause the actual pressure drop to be higher than that calculated above. Calculate the correction factor:

$$F_{correction} = \frac{\Delta P_{calc}}{P} \tag{10-15}$$

If $F_{correction}$ is more than 0.4, choose a larger meter. The actual pressure drop under these conditions can be determined by the following equation:

$$\Delta P = \Delta P_{calc}(1 + 2F_{correction}) \tag{10-16}$$

These calculations *do not* compensate for pressure loss due to piping reduction and expansion that may be required for the meter. This factor should also be calculated to determine the flowmeter system pressure drop.

Velocity Limits

Mass flowmeters are utilized on slurry and dirty liquid applications. The meters should be sized for these applications to limit the velocity to less than 10 feet per second in the flow element. This will minimize erosive wear on the wetted parts by the abrasive fluids.

Accuracy

Coriolis flowmeter accuracy is typically stated as a percentage of rate plus a zero shift error. The zero shift portion of the accuracy statement is a percent of meter capacity, so the accuracy will be more impacted by the zero shift low in the range of a meter than by the percent of rate factor. Figure 10-7 shows this relationship for a typical meter with a stated accuracy of ± 0.2% of rate and a zero shift error of 0.01% of meter capacity. The degradation of the accuracy can be observed low in the range of the meter.

Safety

Hazardous Area Location

Manufacturers of Coriolis mass flowmeters have typically pursued an intrinsically safe system design with the flow element suitable for hazardous area location, while designing the transmitter portion of the system suitable for either general purpose or a Division 2 location. If installation of the entire system in a Division 1 hazardous location is required, provisions must be made for the transmitter and control portion of the system, such as installation in a suitable housing or purging in a manner suitable for the hazard.

Materials

Flow element integrity is an important safety-related issue.

For a vibrating structure such as a Coriolis mass flow sensor, fatigue strength is an important design factor. Fatigue strength refers to the ability of a material to resist failure due to cyclic loading. The endurance characteristics for both 316L stainless steel and Hastelloy C-22 alloy are shown in Figure 10-10. These materials tend to reach what is known as an endurance limit; if the stress applied to the structure in question is below that endurance limit, the structure will not fail due to the vibration.

The flexural stresses seen in mass flowmeters are dependent upon the amplitude of displacement in vibration as well as the geometry of the flow sensor. Most, but not all, manufacturers control the displacement of the sensor tube so that excess stress cannot occur as a result of a component failure in the transmitter or during start-up or shutdown conditions. For the situation where the flow sensor operating stresses do not exceed the fatigue strength curves, fatigue cracking is not a concern. Materials such as stainless steel and high nickel content alloys are typically utilized, because they exhibit endurance limit characteristics shown in the graph. Mechanical conditions in manufacturing that do not result in a smooth surface finish, small grain size, high material toughness, and elimination of stress risers in the tubing material can adversely affect the fatigue charac-

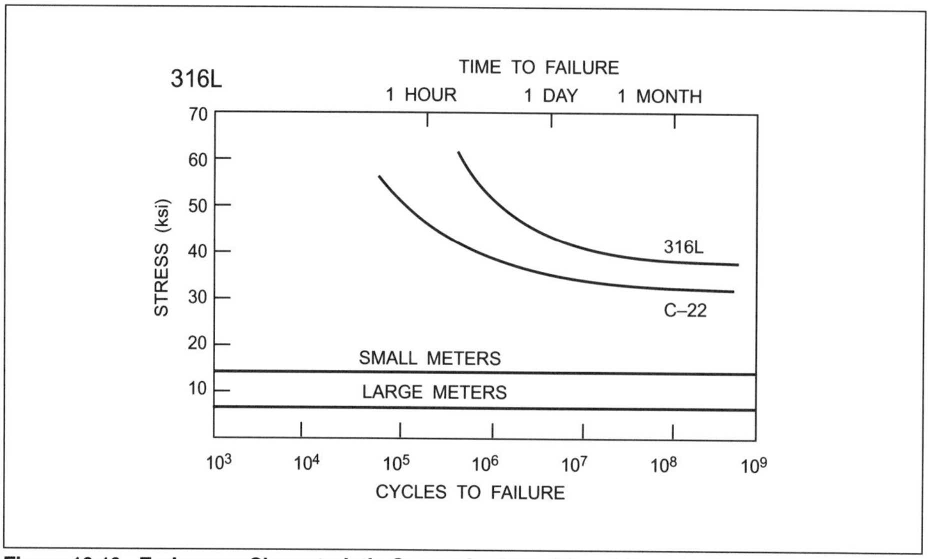

Figure 10-10. Endurance Characteristic Curves for 316L SS and Hasteloy C-22

teristics of a sensor and can result in failure of tubing. It is also interesting to note that experimentation has confirmed that, in the relatively high frequency vibration region seen in mass flowmeters, the endurance limit is independent of wall thickness. This phenomenon is displayed in Figure 10-11, which shows three endurance curves for various wall thicknesses of tubing.

Process conditions can adversely affect the fatigue performance of various materials. While stainless steel is known to have excellent corrosion resistance, the pitting corrosion that can occur in the presence of halogens such as chlorides can result in corrosion-induced stress risers and reduced meter life. For these reasons, it is suggested that the user select a more highly corrosion-resistant alloy for halogen service, such as the Hastelloy alloys, titanium, and tantalum. Table 10-2 gives some examples of some applications and the appropriate metallurgy.

Corrosion guides do <u>not</u> consider corrosion fatigue, but rather weight loss data; therefore, careful evaluation of the metallurgy should be conducted when considering a meter application.

In applications where the integrity of the flow tube cannot be assured, a number of manufacturers have provided containment around the flow element that is suitable for the application pressure rating. This "secondary enclosure" should be provided with fittings for draining any process fluid from the case should it become necessary prior to returning the flowmeter to the manufacturer, as well as for placing attachments such as pressure sensors or other sensing devices that would trip alarms should the enclosure become contaminated with process fluid.

The standard cases on many manufacturers' devices can be provided with rupture discs for venting should the flow tube fail in service.

The standard enclosure can also be provided with purge fittings should they be suitable for detection and venting of process fluids in case of failure.

Calibration

Since Coriolis mass flowmeters are designed to be unaffected by fluid parameters such as density, viscosity, pressure, and temperature, calibration of a meter on one type of fluid is directly transferable to another type. For a dry calibration, the calibration data supplied by the manufacturer must be set and the flowmeter must be zeroed with process fluid at process temperatures under no-flow conditions. For a wet calibration, the most convenient and accurate way to calibrate is using gravimetric proving techniques. Flow calibration is usually performed using water as the process fluid and weigh scales for batching.

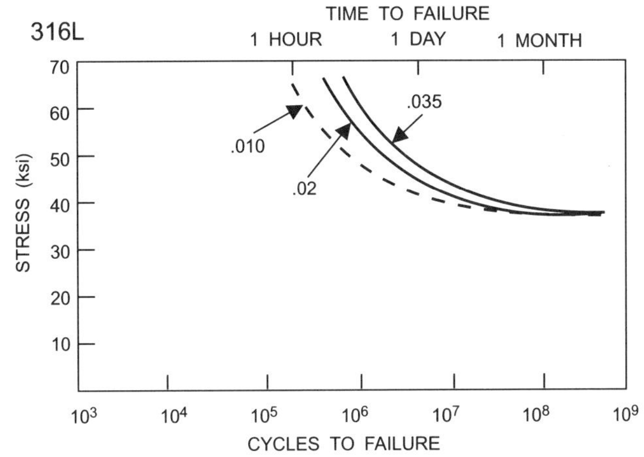

Figure 10-11. Endurance Characteristic Curves for Various Wall Thicknesses of Tubing

Table 10-2. Applications and the Appropriate Metallurgy

Stainless steels are acceptable with:	
Tar sands	Nitric acid
Paint	Isopropanol
Magnetic tape slurries	Lime slurries
Adhesives	Sour crude (H_2S and H_2O)
Polymers	CNG
Urethane	Liquified gases N_2, O_2, LPG
Olefins	Milk cream
Sodium hydroxide	Peanut butter
Potassium hydroxide	Pie fillings
Fuel oils	Fruit juices
Ethylene oxide	Fatty acids
Polypropylene	Beer, beer foam
Asphalt	
Molten sulfur	
Stainless steels are not acceptable with the materials listed below; Hastelloy C-22 should be used with:	
Acetic acid (high concentrations and elevated temperatures)	
Ammonium chloride	
Bromine	
Calcium chloride	
Iodine (other than 100% dry)	
Mustard	
Seawater	
Dyes and inks (if Cl bearing)	

Once the calibration data supplied by the manufacturer has been set and the meter has been zeroed on the test fluid at its operating temperature, a batch run at the desired flow rate can be initiated. The value of totalized flow from the meter can be compared to the mass delivered to the weigh scale, and the calibration can be verified. Verification batches can be run at other flow rates and an accuracy plot generated as shown in Figure 10-7.

Density calibration is performed by filling the sensor with a minimum of two accurately known density calibration fluids and characterizing the natural frequency of the vibrating tube on those fluids. By solving two equations for two unknowns, the transmitter can calculate the slope and offset factors for that particular sensor, from which the density can be calculated. One common technique is to characterize the sensor frequency with no liquid (air) in the flow tube and again with pure water in the flow tube.

Installation

General

Some guidelines need to be followed during installation to ensure optimal performance and ease of use. For liquid applications, it is best to install the meter such that it remains completely full of liquid at all times during the measurement cycle. Mass flowmeters will not provide accurate measurement when the flowmeter is filled only partially with liquid. For liquid applications, this usually means installing the flow element in an orientation that will not allow the trapping of gas bubbles in the measurement portion of the flow element.

In liquid applications, the flow element should not be the highest point in the system, in a location where the liquid can be siphoned out, or in an orientation that allows gas to become trapped in the flow element.

For most liquid flowmeter geometries, it is preferred that the flow element be mounted in a vertical pipe with flow upward because the sensor tends to remain full in this orientation. This orientation is most common for two-phase applications, allowing gas to rise through the flowmeter and allowing the solids to remain easily suspended and not settle out in the flowmeter element. Note that some flowmeter geometries are best installed in other orientations.

In gas or vapor applications, the orientation of the flow element should be such that condensate or other liquids are not trapped in the flow element. This may require the installation of a trap upstream and (or) downstream of the flow element.

Pipe Supports

The piping should support the flowmeter, not vice versa.

Normal vibrations in the piping do not usually interfere with performance of mass flowmeters. Good piping practice dictates the installation of adequate pipe supports on both sides of the flow element. These pipe supports should not be attached to the flow element, but rather to the piping on either side. This will allow the piping to support the flowmeter rather than the flowmeter supporting the process piping. These pipe supports should be mounted to the same reference plane (if possible) so that the shifting of the supporting element does not put stress on the piping system that would be absorbed by the flow element.

In some installations, it may be desirable to install a bypass loop to remove the meter from service while the process is operating. The valves should be installed in locations ensuring that the weight of the valving and bypass loop are not supported by the flow element.

Alignment of the process piping with respect to the meter is crucial to avoid undue stress on the flow element and the potential for leaking process fluid at this connection. On flange fittings, the manufacturer-recommended torquing specifications should be observed. If the meter will operate at more than 25°F above or below ambient temperature, the fit of the couplings or bolts should be checked after operating at temperature to avoid leaking at the process connections. Some manufacturers provide sensors with a corrosion-resistant lining, and, in this case, the manufacturer's torque recommendations on flange bolts should be *carefully* observed. The fittings may not "feel" tight, but the metal face of the flange can cut through the lining if excess pressure is used.

Some manufacturers supply some sensors that are not capable of being supported solely by the process piping. Recommendations for installation and mechanical support of these sensors should be followed closely. These sensors typically address applications with flow rates below 5 pounds per minute.

Downstream Valve

The downstream shutoff valve is recommended to ensure the ability to obtain zero flow conditions at pressure when making the initial zero adjustment. If the flowmetering system is to be used in a batching operation, both the flowmeter and downstream valve should be as close as possible to the receiving vessel to minimize errors.

Wiring

Most manufacturers supply the flowmetering system with the transmitter portion mounted remotely. This allows the installation of the flow element without having the transmitter electronics exposed to process temperature conditions. The manufacturer's recommendations regarding grounding and shielding of the flow element wires should be followed *very* carefully.

Upstream and Downstream Piping Requirements

Most manufacturers state that there are no up or downstream piping requirements such as exist with the most other flowmeter technologies. One manufacturer, however, requires 30 diameters straight run both up and downstream of the meter.

Gas Flow Applications

Due to the low density of gases, the flowmeter size will almost invariably be 1 to 4 pipe sizes less than the process piping. To minimize the pressure drops at the flowmeter inlet and discharge, multiple concentric reducers should be used to reduce and expand the piping in as smooth and gradual a manner as possible.

Special Application Considerations

Temperature extremes should be addressed using specific installation recommendations by the manufacturer. Since most of the meters using Coriolis mass flow technology install the sensor in a sealed enclosure, some steps must be taken to equalize the pressure differential with external barometric pressure if temperature extremes are encountered. In high temperature applications, this involves relieving the pressure due to expansion within the case while in low temperature applications, such as cryogenics, this involves a purge to resupply pressure due to contraction of the gas within the case. It should also be noted that in cryogenic applications, the supply purge should be nitrogen, rather than air. This will avoid the condensation and freezing of water vapor that will occur at cryogenic temperatures.

Flashing can occur in liquid applications and may require that a larger flowmeter be installed to reduce the pressure drop across the flowmeter. This problem usually manifests itself as an unstable measurement, with increasing instability at higher flow rates.

It is possible for meters mounted in series or to the same mounting structure to exhibit "cross-talk" symptoms (vibrations from one flowmeter affecting another flowmeter). This symptom manifests itself as an inability to zero the meter with no flow. For this reason, most manufacturers suggest at least 20 to 30 pipe diameters between meters mounted in series and the installation of isolation brackets around the pipe supports if the meters are to be supported on the same structure. Mechanical isolation of flowmeters mounted in series can often be achieved by locating an expansion joint or suitable flexible hose between individual supports downstream of the first flowmeter and upstream of the second flowmeter.

Maintenance

One of the major benefits of Coriolis mass flowmeters is the lower maintenance requirement. The benefits of no moving parts and low maintenance have, in many cases, been a primary reason for justification of the higher capital expenditure for this technology.

Transmitter

A complete functional check of the transmitter portion of the system should be performed on a semi-annual basis. This can typically be accomplished by utilizing various test points available and performing the tests recommended by the manufacturer. A number of simulator devices are also available for performing a functional checkout of the transmitter. These tests are not designed to replace a flow calibration check that checks the transmitter *and* the flow element.

If a calibration verification must be performed due to contractual or other requirements, a wet calibration check is typically required. If not, a functional verification of the transmitter portion can be performed. In this case, the entire flow system is not "calibrated"—only the electronic portion.

Flow element repair cannot usually be performed in the field and requires replacement of the flow element should some type of failure occur. In applications where the process fluid is suspected of being potentially corrosive or incompatible with the flow element metallurgy, precautions should be taken as noted in the installation guidelines previously stated. A periodic inspection of the flow element and adjacent piping should be conducted. If pitting is occurring in the flow element, steps should be taken to replace that element immediately with a metallurgy that would be more compatible with the process fluid. Pitting on the flow tube can result in premature failure of the flow element and shortened meter life expectancy.

Desired calibration accuracy and end use determine the type of calibration performed.

System Calibration

While the option is always available to return the meter to the manufacturer or to an outside flow laboratory for verification of calibration, many users with a significant installed base of flow devices have constructed their own flow calibration capabilities within their facilities. Among the methods utilized by experienced users of this technology are the following:

(1) Master meter system. In this scheme, a meter is selected for use as a master meter. This meter is submitted to an outside proving or calibration facility for extensive, documented calibration testing. This master meter is then placed in series with the test meter, and a series of tests are performed that require totalizing batches to be run through the test meter at various flow rates. Utilizing this method, a large number of test runs can be completed in a relatively short period of time. If meters of a number of different sizes and flow ranges are used in a facility, the use of several sizes of master meters may be required.

When the master meter is installed in the process piping, this method has the additional benefit of using the process fluid as a calibration medium, but extra precautions need to be taken that the process fluid does not contaminate the master meter system. In the case of hazardous or toxic materials, most users have elected to forego meter proving on the process fluid and have selected water as a calibration medium in an off-line flow calibration facility.

(2) Batch calibration. A portable or stationary tank with weighing capability is required for this method of calibration. Water is usually used to batch liquid into the container on the scale. The installation guidelines mentioned above for batching systems should be carefully observed to avoid calibra-

tion errors. Calibration errors can also result from use of lengths of flexible hose or runs of piping downstream of the batch shutoff valve. The shutoff valve should be a close as possible to the tank being filled. The main advantage of this system is that it can be a direct mass calibration of the metering system. A system of this type may be utilized in manufacturing facilities to calibrate both mass and volumetric meters.

A disadvantage of this type of calibration system is that it can be very large and expensive to design, fabricate, and maintain. Also, calibration can be time-consuming due to the batch and emptying cycles that must occur for each flow rate to be checked.

Since the calibration is only as accurate as the standard to which the meter is being compared, care should be taken that the weighing system is calibrated using NIST traceable standard weights prior to, during, and after the calibration runs are performed. Extensive testing of the calibration stand should be performed periodically to be certain the calibration standard is within specification. The system can provide a highly accurate calibration if the calibration standard is properly calibrated and serviced. Many user installations already have a system of this type in their facility that could be utilized for this purpose in the truck scales used typically for custody transfer of material. Since the resolution of the truck scales are typically only ± 20 pounds, the user must be certain the container has a large enough capacity to result in a valid calibration.

(3) Volumetric standards. Much more effort has been made recently in using either large volume or ballistic provers to conduct mass meter calibrations on a volumetric standard basis. This method has been shown to be valid and quite efficient in conducting provings, particularly in pipeline applications. If the calibration medium is sufficiently well defined, such as most refined petroleum products or water, this calibration can be accurately conducted on a mass basis. It is also possible to measure the density on line and compensate the volume indicated by the prover for this density to get a mass calibration.

The main advantages of utilizing this system are that the techniques are well understood in the pipeline industry, and procedures are well established for doing volumetric proving of the large, installed base of volumetric flowmeters.

(4) Density calibration. A density calibration can be conducted using company standard or API procedures. As in all vibrating tube density measurement systems, a buildup of material or coating on the walls of the vibrating element can adversely impact the calibration accuracy of the system. In applications where this buildup is known to occur, caution should be taken to clean the flow element or conduct frequent calibrations to remove the offsets incurred as a result of this coating buildup.

Applications

The Manufacture of Polypropylene

A number of manufacturers of polypropylene have benefitted from using mass flowmeters to achieve a more accurate material balance and higher product quality control.

Catalyst feed rate is a very critical process variable in achieving high product quality in a polypropylene plant (see Figure 10-12).

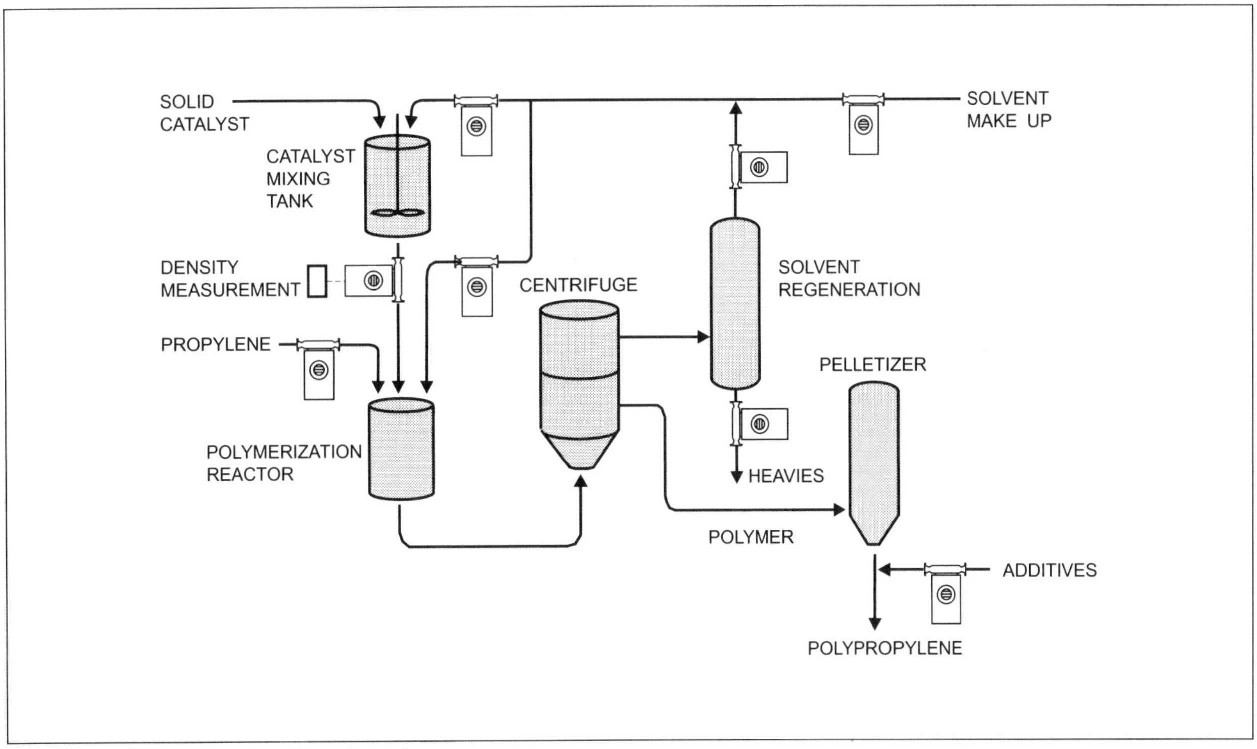

Figure 10-12. Polypropylene Manufacturing
(Courtesy of Micro Motion, Inc.)

Solid catalyst is mixed with solvent and injected into the polymerization reactor. Two critical parameters to be controlled are the catalyst slurry feed rate and the catalyst feed solids content. The metal-based solid catalyst can create maintenance difficulties with turbine or positive displacement flowmeters. The catalyst can also be damaged in some types of positive displacement meters. For these reasons, it is imperative that a flowmeter with no moving parts be used in the application. A magnetic flowmeter is, in many cases, not suitable because the solvent carrier is not conductive. The mass flowmeter is a logical choice, allowing the customer to accurately measure the mass flow rate as well as the solid content of the slurry. Critical process information on the feed rate and solids content can be monitored and controlled through the use of a single sensor.

Hot additives are injected into the polymer to meet the polypropylene manufacturer's customer requirements. This represents a very low flow in ratio to the total process output but is critical to the quality control of the end product. These additives are typically between 100° and 150° Celsius and are viscous materials that tend to coat the piping system. Positive displacement and turbine meters are not suitable for the application due to the interior coating effect as well as the tendency of the material to solidify if the high temperature is not maintained. Velocity-sensing meters, such as the differential pressure types, will not measure accurately when the tube walls become coated. The mass flowmeter will measure accurately even though the interior of the sensor tube becomes coated.

Mass flow rate data on the solvent makeup and solvent regeneration, the propylene feedstock, and the catalyst and additive feed rate are critical to successful plant operation. Accurate calculation of profit and loss on an individual process plant is critical in today's economic environment. Many companies are finding it more difficult to obtain the required accuracy through the use of volumetric flowmeters, even with density compensation. Difficulties in obtaining an accurate measurement through other means have led to greater acceptance of direct mass flow measurement in polypropylene manufacturing.

Fire Retardant Loading[1]

During a typical fire season, the Forest Service-USDA delivers over 11.7 million gallons of fire retardant onto ongoing wildfires; more than 6,000 airtanker missions are flown to deliver the retardant. Volume sight gages on the aircraft tanks have been used to measure the retardant payload being pumped onto an airtanker. These gages are notoriously inaccurate. Engineering estimates indicate these inaccuracies result in non-optimal retardant loadings that vary from 10 percent overload of rated capacity to 40 percent underloads. Increased safety and efficiency results from using an accurate, reliable measurement system to indicate the quantity of fire retardant being loaded into airtankers. Previous attempts, prior to the availability of the system described here, to identify and implement off-the-shelf metering systems met with less than fully successful use and acceptance at various airtanker bases.

The San Dimas Equipment Development Center (SDEDC) began investigating mass flowmeters in 1980, since they appeared to be applicable to the task of measuring retardant quantities being loaded at Forest Service airtanker bases. Field experience has proved the flowmeters to be rugged, dependable, and accurate. A conceptual layout of a retardant measurement system is shown in Figure 10-13.

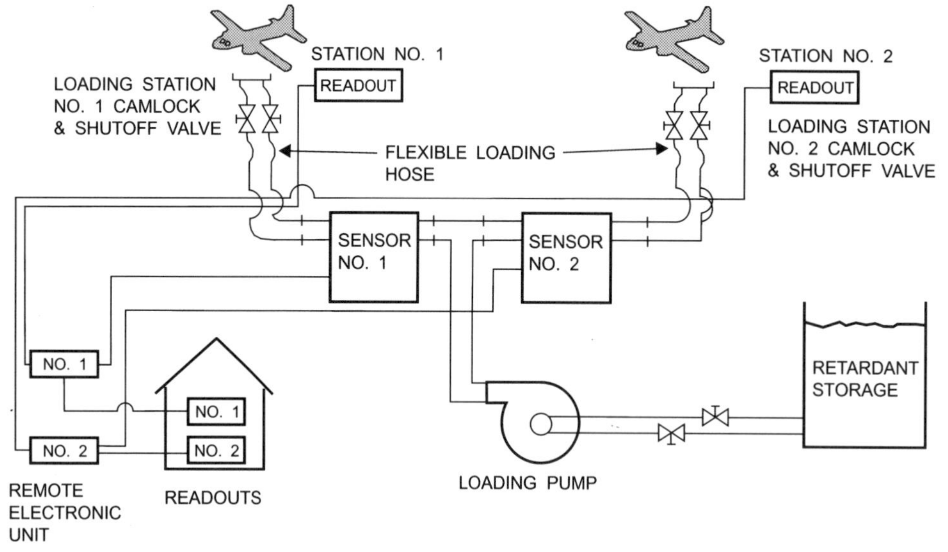

Figure 10-13. Retardant Measurement System Conceptual Layout

---

1 Courtesy of United States Forest Service Equipment Development Center, Special Report 8651-1801.

The use of mass flowmeters at airtanker bases provides not only information that prevents overloading of airtankers but also significant retardant cost savings. Sight gages and turbine meters regularly overstate pumped weights by as much as 10 percent and 5 percent, respectively.

During the operational evaluation, turbine meters utilized for payment were also monitored for error. Errors varied between 1 percent and 8 percent, with a prevailing error of 5 percent.

Airtanker retardant tank gages are very closely calibrated for volume. These do not, however, give accurate indication of volume loaded. Residual retardant is always present in tankers after a drop. The amount depends on retardant type, tank type, and airtanker drop conditions. Loading conditions also vary the amount of retardant loaded. The tank plumbing and sight gage arrangement may be such that the site gage is mounted on the first tank to fill. Before the load has had a chance to level itself, pumps are shut off and the tanks are declared full. Air entrained in retardant tends to rise from the retardant with time. Estimates of entrained air vary from 2 to 10 percent of tank volume. This depends on tanker plumbing, retardant type, and retardant pumping system.

A benefit/cost analysis (a comparison of benefits and the cost of use) of a mass flowmeter, a turbine meter, and sight gages to measure payloads follows. Factors considered and assumptions made in the analysis are shown in Table 10-3. The meters were assumed to have a useful life of 5 years. The annual owning and operating costs were derived by adding annual maintenance cost to ⅕ of the purchase, ancillary equipment, and installation costs. Zero inflation and zero cost of money was assumed. Actual five-year average, base retardant use levels and retardant costs in 1983 were used to calculate projected savings by implementing a turbine meter or a mass flowmeter. Savings and benefit/cost ratios were calculated from Equations (10-17) (mass flowmeter implementation expected annual cost savings) and (10-18) (expected benefit/cost ratios):

Table 10-3. Applicability of Meter Types as Retardant Measurement System

Meter Type	Makes direct measurement of mass flow	Error less than 2% over flow range 800-7,000 lb/min or 3,000-5,000 lb/min	Accuracy significantly affected by retardant type change in viscosity, density, or temperature	Does protrude or significantly block flow stream
Differential pressure transducers	No	No	Yes	Yes
Magnetic	No	No	Yes	Yes
Mass	Yes	Yes	No	No
Positive displacement	No	No	?	Yes
Turbine	No	No	Yes	Yes
Ultrasonic	No	No	Yes	No
Vortex	No	No	Yes	No

Mass flowmeter implementation expected annual cost savings:

$$\text{\$ savings} = \text{(Average annual volume, gal)} \times \text{\%(volume overstatement with current measurement system)} \times \text{(retardant cost \$/gal)]} - \text{[(annual owning and operating cost of new measurement system, \$)]} - \text{[(average annual volume (gal))} \times \text{(\% of overstatement with new measurement system)]} \times \text{(retardant cost \$/gal)]} + \text{annual owning and operating cost of present measurement system \$} \qquad (10\text{-}17)$$

$$\frac{\text{Benefit}}{\text{Costs}} = \frac{\text{Annual cost savings}}{\text{New system cost + annual owning / operating cost} - \text{annual operating cost of old system}} \qquad (10\text{-}18)$$

Applying the equations to data from Table 10-3 yielded Table 10-4. This table shows that implementation of a mass flowmeter results in significant dollar

Table 10-4. Flowmeter Cost/Benefit Analysis

Factor	No. Flowmeter	Option	
		Turbine Meter	Mass Flowmeter (3-in. size)
Purchase cost	—	$1,000	$6,000
Ancillary equipment costs[1]	—	$300[2]	$1,800[3]
Lifetime	—	5 years	5 years
Installation costs	—		
Exterior plumbing (non-recurring capital cost)	—	$300	$300
Manpower[5] (annual cost)	—	$50	$100
Maintenance[4]	—	½ man-day/4 wk $200	½ man-day/season $50
Calibration[4] (per season at ½ man-day each)		4 $200	2 $100
Annual owning and operating costs	—	$770	$1,870
Percent error expected overstatement of volume delivered	10	5	1
Safety	Dependent entirely on sight gages to prevent overloads.	Volume output, w/indication of mass loaded, better than sight gages. No warning if product is denser than specified.	Provides mass flow rate and total. Rate display helps indicate changes in density.

[1] Calibration system cost not included.
[2] Readout
[3] Factory provided D10 RT, and SDEDC LCD readouts, cables, mount.
[4] Assuming 120-day season.
[5] Manpower to set up and calibrate meter costs $100/day.

savings. For example, at Coolidge, AZ, Airtanker Base, approximately 506,000 gallons of retardant are pumped annually. The 10 percent overstatement of volume normally experienced with sight gages results in 50,600 gallons of retardant paid for ($28,800) but not delivered annually. Utilizing the mass flowmeter, with an error of less than one percent, will result in a savings of $24,000 per year (assuming 1.0 percent error—a worst case estimate).

Table 10-5 shows that using a turbine meter is more cost-beneficial than using sight gages and that using a mass flowmeter maximized dollar savings.

A base with average annual retardant costs of greater than $112,200 and utilizing only sight gages for the determination of volume delivered could save enough money in one season, with the mass flowmeter, to pay for the entire meter cost and installation.

Utilizing the mass flowmeter will not only help prevent overloads, it will help identify malfunctioning tank doors. If a sight gage indicates a full load and the meter shows less than a full load has been delivered this suggests that a door is stuck closed or residual retardant is in the tanks. Preventing overloaded takeoffs could prevent potentially disastrous incidents.

Table 10-5. Predicted Cost Savings and Benefit/Cost Ratios

Base, retardant	Retardant use level 5-year average[1] (gal)	Retardant cost 1983 ($/gal)	Annual volume overstatement/ $ cost to FS with no measurement system (10% overstatement)	Utilizing turbine meters in place of sight gages; $ savings/ benefit-cost ratio	Utilizing mass flowmeters in place of turbine meters; $ savings/ benefit-cost ratio	Utilizing mass flowmeters in place of sight gages; $ savings/ benefit-cost ratio
Albuquerque, NM FT-100	112,932[2]	0.57	11,293 / $6,437	$ 2,449 / 3.18	$ 1,475 / 1.34	$ 3,923 / 2.10
Coolidge, AZ FT-100	505,654	0.57	50,565 / 28,822	13,641 / 17.72	10,429 / 9.48	24,070 / 12.87
Fox Field, CA Phos-Chek AF	621,193	0.787	62,119 / 48,888	23,674 / 30.75	18,455 / 16.78	42,129 / 22.53
Hemet Ryan, CA Phos-Chek[4] AF	1,408,597	0.783	140,859 / 110,293	53,606 / 34.813	41,917 / 19.053	95,524 / 25.54[3]
Klamath Falls, AZ M2700	484,081	0.500	48,408 / 24,204	11,395 / 14.80	8,582 / 7.80	21,784 / 11.65
Libby Field, AZ FT-100[5]	98,211	0.721	9,821 / 7,081	2,770 / 3.60	1,732 / 1.58	4,503 / 2.41
Medford, OR FT-93L	484,081	0.470	48,408 / 22,751	10,605 / 13.77	8,001 / 7.27	18,606 / 9.95
Prescott, AZ GTS-R	211,107	0.721	21,111 / 15,220	6,840 / 8.83	4,988 / 4.54	11,830 / 6.33
Redmond, OR FT-93L	464,601	0.470	46,460 / 21,836	10,148 / 13.18	7,634 / 6.94	17,782 / 9.51

[1] For 1977-1981.
[2] For 1979-1981.
[3] Assumed that two new measurement systems were put in.
[4] Phos-Chek registered trademark of Monsanto Co., St. Louis, MO.
[5] Fire-Trol registered trademark of Chemonics Industries, Phoenix, AZ.

Compressed Natural Gas and Flowmetering[2]

The Coriolis mass flowmeter is used almost exclusively by compressed natural gas (CNG) dispensers manufacturers throughout the world.

The metering of CNG is important:

- to demonstrate the cost savings of CNG;
- for testing of new and advanced parts of CNG conversion kits;
- for documentation when paying state taxes on CNG sold;
- for billing and accounting purposes when CNG is sold to another party;
- to help monitor maintenance requirements before major maintenance is required; and
- to automate dispenser systems and for control of cascade sequencing.

The mass flowmeter is a reliable, convenient way of metering CNG. Flowmeters that rely on inferred measurement techniques depend on the accuracy of several critical variables such as volume, temperature, and pressure. Gas volume changes dramatically with changes in pressure and temperature. Pressures range up to 3,000 psi when a CNG-powered vehicle's tanks are full. Gas temperature changes, depending on the rate of the fill. It can be difficult to decide where a CNG temperature measurement should be made. Temperature measurement can be taken inside the storage tank underground, at the nozzles, or inside the automobile's tank. Expert opinion varies about where the best measurement points are located. In addition, temperature and pressure measurements should be made simultaneously with the volumetric measurement.

CNG is chiefly comprised of methane, it has considerable compressibility, and is a non-ideal gas. The Non-Ideal Gas Law states that:

$$PV = nRTZ$$

where:

P = pressure
V = volume
n = mass/molecular weight (i.e., number of moles)
Z = compressibility factor
R = gas constant
T = temperature

As can be seen in Figure 10-14, CNG at 50°F, 2,000 psi, and a specific gravity of 0.7 would have a compressibility factor of about 0.6. If this compressibility factor were not taken into account, the volumetric measurement, when combined with pressure and temperature compensation to obtain the inferred mass, would require a correction of approximately 40%. These types of calculations are cumbersome, and their magnitudes lead to inaccuracies.

Hundreds of mass flowmeters are installed on CNG dispensers in New Zealand, Canada, and the United States. The New Zealand and Canadian governments require that CNG be sold by weight, and the New Zealand Department of Weights and Measures uses portable mass flowmeters to check the flowmeters in CNG dispensers.

In the U.S., most CNG-related sales are made to fleet owners and utility companies. Typically, a utility company sells CNG to fleet owners of either large

---

2 Courtesy: Micro Motion, Inc.

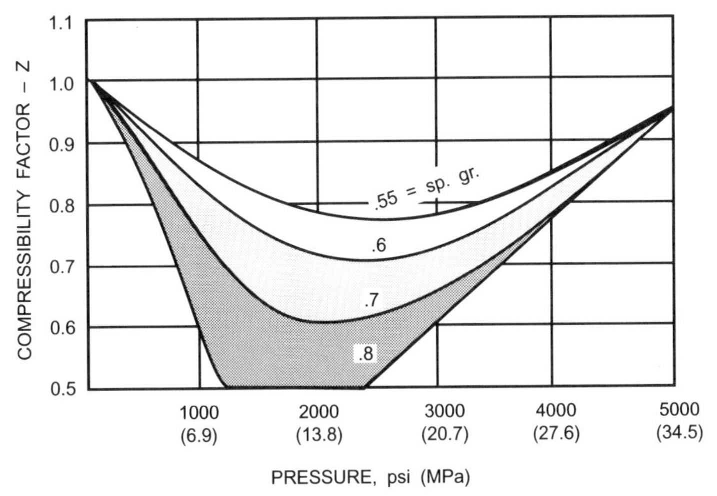

Figure 10-14. Natural Gas Supercompressibility at 50°F (10°C)
(Courtesy of Micro Motion, Inc.)

vehicles or a large number of vehicles. Typical CNG customers include soft drink producers, school bus owners, taxi cab fleets, utility company vehicle fleets, police department vehicles, and the United States Postal Service.

For safe operation of mass flowmeters in CNG applications, it is important to pay close attention to the flowmeter installation procedures. In addition, safety guidelines for CNG application, such as those proposed by the American Gas Association, are recommended for use in the design and installation of CNG-dispensing systems. Other standards, guidelines, or codes may also be applicable.

The accuracy of the mass flowmeter, combined with the independence of temperature, pressure, and compressibility factors, makes this an ideal flowmeter for the measurement of CNG.

About the Author

Lee Smith is the Regional Sales Manager for the western region for Micro Motion, Inc., in Boulder, CO. He has held various sales positions as well as the Director of Marketing position. He has been with Micro Motion since 1981. Prior to joining Micro Motion, Lee worked with The Haltom Co. as an instrumentation manufacturer's representative in Houston, Texas.

James R. Ruesch is the Director of New Product Development at Micro Motion, Inc., Boulder CO, where he has held various engineering positions since 1982. Prior to joining Micro Motion, Jim developed numerous pressure and temperature measurement products for Chrysler Corp., Rosemount, Inc., and the Granville-Phillips Company. Jim received his BSEE from the University of Minnesota in 1971 and is credited with four patents in the fields of flow, density, and pressure measurement.

11

Open Channel Flow Measurement

The majority of industrial liquid flows are carried in closed conduits that flow completely full and under pressure. However, this is not always the case for high volume flows of liquids in industrial waste systems, waterworks, sanitary and storm sewers, and irrigation systems. The flows in these situations are commonly carried in open channels, which are characterized by low system heads and high volumetric flow rates. This chapter will discuss the techniques used for measuring flow in open channels.

Open Channel Flow

There are two basic types of flow systems: flow in closed conduits and flow in open channels. Closed conduit flow is flow in completely filled pipes. Pipes are usually used for fresh water lines or for industrial process lines, and flow through them is often measured by some type of device inserted into the line.

Open channel flow is flow in any channel in which the liquid flows with a free surface. Examples of this are rivers, irrigation ditches, canals, flumes, and other uncovered conduits. Certain closed channels, such as sewers and tunnels when flowing partially full and not under pressure, are also classified as open channels. Open channels are used in most storm and sanitary sewer systems, sewage treatment plants, many industrial waste applications, and some water treatment plants. Most irrigation water is also distributed in open channels.

Methods of Open Channel Flow Measurement

There are many methods of determining the rate of flow in open channels. Some of the more common include the timed gravimetric, dilution, velocity-area, hydraulic structures, and slope-hydraulic radius-area methods.

In the timed gravimetric method, the entire contents of the flow stream are collected in some type of container for a fixed length of time. The weight of the fluid is then determined, and the flow rate calculated. In the dilution method, the flow rate is measured by determining how much the flowing water dilutes an added tracer solution. Although brine tracers have been used, radioactive and fluorescent dye tracers are more commonly used today. In the velocity-area method, the flow rate is calculated by determining the mean flow velocity across a cross section and multiplying this by the flow area at that point. In open channels, this will generally require two separate measurements, one to determine the mean velocity and the other to determine the flow depth.

In the hydraulic structures method, some type of hydraulic structure is introduced into the flow stream. The function of the hydraulic structure is to produce a flow that is characterized by a known relationship between a liquid level measurement at some location and the flow rate of the stream. Thus, by knowing the liquid level at the hydraulic structure, the flow rate in the open channel may be determined. In the slope-hydraulic radius-area method, measurements of water surface slope, cross-sectional area, and wetted perimeter over a length of uniform section channel are used to determine the flow rate, using a resistance equation such as the Manning formula. The Manning formula requires a knowledge of the channel cross section, liquid depth, slope of the water surface, and a roughness factor dependent on the character of the channel.

The timed gravimetric and dilution methods are generally not used for routine industrial flow measurement because they are not suited to provide a continuous record of flow rate. They are more often used for occasional flow rate measurements at a particular time and place, for calibrating some other type of device, or for developing a liquid level-flow rate curve for a particular location.

The velocity-area method is often used like the timed gravimetric and dilution techniques to determine the flow rate in a stream at a particular time for calibration purposes. However, some manufacturers have designed instrument systems that combine depth measurement with a point-velocity sensor. These patented devices typically have been used in temporary situations such as sewer flow measurements for infiltration and inflow. Because of their complexity and expense, these devices are not widely used in industrial flow measurement.

The hydraulic structures method differs from the first three in that, provided a standard type of structure is used and certain installation and application rules are followed, no field calibrations or measurements other than a continuous measurement of liquid level are required to obtain a continuous record of flow rate. Because of this, the hydraulic structures method is widely used for industrial open channel flow measurement.

The slope-hydraulic radius-area method, using the Manning formula, is applied in a manner similar to hydraulic structures in that only a continuous measurement of liquid level is required to obtain a continuous record of flow rate. Because of uncertainties associated with the Manning formula, the accuracies obtainable are not as good as those achieved with hydraulic structures. The slope-hydraulic radius-area technique is not commonly used in industrial open channel flow measurement. It is used only where great accuracy is not required or where the installation of a hydraulic structure is impractical.

Primary Measuring Devices: Weirs and Flumes

The most commonly used method of measuring the rate of flow in industrial open channels is that of hydraulic structures. In this method, flow in an open channel is measured by inserting a hydraulic structure into the channel, which changes the level of liquid in or near the structure. By selecting the shape and dimensions of the hydraulic structure, the rate of flow through or over the restriction will be related to the liquid level in a known manner. Thus, the flow rate through the open channel can be derived from a single measurement of the liquid level in or near the structure.

The hydraulic structures used in measuring flow in open channels are known as primary measuring devices and may be divided into two broad categories — weirs and flumes, shown in Figure 11-1.

A weir is essentially a dam built across an open channel over which the liquid flows, usually through some type of an opening or notch. Weirs are normally classified according to the shape of the notch, the most common types being the triangular (or V-notch) weir, the rectangular weir, and the trapezoidal (or Cipolletti) weir. Each type of weir has an associated equation for determining the flow rate through the weir.

A flume is a specially shaped open channel flow section with an area or slope (or both) that is different from that of the channel. This results in an increased velocity and change in the level of the liquid flowing through the flume. A flume normally consists of three sections, a converging section, a throat section, and a

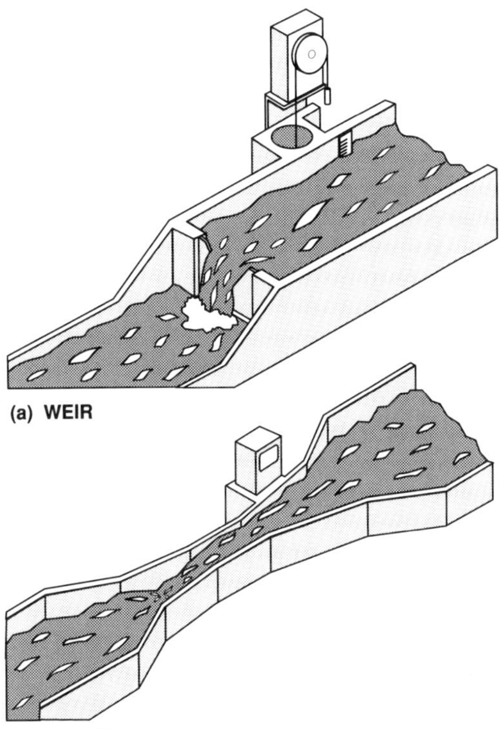

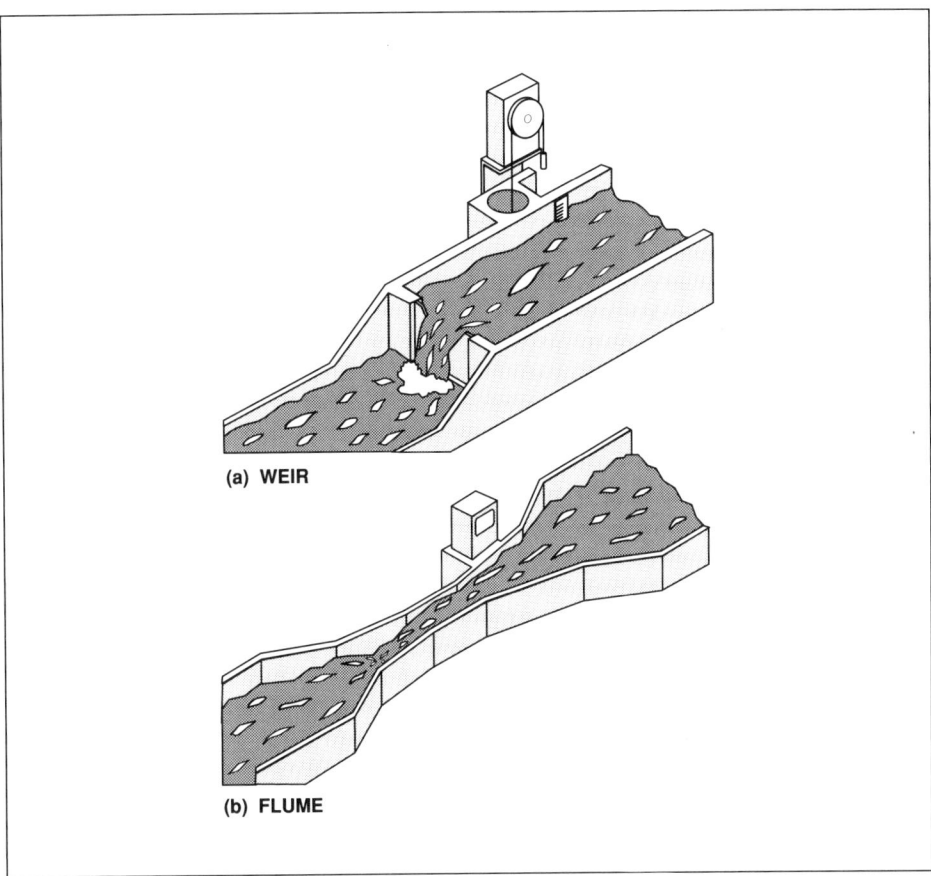

Figure 11-1. Primary Measuring Devices: Weirs and Flumes

diverging section. The flow rate through the flume is a function of the liquid level at some point or points in the flume. The most commonly used types of flumes are the Parshall flume, the Palmer-Bowlus flume, the HS, H, and HL flumes, and the trapezoidal flume, although many other types are available.

Secondary Measuring Devices: Open Channel Flowmeters

The flow rate or discharge through a weir or flume is usually a function of the liquid level in or near the primary measuring device. A secondary measuring device (or open channel flowmeter) is used in conjunction with a primary measuring device to measure the rate of liquid flow in an open channel. The secondary measuring device has two purposes: (1) to measure the liquid level in the primary measuring device, and (2) to convert this liquid level into an appropriate flow rate according to the known liquid level-flow rate relationship of the primary measuring device.

Thus, a combination of a weir or flume (primary measuring device) and an open channel flowmeter (secondary measuring device) is necessary to measure flow in an industrial open channel. The flow measurement system requires both primary and secondary measuring devices to be complete. A weir or a flume (primary device) restricts the flow in a controlled manner and generates a liquid level that is related to the flow rate through the device. An open channel flowmeter (secondary device) measures this level and converts it into a corresponding flow rate according to the known liquid level-flow rate relationship of the primary device.

The first task of an open channel flowmeter is to measure the liquid level at an appropriate point in or near the primary measuring device. This can be measured by a number of different methods, including float, electrical, ultrasonic, bubbler, and submerged probe.

The second task of an open channel flowmeter is to convert the measured liquid level into a corresponding flow rate according to the level-flow rate relationship for the primary measuring device being used. This conversion can be accomplished by a number of different methods, including mechanical cam, electronic analog function generator, electronic digital function generator, electronic memory device, and software.

Any one of the level measurement techniques may be combined with any one of the level-to-flow rate conversion techniques to result in a complete open channel flowmeter. An extensive discussion of the various types of open channel flowmeters is beyond the scope of this chapter.

When designing an open channel flow measuring system, the importance of both the primary and secondary measuring devices should be recognized. In a complete open channel flow measurement system, a proper weir or flume installation can be negated through the use of an inaccurate flowmeter (secondary). Similarly, a very accurate flowmeter (secondary) cannot overcome the inaccuracies of a poorly installed or maintained weir or flume.

As with any measurement system, an open channel flow measurement system can be no more accurate than its least accurate component.

In some instances, a complete open channel flowmeter is not used with a primary device. Instead, a device called a level transmitter is used. A level transmitter simply measures the level associated with the primary device and transmits it to a recording device. The recorded level may then be converted into flow rate, either manually or by a computing device remote from the weir or flume installation.

Weirs

Principles of Operation

Weirs are the simplest, least expensive, and probably most common type of primary measuring device used to measure flow in open channels. A weir is an obstruction or dam built across an open channel over which the liquid flows, often through a specially shaped opening or notch. Weirs are normally classified according to the shape of the notch. The most common types of weirs are the triangular (or V-notch) weir, the rectangular weir, and the trapezoidal (or Cipolletti) weir, shown in Figure 11-2. Each type of weir has an associated equation for determining the flow rate through the weir. The equation is based on the depth of the liquid in the pool formed upstream from the weir.

The edge or surface over which the liquid passes is called the crest of the weir, as shown in Figure 11-3. Generally, the top edge of the weir is thin or beveled with a sharp upstream corner so that the liquid does not contact any part of the weir structure downstream, but, rather, springs past it. Weirs of this type are called sharp-crested weirs.

The stream of water leaving the weir crest is called the nappe. When the water surface downstream from the weir is far enough below the weir crest so that air flows freely beneath the nappe, the nappe is aerated, and the flow is referred to as free or critical. When the downstream water level rises to the point where air does not flow freely beneath the nappe, the nappe is not ventilated, and the discharge rate may be inaccurate because of the low pressure beneath the nappe.

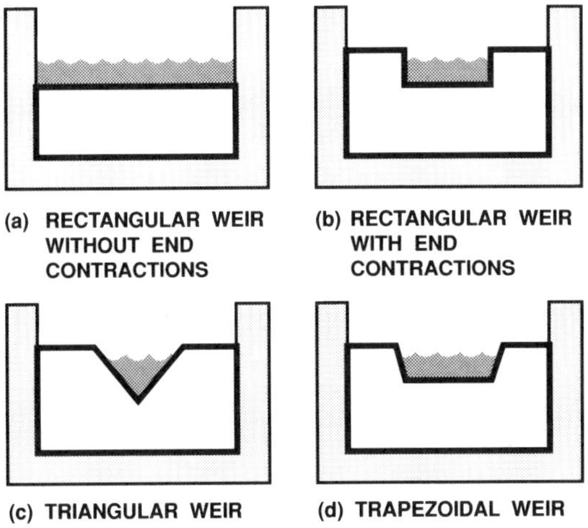

Figure 11-2. Various Sharp-Crested Weir Profiles

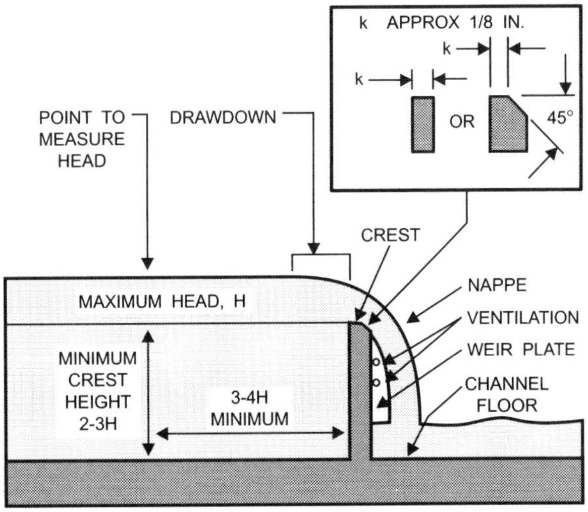

Figure 11-3. Sharp-Crested Weir

When the downstream water level rises above the crest, the flow is referred to as submerged or subcritical. This can affect the discharge rate to a measurable degree, so dependable measurements should not be expected in this range. Submerged, nonventilated flows are undesirable for standard conditions and should usually be avoided. In most cases, weirs should be installed and sized to obtain ventilated and free or critical discharge conditions.

The discharge rate of a weir is determined by measuring the vertical distance from the crest of the weir to the liquid surface in the pool upstream from the crest. This liquid depth is called the head. As shown in Figure 11-3, a slight drop in the liquid surface begins upstream from the weir. This drop occurs at a distance of at least twice the head on the crest and is called the surface contraction or drawdown of the weir. To avoid sensing the effects of drawdown, the head measuring point of the weir should be located upstream of the weir crest a distance of at least three and preferably four times the maximum head expected over the weir, as shown in Figure 11-3. Once the head is known, the flow rate or discharge can be determined using the known head-flow rate relationship of the weir.

TRIANGULAR OR V-NOTCH WEIR

The triangular or V-notch sharp-crested weir (Figure 11-4) consists of an angular notch cut into a bulkhead in the flow channel. The apex of the notch is at the bottom, and the sides are set equally on either side of a vertical line from the apex. The angle of the notch (α) most commonly used is 90°, although triangular weirs with angles of 120°, 60°, 45°, 30°, and 22-½° are also used. As the included angle becomes smaller, the difficulty of producing the geometry of the weir at the apex is increased. A further problem exists in the low head range of the narrow-angled notches, which causes the head/discharge relationship to be unreliable; as the angle narrows, the capillary effect will restrict flow at a surprisingly high head.

The discharge (head vs. flow rate) equation of a free-flowing triangular weir takes the form:

$$Q = KH^{2.5} \qquad (11\text{-}1)$$

where:

Q = flow rate
H = head on the weir
K = a constant, dependent upon angle of notch and units

For flow rate in cubic feet per second, million gallons per day, and gallons per minute, the constant K in the triangular weir equation may be calculated as follows:

For cfs — $$K = 2.50 \tan(\alpha/2) \qquad (11\text{-}2)$$

For mgd — $$K = 1.62 \tan(\alpha/2) \qquad (11\text{-}3)$$

For gpm — $$K = 1120 \tan(\alpha/2) \qquad (11\text{-}4)$$

where:

α = angle of the triangular opening, in degrees

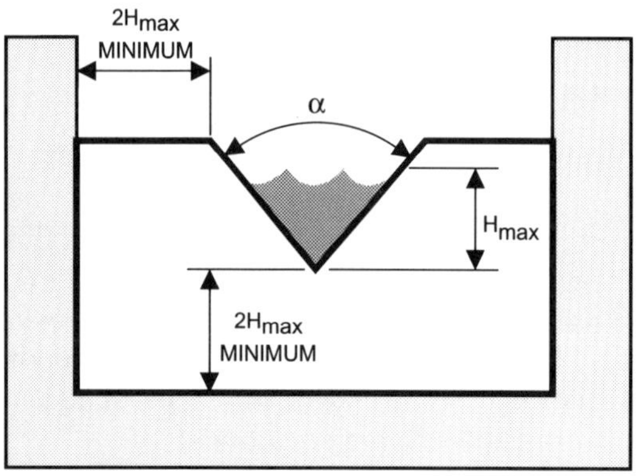

Figure 11-4. Triangular (V-Notch) Sharp Crest Weir

For flow rate in cubic feet per second, million gallons per day, and gallons per minute and head in feet, the discharge equations for the common triangular weirs are as follows:

	cfs	mgd	gpm
120°	$Q = 4.33H^{2.5}$	$Q = 2.80H^{2.5}$	$Q = 1940H^{2.5}$
90°	$Q = 2.50H^{2.5}$	$Q = 1.62H^{2.5}$	$Q = 1120H^{2.5}$
60°	$Q = 1.44H^{2.5}$	$Q = 0.933H^{2.5}$	$Q = 648H^{2.5}$
45°	$Q = 1.04H^{2.5}$	$Q = 0.699H^{2.5}$	$Q = 465H^{2.5}$
30°	$Q = 0.676H^{2.5}$	$Q = 0.437H^{2.5}$	$Q = 303H^{2.5}$
22½°	$Q = 0.497H^{2.5}$	$Q = 0.321H^{2.5}$	$Q = 223H^{2.5}$

RECTANGULAR WEIR

The rectangular sharp-crested weir (Figure 11-5) may be used in one of two configurations. The first configuration (Figure 11-5(a)) consists of a rectangular notch cut into a bulkhead in the flow channel, producing a box-like opening. This configuration is called a contracted rectangular weir, since a curved flow path or contraction results with the nappe forming a jet narrower than the weir opening. The horizontal distances from the end of the weir crest to the side walls of the channel are called the end contractions. These end contractions reduce the width and accelerate the channel flow as it passes over the weir and provide the needed ventilation. Flow through this type of weir is said to be "with end contractions." (Any weir that is narrower than the channel in which it is placed is technically a contracted weir.)

In the second configuration of the rectangular weir (Figure 11-5(b)), the end contractions are completely suppressed by extending the weir across the entire width of the channel. Thus, the sides of the channel also act as the sides of the weir, and there are no lateral contractions. This type of weir is called a rectangular suppressed weir, and flow through it is said to be "without end contractions."

The discharge (head vs. flow rate) equation of a free-flowing rectangular weir with end contractions takes the form:

$$Q = K(L - 0.2H^{1.5}) \qquad (11\text{-}5)$$

where:

Q = flow rate
H = head on the weir
L = crest length of weir
K = constant dependent upon units

For flow rate in cubic feet per second, million gallons per day, and gallons per minute and head in feet, the discharge equation for a rectangular weir with end contractions is as follows:

For cfs — $$Q = 3.33\,(L - 0.2H)H^{1.5} \qquad (11\text{-}6)$$

For mgd — $$Q = 2.15(L - 0.2H)H^{1.5} \qquad (11\text{-}7)$$

For gpm — $$Q = 1500(L - 0.2H)H^{1.5} \qquad (11\text{-}8)$$

where L = crest length of weir in feet.

The discharge equation of a free-flowing rectangular weir without end contractions takes the form:

$$Q = KLH^{1.5} \quad (11\text{-}9)$$

where:

Q = flow rate
H = head on the weir
L = crest length of weir
K = constant dependent upon units

For flow rate in cubic feet per second, million gallons per day, and gallons per minute and head in feet, the discharge equation for a rectangular weir without end contractions is as follows:

For cfs — $$Q = 3.33\,LH^{1.5} \quad (11\text{-}10)$$

For mgd — $$Q = 2.15\,LH^{1.5} \quad (11\text{-}11)$$

For gpm — $$Q = 1500\,LH^{1.5} \quad (11\text{-}12)$$

where L = crest length of weir in feet.

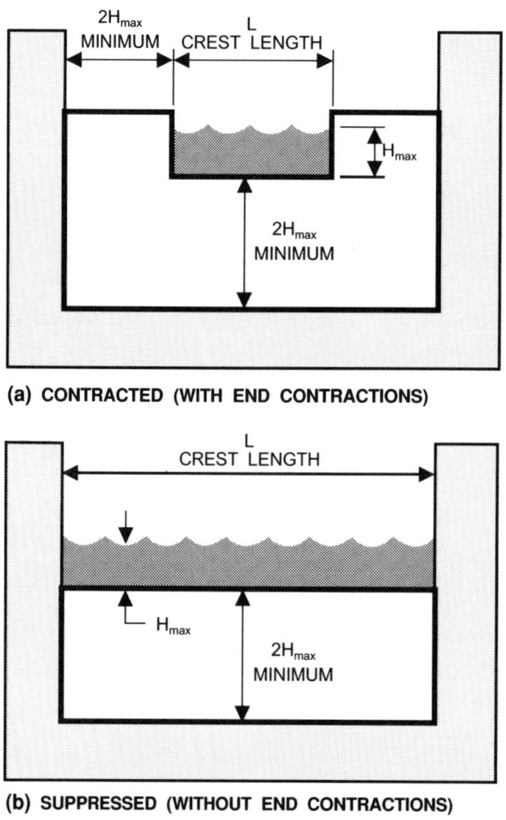

Figure 11-5. Rectangular Sharp-Crested Weir

TRAPEZOIDAL OR CIPOLLETTI WEIR

The trapezoidal sharp-crested weir (Figure 11-6) is similar to a rectangular weir with end contractions except that the sides incline outwardly, producing a trapezoidal opening. When the end inclinations of a trapezoidal weir are in the ratio of 4 vertical to 1 horizontal, the weir is known as a Cipolletti weir (named for the Italian experimenter, Cesare Cipolletti, who first proposed its use). Although the Cipolletti weir is a contracted weir, its discharge occurs essentially as though its end contractions were suppressed. Thus, no correction is necessary for the crest width as in a rectangular contracted weir, resulting in a simpler discharge equation.

The discharge (head vs. flow rate) equation of a free-flowing Cipolletti weir takes the form:

$$Q = KLH^{1.5} \qquad (11\text{-}13)$$

where:

Q = flow rate
H = head on the weir
L = crest length of weir
K = constant dependent upon units

For flow rate in cubic feet per second, million gallons per day, and gallons per minute and head in feet, the discharge equation for a Cipolletti weir is as follows:

For cfs — $$Q = 3.37\,LH^{1.5} \qquad (11\text{-}14)$$

For mgd — $$Q = 2.18\,LH^{1.5} \qquad (11\text{-}15)$$

For gpm — $$Q = 1510\,LH^{1.5} \qquad (11\text{-}16)$$

where L = crest length of weir in feet.

Applications

A weir is low in cost, relatively easy to install, and quite accurate when properly used. Weirs are typically applied to flow measurements in which relatively large head is available to establish free-flow conditions over the weir.

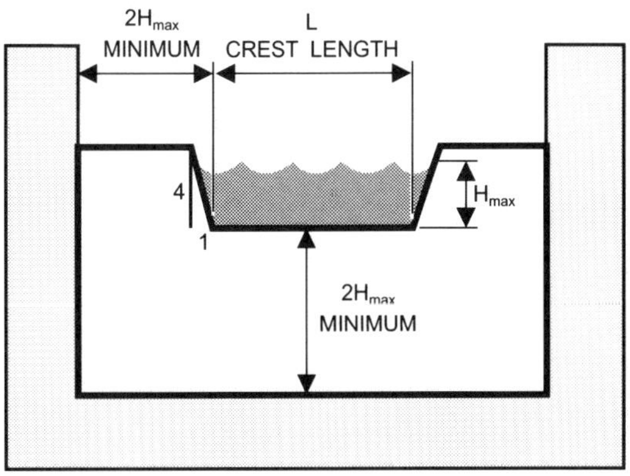

Figure 11-6. Trapezoidal (Cipolletti) Sharp-Crested Weir

TRIANGLE OR V-NOTCH WEIR

The triangular weir is an accurate flow measuring device particularly suited for low flows. Because the triangular weir has no crest length, the head required for a small flow through it is greater than that required with other types of weirs. This is an advantage for small discharges in that the nappe will spring free of the crest, whereas it would cling to the crest of another type of weir and reduce the accuracy of the measurement. The triangular weir is the best weir profile for measuring discharges less than 1 cfs (0.65 mgd or 450 gpm) and has reasonable accuracy for flows up to 10 cfs (6.46 mgd or 4500 gpm).

RECTANGULAR AND CIPOLLETTI WEIR

Rectangular weirs are better suited to measure higher flows than are triangular weirs. The Cipolletti weir offers a slightly wider range than the rectangular weir. However, the measurement accuracy with a Cipolletti weir is inherently less than that obtained with a rectangular or a triangular weir.

Operating Constraints

Remember that accurate flow rate measurements with a weir cannot be expected unless the proper conditions and dimensions are maintained.

Weirs, aside from being operated within their flow limits, must also be operated within the available system head. Typically, most applications are gravity fed, and the amount of permanent head loss (that is, the difference in level before and after the weir) may be limited by physical requirements on the elevation of the inlet and outlet.

Operation of the weir is sensitive to the approach velocity of the liquid, often necessitating a stilling basin or pond upstream of the weir. Such a basin reduces the fluid velocity and provides a place for debris to settle out. Weirs are not suitable for water that carries excessive solid materials or silt, which will deposit in the approach channel behind the weir and destroy the conditions required for accurate discharge measurement. Some silt, sand, or other solid material will inevitably collect in most open channel flow systems. To allow the periodic removal of these deposits, it is suggested that the weir bulkhead be constructed with an opening beneath the notch, through which accumulations can be sluiced as required. A metal plate or plank placed across the upstream side of this opening and securely fixed in place will serve as a cover while the weir is in operation.

Performance

A properly installed weir can typically achieve accuracies of 2 to 5 percent of the rate of flow. However, the overall accuracy obtained from the flow measurement system (the weir and the flowmeter) can be affected by a number of factors, including faulty construction or installation of the weir, improper head measuring location, incorrect zero setting of the flowmeter, flowmeter level measurement inaccuracies, flowmeter level-to-flow conversion inaccuracies, use of the weir outside its proper range, improper maintenance of the weir, and turbulence surges in the approach channel. The reduced accuracy of a level transmitter or flowmeter is most significant in the lower portion of the flow range.

Sizing

TRIANGULAR OR V-NOTCH WEIR

The size of a triangular weir may be estimated using Figure 11-7. It is generally recommended that the minimum head on a triangular weir be at least 0.2 foot to prevent the nappe from clinging to the crest. It is also recommended that the maximum head be limited to 2.0 feet to assure accuracy of the device head-flow rate relationship. Based on these lower and upper head restrictions, Table 11-1 presents the minimum and maximum recommended flow rates for the common triangular weirs.

Table 11-1. Minimum and Maximum Recommended Flow Rates for Triangular Weirs

V-notch angle	Min. head, ft.	Minimum flow rate			Max. head, ft.	Maximum flow rate		
		cfs	mgd	gpm		cfs	mgd	gpm
221/2°	0.2	0.009	0.006	4.04	2.0	2.81	1.82	1260
30°	0.2	0.012	0.008	5.39	2.0	3.82	2.47	1710
45°	0.2	0.019	0.012	8.53	2.0	5.85	3.78	2630
60°	0.2	0.026	0.017	11.7	2.0	8.16	5.28	3660
90°	0.2	0.045	0.029	20.2	2.0	14.1	9.14	6330
120°	0.2	0.077	0.050	34.8	2.0	24.5	15.8	11,000

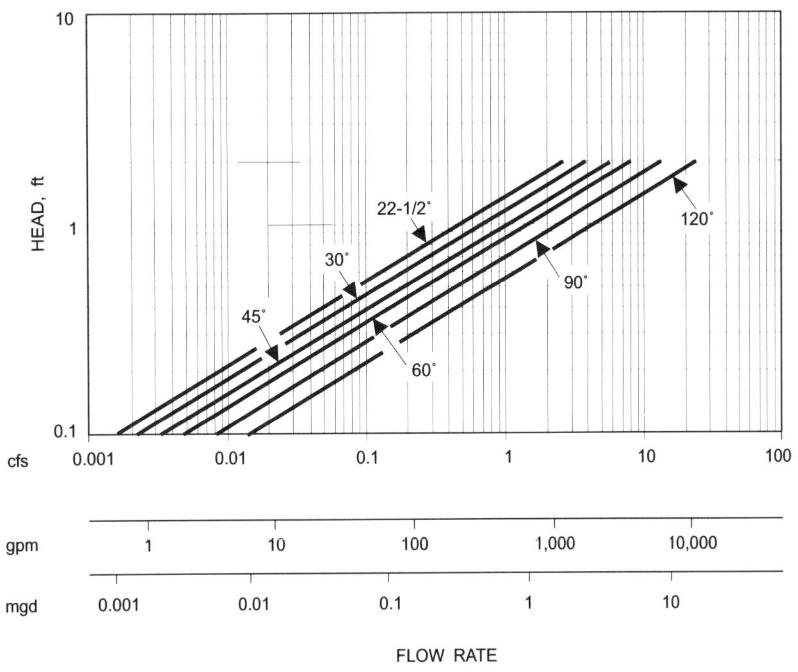

Figure 11-7. Triangular Weir Relationship of Flow Rate and Liquid Head

RECTANGULAR WEIR

The size of rectangular weirs with and without end contractions may be estimated using Figure 11-8. When sizing a rectangular weir, a crest length of 1 foot is the minimum that should be considered, since a triangular weir can more accurately measure the same flow rates as rectangular weirs smaller than 1 foot. It is conventional practice to increase the crest length in increments of 6 inches up to 3 feet, and in 1- foot increments beyond 3 feet, to suit the particular installation. Rectangular weir crest lengths up to 10 feet are common, and, theoretically, there is no maximum crest length. However, beyond 6 or 8 feet, a limit is usually set by economic rather than engineering considerations. In general, it is recommended that the minimum head on a rectangular weir be at least 0.2 foot to prevent the nappe from clinging to the crest. It is also generally accepted practice to limit the maximum head to no more than one half the crest length. However, laboratory experiments have shown that the accuracy of measurement is not impaired to a great extent by exceeding this limit, especially for crest lengths of 1 to 4 feet. But, to ensure strict conformance to the weir head-flow rate relationship, the maximum head limitation of one half the crest length should usually be adhered to. Based on these lower and upper head restrictions, Table 11-2 presents the minimum and maximum recommended flow rates for common rectangular weirs both with and without end contractions.

TRAPEZOIDAL OR CIPOLLETTI WEIR

The size of a Cipolletti weir may be estimated using Figure 11-9. The minimum and maximum recommended heads for a Cipolletti weir are the same as rectangular weirs: minimum head of 0.2 foot and maximum head of no more than one half the crest length. Based on these lower and upper head restrictions, Table 11-3 presents the minimum and maximum recommended flow rates for common Cipolletti weirs.

Construction

A weir consists of a thin plate with a specified opening, as described above. A level sensing device (level transmitter or flowmeter) upstream of the weir is used to sense the liquid head. A variety of materials may be used to construct the weir, depending upon the nature of the liquid in the flow stream. Typically, fiberglass construction with a metal crest, concrete with a metal crest, and all-metal construction are used. Weirs may either be purchased from a weir manufacturer or fabricated locally according to the weir specifications.

To assure accurate discharge measurement, there are certain general weir construction requirements that apply to all types of weirs:

(1) The weir should consist of a thin plate ⅛ to ¼ inch thick with a straight edge or a thicker plate with a downstream chamfered edge. The upstream sharp edge prevents the nappe from adhering to the crest. Knife edges should be avoided because they are difficult to maintain. However, the upstream edge of the weir must be sharp with right-angle corners, since rounded edges will decrease the head for a given flow rate.

(2) The length of the weir crest or the notch angle must be accurately determined, because the percentage error in measured flow rate will be proportional to the error in determining these dimensions.

Installation

When installing a weir, the following general installation requirements apply to all types of weirs:

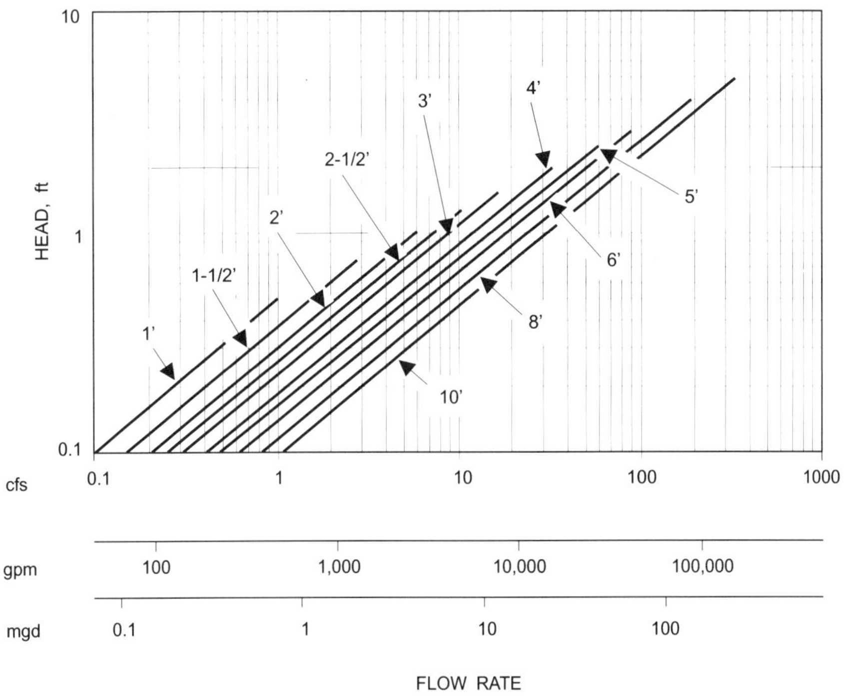

(a) RECTANGULAR WEIR WITH END CONTRACTIONS

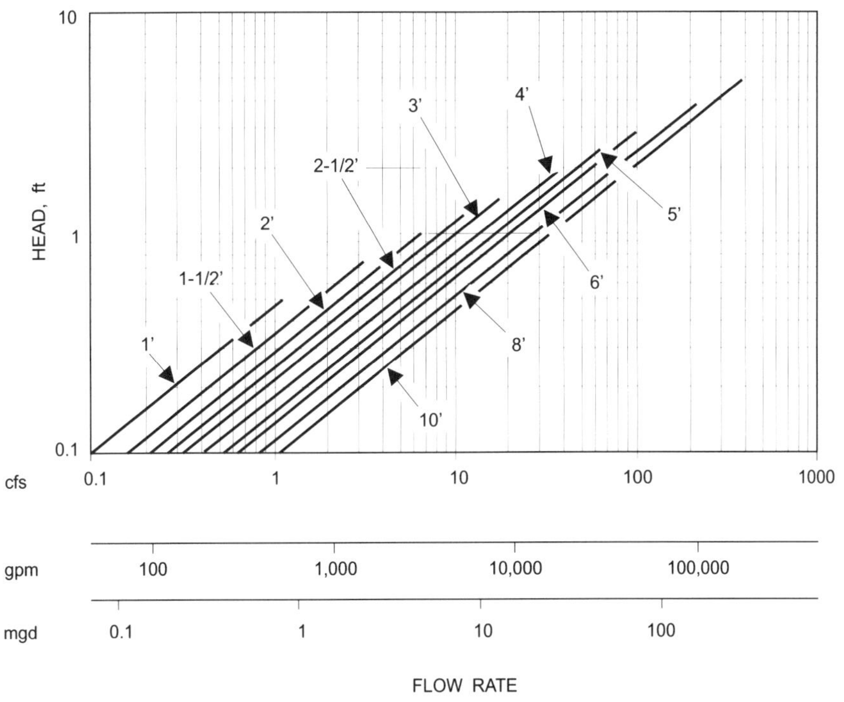

(b) RECTANGULAR WEIR WITHOUT END CONTRACTIONS

Figure 11-8. Rectangular Weir Relationship of Flow Rate and Liquid Head

Table 11-2. Minimum and Maximum Recommended Flow Rates for Rectangular Weirs

(a) Rectangular Weir with End Contractions

Crest length, ft.	Min. head, ft.	Minimum flow rate			Max. head, ft.	Maximum flow rate		
		cfs	mgd	gpm		cfs	mgd	gpm
1	0.2	0.286	0.185	128	0.5	1.06	0.685	476
1½	0.2	0.435	0.281	195	0.75	2.92	1.89	1310
2	0.2	0.584	0.377	262	1.0	5.99	3.87	2690
2½	0.2	0.733	0.474	329	1.25	10.5	6.77	4710
3	0.2	0.882	0.570	396	1.5	16.5	10.7	7410
4	0.2	1.18	0.762	530	2.0	33.9	21.9	15,200
5	0.2	1.48	0.955	664	2.5	59.2	38.3	26,600
6	0.2	1.77	1.15	794	3.0	93.4	60.4	41,900
8	0.2	2.37	1.53	1060	4.0	192	124	86,200
10	0.2	2.97	1.92	1330	5.0	335	217	150,000

(b) Rectangular Weir without End Contractions

Crest length, ft.	Min. head, ft.	Minimum flow rate			Max. head, ft.	Maximum flow rate		
		cfs	mgd	gpm		cfs	mgd	gpm
1	0.2	0.298	0.192	134	0.5	1.18	0.761	530
1½	0.2	0.447	0.289	201	0.75	3.24	2.10	1450
2	0.2	0.596	0.385	267	1.0	56.66	4.30	2990
2½	0.2	0.745	0.481	334	1.25	11.6	7.52	5210
3	0.2	0.894	0.577	401	1.5	18.4	11.9	8260
4	0.2	1.19	0.770	534	2.0	37.7	24.3	16,900
5	0.2	1.49	0.962	669	2.5	65.8	42.5	29,500
6	0.2	1.79	1.16	803	3.0	104	67.1	46,700
8	0.2	2.38	1.54	1070	4.0	213	138	95,600

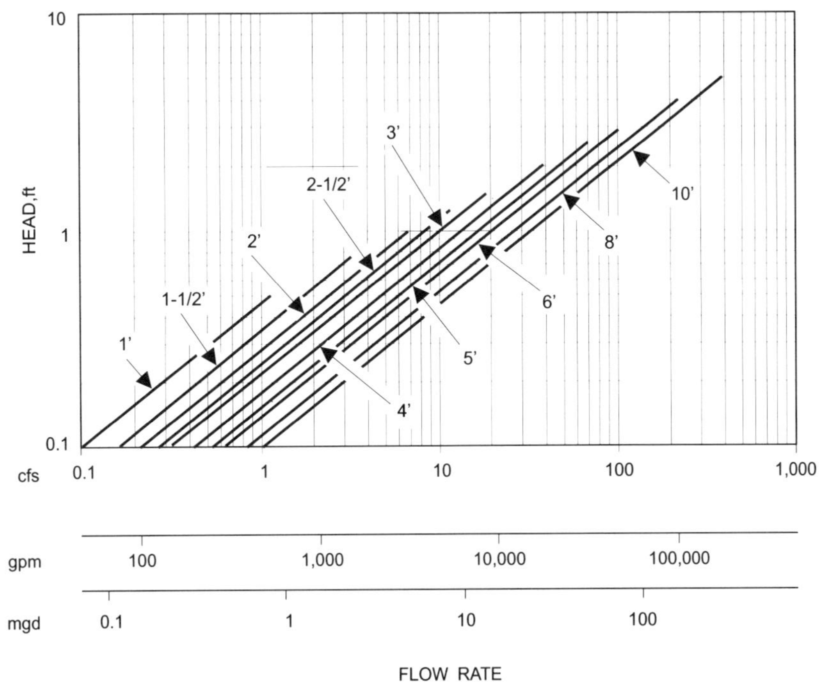

Figure 11-9. Cipolletti Weir Relationship of Flow Rate and Liquid Head

Table 11-3. Minimum and Maximum Recommended Flow Rates for Cipolletti Weirs

Crest length, ft.	Min. head, ft.	Minimum flow rate			Max. head, ft.	Maximum flow rate		
		cfs	mgd	gpm		cfs	mgd	gpm
1	0.2	0.301	0.195	135	0.5	1.19	0.769	534
1½	0.2	0.452	0.292	203	0.75	3.28	2.12	1470
2	0.2	0.602	0.389	270	1.0	6.73	4.35	3020
2½	0.2	0.753	0.487	338	1.25	11.8	7.60	5300
3	0.2	0.903	0.584	405	1.5	18.6	12.0	8350
4	0.2	1.20	0.778	539	2.0	38.1	24.6	17,100
5	0.2	1.51	0.973	678	2.5	66.5	43.0	29,800
6	0.2	1.81	1.17	812	3.0	105	67.8	47,100
8	0.2	2.41	1.56	1080	4.0	214	139	96,000
10	0.2	3.01	1.95	1350	5.0	375	243	168,000

(1) The connection of the weir to the channel should be waterproof. Therefore, the joint between the weir plate the and channel should be packed with chemically inert cement or asphalt-type roofing compound.

(2) The weir should be ventilated, if necessary, to prevent a vacuum from forming on the underside of the nappe.

(3) The height of the weir from the bottom of the channel to the crest should be at least 2 times the maximum expected head of liquid above the crest. This is necessary to lower the velocity of approach. The weir height should never be less than 1 foot.

(4) The approach section should be straight upstream from the weir for a distance of at least 20 times the maximum expected head of liquid and should have little or no slope.

(5) The crest must be set higher than the maximum downstream elevation of the water surface; otherwise, a submerged flow condition will occur instead of the free-flow condition required for reliable flow measurement.

(6) A stilling pond or basin may be required to reduce velocity and the effects of flow turbulence of the liquid upstream of the weir.

(7) The device for measuring the head (level transmitter or flowmeter) should be placed upstream at a distance of at least 3 times the maximum expected head on the weir and should be located in a quiet section of the channel away from all disturbances, preferably in a stilling well. Also, the zero point of the head measuring device must be set exactly level with the weir crest.

TRIANGULAR OR V-NOTCH WEIR

When installing a triangular weir, the minimum distance of the sides of the weir from the channel banks should be at least twice the maximum expected head on the weir. The minimum distance from the crest to the pool bottom should also be at least twice the maximum expected head, as shown in Figure 11-4.

RECTANGULAR WEIR

When installing a rectangular weir with end contractions, the distance from the side of the weir notch to the side of the channel should be at least twice the maximum expected head on the weir. This is necessary to allow the liquid in the channel a free, unconstrained, lateral approach to the weir crest. Special care must also be taken in the installation of rectangular weirs without end contractions to obtain adequate aeration of the nappe. This is usually accomplished by placing vents on both sides of the weir box under the nappe. For rectangular weirs both with and without end contractions, the minimum distance from the crest to the pool bottom should be at least twice the maximum expected head. Refer to Figure 11-5.

TRAPEZOIDAL OR CIPOLLETTI WEIR

All of the installation conditions stated for rectangular weirs with end contractions are also applicable to Cipolletti weirs. Refer to Figure 11-6.

The level transmitter or open channel flowmeter should be installed in accordance with the manufacturer's recommendations.

Be sure that the flowmeter's level-to-flow rate conversion matches the weir being used and that the desired units of measure for flow rate and total flow have been selected.

Calibration

The only calibration required for a weir is to adjust the zero of the level transmitter or flowmeter being used with the weir. The transmitter or flowmeter must be adjusted such that indicated liquid level matches the actual level in the flow stream, relative to the weir's zero point. Adjustment of zero and the span of the level transmitter or flowmeter should be performed per the manufacturer's instructions, which will vary significantly with the manufacturer, as well as with the technology that is used to measure level.

Level transmitters and flowmeters must also be calibrated to compensate for any hydrostatic heads that result from the elevation of the transmitter or flowmeter relative to the crest or the bottom of the weir notch. Non-contact measurements must also be made to account for the elevation of the transducer above the crest level.

Note that the zero point of a weir is the bottom of the weir notch or the elevation of the weir crest.

Safety

Weirs may be located in hazardous locations. The electrical rating of the flowmeter or level transmitter associated with the weir should be consistent with the location.

Weirs also are frequently located in manholes in which there is a possibility of an adverse atmosphere. Before entering a manhole, tests should be made for an explosive atmosphere, the presence of hydrogen sulfide or other dangerous gases, and oxygen deficiency. Whenever an adverse atmosphere is encountered, forced ventilation may be used to create safe conditions, or self-contained breathing apparatus may be used to protect the worker.

Maintenance

In a weir-based open channel flow measurement system, both the weir and the associated flowmeter or level transmitter require periodic maintenance. The operation of the weir is sensitive to any foreign material or debris that may be present upstream of the flowmeter or on the weir plate itself. Therefore, the weir should be periodically inspected and any accumulated debris removed.

The zero and span of the flowmeter or level transmitter should be periodically checked for proper calibration and adjusted per the manufacturer's instructions, if necessary.

Weir size may be estimated by using the graphs of the relationship between flow rate and the liquid head upstream of the weir as shown in Figures 11-7, 11-8, and 11-9 and the tables of the minimum and maximum recommended flow rates (Tables 11-1, 11-2, and 11-3) for triangular, rectangular, and Cipolletti weirs, respectively.

Size a weir for the measurement of 0 to 5000 gallons per minute of water that is flowing in an open channel, assuming that the available head is limited to 1.25 feet.

Referring to the minimum and maximum flow rate tables, it may be seen that the maximum flow rate of 5000 gallons per minute can be accommodated by a 90 degree triangular weir, a 3 foot rectangular weir, or a 2-½ foot Cipolletti weir. However, referring to the head-flow rate tables, it may be seen that the 5000 gallons per minute flow rate generates a 1.82 foot head through the 90 degree triangular weir, which exceeds the available head of 1.25 feet. Therefore, either a rectangular weir or a Cipolletti weir must be used. The 5000 gallons per minute flow rate generates a head of approximately 1.1 feet through a 3 foot rectangular weir and 1.2 feet through a 2½ foot Cipolletti weir, both of which are within the available head. Thus, either the rectangular or Cipolletti weir may be used in this case. Note that for a rectangular weir without end contractions, the crest length must match the channel width. If the channel width is other than 3 feet, the rectangular weir without end contractions may not be used.

Flumes

Principle of Operation

A flume is a specially shaped open channel flow section that restricts the channel area and/or changes the channel slope, resulting in an increased velocity and a change in the level of the liquid flowing through the flume. Normally, a flume (Figure 11-10) consists of a converging section to restrict the flow, a throat section, and a diverging section to assure that the downstream level is less than the level in the converging section. The flume restricts the flow, then expands it again in a definite fashion. The flow rate through the flume may be determined by measuring the head on the flume at a single point, usually at some distance downstream from the inlet. The head-flow rate relationship of a flume may be defined by either test data (calibration curves) or by an empirically derived formula.

Flumes can be categorized as belonging to one of three general families, depending upon the state of flow induced—subcritical, critical, or supercritical. In general, flumes that induce a critical or supercritical state of flow are most commonly used. This is because, when critical or supercritical flow occurs in a channel, one head measurement can indicate the discharge rate if it is made far enough upstream so that the flow depth is not affected by the drawdown of the water surface as it achieves or passes through a critical state of flow. For critical or supercritical states of flow, a definitive head-discharge relationship can be established and measured, based on a single head reading. Thus, most commonly encountered flumes are designed to pass the flow from subcritical through critical at or near the point of measurement.

Most flumes in common use today can be traced to one of three early design sources: rectangular English flumes based upon early work in India around 1908-1914 and the writings of F. V. A. E. Engal; the Parshall flume whose forerunner, a Venturi flume developed by Cone, was extensively modified and tested by Parshall; and flumes of the type first developed by Palmer and Bowlus. The following sections will discuss in detail some of the more popular flume designs currently in use. Included are discussions of the Parshall flume, the Palmer-Bowlus flume, the HS, H, and HL flumes, and the trapezoidal flume.

PARSHALL FLUMES

The Parshall flume (see Figure 11-11) was developed in the 1920s primarily to measure irrigation water, but it is now frequently used in industrial applications. In 1922 Dr. Ralph L. Parshall of the U.S. Soil Conservation Service made some radical changes to the existing Venturi (subcritical) flume design. The essential change introduced by Parshall was a drop in the floor, which produced supercriti-

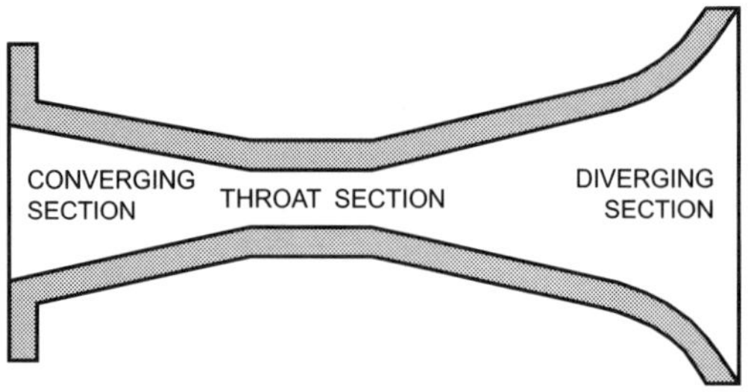

Figure 11-10. General Flume Configuration

cal flow through the throat of the flume. This perfected device was named the Parshall Measuring Flume by the Irrigation Committee of the American Society of Civil Engineers. The flumes are not patented and the discharge tables are not copyrighted.

The constricted throat of the flume produces a head that is related to discharge. The level converging section followed by the downward sloping floor in the throat gives the Parshall flume its ability to withstand relatively high degrees of submergence without affecting the rate of flow. The converging upstream portion of the flume accelerates the entering flow, helping to eliminate deposits of sediment that would otherwise reduce measurement accuracy.

Parshall flume sizes are designated by the throat width, W, as shown in Figure 11-12. Dimensions are available for flumes with throat widths ranging from 1 inch to 50 feet. For convenience, Parshall flumes have been somewhat arbitrarily classified into three main groups: "very small" for 1-, 2- and 3-inch flumes; "small" for 6-inch through 8-foot flumes; and "large" for 10-foot through 50-foot flumes. The flumes cover a range of discharges from 0.01 to 3000 cfs (0.006 to 1940 mgd/4.5 to 1,350,000 gpm) and have overlapping capacities to provide wide latitude in selecting sizes.

The configuration and standard nomenclature for Parshall flumes is given in Figure 11-12. For a given throat width (W), all other dimensions are rigidly prescribed. Since the discharge tables for Parshall flumes are based upon extensive research, faithful adherence to all dimensions is necessary to achieve accurate flow measurement. The flumes must be constructed according to the dimensions given for each flume, because the flumes are not geometrically similar. For example, it cannot be assumed that a dimension in the 12-foot flume will be three times the corresponding dimension in the 4-foot flumes.

Discharge through a Parshall flume can occur for two conditions of flow. The first, free flow, occurs when there is insufficient backwater depth to reduce the discharge rate. For free flow, only the head H_a (refer to Figure 11-12) at the

Figure 11-11. Parshall Flume
(Courtesy of Plasti-Fab, Inc.)

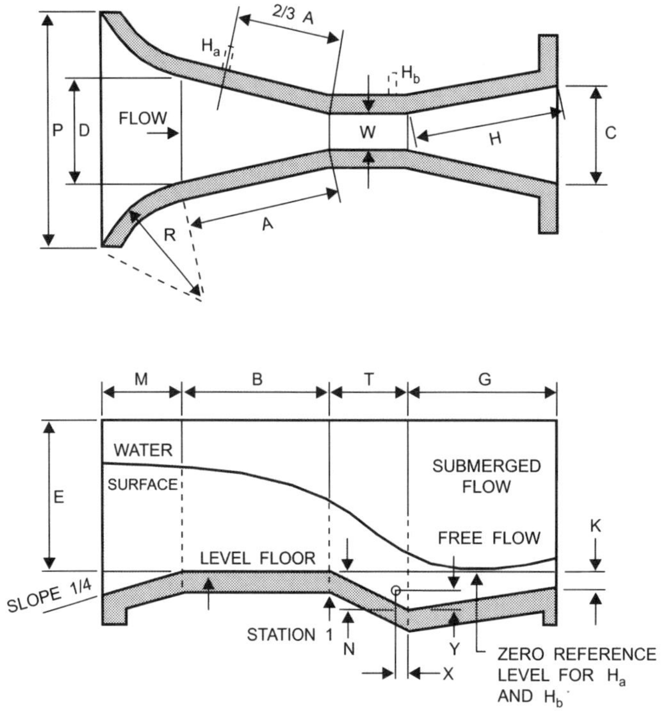

W	A	2/3 A	B	C	D	E	T	G	H	K	M	N	P	R	X	Y
1"	1' 2-9/32"	9-17/32"	1' 2"	3-21/32"	6-19/32"	6" to 9"	3"	8"	8-1/8"	3/4"		1-1/8"			5-16"	1' 2"
2"	1' 4-5/16"	10-7/8"	1' 4"	5-5/16"	8 13/32"	6 to 10'	4-1/2"	10'	10-1/8"	7/8"		1-11/16"			5' 8"	1"
3"	1' 6-3/8"	1' 1/4"	1' 6"	7"	10-3/16"	1 to 1-1/2"	6"	1'	1' 5/32"	1"		2-1/4"			1"	1-1/2"
6"	2' 7/16"	1' 4-5/16"	2'	1' 3-1/2"	1' 3-5/8"	2'	1'	2'		3"	1'	4-1/2"	2' 11-1/2"	1' 4"	2"	3"
9"	2' 10-5/8"	1' 11-1/8"	2' 10"	1' 3"	1' 10-5/8"	2' 6"	1'	2' 6"		3"	1'	4-1/2"	3' 6-1/2"	1' 4"	2"	3"
1'	4' 6"	3'	4' 4-7/8"	2'	2' 9-1/4"	3'	2'	3'		3"	1' 3"	9'	4' 10-3/4"	1' 8"	2"	3"
1' 6"	4' 9"	3' 2"	4' 7-7/8"	2' 6"	3' 4-3/8"	3'	2'	3'		3"	1' 3"	9'	5' 6"	1' 8"	2"	3"
2'	5'	3' 4"	4' 10-7/8"	3'	3' 11-1/2"	3'	2'	3'		3"	1' 3"	9'	6' 1"	1' 8"	2"	3"
3'	5' 6"	3' 8"	5' 4-3/4"	4'	5' 1-7/8"	3'	2'	3'		3"	1' 3"	9'	7' 3-1/2"	1' 8"	2"	3"
4'	6'	4'	5' 10-5/8"	5'	6' 4-1/4"	3'	2'	3'		3"	1' 6"	9'	8' 10-3/4"	2'	2"	3"
5'	6' 6"	4' 4"	6' 4-1/2"	6'	7' 6-5/8"	3'	2'	3'		3"	1' 6"	9'	10' 1-1/4"	2'	2"	3"
6'	7'	4' 8"	6' 10-3/8"	7'	8' 9"	3'	2'	3'		3"	1' 6"	9'	11' 3-1/2"	2'	2"	3"
7'	7' 6"	5'	7' 4-1/4"	8'	9' 11-3/8"	3'	2'	3'		3"	1' 6"	9'	12' 6"	2'	2"	3"
8'	8'	5' 4"	7' 10-1/8"	9'	11' 1-3/4"	3'	2'	3'		3"	1' 6"	9'	13' 8-1/4"	2'	2"	3"
10'		6'	14'	12'	15' 7-1/4"	4'	3'	6'		6"		1' 1-1/2"			1"	9"
12'		6' 8"	16'	14' 8"	18' 4-3/4"	5'	3'	8'		6"		1' 1-1/2"			1"	9"
15'		7' 8"	25'	18' 4"	25'	6'	4'	10'		9"		1' 6"			1"	9"
20'		9' 4"	25'	24'	30'	7'	6'	12'		1"		2' 3"			1"	9"
25'		11'	25'	29' 4"	35'	7'	6'	13'		1"		2' 3"			1"	9"
30'		12' 8"	26'	34' 8"	40' 4-3/4"	7'	6'	14'		1"		2' 3"			1"	9"
40'		16'	27'	45' 4"	50' 9-1/2"	7'	6'	16'		1"		2' 3"			1"	9"
50'		19' 4"	27'	56' 8"	60' 9-1/2"	7'	6'	20'		1"		2' 3"			1"	9"

Figure 11-12. Parshall Flume Dimensions for Various Throat Widths

upstream gage location is needed to determine the discharge from a standard table. Under free-flow conditions a phenomenon known as the hydraulic jump or "standing wave" occurs downstream from the flume. Formation of this is a certain indication of free-flow conditions.

The second condition of flow, submerged flow, occurs when the water surface downstream from the flume is high enough to reduce the discharge. When the discharge is increased above a critical value, the resistance to flow in the downstream channel becomes sufficient to reduce the velocity, increase the flow depth, and cause a backwater effect at the flume. In order to determine the discharge, submerged flow requires the measurement of both an upstream depth, H_a, and a depth in the throat, H_b (Figure 11-12). The ratio of the downstream depth to the upstream depth, H_b/H_a, expressed as a percentage, is referred to as the submergence ratio. Calibration tests show that the discharge of a Parshall flume is not reduced (that is, the flume is operating under free-flow conditions) until the submergence ratio exceeds the following values:

50 percent for flumes 1, 2, and 3 inches wide

60 percent for flumes 6 and 9 inches wide

70 percent for flumes 1 to 8 feet wide

80 percent for flumes 8 to 50 feet wide

When the submergence ratio exceeds the values listed above, the flume is operating under submerged conditions, and submerged discharge tables will have to be used to calculate the discharge. See References 21 and 29 in the Bibliography for a complete discussion of the calculation of discharge under submerged conditions of flow. In general, selecting and installing a Parshall flume so that conditions of free flow exist is desired, since submerged conditions greatly complicate the determination of flow rate.

The discharge (head vs. flow rate) equation of free flow through a Parshall flume takes the form:

$$Q = KH^n \tag{11-17}$$

where:

Q = flow rate
H = head measure at point H_a
K = constant, dependent upon throat width and units
n = constant power, dependent upon throat width

For flow rate in cubic feet per second, million gallons per day, and gallons per minute and head in feet, the discharge equations for a number of common Parshall flumes are as follows:

	cfs	mgd	gpm
1 in.	$Q = 0.338H^{1.55}$	$Q = 0.218H^{1.55}$	$Q = 152H^{1.55}$
2 in.	$Q = 0.676H^{1.55}$	$Q = 0.437H^{1.55}$	$Q = 303H^{1.55}$
3 in.	$Q = 0.992H^{1.547}$	$Q = 0.641H^{1.547}$	$Q = 445H^{1.547}$
6 in.	$Q = 2.06H^{1.58}$	$Q = 1.33H^{1.547}$	$Q = 925H^{1.58}$
9 in.	$Q = 3.07H^{1.53}$	$Q = 1.98H^{1.53}$	$Q = 138H^{1.53}$
1 to 8 ft	$Q = 4WH^{1.522 W^{0.026}}$	$Q = 2.59WH^{1.522 W^{0.026}}$	$Q = 1800WH^{1.55 W^{0.026}}$
10 to 50 ft	$Q = (3.69W + 2.5)H^{1.6}$	$Q = (2.39W + 1.61)H^{1.6}$	$Q = (1660W + 1120)H^{1.6}$

where W = throat width in feet.

Some practitioners have used a simplified version of the Parshall flume that is sometimes referred to as the Montana flume. If a Parshall flume is never to be operated above 70 percent submergence, there is no need to construct the portion of the flume downstream from the end of the flat crest section, shown as Station 1 in Figure 11-12. This configuration, with only the upstream portion of the flume present, is known as the Montana flume. The crest of a Montana flume should be set above the channel bottom. This will assure that the flow profile over the crest section is not modified by backwater from the downstream channel. As long as the 70 percent submergence limit is not exceeded, the standard discharge equations for Parshall flumes may be applied to similarly sized Montana flumes.

PALMER-BOWLUS FLUME

The Palmer-Bowlus flume was developed in the mid 1930s, by Harold V. Palmer and Fred D. Bowlus of the Los Angeles County Sanitation District, as a simple and effective wastewater flow measuring device. The Palmer-Bowlus flume is dependent upon an existing conduit slope and channel contractions (provided by the flume) to produce supercritical flow. This type of flume arose out of a desire to have a primary measuring device that could be inserted into an existing conduit, usually round, with minimal site requirements other than suitable slope.

The Palmer-Bowlus flume is essentially a restriction in the channel designed to produce a higher velocity critical flow in the throat. The flume is most often used in manholes or open round or rectangular bottom channels to measure flow rate. It is also useful as a temporary installation to provide flow data for determining flume size and equipment requirements for permanent installation.

The Palmer-Bowlus flume is a type of Venturi flume characterized by a throat of uniform cross section and a length approximately equal to one diameter of the pipe or conduit in which it is to be installed. A number of different cross-sectional shapes have been proposed, tested, and/or used over the years. Typical shapes of Palmer-Bowlus flumes for installation in round and rectangular conduits are shown in Figure 11-13.

It is important to note that the term "Palmer-Bowlus flume" technically refers only to the general class of flumes that have a form as discussed above. Unlike the Parshall flume, the dimensional configuration of a Palmer-Bowlus flume is *not* rigidly established for each flume size. Much latitude is possible in both the design and construction of Palmer-Bowlus flumes. As shown in Figure 11-13, the cross-sectional configuration of a Palmer-Bowlus flume may assume any of several different shapes: rectangular, trapezoidal (with various slopes and base widths and heights), with or without the bottom slab, etc. Thus, without further qualification, the term "Palmer-Bowlus flume" is not fully definitive of flume configuration and dimensions.

However, there appears to have been a standardization among many commercial manufacturers and users of "Palmer-Bowlus flumes" upon a flume with a trapezoidal throat having the configuration shown in Figure 11-14. This trapezoidal section with a flat bottom is considered to be the preferred design for circular conduits and pipes, since this shape has the least constriction through the critical flow area and provides for minimum head loss through the conduit. Most flumes currently being marketed with the description "Palmer-Bowlus flume" are this particular trapezoidal configuration, with other dimensions as shown in Figure 11-14. Note that all dimensions are proportional to the conduit diameter, D.

Following what appears to be common practice, the term "Palmer-Bowlus flume" will, in the remainder of this chapter, refer to a flume with the configuration shown in Figure 11-15, and the discussion will be limited to flumes of this type. However, it is advisable to remember that the term "Palmer-Bowlus flume" is not definitive and may be applied to flumes with any number of different throat cross-sectional configurations.

Palmer-Bowlus flume sizes are designated by the size of the pipe or conduit into which they fit, not by the throat width as is the case with Parshall flumes. Thus, an 8-inch Palmer-Bowlus flume is designed to be inserted into an 8-inch diameter pipe. Standard Palmer-Bowlus flumes are available from the various manufacturers to fit pipe sizes ranging from 4 inches to 42 inches. Larger sizes are available by special order. The standard sizes appear to be 4, 6, 8, 10, 12, 15, 18, 21, 24, 27, 30, and 42 inches.

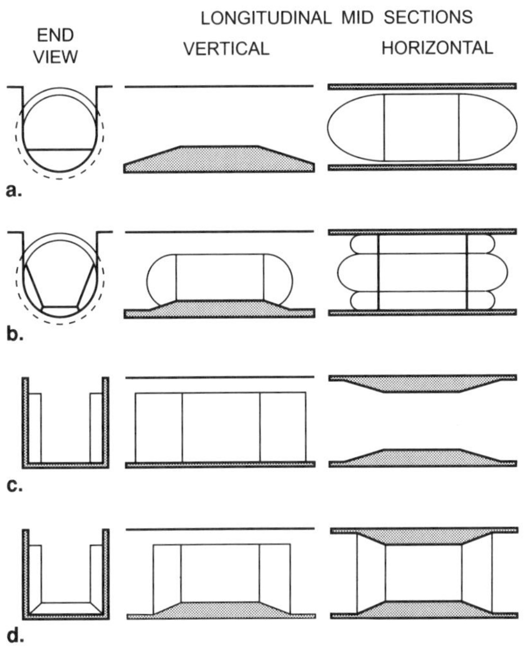

Figure 11-13. Various Cross-Sectional Shapes of Palmer-Bowlus Flumes

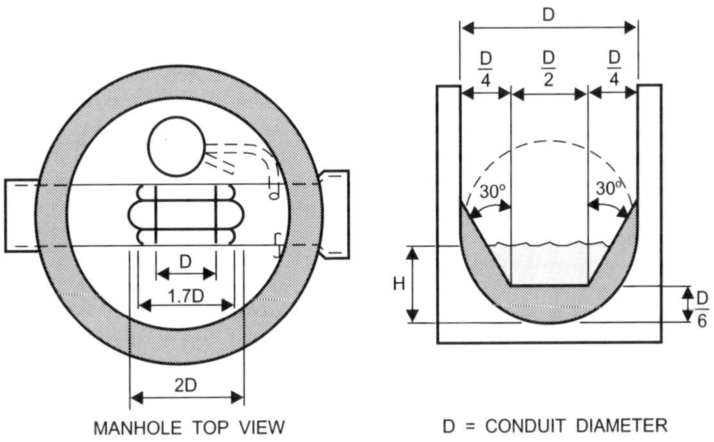

Figure 11-14. Dimensional Configuration of Standardized Palmer-Bowlus Flume Trapezoidal Throat Cross Section

The depth of the liquid above the throat of the flume (not the bottom of the pipe or conduit) is the index of discharge. As shown in Figure 11-16, the ideal location for the level measuring point is at a distance one-half *D* (pipe diameter or channel width) upstream from the entrance of the flume. However, this location is not critical as long as the point of level measurement is located above the upper transition section (Figure 11-16) in a zone where the depth of flow does not change significantly within a range of one pipe diameter.

Tabular data or rating curves, provided by the various manufacturers of Palmer-Bowlus flumes, relate the depth or head of liquid above the flume throat to a corresponding flow rate. Because of the great latitude possible in Palmer-Bowlus flume design mentioned previously, it is important to assure that the rating curve being used is the correct one for the flume in question. The rating curves supplied by two different manufacturers of the same size Palmer-Bowlus flume may have some differences. Therefore, the rating curve used should be the one provided by the manufacturer with the flume.

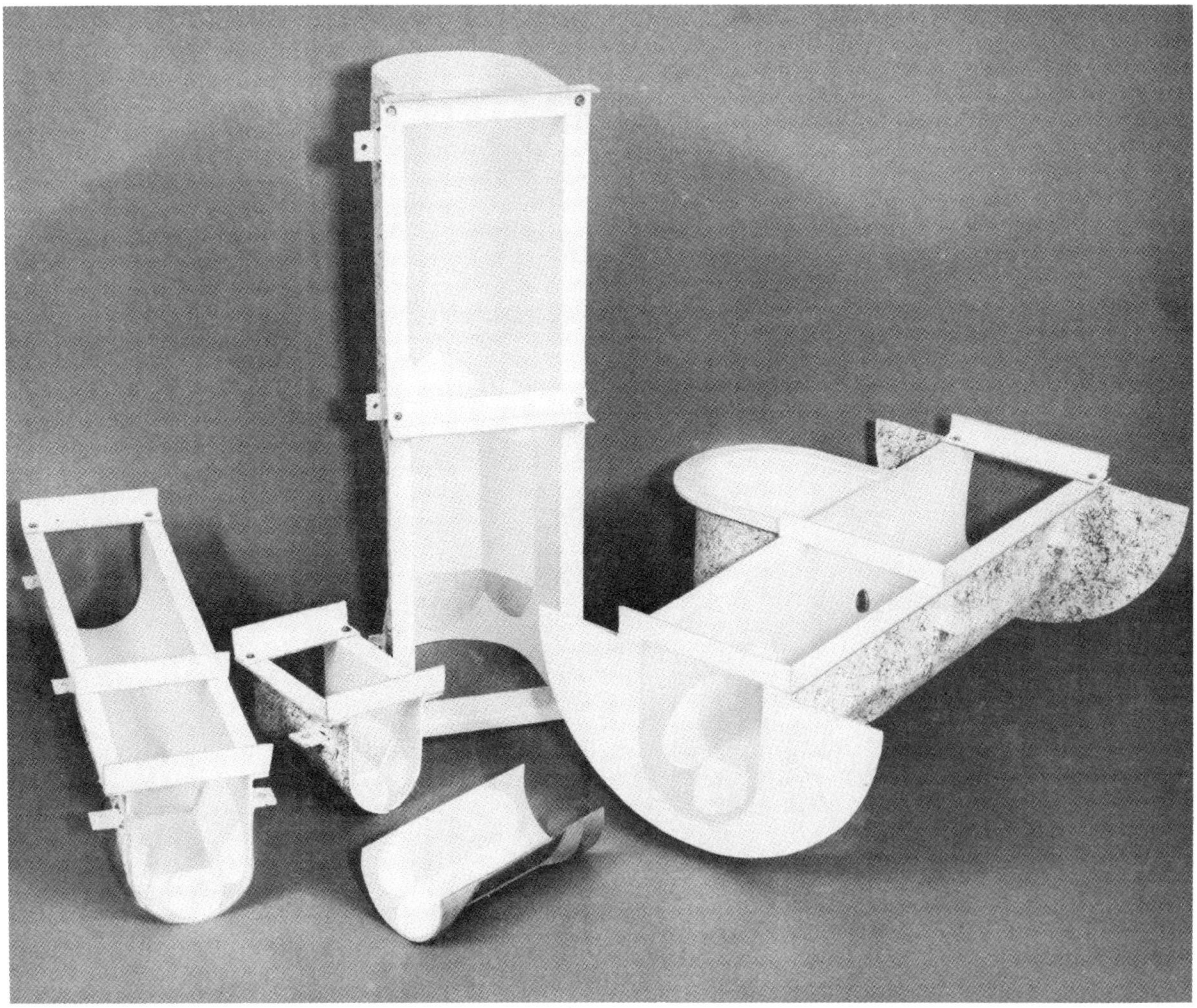

Figure 11-15. Standardized Palmer-Bowlus Flumes
(Courtesy of Plasti-Fab, Inc.)

HS, H, AND HL FLUMES

The HS, H, and HL flumes were developed in the mid 1930s by the U.S. Department of Agriculture (USDA) Soil Conservation Service to measure runoff from small agricultural watersheds and experimental plots. They have served this purpose adequately. Because of their proven performance, H-type flumes are now used to measure runoff from feedlots, runoff from infiltration areas for wastewater disposal, low flows of streams in pollution abatement work, and flow in industrial sewage systems. The H-type flumes are capable of monitoring flow over a wide range with reasonably good accuracy and have the advantage of simple construction. The wide span makes this primary device particularly suitable for measuring drainage water and for portable applications where a wide range of flow rates may be encountered. Other applications may be at installations with normal high flows and very low flows during off hours.

The H-type flumes, shown in Figure 11-17, differ from the flumes discussed above because they are more weir than flume. Their design attempts to combine the sensitivity and accuracy of the sharp-crested weir with the self-cleaning features of a flume. The result is a compromise in both. The flat, unobstructed bottom allows the passage of silt better than a weir. Like the weir, flow control is achieved by discharging through a sharp-edged opening. However, the flow is contracted gently from the sides only, much like the converging section of ordinary flumes. The plane of the exit tilts backward toward the incoming flow.

The forms of the H-type flumes were dictated by the desire for a simple geometric shape and by the character of the flows from small agricultural watersheds and plots. Because these flows often carry vegetal debris and sediment, the sidewalls of the flumes were made to converge gradually to the control opening, reducing the possibility of debris being caught on the flume and providing an increased velocity through the flume to reduce sediment. Also, these flumes must measure the full range of a flow event, so there was a need to provide both sensitivity for low-flow measurement and maximum capacity, with minimum head loss for the design flow, because the available head is usually limited. To meet the requirements for sensitivity and capacity, the tops of the converging vertical sidewalls slope upward from the lip of the outlet. This forms a trapezoidal opening, narrow at the bottom and wide at the top, which acts as the flow control section. The size of the flow control opening is a function of the convergence angle and the sidewall top slope. Varying these parameters also varies the sensitivity and maximum capacity of the flume.

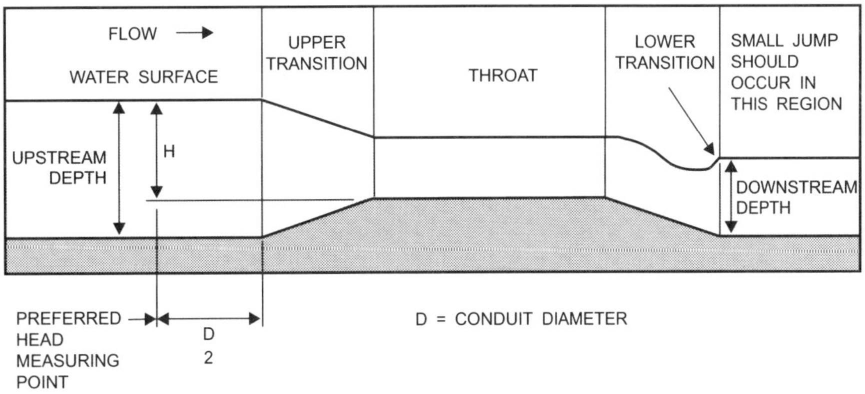

Figure 11-16. Free Flowing Palmer-Bowlus Flume

HS, H, and HL flumes are designated according to the maximum depth attainable in the flume, which is also the depth of the flume at the entrance. Thus, a 1.0-ft H flume has a maximum head of 1.0 foot. Dimensional proportions and maximum capacities of the H-type flumes are shown in Figure 11-18. Note that the dimensions of each type flume are proportional to the maximum depth, *D*. The HS (S for small) flumes were designed to measure relatively small flows, with maximum flow rates ranging from 0.08 cfs to 0.82 cfs (0.05 to 0.53 mgd/36 to 370 gpm). The H flumes were designed to measure medium flows, with maximum flow rates ranging from 0.35 cfs to 31.0 cfs (0.23 to 20 mgd/157 to 430,000 gpm). The HL (L for large) flumes were designed to measure larger flows, with maximum flow rates ranging from 20.7 cfs to 117.0 cfs (13.4 to 75.6 mgd/9300 to 6,140,000 gpm). The head measurement section for each of the flumes is shown in Figure 11-18.

Discharge (head vs. flow rate) equations for free-flow through H-type flumes have been developed by Gwinn and Parsons. However, these equations are quite complex and are not listed here; for further information, see Gwinn and Parsons in the Bibliography.

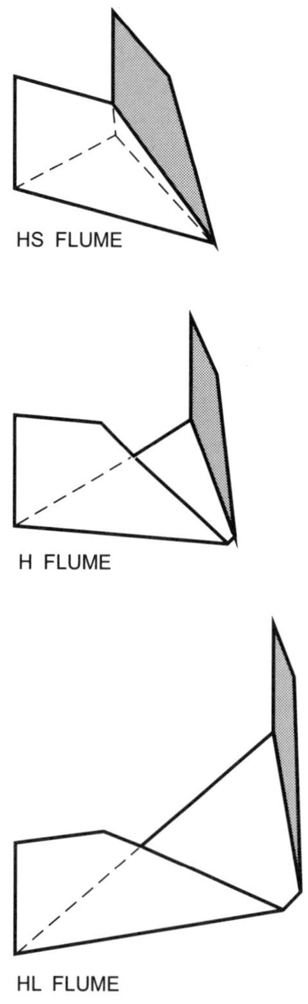

Figure 11-17. HS, H, and HL Flumes

TRAPEZOIDAL FLUME

In an attempt to obtain wider ranges of discharge than those available with Parshall flumes, several investigators have considered the supercritical trapezoidal flume. The general configuration of a trapezoidal flume is shown in Figure 11-19. The outward sloping of the flume walls provides increased sensitivity to low discharge rates for a given size and, hence, increased range.

The trapezoidal flume was developed primarily to measure flow in irrigation channels and has been used for many years by the Agricultural Research Service, U.S. Department of Agriculture. For agricultural applications it is superior to Parshall-type flumes for a number of reasons, particularly for measuring smaller flows. The trapezoidal shape conforms to the normal shape of ditches, especially those that are lined. This minimizes the amount of transition section needed as compared to that required when changing from a trapezoidal shape to a rectangular one and back to the trapezoidal. The trapezoidal shape is also desirable since the sidewalls expand as the depth increases. This means that one structure

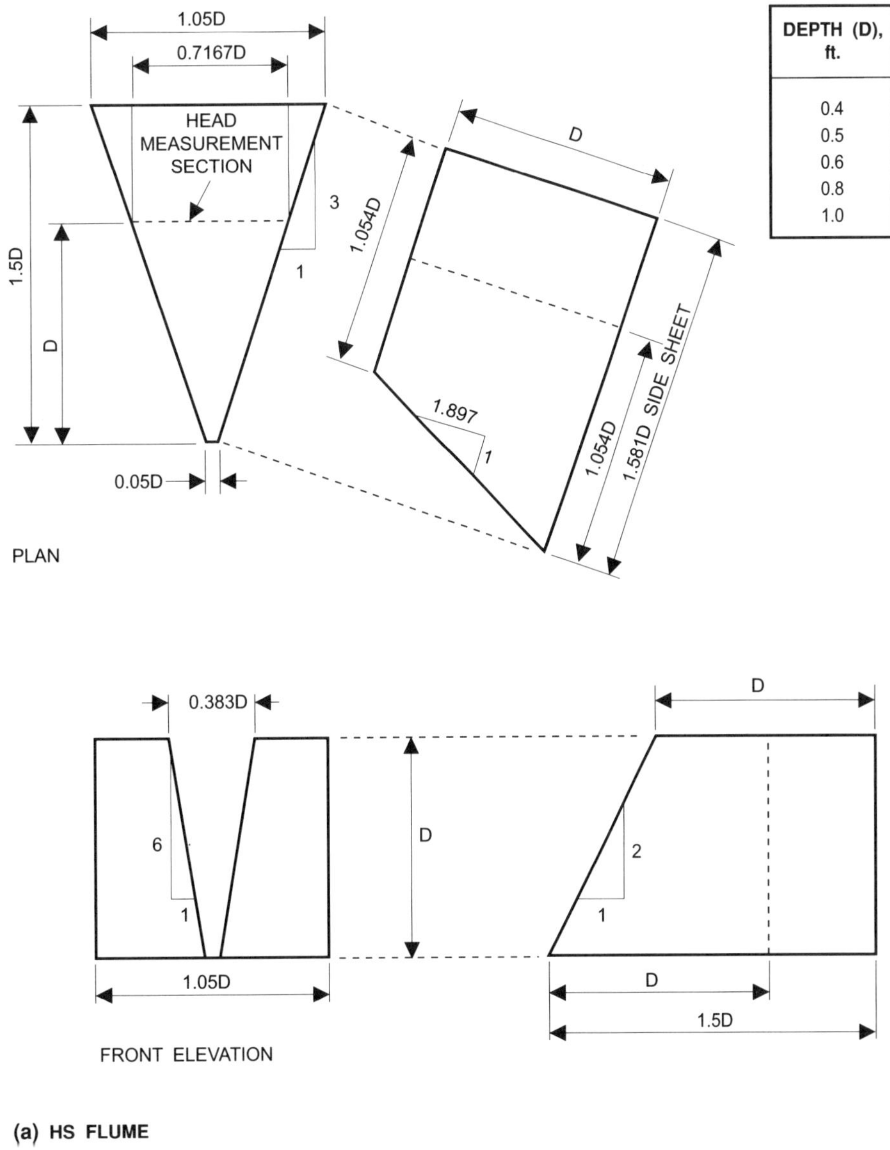

DEPTH (D), ft.	MAX CAPACITY, cfs
0.4	0.085
0.5	0.14
0.6	0.23
0.8	0.47
1.0	0.82

Figure 11-18. Dimensions and Capacities of H-Type Flumes

can convey a larger range of flow and that the entire range of depth for a given range of discharge is smaller. The trapezoidal flume can operate under a higher degree of submergence than the Parshall flume without corrections being necessary. Also, the straight-through bottom of the flume permits the flume to pass trash quite readily and reduces the problem of silt buildup upstream of the flume.

A trapezoidal flume need not include all of the elements shown in Figure 11-19. In some designs, the throat section is absent. Other designs eliminate the diverging and exit sections where channel erosion is not a problem. In yet another variation, the floor slopes slightly towards the center to form a very shallow "V." There are so many variations that no attempt will be made here to describe them all.

Probably the most widely used trapezoidal flumes are those of the various configurations designed and extensively evaluated by Robinson and Chamberlain. These are similar to the flume shown in Figure 11-19 and feature a flat floor throughout the flume that conforms to the general slope of the channel. The

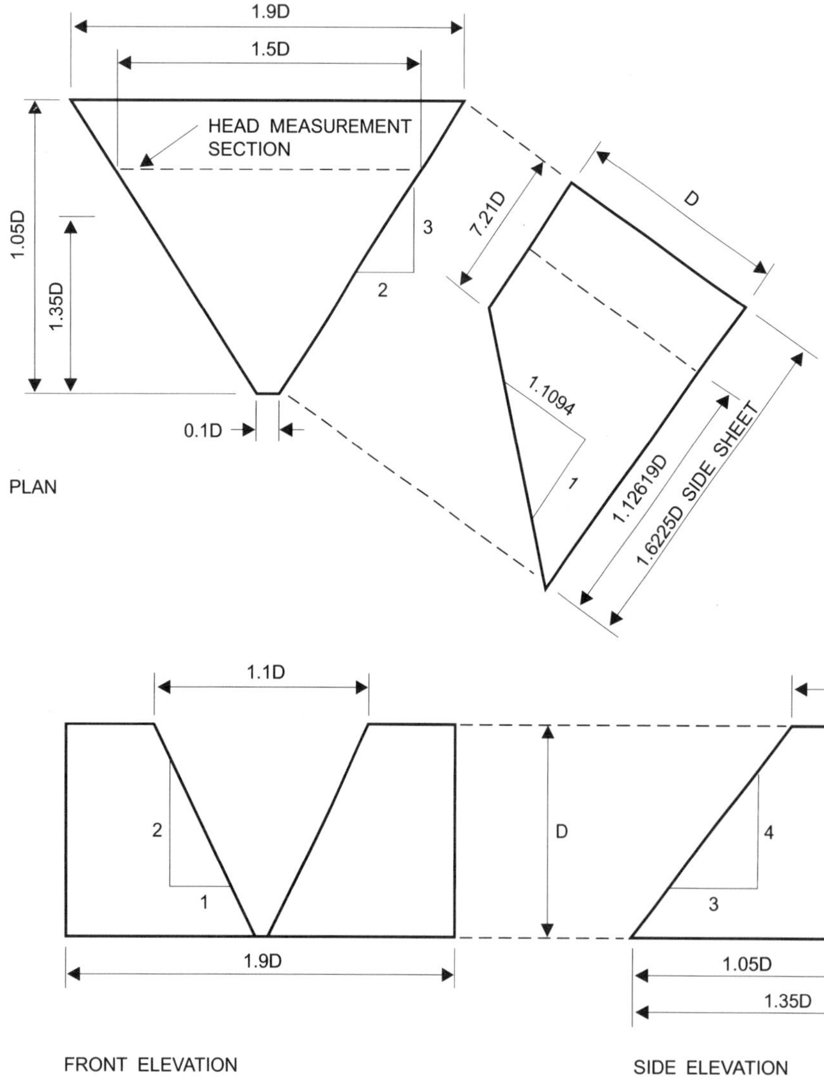

DEPTH (D), ft.	MAX CAPACITY, cfs
0.5	0.347
0.75	0.957
1.0	1.97
1.5	5.42
2.0	11.1
2.5	19.3
3.0	30.7
4.5	84.5

(b) H FLUME

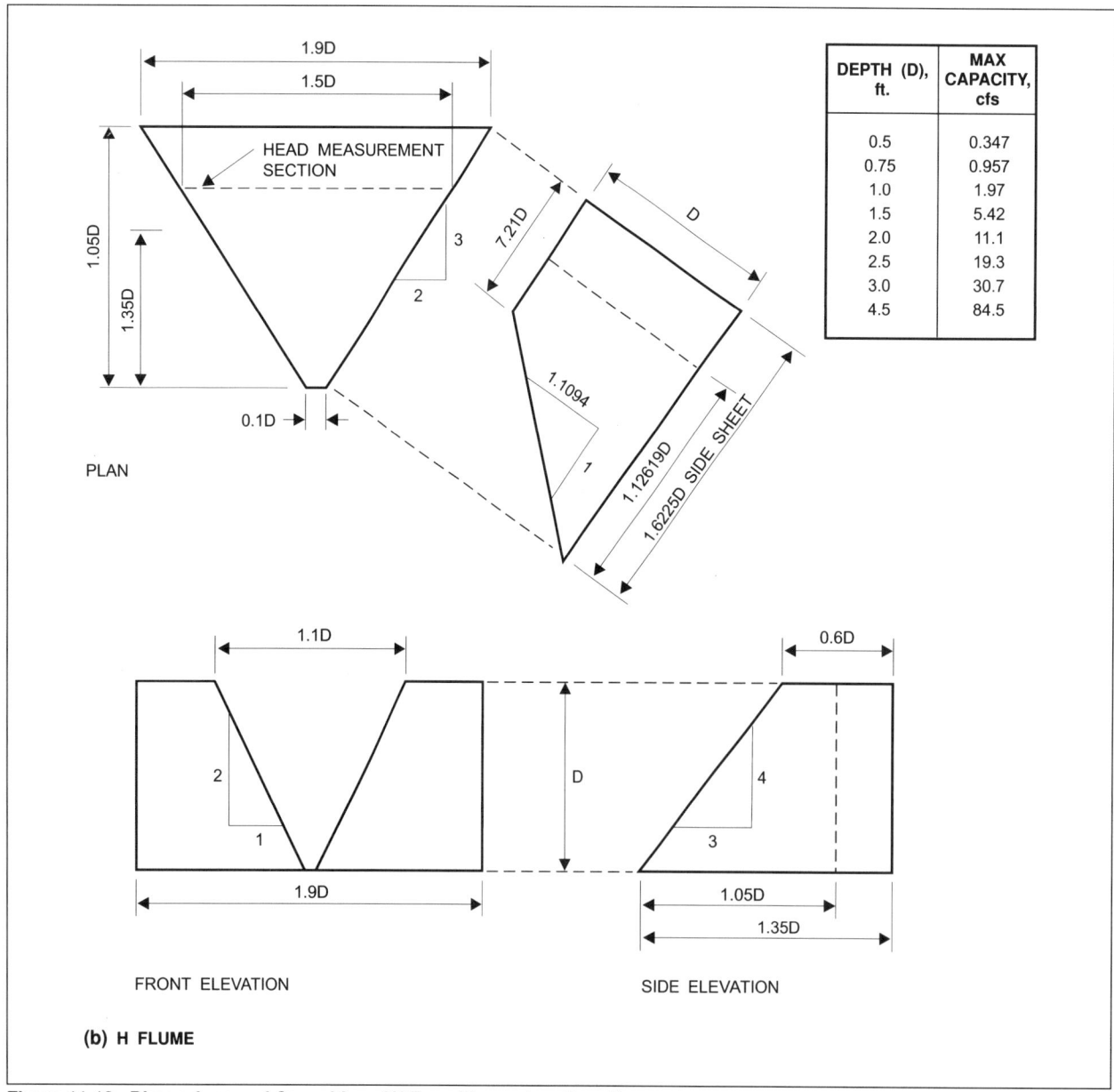

Figure 11-18. Dimensions and Capacities of H-Type Flumes (continued)

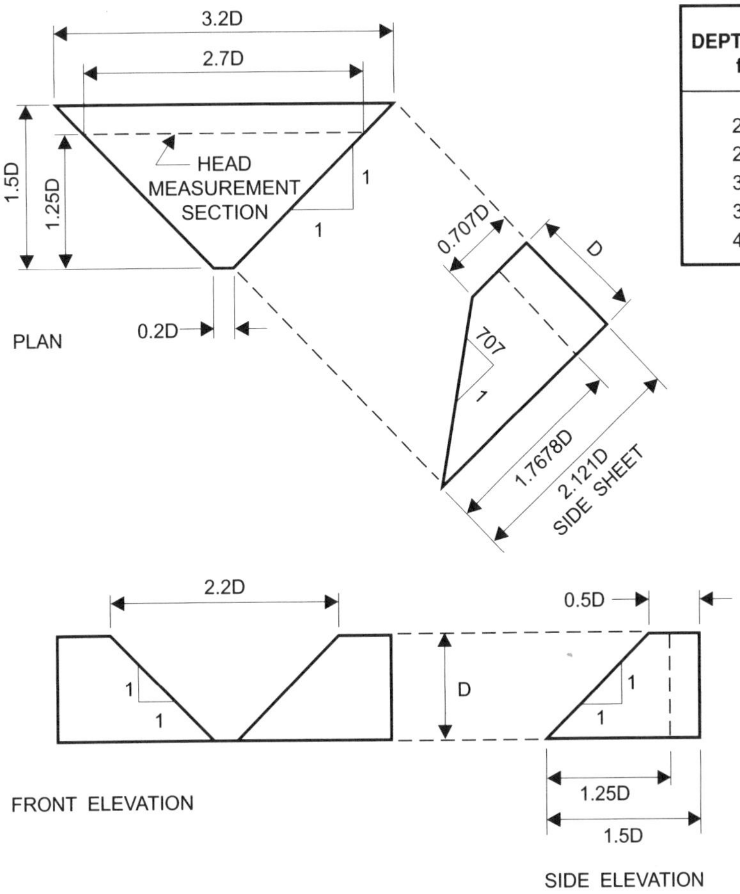

DEPTH (D), ft.	MAX CAPACITY, cfs
2.0	20.7
2.5	36.2
3.0	57.0
3.5	83.9
4.0	117.0

(c) HL FLUME

Figure 11-18. Dimensions and Capacities of H-Type Flumes (continued)

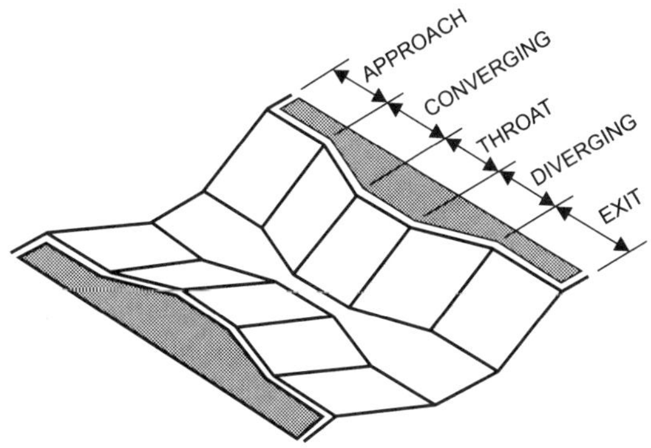

Figure 11-19. Elements of a Trapezoidal Supercritical Flow Flume

flumes were developed primarily for agricultural and stream flow measurement use. However, they are now being used in wastewater and industrial applications.

Of the trapezoidal flumes evaluated by Robinson and Chamberlain, three have been most widely used because they have been made commercially available by flume manufacturers. These, shown in Figure 11-20, are the 12 inch 45°SRCRC flume, the large 60°V flume, and the 2 inch 45°WSC flume. The 12 inch 45°SRCRC flume has a capacity range of 0.16 to 7.1 cfs (0.10 to 4.6 mgd/72 to 3200 gpm). The large 60°V flume has a capacity range of 0.01 to 0.2 cfs (0.006 to 0.13 mgd/4.5 to 90 gpm). The 2 inch 45°WSC flume has a capacity range of 0.025 to 1.8 cfs (0.016 to 1.2 mgd/11 to 810 gpm).

Applications

A flume tends to be self-cleaning since the velocity of flow through it is high and there is no actual "dam" across the channel. Therefore, it is generally more suited to flow channels carrying solids than is the weir. It can also operate with a much smaller loss of head than a weir, which can be important for many applications where the available head is limited. A flume is also not affected nearly as

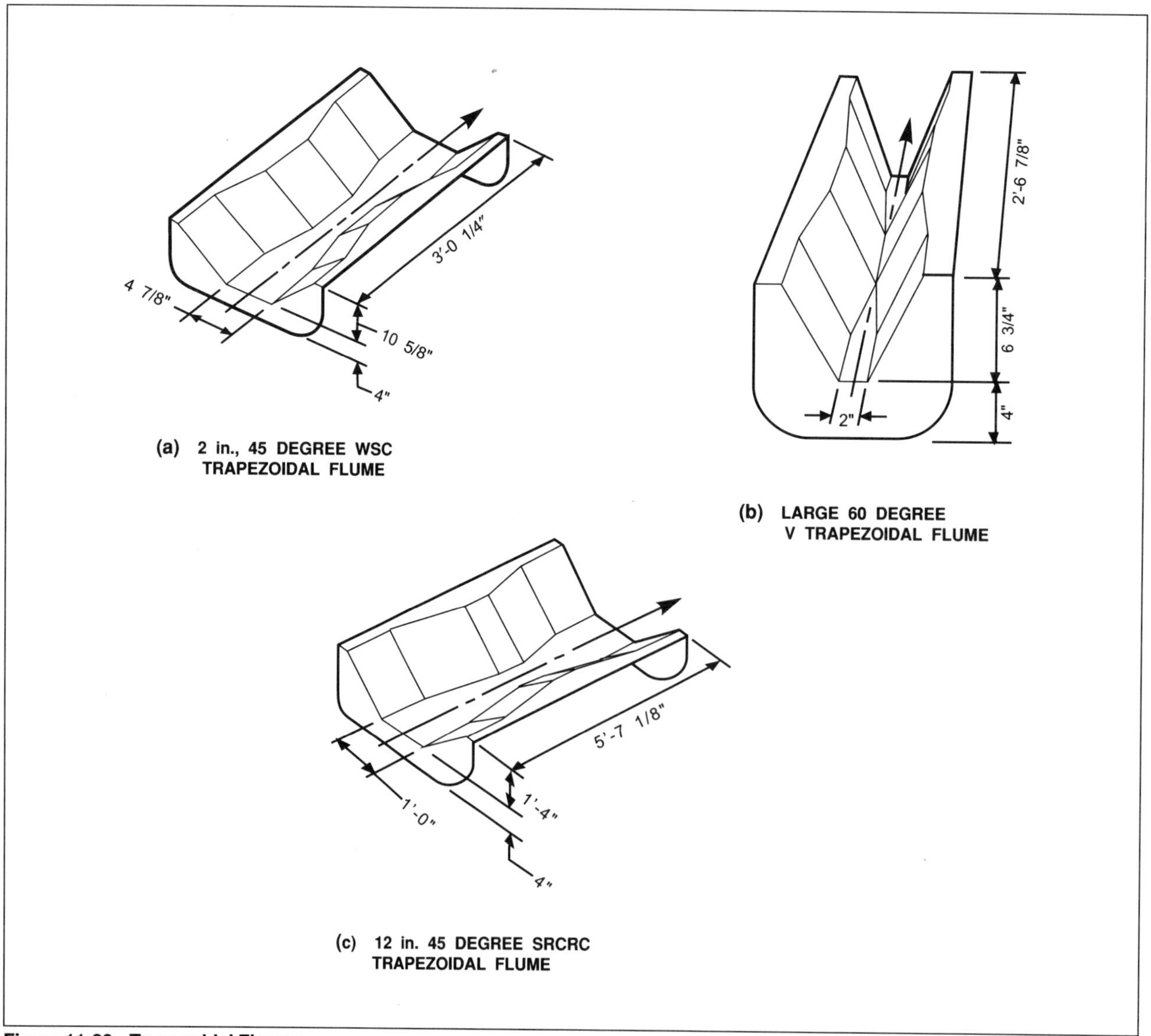

Figure 11-20. Trapezoidal Flumes

much as a weir by varying approach velocities. However, a flume is much more costly than a weir, and the installation is more difficult and time consuming. Flumes are also generally less accurate than weirs.

PARSHALL FLUME

The principal advantages of the Parshall flume are its capabilities for self-cleaning (particularly when compared with sharp-edged weirs), its relatively low head loss, and its ability to function over a wide operating range while requiring only a single head measurement. These characteristics make it particularly suitable for flow measurement in irrigation canals, in certain natural channels, and in industrial sewers.

PALMER-BOWLUS FLUME

The Palmer-Bowlus flume is most often used in manholes or open round or rectangular bottom channels to measure flow rate. It is useful as a temporary installation to provide flow data for determining flume size and equipment requirements for permanent installation. Some of the flume's advantages include accuracy of measurement (comparable to Parshall flumes), low energy loss, and minimal restriction to flow. A principal advantage of the Palmer-Bowlus flume is the comparative ease with which it can be installed in existing conduits, since it does not require a drop in the conduit invert that would be required with a Parshall flume. A disadvantage of Palmer-Bowlus flume is that it has a smaller useful range of flow rates than a Parshall flume, with a range that seldom exceeds twenty to one. Also, the resolution of the Palmer-Bowlus flume is not as good as that of the Parshall flume. For a given change in flow rate, the Parshall flume produces a greater change in head than does the Palmer-Bowlus flume. Thus, more sensitive head measuring instrumentation may be required with the Palmer-Bowlus flume.

HS, H, AND HL FLUMES

The HS, H, and HL flumes are capable of monitoring flow over a wide range with reasonably good accuracy and have the advantage of simple construction. The wide span makes this primary device particularly suitable for measuring drainage water and for portable applications where a wide range of flow rates may be encountered. Other applications may be at industrial installations that have normal high flows and very low flows during off hours.

TRAPEZOIDAL FLUME

The trapezoidal flume was developed primarily to measure flow in irrigation channels and has been used for many years by the Agricultural Research Service, U.S. Department of Agriculture. For agricultural applications it is superior to Parshall-type flumes for a number of reasons, particularly for measuring smaller flows. The trapezoidal shape conforms to the normal shape of ditches, especially those that are lined. This minimizes the amount of transition section needed as compared to that required when changing from a trapezoidal shape to a rectangular one and back to the trapezoidal. The trapezoidal shape is also desirable since the sidewalls expand as the depth increases. This means that one structure can convey a larger range of flow and that the entire range of depth for a given range of discharge is smaller. The trapezoidal flume can operate under a higher degree of submergence than the Parshall flume without corrections being necessary. Also, the straight-through bottom of the flume permits the flume to pass trash quite readily and reduces the problem of silt buildup upstream of the flume.

Operating Constraints

Flumes, aside from being operated within their flow limits, must also be operated within the head range of the associated flowmeter or level transmitter. Flumes are usually operated under free-flow conditions in which the discharge liquid level is low enough that it exerts no backpressure on the liquid in the throat of the flume. Otherwise, the flume will be in a submerged condition, and two head measurements must be made to determine the flow rate.

Operation of a flume is sensitive to turbulence or waves upstream from the entrance to the flume, which can necessitate a section of straight channel upstream of the flume.

Performance

A properly installed flume can typically achieve accuracies of 3 to 10 percent of the rate of flow. However, the overall accuracy obtained from the flow measurement system (the flume and the flowmeter) can be affected by a number of factors, including faulty construction or installation of the flume, improper head measuring location, incorrect zero setting of the flowmeter, flowmeter level measurement inaccuracies, flowmeter level-to-flow conversion inaccuracies, use of the flume outside its proper range, improper installation or maintenance of the flume, and turbulence surges in the approach channel. The reduced accuracy of a level transmitter or flowmeter is more significant in the lower portion of the flow range.

Sizing

PARSHALL FLUME

The size of a Parshall flume may be estimated using Figure 11-21. When selecting a Parshall flume, a number of factors should be considered to assure an accurate flow measurement system. The first consideration is the size of flume to be installed. Because of considerable overlap in flume discharges, it is possible to pass a given discharge through any one of several different standard size flumes. The choice of the proper size also requires consideration of other factors in addition to capacity. For example, a different throat width, *W*, will be required if 20 cfs (8980 gpm/12.9 mgd) is to be discharged with 2.5 feet of depth rather than with 1 foot of depth. In the interests of economy, the smallest practical size should usually be selected.

In selecting a Parshall flume size, it is usually necessary to use a "trial-and-error" system on several sizes believed adequate. The final selection is normally made on the basis of the original channel dimensions. Thus, if a 2-foot flume can accommodate the discharge without overrunning the upstream channel banks or flooding other outlets and facilities, it would be preferred over a 3- or 4-foot flume. However, when the width of the channel is considered, it may be just as economical to use a 3- or 4-foot flume because longer and more costly wingwalls may be needed to span the channel when using the narrower flume.

The minimum and maximum recommended flow rates for free flow through Parshall flumes have been experimentally determined. Table 11-4 presents the minimum and maximum recommended flow rates for a number of common sizes of Parshall flumes.

PALMER-BOWLUS FLUME

The size of a Palmer-Bowlus flume may be estimated using Figure 11-22. When choosing the size of Palmer-Bowlus flume to be installed at a particular location, both the diameter of the conduit or pipe in which the flume is to be installed and the range of expected flow rates should be considered. For maximum accuracy, the size of the flume should be determined by the expected volume of flow, rather than strictly by the pipe diameter. Many large sewer lines have very low flows and, therefore, require only small size Palmer-Bowlus flumes; end bulkheads can be installed to match the smaller flume to the larger pipe inside diameter. Sometimes, a slightly larger than required flume, which will provide adequate accuracy without the difficulty of bulkheads and with the capacity to accommodate possible future increases in flow rates, can be economically justified for use.

Table 11-5 presents recommended minimum and maximum flow rates through a number of Palmer-Bowlus flumes.

HS, H, AND HL FLUMES

The HS flumes were designed to measure relatively small flows, with maximum flow rates ranging from 0.08 to 0.82 cfs (0.05 to 0.53 mgd/36 to 370 gpm). The H flumes were designed to measure medium flows, with maximum flow rates ranging from 0.35 to 31.0 cfs (0.23 to 20 mgd/160 to 14,000 gpm). The HL flumes were designed to measure larger flows, with maximum flow rates ranging from 20.7 to 117.0 cfs (13.4 to 75.6/9290 to 52,500 gpm).

Of the three H-type flumes, H flumes appear to be the most widely used. The size of H flumes may be estimated using Figure 11-23.

Table 11-6 presents the minimum and maximum recommended flow rates for free-flow through a number of common sizes of H flumes.

TRAPEZOIDAL FLUME

The size of a trapezoidal flume may be estimated using Figure 11-24. The 12 inch, 45°SRCRC flume has a capacity range of 0.16 to 7.1 cfs (0.10 to 4.6 mgd/72 to 3200 gpm). The large 60°V flume has a capacity range of 0.01 to 0.2

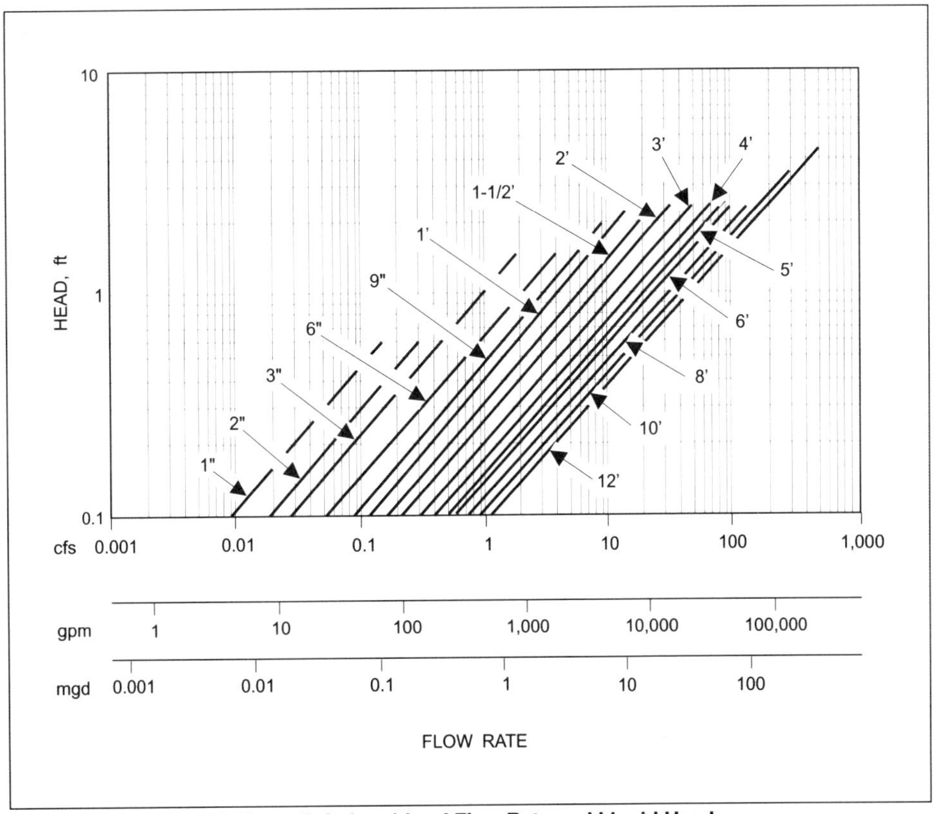

Figure 11-21. Parshall Flume Relationship of Flow Rate and Liquid Head

cfs (0.006 to 0.13 mgd/4.5 to 90 gpm). The 2 inch, 45°WSC flume has a capacity range of 0.025 to 1.8 cfs (0.016 to 1.2 mgd/11 to 810 gpm).

Table 11-7 presents the minimum and maximum recommended flow rates for free flow through these three commonly used trapezoidal flumes.

Construction

A flume is a three-dimensional structure of a specified configuration, as described previously. Because of their complex configuration, flumes are usually purchased prefabricated from one of a number of flume manufacturers. A variety of materials may be used to construct the flume, depending upon the nature of the liquid in the flow stream. Typical materials of construction include fiberglass, reinforced plastic, and stainless steel. Larger flumes are usually constructed on site from wood, concrete, galvanized sheet metal, or fiberglass.

Installation

When installing a flume, the following general installation requirements apply to all types of flumes:

Table 11-4. Minimum and Maximum Recommended Flow Rates for Parshall Flumes

Throat width, W	Min. head, ft.	Minimum flow rate			Max. head, ft.	Maximum flow rate		
		cfs	mgd	gpm		cfs	mgd	gpm
1 in.	0.07	0.005	0.003	2.24	0.60	0.153	0.099	68.7
2 in.	0.07	0.011	0.007	4.94	0.60	0.306	0.198	137
3 in.	0.10	0.028	0.018	12.6	1.5	1.86	1.20	835
6 in.	0.10	0.054	0.035	24.2	1.5	3.91	2.53	1750
9 in.	0.10	0.091	0.059	40.8	2.0	8.87	5.73	3980
1 ft.	0.10	0.120	0.078	53.9	2.5	16.1	10.4	7220
1½ ft.	0.10	0.174	0.112	78.1	2.5	24.6	15.9	11,000
2 ft.	0.15	0.423	0.273	190	2.5	33.1	21.4	14,900
3 ft.	0.15	0.615	0.397	276	2.5	50.4	32.6	22,600
4 ft.	0.20	1.26	0.816	565	2.5	67.9	43.9	30,500
5 ft.	0.20	1.55	1.00	696	2.5	85.6	55.3	38,400
6 ft.	0.25	2.63	1.70	1180	2.5	103	66.9	46,200
8 ft.	0.25	3.45	2.23	1550	2.5	139	90.1	62,400
10 ft.	0.30	5.74	3.71	2580	3.5	292	189	131,000
12 ft.	0.33	7.93	5.13	3560	4.5	519	335	233,000

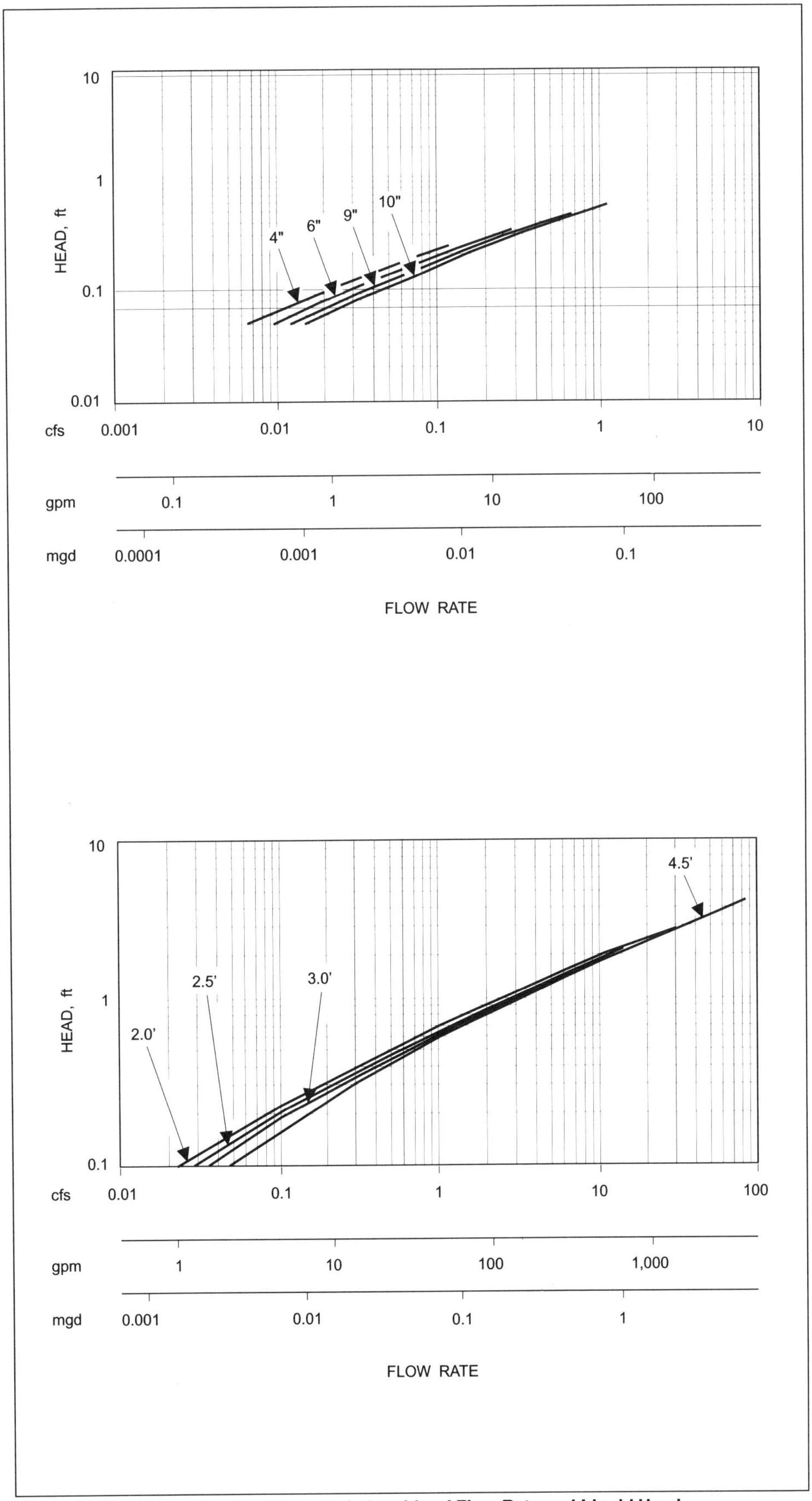

Figure 11-22. Palmer-Bowlus Flume Relationship of Flow Rate and Liquid Head

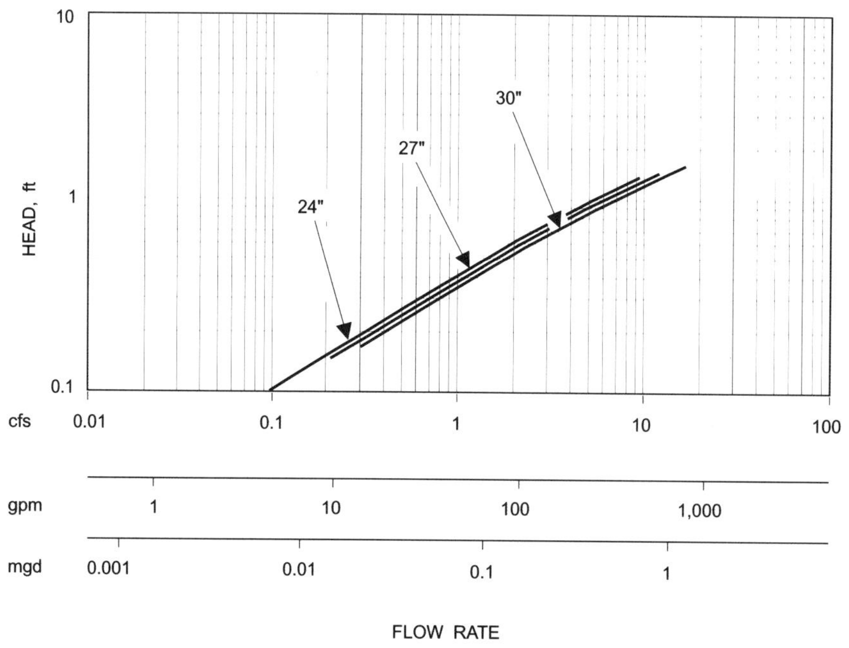

Figure 11-22. Palmer-Bowlus Flume Relationship of Flow Rate and Liquid Head (continued)

Table 11-5. Minimum and Maximum Recommended Flow Rates for Palmer-Bowlus Flumes
(Courtesy of Plasti-Fab, Inc.)

Flume size, in.	Max. slope for up-stream, percent	Min. head, ft.	Minimum flow rate			Max. head, ft.	Maximum flow rate		
			cfs	mgd	gpm		cfs	mgd	gpm
4	2.2	0.02	0.001	0.001	0.563	0.25	0.121	0.078	54.5
6	2.2	0.03	0.004	0.002	1.68	0.35	0.295	0.190	132
8	2.0	0.03	0.008	0.005	3.43	0.50	0.690	0.445	310
10	1.8	0.05	0.014	0.009	6.17	0.60	1.12	0.722	502
12	1.6	0.05	0.016	0.011	7.35	0.70	1.68	1.08	752
15	1.5	0.06	0.027	0.017	12.1	0.90	3.09	1.99	1385
18	1.4	0.08	0.051	0.033	22.7	1.05	4.61	2.98	2070
21	1.4	0.10	0.081	0.052	36.5	1.25	7.04	4.54	3160
24	1.3	0.10	0.096	0.062	43.0	1.40	9.47	6.11	4250
27	1.3	0.11	0.126	0.081	56.4	1.60	13.1	8.44	5870

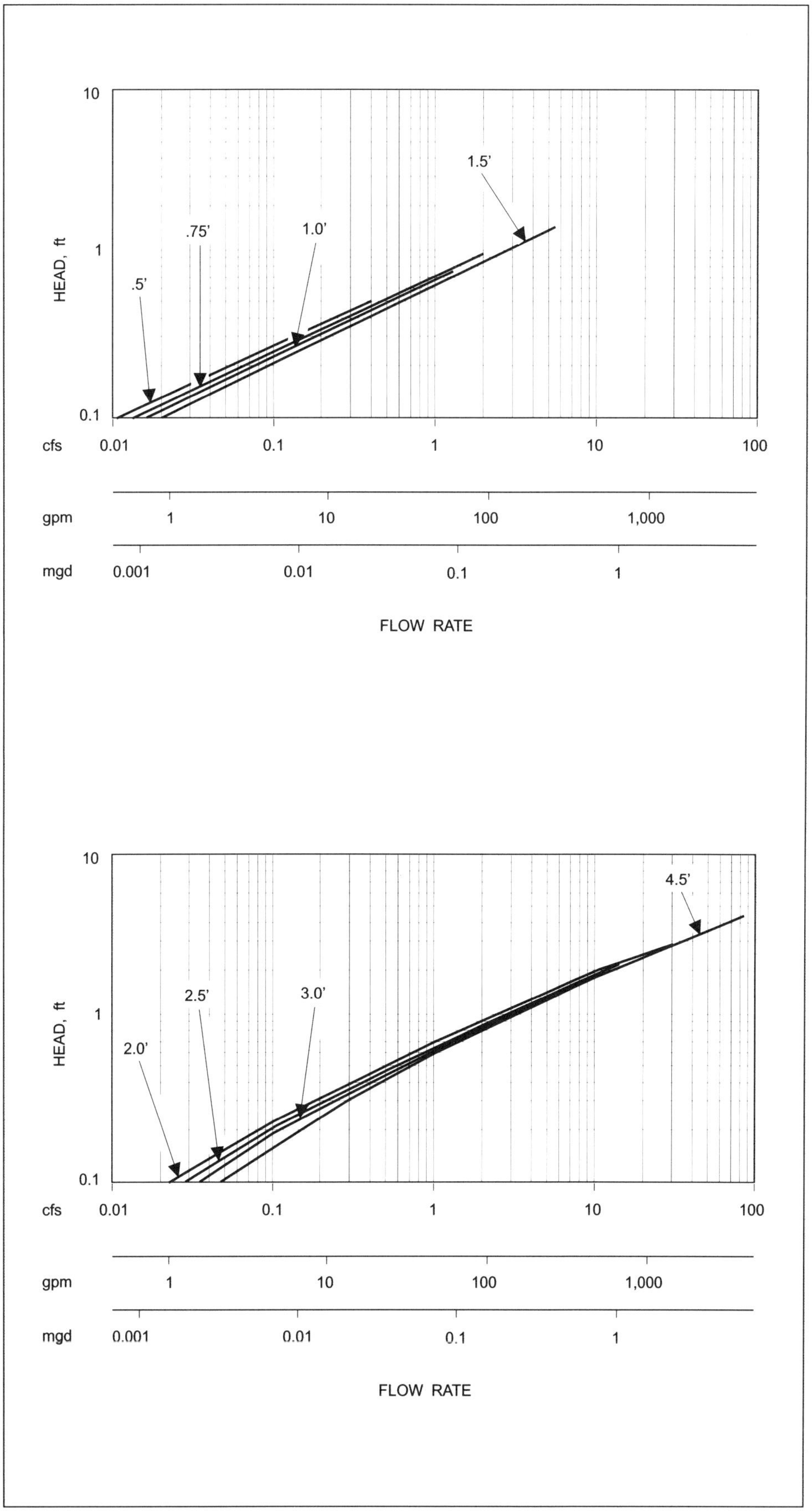

Figure 11-23. H Flume Relationship of Flow Rate and Liquid Head

Table 11-6. Minimum and Maximum Recommended Flow Rates for H Flumes

H flume size, ft.	Min. head, ft.	Minimum flow rate			Max. head, ft.	Maximum flow rate		
		cfs	mgd	gpm		cfs	mgd	gpm
0.50	0.02	0.0004	0.0003	0.180	0.50	0.347	0.224	156
0.75	0.02	0.0006	0.0004	0.269	0.75	0.957	0.619	430
1.0	0.02	0.0007	0.0005	0.314	1.0	1.97	1.27	884
1.5	0.02	0.0011	0.0007	0.494	1.5	5.42	3.50	2430
2.0	0.02	0.0014	0.0009	0.628	2.0	11.1	7.17	4980
2.5	0.02	0.0018	0.0012	0.808	2.5	19.3	12.5	8660
3.0	0.02	0.0021	0.0014	0.942	3.0	30.7	19.8	13,800
4.5	0.02	0.0031	0.0020	1.39	4.5	84.5	54.6	37,900

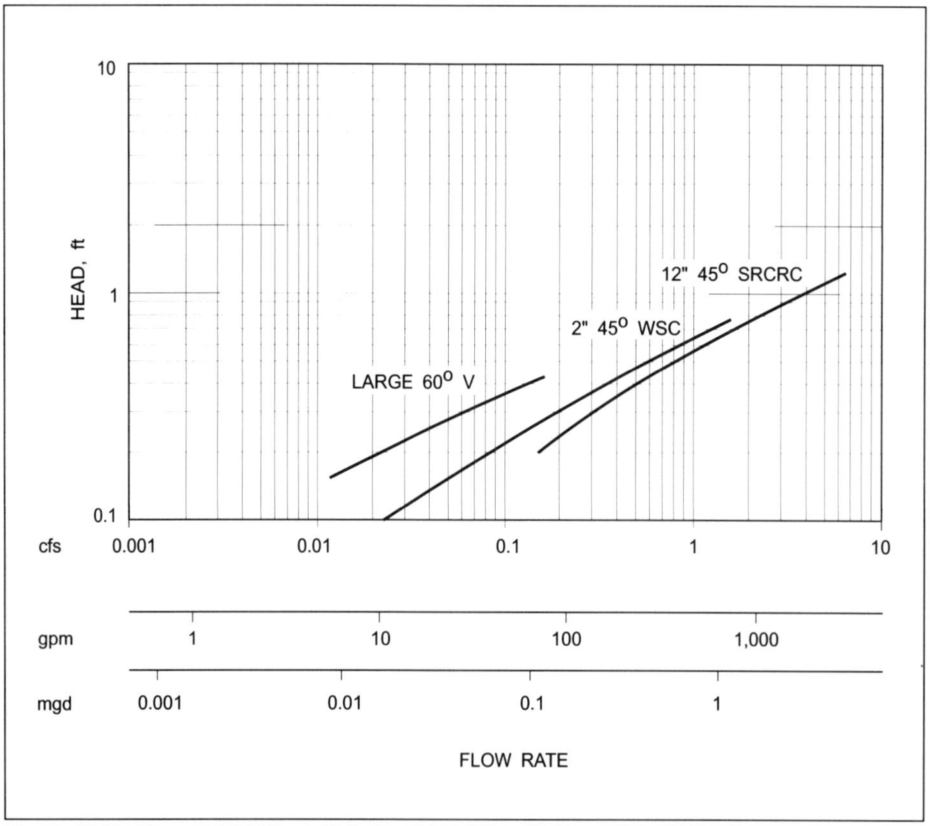

Figure 11-24. Trapezoidal Flume Relationship of Flow Rate and Liquid Head

Table 11-7. Minimum and Maximum Recommended Flow Rates for Trapezoidal Flumes

Flume type	Min. head, ft.	Minimum flow rate			Max. head, ft.	Maximum flow rate		
		cfs	mgd	gpm		cfs	mgd	gpm
Large 60° V	0.14	0.010	0.006	4.37	0.45	0.198	0.128	88.8
2" 45° WSC	0.10	0.023	0.015	10.3	0.77	1.82	1.18	817
12" 45° SRCRC	0.20	0.160	0.103	71.8	1.29	7.08	4.58	3180

(1) A flume should be located in a straight section of the open channel, without bends immediately upstream.

(2) The approaching flow should be well distributed across the channel and relatively free of turbulence and waves.

(3) Generally, a site with high velocity of approach should not be selected for a flume installation. However, if the water surface just upstream is smooth with no surface boils, waves, or high velocity current concentrations, accuracy may not be greatly affected by velocity of approach.

(4) Consideration should be given to the height of the upstream channel, with regard to its ability to sustain the increased depth caused by the flume installation.

(5) Although less head is lost through flumes than over weirs, it should be noted that significant losses may occur with large installations.

(6) The possibility of submergence of the flume due to backwater from downstream should also be considered, although the effect of submergence upon the accuracy of most flumes is much less than is the case with weirs.

PARSHALL FLUME

When installing a Parshall flume, particularly in the very small sizes, the crest should be used as an index. Careful leveling is necessary in both longitudinal and transverse directions if standard discharge tables are to be used. The flume should be set on a solid foundation to prevent settlement or heaving. Collars should be attached to either or both the upstream and downstream flanges of the flume and should extend well out into the channel banks and invert to prevent flow from bypassing the structure and eroding the foundation.

To assure accurate discharge measurement, the approach flow conditions of the Parshall flume should be considered. The approaching flow should enter the converging section reasonably well distributed across the entrance width, and flowlines should be essentially parallel to the flume centerline. Surges and waves of any appreciable size should be eliminated. Also, the flow at the flume entrance should be free of "white" water and free from turbulence in the form of visible surface boils.

Experience has shown that a Parshall flume should not be placed at right angles to flowing streams unless the flow is effectively straightened and uniformly redistributed before it enters the flume.

PALMER-BOWLUS FLUME

Palmer-Bowlus flumes are available in a number of installation configurations. The permanent-type flume is intended to be embedded in poured concrete for new construction and has the same inside radius as the pipe to which it is joined. The permanent type of flume is usually available with or without an integral approach section; the approach section is sometimes necessary to assure a smooth approach to reduce the turbulence present in a poorly defined channel. The invert or insert type of flume is used for temporary measurements or for permanent installation in an existing pipe. The flume is inserted and seated into the existing half section of pipe, with the outside radius of the flume being the same as the inside radius of the pipe. A final type is the cutback or exit type of flume. This special version of the Palmer-Bowlus flume is intended for temporary or permanent installation in the downstream exit pipe of a manhole, allowing space for upstream monitoring and sampling installations.

When installing a Palmer-Bowlus flume it is important to consider the conduit or channel slope. A minimum channel slope (which applies only to the downstream section) is necessary to maintain critical flow through the throat of the flume and prevent the flume from becoming submerged. It has been determined that the required critical flow will occur if the downstream depth of flow is not greater than 85% of the upstream depth (Figure 11-16), that is, if the submergence ratio is less than 85%. This is considered to be the upper limit of submergence allowable for proper flume operation. In new installations, effective operation can be assured by building in a slight drop on the downstream end of the installation. A small jump or rise in the water surface just below the throat of the flume is evidence that the required critical flow is occurring through the flume, as shown in Figure 11-16.

There is also a maximum allowable upstream slope, which is necessary to assure that the upstream flow is subcritical (lower in velocity than the throat of the flume) and is not turbulent. Otherwise, the flow at the point of measurement would be too choppy and rough to make meaningful head determinations. Also, the flow through the flume and at the measuring point above the throat must be straight and parallel. In order to obtain proper head readings, corrective measures must be taken to quiet any turbulence so that the upstream flow is tranquil and the flow is uniformly parallel through the flume. The upper limit of channel conduit slope necessary to maintain subcritical flow in the upstream section is normally on the order of a 2% slope for smaller flume sizes and a lesser slope for larger flume sizes.

When installing a Palmer-Bowlus flume, the downstream outlet pipe slope should be greater than or equal to the upstream pipe slope. The flume itself should be level, although a small slope will not significantly affect the accuracy of the flume. The downstream outlet pipe should be free of obstructions that would contribute to submergence. Upstream turbulence or obstructions to the flow should be avoided. There should be no bends, drop manholes, flow junctions, etc., within 25 pipe diameters (D) upstream of the flume location.

Wells and Gotaas [Ref. 30] conducted detailed studies on Palmer-Bowlus type flumes. Based on these studies, they summarized the characteristics of a freely discharging (nonsubmerged), properly functioning Palmer-Bowlus flume as follows:

(1) Upstream channel: The flow in the upstream channel is smooth, tranquil flow. There should be no aeration or prominent surface waves, especially at the point of upstream depth measurement.

(2) At the flume: The water enters and passes through the flume smoothly and with little turbulence. The surface profile should drop throughout the length of the flume. Streamlines are evident in the flow even after the point of critical depth has been reached.

(3) Downstream channel: A shooting flow is evident on the downstream side of the flume, indicating that the free discharge necessary for proper functioning of the flume prevails. In no case should the flow merely "neck down," i.e., show a smooth surface depression at the flume. In some instances, a hydraulic jump may form immediately downstream from the flume. Operation will be satisfactory if the upstream edge of the jump remains below the throat of the flume.

HS, H, AND HL FLUMES

When installing an H-type flume, the preferred configuration of the approach channel is rectangular, having the same depth and width as the flume and a length 3 to 5 times the depth of the flume. An H-type flume is generally installed with a free spill off the downstream end, as shown in Figure 11-25. Although H-type flumes were not originally intended to be used with a pipe at either the inlet or outlet end, they can often be installed quite satisfactorily in this manner. An H-type flume may be used in conjunction with an outlet pipe, as long as the water is permitted to flow or spill away from the flume in such a way that it does not slow down the flow through the flume notch. The water must spill from the flume or flow unimpeded in a manner comparable to spilling off the end.

An H-type flume must be installed so it is level, with the bottom of the flume at the same elevation as the inlet channel. The velocity of the water entering the flume must be subcritical and not turbulent. Generally, H-type flumes should be installed so there is free discharge, that is, the flume is not operating in a submerged condition. This may require enlargement and/or regrading of the outfall channel. Submergence of an H-type flume is defined as the ratio, expressed as a percentage, of the downstream water depth to the depth in the head measuring section. Tests have shown that submergence of 30 percent has less than a 1 percent effect on the calibration, and a 50 percent submergence has less than a 3 percent effect.

TRAPEZOIDAL FLUME

The trapezoidal flume, as with most flumes, should be installed so the flow through it is critical, not submerged. Submergence is defined as the ratio, expressed as a percent, of the downstream depth, h_4 (as shown in Figure 11-20), to the upstream depth. Test results have shown that corrections for submerged flow will have to be made for submergence ratios of greater than 80%.

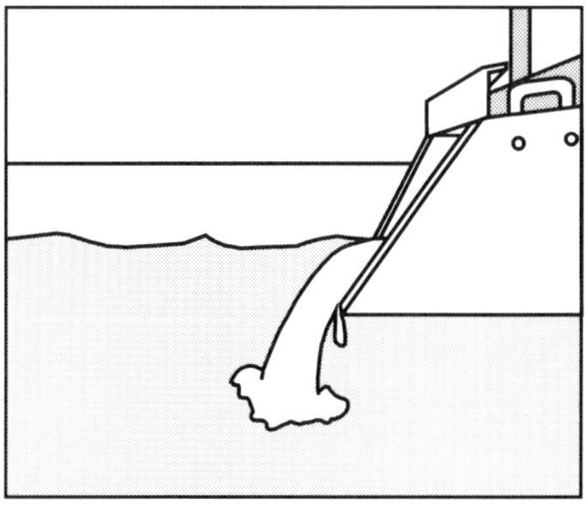

Figure 11-25. Discharge from H-Type Flumes

A trapezoidal flume should be installed level. Occasionally a flume is installed with a slight slope, which necessitates the adjustment of the flowmeter or level transmitter zero level so that it is at the same elevation as the flume throat. If the flume is installed in an earth ditch, the flume bottom should always be placed higher than the ditch bottom. If the flume is installed in a concrete ditch having a flat slope, the flume may become submerged; if this is the case, the flume should also be raised above the bottom of the channel.

The level transmitter or open channel flowmeter should be installed in accordance with the manufacturer's recommendations.

Be sure that the flowmeter's level-to-flow rate conversion matches the flume being used and that the desired units of measure for flow rate and total flow have been selected.

Calibration

The only calibration required for a flume is to adjust the zero and span of the level transmitter or flowmeter being used with the flume. The transmitter or flowmeter must be adjusted such that the indicated liquid level matches the actual level in the flow stream, relative to the flume's zero point. Note that the zero point of a flume is defined for each particular type of device and is not always obvious. For example, the zero point of a Palmer-Bowlus flume is the floor of the flume's throat, not the bottom of the pipe at the point where the level measurement is made, as might be expected. Adjustment of zero and span of the level transmitter or flowmeter should be performed per the manufacturer's instructions, which will vary significantly with the manufacturer, as well as with the technology that is used to measure level.

Level transmitters and flowmeters must also be calibrated to compensate for any hydrostatic heads resulting from the elevation of the transmitter or flowmeter relative to the zero point of the flume. Non-contact measurements must also be made to account for the elevation of the transducer above the crest level.

Safety

Flumes may be located in hazardous locations. The electrical rating of the flowmeter or level transmitter associated with the flume should be consistent with the location.

Flumes also are frequently located in manholes in which there is a possibility of an adverse atmosphere. Before entering a manhole, tests should be made for an explosive atmosphere, the presence of hydrogen sulfide or other dangerous gases, and oxygen deficiency. Whenever an adverse atmosphere is encountered, forced ventilation may be used to create safe conditions, or self-contained breathing apparatus may be used to protect the worker.

Maintenance

In a flume-based open channel flow measurement system, both the flume and the associated flowmeter or level transmitter require periodic maintenance. The operation of the flume is sensitive to any foreign material or debris that may be present upstream of the flume. Therefore, the flume should be periodically inspected and any accumulated debris removed.

The zero and span of the flowmeter or level transmitter should be periodically checked for proper calibration and adjusted per the manufacturer's instructions, if necessary.

Flume size may be estimated by using the graphs of the relationship between flow rate and the liquid head associated with the flume (Figures 11-21, 11-22, 11-23, and 11-24), and the tables of the minimum and maximum recommended flow rates (Tables 11-4, 11-5, 11-6, and 11-7), for Parshall, Palmer-Bowlus, H-type, and trapezoidal flumes, repectively.

Select and size a flume for the following conditions. The flow channel is an existing 18-inch diameter round pipe. Based on rough measurements, the normal minimum flow rate is approximately 350 gallons per minute and the normal maximum flow rate is approximately 1200 gallons per minute. The available head is limited to 0.95 foot.

Of the commonly available types of flumes, the Palmer-Bowlus flume is probably most suitable for this type of application because of its ease of installation in an existing round pipe. This, of course, assumes that a suitable manhole can be located, one that has a "U" channel section in which the flume may be installed. The logical size of Palmer-Bowlus flume to be installed in an 18-inch sewer line is, of course, an 18-inch flume. Referring to Table 11-5, it may be seen that the minimum and maximum flow rate of the site in question falls within the useful flow rate range of an 18-inch Palmer-Bowlus flume. Referring to Figure 11-22, it may be seen that an 18-inch Palmer-Bowlus flume generates a head of approximately 0.80 foot at the maximum expected flow rate of 1200 gallons per minute. This is acceptable since it is within the available system head of 0.95 foot. Thus, an 18-inch Palmer-Bowlus flume may be used for this installation.

Bibliography

1. Ackers, P.; White, W. R.; Perkins, J. A.; and Harrison, A. J. M., *Weirs and Flumes for Flow Measurement*, New York, NY: John Wiley & Sons, 1978.
2. *Annual Book of ASTM Standards, Part 31, Water*, Philadelphia, PA: American Society for Testing and Materials, 1974.
3. Associated Water & Air Resources Engineers, *Handbook for Monitoring Industrial Wastewater*, Environmental Protection Agency Technology Transfer Publication, August, 1973.
4. Chow, Ven Te, *Open-Channel Hydraulics*, New York, NY: McGraw-Hill Book Company, 1959.
5. Cone, V. M., "The Venturi Flume," *Journal of Agricultural Research*, pp 115-129, Vol. 9, No. 4, April 12, 1917.
6. Cone, V. M., "28th Annual Report," Colorado Agricultural Experiment Station, 1915.
7. Debevoise, N. T., and Fernandez, R. B., "Recent Observations and New Developments in the Calibration of Open Channel Wastewater Monitors," *Journal Water Pollution Control Federation*, pp 1188-1191, Volume 56, Number 11, November 1984.
8. Engal, F. V. A. E., "Non-Uniform Flow of Water: Problems and Phenomena in Open Channels with Side Contractions," *The Engineer*, pp 392-394, 429-430, 456-457, No. 155, 1933.
9. *Field Manual for Research in Agricultural Hydrology*, Agriculture Handbook No. 224, Washington, D.C.: Agricultural Research Service, Soil and Water Conservation Research Division, U.S. Department of Agriculture, 1962.

10. Grant, Douglas M., *Isco Open Channel Flow Measurement Handbook*, 3rd Edition, Lincoln, NE: Isco, Inc., 1989.

11. Gwinn, W. R., and Parson, D. A., "Discharge Equations for HS, H, and HL Flumes," *Journal of the Hydraulics Division, ASCE*, pp 73-88, Vol. 102, No. HY1, Proc. Paper 11874, January 1976.

12. Harris, James P.; Kacman, Stephen A.; Grant, Forest; and Tomcik, John, "Flow Monitoring Techniques in Sanitary Sewers," *Deeds & Data*, Washington, D.C.: Water Pollution Control Federation, July 1974.

13. Kilpatrick, F. A., "Use of Flumes in Measuring Discharges at Gaging Stations," *Surface Water Techniques*, Book 1, Chapter 16, U.S. Geological Survey, Washington, D.C.: United States Department of the Interior, 1965.

14. Kirkpatrick, George A., and Shelley, Philip E., *Sewer Flow Measurement —A State-of-the-Art-Assessment*, EPA Environmental Protection Technology Series, EPA-600/2-75-027, 1975.

15. Kulin, Gershon, and Compton, Philip R., *A Guide to Methods and Standards for the Measurement of Water Flow*, Springfield, VA: National Technical Information Service, 1975.

16. Ludwig, John L., and Ludwig, Russell G., "Design of Palmer-Bowlus Flumes," *Sewage and Industrial Wastes*, pp 1096-1107, Volume 23, Number 9, September, 1951.

17. Ludwig, Russell G., and Parkhurst, John D., "Simplified Application of Palmer-Bowlus Flowmeters," *Journal, Water Pollution Control Federation*, pp 2764-2769, Volume 46, Number 12, December, 1974.

18. Mougenot, G., "Measuring Sewage Flow Using Weirs and Flumes," *Water & Sewage Works*, pp 78-81, July 1974.

19. Palmer, Harold K., and Bowlus, Fred D., "Adaptation of Venturi Flumes to Flow Measurements in Conduits," *Transactions, American Society of Civil Engineers*, pp 1195-1216, Volume 101, 1936.

20. Parr, A. David; Judkins, Joseph F.; and Jones, Thomas E., "Point-Velocity Discharge Measurement Method for Sewers," *Journal Water Pollution Control Federation*, pp 113-117, Volume 53, Number 1, January 1981.

21. Parshall, R. L., "Measuring Water in Irrigation Channels with Parshall Flumes and Small Weirs," U.S. Soil Conservation Service, Circular 843, May 1950.

22. Robinson, A. R., "Trapezoidal Flumes for Measuring Flow in Irrigation Channels," ARS 41-140, Agricultural Research Service, U.S. Department of Agriculture, March 1968.

23. Robinson, A. R., "Water Measurement in Small Irrigation Channels Using Trapezoidal Flumes," *Transactions of the American Society of Agricultural Engineers*, pp 382-385, Volume 9, Number 3, 1966.

24. Robinson, A. R., and Chamberlain, A. R., "Trapezoidal Flumes for Open-Channel Flow Measurement," *Transactions of The American Society of Agricultural Engineers*, pp 120-128, Volume 3, Number 2, 1960.

25. Smoot, G. F., "A Review of Velocity Measuring Devices," United States Department of the Interior, Geological Survey Open File Report, April 1974.

26. Spitzer, David W., *Industrial Flow Measurement*, Research Triangle Park, NC: Instrument Society of America, 1990.

27. *Stevens Water Resources Data Book*, 2nd Edition, Beaverton, OR: Leupold & Stevens, Inc., 1974.

28. Vennard, John K., *Elementary Fluid Mechanics*, New York, NY: John Wiley & Sons, Inc., 1961.

29. *Water Measurement Manual*, United States Department of the Interior, Denver, CO: Bureau of Reclamation, 1967.

30. Wells, Jr., Edwin A., and Gotaas, Harold B., "Design of Venturi Flumes in Circular Conduits," *Transactions, American Society of Civil Engineers*, pp 749-771, Volume 123, 1958.

About the Author

Douglas M. Grant graduated in Mechanical Engineering from the University of Nebraska. He has been with Isco, Inc., of Lincoln, Nebraska, since 1969 and has served as Vice President and General Manager of its Environmental Division since 1987. Mr. Grant is the author of *The Isco Open Channel Flow Measurement Handbook* and numerous trade journal articles and technical papers on open channel flow measurement.

12

Oscillatory Flowmeters

Oscillatory flowmeters employ vortex formation and the Coanda effect, both of which produce a digital or pulse output that comes from the natural dynamics and physics of the fluid. This natural pulse relates frequency to flow rate and is produced without moving parts, which makes these flowmeters quite reliable.

The Vortex Shedding Flowmeter

The vortex shedding phenomenon is nothing new. It occurs in nature. The first recorded observation was by Leonardo da Vinci more than 400 years ago when he noted the formation of vortex swirls downstream of a rock in a stream of water. At that time, while interesting to observe, the phenomenon was of no practical value. It required modern electronics to make some use of this phenomenon.

Operating Principle

When a flowing medium strikes a non-streamlined object or obstruction, it separates and moves around the object and passes on downstream. At the point of the contact with the object, vortex swirls or eddy currents separate from the object on alternating sides. When this occurs, the separation or shedding causes a local increase in pressure and a decrease in velocity on one side of the object, and a local decrease in pressure with corresponding increase in velocity on the other side of the object (see Figure 12-1). After shedding from one side, the process is reversed and a swirl or vortex is shed on the other side of the object. In this way the vortex swirls are shed continuously—180 degrees out of phase with each other. The frequency of the shedding process is proportional to the velocity of the material flowing past the object known as the bluff body.

An illustration of the vortex shedding phenomenon is a flag waving in the wind. As the wind passes the flag pole, the pole acts as the bluff body, causing vortices to be shed. As the vortices are shed by the pole, the vortex swirls past the flag with the alternating high and low pressure areas on either side of the flag. These alternating high and low pressure areas cause the flag to wave in the wind.

Another illustration that demonstrates the principle is a canoe paddle being pulled through a body of water. As the paddle is pulled through the water, a series of vortices (known as a vortex street) is generated. Vortex swirls being shed on alternate sides from the edges of the canoe paddle can be seen in the water. The paddle trying to twist can be felt—first toward one side, then the other—as the vortex being shed generates the high and low pressure areas at the edge of the paddle blade. As the vortex street moves downstream, the swirl increases in pressure and decreases in velocity until eventually the swirl ends and normal flow resumes.

Early studies were conducted by Theodor von Karman in 1912 using a right circular cylinder as the bluff body. More recent work has proved that the use of a non-streamlined object having a sharp edge to define the point of vortex separation yields a more consistent vortex swirl. Due to the pioneering work by von Karman, the series of vortex swirls have been called the von Karman vortex street (see Figure 12-2).

The vortex shedding phenomenon not only occurs in streams around rocks, but also around bridge piers and pilings, around offshore drilling platform supports, and by the wind around tall buildings. Engineers must take the vortex shedding phenomenon into account when designing these structures. If the vortex swirl, while moving downstream, strikes another object directly in its path, damage may occur. Offshore platform legs have been damaged and tall buildings have had problems with window glass being broken due to the vortices shed by an upstream leg or another tall building in the flow path. In a pipeline, this effect is normally dissipated within a few pipe diameters downstream of the bluff body. It is not likely to cause any problem within the piping system.

The Bluff Body

The frequency of the shedding is directly proportional to the velocity of the flow in the pipe, which permits measurement of the flow rate by sensing the frequency of the vortex formation on the alternate sides of the bluff body. The fluid velocity has been determined to be proportional to the vortex frequency and width of the bluff body. The proportionality ratio is called the Strouhal number, which is dimensionless.

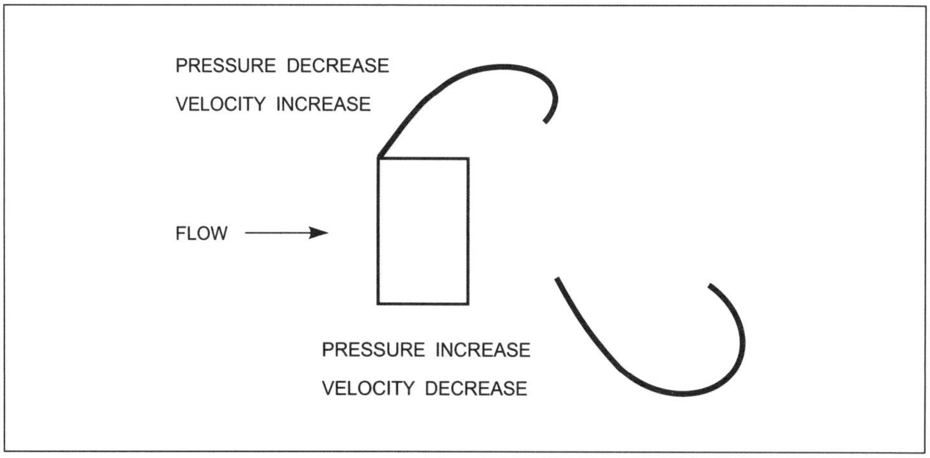

Figure 12-1. Operating Principle

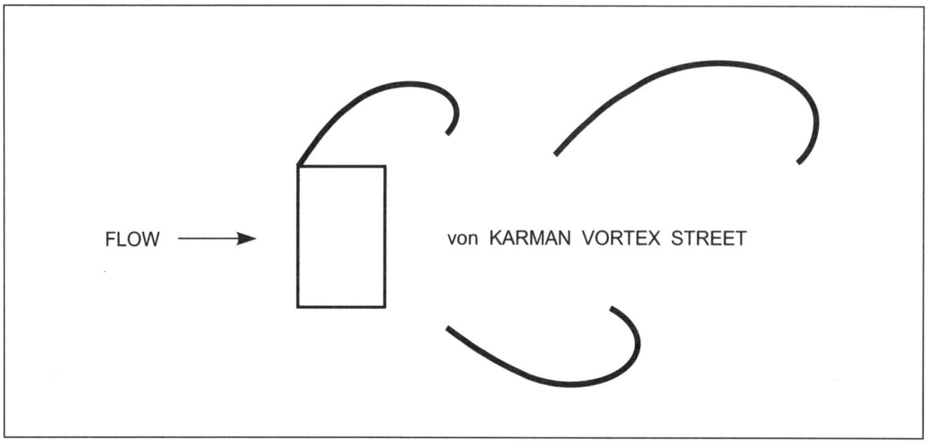

Figure 12-2. von Karman Vortex Street

$$\text{St} = \frac{fd}{\bar{v}}$$

where St is the Strouhal number, f is the frequency of the shedding, d is the width of the bluff body, and $\bar{v}$ is the average fluid velocity.

The width of the bluff body within a given meter is fixed, therefore, a constant. The frequency of shedding then is linearly proportional to the average flowing velocity over a wide range of Reynolds numbers. Typical vortex frequencies range from as low as one or two cycles per second to thousands of cycles per second depending on flowing velocity, flowing medium, and meter size. Meters measuring gas normally have frequencies typically about 10 times the frequency found on meters in liquid applications. This is due to the fact that gas flowing velocities are normally much higher than liquid velocities in the same size pipeline.

Since proper vortex formation is essential, bluff body shapes and dimensions have been experimentally determined to achieve the desired balance of characteristics. This has resulted in many bluff body configurations being supplied by the various vendors of vortex meters. The one feature that has become universal is that the bluff body has a sharp edge on its upstream edge (the edge that faces into the flow). It is this sharp edge that has improved the strength and regularity of vortex shedding. It has improved both the accuracy and rangeability of vortex shedding necessary to meet industrial standards. Most vortex meters today will function within the manufacturer's stated accuracy at all Reynolds numbers from 10,000 or 20,000 to as high as 10,000,000. Many vendors will also permit measurement at very low Reynolds numbers at a somewhat reduced accuracy. In many cases, if the meter is to be used for process control applications, this reduced accuracy may not cause any significant problem to the user.

To look at the mathematics of vortex metering, the relationship of volumetric flow is:

$$Q = A\bar{v}$$

where Q equals the volumetric flow rate, v equals the average velocity, and A equals the cross-sectional area of the meter body.

If the average velocity in terms of Strouhal number is substituted, the relationship becomes:

$$Q = \frac{fdA}{\text{St}}$$

Since the Strouhal number, which is constant over a wide range of Reynolds numbers, and the d (bluff body width), and meter cross section are constants (which is defined as K), the equation can be rewritten as:

$$Q = \frac{f}{K}$$

As with other frequency producing flowmeters such as turbine meters, the K factor can be defined as pulses per unit volume—pulses per gallon or pulses per cubic foot, etc. This being the case, all that is needed is to determine how many pulses per unit time are being generated to find the flow rate.

Conversely, it is necessary only to count pulses to know how much flow has been accumulated.

The Vortex Swirl

Many methods are used to detect the vortex swirl. Early designs incorporated a magnetic shuttle ball that moved back and forth within the bluff body as the vortices caused the pressure to change from one side of the bluff body to the other. This had a tendency to cause problems, because the small passage that the ball moved in tended to plug if any foreign matter was present. Another early design incorporated thermistors that would detect the small differences in temperature on the bluff body as the vortex swirls separated from the bluff body. In some instances this design was a problem due to interference from the process temperature. Many of the later designs utilize piezoelectric crystals, which generate an electrical pulse when a vortex swirl passes. Others rely on the torquing motion on the bluff body caused by the alternating high and low pressure areas generated during the vortex shedding process. The stresses on the bluff body may be detected by piezoelectric crystals or by a differential switched capacitance. Still another method is to direct a sonic beam across the flow diagonally. This beam is deflected by the movement of the vortex swirl, thereby generating a pulse output. The vendors of vortex meters have spent considerable time and effort to develop the best method of sensing the vortex swirls. Each has reason to claim technical superiority. It is for the user to decide, often from experience, which is the best for the particular application.

When To Use the Vortex Meter

Since there is no universal flowmeter that is applicable for all flow measurements, it is important to determine where each type of meter has its advantages and limitations. The satisfaction or lack of satisfaction with any given flowmeter installation depends on how well the flowmeter's characteristics meet the needs of the application.

The vortex meter works well on relatively clean liquids, gas, and steam that do not contain a significant amount of solids. The pipe Reynolds number should be above the minimum specified by the manufacturer (typically 10 - 20,000) to obtain desired accuracy. Therefore, liquids usually have a relatively low viscosity. The vortex shedding meter usually has at least a 10 to 1 rangeability (10% to 100%) and frequently ranges much greater than that. With a vortex meter, useful ranges up to 40 to 1 are possible. However, the measurement range is fixed by the meter itself. The typical vortex shedding flowmeter with a wet factory calibration has a typical accuracy of ± 0.75% of reading or better for liquids and ± 1.5% of reading for gases and steam.

The vortex shedding flowmeter is easy to install, requiring few components and hardware, thereby saving installation time and cost. Installation requires mounting the flowmeter body between appropriate flanges and making the appropriate electrical connections.

On liquids where the Reynolds number is less than 20,000, a vortex shedding flowmeter should be used with caution. With low Reynolds number liquids, the shedding of the vortex swirl tends to become less regular. This irregular shedding of the vortex causes the output to become nonlinear. To determine just how low a Reynolds number may be permitted and still have an effective meter, refer to the literature of the vendor in question. Some vortex flowmeter designs are able to operate at lower Reynolds numbers than other meter designs. Some are able to operate at low Reynolds numbers with some loss of accuracy. In some control applications, reduced accuracy at low flows may not preclude the use of a vortex meter. Note that vortex shedding (and, hence, flowmeter output) ceases when Reynolds numbers fall below 3000 to 5000.

Another questionable area for the vortex shedding flowmeter is measurement of gases that have a low density at the operating pressure. The problem with low density gases is that the vortex swirl does not have a strong enough pressure pulse to enable the sensor to distinguish between a vortex swirl and pipeline noise. When this occurs, the vortex meter will not be able to measure flows at the low end of the flow range, and the output will go to zero. Since the actual usable Reynolds number range and the usable gas density vary from one vendor to another, it is recommended the vendor's literature be consulted if the operating parameters fit into either of these conditions.

Vortex shedding flowmeters have had great success in applications where clean liquids and gases are being metered. It may still be used if small amounts of foreign matter are present. Since there are no moving parts in this flowmeter, there is little concern for erosion or physical damage caused by the foreign matter. In some designs there are, however, pockets where small particles may collect and interfere with proper operation of the meter. Small solids, particularly if they are slightly abrasive, can be a major source of trouble to many flowmeters and can be destructive to meter internals. With a vortex shedding flowmeter, the flow strikes the bluff body and separates at the sharp edge. The effect of erosion of the sharp edge is small and often causes no significant accuracy problem (see Figure 12-3).

Installation

The vortex shedding flowmeter is not attitude sensitive; flow can be upward or horizontal without affecting performance. The pipe must be kept full of liquid at all times when metering is required. The optimum installation remains with the flow in a vertically upward direction since this assures that the pipe will be full. Operating with less than a full pipe will result in an error in flow measurement.

Straight run requirements vary from one vendor to another (see Figure 12-4). While most vendors use the same recommendations that are published for orifice meters, the design of some vortex meters will permit shorter straight runs. These reduced straight runs sometimes carry a penalty of reduced accuracy. In some applications where piping is very tight, the reduced accuracy may be an acceptable trade-off, depending on the purpose of the meter (see Figure 12-5). This trade-off gives the user fewer constraints in locating the meter in the piping system. A reasonable loss of accuracy will often permit a vortex shedding flowmeter to be installed where other meter types cannot fit.

When mounting this meter in the pipeline, correct installation is important. The mating pipe should ideally be the same diameter as the meter bore. Most vendors' literature indicates the size pipe that should be used to mate up with their meter (see Table 12-1). Either schedule 40 or schedule 80 pipe is normally specified. Some vortex shedding flowmeters may be able to function in other pipe schedules with or without an adjustment to the meter K factor (see Table 12-2). However, other meters will not function with any acceptable level of accuracy in any pipe schedule other than that specified. It may be desirable to reduce the

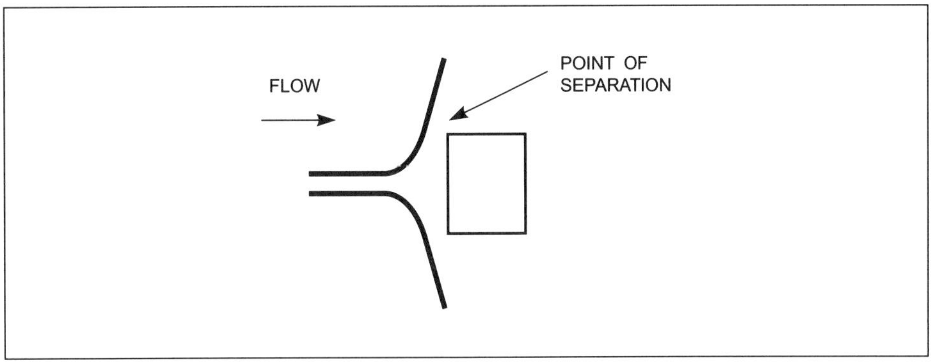

Figure 12-3.

pipe size, but the pipe comprising the straight run upstream and downstream of the flowmeter must be of the specified schedule for the meter accuracy to be met.

It is important also that the alignment procedure specified by the manufacturer be followed closely. If a vortex shedding flowmeter is not mounted and aligned properly in the pipe, turbulence caused by the misalignment will interfere with the proper formation of the vortex street. This will cause the signal to be "noisy," because the turbulence occurs just at the inlet of the meter where the vortex swirl is being generated. The vortex swirl becomes irregular, resulting from the hydraulic noise. Poor meter performance is the penalty. Vortex shedding flowmeter manufacturers provide tools or other alignment gages with their meters to allow the user to install the meter correctly (see Figure 12-6).

Along with the meter alignment, it is important that the gaskets installed with the meter not protrude into the flow stream. Protruding gaskets will also produce hydraulic noise, which will affect meter performance. The gaskets used should have an I.D. no smaller than the meter bore; they should be self-centering; and they should be cut from thin sheet gasket material. Thick gaskets such as spiral wound gaskets (Flexitalic™, for example) will leave a gap between the end of the pipe (flange face) and the meter body face. This gap can also be a source of turbulence and contribute to meter problems. To help install the meter with the correct gaskets, some vendors provide these gaskets. When possible, it is best to use the gaskets supplied.

How To Select the Correct Size Vortex Shedding Flowmeter

Consult the vendor about computer software for automatic size selection.

All vendors of vortex meters have their own calculation method to select the correct size meter. In fact, many vendors have a computer program that will take inputs of the process operating conditions and automatically select the best size meter for the application (see Figure 12-7). Often, these programs are offered to users either free or at a nominal charge. Additionally, all have printed instructions to enable the user to manually determine the correct meter size.

As a general rule, the best size selection often is one pipe size smaller than the pipe into which the meter will be installed. The reason for this is that the vortex

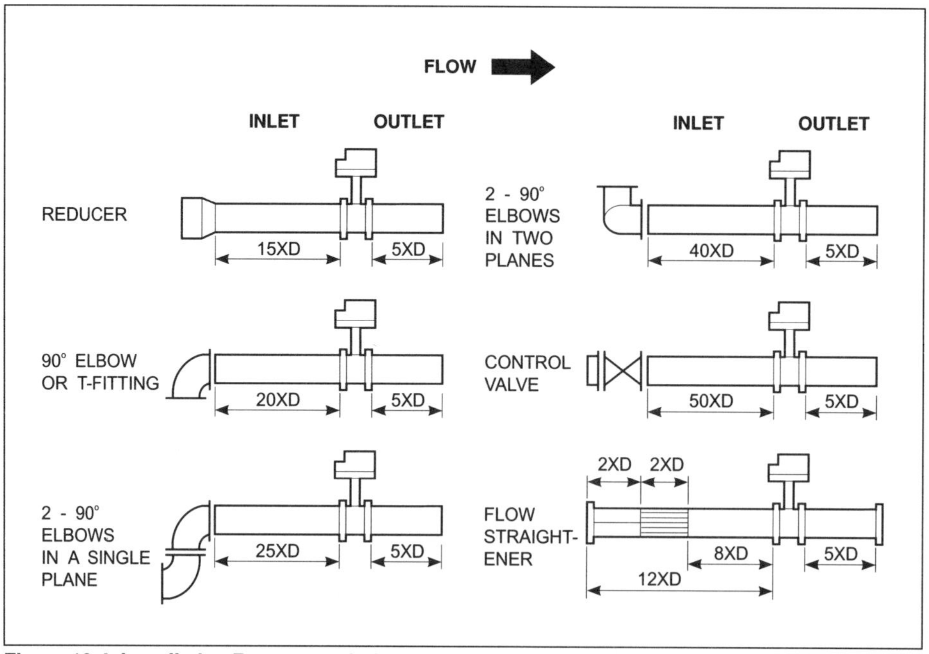

Figure 12-4. Installation Recommendations

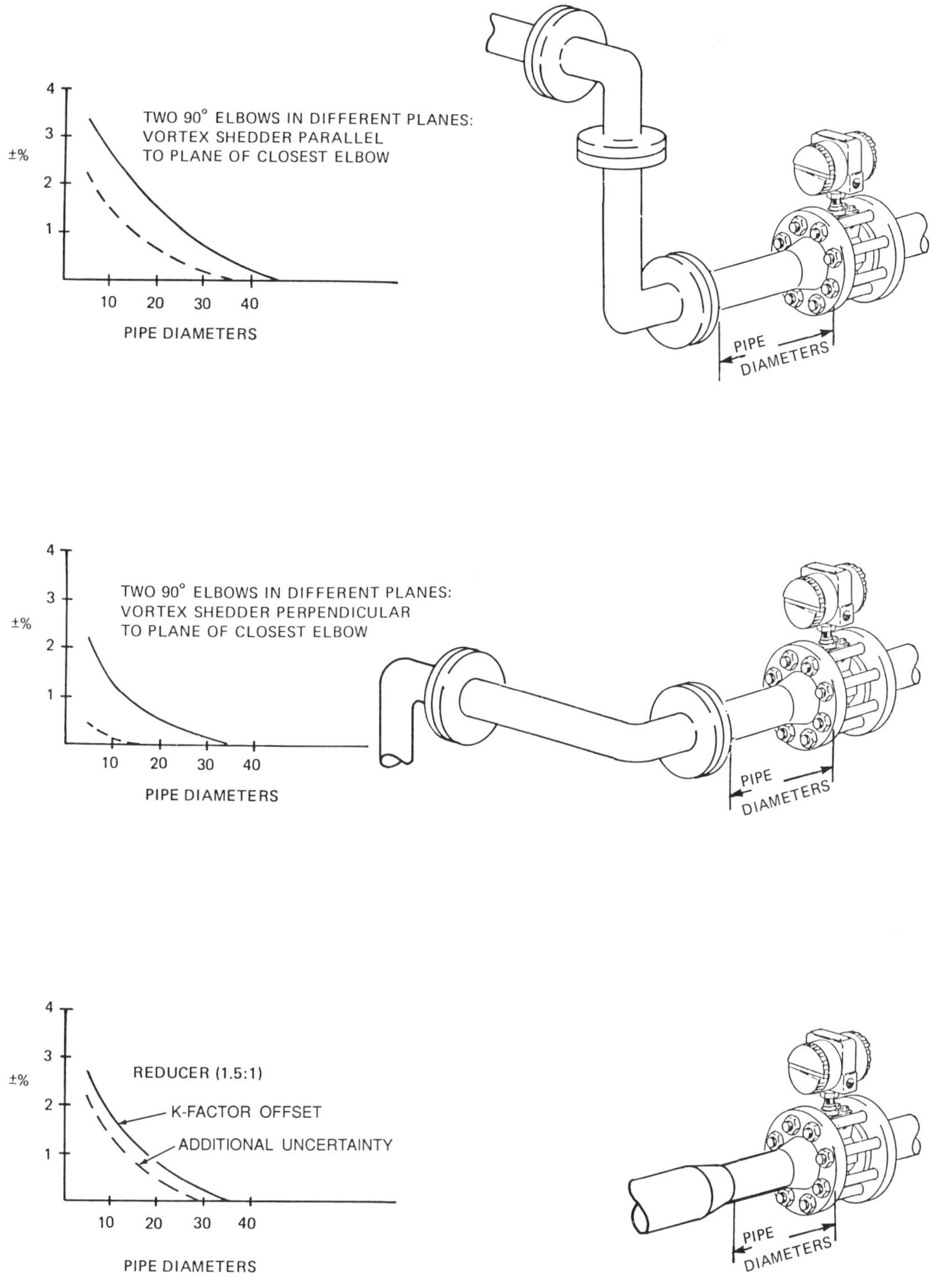

Figure 12-5. Straight Runs
(Courtesy of The Foxboro Company)

Table 12-1.
(Courtesy of Fischer & Porter Co.)

Piping Configuration		Additional Inaccuracy with Upstream Piping Length of				
		2 Diameters	4 Diameters	6 Diameters	8 Diameters	10 Diameters
Single Elbow in Plane of Shedder	A	DO NOT USE	- 1.6% of rate	- 0.9% of rate	- 0.2% of rate	0
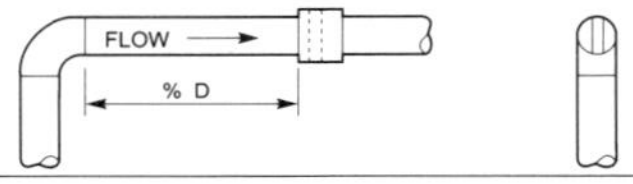	B	DO NOT USE	± 0.3% of rate	± 0.2% of rate	± 0.1% of rate	0
Single Elbow in Plane Perpendicular to Shedder Plane	A	+ 0.1% of rate	- 0.6% of rate	- 0.2% of rate	+ 0.4% of rate	+ 0.5% of rate
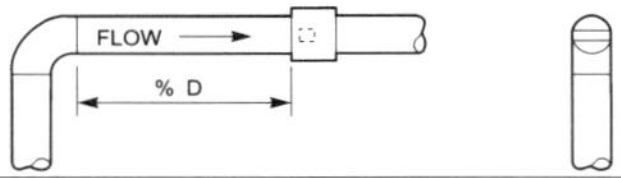	B	± 0.1% of rate	± 0.2% of rate	± 0.1% of rate	0	0
Two Elbows Close-Coupled, with Close Elbow in Plane of Shedder	A	- 0.8% of rate	- 0.3% of rate	0	+ 0.3% of rate	+ 0.6% of rate
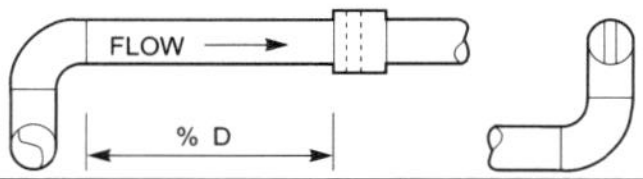	B	± 0.2% of rate	± 0.1% of rate	± 0.1% of rate	± 0.1% of rate	0
Two Close-Coupled Elbows with Close Elbow Perpendicular to Shedder	A	+ 0.5% of rate	- 0.2% of rate	- 0.1% of rate	0	+ 0.2% of rate
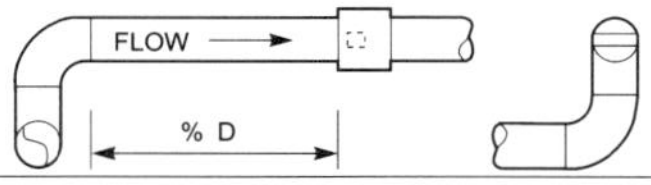	B	± 0.3% of rate	± 0.2% of rate	± 0.2% of rate	± 0.1% of rate	± 0.1% of rate
1.3 : 1 Reducer	A	DO NOT USE	+ 0.7% of rate	+ 0.4% of rate	+ 0.2% of rate	0
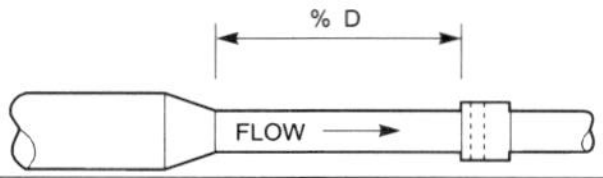	B	DO NOT USE	0	0	0	0
Expander	A	← DO NOT USE →				
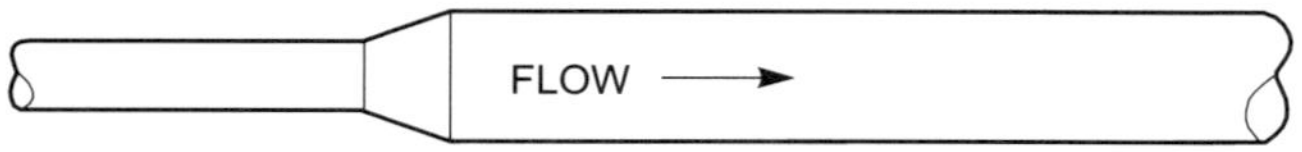	B					

A = Percent Change in $\overline{K}$
B = Additional Linearity to be Added to Standard Linearity

Table 12-2. K Factor Offest
(Courtesy of The Foxboro Company)

Flowmeter Size		Liquid Flow			Gas and Steam Flow		
mm	**in.**	**Sched 10**	**Sched 40***	**Sched 80**	**Sched 10**	**Sched 40**	**Sched 80**
50	2	+ 0.4%	ref	- 0.8%	+ 0.4%	ref	- 0.8%
80	3	+ 1%	ref	- 0.8%	—	ref	- 0.8%
100	4	+ 1%	ref	- 0.2%	—	ref	- 0.2%

* K Factor stamped on data plate is based on calibrating with schedule 40 pipe. The flowmeter bore diameter is equivalent to the inside diameter of a schedule 80 pipe.

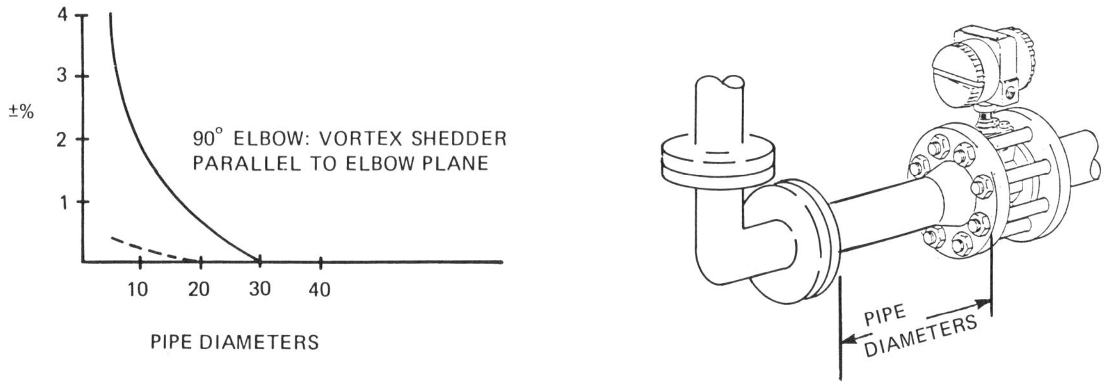

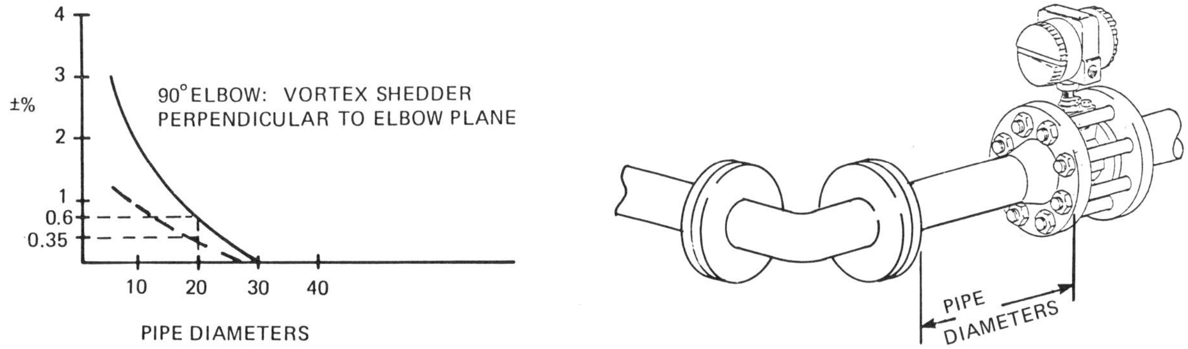

shedding flowmeter works best when the flowing velocity is high. Pipe sizes are selected so that the flowing velocity is relatively low. This lower velocity in the pipe creates less pressure drop and less wear on the pipes themselves. The flowmeter can operate at high flowing velocities without adverse effects on the meter. Vortex shedding flowmeters can often be overranged by as much as 50% without adverse effects. Since there are no moving parts in it, overranging does not endanger most meters.

The area where the vortex meter is vulnerable to loss of performance is at the low end of the flow range—at the bottom of the flow scale. Flow rates that are too low will cause the signal to be lost or to become nonlinear.

OPERATING DENSITY

If the meter is measuring gas flow, the operating density becomes important. At some point the density/flowing velocity relationship becomes so low that the meter sensors are not able to distinguish the flow signal from pipeline noise. When this occurs, the electronics are usually designed to indicate a no-flow condition. Just where in the flow velocity/density this occurs varies from one vendor to another due to the varying types of sensors and their sensitivity.

REYNOLDS NUMBER

On liquid applications, at the low end of the flow range, the problem is likely to be related to the Reynolds number. Each vendor provides information as to just how low a Reynolds number for a particular meter will be linear. Below that value a vortex shedding flowmeter does not shed vortices that are regular enough to meet reasonable accuracy requirements. Some vendors will specify only a minimum usable Reynolds number that will meet the stated accuracy of their meter.

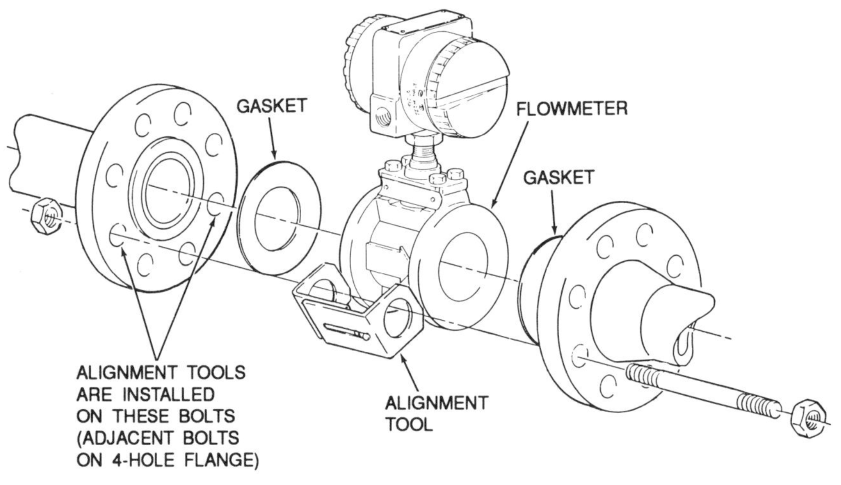

Figure 12-6. Tool for Proper Alignment

CUSTOMER NAME	:	ABC CHEMICALS	TEMPERATURE	:	120.0˚F
PROJECT NAME	:	DISTILLATION EXPAND.	PRESSURE	:	40.0 psig
TAG NAME	:	FT-123	DENSITY	:	0.850 S.G.
FLUID NAME	:	SOLVENT			
FLUID TYPE	:	LIQUID	VISCOSITY	:	0.6 cSt
CONNECTION	:	A N S I			
PRESS. RATING	:	150 lbs	FULL SCALE	:	175 gpm

METER SIZE	:	2" ANSI
METER INNER DIAM.	:	2.07 in.
METER LENGTH	:	2.56 in. WAFER
MINIMUM FLOW	:	8.24 gpm
.75% OF RATE FROM	:	8.50 gpm
MAXIMUM FLOW	:	308 gpm
PRESS. LOSS (at fs)	:	4.6 psi

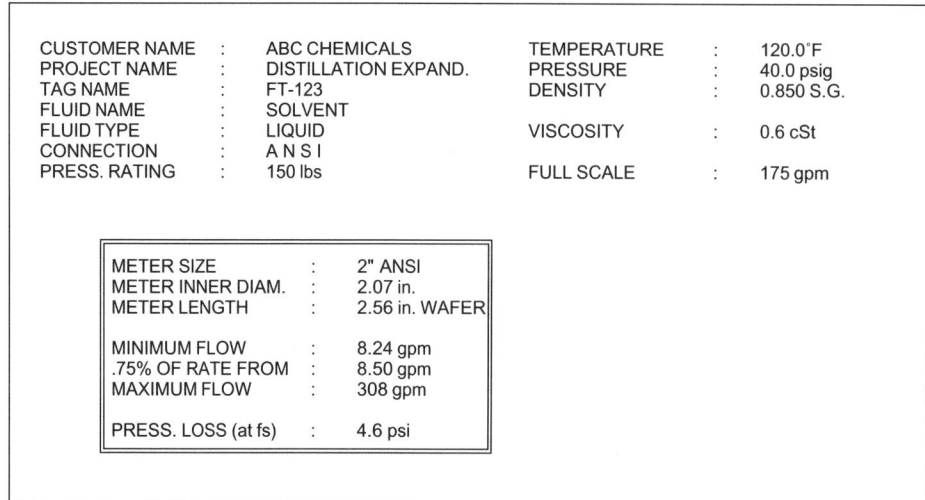

Figure 12-7. Software for Meter Size Selection

Other vendors will identify a minimum Reynolds number that will maintain the stated accuracy of the meter, plus a lower Reynolds number above which the meter will indicate flow at some reduced accuracy (see Table 12-3). The meter may be used in this Reynolds number range provided the reduction in accuracy does not adversely affect the application. The reduced accuracy may or not be quantified in the literature. Depending on the requirements of the application, it may be acceptable to operate at the low Reynolds number range. The main concern in many applications is that the meter be repeatable rather than absolutely accurate. This can be the case when the meter is being used for process control applications. If the application is for accounting or custody transfer, absolute accuracy is of prime importance.

FLUID VELOCITY

The procedure to select the correct size meter will vary, but all vortex shedding flowmeter sizing is based on the acceptable flowing velocity within the meter. Generally, the minimum velocity on gas service is between 10 and 20 feet per second (fps) with a maximum flowing velocity of between 200 and 250 fps. Some vendors will permit flowing velocities up to 650 fps, but this is normally considered too high by most piping standards. For liquid applications the minimum flowing velocity may be in the 1 to 2 fps range with a maximum velocity between 20 and 25 fps. These values are dependent upon the flowing density of the gas or the Reynolds number in liquid applications.

PRESSURE DROP

Another consideration in meter size selection is pressure drop. If the larger meter size—the same size as the pipeline—is able to match the flow range requirements of the application, then that size meter will normally offer an ad-

Table 12-3. Capacity Table for Liquids
(Courtesy of Fischer & Porter Co.)

Meter Size		Overall Linear Flow Range[1, 2]		Minimum Linear Flow Rate[3]		Accuracy (including Non-linearity)		Minimum Operating Nonlinear Q/v[4]	
in.	DN	gpm	l/s	gpm	l/s	Dry Calibration	Wet Calibration	gpm/cSt	mm^2/s
1-1/2"	40	10 - 110	0.63 - 6.9	cSt x 10	mm^2/s x 0.63	± 1.5%	± 0.75%	2.4	0.15
		7.5 - 110	0.47 - 6.9	cSt x 7.5	mm^2/s x 0.47	± 2.0%	± 1.0%	2.4	0.15
2"	50	13.5 - 220	0.85 - 13.9	cSt x 13.5	mm^2/s x 0.85	± 1.5%	± 0.75%	3.1	0.20
		12 - 220	0.76 - 13.9	cSt x 12	mm^2/s x 0.76	± 2.0%	± 1.0%	3.1	0.20
3"	80	29 - 440	1.8 - 27.8	cSt x 29	mm^2/s x 1.8	± 1.25%	± 0.75%	4.6	0.29
		22 - 440	1.4 - 27.8	cSt x 22	mm^2/s x 1.4	± 1.5%	± 1.0%	4.6	0.29
4"	100	47 - 750	3.0 - 47.3	cSt x 47	mm^2/s x 3.0	± 1.25%	± 0.5%	6.1	0.38
		37 - 750	2.3 - 47.3	cSt x 37	mm^2/s x 2.3	± 1.5%	± 0.75%	6.1	0.38
6"	150	105 - 1650	6.6 - 104	cSt x 105	mm^2/s x 6.6	± 1.25%	± 0.5%	9.1	0.57
		82 - 1650	5.2 - 104	cSt x 82	mm^2/s x 5.2	± 1.5%	± 0.75%	9.1	0.57
8"	200	180 - 2850	11.4 - 180	cSt x 180	mm^2/s x 11.4	± 1.25%	± 0.5%	12.1	0.76
		142 - 2850	9.0 - 180	cSt x 142	mm^2/s x 9.0	± 1.5%	± 0.75%	12.1	0.76

1 For liquids with viscosity of 1 cSt, or lower, at 70°F (21°C).
2 For specific gravities above 1.2, maximum flow will be reduced.
3 LV-3 is affected by viscosity. Performance below the calculated minimum is not guaranteed.
4 Flow rate divided by kinematic viscosity mm^2/s (=cSt) must not be lower than the factor listed, and min. operating flow can never be lower than for water.

vantage of lower pressure drop. It also has the advantage of not having to reduce the size of the pipeline when installing the meter in an existing line. On liquid applications where the flowing pressure is near the vapor pressure, pressure drop considerations can be very important. The internal pressure drop may be high enough to cause the liquid to flash or cavitate within the pipe. This has an adverse effect on meter accuracy, but, more importantly, flashing and cavitation can cause internal damage to the meter or the pipeline.

TWO-PHASE FLOW

Measurement of two-phase flow (gas and liquid) within the same meter is difficult. If two-phase flow is present, the vortex shedding flowmeter will not be accurate. A vortex meter is a volumetric device; it cannot distinguish what portion of the flow is liquid and what portion of the flow is gas. The meter will report all the flow as gas or all as liquid depending on the original calibration of the meter.

The vortex shedding flowmeter may be used to measure wet steam, however, if a somewhat reduced accuracy can be tolerated. In wet steam, where both water droplets and gaseous steam are both present, it is difficult to achieve an accurate measurement. Volumetric meters cannot differentiate between gas and liquid. The distribution of the liquid drops within the steam is not homogeneous. In vertical pipes, they tend toward slug flow where there is all gas, all liquid, or somewhere in between. In horizontal pipes, the liquid tends to flow along the bottom of the pipe. Temperature measurements cannot be used to differentiate between liquid and vapor since boiling water and saturated steam are at the same temperature. Under this dual-phase condition, the measurement of one phase cannot be accurate with any volumetric flowmeter. Since the vortex meter has only a small obstruction (the bluff body), there is no place for the water to collect, and it is free to pass on downstream, especially when the bluff body is oriented in the horizontal plane.

Design

Vendors of vortex shedding flowmeters have done considerable testing and development to determine what is, in their opinion, the optimum bluff body configuration (see Figure 12-8). Most have been issued patents for their designs. The one point almost all have in common is the sharp edge on the upstream side of the bluff body to improve both the accuracy and the regularity of the vortex shedding. Many claims are made as to the best bluff body shape. All have valid points, and as in many other areas, some are the advertising person's effort to make a product appear superior. The main point to bear in mind is that all meters work when properly applied—combining the correct meter to the application requirements. While one vendor's product will have a wider rangeability, another's may have better performance at some other point that may be more important to the application requirements.

CONNECTIONS

Shedding vortex flowmeters are available with different types of connections, the most common of which is what has become known as "wafer body" construction. That is, the meter body is clamped between the mating pipeline flanges, using either long bolts or tie rods to hold the meter in place. Meter bodies with integral flanges are also available in many sizes and flange ratings; however, flanged body meters tend to be more costly than the wafer body designs. Some vendors also offer threaded connections in the smaller sizes. In addition to the savings of not furnishing flanges on the wafer body, the wafer body can be mated up to more than one flange rating, which adds flexibility.

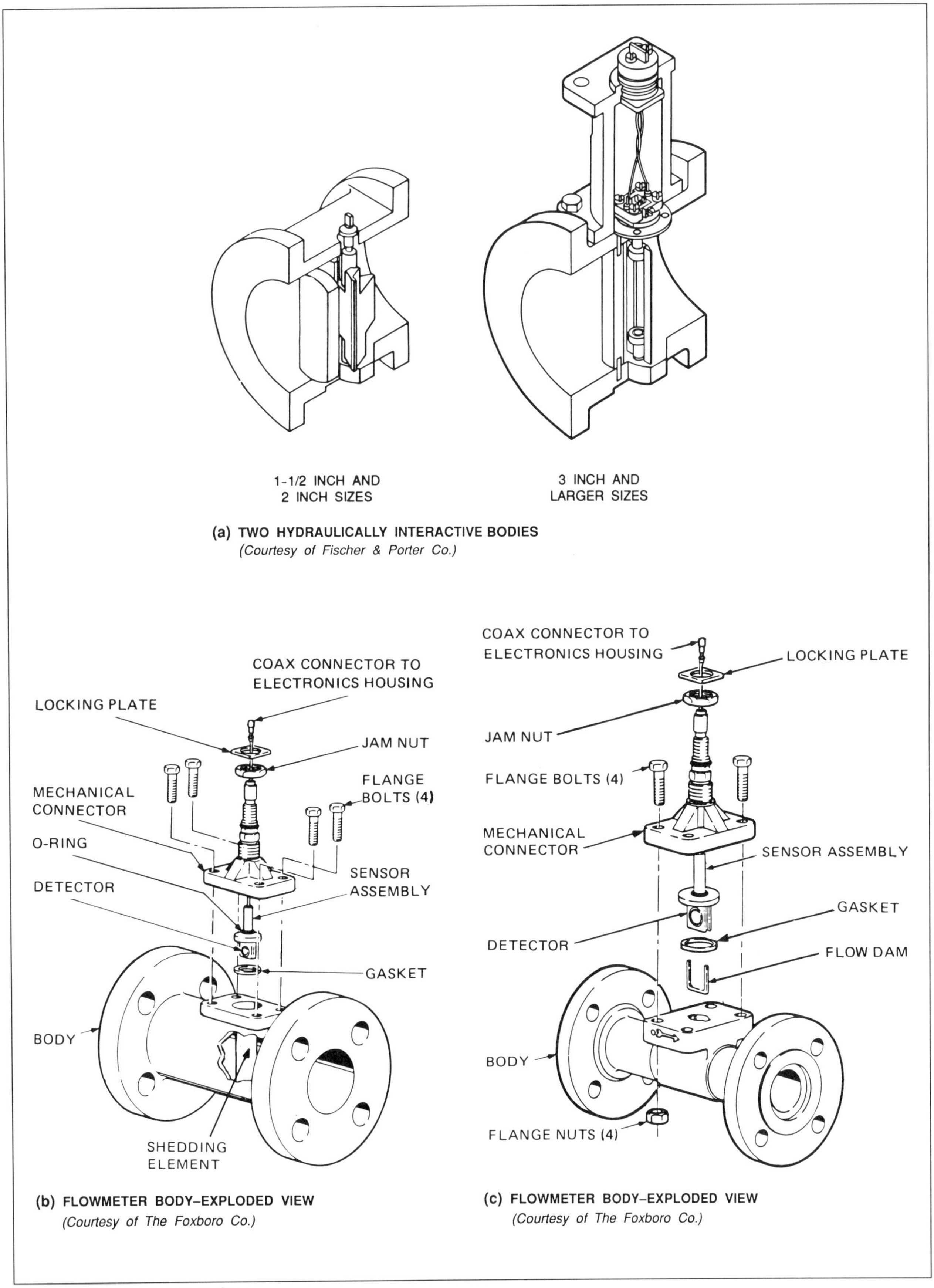

(a) TWO HYDRAULICALLY INTERACTIVE BODIES
(Courtesy of Fischer & Porter Co.)

(b) FLOWMETER BODY–EXPLODED VIEW
(Courtesy of The Foxboro Co.)

(c) FLOWMETER BODY–EXPLODED VIEW
(Courtesy of The Foxboro Co.)

Figure 12-8. Body Configurations

Materials of Construction

The standard material of construction is 316 stainless steel. Some manufacturers offer titanium, Hastelloy C™, or plastics as alternate materials. For most of the applications where the vortex shedding flowmeter is applied, one of these materials is satisfactory. Cost considerations often make the use of other materials of construction prohibitive.

Safety

Since the vortex shedding flowmeter is an electronic device that is mounted in the process piping, it frequently must be installed in a hazardous atmosphere. As a result, most manufacturers have designs that have been approved by a third party agency such as Factory Mutual (FM) or the Canadian Standards Association (CSA) so that meters are available in non-incendive, explosion-proof, or intrinsically safe designs.

Third party agencies will review and test the meters to see that they are designed and manufactured in accordance with their safety standards. For details on which agency and what standards are met, consult the manufacturer.

When a vortex shedding flowmeter is used to measure a lethal material, it may be well to consider a totally welded design.

When a vortex shedding flowmeter is being used to measure very hazardous material, it may be well to consider the flanged body instead of the wafer body. With two direct bolting flanges instead of long tie rods to connect the meter into the pipeline, many consider this to be an extra margin of safety.

Maintenance

Since the vortex shedding flowmeter (as it is currently being built) has no moving parts, little regular maintenance is required. The internals (meter body and bluff body) are very rugged. The sharp edge of the bluff body is subject to very little wear since this edge is the point of separation of the vortex swirl.

Mechanically, very little maintenance is required once the meter is installed correctly. Unless the application parameters are changed, recalibration is seldom required.

Many designs permit the sensor to be changed without removing the meter from the pipeline. Some vendors provide for isolation of the sensor from the process fluid so that it is not necessary to shut down the flow to service the sensor (see Figures 12-8 and 12-9). On the other hand, with some designs it is not possible to change the sensor in the field. Consult the instruction manual for the particular meter.

The ease of changing the electronics also varies with design. In most meters it will be necessary to check and recalibrate the meter electronics after changing the electronics. It is rare that a vortex shedding flowmeter will have to be wet flow calibrated again unless the sensor has been changed. In many designs it is not necessary to wet flow calibrate the meter even then.

If it is suspected that the meter is not reading correctly, the installation should be checked first. Experience has shown an incorrectly installed meter is more likely to be the culprit than an actual problem with the meter itself. Unanticipated process conditions also can adversely affect performance.

Calibration

If the user provides the correct operating information about how the meter will be used, the manufacturer will calibrate the meter to (but not at) those conditions. In most cases, the factory calibration includes wet flow calibration where the meter is installed in a calibration standard where water or air can be used to calibrate the flowmeter element. The electronics are then adjusted so that the output signal is correct for the application.

In many standard applications, vortex meters are being used with great success. These include metering water, compressed air, natural gas, steam and chemicals.

One unusual application that proved very successful was on unloading ammonia tank cars. The liquid ammonia was pumped out of the tank car by pressurizing the car with ammonia gas. Since the liquid was drawn off at irregular intervals in varying amounts, it was difficult to tell when the car was empty. Other types of flowmeters were tried but none could tell when the line changed from liquid to gas, indicating that the tank car was empty. A vortex shedding flowmeter was installed and calibrated to measure liquid ammonia. Since the frequency of vortex shedding is normally about ten times higher for gas than liquid, but at a lower amplitude of pressure pulse, the vortex meter electronics could measure the gas flow as outside its calibration range. When this occurred, the meter indicated a zero flow condition, which resulted in considerable savings of time and liquid ammonia, which frequently was found in the bottom of tank cars thought to be empty.

Another application for a vortex meter is dirty steam. In this instance, the steam was recovered after serving its primary purpose, but it still contained enough heat to be useful. The previous meter was an orifice plate with a differential pressure transmitter. The foreign matter picked up by the process caused clogging of the sensing lines, which required a weekly cleaning. The orifice flowmeter was replaced by a vortex shedding flowmeter, which has no sensing lines to clog. The user estimated a savings of four to six man hours per month since installing the vortex meter.

Fluidic Flowmeters

Operating Principle

As the flow enters the meter body, the flow stream attaches itself to one wall of the passage (see Figure 12-10) (the Coanda effect). A small portion of the flow is channeled through the feedback passage. This causes the flow stream to separate from that side of the meter body and attach itself to the opposite side of the meter body. Another feedback passage on the other side of the meter then causes the flow to flip over to the other side of the meter body where the process is repeated continuously. The frequency of the oscillation is directly proportional to the flow velocity through the meter. By locating a sensor in one of the feedback passages, the oscillation can be detected.

Sensors

The meter currently available has a choice of sensor types. One sensor is a strain gage; the other uses a thermistor. The strain gage will detect a pressure surge as the flow passes through the feedback channel. The thermistor will detect the instantaneous change in temperature caused by the feedback flow. The output from the sensor is fed into either an electronic or a pneumatic transmitter. The output, therefore, is either a 3-15 psi pneumatic signal, a 4-20 mA linear signal, or a pulsed output signal.

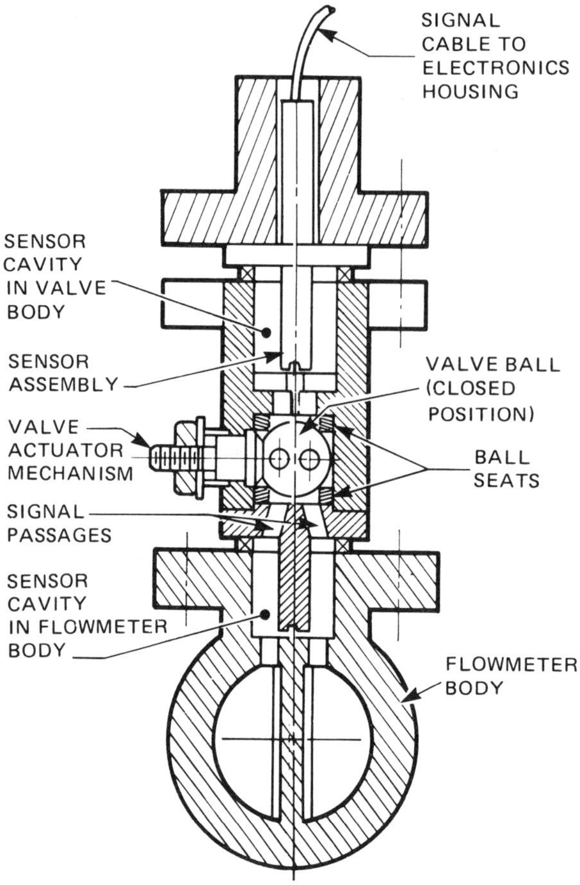

Figure 12-9. Cross-Sectional View of Optional Isolation Manifold
(Courtesy of The Foxboro Co.)

The fluidic meter is designed to be used only on liquid flow applications; it cannot be used on gas or steam. This type of meter should not be used to measure slurry flow; however, small solids can be tolerated as long as the solids do not exceed 2% by volume.

Fluidic meters are available in line sizes from 1 in. to 4 in. (see Figure 12-11 and Table 12-4). In larger line sizes, the 1-in. unit can be used in a bypass arrangement in conjunction with an orifice plate. In the bypass mode, an orifice plate creates a pressure drop in the main line. This pressure drop causes a small amount of the flow to pass through the bypass piping. This bypass flow is in direct proportion to the main line flow. By metering the bypass flow and applying the correct scaling value, the main line flow is determined. Since the fluidic meter has an output linear with flow, the measurement output from the bypass meter remains linear. Because the meter is not measuring the amount of pressure drop across the orifice plate, no square root extractor is required.

The rangeability of the fluidic meter can be up to 30 to 1, with an accuracy of ± 1% of reading, which is typical of other types of oscillatory flowmeters. The choice of sensors—strain gage or thermistor—is dependent upon the range of

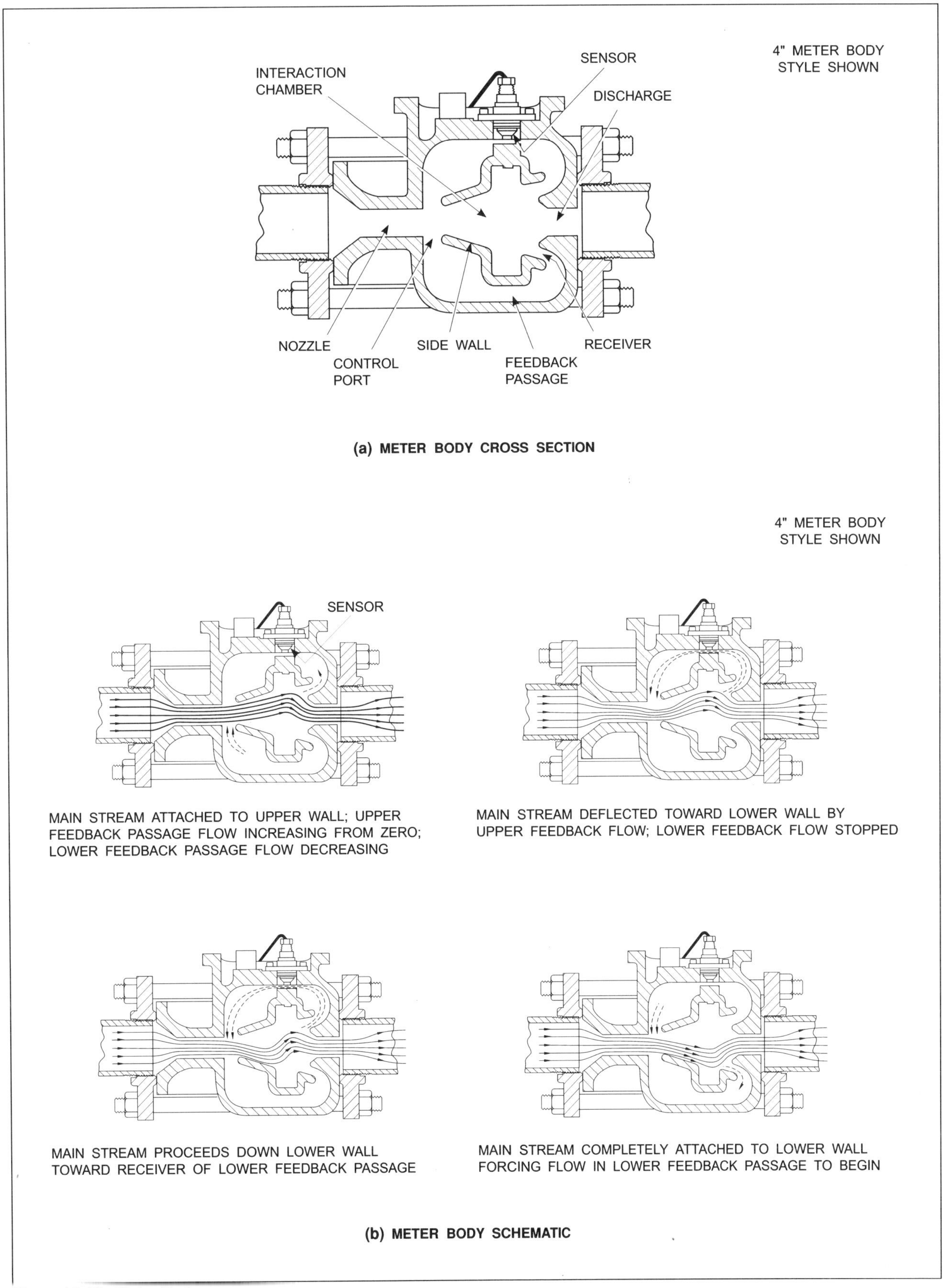
INTERACTION CHAMBER
SENSOR
DISCHARGE
4" METER BODY STYLE SHOWN
NOZZLE
CONTROL PORT
SIDE WALL
FEEDBACK PASSAGE
RECEIVER
(a) METER BODY CROSS SECTION
4" METER BODY STYLE SHOWN
SENSOR
MAIN STREAM ATTACHED TO UPPER WALL; UPPER FEEDBACK PASSAGE FLOW INCREASING FROM ZERO; LOWER FEEDBACK PASSAGE FLOW DECREASING
MAIN STREAM DEFLECTED TOWARD LOWER WALL BY UPPER FEEDBACK FLOW; LOWER FEEDBACK FLOW STOPPED
MAIN STREAM PROCEEDS DOWN LOWER WALL TOWARD RECEIVER OF LOWER FEEDBACK PASSAGE
MAIN STREAM COMPLETELY ATTACHED TO LOWER WALL FORCING FLOW IN LOWER FEEDBACK PASSAGE TO BEGIN
(b) METER BODY SCHEMATIC

Figure 12-10. Meter Body
(Courtesy of Moore Products Co.)

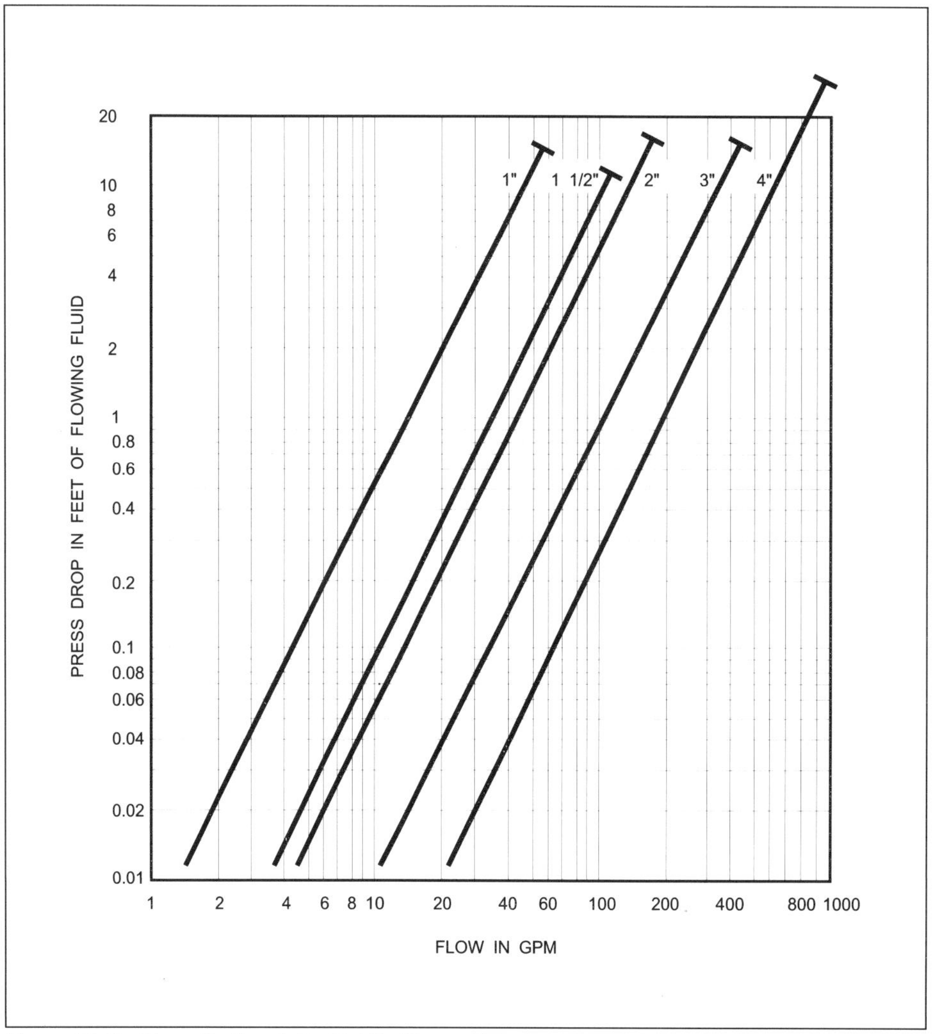

Figure 12-11. Capacity Graph for Fluidic Flowmeter
(Courtesy of Moore Products Co.)

The meter can be installed with the flow either horizontally or vertically upward. Vertical downward flow should be avoided as with most types of flowmeters.

flow required and the fluid coating properties. The strain gage sensor is less costly, but it will not measure flow rates as low as the thermistor can measure. As with the vortex shedding flowmeter, there are no moving parts, so maintenance is low. The fluidic meter may be overranged by as much as 400% without adversely effecting the meter or its calibration. Installation is easily accomplished since no external hardware is required. The meter is mounted between pipeline flanges (with appropriate upstream and downstream straight run), the wiring or pneumatic lines are connected, and the meter is ready to perform.

Table 12-4.
(Courtesy of Moore Products Co.)

METER SIZE	MINIMUM FLOW RATE (USGPM)			
	ANALOG SERVICE		TOTALIZING/BATCH SERVICE	
	THERMAL SENSOR	STRAIN GAGE SENSOR	LINEAR RANGE	EXTENDED RANGE
1	1	3.25	7	4
1-1/2	1.5	7.60	12	8
2	2	10.5	40	10
3	3	23.0	200	10
4	4	40.0	47	10

NOTE: All values shown are for water @ 68° F (1 cSt). For other fluids, refer to the following:

1. For a meter with a *thermal sensor*, multiply the value shown by the viscosity of the fluid in centistokes (cSt).
2. For a meter with a *strain gage sensor*, divide the value shown by the square root of the specific gravity.

METER SIZE	CAPACITY LIMITS		
	FULL SCALE CALIBRATION RANGES *		
	MINIMUM RANGE (USGPM)		MAXIMUM RANGE (USGPM)
1	0-15	Thru	0-50
1-1/2	0-45	Thru	0-120
2	0-75	Thru	0-180
3	0-150	Thru	0-450
4	0-350	Thru	0-1000

* Ranges shown correspond to 4-20 mA or 3-15 psig outputs

About the Author

William Gotthardt attended Juniata College and Temple University. He worked as an Applications Engineer in variable area meters at Schutte & Koerting for six years, as Applications Engineer and Product Manager for variable area meters at Brooks Instruments for seven years, and as Product Manager and Applications Specialist for vortex meters at Fisher Controls for nine years. He is presently Product Manager and Flow Consultant for Endress + Hauser.

13

Positive Displacement Flowmeters for Liquid Measurement

Positive displacement liquid flowmeters have long been used to measure liquid products. Over the years, numerous design improvements have resulted in an expanded product line that now serves industrial as well as petroleum and domestic applications.

Principles of Operation

A liquid flowmeter is, in essence, a hydraulic motor with high volumetric efficiency that absorbs a small amount of energy from the flowing stream. This energy is used to overcome internal friction in driving the flowmeter and its accessories and is reflected as a pressure drop across the flowmeter. Pressure drop is regarded as a necessary evil that must be minimized. It is the pressure drop across the internals of a positive displacement flowmeter that actually creates a hydraulically unbalanced rotor, which causes rotation.

A positive displacement flowmeter is one that continuously divides the flowing stream into known volumetric segments, isolates the segments momentarily, and returns them to the flowing stream while counting the number of displacements.

Elements of Construction

A positive displacement flowmeter can be broken down into three basic components: the external housing, the measuring unit, and the counter drive train.

The external housing is the pressure vessel that contains the product being measured. It can be of single- or double-case construction, with single-case having the housing and the measuring chamber walls as one integral unit. With double-case construction, the external housing is separate from the measuring unit and serves only as a pressure vessel. This type of construction has two major advantages: (1) The measuring chamber walls sense only the pressure difference between the flowmeter inlet and outlet, which allows for thinner chamber walls with less distortion, and (2) system piping stresses are absorbed in the external housing and are not transmitted to the precision measuring element.

The measuring unit is a precision metering element and is made up of the measuring chamber and the displacement mechanism. In today's market, the oscillating piston, sliding vane, oval gear, trirotor, birotor, and nutating disc types of positive displacement flowmeters are the most common (see Figure 13-1).

The counter drive train is used to transmit the internal motion of the measuring unit into a usable output signal. Many positive displacement flowmeters use a mechanical gear train that requires a rotary shaft seal or packing gland where the shaft penetrates the external housing. Other flowmeters may use reed switch outputs, magnetic drive couplings (contact closure), and differential inductance pickoffs. These last three offer the advantages of progressively lower driving torque (and, hence, less drag at low flows) and isolated product with no seals to leak.

Design Considerations

In capillary seal positive displacement meters, the capillary action of the metered product forms a liquid seal between the moving and the stationary parts. This type of flowmeter requires very close clearance dimensions and is sensitive to differential pressure.

Product slippage is the most crucial problem affecting the accuracy of a capillary seal positive displacement flowmeter. All capillary seal positive displacement flowmeters have some clearance between moving and stationary parts and some differential pressure across these clearances. For this reason, some product will always be allowed to bypass the measuring chamber by "slippage" through these clearances.

If slippage were a constant at all operating conditions, it could be corrected for by the counter drive train gearing and would cause no inaccuracy. However, it is not constant and does vary with flow rate, pressure drop, temperature, viscosity, and clearance dimensions.

Applications

The positive displacement flowmeter can offer excellent accuracy, repeatability, and reliability in many applications. It can measure varying and high viscosity products over a broad range of flow rates. Positive displacement flowmeters typically have limitations on applications with extremely dry (non-lubricating) liquids and liquids that contain solid particles.

Many system and flowmeter parameters determine the suitability of a flowmeter for a specific application. A full understanding of each is essential in selecting the best flowmeter for the application in question.

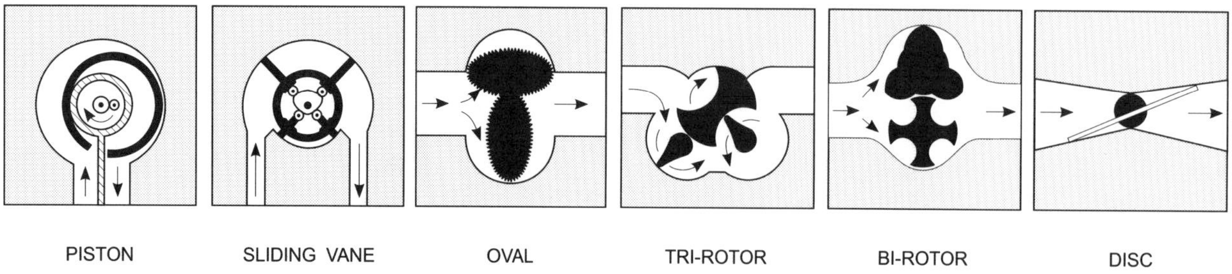

Figure 13-1. Six Common Positive Displacement Meter Principles

System Parameters

Viscosity

Viscosity is defined as the property of a liquid that presents a resistance to flow or the ability of the fluid to flow over itself. It is this property that seals the clearances through capillary action, thus blocking or partially blocking a potential leak. Increasing viscosity improves the performance of a positive displacement flowmeter and often extends the low flow limit significantly.

Unfortunately, an increase in viscosity is also accompanied by an increase in pressure drop as the moving parts expend more energy in shearing the heavier product. Since high differential pressures increase wear, most flowmeter manufacturers specify a maximum allowable differential pressure and derate the flowmeter's flow capacity as viscosity increases. While the maximum flow coefficient varies from flowmeter to flowmeter, Figure 13-2 represents a typical curve for derating flow capacity as a function of viscosity.

In an effort to regain rated flow capacities, it is common to compensate for viscous products by increasing clearance dimensions with "extra-clearance rotors." This effectively decreases the shearing forces, which, in turn, reduce the differential pressure and allow the flow range to be extended on viscous products. The curve shown in Figure 13-2 would shift to the right with extra-clearance rotors.

A primary advantage of the positive displacement flowmeter is its ability to measure varying and high viscosity products without appreciable shift in accuracy.

Figure 13-3 is a typical positive displacement flowmeter curve for a flowmeter initially calibrated on 1 centipoise (cP) product and then applied to 100 cP product. Note that not only is there a maximum shift of 1.2%, but above 100 cP, no shift is evident. Typically a positive displacement flowmeter's accuracy will lose its sensitivity to viscosity changes at about 100 cP.

Temperature

Temperature is a system parameter that must always be considered in flowmeter selection. Its effect upon viscosity causes a shift in the performance curves, as discussed in the previous section. Accompanying a temperature change are corresponding changes in product volume (density), measuring chamber volume, and clearance dimensions.

A change in product volume can be corrected to an accepted base temperature, typically 60°F in the U.S., with an automatic temperature compensator. A change

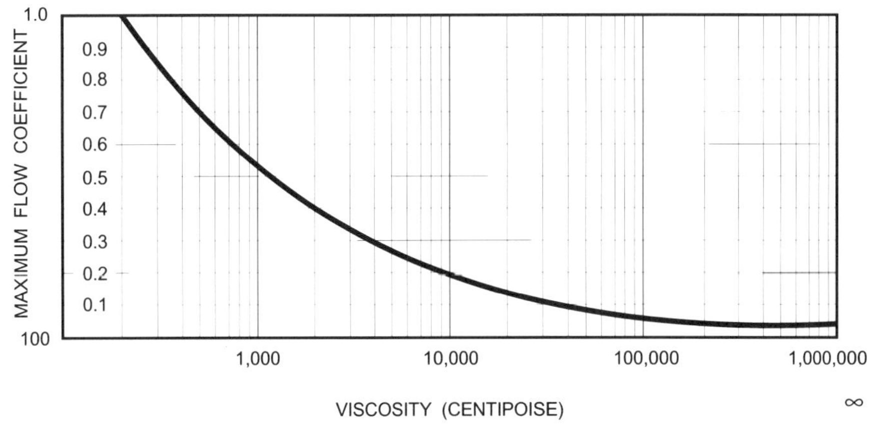

Figure 13-2. Maximum Flow Coefficient vs. Viscosity

in measuring chamber volume, if necessary, can be corrected with a minor adjustor setting that changes the output gear ratio. Changes in clearance dimensions with elevated temperature often necessitate the use of extra-clearance rotors, referred to as "high temperature trim." This extra clearance takes into account a difference in the linear coefficient of thermal expansion of the rotors and housing and prevents flowmeter lockup.

Temperature extremes often require the use of jacketing c tracing to maintain temperature equilibrium throughout the flowmeter. cold flowmeter were shocked with high temperature product and allowed to operate prior to reaching thermal equilibrium, it is likely that the rotors, being of less mass than the housing, would expand faster and score the measuring chamber wall. For this reason, manufacturers specify a minimum time for "warm-up" prior to use that will ensure temperature equilibrium throughout the flowmeter.

Flowmeter accessories must often be protected from high temperature by the use of cooling fins and/or counter extensions. The manufacturer's specifications will dictate such requirements if they are necessary.

Pressure

In addition to the working pressure, the user should be aware that shock pressures and expansion pressures can be significant and should be minimized.

System pressure must be contained by the flowmeter housing. While housings are available in many pressure ratings, they usually will correspond to flange rating if applicable. As with the flange ratings, the maximum working pressure must be derated at elevated temperatures.

Flow Rate

System flow rate limits should be known to ensure a flowmeter with sufficient rangeability to handle both maximum and minimum flow rates. Excessively low rates tend to under-register flow as slippage increases, and excessively high flow rates increase wear. A flowmeter should normally operate around the midpoint of its related flow range for optimum performance.

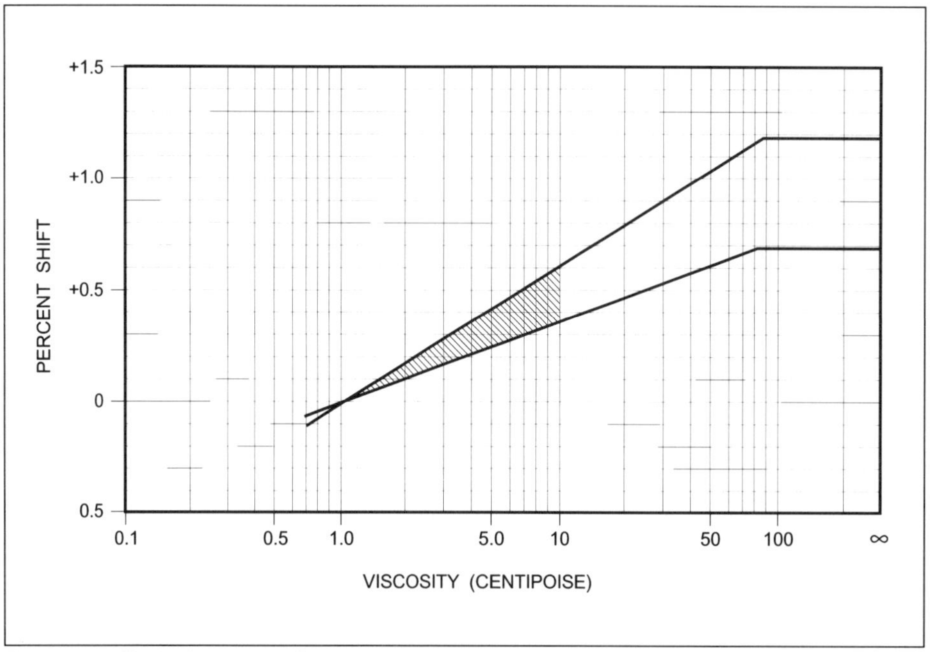

Figure 13-3. Typical Accuracy Shift vs. Viscosity

Product

The product to be measured is most important in flowmeter selection. Chemical compatibility must be considered in the selection of materials of construction of all internal wetted surfaces. Dry, abrasive products are best handled by using an automatic pressure lubricated system that isolates bearings and gears from the product and offers positive, self-contained lubrication. Because of the close clearances inherent in a positive displacement flowmeter, entrained solids are not readily passed and should be removed by an appropriately meshed strainer upstream of the flowmeter.

A positive displacement flowmeter is volumetric and will yield gross inaccuracies with a product that contains either free or entrained air. Removal of this air with an appropriately sized air eliminator is essential. Large volumes of free air not only impair accuracy but can also overspeed and destroy a meter.

Consider a large section of pipeline filled with air that is contained on both ends with the liquid product. Two problems will arise. As the air mass passes through the flowmeter, its lack of viscous drag will increase the speed of the rotors. Immediately after the large air mass passes through the flowmeter, the rotors are struck with a wall of liquid product than can severely jar the system. Second, and more importantly, as the air mass passes through a control valve downstream of the flowmeter, the entire flowing stream accelerates, which can excessively overspeed the flowmeter, causing damage to the bearings.

Flowmeter Parameters

Accuracy

Accuracy is an indication of a flowmeter's ability to correctly measure product and is usually expressed as either percent registration (see Figure 13-4) or percent variation of the flowmeter factor. Accuracy can be expressed at any given flow rate as:

$$\text{Percent Registration} = \frac{\text{Actual Quantity}}{\text{Metered Quantity}} \times 100$$

A typical percent registration curve for a positive displacement flowmeter will peak at approximately 25% of the rated flow capacity. At high flow rates the exponential increase in differential pressure increases the slippage flow rate, which reduces percent registration (increases flowmeter factor). At low flow rates there is a shortage of energy (differential pressure) to extract from the flowing stream to drive the flowmeter and its accessories. This increases the relative significance of the flowmeter's driving torque, which again results in an increase in slippage and a decrease in percent registration. Figure 13-4 graphically illustrates a typical capillary seal positive displacement flowmeter's performance as described above. Notice how the percentage registration falls as the product viscosity decreases.

A change in accuracy is often the first symptom of a mechanical defect. Consecutive logs of flowmeter provings can be used to identify a problem at an early stage.

All positive displacement flowmeters have a primary accuracy expressed as a percent *of rate*. This means that within the recommended operating range, regardless of flow rate, the primary flowmeter output (typically, shaft rotation or pulses) will be within the specified percentage of the actual volume (see Figure 13-5).

Rangeability

Rangeability is a flowmeter's ability to cover a range of flow rates without exceeding the accuracy limits specified. The flow range of a flowmeter is specified as a maximum and minimum and is directly related to the flowmeter design and the viscosity of the product. A typical flow range is from 20% to 100% of maximum flow (5:1 flow range) although 10:1 and greater flow ranges are not uncommon for positive displacement flowmeters. The maximum rated flow capacity of a flowmeter is usually applicable to intermittent flow and should be derated if continuous flow is expected. While most flowmeters can withstand short duration

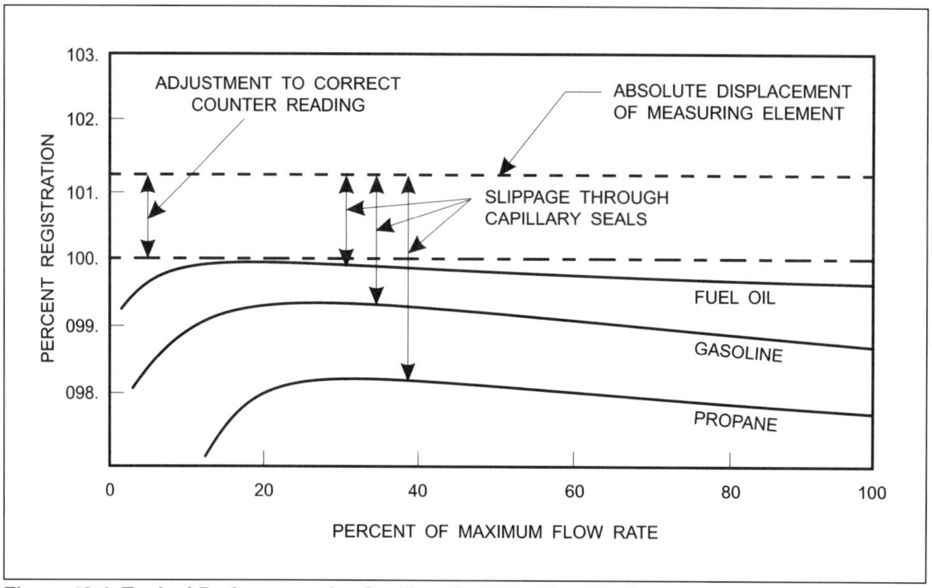

Figure 13-4. Typical Performance for Capillary Seal Positive Displacement Flowmeter

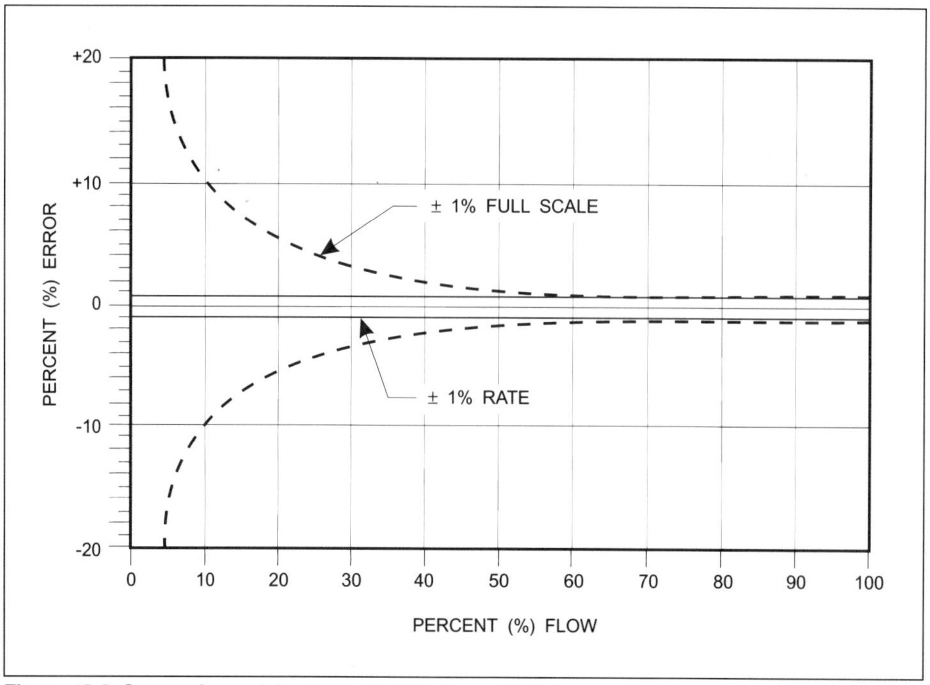

Figure 13-5. Comparison of Accuracy Full Scale vs. % Rate

overspeeding, a rule of thumb is that wear increases to the third power as the flow rate exceeds the maximum rated capacity.

Repeatability

Repeatability is a flowmeter's ability to repeat its accuracy at a given flow rate. Typical repeatabilities for positive displacement flowmeters will be plus or minus ± 0.05% or better.

Pressure Drop

Pressure drop is a reflection of the energy lost in driving the flowmeter and its accessories. As discussed earlier, it is the pressure drop across the internals of a positive displacement flowmeter that actually creates a hydraulically unbalanced rotor, which causes rotation.

Sizing

The size of positive displacement flowmeters varies from ¼ inch through 16 inches and in flow capacities from fractions of a gallon per hour to 13,000 barrels per hour. A positive displacement flowmeter should be sized based upon product viscosity as well as flow rate and should not be operated continuously at the maximum rated flow capacity.

Output Signal

The output signal is obtained through the counter drive train and is available in either mechanical or electrical form with a digital or analog signal. The rotary motion of the flowmeter internals can be converted into an output signal that is capable of driving a broad line of accessories. Before one can finalize the specifications for a flowmeter, however, the accessories and the required output signal to drive these accessories must be known.

Cost

The cost of one type of flowmeter relative to another will vary based upon size and manufacturer. The initial cost, however, is only one of several costs that should be considered. For example:

(a) Accuracy: Consider a 16 inch pipeline flowmeter flowing at 12,000 barrels per hour. With oil priced at $20 per barrel, an improved accuracy of only 0.05% could result in savings of $120 for every hour of operation.

(b) Maintenance: Recurring costs in maintaining a flowmeter can be a significant factor in overall flowmeter cost.

Conclusions

A number of system and flowmeter parameters affect the performance of any given flowmeter. By understanding the effects and interrelationships of these parameters, a user can select the optimum flowmeter for an application and expect the best performance possible from the entire system.

Flowmeter manufacturers are continuing to make improvements to existing designs as well as developing new technology aimed at better satisfying the needs of the market. The positive displacement flowmeter has satisfied many needs in the past and will play a vital role in serving the future needs of industry.

Bibliography

1. ASME Research Committee on Fluid Meters, "Fluid Meters, Their Theory and Application," Sixth Edition, New York, NY:ASME, (1971).

About the Author

R. Gary Barnes has been involved in the flow measurement industry for more than twenty years. He is Product Manager/Industrial Products for Brooks Instrument Division/Emerson Electric Company. A graduate of Georgia Southern University, Gary has papers published by ISA and the International School of Hydrocarbon Measurement.

14

Target Flowmeters

Versatility and low installation cost make the target flowmeter a viable candidate for many difficult flow measurement applications. Full bore, in-line meters are available in sizes from ½ to 6 inches. Insertion versions for larger line sizes are also available. Target flowmeters are used on liquids, gases, vapors, dirty fluids, light slurries, and high viscosity fluids, particularly where fluid characteristics preclude the use of flowmeters with pressure taps or moving parts. Process temperatures vary from cryogenic to low temperature steam.

Advantages:
Low cost
Ease of installation, service
Simple construction, high reliability
No pressure ports to plug
Can be used on most fluids, clean or dirty

Operating Principle

Whenever there is flow past an obstacle in a pipe, a force, commonly referred to as drag, is generated to push or drag the obstacle in the direction of flow. Such an obstacle left unsupported would be carried away with the fluid. If, on the other hand, the obstacle was constrained by a force equal and opposite to the drag, the magnitude of that force could be used to determine the rate of flow. This is the underlying principle behind the target flowmeter.

There are two primary contributors to drag. One results from the force generated by the fluid viscosity as it slides by the obstacle. This is referred to as friction drag and has its major influence when the flowmeter is operated in the laminar flow regime. The second contributor is the so-called pressure drag. Pressure drag is the force resulting from the difference between the pressures immediately upstream and immediately downstream of the obstacle. For turbulent flows, the pressure drag is the prime contributor to the total drag on the obstacle.

The obstacle or target typically used in practice is a circular disc mounted concentrically in a pipe. The idealized flow past such a target is shown in Figure14-1. The upstream face sees a relatively high pressure since the forward motion of a large percentage of the fluid is abruptly stopped before turning and traveling around the target. As the fluid passes through the annulus around the target, it sees an increase in velocity and, as a result, a decrease in pressure. At the downstream edge of the plate, this high velocity, low pressure flow separates from the target surface, setting up turbulence downstream of the target. This results in a relatively low pressure region near the downstream face. In the simplest terms, the pressure drag on the target is given by:

$$F_\tau = (\text{constant}) \times \left(\frac{1}{2}\rho\bar{v}^2\right) \times A_\tau \tag{14-1}$$

where:

A_τ = target area
F_τ = target force
ρ = fluid density at process temperature and pressure
$\bar{v}$ = average velocity in the annulus
$\frac{1}{2}\rho\bar{v}^2$ = dynamic pressure

Mass rate of flow, $\dot{m}$ is given by:

$$\dot{m} = \rho A_A \bar{v} = \rho(A_p - A_\tau)\bar{v}$$

where:

A_A , A_p = annular and pipe area, respectively

The mass flow rate in terms of target force is given by:

$$\dot{m} = (\text{constant}) \times \sqrt{\frac{\pi}{2}\rho F_\tau} \times \left(\frac{D - d_\tau^2}{d}\right) \quad (14\text{-}1a)$$

$$= (\text{constant}) \times \sqrt{\frac{\pi}{2}\rho F_\tau} \times D\left(\frac{1 - \beta_\tau^2}{\beta_\tau}\right)$$

$$= KD\sqrt{\rho F_\tau}$$

where:

D, d_τ = pipe and target diameters, respectively

$\beta_\tau = \frac{d_\tau}{D}$

K = constant that includes target blockage

Since K is a dimensionless constant, Equation (14-1a) holds true for any consistent system of units. In the SI system, D is in meters, ρ is in kilograms per cubic meter, F_τ is in newtons, and $\dot{m}$ is in kilograms per second. In the US it is customary not to use a consistent system of units, particularly when units of mass are involved. The most common mass unit in the US is the pound mass (lbm). In order to allow the use of lbm and at the same time keep K dimensionless, Equation (14-1a) has to be modified slightly to include the conversion factor g_c (32.17) to become:

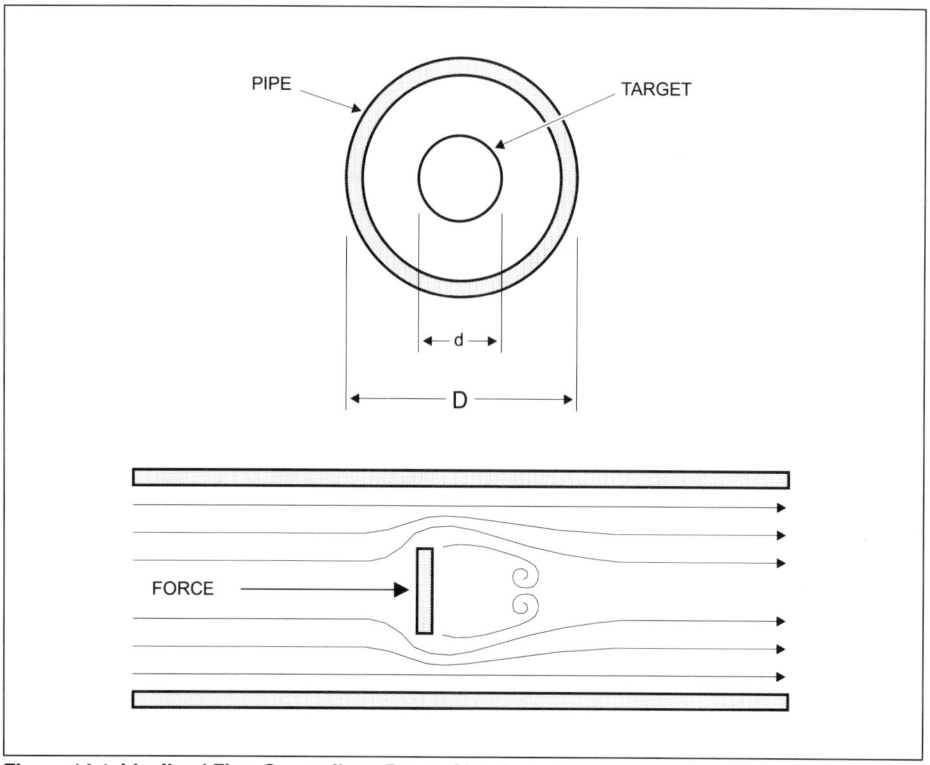

Figure 14-1. Idealized Flow Streamlines Past a Circular Disc

$$\dot{m} = KD\sqrt{\rho g_c F_\tau} \qquad \text{(U.S. Customary Units)} \qquad (14\text{-}1b)$$

For Equation (14-1b), D is in feet, ρ is in lbm per cubic foot, F_τ is in lbs, and $\dot{m}$ is in lbm per second.

To obtain the relationship between volumetric flow rate, Q, and force, divide Equations (14-1a) and (14-1b) by fluid density to get:

$$Q = (\text{constant}) \times \sqrt{\frac{\pi F_\tau}{2\,\rho}} \times D\left(\frac{1-\beta_\tau^2}{\beta_\tau}\right) \qquad (14\text{-}2a)$$

$$= KD\sqrt{\frac{F_\tau}{\rho}} \qquad \text{(SI Units)}$$

where Q is in cubic meters per second, and

$$Q = KD\sqrt{\frac{F_\tau g_c}{\rho}} \qquad \text{(U.S. Customary Units)} \qquad (14\text{-}2b)$$

where Q is in cubic feet per second.

Equations (14-1) and (14-2) represent the equations for an idealized target that is freely suspended in a pipe. The true working equations must take into account the additional forces introduced by the target support(s). In most cases this can be dealt with by adjusting K, which is typically determined experimentally and will vary with support design.

Accuracy/Turndown

Accuracy for a target flowmeter is typically stated as a percent of full scale and, depending on design, is from ± 0.5 to ± 2.0%. When stated in this manner the accuracy at a flow rate lower than the full scale value will be proportionately less. The rate accuracy at a given flow rate can be calculated as:

$$\text{rate accuracy} = \frac{\text{full scale}}{\text{actual flow}} \times \text{full scale accuracy statement}$$

For example, a meter with a ± 0.5% of full scale accuracy statement and a full scale set at 100 gpm, operating at a flow rate of 40 gpm, would have a rate accuracy of:

$$\textbf{rate accuracy} = \frac{\textbf{100 gpm}}{\textbf{40 gpm}} \times \textbf{0.5\%}$$

$$= \pm\,\textbf{1.25\%}$$

Likewise, at 20 gpm the accuracy would be ± 2.5%.

The reason the accuracy of the device falls off in this manner is explained by considering the relationship between flow and target force. Since the force is proportional to the square of the flow, a force sensor suitable for a 5:1 range of flow would have to have a dynamic range of 25:1. For a 10:1 flow range the sensor would have to be capable of measuring force over a 100:1 range. Although some force sensors are capable of extremely wide ranges, influences that may introduce noise to the measurement and limit useful range are prevalent in most applications. It is best to operate within a 4 to 5:1 range and to select the full scale flow such that the normal operating rate is approximately 70% of the full scale flow.

There are two ways to adjust full scale. One is to vary the target force by changing the diameter of the drag disc. The other is to adjust the output sensitivity of the transmitter or indicator that converts the force to a useful output.

Temperature Effects

Since the target force is dependent on geometry, as the temperature varies from the reference or calibration temperature an error will be introduced due to thermal expansion of the meter. Assuming the target disc and meter body are of the same material, a correction can be applied to the flowmeter equations as follows:

$$\dot{m} = (1 + 2\alpha\,\Delta T) \times KD\sqrt{\rho F_\tau} \quad (14\text{-}3)$$

and

$$Q = (1 + 2\alpha\,\Delta T) \times KD\sqrt{\frac{F_\tau}{\rho}} \quad (14\text{-}4)$$

where:

α = linear coefficient of thermal expansion for target flowmeter material (assumes target and body material are the same)

ΔT = process temperature minus reference or calibration temperature

Remember: the geometry and, hence, the accuracy of a target flowmeter will depend on temperature. For operation at temperatures far from the reference or calibration temperature, the effect of thermal expansion should be considered. For example, for 300 series stainless steel an error of approximately 0.5% per 100°C will result if no temperature correction is applied.

Reynolds Number

A typical curve of Reynolds number versus flow coefficient [Ref. 1] is shown in Figure 14-2. Above some Reynolds number where the flow is predominately turbulent, the flow coefficient is reasonably constant and suitable for flow mea-

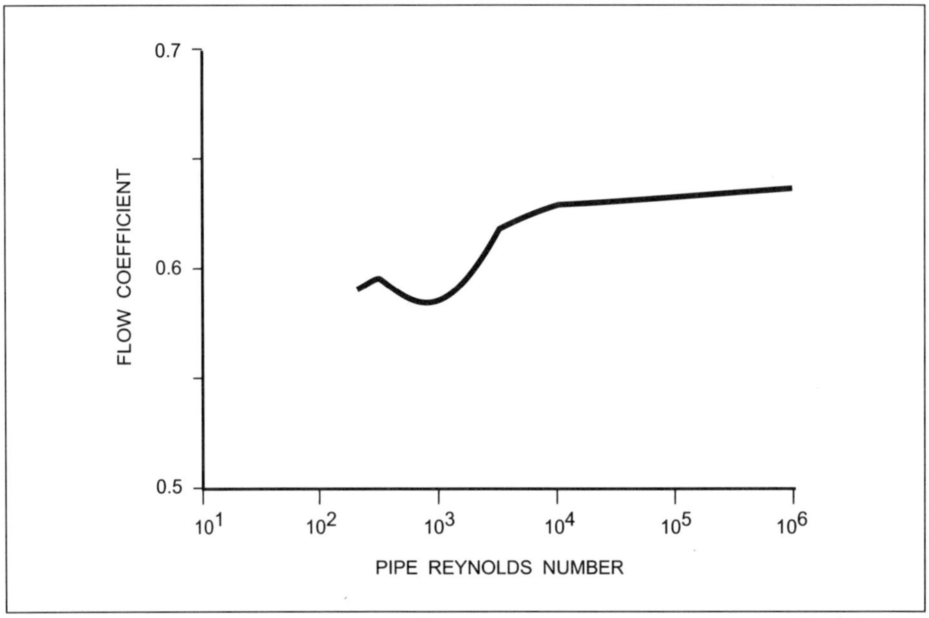

Figure 14-2. Example of Reynolds Number Characteristics for a Target Flowmeter [Ref. 1]

surement. This Reynolds number varies with design from 1000 to 4000. For lower Reynolds numbers, where the flow is in either the laminar or transitional flow regime, care must be exercised to size the meter so the Reynolds number falls on a local flat part of the curve. If this is not possible, some provision such as a flow computer can be added to correct for the non-constant coefficient.

Design Considerations

To sense a moderate to low force while providing a positive seal against a relatively high process pressure poses a design challenge. Substantial zero offsets due to process pressure variations can be introduced if the seal is not properly designed.

The deflection of the target must be effectively zero to avoid the nonlinearities that would result from changes in the target orientation.

The strut used to support the target and transmit the force will, by its presence, alter the effective target geometry from the idealized one. For this reason the designer must carefully consider the shape and location of any support members.

Construction

Although many alternative target shapes have been suggested in the literature [Refs. 2, 3], a thin circular disc mounted perpendicular to the flow direction still remains the most commonly used target geometry for fully developed turbulent flow. In some flowmeter designs the target is field replaceable, enabling ease of maintenance or range change. The connection between the target and the force transducer is typically a circular rod that extends from the target through a shaft seal or flexure. The methods employed to transform the force to a useful output differ considerably with manufacturer and application.

Force Balance

One approach uses force balance as the sensing means. With force balance the shaft pivots about a flexure much like a lever and fulcrum (Figure 14-3). The

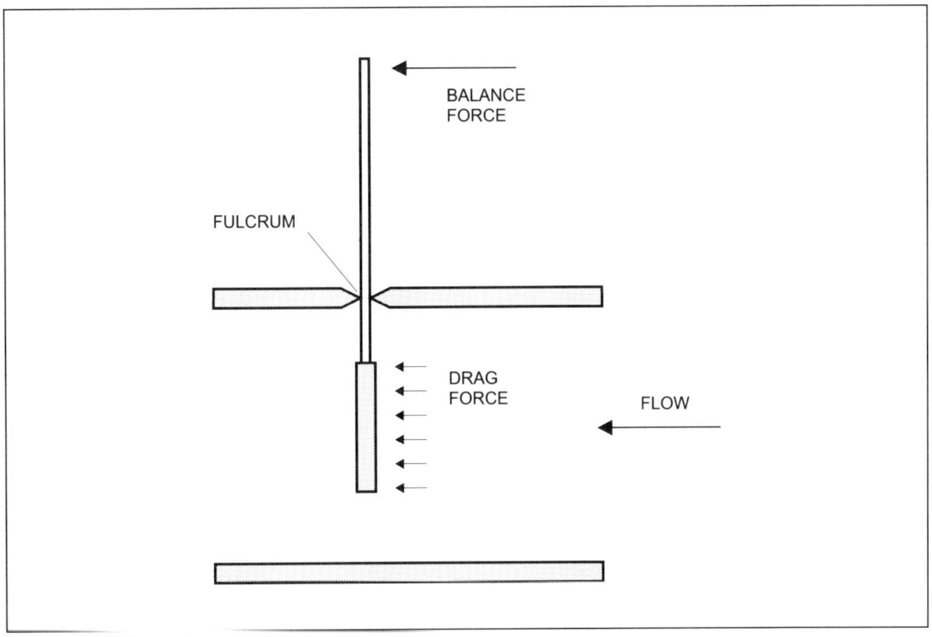

Figure 14-3. Target as a Lever

drag force acts on one end of the shaft, causing it to rotate about the flexure. A force is applied at the other end of the shaft to null out the drag-induced rotation. The amount of force required to provide the nulling is thereby a measure of the target force. Force balance can be implemented magnetically as shown in Figure 14-4(a). The force acting on the target would tend to rotate the force bar in a clockwise direction. This motion is transmitted to an electromagnetic detector, which generates a current proportional to that motion. The current is amplified and applied to a feedback coil to generate a force to null the target motion. The feedback current is converted to a 4-20 mA signal that is proportional to the square of the flow. For installations where it is undesirable to transmit signals via current, force balance can be implemented entirely by pneumatics (Figure 14-4(b)). For this case a pneumatic relay is used to null out the target motion and provide a 3-15 psi output pressure that also is proportional to the square of flow.

Resistive Strain Gage

This approach measures the strain produced by the force transmitted through the shaft (Figure 14-5). Depending on the flow ranges to be measured, either foil or semiconductor gages are used. Semiconductor gages provide 100-200 times more sensitivity than foil gages but are typically limited to more moderate temperatures. Also, for maximum accuracy, Wheatstone bridges with temperature compensation are often used. The output is generally a DC voltage proportional to the square of velocity with level in the range 2-4 mV per volt of supply voltage at full scale value.

Mechanical Force Gage

Still another variation directly couples the target to a mechanical gage (Figure 14-6). The motion of the target is amplified by means of gears. The mechanism can be supplied with a transmitter to convert the motion to a 4-20 mA output.

Materials of Construction

The most common material used for the wetted parts is 300 series stainless steel. Optional materials including Inconel™, Hastelloy C™, and Monel™ are available. Special coatings such as Teflon™ are also available. For abrasive fluids, Stellite™ or boron-coated stainless steel can offer the abrasion resistance needed for a reasonably long life.

Difficult Fluids

Although the target flowmeter is not considered on a par, in terms of accuracy, with some available devices, it does tend to be repeatable and reliable. Its reliability is no surprise since there are no moving parts or bearings to degrade, no pressure lines to plug, and a minimum of components. If care is taken with respect to material selection, the target flowmeter will provide a reliable measurement of most liquids and many gases, particularly for control applications. It has been used successfully with slurries, paints, pulp stock, emulsions, and low grade petroleum products; consider, for example, that the target flowmeter has been successfully used to measure coal-oil slurries, which are abrasive, non-Newtonian and have a tendency to plug.

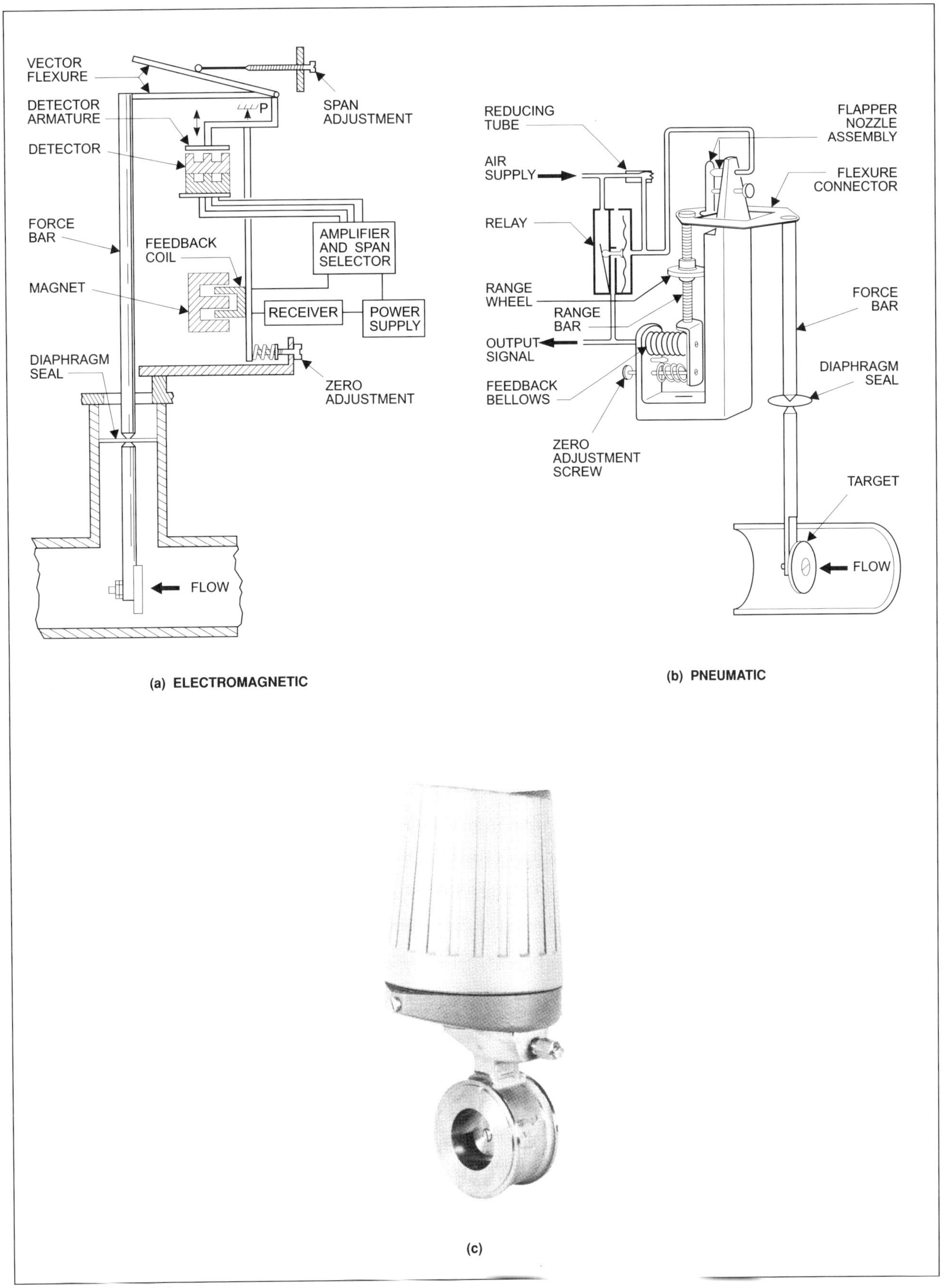

Figure 14-4. Force Balance
(Courtesy of the Foxboro Company)

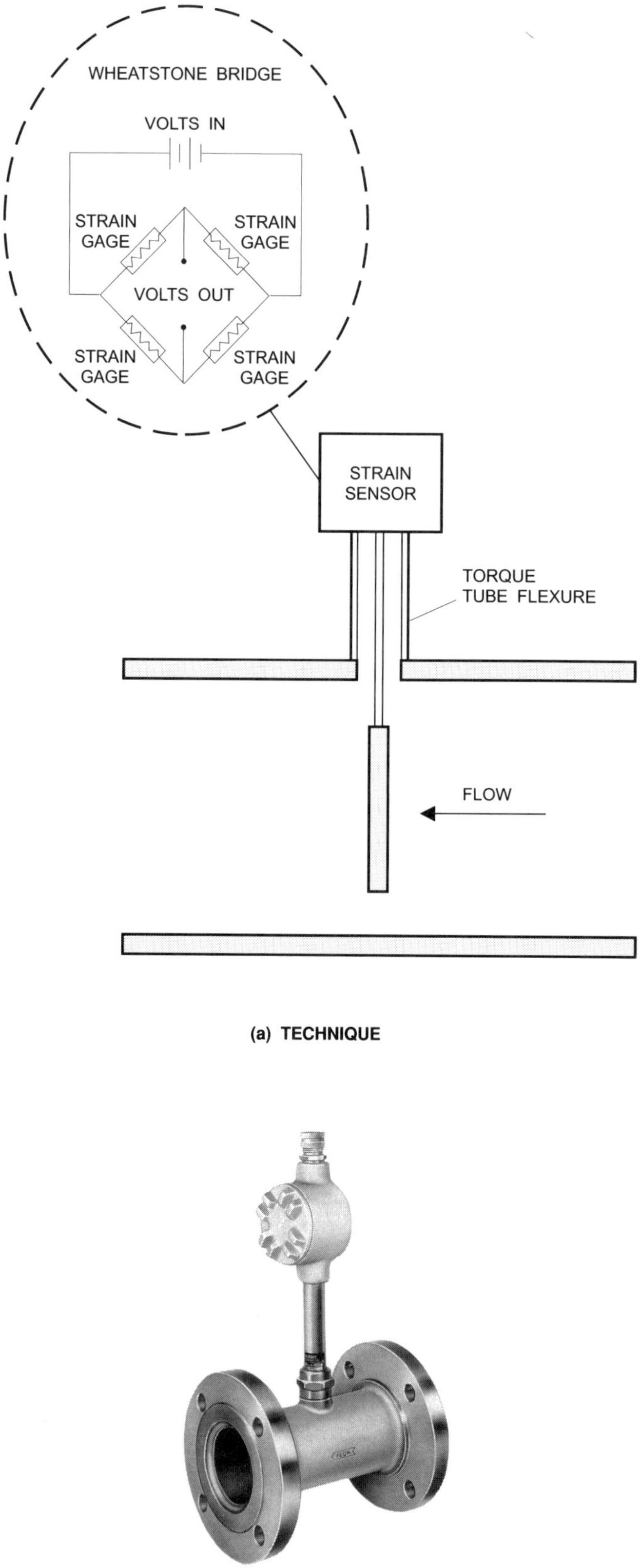

(a) TECHNIQUE

(b)
(Courtesy of Hersey)

Figure 14-5. Strain Gage

Figure 14-6. Mechanical Force Gagé

Sizing Considerations

As with any head-type flowmeter, sizing of a target flowmeter involves the balancing of two contradictory criteria. The target force must be large enough to provide the necessary sensitivity for the force sensor, but the unrecoverable loss in pressure, which will increase as target force increases, must not exceed acceptable limits.

To ensure that target force levels are within the operating limits of the sensor, it is necessary to calculate the force generated at the flows of interest. Operating limits depend on sensor design.

The pressure drop, like the target force, can be expressed as:

$$\Delta P = (\text{constant}) \times (\rho \bar{v}^2)$$

For simplicity it is often presented as a ratio or percentage of target force, thus:

$$\Delta P = K_{\text{loss}} \times F_\tau \qquad (14\text{-}5)$$

Note K_{loss} in Equation (14-5) is not a dimensionless constant. It has units of (area)$^{-2}$ and must be applied accordingly.

A 4-inch schedule 40 (D = 4.026 inch = 0.336 ft) target flowmeter with a β_τ = 0.665 is used to measure water flow at a temperature of 68°F (ρ = 62.32 lb/ft^3). At 5 feet per second (27.55 pounds per second) calculate the force on the target, assuming K for these conditions is 0.676:

Rearrange Equation (14-1b) in terms of force to yield:

$$F_\tau = \left(\frac{\dot{m}}{K}D\right)^2\left(\frac{1}{\rho g_c}\right)$$

$$= \left(\frac{27.55}{0.672 \times 0.336}\right)^2\left(\frac{1}{62.32 \times 32.17}\right)$$

$$= 7.42 \text{ lb}$$

Applying the same flowmeter, what would the force be for nitrogen at 145 psia, 68°F (ρ = 0.720 lbs/ft^3) and flowing at 60 feet per second (3.82 pounds per second)?

$$F_\tau = \left(\frac{3.82}{0.672 \times 0.336}\right)^2\left(\frac{1}{0.720 \times 32.17}\right)$$

$$= 12.4 \text{ lb}$$

K_{loss} for this meter is given in [Ref. 4] as 0.09 in.$^{-2}$ Determine the pressure losses for the water flow and the nitrogen flow. From Equation (14-5):

For water —

$$\Delta P_{loss} = 0.09 \times 7.42 = 0.67\text{psid}$$

For nitrogen —

$$\Delta P_{loss} = 0.09 \times 12.4 = 1.12\text{psid}$$

Note that, although the density and the mass flow rate of the nitrogen are less than those of water, both the target force and pressure loss for the nitrogen are higher in the example. This is due to the higher nitrogen velocity and demonstrates the dominance the squaring of the velocity can have on these forces.

Calibration

The calibration procedure for target flowmeters depends on the accuracy required. Highest accuracies can be achieved by performing a wet flow calibration of the target flowmeter installed in the meter run. Since target flowmeters are generally used where accuracy is not the primary concern, a lower accuracy bench calibration is more often used.

Using procedures outlined above, calculate the force that acts on the target at the desired full scale flow. After zeroing the transmitter, a weight equal to that calculated force is suspended from the target support (Figure 14-7). The span is then adjusted to give the correct full scale output. Since the calculated force is generally not an even number, it may be difficult to obtain a weight of exactly the correct value. As an alternative, a weight that is close to, but less than, the full scale value can be substituted, and the output can be calculated based on the ratio of the lower weight to the full scale weight.

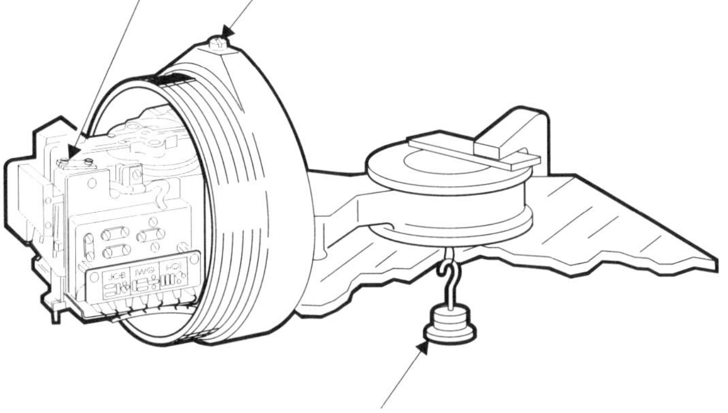

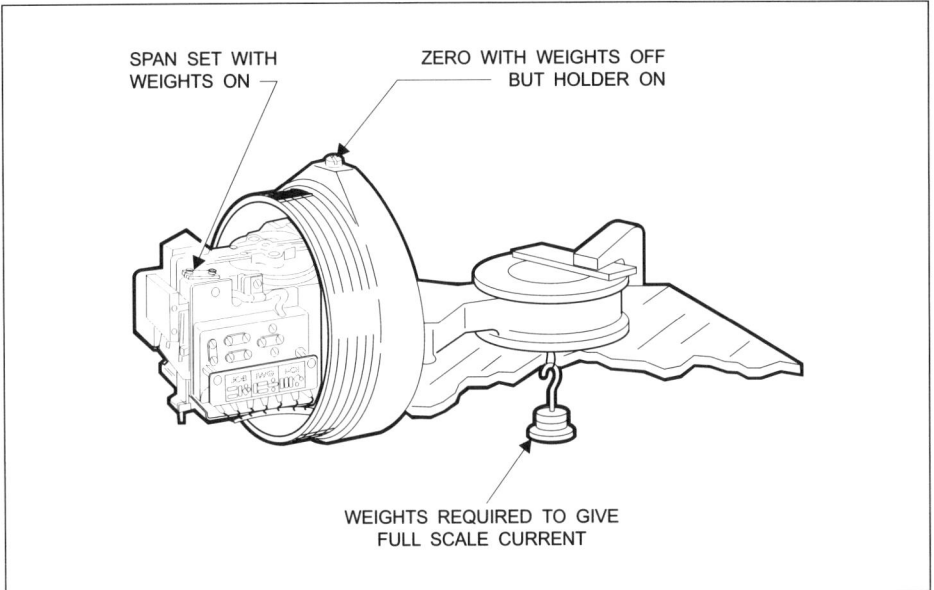

Figure 14-7. Bench Calibration
(Courtesy of The Foxboro Company)

For example, if the actual full scale force is 15.45 lb, a 15 lb weight could be used if the transmitter were spanned at 19.53 mA.

$$\text{span output} = (16\ \text{mA}) \times \left(\frac{\text{actual weight}}{\text{full scale force}}\right) + 4\ \text{mA}$$

$$= (16\ \text{mA}) \times \left(\frac{15.00\ \text{lb}}{15.45\ \text{lb}}\right) + 4\ \text{mA}$$

$$= 19.53\ \text{mA}$$

After the bench calibration and final installation, the installed flowmeter must be re-zeroed under zero flow with the process at as close to flowing temperature and pressure as possible.

Installation

To obtain the best performance from a target flowmeter, straight runs of upstream and downstream pipe having the same diameter as the meter bore are required. The lengths of straight run will depend on the type of flow disturbance (i.e., elbow, reducer, etc.) that precedes the flowmeter. The lengths can be reduced by the use of a flow conditioner. Typically, straight run requirements are the same as for an orifice plate flowmeter having the same blockage. Installation requirements for the orifice plate are well documented in the literature [Refs. 4, 5]. Assuming the blockage due to the target support strut can be neglected in con-

A 0.8 beta target would correspond to a 0.6 beta orifice plate, thus:

$$\beta_e = \sqrt{1 - 0.8^2}$$
$$= \sqrt{1 - 0.64}$$
$$= 0.6$$

sideration of installation requirements, the following equation can be used to calculate effective orifice beta:

$$\beta_e = \sqrt{1 - \beta_\tau^2}$$

where β_e = effective orifice β ratio.

Maintenance

Since meter factor depends on target geometry, the target and pipe should be periodically checked for coating or erosion. Any buildup should be cleaned. Do not scour the surface or edges. One or two nicks on the target will not greatly affect performance, and it is better to leave them than attempt to file them off.

Target flow coefficients can vary considerably with the edge sharpness. Therefore, any noticeable rounding of the edges due to erosive wear can cause errors and reduce repeatability. Generally, this cannot be handled simply by re-calibration. Replacement of the target is recommended if the edges become rounded.

References

1. Curran, D. E., "Laboratory Determination of Flow Coefficient Values for the Target-Type Flowmeter at Low Reynolds Numbers," *Flow, Its Measurement and Control in Science and Industry*, Vol. 2, Research Triangle Park, NC: Instrument Society of America, 1981.
2. Miller, R. W., "Flowmeter Having Uniform Response under Both Laminar and Turbulent Flow Conditions," Patent, #4517847, US, 1985.
3. Scarpa, T. J., "Flowmeter," Patent, #4604906, US, 1986.
4. Miller, R. W., *Flow Measurement Engineering Handbook*, 2nd edition, NY: McGraw-Hill, 1989.
5. ASME, *Fluid Meters*, 6th edition, NY: ASME, 1971.

About the Author

Wade Mattar is a Principal Engineer at Foxboro Company and has over 16 years experience at Foxboro doing Research and Development into new and improved flow measurement techniques. He received a B.S. in Aeronautical Engineering from Boston University and an M.S. in Mechanical Engineering from Northeastern University.

15

Thermal Mass Flowmeters and Controllers

There have been numerous applications for the thermal mass flowmeter since they were first introduced commercially. Early development was spurred by the space program, which required a small, low-powered mass flowmeter to measure air flow in space suits. The rise of air pollution concerns demanded accurate, electronic mass flowmeters and controllers for a new generation of pollution-monitoring equipment and gas blending apparatus. Unfortunately, this market shrank when funding for environmental concerns was cut.

Thermal mass flowmeters came of age when the semiconductor manufacturing industry required gas flowmeters that would accurately measure and control small flows of all types of clean gases being used in their diffusion process. Additional applications include leak testing of automotive parts and medical analytical equipment. There are also numerous special niches such as measuring uranium hexafluoride in the nuclear industry or gas velocity in flare stacks.

Principles of Operation

The thermal mass flowmeter, as its name implies, depends on the variation of one or more of the heat characteristics of the fluid as a function of flow. While in theory such flowmeters will measure liquids, in practice the commercial versions are limited to measuring gases. There are two types of thermal mass flowmeters: those that measure the "rate of heat loss to the flow stream" and those that measure the "temperature rise of the flow stream."

Rate of Heat Loss Flowmeter

This type of flowmeter measures the "rate of heat loss" to the flow stream from a heated element such as a resistance wire, thermistor, thermocouple, or thin film sensor (see Figure 15-1). These flowmeters are characterized by the classic King equation for a hot wire:

$$qt = \Delta T\,[k + 2(kC_v\rho\pi d v)^{1/2}] \qquad (15\text{-}1)$$

where:

qt = rate of heat loss per unit time
ΔT = mean temperature elevation of wire
d = diameter of wire
k = thermal conductivity of the fluid stream
C_v = specific heat of the fluid stream at constant volume
ρ = density of fluid stream
$\bar{v}$ = average velocity of the fluid stream

Note that the rate of heat loss is proportional to the square root of "the density times area, times velocity" to give mass flow. Note also that it is dependent upon three thermal characteristics of the gas: temperature, thermal conductivity, and specific heat.

In practice, some sensors are heated with a constant power, and the flow is measured as a function of the change in temperature. Others maintain the sensor at a constant temperature, and the flow is measured as a function of the required power. The constant temperature method usually results in faster response.

This principle is widely used to measure air velocity, and when the sensor is confined in a pipe, it becomes a flowmeter. The sensors normally are very small, require very little power, and have very fast response times. The main disadvantages are nonlinear response, sensitivity to the thermal conductivity of the gas, and dependency on the velocity profile due to the single point of measurement.

Temperature Rise Flowmeter

The second type measures the "temperature rise" of the flow stream. Typically, the temperature of the flow stream changes as it passes over a heated grid or through a heated tube (see Equation (15-2)).

$$W = \frac{H}{(\Delta T * C_p)} \tag{15-2}$$

where:

W = mass flow
H = heat (power) input
ΔT = temperature change
C_p = specific heat at constant pressure

Historically, the Thomas flowmeter shown in Figure 15-2 consisted of a large pipe, a heated grid in the center of the pipe, and thermometers upstream and downstream of the heated grid. A lot of power was required to heat the entire gas stream, but the flowmeter worked. Unfortunately, Thomas was employed by the natural gas industry, and they were just not going to install any electrically heated element in their gas lines.

Later work by Laub produced a design, shown in Figure 15-3, that placed the heater and thermometers on the outside surface of the pipe. Heat was transferred through the wall of the pipe and heated only the thin boundary layer of gas next to the wall. The design was safer because the heaters and thermometers were not in direct contact with the gas. The response time was slow because of the thick-

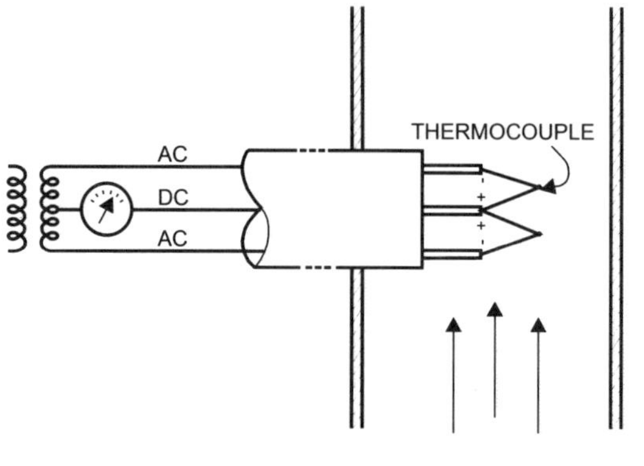

Figure 15-1. Rate of Heat Loss Flowmeter

ness of the wall of the pipe, and the output was unreliable because only the boundary layer was heated; this was affected by the Reynolds number, velocity profile, and the viscosity of the gas.

Benson and others, expanding on the work of Thomas and Laub, designed a flowmeter with the heater and thermometers on the outside surface of the pipe, but they reduced the pipe to a thin wall capillary tube so that the entire flow stream was heated and response time was substantially improved.

Today most commercially available thermal mass flowmeters are a variation of this design. They differ in how they heat the capillary tube. Some wrap an insulated resistive heater wire around the tube, others inductively heat the tube. They also differ in how they measure the temperature change in the tube. Some use temperature-sensitive resistance wire, others use thermocouples or thermistors.

Figure 15-4 shows schematically a typical thermal flowmeter. The capillary tube is heated by an insulated resistive wire wrapped around its center. The ends of the capillary tube are held at ambient temperature by the base of the transducer, which acts as a heatsink. The thermometers are thermocouples attached upstream and downstream of the heater. When power is applied to the heater and no flow is present, a symmetrical temperature gradient exists across the tube. When flow passes through the tube, heat is transferred from the tube to the gas stream on the inlet side, reducing the temperature at TC-1. As the gas continues through the tube it begins to return the heat to the tube on the outlet side, increasing the temperature at TC-2. The difference in temperature, measured in millivolts by the thermocouples, is proportional to the mass flow through the tube.

A flowmeter with a single heated capillary tube typically has a limited full scale flow capacity of about 10 to 50 sccm. When the single heated capillary tube is combined in parallel with similar unheated capillary tubes or other suitable bypass arrangements, flows up to 15,000 slpm can be routinely measured.

An innovative version of this type of flowmeter is produced on an integrated chip. The capillary is etched in the chip, and the temperature sensors are semiconductor materials. The device is very small and has very fast response time.

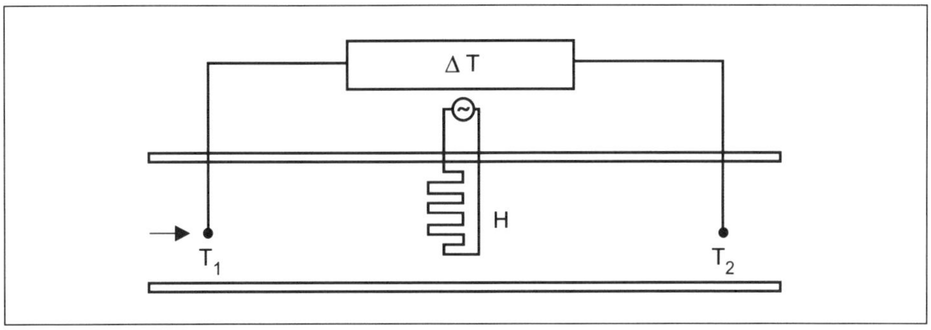

Figure 15-2. Thomas Flowmeter

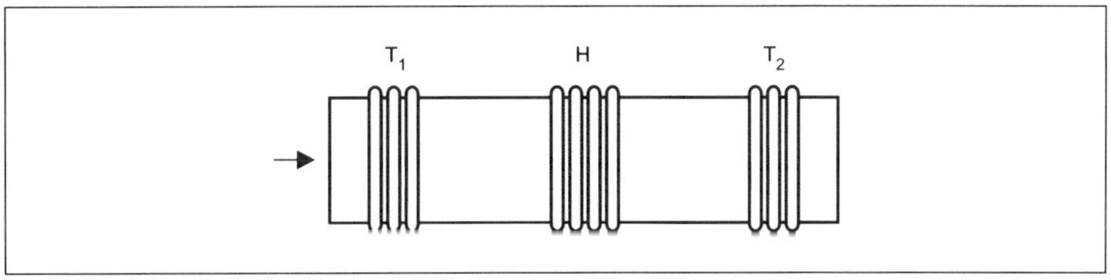

Figure 15-3. Laub Flowmeter

Design Features

Most commercial versions of the thermal mass flowmeter share similar design features. A typical mass flowmeter/controller, as shown in Figure 15-5, is about 3 in. long by 1 in. wide by 5 in. high. The body is constructed of 316SS with Viton O-rings used as seals. The 316SS capillary tube typically has an O.D. of 0.01 inch and is 0.002 inch thick. The 316SS bypass element is contained in the base, and it determines the flow range of the flowmeter. The inlet and outlet fittings are ¼ inch or ½ inch Swagelok™, although most offer VCR or VCO fittings as an option. A few manufacturers offer Monel™ and other O-ring materials or other materials instead of 316SS as an option.

The upper part of the flowmeter contains the electronic circuit card, which controls the power to heat the capillary tube. It also amplifies and linearizes the millivolt output signal from the sensors. Some manufacturers make electrical connections directly to the PC board; others use the type "D" electrical connector, although there does not seem to be any standard pin configuration; some can provide terminal connections.

Input power is usually ± 15 V DC at 50 to 200 mA, although a few models require a single-ended 24 V DC supply. Output is usually 0-5 V DC and is linearly

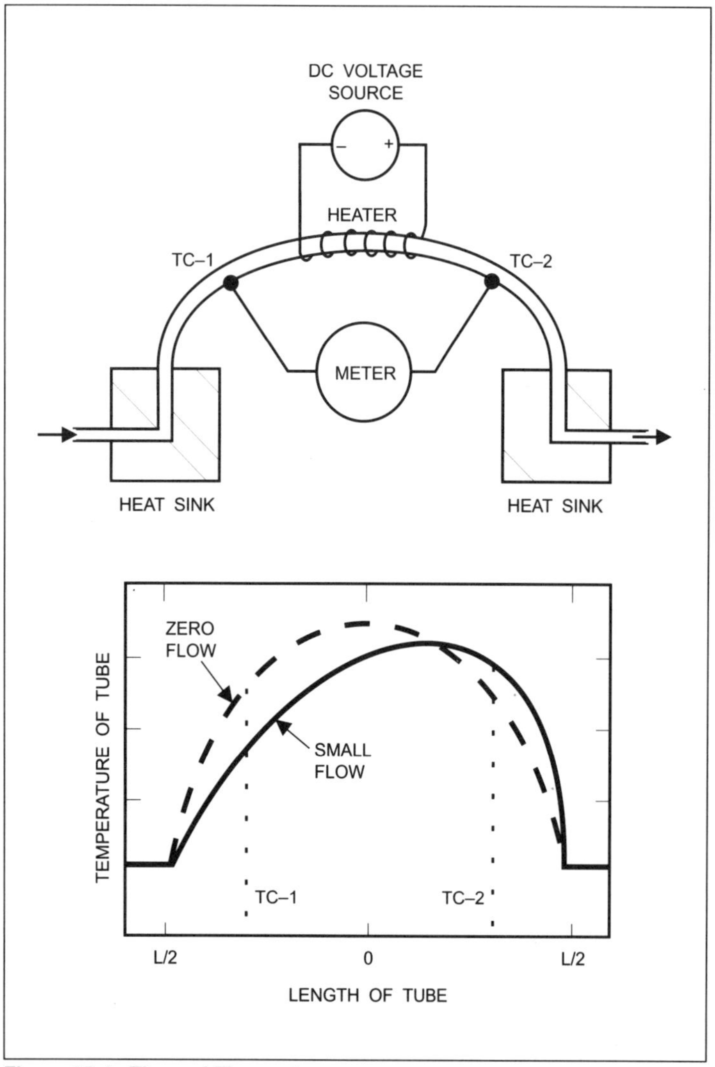

Figure 15-4. Thermal Flowmeter

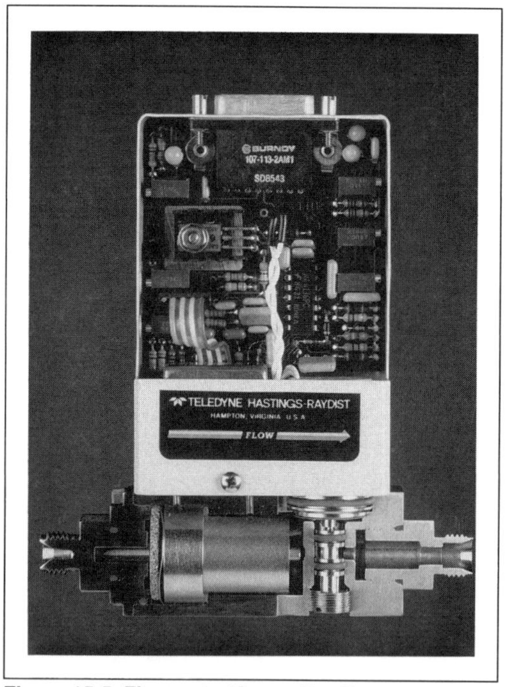

Figure 15-5. Flowmeter/Controller Cutaway View
(Courtesy of Teledyne Hastings-Raydist)

proportional to flow; however, 4-20 mA outputs are also available. Scaling for special gases or special flow units can be performed at the flowmeter or at the readout device.

Most manufacturers keep the power (heat input) constant and measure the mass flow as a function of temperature change in the capillary tube. Since Equation 15-2 shows that mass flow is inversely proportional to temperature change, the output will be nonlinear. Some manufacturers linearize the output electronically, while others use only a small segment of the output that is "approximately" linear.

Controllers

Approximately half of all thermal mass flowmeters are flow controllers; that is, they contain a flow control valve and a feedback circuit that compares the actual flow with the desired flow and adjusts the valve until the two values are equal.

Early flow controllers used motor-driven metering valves. They were effective and reliable but bulky and very slow. A major innovation was the thermal expansion valve, shown in Figure 15-6. The valve consists of a cone suspended just below an orifice by a wire having a high coefficient of expansion. When voltage is applied, the wire heats up and expands, forcing the cone away from the seat,

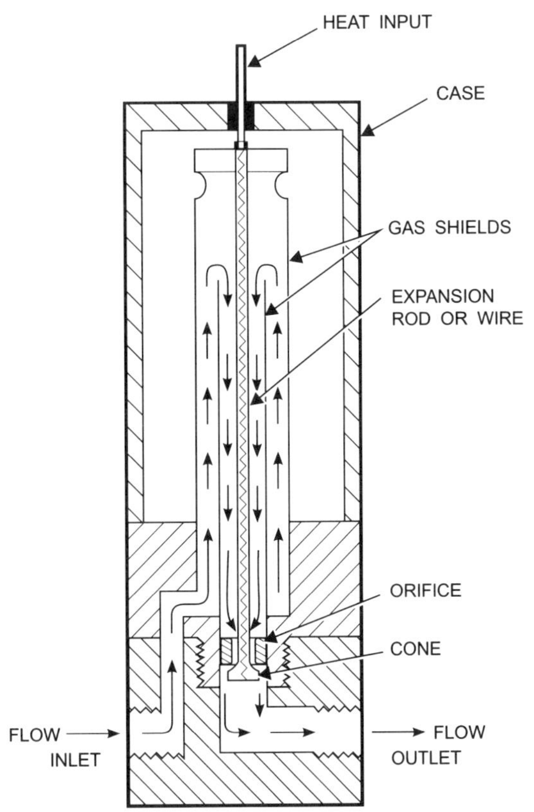

Figure 15-6. Thermal Expansion Valve

thus allowing gas to flow through the orifice. The feedback circuit compares the actual flow signal with the desired flow signal and adjusts the temperature of the wire (hence, its expansion) to keep these two signals equal. Thousands of these controllers are still in use, particularly in the semiconductor industry.

Most flow controllers currently use the floating or proportional solenoid type of valve shown in Figure 15-7. In a typical ON-OFF solenoid valve, a spring pushes the plunger down against the orifice, creating the off or closed position. When full voltage is applied to the valve coil, an electromagnetic force pulls the plunger away from the orifice, creating the on or open position. If less than full voltage is applied to the coil, the plunger is still pulled away from the orifice, but it does not completely open the valve. It "floats" above the orifice, controlling or metering the flow of gas through the valve. The feedback circuit controls the voltage applied to the coil (hence, the position of the plunger) until the actual flow signal is equal to the desired flow signal. Larger flows can be controlled by using the floating solenoid valve as a pilot valve to control a much larger plunger and seat, as shown in Figure 15-8.

It must be remembered that control valves require a pressure differential to operate—typically 5 to 50 psid for the floating solenoid type. It must also be remembered that the valve is not independent of gas temperature and pressure like the mass flowmeter. While a mass flowmeter may function equally well at 0 or 250 psig, the control valve may require different sizes of orifices.

The data in Table 15-1 is a composite taken from the spec sheets of 7 leading manufacturers of mass flowmeters and controllers. The data attempts to represent a typical 1 slpm of air mass flowmeter and mass flow controller. Since performance specifications differ for each manufacturer, the reader is urged to review the individual spec sheet for any mass flowmeter or controller of interest.

Sizing

Sizing a mass flowmeter can be confusing. Consider the ranges of flow—from grams per second of carbon dioxide to ranges in slpm of air. The mass of solids and liquids is usually measured in grams or pounds, but "traditionally" the mass

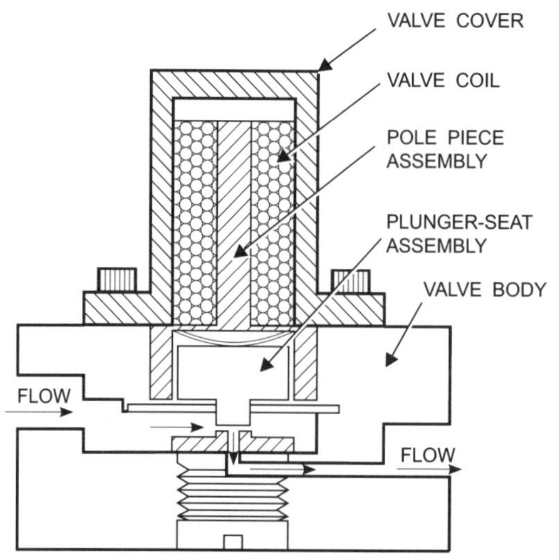

Figure 15-7. Porportional Solenoid Valve

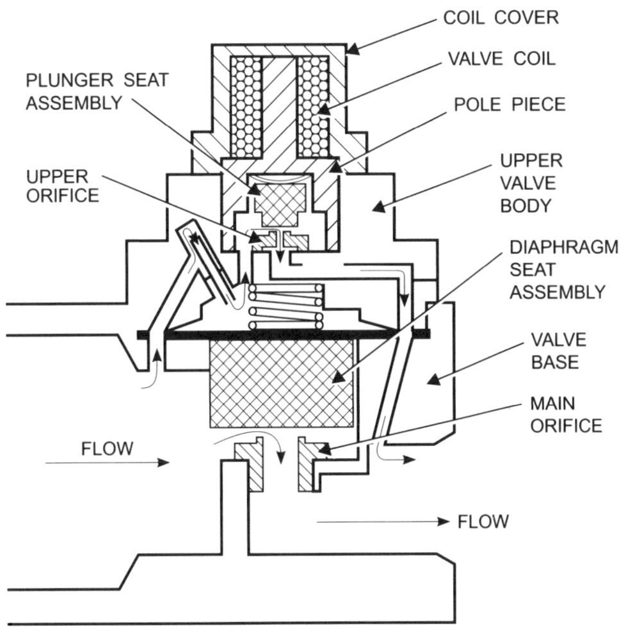

Figure 15-8. Two-Stage Proportional Solenoid Valve

of gases has been measured in terms of the volume it occupies at an agreed upon temperature and pressure, known as standard conditions. While the thermal mass flowmeter is one of the few gas flowmeters that could be calibrated directly in mass units of grams/second, etc., tradition dictates using the more conventional standardized volume units. Most manufacturers agree that the standard pressure should be one atmosphere (760 mmHg); however, they do not agree on a standard temperature. The most common standard temperatures used are 0°C, 20°C, and 25°C.

Know to which standard temperature your flowmeter has been calibrated, because there is over an 8% difference in calibration between 0°C and 25°C.

Gas correction factors are based on the ratio of the "standardized volume flow rate" of the gas to that of air. Do not apply them directly to flow rates given in mass units such as grams/second or lbs/hour. Be sure to convert the flow rates to volumetric units such as slpm or sccm. Note also that the gas correction factor for some gases is dependent upon the operating temperature of the sensor. Be sure to use the factor given by the manufacture of the flowmeter.

How would a flowmeter be selected to measure 1 gram/second of carbon dioxide if the manufacturer calibrates to 760 mmHg and 20°C?

The density of carbon dioxide is 1.84 grams/liter at 760 mmHg and 20°C. (Most manufacturers publish a list of gas densities in their catalogs.)

$$Q_{CO_2} = (1\ \text{gr/sec}) * \frac{(60\ \text{sec/min})}{1.84\ \text{gr/liter}}$$
$$= 32.61\ \text{standard liters/min of}\ CO_2$$

Since the thermal flowmeter is sensitive to the heat capacity of the gas, one additional factor must be considered: the gas correction factor. For carbon dioxide this factor is 0.73; that is, a flowmeter that reads full scale with 1.00 slpm of air would read full scale with only 0.73 slpm of CO_2. (Again, most

(continued)

(continued)

manufacturers publish a list of gas correction factors that apply to their flowmeters.) To obtain the equivalent air flow rate, the gas flow rate is divided by the gas correction factor.

$$Q_{\text{air equivalent}} = \frac{(32.61 \text{ slpm of } CO_2)}{0.73}$$
$$= 44.67 \text{ slpm of air equivalent}$$

In this case, a flowmeter having a full scale air range of 50 slpm would be an appropriate choice.

Table 15-1. Performance Table

	Typical Flowmeter	Typical Flow Controller
Minimum FS Air Range	5 sccm	10 sccm
Maximum FS Air Range	200-1500 slpm	30-750 slpm
Accuracy % FS	± 1.0%	± 1.0%
Linearity % FS	± 0.5%	± 0.5%
Repeatability % FS	± 0.2%	± 0.2%
Response Time 98%	< 1.0-9.9 sec	< 1.0-9.9 sec
Pressure Correction	0.01%/psi	0.01%/psi
Max Pressure	1000-2500 psi	500-1500 psi
Temperature Range	–20°C-60°C	–20°C-60°C
Temperature Correction	0.02-0.20%/°C	0.02-0.20%/°C
Pressure Drop	0.05-12" H_2O	5-50 psi
Power Supply		
Voltage	± 15	± 15
Current	30 mA	30-300 mA
Dimensions with Fittings		
Height	76-160 mm	76-160 mm
Width	32-50 mm	32-52 mm
Length	60-150 mm	60-160 mm
Weight	300-2000 gr	500-2500 gr
Materials of Construction	316SS/Monel	316SS/Monel
Seals (O-rings)	Viton/Other	Viton/Other
End Connections	¼" Swagelok	¼" Swagelok
Optional	VCR, VCO	VCR, VCO

Safety

The thermal mass flowmeter has an excellent safety record. The electrical heaters are attached to the outside wall of the sensor tube and do not come in direct contact with the gas. Most units are rated to 500 psig, and many can be rated to over 1000 psig. Most manufacturers helium leak test their flowmeters to better than 10^{-8} sccs.

The primary safety concern associated with the thermal flowmeter comes from the use of hazardous or corrosive gases. The material selected for the elastomer O-ring seals must be considered carefully to ensure it will stand up to the gases and operating conditions it will encounter. There is a trend toward the use of all-metal seals instead of elastomers when dealing with corrosive gases. Moisture in the flowmeter or in the connecting lines can react with some gases to form compounds more corrosive than the gas itself. Failure to properly purge the flowmeter when changing gases can sometimes result in corrosive or explosive mixtures being formed.

Calibration

Most thermal flowmeters can trace their calibration back to NIST through some type of volumetric calibrator. The most frequent type used below 50 slpm is the precision-bore glass tube with mercury-sealed piston. This device accurately measures the volume of gas discharged from the flowmeter in a given period of time. By also measuring the temperature and pressure of the gas, the volume flow rate can be converted to mass flow rate. This is a sampling-type calibrator as opposed to a continuous reading calibrator.

Above 50 slpm some use a bell prover, whose operation is similar to the piston calibrator described above. However, the preferred method is the sonic nozzle. As long as the downstream pressure is less than half the absolute upstream pressure, sonic flow exists through the nozzle, and changes in the downstream pressure have little or no effect on the flow rate. Hence, the flow rate becomes dependent only on the upstream pressure and not on the differential pressure across the nozzle. It is still necessary to measure the temperature to determine mass flow.

Most manufacturers offer air or nitrogen calibrations as standard. They will adjust the calibration for other gases using the gas correction factors published for their flowmeter. Some will calibrate the flowmeter with a specific gas.

A rather unique but simple method of calibrating a thermal flowmeter for a gas other than air is occasionally used, typically by the user in the field (see Figure 15-9). A gas with an unknown calibration factor is connected to a flowmeter calibrated for air by means of a regulator, a metering valve, and a coil of small diameter tubing. The tubing is the heart of the method. It must have a volume at least twice that of the full scale range of the flowmeter, its length must be at least 50 to 100 times its diameter, and it must be filled with air.

The regulator should be set at about 15 psig and the metering valve opened quickly to obtain a reading on the flowmeter anywhere in its midrange. At first, the flowmeter will measure only air and within 10 to 30 seconds should reach a stable reading that should be recorded. Soon the measurement will begin to drift

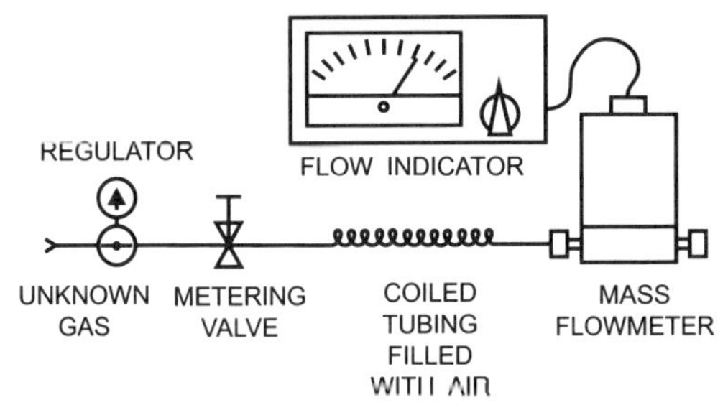

Figure 15-9. Calibration Method

as the mixture of air and the unknown gas enter the flowmeter. After a few minutes, depending on the size and length of the tubing, the reading will stabilize again when it measures only the unknown gas. This reading should be recorded. It should be understood that the flow rate has not changed, because it was set by the regulator and metering valve, which have remained untouched. The ratio of the final reading to the initial reading is the gas correction factor for the unknown gas. This can be used to calibrate this or other similar flowmeters. This method can be used to obtain calibration factors for corrosive or hazardous gases that cannot be measured with primary calibration standards.

Installation

Be sure to read the flowmeter manual before installing the flowmeter.

Most thermal flowmeters can be mounted in any position, but there is usually a preferred orientation, because drift and pressure effect are somewhat dependent upon position.

Unless the gas and piping are very clean, a filter upstream of the flowmeter is recommended. Particles, moisture, and oil vapor are major culprits in plugging a flowmeter or shifting its calibration. Sintered metal filters are built into some flowmeters. They do a good job of filtering out particles but do not always filter out moisture and oil vapors.

Be sure to examine the volume of the system. For a 10 sccm flowmeter with downstream tubing and fittings having a volume of 10 liters, it could take 20 minutes for the flow to stop after the valve is shut off. This is not the response time of the flowmeter; it is the time required for gas trapped in the system to pass through the flowmeter as the pressures equalize. Many hours have been spent looking for leaks in a closed system, only to discover that normal expansion and contraction of the gas in the system caused a measurable amount of flow to pass through the flowmeter as the pressures equalized.

Maintenance

Thermal mass flowmeters have no moving parts, and when used on clean, noncorrosive gases, they can give years of dependable service. The bypass element should be examined to make sure it is clean. The O-rings should be checked and replaced if they show any sign of deterioration, nicks, or loss of flexibility. It is a good idea to have the calibration checked annually. The valve in the flow controller does have moving parts. Springs may change their tension or valve seats may wear, causing the valve to leak in the closed position.

If a sensor plugs, it may be possible to clean it by running a fine wire (< 0.008 in. diameter) through it. Always run the wire from the outlet to the inlet if possible, in case the blockage is confined to the entrance of the sensor. Many flowmeters now have replaceable sensors, and it is much quicker to install a new sensor. Similarly, the electronic circuit card is replaceable.

References

1. Thomas, C. C., "The Measurement of Gases," *Journal of the Franklin Institute*, November 1911.
2. Hastings, C. E., and Weislo, C. R., "A Compensated Thermal Anemometer and Flowmeter," *American Institute of Electrical Engineers*, March 1951.
3. Laub, J. H., "An Electrical Flowmeter," *Electrical Engineering*, December 1969.

4. Benson, J. M.; Baker, W. C.; and Easter, E., "Thermal Mass Flowmeters," *Instruments and Control Systems*, February 1970.

5. Baker, W. C., and Pouchot, J. F., "Pitot/Thermal Gas Flow Probe with Purge Measures Low, Dirty Flow without Fouling," *Measurements & Data*, November-December 1974.

6. Benson, J. M., "A Survey of Thermal Devices for Measuring Flow," *Flow: Its Measurement and Control in Science and Industry*, ISA, 1974.

7. Baker, W. C., and Pouchot, J. F., "The Measurement of Gas Flow," *The Journal of the Air Pollution Control Association*, Volume 33, Numbers 1 & 2, January-February 1983.

8. Bray, B. J., "Proving the Accuracy of Gas Flow Calibration Systems," *Flow Measurement Symposium*, page 165, ASME 1966.

9. Olin, J. G., "Process Gas Mass Flow Controllers: An Overview," *Solid State Technology*, April 1988.

10. Sullivan, J. J.; Ewing, J. H.; and Jacobs, R. P., "Calibration Techniques for Thermal-Mass Flowmeters," *Solid State Technology*, April 1985.

About the Author

William C. Baker is Vice President of Marketing for Teledyne Hastings-Raydist and has been with that company for 30 years. He obtained his B.S. in Electrical Engineering from VPI and his M.S. in Administration from George Washington University. He holds five patents on the thermal mass flowmeter and has written several technical papers on mass flow and vacuum-related topics. He is a member of Tau Beta Pi, Eta Kappa Nu, ISA, AVS, and ASM.

16

Tracer Dilution Measurement of Flow

Measurement of flow by dilution methods depends on the determination of the degree of dilution of an added tracer solution by the flow. Dilution methods of measuring discharge have been known since at least 1863 [Ref. 11]. Until recently, chemical salts [Ref. 8] were the most commonly used tracers. Radioactive tracers have been used successfully, but handling problems have limited widespread use [Ref. 9]. The development of stable fluorescent dyes and fluorometers that can measure them at very low concentrations has greatly enhanced the use of dilution methods [Refs. 5, 7]. Hence, the use of fluorescent dyes as the tracer is addressed, although the principles discussed apply to any tracer.

Typical examples of situations where the dilution methods might be used are turbulent mountain streams, pipes, canals, sewers, ice-covered streams, and sand-channel streams.

Dilution methods are useful under the following flow conditions:

- **Open or closed channel flow**
- **Where high velocities, turbulence, or debris preclude other measuring devices**
- **Where, for physical reasons, the flow is inaccessible to a current meter or other measuring device**
- **Where the physical properties, such as cross-sectional area and/or velocity profile, cannot be accurately measured as part of the discharge measurement or are changing during the measurement**

Theory

Measurement of the degree of dilution of a known quantity of tracer after its mixing in a flowing stream of water is the basis of dilution gaging. There are two main approaches: (1) the slug-injection of a known amount of tracer into the flow, which requires that the dilution of the tracer be accounted for by the complete measurement of its mass downstream (for this reason, it is sometimes referred to as the total recovery method) and (2) the constant-rate injection of a tracer solution into the flow, which requires only the measurement of the plateau level of concentration that results downstream after equilibrium has been reached. The principles are simple, yet their successful application in streams, canals, pipes, and elsewhere requires a good understanding of the dispersion process. No elaborate theoretical treatment is used in explaining these processes; however, the reader is urged to heed the following principles, for they can eliminate many of the problems others have had in performing such measurements.

Slug-Injection

Soluble tracers injected into a stream behave in the same manner as the water particles themselves. The slug-injection or instantaneous "dumping" of a quantity of tracer into a flowing stream is the simplest of all methods from the standpoint of equipment needs. Where radioactive tracers are employed, handling problems make slug injection virtually the only feasible method. The dispersion of the tracer in the receiving stream takes place in all three dimensions of the channel. Vertical dispersion is normally completed first, lateral later, depending upon the width of the stream and velocity variations. Longitudinal dispersion, having no boundaries, continues indefinitely. The solid curves in Figure 16-1 show the resulting time-concentration response curves at different distances downstream that may result from a single midchannel slug injection of tracer.

The discharge as measured by the slug-injection tracer-dilution technique is:

$$Q = \frac{m}{A_c} \tag{16-1}$$

where:

Q = the volume of flow of the stream
m = the mass of tracer injected
A_c = the area under the response curve obtained after adequate mixing of the tracer in the flow

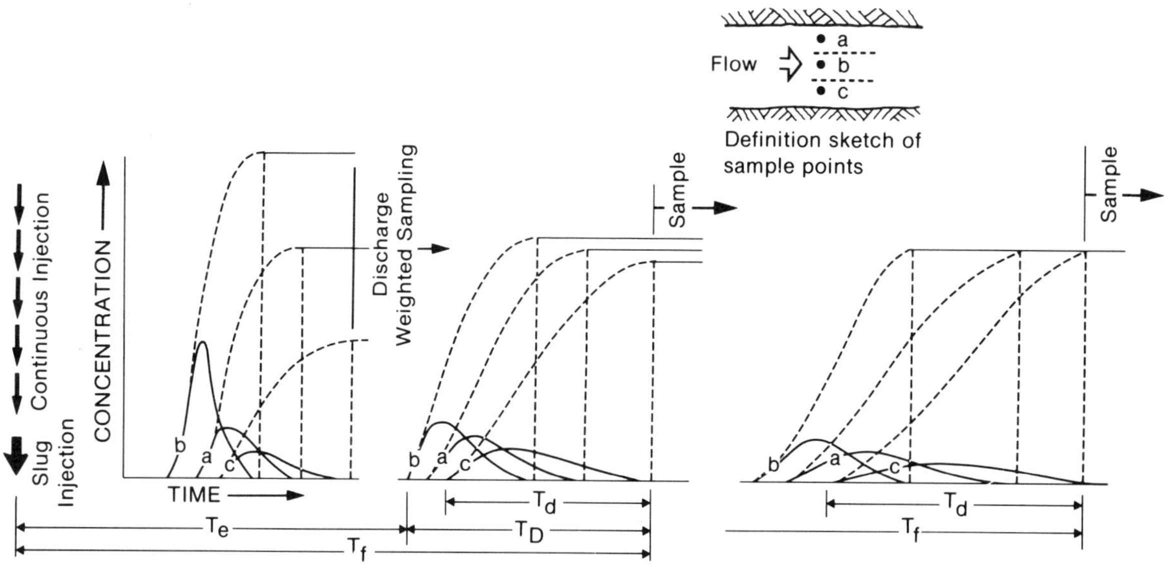

SHORT DISTANCE

1. Curve areas not the same, lateral mixing incomplete.
2. Rapid buildup to plateaus of different levels.
3. Dilution-discharge measurement can be obtained by using discharge proportional sampling, or by discharge weighting the concentration data.

OPTIMUM DISTANCE

1. Curve areas about the same, mixing nearly complete.
2. Fast buildup to plateaus nearly the same in concentration, mixing adequate.
3. Good dilution discharge measurement can be made with moderate length injection, T_D.

LONG DISTANCE

1. Curve areas identical, perfect mixing.
2. Slow buildup to plateau concentrations exactly the same but only after long period of injection.
3. Excellent dilution discharge measurement can be made if injection is long enough to allow sampling after slowest point in section has reached an equilibrium plateau.

GENERAL: Minimum length of constant injection required, T_D, is equal to longest time slug dye cloud would be present at any location in the stream. Danger is in not injecting long enough and not sampling late enough; samples taken prior to T_f would give impression mixing was poor and (or) yield data suggesting, incorrectly, a tracer loss.

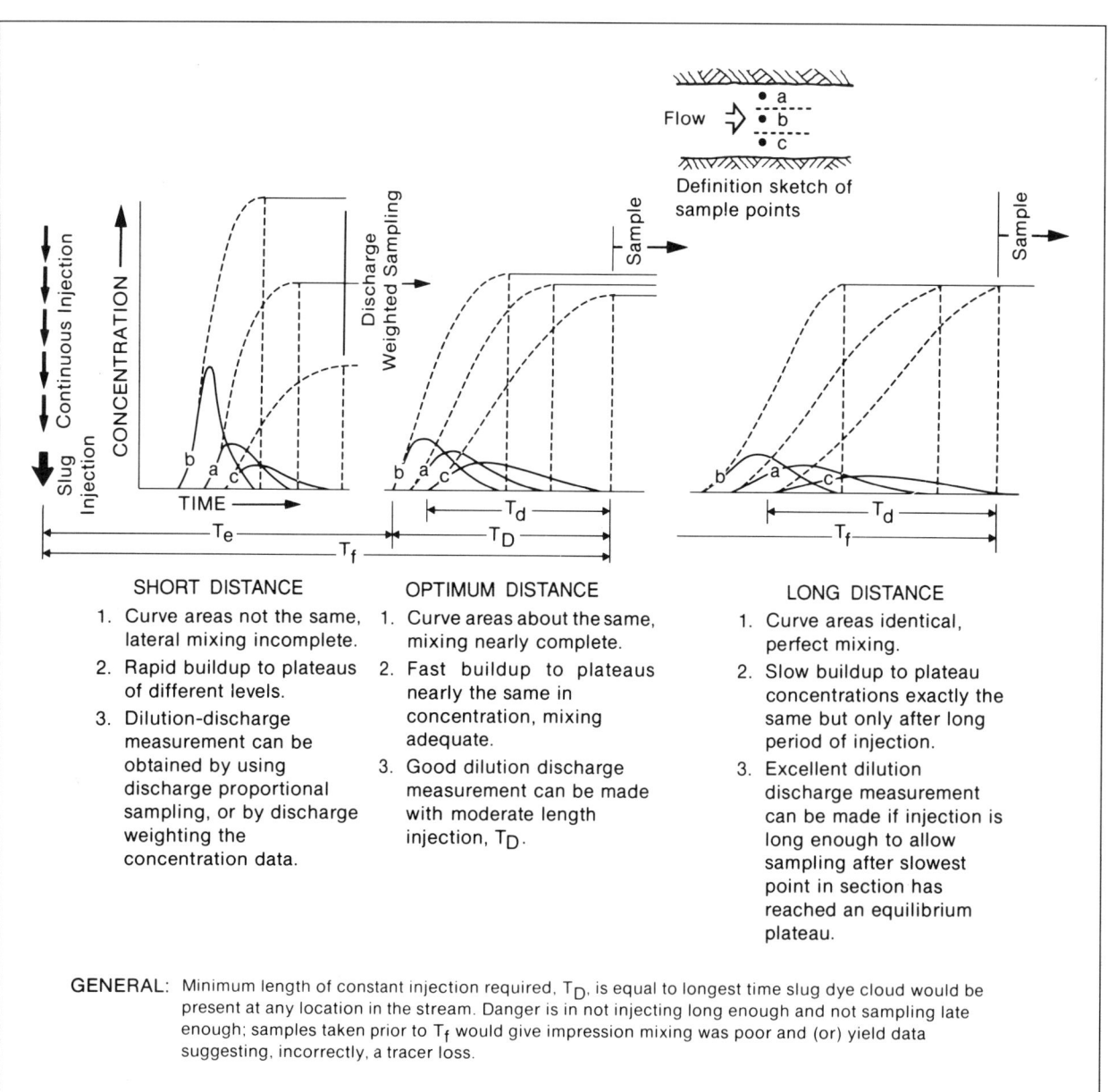

Figure 16-1. Time-Concentration Curves for Slug and Constant Injections Observed at Three Points Laterally across the Channel at Three Different Distances Downstream from the Injection Point
(Courtesy of U.S. Geological Survey)

At short distances downstream from a slug injection, the tracer is not fully mixed in the total flow of the stream, being more in the center than along the banks. Furthermore, the response curve measured in the center may be much shorter in duration, T_d , than for those along the banks. This is a common occurrence, because flow along the banks is usually slower and the channel banks tend to slow and elongate the tracer cloud. At such a short distance, an accurate measurement of discharge by dilution cannot be made by ordinary methods. When uniform mixing is reached, the areas under the time-concentration curves are essentially the same regardless of shape; at too short a distance, they are not.

Most important, therefore, is a measurement of the response curve(s) far enough downstream that mixing is almost complete in a cross section.

From a practical standpoint, complete mixing does not need to be attained. A good dilution-discharge measurement can be made at what is defined here as an optimum distance, L_o, downstream. The optimum distance is usually where mixing is about 95 percent complete. The distance is optimum because T_D is not too long, and thus, sampling of the complete response curves at several points laterally across a section is feasible. Note that the peaks of the response curves are not the same and that their lengths, or durations, and arrival and departure times are different. Nevertheless, the areas under the individual response curves are nearly the same, which indicates good mixing and allows a good dilution-discharge measurement.

If the response to the tracer slug is measured farther downstream, mixing will be nearly perfect, and the individual time-concentration curves will be nearly identical in area; therefore, a very accurate measurement of discharge may be obtained. That is true, however, only if sampling has been performed for a sufficient time, particularly of the drawn-out tails of the clouds, and if loss of tracer has not occurred because of excessive time of exposure. Frequently, failure to measure the low-level concentrations of the trailing edge of the tracer is misinterpreted as a loss of tracer. It is not a real loss but merely a sampling or measurement failure.

As may be seen, the chief advantage of the slug-injection method is simplicity of injection. The chief disadvantage is the requirement of extensive sampling and complete measurement of the downstream response curves.

Constant-Rate Injection

As illustrated in Figure 16-1 by the dashed lines, a continuous, constant-rate injection of tracer will produce, after sufficient duration of injection, plateaus of concentration. These plateaus will vary laterally if too short a mixing distance is employed. At an optimum distance they will have nearly the same concentrations, and a good measurement of discharge can be obtained as the plateau(s) provide the measure of dilution needed so that the continuity equation

$$qC = Qc \tag{16-2}$$

applies, whereby the amount of tracer being injection, qC, equals that passing the sampling section, Qc,

where:

q = the constant rate of tracer injection and is assumed to be very small relative to the flow being measured, Q
C = the concentration of the tracer being injected
c = the resulting plateau concentration after dilution by Q

At a long distance the resulting plateau concentrations are virtually identical *if the constant injection is made over sufficient time and if sufficient time is allowed* for buildup across the entire channel before sampling. Depending on the nature of the channel and the mixing distance selected, the injection duration and the lapsed time necessary before sampling the correct plateau concentration may be greater than anticipated.

Certain conclusions can be drawn from examination of Figure 16-1 for a given stream and flow:

(1) Sampling of the response curves from a slug injection must be for a period of time, T_D, and until time T_f; T_f is also the earliest time at which the plateau resulting from a constant injection can be sampled. Thus, the effective elapsed time required to make a discharge measurement is essentially the same by either method.

(2) The duration of a continuous injection needed to reach equilibrium plateau concentrations can be determined accurately by examination of the slug-response curves for a given site. The injection time must be at least equal to the time at which tracer is last present in the stream minus the earliest time of arrival of the leading edge of the tracer for that section, T_D. For practical purposes, the injection time must be slightly longer than T_D to allow for sampling.

(3) For the constant-injection method, plateaus develop earlier in the main flow than they do close to the streambanks, where the flow is slower.

Tracer and Instrumentation Requirements

Tracers

After World War II, radioisotopes such as tritium (heavy hydrogen) gained favor as tracers, but their use was severely limited by handling problems, the special training required, and a general lack of understanding by the public.

A search for a suitable substitute for radioisotopes led to the use of several of the more conservative fluorescent dyes. Two dyes, variations of the same basic organic structure (xanthene), are preferred for use as water tracers: rhodamine WT and pontacyl pink (also known as intracid rhodamine B, pontacyl brilliant pink B, and acid red 52). These dyes are generally good tracers because they are (1) water soluble, (2) highly detectable—strongly fluorescent, (3) fluorescent in a part of the spectrum not common to materials generally found in water, thereby reducing the problem of background fluorescence, (4) harmless in low concentrations, (5) inexpensive, and (6) reasonably stable in a normal water environment.

Rhodamine WT is preferred for most water-tracing uses and will be used in all examples.

With a good fluorometer, rhodamine dyes can be measured at concentrations as low as 0.05 µg/l; accurate dilution-type flow measurements are best made at concentrations of about 10 µg/l. Costs for rhodamine WT dye will vary depending on the quantities ordered, but, as a rule of thumb, $1.00 will measure about 10 ft^3/s or 4,000 gal/min of water flow using the constant-injection method and about double these discharges using the slug-injection method. The amount of dye and its cost is not a limiting factor in the size of flows that can be measured; rather, in natural streams, limitations are the excessive length of channel and the customary time required to accomplish mixing when large flows are to be measured.

Mixing Lengths

The performance of either the slug or constant-rate injection type of discharge measurements requires an estimate of the mixing length required; both methods will have the same mixing length for like physical and hydraulic conditions. The optimum distance required for adequate lateral mixing of a tracer injected in the middle of the streamflow may be estimated by the following equation (adapted from Yotsukura and Cobb, Equation (16-29) [Ref. 13], and Fischer and others, Equations (16-5) and (16-10) [Ref. 3]):

$$L_o = K\frac{vB^2}{E_z} \tag{16-3}$$

where E_z is the transverse mixing coefficient to be selected from Table 16-1; L_o is the distance required for optimum mixing, in feet; K is a coefficient to allow for different degrees of mixing and numbers of locations of injection points, to be selected from Table 16-2; v is the mean stream velocity, in feet per second; and B is the average stream width, in feet.

Thus, for optimum mixing (95 percent) and a center injection, $K = 0.1$. The width of the stream is the most important factor in determining the mixing length. Thus, in man-made canals and conduits, where width-to-depth ratios are normally smaller than for a natural stream, mixing lengths are shorter and excellent dilution measurements can be made.

A common fallacy is that shallow, turbulent white water streams produce rapid mixing. Turbulence is effective in yielding rapid vertical mixing, but unless there is substantial depth lateral dispersion occurs gradually and the mixing length may be quite long.

Inspection of Table 16-2 reveals that two injection points will reduce the required mixing length by a factor of 4, whereas a single-side injection will likely increase it by a factor of 4.

Table 16-1. Values of the Transverse Mixing Coefficient, E_z, for Selected Average Flow Depths and Slopes
(Courtesy of U.S. Geological Survey)

Depth, d (ft)	Slope, s, (ft/ft)				
	0.001	0.005	0.010	0.050	0.100
1.0	0.04	0.08	0.1	0.3	0.4
2.0	0.1	0.2	0.3	0.7	1.0
3.0	0.2	0.4	0.6	1.3	1.9
4.0	0.3	0.6	0.9	2.0	2.9
5.0	0.4	0.9	1.3	2.8	4.0

Dilution-type measurements are usually very readily performed in man-made hydraulic structures because advantage can be taken of valves, pumps, turnouts, drops, and expansions, which expedite mixing. Figure 16-2 shows the multiple injection of dye as a line source in the approach to a Parshall flume with sampling just downstream of the tailwater jump.

The literature is replete with descriptions of flow measurements in pipes. In straight uniform pipes, mixing lengths to produce accurate dilution flow measurements are reported as varying from 40 to 200 pipe diameters.

Tests in pipes by Filmer [Ref. 2] using dye to determine probable errors for various pipe lengths to be expected when using dilution methods showed probable errors of about 6 percent to 0.56 percent for mixing lengths of 100 and 200 pipe diameters, respectively. In almost all cases, effective mixing lengths can be drastically reduced by multiple injection schemes and (or) by utilizing existing or

Table 16-2. Values for Coefficient, K, for Different Degrees of Mixing and Numbers and Location of Injection
(Courtesy of U.S. Geological Survey)

Number and location of injection points	Coefficient, K		
	Percentage mixing		
	90	95[3]	98
One center injection	0.070	0.100	0.140
Two injection points[1]	0.018	0.025	0.035
Three injection points[2]	0.008	0.011	0.016
One side injection point	0.280	0.400	0.560

[1]For an injection made at the center of each half of flow.
[2]For an injection made at the center of each third of flow.
[3]Optimum mixing.

Not visible is dye injection pump, which preinjects dye into a hose line from a domestic water supply. Eight samples taken across the channel about 10 ft below the hydraulic jump in the bottom of the picture yielded good results.

Figure 16-2. Multiple Dye Injection in the Approach to a Parshall Flume
(Courtesy of U.S. Geological Survey)

added physical roughness or turbulence producing mechanisms. Pumps, for example, have been reported to produce almost instant mixing [Ref. 1].

It may be concluded that for closed conduit flow, each situation must be evaluated in selecting a mixing length that will produce adequate mixing for accurate dilution flow measurements.

Slug-Injection

The volume of tracer required for a slug-injection discharge measurement is a function of the stream discharge, the measurement-reach length, the stream velocity, and the peak concentration to be achieved at the sampling site. The dosage requirements of rhodamine WT 20 percent dye for natural streams may be estimated from the empirical equation

$$V_s = 3.79 \times 10^{-5} \frac{QL}{v} C_p \qquad (16\text{-}4)$$

or from the graph in Figure 16-3

where:

C_p = the peak concentration at the sampling site, in micrograms per liter
L = the length of the measurement reach, in feet
Q = the stream discharge, in cubic feet per second
V_s = the volume of rhodamine WT 20–percent dye, in milliliters
v = the mean–stream velocity, in feet per second

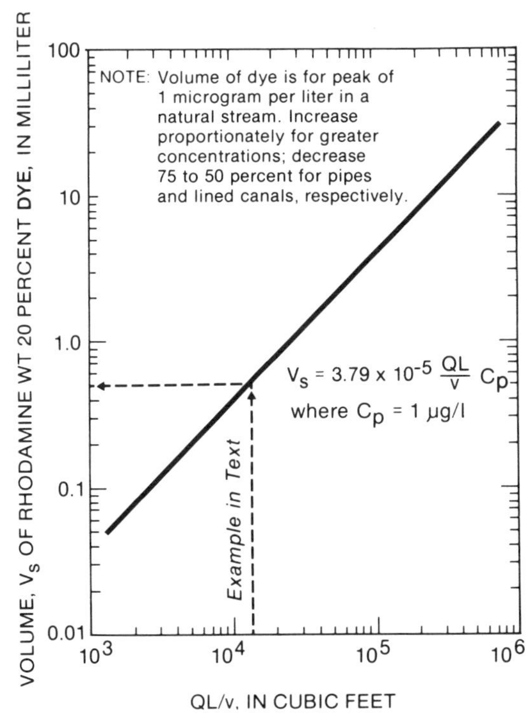

Figure 16-3. Quantity of Rhodamine WT 20 Percent Dye Required for Slug Injection to Produce a Peak Concentration of 1 Microgram per Liter at a Distance Downstream, *L*, at a Mean Velocity, *v*, and with a Dishcarge, *Q*, in the Reach
(Courtesy of U.S. Geological Survey)

The graph or Equation (16-4) is for a peak of 1 μg/l. For most discharge measurements, a peak concentration of 10 to 20 μg/l is recommended; hence, volumes should be increased accordingly. Experience indicates that in lined channels and smooth pipes flowing full, the dosages may be reduced to one half to one fourth, respectively, of the amount needed in a natural stream.

Rhodamine WT 20 percent dye is rather viscous and tends to cling to the sides of graduated cylinders and other laboratory glassware; mixing with a larger quantity of water before injection will help to obtain an accurate injection of the amount measured.

Just before withdrawing the concentrated dye, 10 liters or so of river water should be measured accurately into a container (most buckets will readily contain 10 to 12 liters). The exact amount of dye withdrawn (it will probably be slightly different from the amount computed) should be measured and delivered to the larger container. When using a "To Contain" pipet, the contents should be blown out into the container, care being taken not to allow droplets of dye on the outside of the pipet to spill into the container. The exact volume of rhodamine WT 20 percent dye and water mixture to be injected into the stream must be recorded.

A sample of the mixed tracer solution must be retained in the event standards are to be prepared from it; 50 ml is a desirable sample amount. The removal of the sample from the total amount injected must be noted.

In the case of open channel flow, the dye solution should be emptied instantaneously into the center of the flow. It is advisable to quickly rinse the bucket with clean water and also empty this into the flow. The 30-second delay will be insignificant to the measured response curve(s) downstream.

It is important to remember that the total recovery method can be applied to any type of injection as long as A_c in Equation (16-1) can be measured. Thus, slug injections in pipes or other closed conduits can be accomplished with injection pumps using a sufficiently concentrated dye solution and a short burst of tracer.

Constant-Rate Injection

In most cases the constant-rate injection method of measuring discharge by dilution is preferred over the slug-injection method because of the less stringent sampling requirements.

Equation (16-2) is in a more usable form as

$$Q = 5.89 \times 10^{-7} q \frac{C}{\overline{c}} \qquad (16\text{-}5)$$

where:

Q = the discharge of the stream, in cubic feet per second
q = the rate of injection of the tracer, in milliliters per minute
C = the concentration of the dye solution injected into the stream, in micrograms per liter
$\overline{c}$ = the equilibrium, or plateau concentration, in micrograms per liter, averaged or weighted across the sampling section

This equation not only is used to compute the discharge measurement but also, by solving for q for a given c, may be used to estimate dye quantities for preparing injection solutions.

The curves in Figure 16-4 were computed from Equation (16-5) for a plateau concentration, $\overline{c}$, of 10 μg/l. A dye concentration of 10 μg/l in the stream is well above background levels commonly experienced. In most cases a good dye-dilution discharge measurement can be made with concentrations as low as 2 μg/l. Figure 16-4 is used only to estimate the injection rate needed; the allowable latitude in $\overline{c}$ virtually guarantees a successful measurement, no matter how poor the estimate may be. Furthermore, if 2 μg/l is viewed as a lower limit for $\bar{c}$, discharges five times those in Figure 16-4 may be measured, using the concentrations and injection rates shown.

Table 16-3 provides convenient volumes of water and dye to mix and obtain the concentrations shown in Figure 16-4.

The constant-injection dye dilution method requires apparatus capable of injecting dye at a small, constant rate reliably and accurately. Equipment that can be purchased or fabricated that will produce constant-rate injection includes battery-driven pumps, mariotte vessels, and various chemical-feed devices.

Figure 16-5 shows a small fluid metering pump that operates on a 12-volt DC battery withdrawing a dye solution from a graduated cylinder. It is a valveless, variable, positive-displacement pump that can be set for a rate up to 48 ml per minute. Different models are available in a range of capacities and produce accuracies on the order of one percent. The valveless feature is desirable because it is cleaner and can handle dirt and foreign material in the dye solution. The rate settings have been found to be reproducible within a fair degree of accuracy; nevertheless, the actual injection rate should be independently measured. This is done by measuring and plotting the change in volume in the graduated cylinder with time and calculating the injection rate from the slope of the resulting line. The advantages of this method of measuring the injection rate are that

(1) any change in the injection rate during the test will be revealed by a change in the slope of the line;

(2) the injection is not interrupted during the discharge measurement and may be observed throughout;

(3) sufficient data are obtained to guarantee an accurate measurement of the injection rate; and

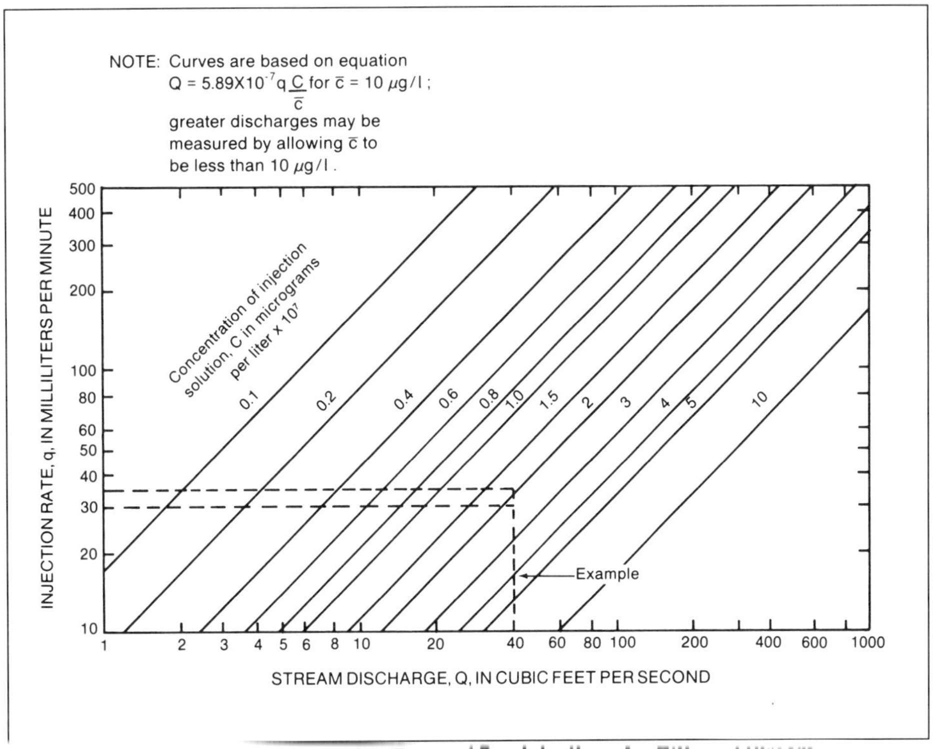

Figure 16-4. Graph Used for Estimating Rates of Dye Injections for Different Stream Discharges and Concentrations of Injection Solutions
(Courtesy of U.S. Geological Survey)

(4) separate sampling of the dye-injection rate is not necessary, thus reducing the chance of contamination; this is particularly important if only one person is performing the discharge measurement.

Regardless of manufacturers' ratings, all pumps and injection devices should be bench tested to check rates and stability of each rate.

Thin-wall, ⅛-inch-diameter brass tubing is used in the cylinder as a withdrawal line to prevent an error in the measured volumes. The pump shown in Figure 16-5 is designed for pumping against a relatively low pressure of 5 psig. For pumping into pipes and under ice, and for multiple-line injections, pumps capable of injecting against greater pressures are available. A similar approach may be taken by weighing the injection container over selected time increments. This approach is shown in Figure16-6, using a laboratory balance to weigh the change in contents of a chemical feed tank injecting into a pipeline.

Sampling

Normally, sampling of the flow containing the diluted dye is performed manually, obtaining grab samples at selected times, depending on the method. Bottles, either glass or plastic of at least 1 oz size, are required. Samples of 100 ml allow repeat fluorometric analysis, and yet are not too large to limit their placement in temperature control baths, etc.

Some fluorometers are designed for flow-through performance, displaying fluorescence on a convenient graph. This allows better control of the grab sampling that should still be performed. Follow-up fluorometric analysis in the laboratory is desirable as *all* samples are best brought to one temperature for accurate analysis. Furthermore, samples will normally be sealed and stored out of sunlight overnight in a temperature bath to permit suspended sediment to settle and air bubbles to escape.

Table 16-3. Convenient Volumes of Water and Dye to Mix for the Preparation of Bulk Solutions of Selected Dye Concentrations
(Courtesy of U.S. Geological Survey)

	Desired concentration, C in µg/l x 10^7	Volumes to mix together		Dilution factor D_C x 10^{-2}	Volume of mixture in gallons
		V_w, water, ml	V_d, 20% rhodamine WT dye, ml		
	(1)	(2)	(3)	(4)	(5)
1	10	10,700	7,750	42.0	4.87
2	10	11,000	8,000	42.1	5.02*
3	5	14,100	3,750	21.0	4.72
4	5	15,000	4,000	21.1	5.02*
5	4	14,850	3,000	16.8	4.72
6	3	15,600	2,250	12.6	4.72
7	2	16,350	1,500	8.40	4.72**
8	1.5	16,350	1,100	6.30	4.61
9	1.5	17,850	1,200	6.30	5.03*
10	1.0	17,100	750	4.20	4.72
11	0.8	17,250	600	3.36	4.72
12	0.6	17,400	450	2.52	4.72
13	0.4	17,550	300	1.68	4.72
14	0.2	17,700	150	0.80	4.72
15	0.1	17,775	75	0.42	4.72

*In many instances, commercial 5-gallon containers are slightly greater in actual volume; the smaller volumes allow for easier mixing.
**Used in example.

Aside from sampling the dye cloud, one or more background samples should be taken upstream of the test reach or prior to the test. These are handled identically to the stream dye samples.

Important, too, is the retention of about 50 ml of the dye mixture *actually injected*. It may be used to calibrate the fluorometer, as will be explained later. A dilution flow measurement that uses a fluorometric calibration based on this sample is more accurate than one based on the original 20 percent rhodamine dye supplied by the manufacturer. Every effort should be taken to avoid contaminating the other samples and equipment with this concentrated dye sample; collection by a person not involved in the rest of the measurement is advisable.

The passage of the entire tracer cloud must be monitored by the slug-injection method of discharge measurement. Sampling should begin before the tracer arrives at the section and continue until it has entirely passed. Large numbers of samples are normally required to make certain the entire cloud is measured. For a stream or canal, samples should be taken from at least three points in the section in what is estimated to be equal increments of flow. Samples may be given a preliminary fluorometric analysis in the field to ascertain if the dye has arrived and (or) left and, hence, dictate frequency and duration of sampling.

If a fluorometer is not used to concurrently analyze samples, the flow must be oversampled to make certain the response curve(s) is adequately defined. As a

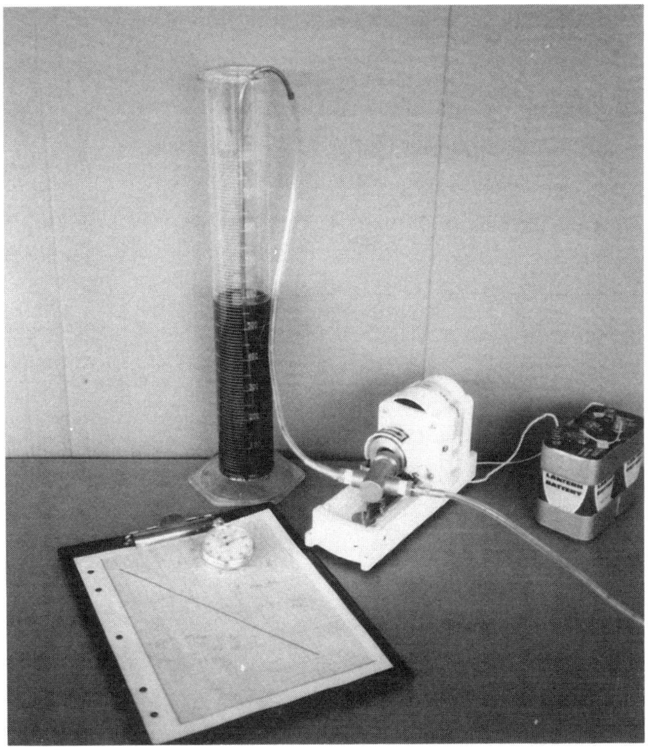

Figure 16-5. Pump and Graduated Cylinder Used for Constant-Rate Dye Injection
(Courtesy of U.S. Geological Survey)

rough approximation, sampling past the peak should continue for two to four times as long as it takes for the peak to arrive. In some instances, the arrival of the peak can be ascertained visually. On small streams, samples must be taken at very short time intervals; therefore, two people are needed, one sampling and one noting the exact time and recording the data. Sampling may be less frequent as the dye cloud recedes.

In contrast to the slug injection method, when using the constant-rate injection method, immediate and complete sampling of the dye downstream is *neither necessary nor desirable*. The injection must be maintained at a constant rate long enough for plateau concentrations to be *fully* established laterally across the channel and with time.

The best practice is to inject longer and sample later than is likely needed.

Sampling should, of course, be done toward the end of the injection but before its termination. Regardless of method, all samples are sealed and stored out of direct sunlight and forwarded to the laboratory for more precise fluorometric analysis.

Flow in pipes or other closed conduits is typically at a much higher velocity than in a stream. Furthermore, utilization of pumps, bends, orifices, etc., to improve mixing may result in the use of short pipe lengths to perform the dilution flow measurements. In such instances the high level of turbulence may cause pulsating dye concentrations, which should not be sampled instantaneously. For example, when using the constant-rate method in a pipe, the plateau is more accurately measured if oversize samples are collected over 15- to 30-second periods and a 100-ml mixed sample of each is retained for analysis.

Figure 16-6. Pressurized Chemical-Feed Tank Being Used to Inject Dye into a Pipeline—A Laboratory Balance Measures the Injection Rate
(Courtesy of U.S. Geological Survey)

Fluorometric Analysis

A fluorometric analysis must be performed to measure the magnitude of the dilution of the dye in the flow.

The two fundamental types of fluorometers are: (1) fluorescence spectrometers or spectrofluorometers that are used for spectral analyses of fluorescent substances and (2) filter fluorometers that are more readily used for selective analysis of large numbers of water samples.

A filter fluorometer, or fluorometer, is an instrument that gives a relative measure of the intensity of light emitted by a sample containing a fluorescent substance; the intensity of fluorescent light is proportional to the amount of fluorescent substance present. However, a fluorometer reading by itself is a number having little meaning until it is compared with readings for samples of known concentrations (standards) on the same fluorometer under the same instrument and temperature conditions. Generally, a reading for a given sample on one fluorometer cannot be compared directly with a reading for the same sample on a different fluorometer. Every fluorometer is different and must be individually calibrated.

A filter fluorometer consists of six basic components, shown in Figure 16-7.

A number of companies market fluorometers that can be used; reference to a specific fluorometer is for the purpose of illustration and should not be regarded as an endorsement of a particular brand of equipment.

One of the most commonly used fluorometers is the Turner Designs® model 10 (shown in Figure 16-8), which, because it is solid state, can be readily used in the field as well as for laboratory analysis.

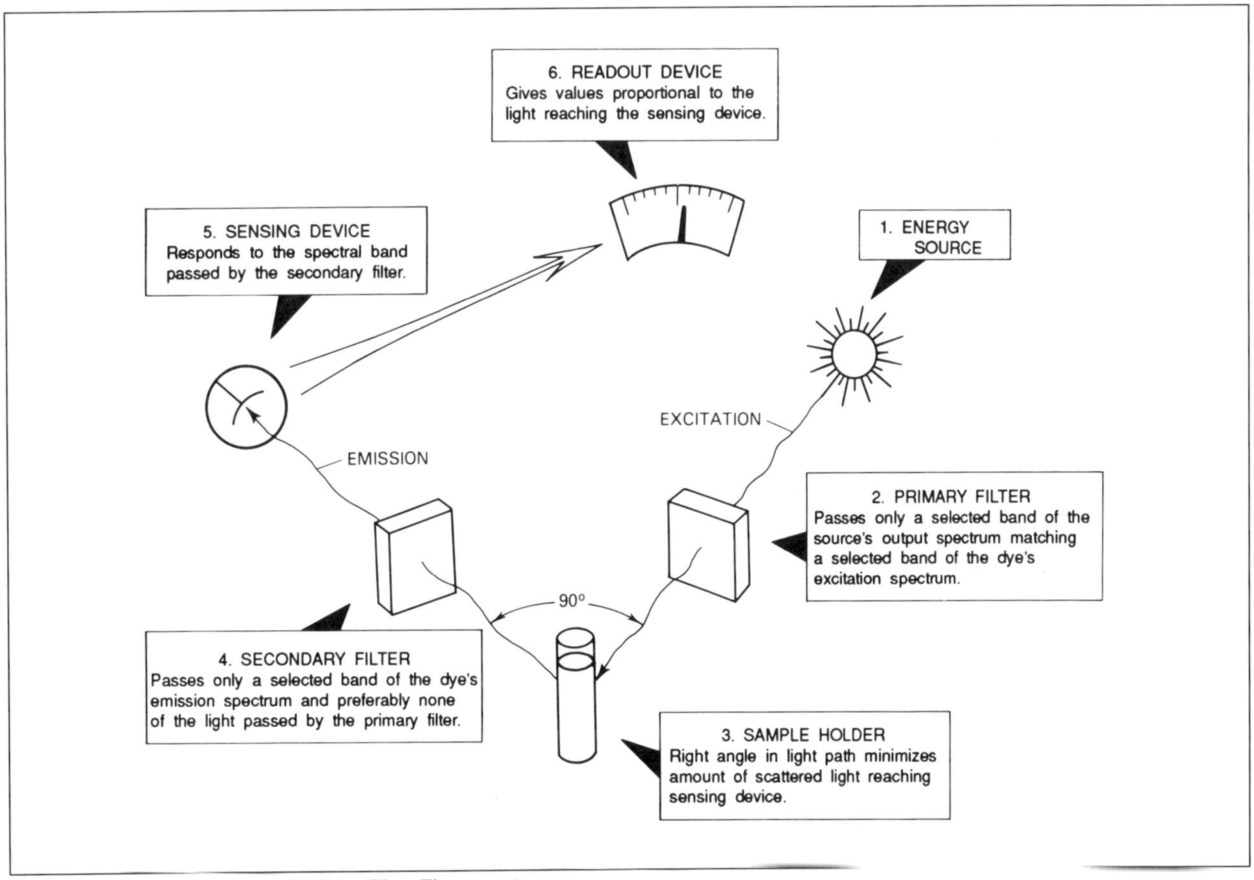

Figure 16-7. Basic Structure of Most Filter Fluorometers
(Courtesy of U.S. Geological Survey)

All filter fluorometers require calibration using standards prepared from the dye used. These standards (samples) are prepared using a serial dilution process by which the concentrated dye is *accurately* reduced to a series of samples, which can be analyzed on the fluorometer for the range in concentrations expected of the test samples.

Rhodamine WT, as supplied by the manufacturer, has a concentration of 20 percent by weight. At least a four-step serial dilution is required to obtain standard concentrations in the range needed. Thus, the concentrations of final standards obtained by a four-step serial dilution may be computed by:

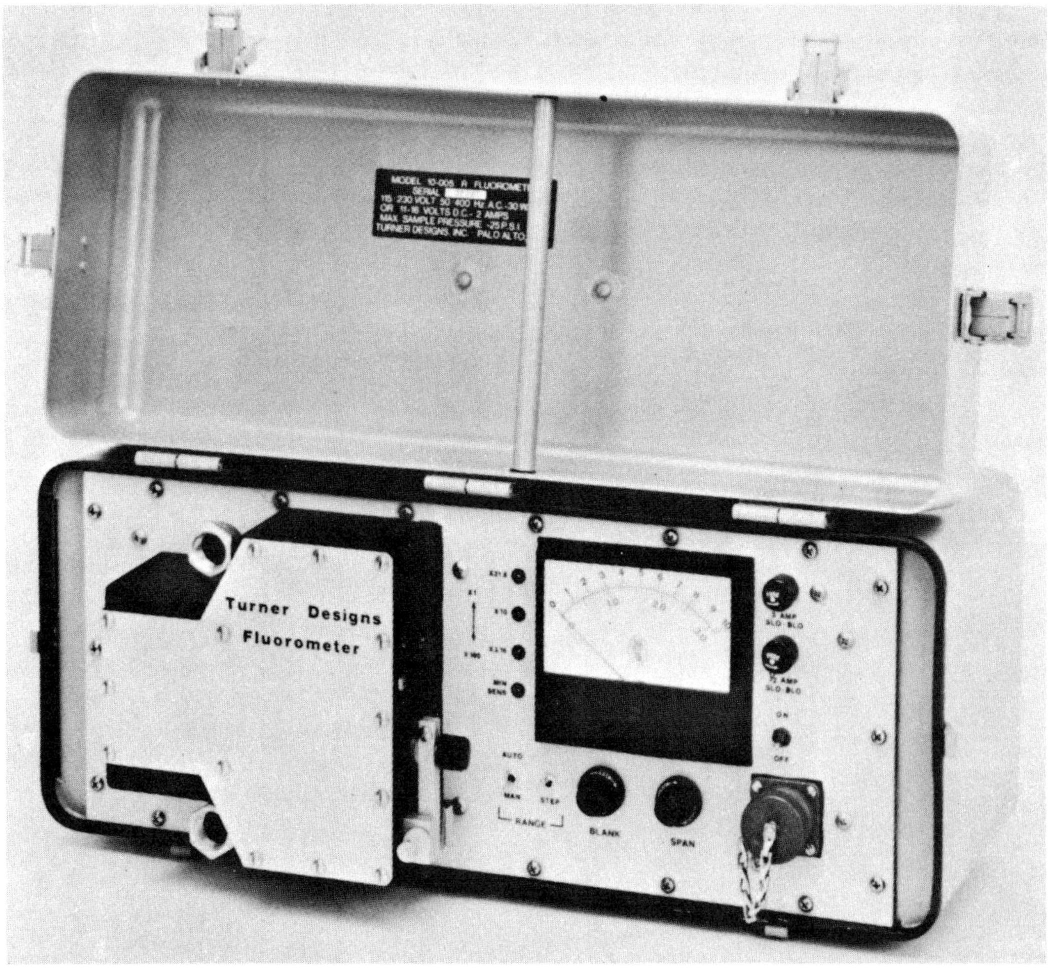

Figure 16-8. A Fluorometer Suitable for Field Use, Having Both Flow-Through and Individual-Sample Analysis Capability
(Courtesy of U.S. Geological Survey)

$$C_f = C_s\mathrm{SG}\left[\frac{V_d}{V_w + V_d}\right]_1 \times \left[\frac{V_d}{V_w + V_d}\right]_2 \times \left[\frac{V_d}{V_w + V_d}\right]_3 \times \left[\frac{V_d}{V_w + V_d}\right]_4 \quad (16\text{-}6)$$

or

$$C_f = C_s\mathrm{SG}\, D_1 \times D_2 \times D_3 \times D_4 \quad (16\text{-}7)$$

where the terms are as previously defined except that:

C_f = final concentration of the standard obtained after the fourth step;
C_s = concentration of the dye solution, usually as obtained from the manufacturer
$D_{1,2,3,4}$ = the dilution factor for each step
SG = specific gravity of the initial dye solution
V_d = pipet volume of the dye solution taken from the previous step dilution
V_w = volume of the added diluent, usually distilled water

For rhodamine WT, Table 16-4 provides a range of convenient pipet and dilutent volumes to obtain a "working solution," which is 100 µg/l at the end of three dilutions. It will be noted that volumes, V_w, for the first dilution are uneven, in order to compensate for the specific gravity of 20 percent rhodamine WT dye, which is 1.19. Such a procedure simplifies subsequent dilution computations and lessens chances of errors. This "working solution" is used for all fourth or final standards. It may be retained for future use where the same dye lot is to be used. Thus, the first three serial dilutions need not be repeated every time a dye test is performed from the same dye lot. This "working solution" should be sealed and stored out of direct light.

Table 16-4. Convenient Three-Step Serial Dilutions for Preparation of Working Solution
(Courtesy of U.S. Geological Survey)

Dye used in test		First		Second		Third		Working solution in µg/l
		V_d (ml)	V_w (ml)	V_d (ml)	V_w (ml)	V_d (ml)	V_w (ml)	
Rhodamine WT (20 percent; SG 1.19)	(a)	50	3,792	20	3,500	20	3,500	100
	(b)	25	2,585	20	3,000	20	3,000	100
	(c)	20	2,068	20	3,000	20	3,000	100
	(d)	20	1,158	10	2,000	10	2,000	100

Because rhodamine WT, as supplied by the manufacturer, is quite viscous, small-volume measurements are apt to be in error. For this reason, the smallest pipet volume suggested for the first dilution is 20 ml; use a "to contain" type of pipet if available.

It is customary practice to vary the pipet volume, V_d, and diluent volumes, V_w, in step 4 sufficiently to obtain a range in final standard concentrations. Table 16-5 provides a range of convenient pipet and diluent volumes for the fourth dilution step to provide a complete range of final standard concentrations. Even volumes were employed to the extent possible both for convenience and to lessen potential measurement errors. Normally, all concentrations provided would not be needed; those desired would merely be chosen from Table 16-5. Judicious use of Table 16-5 will often permit the use of only one or two pipets for the fourth-step dilutions, lessening cleaning and handling problems as well as potential errors. For convenience, Table 16-5 also provides for recording fluorometer readings during the calibration process.

The application of Tables 16-4 and 16-5 in preparing dye standards is illustrated by the example in Figure 16-9. Figure 16-10 shows the four calibration curves obtained for standards analyzed at 70°; a laboratory calibration such as this one for each fluorometer is helpful in showing its sensitivity and later in selecting a scale for analyzing a given set of river samples.

Space precludes an elaborate explanation of fluorometric analysis procedures; the report by Wilson and others [Ref. 12] provides detailed instructions on the use of fluorometers and analysis procedures.

Table 16-5. Convenient Fourth-Step Dilutions for Preparing Dye Standards Using a 100-micrograms-per-liter Working Solution
(Courtesy of U.S. Geological Survey)

Final standard number	V_d (ml)	V_w (ml)	Final standard concentration (μg/l)	Fluorometer scale			
Background	—	—	—				
1	300	100	75				
2a	200	200	50				
b	250	250					
3a	100	150	40				
b	200	300					
4	100	233	30				
5a	50	150	25				
b	100	300					
6a	50	200	20				
b	100	400					
c	125	500					
7a	50	283	15				
b	100	566					
8a	20	180	10				
b	25	225					
c	50	450					
d	100	900					
9a	20	230	8				
b	25	288					
c	50	575					
10a	20	313	6				
b	25	392					
c	50	784					
11a	20	380	5				
b	25	475					
c	50	950					
12a	20	480	4				
b	25	600					
c	50	1,200					
13a	20	647	3				
b	25	808					
c	50	1,617					
14a	10	490	2				
b	20	980					
c	25	1,225					
d	50	2,450					
15a	10	990	1.0				
b	20	1,980					
c	25	2,475					

In brief, fluorometric analysis consists of the following:

(1) Bring *all* samples, background, dye water, and standards to the same temperature.

(2) Allow the fluorometer to warm up adequately.

(3) Analyze all standards in the fluorometer in an identical manner: same cuvette and same rinsing procedures.

(4) Plot the fluorometer calibration on rectilinear graph paper, and reconcile any data points that do not conform to the curve by reanalysis of fresh samples on the fluorometer and, if necessary, by preparation and analysis of new standards.

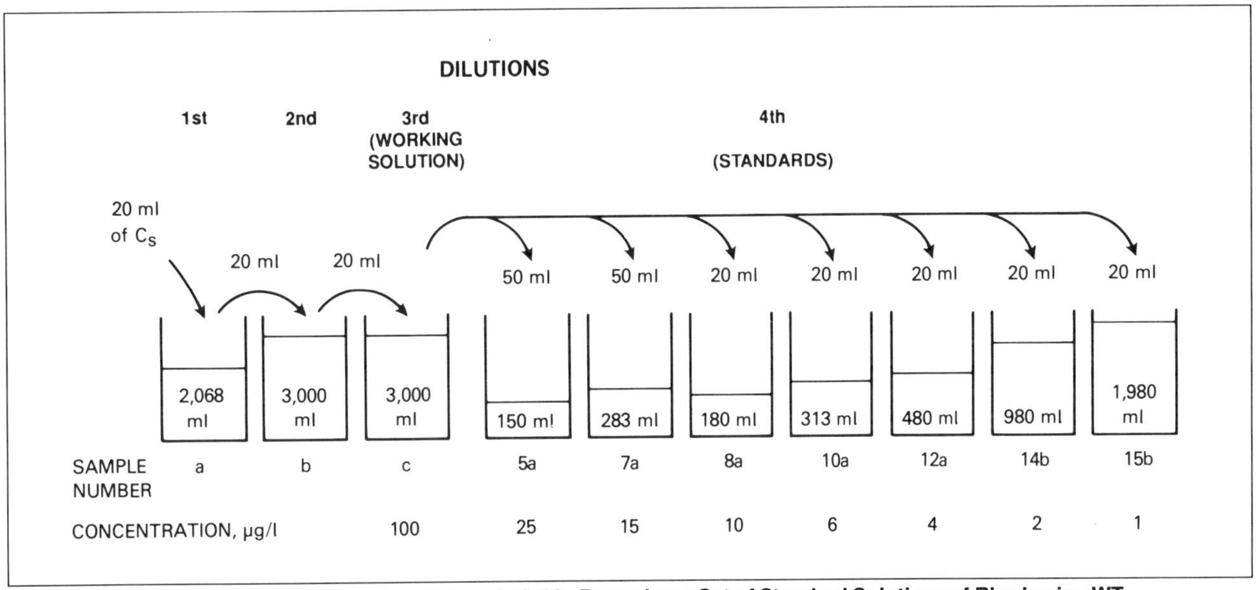

Figure 16-9. Example of the Use of Tables 16-4 and 16-5 in Preparing a Set of Standard Solutions of Rhodamine WT
(Courtesy of U.S. Geological Survey)

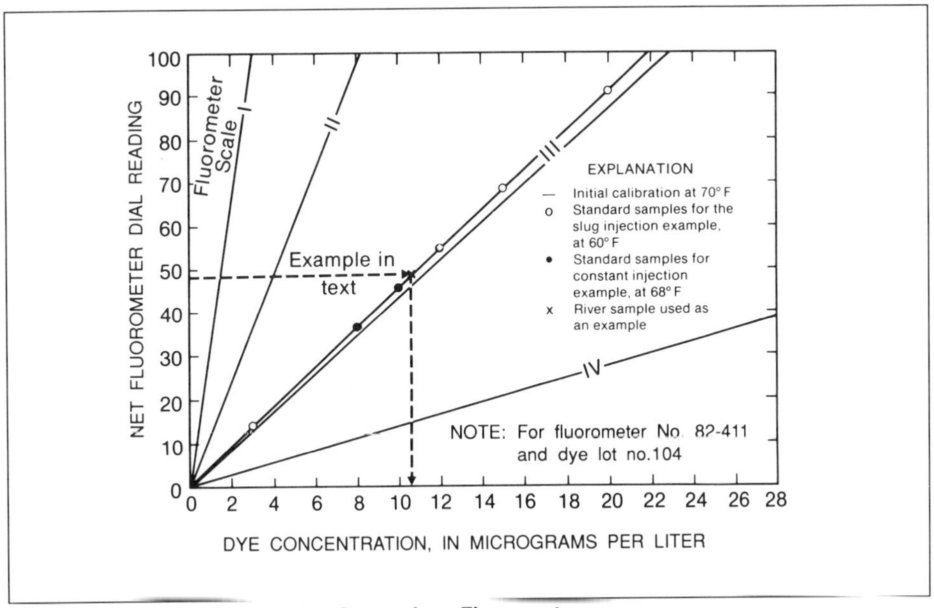

Figure 16-10. Typical Calibration Curves for a Fluorometer
(Courtesy of U.S. Geological Survey)

Performance of Slug-Injection Type of Flow Measurement

The slug-injection method is sometimes preferred because of the simplicity of injection and because less tracer is needed. Equation (16-1) may be written as:

$$Q = 5.89 \times 10^{-7} \frac{SGV_I C}{A_c} \tag{16-8}$$

where:

Q = the discharge of the stream, in cubic feet per second
V_I = the volume of concentrated dye solution injected into the stream, in milliliters
C = the concentration of the dye solution injected into the stream, in micrograms per liter
A_c = the area under the time–concentration curve, in units of minutes time micrograms per liter

Successful application of Equation (16-8) requires that the mass of tracer injected be fully recovered at the sampling point.

A schematic diagram of the step-by-step performance of a slug-injection type of dye-dilution measurement on a small stream is shown in Figure 16-11. The steps in performing such a test use the example shown.

(1) Selection of measurement reach. The stream discharge and the other stream characteristics are estimated; using Equation (16-3), compute an optimum stream reach length of about 700 feet.

(2) Background samples. Samples of the stream water are needed before dye injection in order to obtain a reading of the background (see sample Number 202 on the note form in Figure 16-11).

(3) Dye injection. For the channel and flow conditions shown, it was decided to try for a peak concentration of about 20 µg/l, using rhodamine WT 20 percent. From Figure 16-3, a volume of 0.5 ml is obtained for a peak of 1 µg/l; thus, for a peak of 20 µg/l, 10 ml of rhodamine WT dye is required.

Next, 10 liters of river water is carefully measured into a clean bucket to which the 10 ml of dye is added and mixed. A 50-ml sample of this mixture is retained for future use. The entire contents (9,960 ml) are dumped as a slug into what is judged to be the centroid of the flow. The time is noted, and the person at the sampling site is notified. A good practice is to start a stopwatch at the instant of injection so that all samples are recorded with respect to lapsed time from injection (see the notes and data plots of Figure 16-11).

(4) Sampling the dye cloud. The passage of the entire tracer cloud must be monitored. Sampling should begin before the tracer arrives at the section and continue until it has entirely passed. Large numbers of samples are normally required to make certain the entire cloud is measured. Samples should be taken from at least three points in the section in what is estimated to be equal increments of flow. Samples 203 through 241 are those taken to define the dye cloud at points a, b, and c across the channel. All samples are sealed and stored out of direct sunlight. Notes describing the measurement accompany the samples to the laboratory.

(5) Fluorometric analysis. The quantity of dye injected in step 3 was predicated on obtaining a peak of about 20 µg/l. Examining Table 16-5, four standards—20, 15, 12, and 3 µg/l—are prepared from the 100-µg/l work-

ing solution retained for this purpose. Those standards and all river and background samples are brought to a single temperature of 68°F. The calibration for scale III is plotted as shown in Figure 16-10. Note that net readings are used, the background for distilled water being subtracted first.

All river samples including the background sample, 202, are analyzed on the fluorometer, and, using the above calibration, absolute concentrations are determined for each; the stream background reading is subtracted from each stream sample before entering the calibration curve.

(6) Data analysis. The time-concentration curves for the three sample points, a, b, and c, are plotted and their areas determined. Comparison of the individual curve areas indicates that adequate mixing has been achieved. An average curve area, A_c, of 34.2 µg/l times minutes is obtained.

(7) Computation of discharge. Using Equation (16-8) with V_s substituted for V_I and C_s substituted for C, the discharge may be computed as:

$$Q = 5.89 \times 10^{-7}\left[\frac{1.19 \times 10.0 \times 20 \times 10^{7}}{34.2}\right] = 41.0 \text{ ft}^3\text{/s}$$

During the preparation of the solution for injection, 10 ml of dye was mixed with 10 liters of river water, and a 50-ml sample of the mixture was retained (step 3, Figure 16-11). The concentration used in the earlier computation was for the manufacturer's stock solutions: 20×10^{-7} µg/l for rhodamine WT. Using Equation (16-6) for only the first dilution, it can be seen that the 10 liters had a concentration of:

$$C = 1.19 \times 20 \times 10^{7}\left[\frac{10.0}{10{,}000 + 10.0}\right] = 23.8 \times 10^{4} \text{ µg/l}$$

Hence, a major dilution was made at this point. Thus, the 50-ml sample may be used in preparing standards for this particular measurement instead of starting with the stock solution.

In preparing standards from C (above), note that, according to Equation (16-6), 20 ml of dye solution added to 1,980 ml of water is a dilution of 10^{-2}. If repeated, a total dilution of 10^{-4} is obtained, because the final dilution in any serial dilution is the product of the individual dilutions. For any dilution after the first, the specific gravity term may be ignored. Thus, if C (above) was diluted twice, as described, for a total dilution of 10^{-4}, the resulting concentration would be $23.8 \times 10^4 \times 10^{-4} = 23.8$ µg/l. While this would be off scale III in Figure 16-10, other standards may be produced by judicious selection of dye and water volumes. The analysis in Figure 16-10 would then be based on the fluorometer calibration thus obtained, and computations would be based on the new volume and concentration injected such that the discharge (using Equation (16-8)) would be:

$$Q = 5.89 \times 10^{-7}\left[\frac{(10{,}000 + 10 - 50) \times 23.8 \times 10^{4}}{34.2}\right] = 40.8 \text{ ft}^3\text{/s}$$

This alternative method of analysis is more accurate than the first, because any errors that occur when measuring the 10 ml of dye for use in the solution to be injected are canceled. Similarly, in preparing standards, only two dilutions are necessary, instead of four as in the previous example. Furthermore, if the exact concentration of the solution injected is unknown, a measurement is still possible if a sample of the solution is retained and used in the laboratory to develop a calibration in the manner previously described. The importance of always retaining a sample of the actual dye solution that was injected becomes apparent.

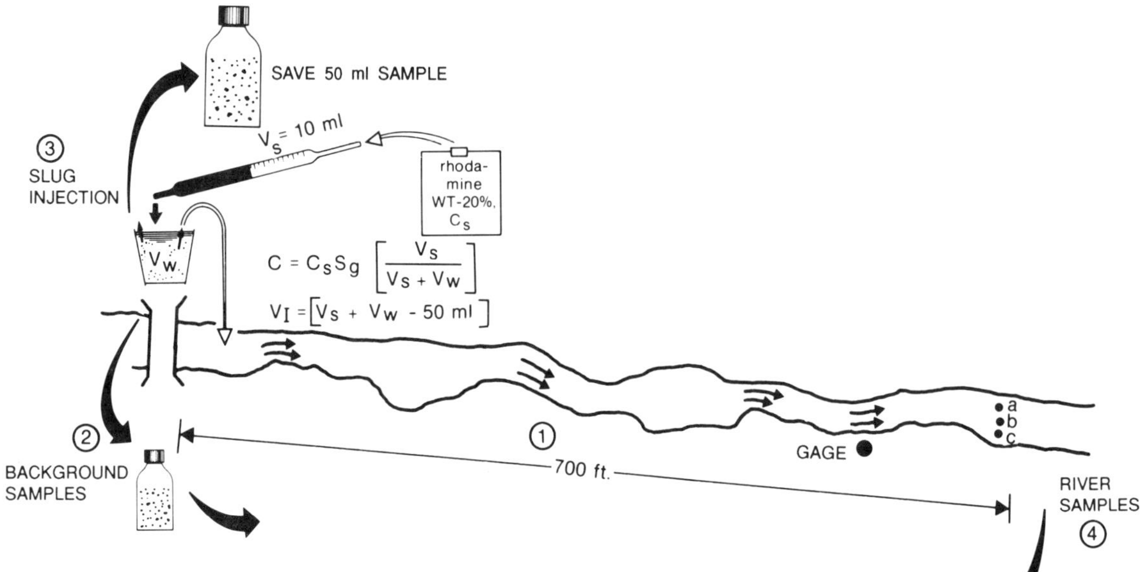

⑤ FLUOROMETRIC ANALYSIS

Sample number	Location in section	Lapsed time (min-s)	Fluorometer reading Scale I	II	III	IV	Net reading	Concentration in (µg/l)
250	Distilled water			0.1	0.1			0
251	Standard				91.1		91.0	20
252	Standard				68.4		68.3	15
253	Standard				54.7		54.6	12
254	Standard				13.9		13.8	3
202	Background	0			02			
203	a	4-40			0.2		0	0
204	b	4-46			0.2		0	0
205	c	4-52			0.2		0	0
206	a	5-0			2.4		2.2	0.4
207	b	5-08			0.9		0.7	0.1
208	c	5-13			0.2		0	0
209	a	5-20			31.5		31.3	6.8
210	b	5-26			11.0		11.8	3.0
211	c	5-32			15.2		15.0	3.2
212	a	5-40			69.2		69.0	15.2
213	b	5-46			43.5		43.3	9.4
214	c	5-53			42.2		42.0	9.2
215	a	6-0			86.6		86.4	19.0
216	b	6-07			80.7		80.5	17.7
217	c	6-12			70-2		70.0	15.4
218	a	6-20			79.5		79.3	17.4
219	b	6-26			81.4		81.2	17.8
220	c	6-31			76.5		76.3	16.7
230	a	8-0			6.5		6.3	1.3
231	b	8-10			13.7		13.5	2.9
232	c	8-20			16.7		16.5	3.6
233	a	8-30			0		0	0
234	b	8-40			6.9		6.7	1.4
235	c	8-50			9.5		9.3	2.0
236	a	9-0			0.2		0	0
237	b	9-10			1.7		1.5	0.3
238	c	9-20			6.2		6.0	1.2
239	a	10-0			0.2		0	0
240	b	10-10			0.2		0	0
241	c	10-20			0.4		2.2	0.4

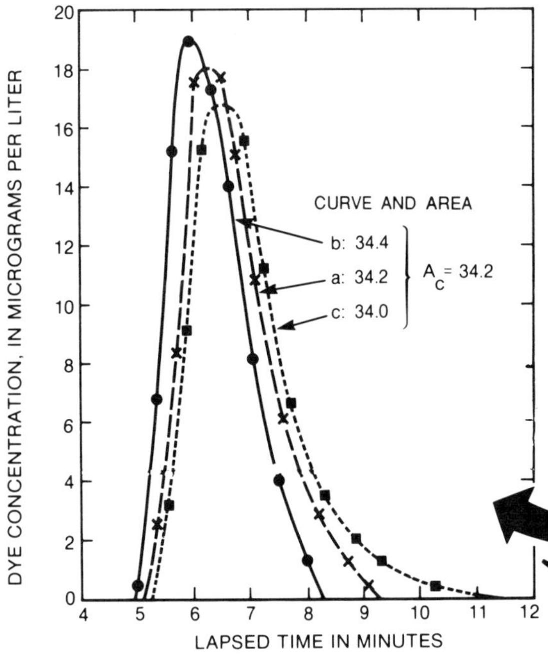

⑥ OBSERVED TIME-CONCENTRATION CURVES

⑦ COMPUTATION OF DISCHARGE

$$Q = 5.89\times10^{-7}\ \frac{S_G V_s C_s}{A_c}$$

$$Q = 5.89\times10^{-7}\left[\frac{1.19\times10\times20\times10^{7}}{34.2}\right] = 41.0\ \text{ft}^3/\text{s}$$

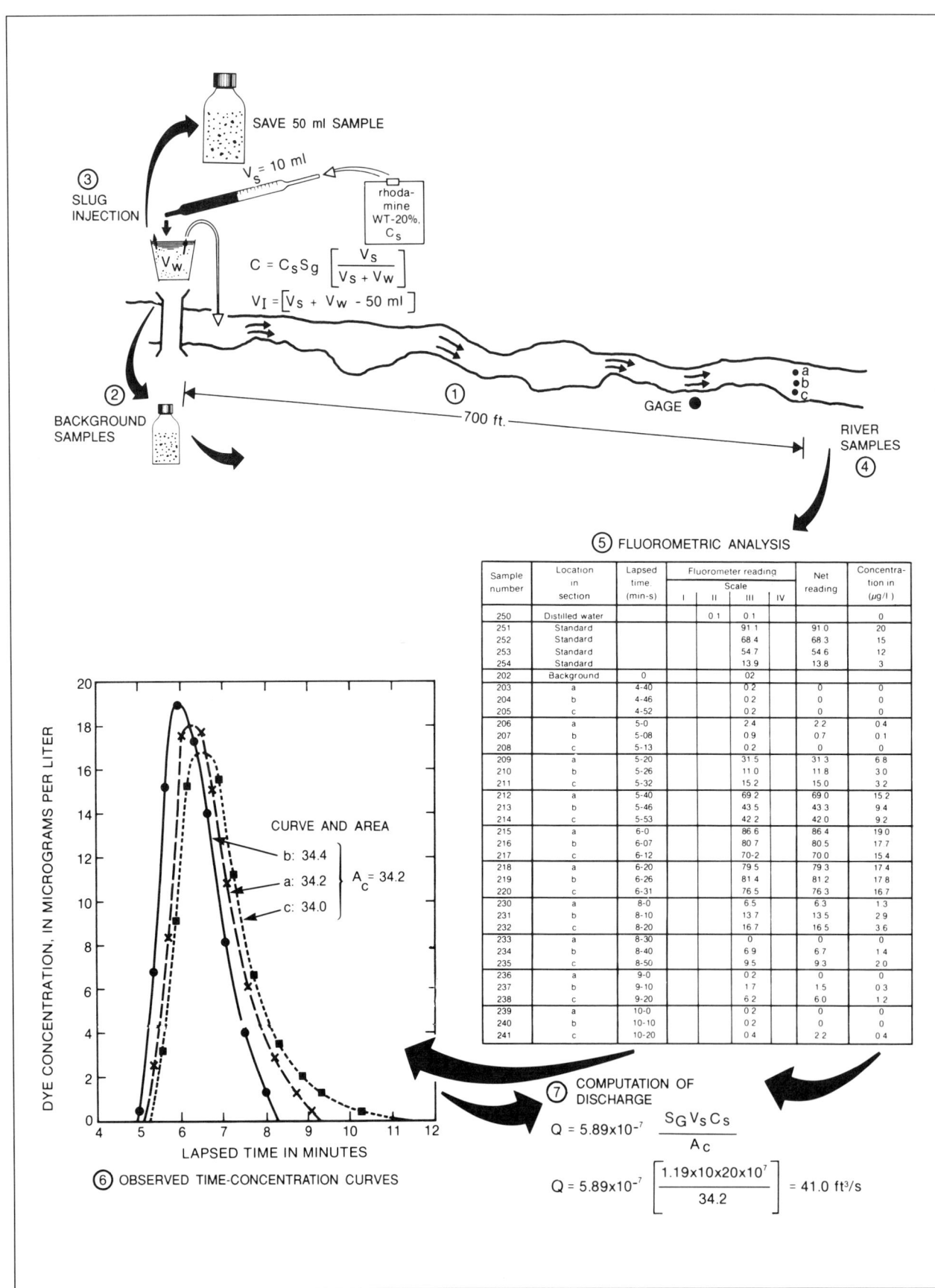

Figure 16-11. Schematic Diagram Depicting the Performance of Slug-Injection Type Dye-Dilution Discharge Measurement
(Courtesy of U.S. Geological Survey)

Performance of Constant-Rate Type of Flow Measurement

A schematic diagram of the performance of a constant-rate dye-dilution measurement is shown in Figure 16-12. The steps in performing such a measurement use the example shown.

(1) Selection of measurement reach. The stream discharge, mean velocity, and geometry are estimated; using Equation (16-3), an optimum reach length of 700 feet is computed.

(2) Background samples. Stream background samples are taken before handling dyes and injection equipment.

(3) Dye preparation, injection, and sampling. For an estimated streamflow of 40 ft^3/s, Figure 16-4 shows that an injection rate of about 35 ml per minute of a 2.0×10^7 µg/l (2 percent) solution should yield a plateau concentration of about 10 µg/l. Referring to Table 16-3, a 2 percent bulk solution could be prepared by mixing those quantities of water and dye shown in columns 2 and 3, line 7. The 2 percent dye mixture is placed in a 1,000 ml graduated cylinder.

The intake line to the pump is secured inside the graduated cylinder to reach nearly to the bottom. The plastic discharge line leading from the pump should be positioned and secured where the injection will enter what is judged to be the centroid of flow. The pump is turned on and set to a rate of about 35 ml/min. At the end of the injection—after steps 4 and 5—a sample of injected solution should be bottled, labeled, and stored separately from the stream samples.

(4) Measurement of injection rate. As soon as the injection rate has stabilized, the change in the volume of dye solution level in the cylinder should be timed with a stopwatch. The watch is not stopped, but the time and volume are observed simultaneously several times before and after stream samples are collected downstream. These volumes are plotted versus time and the rate determined from the slope of the line as 31.2 ml/min.

(5) Sampling the plateau. In contrast to the slug-injection method, immediate and complete sampling of the entire dye response downstream is not necessary. The injection must be maintained at a constant rate long enough for plateau concentrations to be fully established laterally across the channel and with time. The best practice is to inject longer and sample later than is likely needed. Sampling should, of course, be done toward the end of the injection but before its termination.

Stream samples containing the diluted dye should be taken at a minimum of three points across the flow at the sampling section (labeled a, b, and c in Figure 16-12).

(6) Fluorometric analysis. Assuming that c values at sampling points a, b, and c will be on the order of 10 µg/l, standards in that concentration range are required. A working solution of 100 µg/l has been prepared from the dye lot used and is stored in the laboratory. Thus, according to lines 9c and 8c of Table 16-5, standards of 8 and 10 µg/l are prepared by mixing 50 ml of 100-µg/l solution in each of two containers containing 575 ml and 450 ml of distilled water, respectively. The 8- and 10-µg/l standards and a background sample of the distilled water used in their preparation are placed in sample bottles similar to those used for collecting field samples. All stream samples, stream background, standards, and distilled water background

samples are brought to a common temperature—68°F. All samples are analyzed on one fluorometer scale.

The net dial readings for the 8- and 10-μg/l standards are plotted on the existing fluorometer calibration after subtracting the dial readings for the distilled water sample. If the readings do not plot exactly on the established calibration curve, a new curve is drawn through the two points for that scale parallel to the original calibration (see Figure 16-10). The net dial readings for the stream samples are averaged, and, using the new calibration curve, the net concentrations of the stream samples for each set are determined, as shown in column 9 of Figure 16-12. In this case, an average plateau concentration, $\bar{c}$, of 10.62 and 10.63 μg/l was obtained for the sets taken at 20 and 25 minutes.

(7) Computation of discharge. Based on the data presented and using Equation (16-5), the discharge may be computed as:

$$Q = 5.89 \times 10^{-7}\,(31.2)\left[\frac{2.00 \times 10^{7}}{10.63}\right] = 34.6 \text{ ft}^3/\text{s}$$

An alternate method of analysis and computation is more accurate than that described above and involves using relative concentrations. This method has the advantage of avoiding errors that might occur if the manufacturer's solution was not mixed each time dye was extracted or if an error was made in measuring the quantities of dye and water in preparing the injection solution. Thus, the stream discharge could be determined if the absolute value of C was unknown but a sample of C was retained from each test. This is good practice regardless of the method of analysis. The fact that $C/\bar{c}$ in Equation (16-5) is a ratio allows the comparison and measurement of $\bar{c}$ relative to C using dial readings and the total dilution factor from the laboratory serial dilution. By this method, *a three-step serial dilution of C must be done for each test* where a different injection solution is used. This emphasizes the advantages of preparing a bulk quantity of dye solution of a single concentration, such as one of those in Table 16-3, for a series of discharge measurements. By starting with the injected solution, C, only a three-step serial dilution need be prepared to obtain a standard about equal to that expected in the stream for $\bar{c}$. Table 16-6 provides convenient combinations of dye and water to yield a range of standards from about 7 to 13 μg/l (column 8), which should be the range obtained in the field samples. Column 9 in Table 16-6 provides the total dilution factor, D_T, for whatever concentration is used. Based on Equation (16-6), it can be seen that the standard, C_3, is equal to $SGCD_T$. Thus, $C = C_3/SGD_T$ and the discharge equation, substituting net dial readings for concentrations, is:

$$Q = 5.89 \times 10^{-7}\, q\left[\frac{R}{\bar{r}} \times \frac{1}{SGD_T}\right] \qquad (16\text{-}9)$$

As an example, the same test conditions as those previously used and presented in Figure 16-12 are assumed, except that now a three-step serial dilution of C must be performed. Referring to Table 16-6, a convenient serial dilution combination for the case where $C = 2.00 \times 10^{7}$ μg/l is found on line 23c (for $C_3 = 10.60$), since the injection rate for the estimated stream discharge was chosen to produce a plateau concentration of about 10 μg/l. The standard, C_3, is analyzed on the fluorometer on scale III at the same time the field samples are analyzed. This sample gives a dial reading of 49.6 for a net R of 49.4 (the stream background was 0.2). Thus, for the stream samples taken at 25 minutes after the start of injection and having an average dial reading, $\bar{r}$, of 48.6, the discharge may be computed as:

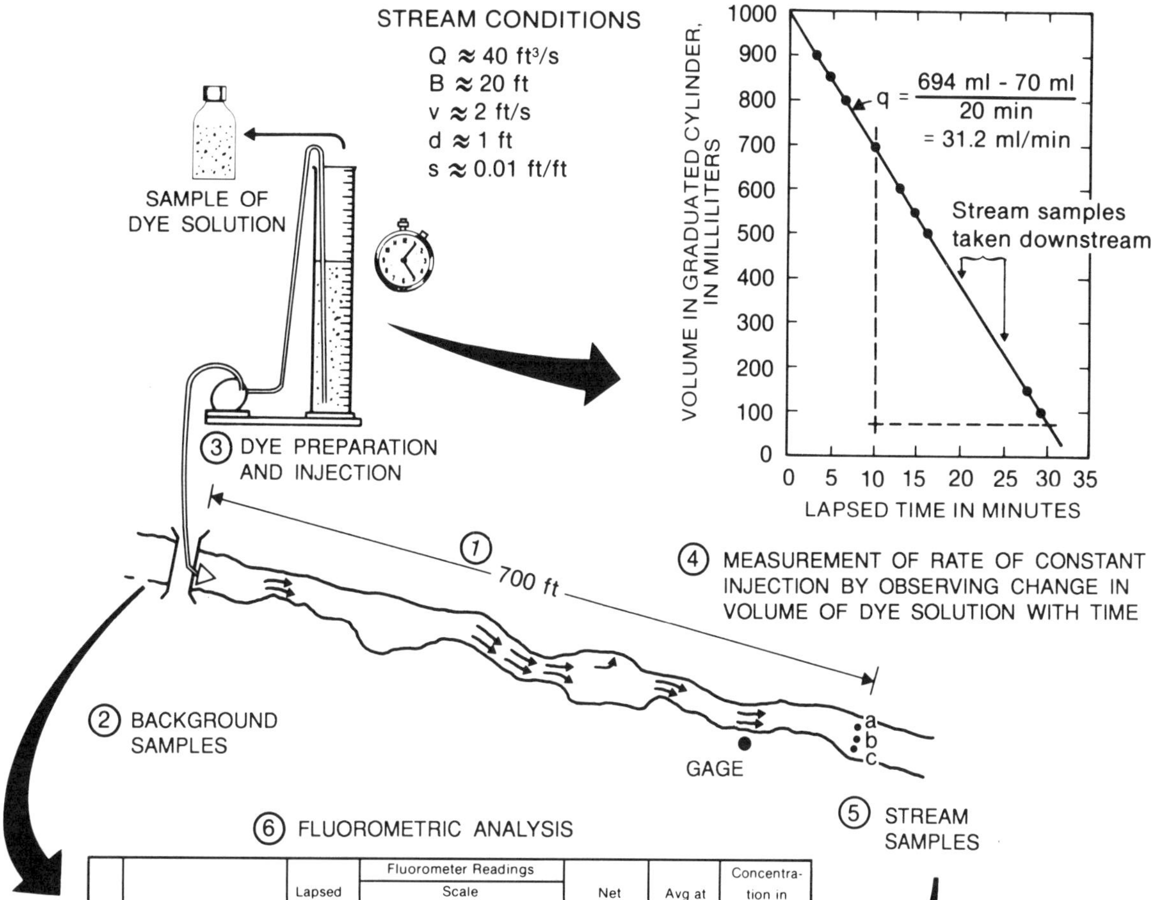

(6) FLUOROMETRIC ANALYSIS

	Samples, type, and location (1)	Lapsed time (min.) (2)	Fluorometer Readings Scale I (3)	II (4)	III (5)	IV (6)	Net readings r (7)	Avg at section $\bar{r}$ (8)	Concentration in stream, $\bar{c}$ (μg/l) (9)
1.	Stream Background	0			0.2				
2.	Stream Background	0			0.2				
3.	Pt a	20			48.7		48.5		
4.	Pt b	20			49.0		48.8	48.5	10.62
5.	Pt c	20			48.5		48.3		
6.	Pt a	25			48.6		48.4		
7.	Pt b	25			49.1		48.9	48.6	10.63
8.	Pt c	25			48.6		48.4		
9.	Standard 8 μg/l	Lab			36.6		36.5		
10.	Standard 10 μg/l	Lab			45.6		45.5		
11.	Distilled Water	Lab			0.1				

(7) COMPUTATION OF DISCHARGE

$$0 = 5.89 \times 10^{-7}\, q\, \frac{C}{\bar{c}}$$

At 20 minutes : $Q = 5.89 \times 10^{-7} (31.2) \left[\frac{2.00 \times 10^{7}}{10.62}\right] = 34.7\ \text{ft}^3/\text{s}$

At 25 minutes : $Q = 5.89 \times 10^{-7} (31.2) \left[\frac{2.00 \times 10^{7}}{10.63}\right] = 34.6\ \text{ft}^3/\text{s}$

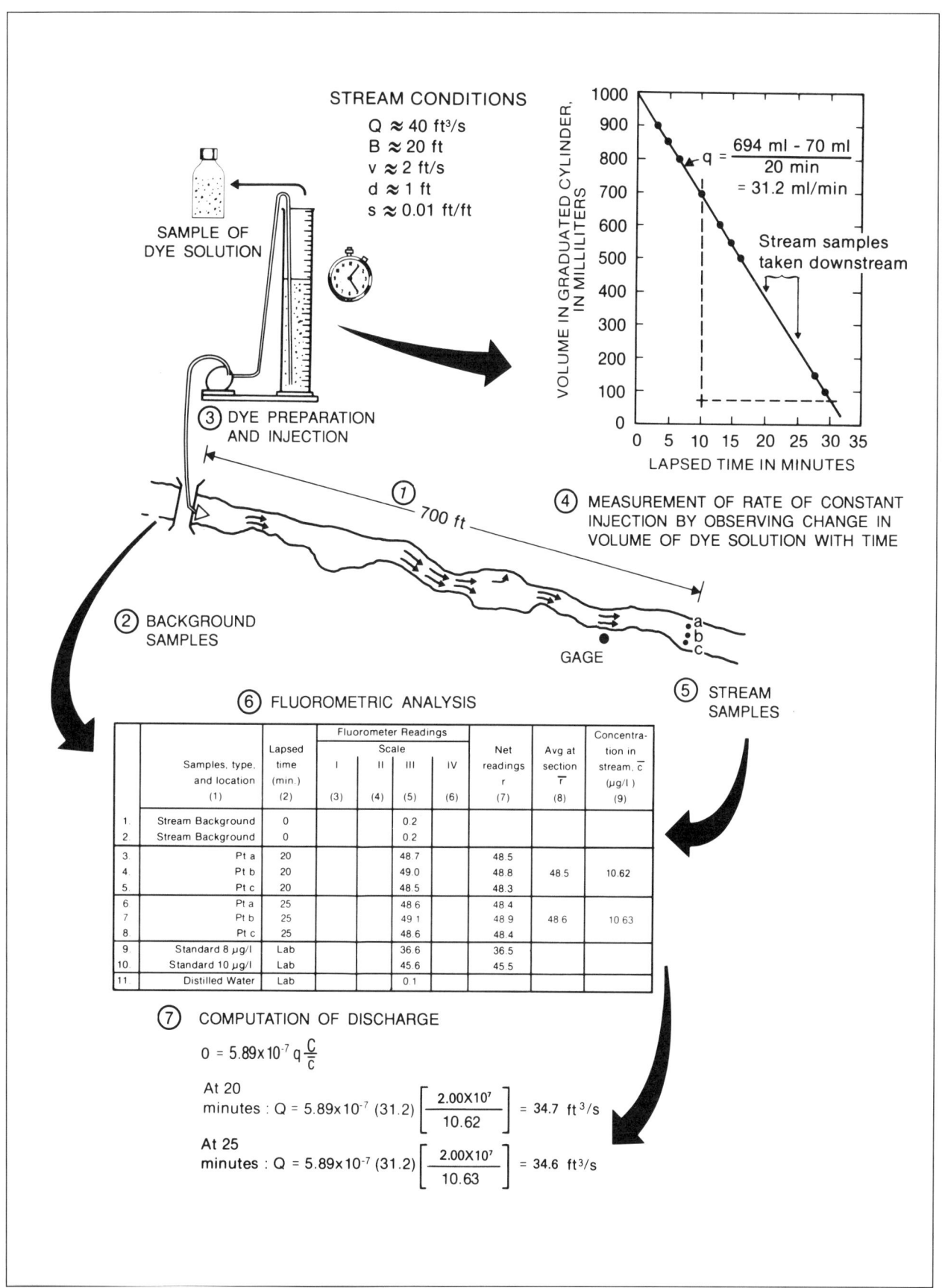

Figure 16-12. Schematic Diagram of the Performance of a Constant-Rate Dye Dilution Discharge Measurement
(Courtesy of U.S. Geological Survey)

Table 16-6. Convenient Three-Step Serial Dilution Combinations to Prepare Selected Standards for Selected Dye-Injection Solutions; Bulk Solutions
(Courtesy of U.S. Geological Survey)

Injection solution C in µg/l x 10^7		Dilutions: First		Second		Third		Resulting conc. C_3 in µg/l	Resulting dilution factor, D_T x 10^{-7}
		V_d	V_w	V_d	V_w	V_d	V_w		
(1)		(2)	(3)	(4)	(5)	(6)	(7)	(8)	(9)
16a	0.1	50	3000	50	1000	20	2000	7.73	77.29
b		50	3000	50	1000	20	1750	8.82	88.21
c		50	3000	50	1000	20	1500	10.27	102.71
d		50	3000	50	1000	20	1000	15.31	153.07
17a	0.2	50	3000	50	2000	20	2000	7.92	39.59
b		50	3000	50	2000	20	1750	9.04	45.18
c		50	3000	50	2000	20	1500	10.52	52.61
d		50	3000	50	2000	20	1250	12.59	62.97
18a	0.4	50	3000	50	3000	20	1500	14.14	35.36
b		50	3000	50	3000	20	1750	12.15	30.37
c		50	3000	50	3000	20	2000	10.64	26.61
d		50	3000	50	3000	20	2500	8.53	21.33
19a	0.6	50	3000	20	2000	20	3000	6.45	10.75
b		50	3000	20	2000	20	2500	7.73	12.88
c		50	3000	20	2000	20	2000	9.64	16.07
d		50	3000	20	2000	20	1500	12.82	21.36
20a	0.8	50	3000	20	2000	20	3000	8.60	10.75
b		50	3000	20	2000	20	2750	9.38	11.72
c		50	3000	20	2000	20	2500	10.31	12.88
d		50	3000	20	2000	20	2000	12.86	16.07
21a	1.0	50	3500	20	3000	20	2500	7.40	7.40
b		50	3500	20	3000	20	2000	9.24	9.24
c		50	3500	20	3000	20	1750	10.54	10.54
d		50	3500	20	3000	20	1500	12.27	12.27
22a	1.5	50	3500	20	3000	20	3500	7.95	5.30
b		50	3500	20	3000	20	3000	9.27	6.18
c		50	3500	20	3000	20	2500	11.10	7.40
d		50	3500	20	3000	20	2000	13.85	9.24
23a	2.0	50	3500	20	3500	10	2000	7.96	3.98
b		50	3500	20	3500	20	3500	9.09	4.55
c		50	3500	20	3500	20	3000	10.60	5.30
d		50	3500	20	3500	20	2500	12.70	6.35
24a	3.0	25	3500	20	3000	20	3500	8.01	2.67
b		25	3500	20	3000	20	3000	9.33	3.11
c		25	3500	20	3000	20	2500	11.18	3.73
d		25	3500	20	3000	20	2000	13.95	4.65
25a	4.0	25	3500	20	3500	10	2000	8.02	2.00
b		25	3500	20	3500	20	3500	9.16	2.29
c		25	3500	20	3500	20	3000	10.67	2.67
d		25	3500	20	3500	20	2500	12.79	3.20
26a	5.0	20	3500	20	3500	10	2000	8.03	1.61
b		20	3500	20	3500	20	3500	9.17	1.83
c		20	3500	20	3500	20	3000	10.69	2.14
d		20	3500	20	3500	20	2500	12.81	2.56
27a	10.0	20	3500	20	3500	10	3750	8.59	0.859
b		20	3500	20	3500	10	3500	9.20	0.920
c		20	3500	20	3500	10	3000	10.73	1.07
d		20	3500	20	3500	10	2500	12.86	1.29

$$Q = 5.89 \times 10^{-7} \times 31.2 \left[\frac{49.4}{48.6} \times \frac{1}{1.02 \times 5.30 \times 10^{-7}} \right] = 34.6 \text{ ft}^3/\text{s}$$

The dilution factor, D_T, was picked from column 9, line 23c of Table 16-6 for the standard chosen. Note that in Table 16-6, the first two dilutions for any one concentration, C, are the same. Thus, the second dilution may be treated as a working solution. If the 10.60-µg/l standard was not close to that experienced in the stream (R did not approximately equal $\bar{r}$, another standard could be prepared quickly, starting with the second dilution.

This method must be considered the most accurate of those presented, because only a three-step serial dilution of C is involved. Any error in preparing the injection solution is eliminated, and fluorometer dial readings are used directly without reçourse to calibration curves.

Acknowledgments

The techniques and methodology described in this report have been largely developed by the author and his colleagues in the Water Resources Division of the U.S. Geological Survey; more specifically, E. D. Cobb, J. F. Wilson, Jr., and J. F. Bailey. Furthermore, much of the material presented herein is taken from publications of the U.S. Geological Survey, particularly their "Technique" series of Water-Resources Investigations.

References

1. Clayton, C. G., "The use of a pump to reduce mixing length in the dilution method of flow measurement," *United Kingdom Atomic Energy Authority Research Group Report*, 1964.

2. Filmer, R. W., and Yevdjevich, V. M., "The use of tracers in making accurate discharge measurements in pipelines," *CER 66RWF-EMY*, Colorado State University, 1966.

3. Fischer, H. B., and others, *Mixing in Inland and Coastal Waters*, New York: Academic Press, Inc., 1979.

4. Holley, E. R., "Dilution method of discharge measurement in pipes," National Bureau of Standards Flow Measurement Symposium. *Special Publication 484, Proceedings*, Gaithersburg, MD., pp 395-421, 1977.

5. Kilpatrick, F. A., "Flow calibration by dye-dilution measurement," *Civil Engineering*, pp 74-76, 1968.

   ____ "A comprehensive evaluation of dye-dilution discharge measurement applications," *ASCE Hydraulics Division 17th Annual Specialty Conference*, Utah State University at Logan, 1969.

   ____ "Dosage requirements for slug injection of rhodamine BA and WT dyes," *U.S. Geological Survey Professional Paper 700-B*, pp B250-253, 1970.

6. Kilpatrick, F. A., and Cobb, E. D., "Measurement of discharge of using tracers," *U.S. Geological Survey Techniques of Water-Resources Investigations*, Book 3, Chapter A16, 1985.

7. Morgan, W. H.; Kempf, D.; and Phillips, R. E., "Validation of use of dye dilution method for flow measurement in large open and closed channel

flows," *National Bureau of Standards Flow Measurement Symposium*, Gaithersburg, MD, 1977.

8. Ostrem, G., "A method of measuring water discharge in turbulent streams," *Geographical Bulletin* no. 21: Geographical Branch, Ottawa, Canada, 1964.
9. Schuster, J. C., and Hansen, R. L., "Discharge measurements using radioisotopes in high-head turbines and pumps at flatiron power and pumping plant: Colorado-Big Thompson Project TID-25177," *Bureau of Reclamation Report* no. 40, Denver, CO,1968.
10. Smart, P. L., and Laidlaw, I. M. S., "An evaluation of some fluorescent dyes for water tracing," *Water Resources Research* 13, 1: 15-33, 1977.
11. Spencer, E. A., and Tudhope, J. R., "A literature survey of the salt-dilution method of flow measurement," *Institute of Water Engineers Journal* 12, pp 127-138, 1958.
12. Wilson, Jr., J. F.; Cobb, E. D.; and Kilpatrick, F. A., "Fluorometric procedures for dye tracing," *U.S. Geological Survey Techniques of Water-Resources Investigations*, Book 3, Chapter A12, 1984.
13. Yotsukura, N., and Cobb, E. D., "Transverse diffusion of solutes in natural streams," *U.S. Geological Professional Paper* 582-C, 1972.

About the Author

Frederick A. Kilpatrick is Hydrologist, U.S. Geological Survey, Water Resources Division, Office of Surface Water. He earned his Bachelor's and Master's degrees at the Georgia Institute of Technology in Atlanta.

Mr. Kilpatrick graduated from Georgia Tech in 1953 with a B.S. in Civil Engineering, after which he served as Sanitary Engineer in the Army Medical Service until 1956. He obtained M.S. degree from Georgia Tech in 1958.

As a co-op student during his college years, Mr. Kilpatrick worked for the Water Resources Division of the U.S. Geological Survey and remained with the Division full-time from 1958 until his retirement in 1986. During these 28 years with the USGS, he worked in the Research Section in Atlanta, GA, Fort Collins, CO, and Washington, D.C.; in the Radiohydrology Section, consulting to the Atomic Energy Commission; and as Assistant Energy Program Coordinator in Reston, VA.

His principal interests have been the use of tracers in hydrologic studies, flow measurement techniques, and hydrologic instrumentation development and application.

Mr. Kilpatrick is the author of about 30 technical papers on the use and evaluation of flumes, hydrology, water monitoring, water demand, tracer simulation, and dye techniques.

17

Turbine Flowmeters

Turbine flowmeters are designed to accurately measure the flow of liquids and gases in pipes. They are volumetric flow measuring devices and have been commercially available since the late 1940s. Sizes exist from a variety of manufacturers to cover the flow range from 0.001 gpm to over 25,000 gpm for liquid service and 0.001 acfm to over 25,000 acfm for gas service. End connections are available to meet the various piping systems. The flowmeters are typically manufactured from austenitic stainless steel but are also available in a variety of materials, including plastic. Turbine meters are applicable to all clean fluids over a pressure range from subatmospheric to over sixty thousand psi and temperature ranges from cryogenic to about 800°C (1500°F).

The turbine flowmeter is perhaps the most accurate type of meter available. It is capable of repeating to 0.025% of reading with accuracy and traceability to 0.05% of reading for liquid service; it is also capable of repeating to 0.1% of reading with accuracy and traceability to 0.3% of reading for gas service. A turbine meter has only one moving part — the rotor. Components can be selected for compatibility with most fluids, such as corrosive chemicals, dairy products, cryogenic liquids, steam, fuels, and water (including de-ionized water). The output signal from the flowmeter and sensor is an electronic pulse, but other output signals such as analog, visual, or digital are available.

Advantages:
Highly accurate
Corrosion-resistant materials
Long-term stability
Liquid or gas operation
Analog or pulse output
Wide operating range
Low pressure drop
Wide temperature and pressure limits
High shock capability
Wide variety of electronics available

Principles of Operation

A turbine flowmeter consists of a rotor mounted on a bearing and shaft in a housing. The fluid to be measured in passed through the housing, causing the rotor to spin with a rotational speed proportional to the velocity of the flowing fluid within the meter. A device to measure the speed of the rotor is employed to make the actual flow measurement. The sensor can be a mechanically gear-driven shaft to a meter or an electronic sensor that detects the passage of each rotor blade generating a pulse. The rotational speed of the sensor shaft and the frequency of the pulses is proportional to the volumetric flow rate through the meter.

Figure 17-1 shows a vector diagram of the forces involved. Vector V has an axial velocity v and no velocity component in the radial or tangential direction. The lack of a velocity component in the tangential direction is an important element in the system and will be discussed in more detail. The rotor blade is oriented at an incidence angle α to the flow stream. The momentum of the flowing fluid imparts a rotational velocity on the rotor, causing the flow to change direction and depart from the rotor in a swirling direction. The rotational velocity is nearly directly proportional to the velocity or flow rate through the flowmeter.

While the blade in the turbine meter works much like a typical turbine blade, there are some significant differences. A standard turbine blade is designed to do work by turning a shaft from which a major portion of the energy in the system is extracted. It is therefore necessary to design the most efficient blade possible.

The rotor of a conventional power-producing turbine is usually designed to operate at a specific speed or over a limited range of speeds. Seldom is a power-producing turbine allowed to operate at its "free-spin speed." The rotor in a turbine meter is not attached to a shaft to produce useful work and is always operating at its "free-spin speed." The range in speed over which the turbine flowmeter rotor operates is usually quite wide. The most important design characteristic of the turbine rotor is stability, not efficiency as with a drive turbine. A good turbine flowmeter must transform an axial velocity into a rotational velocity as precisely and repeatably as possible. It is not necessary to produce the highest free-spin speed possible, but to produce the most stable representation of the axial velocity and one that is the least affected by secondary influences.

Examination of Figure 17-2 shows that if the radial velocity profile were "flat," the tangent of the incidence angle α must vary as the radius changes in order to maintain the tangential velocity component in proportion to its relative blade velocity component. If the radial profile is not "flat," which is in reality the case, then in order to maintain a proportionally relative tangential velocity relationship, the radial twist of the blade would be proportionally biased. These factors would be taken into account if one were designing an efficient blade. Another factor that would be taken into account in designing an efficient blade is the shape of the blade in terms of the airfoil shape to produce the least drag (resulting in the lowest axial force component) and the highest tangential force component (resulting in the highest torque or rotational speed). Again, for a turbine meter, the most important characteristic is stability, and the design parameters that make an efficient blade also make the blade sensitive to minor changes in environmental conditions such as velocity profile. Another desirable characteristic of a turbine meter is to maintain the meter constant (*K* factor) at the

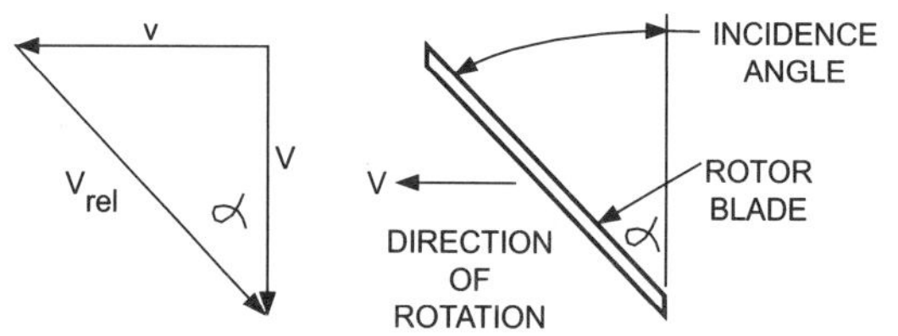

Figure 17-1. Vector Diagram

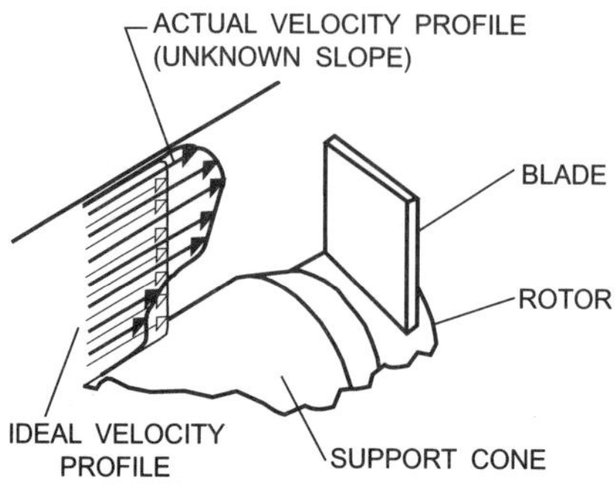

Figure 17-2. Velocity Profile

same value over as wide a range in flow as possible. A highly efficient blade tends to vary in efficiency as a function of speed (flow rate), resulting in a significant decrease in meter constant as a function of flow rate.

In reality, the inefficient blade designs are better for a turbine meter. A blade design that does not change its efficiency with operating conditions, such as velocity, velocity profile, temperature, or density, is best. In other words, a design that is minimally sensitive to changes in Reynolds number is optimum.

While the design details can be tailored to minimize the effects of Reynolds number on a turbine meter, it is more difficult to minimize the effects of a swirling component to flow. Since a turbine meter contains a rotating component, it is sensitive to a swirling component of the flow velocity. The upstream support of the meter is used by most manufacturers to "help" reduce any swirl, but it cannot be used to completely eliminate swirl due to flowmeter size limitations.

The drag imposed on the rotor by the other components in the meter greatly affect the operation of the flowmeter, especially at the lower end of its operating limits. The bearings are one of the largest contributors to drag. If the meter has a mechanical-type sensing device, it will contribute to the drag on a rotor. The electronic-type pickoff can impart drag on the rotor as well. The larger the meter, however, the less sensitive it is to these drag forces because there is proportionally more blade area, generating more torque to drive the rotor. However, in small meters, especially in gas applications, bearing and sensor drag is of prime importance at lower flow rates.

Sensors

Mechanical sensors are installed in some turbine flowmeters. A mechanical sensor usually consists of a shaft that is driven off a gear on the rotor shaft. This shaft is connected to a device much like the speedometer in an automobile. Some of these indicators are capable of displaying both flow rate and totalized flow (much like a speedometer and odometer), while others are set up only to totalize the flow (much like the domestic water meter).

Electronic sensors, usually called "pickoffs," are usually supplied by the manufacturers with the turbine meter. There are three basic types of pickoffs: magnetic inductive, magnetic reluctance, and modulated carrier frequency. They are mounted to the housing or body of the turbine and spaced a short distance away from the tip of the rotor through a thin membrane or diaphragm.

The end of the pickoff should NEVER be in contact with the fluid. No reputable manufacturer would produce a meter without a diaphragm for safety.

The magnetic inductive pickoff requires that a magnet be imbedded in each turbine rotor blade. A simple coil and bobbin arrangement in the end of the pickoff is used to produce an electric pulse at the passage of each blade (magnet) under the pickoff. The signal produced by this type of pickoff is usually a sinusoidal wave that has a frequency of one pulse per blade passage in which the amplitude varies from a few millivolts at the lowest frequency to a few hundred millivolts at the highest frequency. No amplification of this signal is required if the signal is only transmitted only a short distance (typically fifty feet or less) and the wire used is a twisted pair of conductors.

The magnetic reluctance pickoff contains a magnet within the pickoff itself, usually with a bobbin and coil around it. This type requires the rotor to be manufactured of a material that has some magnetic permeability. It does not have to be a highly magnetic material but has to be of sufficient magnetic permeability for the pickoff to "see" it. The signal characteristics produced by this type of pickoff are the same as those of the magnetic inductive type.

The modulated carrier frequency pickoff contains a ferritic bobbin and core with a coil around it. An accompanying amplifier contains a circuit to electronically oscillate with the inductance of the pickoff coil. The frequency of oscillation (commonly about 50 kHz) is much higher than the frequency of the passing blades (commonly about 2 kHz) and acts as a carrier, which is modulated by the passing of the rotor blades. The carrier frequency is filtered out in the amplifier (demodulated), leaving only the blade-passing frequency. Since the resulting frequency is produced in an amplifier, the amplifier output is usually transformed into a square wave pulse at a constant voltage amplitude, usually 5 or 10 volts peak-to-peak. External power, usually 24 V DC or 110 V AC, is used to operate the amplifier. The rotor must still be made of a material that contains a minimal amount of magnetic permeability for the pickoff to "see" it, but the material choices are much wider than with the magnetic pickoffs. Since there is no magnet involved in this configuration, no drag is imparted on the rotor by this type of pickoff.

Readout

The balance of the components in the flowmeter system are associated with "reading" and are widely diverse in nature. Frequency sensing devices can be used to transform the signal into a visual display or to condition the signal to be compatible with other devices.

Most turbine flowmeter manufacturers offer a wide array of devices and electronic packages to fit most applications. These include: amplifiers for the magnetic types of pickoffs that transform the low-level sine wave into a higher-level square wave pulse for transmission over greater distances and help isolate the signal in "noisy" environments; frequency-to-analog devices to convert the frequency signal to a 4 to 20 mA (or other desired analog) signal; linearizers that are tailored to each meter to compensate for any nonlinearities in the meter and produce a linear output signal over a wide flow range; displays of either an analog or a digital type in terms of either rate or total accumulated flow or both; microprocessor-based devices (flow computers) that can be programmed to take into account changes in meter performance with changing fluid properties, to name a few.

The various types of amplifiers and displays may be mounted on the meter or are available in a variety of packages for field mounting or panel mounting. They are available in NEMA-rated enclosures or so-called "explosion-proof" or hazardous location-rated enclosures. Most manufacturers offer a variety of intrinsically safe devices as well.

Each system is custom-tailored to an application. All reputable manufacturers have an applications engineering department to work with specific needs and to advise what equipment will fit a particular application. While the turbine flowmeter itself has been around for decades, little development is underway to improve the meter; however, the electronic devices associated with the turbine flowmeter are undergoing changes that will result in more cost-effective performance.

Flowmeter Design and Materials of Construction

The turbine flowmeter consists of a housing or body, a rotor supported by a bearing(s), a shaft, an upstream support structure, usually a downstream support structure, a sensing device, and a readout device(s). Some meters have two rotors or other unique features. As with any flowmeter, the design of a turbine flowmeter is a compilation of compromises that result in the best overall design in terms of performance and cost. Each designer may weigh the various factors differently in terms of goals and experience. There is no "correct" design. All will contain certain advantages and work well within certain limitations, given an associated level of cost. It is the understanding of these advantages and limitations and the relative cost of each feature that make the selection of the best meter system possible.

Rotor

The most common rotor material is a 400 series martensitic stainless steel, typically 416 or 430 ss. The most common alternative material is 17-4PH stainless steel, which is used when the other materials are not suitable. One of the biggest misconceptions regarding a rotor is the selection of its material of construction. Frequently, an "all 316 stainless" construction for the flowmeter is specified. However, as previously discussed, the rotor material must contain some degree of magnetic permeability if a magnetic reluctance or modulated carrier frequency type of pickoff is to be used. 300 series stainless steels do not have any magnetic properties and cannot be used for a rotor material unless it is "slugged" with a magnet and a magnetic inductive pickoff is used, or it is "slugged" with a piece of ferritic material and one of the other type of pickoffs is used. Some (more expensive) exotic nickel alloys, such as Monel™ can be used for the rotor material if they contain some iron in their chemistry. In general, the more iron in the material, the better chance of its use. Other low-ferrous metals such as Hastelloy B™, which contains 2% iron, can be used successfully (with a modulated carrier frequency pickoff only), while Hastelloy C™ contains 4% to 7% iron and is harder to detect.

Some manufacturers offer custom fabrication of flowmeters using selected materials, while others do not. One should contact the various manufacturers to determine the availability of various material choices. The manufacturers offering a wide variety of materials usually have a considerable amount of experience in solving a material selection problem. If a user has a strong preference toward a particular material for the rotor, the manufacturers will usually know whether or not it is a viable choice.

Slugging the rotor with a magnet or other ferritic material complicates the rotor fabrication significantly, increasing the cost and creating several other limitations. Figure 17-3 shows two common methods of slugging a rotor. Either method may be used for the installation of a magnet, assuming that the magnet material is compatible with the fluid in which the meter will be used. However, the radial tip method is the only practical consideration when operating in an environment that is corrosive to ferritic materials or magnet materials. A small closing weld may be placed over the slug to protect it from the fluid. The disadvantage of this type of construction is that, in order to use any reasonable sized pin or other slug, the blade must be thick enough to contain the pin or slug and still have some reasonable wall thickness. Typically ⅛ inch overall blade thickness would be required. This characteristic limits the design to larger meters, which would typically have that blade thickness, or to upsizing the meter one size and using thicker blades. This is commonly done when fabricating a meter from an exotic material such a tantalum for use in very corrosive environments such as chlorinated or brominated compounds.

The ratio of the blade tip radius to the blade hub radius is defined as the radius ratio of the rotor. Some manufacturers use a low radius ratio, meaning that the blade hub is large with respect to the blade height. More visually, the blades are short with respect to the hub diameter. Figure 17-4 shows a meter of this type. Other manufacturers offer meters of a high radius ratio, where the blade height is larger than the diameter of the hub. Figure 17-5 illustrates a design utilizing a high radius ratio. Large radius ratio designs will make the overall size of the meter smaller and usually at less cost for a certain flow range, but the small radius ratio designs will result in a meter that has wider rangeability at low densities, such as in gas applications, and usually will have a lower differential pressure.

The blade incidence angle α (see Figure 17-1) is selected to set the rotor speed and resulting output frequency for a specific flow rate. For blades containing no twist, the maximum practical limit is 45 degrees. Incidence angles greater than 45° will produce more downstream thrust on the rotor than driving torque, increasing the required size of the bearing, which results in decreased performance at the low end of the meter. If a twisted blade is used, the incidence angle at the tip may be greater than 45°, but the incidence angle at the hub will be less than 45 degrees. The mean incidence angle will normally be less than 45 degrees. Factors that determine the selection of the incidence angle include allowable bearing speed, bearing life, overspeed characteristics, maximum to minimum flow range (turndown) of the flowmeter, and differential pressure across the meter. In general, the higher the incidence angle, the higher the rotor speed, the higher the resolution of the meter, the higher the bearing loads, the greater the turndown, and the higher the differential pressure. Most of these characteristics conflict with the others, and any viable design is a compromise of these characteristics. While the rotor incidence angle used in a flowmeter design is not usually of concern to a user, many of the characteristics that result from the incidence angle selection are of concern.

At low densities, such as in most gas applications, the allowable velocity through the meter can be much greater than for liquids. The resulting blade incidence angle can, therefore, be much lower. Some manufacturers use incidence angles as low a 5 degrees for low-density gas applications, while others maintain larger incidence angles typical of those for liquid applications.

An interesting phenomenon results when using low blade incidence angles for gases and high incidence angles for liquids. The same numerical value for the usable flow range for a given size meter can be specified when the liquid flow rate is quoted in gallons per minute (gpm) and the gas flow rate is quoted in actual cubic feet per minute (acfm). This is due to the fact that there are approximately 7.5 gallons per cubic foot and the velocity through the meter for low density gases is approximately 7.5 times that allowed for liquids. The blade incidence angle is appropriately reduced to maintain the same maximum frequency

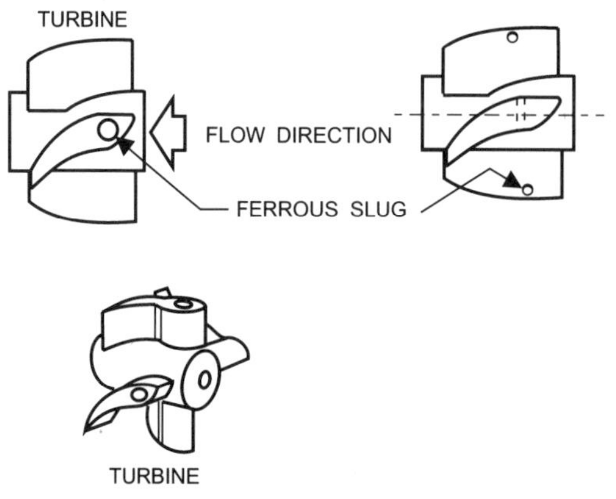

Figure 17-3. Methods of Slugging Rotor
(Courtesy of Great Plains Industries)

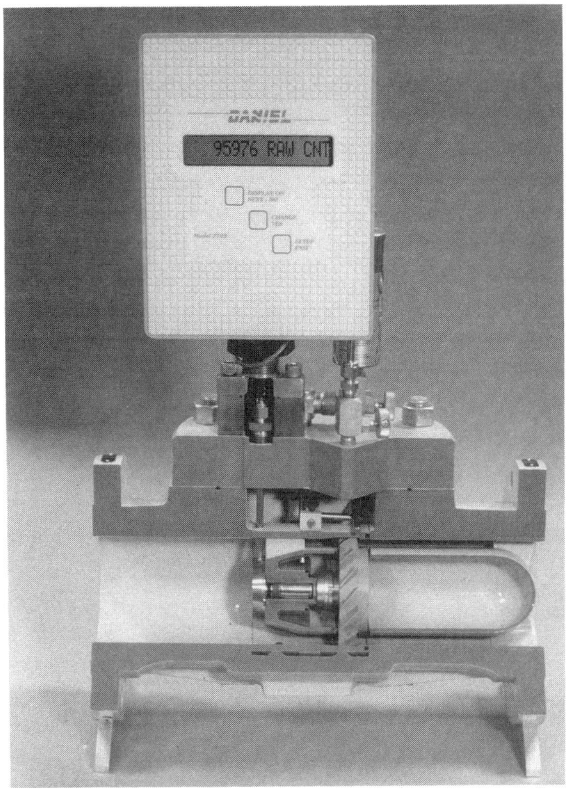

Figure 17-4. Turbine Meter with Low Radius Ratio and High Blade Incidence
(Courtesy of Daniel Industries)

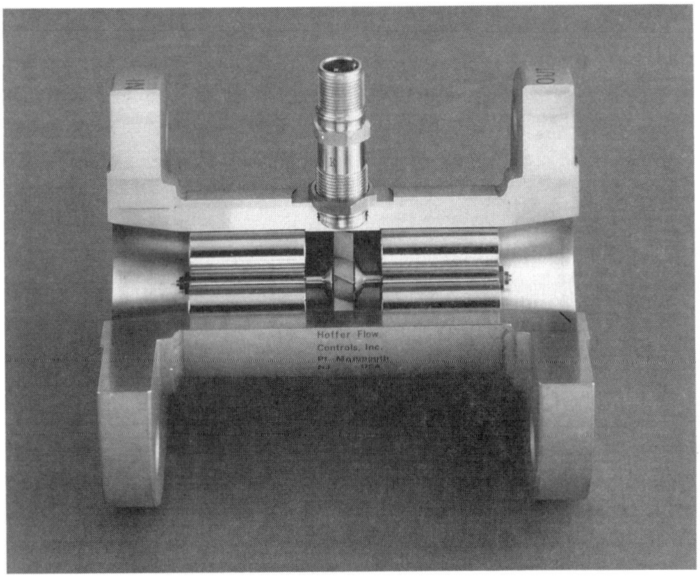

Figure 17-5. Turbine Meter with High Radius Ratio
(Courtesy of Hoffer Controls)

for both fluids. Some manufacturers use a range of intermediate incidence angles for higher density gases and limit the flow range appropriately in order to keep the thrust load on the bearings somewhat constant.

Bearings

The bore of the rotor is designed to suit the type of bearing being used. Since there are a wide variety of bearings used, there are as many bore configurations. The bearing used must (1) allow the rotor to spin freely with a minimum amount of friction and (2) position the rotor in its correct axial position while overcoming the dynamic pressure of the flowing fluid that is trying to drive the rotor downstream. The only work the rotor must produce is to overcome the friction from the bearings, mechanical readout, and the pickoff drag (if a magnetic-type pickoff is used). The lower the friction in the bearings and other components, the better the meter will respond to the flowing fluid.

Three types of bearings are used in turbine flowmeters: the ball bearing, the journal bearing, and the pivot bearing. There are many variations to each of these types, but almost all designs can be categorized into these types. Figure 17-6 shows each of these types.

BALL BEARINGS

The ball bearing is perhaps the most common type used. This type of bearing provides a relatively sturdy, low drag arrangement. They may be used by mounting the bearing(s) in the rotor with the shaft fixed in the end supports or by mounting the bearings in the supports with a rotating shaft. While the latter offers better rotor stability, the former eliminates drag due to shaft misalignment and is more commonly used. Some manufacturers use only one rotor-mounted bearing to further reduce friction, while others use two rotor-mounted bearings to increase stability and life, usually at the sacrifice of low end performance but at the gain of higher allowable frequencies. Both designs work well within their rated specifications.

The greatest limitation of the ball bearing is that they are available essentially in a single material for the balls and races—440C stainless steel. The retainers are available in a variety of materials including 303 or 410 stainless steel, phenolic, or varieties of Teflon™, or with no retainer at all. The bearings rely on the metered fluid for their lubrication and cannot be used in fluids that will chemically react with either the ball or retainer materials. While 440C is classed as a stainless steel, it is not suitable for extended service in water or acids and is not suitable for fluids containing particulate contamination. It is also advisable to use fine filtration upstream of the flowmeter. Usually 10 micron filtration is typically specified, but each manufacturer should be consulted for specific requirements.

Ball bearings will provide excellent service in fuels, most oils, alcohol, Freon™, and cryogenic liquids. They are not suited for crude oils containing hydrogen sulfides. When used at cryogenic temperatures, it is necessary to use a shaft and rotor material that has a very similar thermal expansion coefficient to that of the bearing in order to prevent binding or loosening of the bearing. Ball bearings will provide excellent service in clean, noncorrosive gases including air, argon, helium, hydrogen, oxygen, ammonia, methane, nitrogen, etc. Gas applications provide little if any lubricity for the ball bearings. Either sealed and lubricated bearings, or bearings specifically suited for gas service, must be used. Ball bearings designed for liquid service will not work well in gas applications. While many manufacturers offer turbine meters for gas applications, not all use the best bearings for that service.

Standard ball bearings may be used up to 230°C (450°F). Ball bearings may be used up to about 400°C (750°F) with a higher temperature heat treat processing.

These bearings are available on special order only and will result in long lead times when ordering meters using them.

Some bearing manufacturers offer impregnated coatings, usually of a molybdenum or tungsten disulfide material to provide lubricity at elevated temperatures. Ceramic ball bearings made of silicon nitride and silicon carbide are recent developments in the bearing industry. Limited testing has been performed on these bearings in turbine meters with little or no success. Perhaps future development will add ceramic ball bearings to the list of potential choices.

JOURNAL BEARINGS

Journal bearings, or "sleeve bearings" as they are often called, consist of a bushing rotating on a stationary shaft or journal. Sliding friction is characteristic of this type of bearing as opposed to rolling friction in the ball bearing. As a result, the relatively high drag of the journal bearing limits the rangeability and linearity of the meter. Journal bearings can be made of a wide variety of materials and, therefore, can operate satisfactorily in many fluids in which a ball bearing cannot. Bonded graphite sleeves on stainless steel journals, tungsten carbide sleeves on tungsten carbide journals, Teflon sleeves on a variety of metal jour-

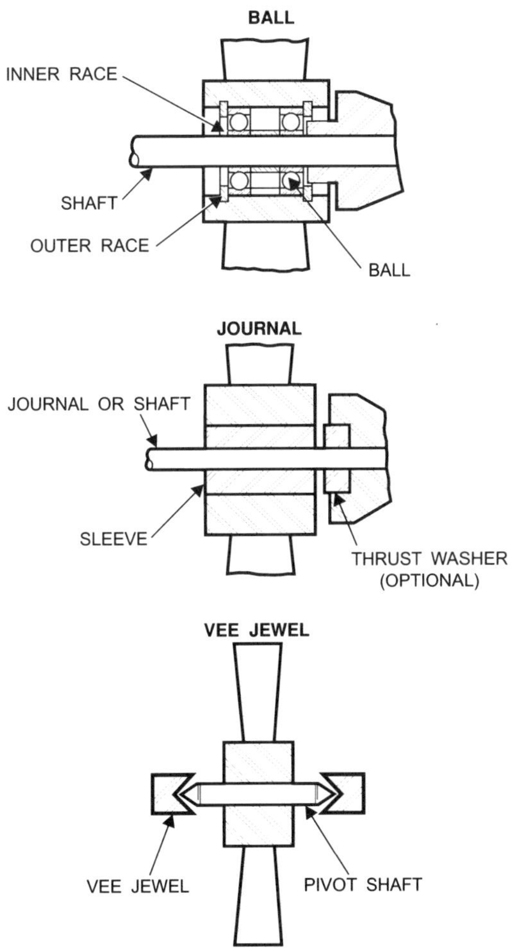

Figure 17-6. Types of Bearings Used in Turbine Meters
(Courtesy of EG&G Flow Technology)

nals, ceramic sleeves on ceramic journals, and stellite sleeves on stellite journals are all commonly used as bearing materials in turbine meters.

The journal bearings provide only radial positioning of the rotor; therefore, separate thrust bearings must be utilized to position the rotor. Some manufacturers utilize a thrust-balancing scheme based upon local differential pressures across the rotor to minimize the loads. Since most of the load is caused by viscous drag and the meter must operate over a wide range of flows and viscosities, it is not possible to completely compensate for the loads in this manner at all operating conditions. If too much thrust compensation is provided, it is possible for the rotor to actually be pushed forward against the upstream support, resulting in a large change in the friction. A step change in the meter constant (*K* factor) will be observed if this occurs, an undesirable characteristic. Commonly, a thrust bearing is mounted in the upstream and downstream supports, which react against the end of the rotor or the sleeve. This type of thrust bearing works much like the radial-type journal bearing, and does add to the bearing friction. The same list of materials used for the journal bearings may be used for the thrust bearings.

Tungsten carbide is a very hard material, exceeded in hardness only by diamonds and a small list of other ceramics. It is, therefore, suitable for fluids that contain some abrasive particles. It will provide excellent service in most types of water and non-oxidizing acids. It has excellent wear properties, and if properly used will almost never wear out. Stellite and aluminum oxide ceramic offer similar properties to abrasive fluids and are chemically resistant to some fluids that the tungsten carbide is not. Neither have as good a wear resistance as the tungsten carbide. Aluminum oxide is resistant to almost all acids and is often the material of choice in oxidizing acids. Tungsten carbide and aluminum oxide bearings are significantly more expensive than graphite. Tungsten carbide may be used up to 650°C (1200°F). Above that temperature the binder breaks down and the base material is free to migrate from the matrix.

Graphite bonded with a variety of materials is commonly used as a sleeve material conjunction with a metal journal. Graphite is not very hard and is not suitable in abrasive environments. Graphite is compatible with a wide variety of fluids and may be used with journal materials such as Hastelloy™ to provide a very corrosion-resistant bearing configuration. It is often used with a 304 stainless steel shaft in water service. Graphite is limited to a maximum of 540°C (1000°F), depending on the binder material. The limiting temperature usually occurs due to breakdown of the binder material.

Journal bearings are not normally suited for operation in gas applications since the gas properties lack sufficient viscosity to generate a film on which the journal may slip. Bearing "ringing" usually occurs in gaseous fluids. The one exception to this rule is graphite sleeves running on a metal journal. If the bearing clearances are properly controlled, satisfactory operation can be achieved in steam and air service. The lubricating properties of the graphite are sufficient to provide adequate lubricity.

Table 17-1 illustrates a list of fluids commonly metered with graphite bearings. These lists are by no means all inclusive. Many other fluids may successfully be used with journal bearings. Consult the manufacturer or other material compatibility charts for other fluids.

PIVOT BEARINGS

The third type of bearings used in turbine flowmeters are pivot bearings, which are constructed of a tapered shaft spinning in a cupped conical support. The support may be spring loaded against the shaft to eliminate end play, or the end clearance is carefully controlled. The shaft may spin or roll in the end support, depending upon the endload, clearance, spin rate, and bearing loads. Both the

Table 17-1. Fluids and Solutions Compatible with Graphitar 30A™ Journal Bearings

Acetic Acid (dilute)	Ethane	Naphtha	Iric Acid
Acetic Acid (glacial)	Ethyl Acetate	Naphthalene	Varnish
Acetone	Ethyl Alcohol	Nickel Chloride	Vegetable Oil
Alcohol	Ethyl Chloride	Nickel Sulphate	Vinegar
Alum	Ethylene	*Nitric Acid	Vinyl Acetic Acid
Amidol	Ethylene Dichloride	Nitro Benzoic Acid	Water
Ammonia	Ferrous Sulphate	Nitrogen	Wine
Amyl Alcohol	Fluroroform	Olive Oil	Zinc Chloride
Aniline	Formic Acid	Oxalic Acid	Zinc Sulphate
Aqua Regia	Freon (11)	Pentane	
Arsenic Acid	Freon (12)	Petroleum	
Beer	Furfuryl Alcohol	Phenol	
Beet Juice	Gasoline	Phosphoric Acid	
Benzaldeyhne	Glycerine	Poric Acid	
Benzilic Acid	Hexane	Potassium Chloride	
Benzine	Hydrbromic Acid	Potassium Hydroxide	
Bleach Solutions	Hydrocarbon Gases	Potassium Nitrate	
Boric Acid	Hydrochloric Acid	Propyl Alcohol	
Brine	Hydrocyanic Acid	Salicylic Acid	
Bromine Water	Hydrofluoric Acid	Salt Water	
Butane	Hydrogen Peroxide	Sea Water	
Butyl Alcohol	Hydrogensulphide Water	Sewage	
Calcium Carbonate	Hypo	Soap	
Calcium Chloride	Iodine-Water	Sodium Bicarbonate	
Calcium Hydoxide	Isopropyl Alcohol	Sodium Carbonate	
Calcium Oxide	Isopropyl Ether	Sodium Chloride	
Carbolic Acid	Kerosene	Sodium Hydroxide	
Carbon Dioxide	Lactic Acid	Sodium Hydrosulphate	
Carbon Dioxide-Water	Lime Water	Sodium Hypochlorite	
Carbon Disulphide	Lithium Chloride	Sodium Nitrate	
Carbon Tetra-Chloride	Lithium Hydroxide	Sodium Silicate in Water	
Chlorine	Lubricating Oil	Sodium Sulphate	
Chlorine-Water	Lye	Sodium Tetrobirate	
Chloroform	Maleac Acid	Sodium Thiosulphate	
Citric Acid	Magnesum Sulphate	Sorbic Acid	
Copper Sulphate	Menthol	Starch	
Cottonseed Oil	Mercury	Stearic Acid	
Creosole	Methane	Sugar	
Cupric Chloride	Methyl Alcohol	Sulphur	
Cyanic Acid	Methyl-Ethyl Ether	Sulphuric Acid	
Destrose	Milk	Toluene	
Dibenzyl	Mine Water	Tributyl Alcohol	
Dichloroether	Mineral Oil	Turpentine	

shaft and support are made from hard materials, usually tungsten carbide or sapphire jewel. The friction generated by pivot bearings is considerably less than that of ball bearings, although the load-carrying capacity is also considerably less than ball bearings. The materials used are impervious to attack by most chemicals and are not subject to abrasion due to particulate contamination because of the material hardness. The allowable speeds are low compared to ball bearings. The characteristics of pivot bearings make them an ideal choice for low flow rates and low density operation at low flow rates.

Pivot bearings are frequently used in insertion meters and are well suited for operation in small turbine meters of both the axial and tangential designs. Rotors much over 1 inch in diameter usually generate radial and axial loads too high for pivot bearings, although they are used in flowmeters up to 2 inches in diameter in low-flow gas applications.

While most bearings used in turbine flowmeters fall into the above three categories, variations of these principles are also used. A combination of the ball and pivot bearing is used in which the conical end of the shaft is run directly against the balls in a ball bearing having no inner race. The friction is quite low in this arrangement, but it is subject to misalignment loads and the resulting friction. Ball end pivot bearings use a spherical ended shaft sliding in a cup. This arrangement is capable of somewhat higher loads than the pivot bearing, but the friction is also higher.

Supports

The supports have the dual function of supporting the rotor shaft, positioning it properly in the housing, and at the same time straightening the flow streamlines and directing it into the rotor. The designs of the supports take a wide variety of configurations. They are fabricated from machined bar stock, castings, welded fabrications, stacked tube bundles much like a flow tube type of straightener, and formed sheet metal. Each design is intended to emphasize one feature or another in terms of flow conditioning, support of the rotor, shaft alignment, and cost. All of the designs are functional within certain limitations. The longer upstream support provides better flow straightening but is typically associated with a shorter downstream support, sacrificing bi-directional operation.

Some designers use no downstream support at all by cantilevering the shaft from the upstream support, making installation of the rotor easier, but no thrust reduction by pressure balancing is possible. The machined configurations, perhaps, provide better alignment but usually at a higher cost. Castings may reduce the cost of machining but the manufacturing volume must be high enough to recover the cost of the tooling. With the vast array of sizes, bearing configurations, and material selections, it is not always possible to reduce the cost of a machined part with a casting. The tube bundle and the formed sheet metal designs are lower cost approaches but may not support and align the rotor shaft as well as the stiffer and machined designs without additional procedures or processes being accomplished during the fabrication to assure that the alignment is correct. Most manufacturers bend or form the trailing edge of the upstream support as a method of "trimming" the meter to set the desired meter constant and to adjust the linearity of the meter over its operating range. The closer to the rotor the trailing edge of the support, the more effective the bend will be in trimming the meter. Figure 17-7 shows this feature.

Most designs allow assembly of the supports, shaft, bearings, and rotor as a subassembly, commonly called the "internals." This assembly is then slid into the housing. Since the meters are trimmed during calibration, it may be necessary to test and remove the internals several times during the initial calibration in order to meet the overall specifications. It is, therefore, desirable to make this installation as simple as practical. Once a meter has been trimmed at the factory, it is usually not necessary to retrim it again during subsequent recalibrations, even if the bearings have been replaced, so long as the meter is run over the same Reynolds number range as it was during the initial calibration.

Retention

The method of retention of the internals into the housing is an area of much concern for meter designers. Most designs require some sort of a step or retaining ring protruding into the flow stream to provide the desired retention. The greater the protrusion, the greater the disturbance to the flow stream and the greater the differential pressure loss to the meter. The smaller the protrusion, the lesser the strength of the retention. Round wire cross-section retaining rings provide minimal flow disturbance but are not as strong as a stiffer cross-section type of retaining ring, and the round section retainers can be dislodged from the groove if the meter is used in extreme pulsating service. Designs designated as "high shock" meters usually employ the stiffer cross-section retainers or a machined step in the flow path on one end and a retainer ring on the other end. The method of retaining the bearings in the rotor may also be stiffened in a high shock design. This seemingly insignificant detail in the design of a flowmeter can double the differential pressure loss across the meter. If low differential pressure is desired in an application, it will generally by achieved at the sacrifice of ruggedness.

Housings

The housing provides (1) for the positioning of the internals, (2) for the relative positioning of the pickoff, (3) the outer flow path for the fluid, (4) the containment for the internal pressure of the fluid, and (5) for connection of the flowmeter to the mating piping.

Flowmeters are manufactured with every conceivable end connection. The more common connections include: National Pipe Threads (NPT), ANSI pipe flanges, AN or MS connections, and metric flanges. Other fittings commonly used for special purposes include high pressure clamp or "Greyloc" fittings and Tri-Clover™ fittings in the food processing industry. Most of the flowmeter sizes are available in all end fitting types.

The material used for the housing must be a nonmagnetic material in order not to affect the operation of the pickoff. If the housing were made of steel, the pickoff would not be able to detect the rotor pulses. Commonly, the housings are

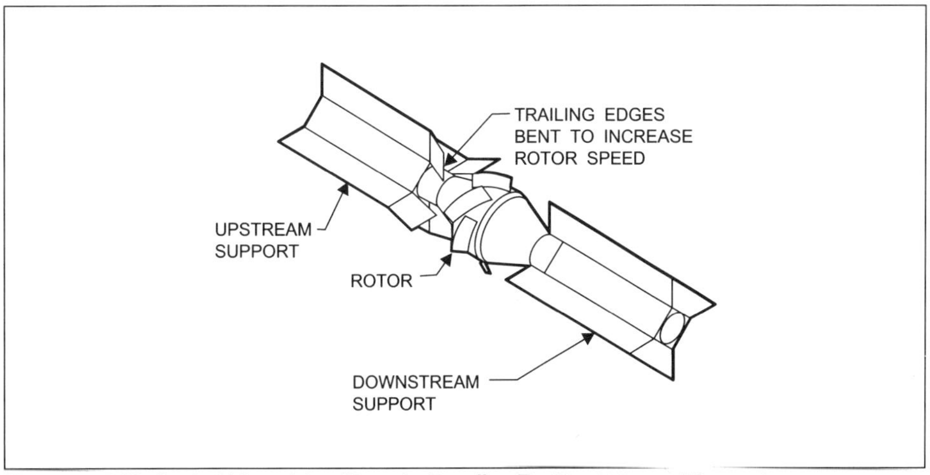

Figure 17-7. Method of Linearizing Meter by Bending Trailing Edges of Support
(Courtesy of EG&G Flow Technology)

made of stainless steel, usually 303, 304 or 316. Materials such as aluminum, Hastelloy, bronze, various plastics, and other nonmagnetic materials can be used. It is common practice of some manufacturers to use steel flanges welded to a stainless steel body, especially in larger meters.

One of the most critical areas in a flowmeter housing is the thin membrane between the pickoff and the inside diameter of the housing over the rotor. This membrane must be thin enough to allow the pickoff to sense the passage of the rotor blades, yet thick enough to withstand the internal pressure of the meter. The end of the pickoff should never be in contact with the fluid. It may be in direct contact with the membrane, but the membrane should be of sufficient thickness to withstand the internal pressure without relying on any support from the pickoff. This is usually not a problem except in very high pressure meters.

Pickoffs

The last component in the standard turbine flowmeter is the pickoff. As outlined above there are three basic types: magnetic inductive, magnetic reluctance, and modulated carrier frequency. All three types are used by the various manufacturers of turbine meters, and some manufacturers offer all three types.

The magnetic reluctance type is perhaps the most common. Figure 17-8 shows a cross section of a typical pickoff of this type. A small bar magnet is wrapped with a coil over a bobbin. The wires from the coil are routed to the connector on top of the pickoff and potted in place. As a rotor blade is passed through the magnetic field, an alternating current voltage is generated in the coil at a frequency equal to the blade passage frequency. The amplitude of the voltage is also a function of the frequency. At the lowest frequency at which the meter will operate, the voltage may be only a few millivolts (typically, about 10 millivolts). The voltage increases as the frequency of the meter increases to a peak level of a few hundred millivolts at the upper end of the meter's range. The waveform is a pure sinewave. Since the signal is produced within the pickoff, no external amplifier is required to measure the signal. Any instrument capable of measuring a low level sine wave may be used to measure the frequency.

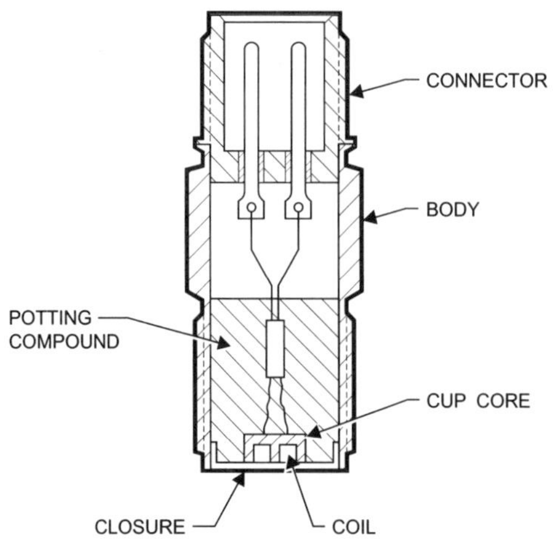

Figure 17-8. Cross Section, Modulated Carrier Pickoff
(Courtesy of EG&G Flow Technology)

The magnet in the pickoff does exert a force on the blades of the rotor, causing a drag that must be overcome by the fluid acting on the blades. This drag can be of sufficient force to affect the linearity of the meter at the lower flow rates. It also will stop the rotor from turning at all, limiting the range of the meter. The drag is of sufficient force to severely affect a small meter in gas service and, therefore, is generally not used in gas service. Some manufacturers offer a "low drag" magnetic pickoff, which minimizes the drag on the rotor by reducing the magnetism in the magnet. This reduces the sensitivity of the pickoff but renders it more useful in gas service.

The magnetic inductive pickoff requires the installation of a magnet (or ferrite core) in the blades of the rotor. This concept offers less drag but complicates the fabrication of the rotor. A coil in the pickoff is used to generate an alternating current at a frequency equal to the blade passing frequency, much in the same manner as the magnetic reluctance type. An amplifier is also required with this type of pickoff. The two types of pickoffs are not interchangeable.

The modulated carrier pickoff uses a high frequency carrier frequency, usually 40-50 kHz, which is modulated by the passing turbine rotor blades. With this design, no significant drag is imposed on the rotor; thus, its rotational velocity can be sensed without influencing its ability to follow the flow rate. The performance of the meter at low flow rates or in low density fluids is greatly enhanced. Construction of this type of pickoff is similar to the other pickoffs. It uses a small coil and cup core at the base. The leads from the coil are routed to the connector on the top of the pickoff.

The carrier type of pickoff can be used from cryogenic temperatures up to 267°C (450°F). A high temperature version is also available from some manufacturers that can be used up to 400°C (750°F). The carrier frequency pickoffs must be used in conjunction with an amplifier that, when coupled to the pickoff, oscillates to create the carrier frequency. Therefore, no signal is observed from the pickoff alone. Likewise, a carrier frequency amplifier cannot be tested without a pickoff attached. The length of the leads between the pickoff and the amplifier do affect the oscillation frequency. If long leads are to be used, the amplifier must be tuned with the leads in the system. The carrier frequency pickoff should be selected for all gas applications in meters 2 inches and smaller and for liquid meters smaller than 1 inch if an extended range is desired.

A common error in troubleshooting a system is made by trying to detect with an oscilloscope a signal from an unconnected carrier pickoff.

Performance/Limitations

The "accuracy" of an instrument is an often misunderstood and often misused term. It implies how well the reading presented by the meter represents the absolute truth. However, often the user does not use the instrument within its limitations and the measurement is not representative of the instrument's true capability. The manufacturer advertises the potential accuracy of the instrument under ideal conditions. While the instrument may have a certain ideal potential, it may be unrealistic to assume that any given user can achieve that potential in the real world.

The accuracy of a turbine flowmeter in general is a combination of three elements: repeatability, calibration, and application effects. The easiest of these to evaluate is repeatability, which can be determined by any user by comparing the repeated results of a meter to a standard. It must be kept in mind that the repeatability demonstrated is the combined repeatability of the meter and the standard. For instance, if the standard by which a turbine flowmeter is being compared is the proverbial "bucket and stop watch," the accuracy of what constitutes a "full bucket" and how accurately the stopwatch is started and stopped is just as much a part of the repeatability as the flow reading from the meter. If the flowmeter is being compared to another flowmeter, the repeatability of each contributes to the demonstrated repeatability. Many good calibration systems are sophisticated versions of the bucket and stopwatch or the master flowmeter.

Repeatability

In order to evaluate the repeatability of a flowmeter, the standard by which it is compared should be a minimum of 4 times more repeatable than the expected repeatability of the meter being evaluated. A standard that is 10 times more repeatable would be much more desirable but is often hard to find, especially when trying to evaluate a flowmeter with the good repeatability characteristics of the turbine flowmeter. If the standard is 4 times more repeatable than the meter being tested, then 20% of the demonstrated repeatability is attributable to the standard. If the standard is 10 times more repeatable than the meter being evaluated, then 9% of the demonstrated repeatability is attributable to the standard.

Most manufacturers of turbine flowmeters accurately or conservatively specify the repeatability of their meters. For precision in-line turbine meters in liquid service, repeatabilities of 0.02% of reading can be expected. When the meters are used over an extended range, especially small flowmeters, the repeatability may not be quite that good at the lower flow rates. Once the meter becomes nonlinear (the K factor is increasing rapidly as a function of frequency or flow rate), the repeatability generally increases. Tangential flowmeters are typically not as repeatable as the axial type. Typically, 0.1% of reading can be expected for this type of meter. In gas service, repeatability for turbine flowmeters is not as good as with liquids. Repeatability of 0.25% of reading is common for axial turbine meters. Repeatabilities of 0.1% of reading are attainable in some meters.

Accuracy

The calibration accuracy of a turbine meter is not at all a function of the meter itself; it is simply a function of the traceability of the standard against which it is being calibrated. Most manufacturers and metrology laboratories want to believe that their techniques and traceability are extremely accurate and, therefore, little if any bias is added to the repeatability to determine the accuracy. This is often not the case, however. It must be kept in mind that the National Institute of Standards and Technology (NIST) considers itself to be accurate only to within ± 0.15% of reading for liquid calibrations. While this estimate of the NIST accuracy is considered by many experienced members of the flow community to be very conservative, it is still an estimate to be reckoned with. Any standards laboratory that claims a flow traceability of ± 0.1% of reading to NIST is actually claiming ± 0.18% of reading (the root sum square of 0.1 and 0.15) traceability to the absolute truth.

There are other ways to maintain traceability to NIST other than on a direct flow basis. A mass calibrator can maintain its traceability by the certification of its weights and timing devices, and a volume calibrator can maintain its traceability by the certification of its volume and timing devices. Both of these methods can show their accuracies to be significantly better than the accuracies listed in the previous paragraph for traceability on a flow basis. Many manufacturers and laboratories can show traceabilities of 0.02% on these bases. However, the real proof is in a flow traceability using a suitable artifact, not in a mass and time traceability or a volume and time traceability. A single turbine flowmeter calibrated by NIST is not a suitable artifact. When traceability is demonstrated on a flow basis, it is difficult to maintain a traceability of better than ± 0.15% of reading to the absolute for liquid calibrations. Many laboratories that show a traceability on a mass time or a volume time basis of ± 0.02% cannot show a direct flow traceability of even ± 0.15% of reading. While this fact may seem somewhat paradoxical, it is true.

Accurate traceability in gas measurement is more difficult to attain than with liquids since compressibility effects must be taken into account. Volume time calibrators can be accurately measured much as in liquid calibrators, and sonic Venturis can be accurately calibrated on a pressure-volume-time-temperature basis. However, the true traceability should be on a direct flow basis using a suitable artifact. To date, little direct flow traceability has been accomplished, although it is currently in the planning stages. While, as in the liquid calibration traceability, some methods can claim highly accurate traceability, expectations of better traceability of less than ± 0.5% of reading on a flow basis may be optimistic. Some meter types and some Reynolds number ranges may be more realistically in the ± 1% of reading range. Turbine meter *K* factors are less sensitive to changes in Reynolds number at higher Reynolds numbers. It may, therefore, follow that at high Reynolds numbers, better traceability may be expected.

Application

The third aspect of the accuracy of a turbine meter is the effect of the actual application. Effects such as improper or no straightening sections, operation of the meter at Reynolds numbers different from those of calibration, not compensating for extremes in operating temperatures, and use of fixed *K* factor electronics on a meter used in extended ranges can introduce percent errors into a meter. It doesn't make much economic sense to calibrate a turbine meter to an accuracy of ± 0.15% of reading and then introduce a 3% bias error due to misapplication, although it is commonly done.

The most accurately calibrated flowmeter can be completely defeated by misapplication.

The calibration constant for a turbine flowmeter, generally called the *K* factor, is expressed as the number of blade pulses transmitted by unit volume of flow through the meter, i.e., pulses per gallon, pulses per cubic meter, etc. It is sometimes expressed as the reciprocal or gallons per pulse and is then typically called the "meter factor." A plot of the *K* factor versus the frequency output or the flow rate is a dimensionalized expression of meter linear performance. Properly scaled, this presentation clearly shows the nonlinearity of the meter as a function of frequency or flow rate. This presentation does not illustrate the total performance of the meter. It simply illustrates the performance of the meter at one temperature and one fluid viscosity.

Linearity of a turbine flowmeter is defined by the ISA-RP31.1 as the maximum deviation in *K* factor from the average *K* factor over a specific range. Expressed algebraically as:

$$\%\ \text{Linearity} = \frac{(K_{\text{max}} - K_{\text{ave}}) \times 100}{K_{\text{ave}}} \qquad (17\text{-}1)$$

The linearity of a turbine flowmeter in liquid service is typically quoted by most manufacturers as ± 0.5% of reading over a 10:1 range at constant viscosity and temperature. Most manufacturers can improve the linearity over the same flow range to ± 0.25% of reading with a little extra "tweaking" of the upstream support and/or trimming of the blade tips. When operating over a wider flow range, the linearity will increase.

Linearity is also dependent upon the viscosity of the fluid. Most turbine flowmeters may be linearized at any viscosity up to approximately 30 centistokes over a 10:1 range but will become nonlinear at other viscosities. The deviation in linearity can change as much as 30% over a viscosity change from 1 centistoke to 100 centistokes at the lowest operating frequencies but may be as little as only 1% at the highest operating frequencies. Figure 17-9 illustrates this characteristic for a typical turbine flowmeter. It must be kept in mind that even though the meter may become nonlinear, it is still repeatable.

Turbine flowmeters in gas service are significantly more nonlinear than in liquid service. The linearity in gas service is commonly expressed as a percentage of full scale as opposed to an absolute percentage. This means that a meter having a linearity of ± 0.5% of full scale can be 5% nonlinear at 1/10th the maximum flow rate and 10% nonlinear at 1/20th the maximum flow rate.

Tangential turbine flowmeters in gas service can be extremely nonlinear. The degree of nonlinearity is dependent upon the range of the meter, with the lower flow rate meters having a linearity as high as 80%.

The linearity of a turbine flowmeter is affected by factors other than the viscosity of the fluid, such as the type of pickoff and the type of bearings used. Modulated carrier frequency pickoffs impart no drag on the rotor and, thus, do not affect linearity. Both types of magnetic pickoffs impart a drag on the rotor and increase the linearity of the meter. Likewise, any bearing type that causes increased drag on the rotor will increase the linearity, journal bearings having the greatest effect and jewel pivot bearings having the least effect. Drag on the rotor from the bearings or from the pickoff will cause the rotor to decrease its frequency and, thus, limit the range of the meter as well as increase the linearity.

Turbine flowmeters are sensitive to Reynolds number, which can be expressed as a ratio of velocity over kinematic viscosity. Since the frequency output is almost proportional to velocity, Reynolds number sensitivity can be expressed as frequency/kinematic viscosity. Frequency/kinematic viscosity is a simplified and approximated form of Reynolds number and is commonly used in conjunction with turbine flowmeters to express operating conditions at constant temperature.

Temperature changes affect the diameter of the turbine meter as well as the viscosity and density of the fluid. To account for the temperature changes in the meter and the fluid, Strouhal number and Reynolds number are used to characterize the performance of a turbine flowmeter. Strouhal number incorporates the *K* factor as well as the change in diameter of the meter as a function of temperature (which, in turn, changes the *K* factor). Reynolds number includes changes in flow rate (velocity), kinematic viscosity, and temperature as it affects the meter diameter (velocity) and fluid viscosity.

Strouhal number as a function of Reynolds number is the only completely correct method of characterizing the performance of the turbine flowmeter. All of the more commonly used methods are only approximately correct because they eliminate one or more of the secondary variables.

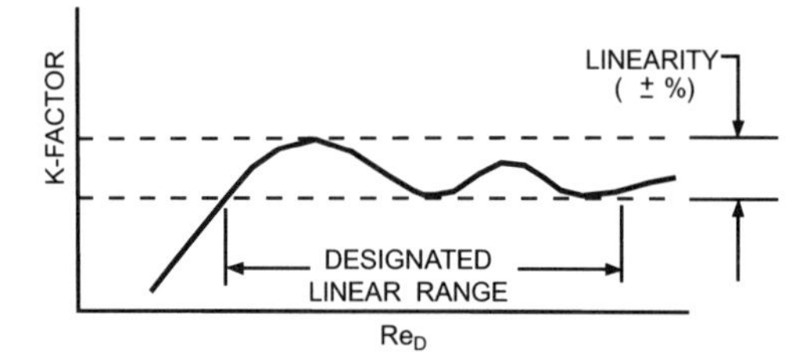

Figure 17-9. K Factor vs. Reynolds Number for a Typical Turbine Flowmeter

Sizing

Turbine flowmeters are manufactured in a wide variety of sizes and flow ranges. The smallest common meters will measure liquid flow as low as 0.001 gpm or gas flows as low as 0.001 cfm. Meters designed to measure tens of thousands of gallons per minute are also available. Most turbine flowmeters have a linear flow range of at least 10:1 from maximum rated flow, with many being linear to 20:1. Most meters are repeatable over flow ranges of 100:1 but at a greatly increased nonlinearity. Some manufacturers claim operating ranges in excess of these guidelines, but it would be advisable to investigate such claims extensively before accepting them. Extended ranges beyond these limits may be possible at some operating conditions but probably not at the wide variety of conditions common for the above ranges.

The most common error in sizing a flowmeter is to select a 2-inch meter for use in a 2-inch line or other equivalent sizing.

Meters are designed to operate in a reasonably high velocity flow stream. The velocities best suited for the meter may not be best suited to transport the media. With properly sized piping, if the piping runs are short, the pipe size can be the same as the meter size; however, if the piping runs are long, the pipe size may be larger than the meter size. In certain aerospace environments, piping velocities selected to minimize weight may be too high for proper meter operation. In any event, the flow ranges expected in the piping system must be matched to the meter's flow range whether or not the line size and the resulting meter size are equal.

The second most important consideration in selection of a meter size is the resulting pressure differential. Meter pressure differential is often not considered in selecting a flowmeter until after the meter is installed and the system performance is not satisfactory.

Most manufacturers rate the pressure differential at maximum normal rated flow for water or light hydrocarbon service. The differential pressure is a function of the square of the flow; thus, if the meter is used in an extended range, the resulting differential pressure loss will be significantly greater at the higher flows. The density and viscosity of the fluid also will affect the resulting pressure drop across the meter. If data from a specific manufacturer is not available at conditions other than the one specified, the following formula will yield a reasonable approximation of the pressure loss at other conditions:

$$\Delta P_1 = 1.2\, \Delta P\, \nu^{0.25}\, \mathrm{SG} \left(\frac{Q_1}{Q} \right)^2 \qquad (17\text{-}2)$$

where:

ΔP_1 = pressure loss with new fluid in psid
ΔP = pressure loss in MIL–C–7024B (from mfg. data)
ν = kinematic viscosity of fluid in centistokes
SG = specific gravity of new fluid
Q = flow rate at pressure loss ΔP
Q_1 = flow rate at which new pressure loss is required

It is often found that the resulting differential pressure at a particular flow condition cannot be tolerated in a system. Selecting the next larger size meter will lower the resulting differential pressure and may still function. However, the larger meter will then probably be operating at the lower end of its flow range and may be somewhat more nonlinear.

The operating pressure and temperature for the meter must be considered when sizing or selecting a meter (see Figure 17-10). Turbine flowmeters work well at both low and high pressures, but the end connection must be suitable for the pressure and temperature service. Turbine meters are suitable for temperature ranges from cryogenic service to about 650°C (1200°F) depending on the bearing selection.

The type of pickoff selected must also be considered. For most applications the magnetic pickoff is suitable. However, for meters below 1½ inch in extended low ranges or for gas service, a carrier frequency pickoff may be desirable to improve the linearity and range of the meter. High temperature pickoffs are available for service above about 200°C (350°F), the typical upper limit for a standard pickoff.

Bearing selection is the next consideration that should be made when sizing and selecting a meter. Ball bearings are suitable for clean liquid or dry gas service compatible with 440C stainless steel. High temperature ball bearings are available for service up to 400°C (750°F) for both gas and liquid. Journal bearings of a variety of materials are available for corrosive fluids or fluids containing some particulates, but flow range and linearity are sacrificed when using journal bearings. Pivot bearings or other specialty bearings are available for small meters where extremely low drag is required.

Calibration

The best of meter designs, the most appropriately sized meter, and the most properly installed meter are all but worthless if the calibration of the meter is not also appropriate and correct.

Most users familiar with turbine flowmeters are aware of the meter designs that suit their particular service and are equally aware of the general guidelines for the meter's installation. However, most users are unaware of the proper calibration method for the particular application.

The ideal calibration for any meter would be traceable to a NIST primary calibration system at the exact temperature and pressure and on the exact fluid

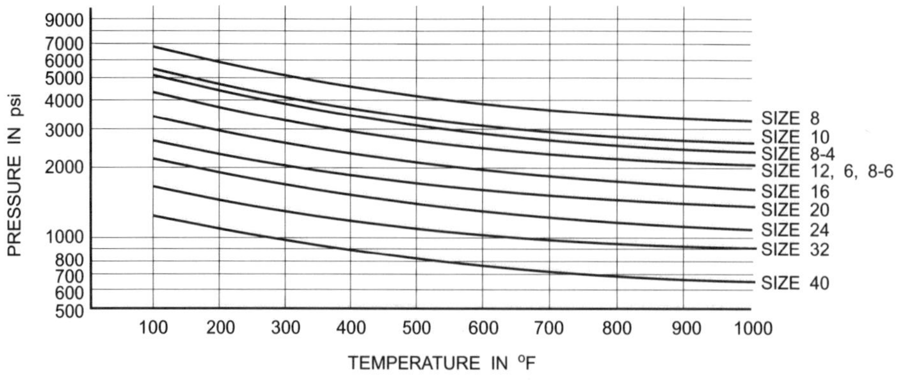

NOTE: The pressure ratings shown reflect a 6:1 factor of safety. For operation at pressure approaching the upper limits shown, care should be exercised in the selection of adequately pressure rated mating female connectors.

Figure 17-10. Pressure-Temperature Ratings of a Turbine Flowmeter
(Courtesy of Shute & Koerting Div., Ketema, Cox Instruments)

that the meter will be used to measure. Unfortunately, this is not always possible or practical. In fact, most of the time it is not practical.

Also, unfortunately, the turbine flowmeter, like all other meters, is not a perfect device; therefore, it is subject to influences other than flow. Every type of meter has a set of parameters that can be used to correlate these influences. The correct correlation parameter for the turbine flowmeter is Strouhal number vs. Reynolds number.

Frequency/Flow Rate

Turbine flowmeter calibration data are typically displayed or presented as flow rate as a function of frequency. This presentation is satisfactory as long as the meter is calibrated in exactly the same fluid and at the same temperature (and pressure for gas applications) as the application conditions. It does, however, severely limit the resolution to which the data can be determined.

K Factor/Frequency

A better and more commonly used method to present the data is *K* factor as a function of frequency. *K* factor is defined as frequency/flow rate (pulses/gallon). It is sometimes presented as the meter factor, flow rate/frequency (gallons/pulse). This dimensionalized presentation increases the resolution of the data, but it is still limited to the kinematic viscosity (and temperature and pressure) as in the calibration. However, for applications in which the meter is calibrated at the same conditions as the application, it is a good presentation of the data.

Universal Viscosity Curve

A commonly used method of presenting the data when the meter is to be used over a wide range of viscosity is called the "universal viscosity curve." The universal viscosity calibration consists of a series of calibrations at various kinematic viscosities covering the range of interest. These various viscosities are usually achieved by changing the fluid in which each separate calibration is accomplished to a fluid of the desired kinematic viscosity. If the data from all of the calibrations is presented as *K* factor as a function of frequency/kinematic viscosity (an abbreviated form of Reynolds number), a continuum of the data will follow a single line if certain limitations of the range are observed.

Figure 17-11 shows a typical calibration curve for a turbine flowmeter at 0.8 centistoke viscosity over a flow range of 100:1. The data is plotted as *K* factor vs. frequency/viscosity (Reynolds number). Superimposed is a second calibration of the same meter at a viscosity of 9.6 centistokes over the same flow range. Note that the top 20:1 flow range of the higher viscosity follows a continuum of the data from the top 20:1 of the lower viscosity. A third calibration of the same meter calibrated at 94.08 centistokes is added to the other calibrations. Although it can't be seen by the curve, only the upper 20:1 of the flow range follows a continuum from the other data. The lower portion the extended range in each case tails off below the continuum. It is possible to express the normal range of this data as a polynomial equation. In this manner, the meter can be used over a wide range of viscosities.

The conditions over which the universal viscosity calibration are valid are within the normal operating range of the meter and at viscosities below 100 centistokes. They cannot be used at higher viscosities and over flow ranges exceeding the normal range. These calibrations are also limited to operating at the temperature of the calibration.

A word of caution is in order. For most meters the normal range is 10:1. It is a common error to try to use the meter in a "universal viscosity" manner over a flow range of 100:1. This does not work and should be avoided! Extended range calibrations at various viscosities that will not follow the universal viscosity curve concept are called "multiple viscosity calibrations" and are valid only at the discrete viscosities used in the calibration. Manufacturers will sell calibrations with the meters at multiple viscosities over a 100:1 flow range if requested. The data for the extended portion of the range is valid only at the exact viscosity and at the exact temperature of the calibration. Interpolation of the data is not very practical and should be avoided.

The fallacy in universal and multiple viscosity calibrations is that the various calibrations at different viscosities are usually achieved by changing the fluid to another that has a different viscosity. The calibration is usually accomplished at room temperature for all viscosities. What is ultimately achieved is a series of calibrations for a fluid that has various viscosities at the same temperature. A fluid with such a characteristic is not found in the real world.

All known fluids have only one viscosity at one temperature (and pressure). The effect of temperature has been ignored in this type of data presentation. The resulting error for a typical turbine meter amounts to about 0.3%/100°F. Those who think they have obtained a 0.05% calibration are mistaken. They may have achieved this performance over a very limited temperature range, but at all other temperatures they have not achieved the desired performance.

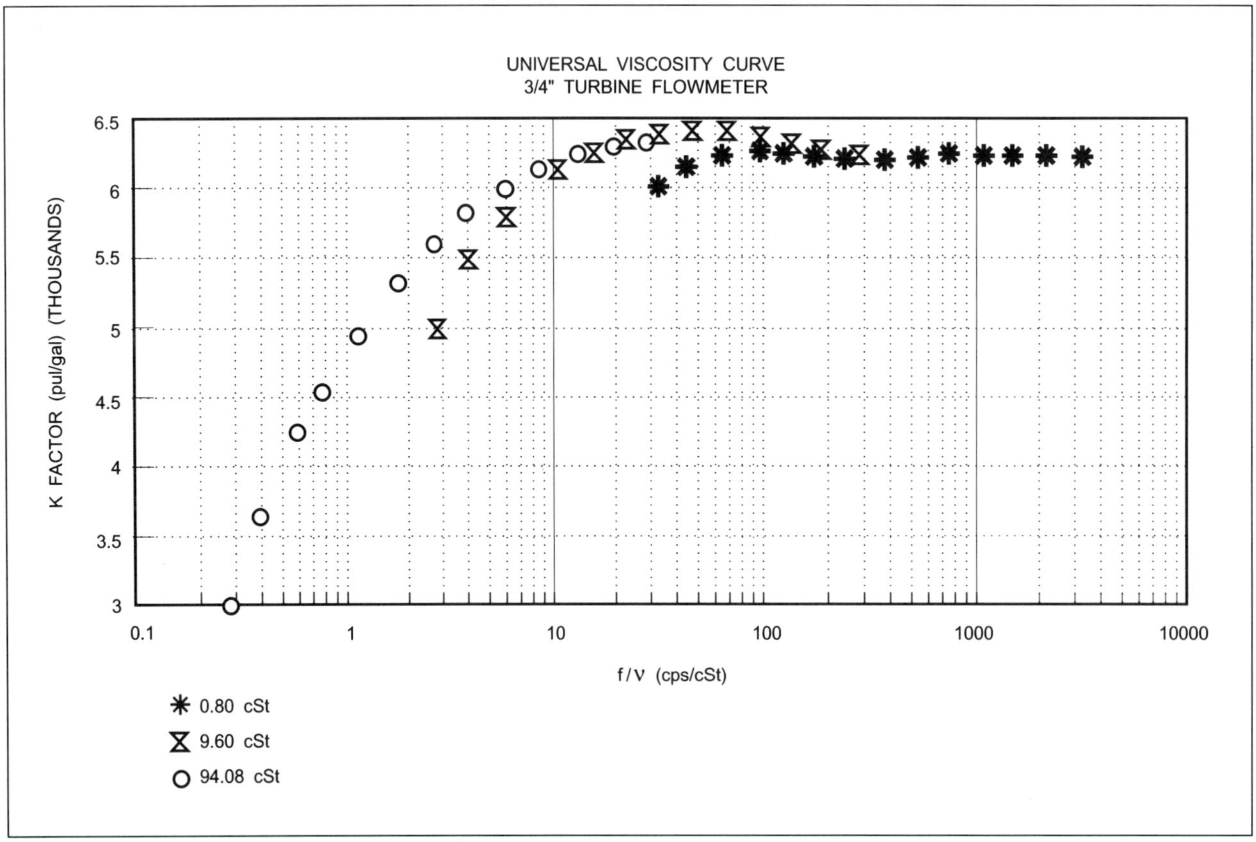

Figure 17-11. Three Fluid Universal Viscosity Calibration
(Courtesy of EG&G Flow Technology)

K Factor/Reynolds Number

A still better way to present the data is with the *K* factor as a function of Reynolds number. This method of correlation incorporates the effect of temperature on the meter body as it effects the velocity of the fluid but does not incorporate the effects of temperature on the meter body as it effects the *K* factor. This correlation is perhaps the most commonly used method of presenting turbine flowmeter performance.

Strouhal Number/Reynolds Number

The best way to present the data for a turbine flowmeter is Strouhal number as a function of Reynolds number, i.e., through the use of two dimensionless parameters. It is by the use of these two parameters that potential error is left in the correlation. The St vs. Re presentation takes into account all of the secondary effects to which the meter is sensitive. This presentation or correlation is correct for both liquids and gases. It is especially important for gas calibrations, since the density and kinematic viscosity are a function of both temperature and pressure. As stated above, the use of frequency/kinematic viscosity is an abbreviated form of Reynolds number that ignores the effect of temperature on the meter body. *K* factor also ignores the effects of changing temperature on the meter body, since the meter will change diameter when the temperature changes. The use of Strouhal number instead of simple *K* factor will account for this temperature effect. The same limitations that should be applied to the universal viscosity curves (UVC) must be observed with a Strouhal vs. Reynolds correlation, as the meters will follow this correlation only within those limitations.

Strouhal number and Reynolds number in their basic forms are difficult to use since they both contain a form of the variable being determined by the measurement. Strouhal number is defined as:

$$St = \frac{fD}{\bar{v}} \qquad (17\text{-}3)$$

where:

f = meter output frequency
D = reference diameter of meter
$\bar{v}$ = velocity of the fluid through the reference diameter

and Reynolds number is defined as:

$$Re_D = \frac{D\bar{v}\rho}{\mu} \qquad (17\text{-}4)$$

where:

ρ = density of the fluid
μ = absolute viscosity of the fluid

Kinematic viscosity, the ratio of the absolute viscosity to density, is:

$$\nu = \frac{\mu}{\rho} \qquad (17\text{-}5)$$

Thus:

$$Re_D = \frac{D\bar{v}}{\nu} \qquad (17\text{-}6)$$

where ν = kinematic viscosity.

Since both Re_D and St contain velocity as a term and the velocity is directly proportional to flow rate, it is difficult to use these two forms of these dimensionless parameters. More useable forms of these parameters can be obtained by the following means.

K factor was earlier defined as:

$$K = \frac{f}{Q} \qquad \left(\frac{\text{Frequency}}{\text{Flow rate}}\right)$$

but $Q = A\bar{v}$, so Q is proportional to $D^2\bar{v}$. Then,

$$K = \frac{f}{\bar{v}D^2} \tag{17-7}$$

multiplying by D^3 will yield the dimensionless parameter Strouhal number:

$$\text{St} = \frac{f \cdot D^3}{\bar{v}D^2} = \frac{fD}{\bar{v}} = KD^3 \tag{17-8}$$

The diameter of the meter will change as a function of temperature as:

$$D = D_0\,[1 + \alpha\,(T - T_0)\,] \tag{17-9}$$

where:

D_0 = meter diameter at reference temperature
α = linear coefficient of expansion of meter body
T_0 = reference temperature

This equation makes the assumption that the materials of construction of the rotor and the housing are identical, which for almost all turbine flowmeters is not the case. However, for most applications this equation will yield acceptable results. The only way to improve on this equation (other than to perform a calibration at the same temperature) is to perform extensive calibrations on similar meters over a wide temperature range. It is not practical to perform calibrations at small temperature differences and extrapolate the results. The accuracy of even the best calibrations do not permit this extrapolation with better certainty than the use of the Equation (17-9).

Most turbine flowmeter housings are manufactured from austenitic stainless steel—303, 304, or 316 ss. The coefficient of linear thermal expansion for these materials is approximately 9.6×10^{-6} %/°F. The resulting correction is therefore 0.003%/°F, 0.03%/10°C, or 0.3%/100°F. It can be seen that the error (for not incorporating the correction) for a few degrees in temperature may not be significant. However, to ignore this correction for a 100°F difference in temperature introduces significant bias error into the calibration.

Substituting Equation (17-9) into Equation (17-8) yields:

$$\text{St} = \frac{fD}{\bar{v}} = K\{D_0\,[1 + \alpha\,(T - T_0)\,]\}^3 \cong KD_0^3\,[1 + 3\alpha\,(T - T_0)\,] \tag{17-10}$$

Any of the above forms or proportionalities of St may be used. The author prefers the KD^3 form for its simplicity in the presentation, keeping in mind that the diameter must be corrected by the $[1 + \alpha\,(T - T_0)]$ term.

If the dimensionless parameter Reynolds number is multiplied by the other dimensionless parameter St, a very usable form of a dimensionalized dimensionless term is obtained:

$$\text{Re}_\text{D} \cdot \text{St} = \frac{vD}{\nu} \cdot \frac{fD}{v} = \frac{fD^2}{\nu} \tag{17-11}$$

This dimensionless parameter is equivalent to the UVC parameter $\frac{f}{\nu} \cdot D^2$. The addition of the D^2 term transforms the erroneous $\frac{f}{\nu}$ correlation into a correct correlation. The correct correlation parameter may be used in any of the following forms:

$$Re_D \cdot St = \frac{fD^2}{\nu} = \frac{f}{\nu}\left\{D_0\left[1 + \alpha\left(T - T_0\right)\right]\right\}^2 \cong \frac{fD_0^2}{\nu}\left[1 + 2\alpha\left(T - T_0\right)\right] \qquad (17\text{-}12)$$

Here again, the author prefers the $\frac{fD^2}{\nu}$ form for its simplicity in presentation. The diameter must again be corrected by the temperature term.

In addition to the temperature correction terms presented here, there are similar pressure correction terms that have to do with the deformation of the meter body due to pressure. However, most meter bodies are very thick walled structures and this term is insignificant. The effect of pressure on the bulk modulus of the liquid is likewise ignored.

Figure 17-12 shows the results of a meter calibrated on a specific fluid (in this case, JP-4) at temperatures of 50, 80, and 110°F. The viscosity of the fuel changed from 0.92 to 1.36 centistokes. Note that all of the calibrations form a continuum of data that could be expressed in a polynomial form to be used at any conditions in that range. Note the very expanded scale of the *K* factor term. The data scatter over most of the plot is less than ± 0.04% of reading. Also note that, below the normal 10:1 turndown of the meter, the data becomes a little more ragged, indicating that the meter does not follow this concept over a larger turndown. This same data, when plotted in the conventional *K* factor vs. frequency/viscosity basis, showed as much as ± 0.1% of reading scatter due to the fact that the temperature growth of the meter body is ignored in that type of correlation.

As can be seen in the above equations, the term that most significantly affects the calibration and operation of the turbine flowmeter is kinematic viscosity (not to be confused with absolute viscosity).

A turbine flowmeter should be calibrated at the same kinematic viscosity at which it will be operated. This is true for both fluid states, liquid and gas.

It is common practice to ignore the viscosity of gases and equate the density during gas calibrations, most likely because the absolute viscosities of gases are low, but it must be kept in mind that the densities of gases are equally low; therefore, the kinematic viscosities of gases may not be low.

It is desirable but not necessary to calibrate a meter at the same temperature at which it will be operated in service, as the temperature effects may be compensated for analytically as shown above. If the kinematic viscosity and temperature are the same during the calibration and in service, the correct flow rate will be measured (assuming a properly sized and installed meter). If they are not calibrated at the correct kinematic viscosity, the meter is then calibrated in a Reynolds number range that is not correct for the application. This change in Reynolds number range can create significant errors in the measurement.

If the meter has been calibrated in the proper Reynolds number range, i.e., at the correct kinematic viscosity, the calibration may be adjusted or corrected to other temperatures of operation. They may not, however, be adjusted to other Reynolds number ranges (other kinematic viscosities).

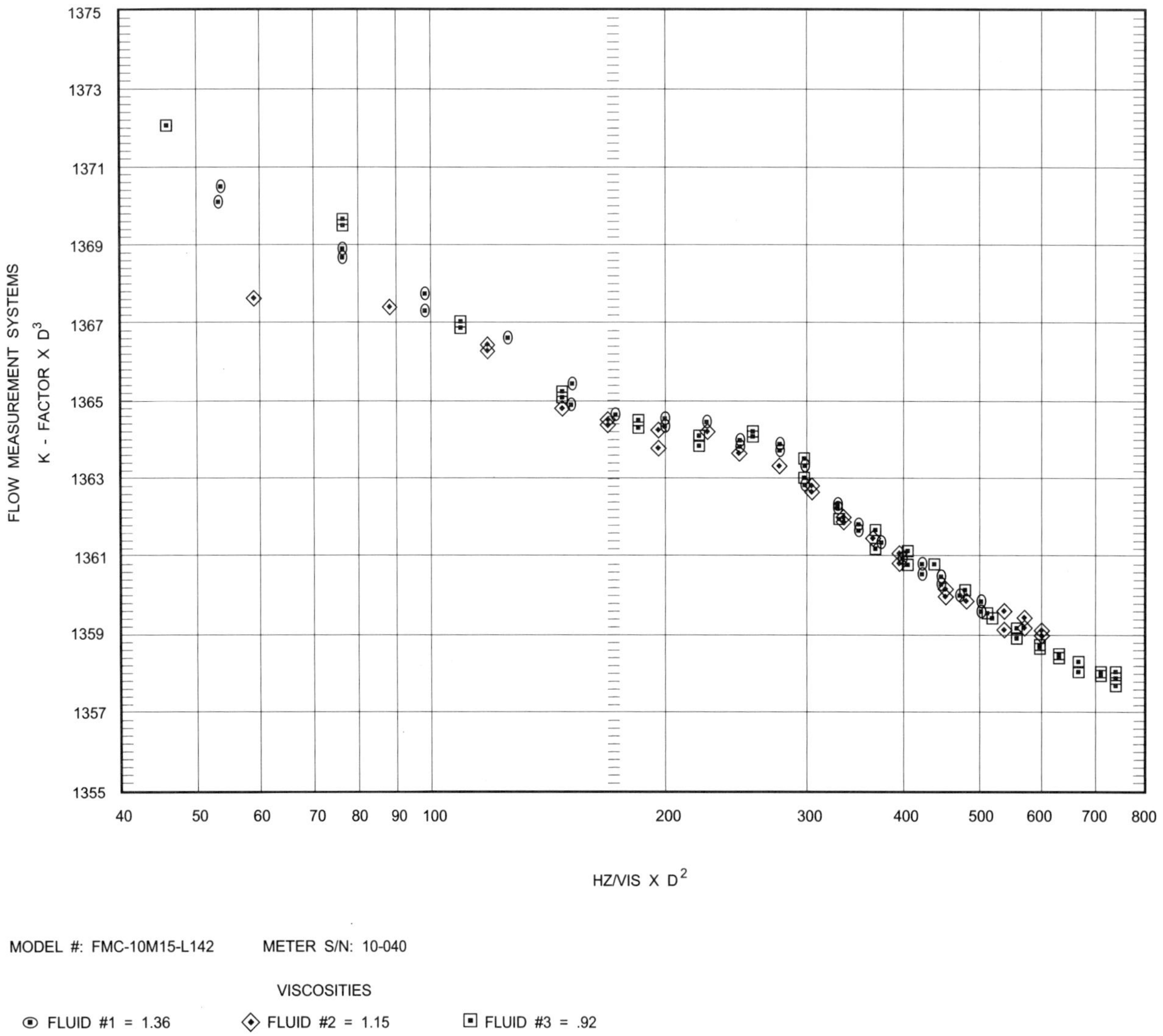

Figure 17-12. Meter Calibration Results
(Courtesy of NIST)

Viscosity

The absolute viscosity of gases is a function of temperature and only a weak function of pressure. The density of liquids is only a weak function of pressure, but the density of gas is a strong function of pressure. These facts can be used to obtain a desired kinematic viscosity of a gas by varying the density of the gas.

The following computations will illustrate this concept. The absolute viscosity of air at standard conditions is 1.21 $(10)^{-5}$lb/ft-sec and the density is well known to be 0.0760 lb/ft^3, yielding a kinematic viscosity of 1.59$(10)^{-4}$ft^2/sec (14.8 centistokes). The absolute viscosity of water at 68°F is 6.51$(10)^{-4}$lb/ft-sec, and the density is 62.1 lb/ft^3 yielding a kinematic viscosity of 1.04$(10)^{-5}$ft^2/sec (0.97 centistoke) or 15.2 times less (yes, less!) than air at STP. If the density of the air is increased to 15.2 atmospheres and the temperature is maintained at standard temperature (68°F), the density is increased about 15.2 times, and the resulting kinematic viscosity is decreased 15.2 times to the same viscosity as water. That's correct: the kinematic viscosity of air at about 15.2 atmospheres (compressibility has been ignored to make a point) is the same as water.

The conclusion from these facts and the above statement that, if the Reynolds number is correct, a valid calibration result would lead one to the conclusion that a water calibration could be simulated with air at about 15 atmospheres. That is exactly what could be done! Empirical verification of this concept has been accomplished. The only difference is, since the density of the water and air are so different at the conditions where the kinematic viscosities are equal, the calibration at the very low end of the meter operating range may be lower for the air than that of the water, since insufficient momentum exists with the air to totally overcome the bearing drag. One would not normally need to simulate a water calibration with air, but this concept is a powerful tool in setting up other calibration schemes. It is not always possible to calibrate a flowmeter in the exact service conditions. The use of surrogate fluids may be used so long as the above described conditions of St and Re_D are maintained.

Calibration Procedures

One of the following procedures should be used in order to best simulate the correct fluid characteristics.

(1) Find a fluid that can be used in the calibration process that has the same kinematic viscosity at the operating temperature, and perform the calibration at the operating temperature. No corrections to the data are required when using this procedure, and the calibration will be correct. A good technique that can be used to obtain the desired kinematic viscosity is to blend soluble fluids; Stoddard solvent and a very viscous oil are commonly used.

A turbine flowmeter used on hydrogen gas at 200 psia and 32°C (90°F) is to be calibrated on a bell prover. Determine the best calibration scheme.

The absolute viscosity of hydrogen is 0.0008936 centipoise at 32°C (90°F), and the density at the operating condition is 0.001100 gr/cc. The resulting kinematic is then 0.8122 centistoke. Air at 21°C (70°F) and 275 psia will have this same kinematic viscosity. A throttling valve must be placed downstream of the meter to keep the meter at 275 psia and the bell at atmospheric pressure. Since the calibration temperature and operating temperature are so close together, the temperature correction does not have to be made for most practical applications. To be completely correct, a correction reducing the *K* factors (St) by 0.057% and increasing the frequencies (Re_D) 0.038% should be made.

(2) If a calibration at the correct operating temperature cannot be performed, then find (or blend) a fluid that has the correct kinematic viscosity at the temperature at which the calibration will be performed. Perform the calibration at these conditions and then correct the calibration data to the operating temperature conditions using any form of the St and Re_D equations above. Keep in mind that a turbine flowmeter is a volumetric device and that "a gallon is a gallon" regardless of the temperature. If performing the calibration on a mass calibrator, then correction for the change in density of the fluid between the two temperatures must be performed.

The flow rate of an oil is to be measured at 180°F where the kinematic viscosity of the fluid is 2.5 centistokes. The calibration laboratory does not have the capability to calibrate the meter at elevated temperatures.

A calibration can be achieved by simply blending a fluid to have a kinematic viscosity of 2.5 centistokes at room temperature, calibrate the meter in this surrogate fluid, and then correct the data to the desired temperature using the terms:

$$\mathrm{St} = KD_o^3\,[1 + 3\alpha\,(T - T_o)\,] \qquad (17\text{-}10)$$

and

$$\mathrm{Re_D}\,\frac{f}{\nu}\cdot D_o^2\,[1 + 2\alpha\,(T - T_o)\,] \qquad (17\text{-}12)$$

Thus, a correction of 1.008064, or 0.8064% reduction is appropriate for the *K* factor (St) and 1.002112 or 0.2112% increase for frequency (Re_D).

If the same meter were to be used in the same fluid over a temperature range of ambient to 82°C (180°F) where the viscosity of the fluid changes from 100 centistokes at 21°C (70°F) to the above conditions, then the meter may be calibrated at 100 centistokes, 30 centistokes, 10 centistokes, and 2.5 centistokes. The selection of these intermediate viscosities is somewhat arbitrary, but it should be done in a geometric fashion, allowing sufficient overlap of the data. In this case, approximately a 3:1 ratio in viscosity was chosen each time, allowing for plenty of overlap of each data set. A 4:1 ratio was allowed at the upper end of the Reynolds number range where the meter will be most linear. After each individual calibration at room temperature, each set of data is independently adjusted in the above manner and all sets then jointly presented in a St vs. Re, (KD^3 vs. $\frac{fD^2}{\nu}$) form. The meter is now capable of correctly measuring flow rates within its normal flow range at any viscosity between 100 and 2.5 centistokes for the desired fluid. It will not accurately measure flow rates for any fluid within that viscosity range; only a fluid that has the same temperature vs. kinematic viscosity characteristic as the desired intended fluid can be accurately measured using this calibration.

(3) Liquids that have kinematic viscosities below 0.6 centistoke that don't also have a vapor pressure above atmospheric pressure are difficult to find. Some calibration systems are equipped to handle such fluids. However, it may be possible to simulate the same calibration using air.

Refrigerant freons have kinematic viscosities in the range of 0.2 to 0.4 centistoke at room temperature; however, the vapor pressure of these fluids is in the range of 100 psig. Some calibration systems are equipped to maintain the required pressure in order to maintain the fluid in the liquid state. These calibrators are more difficult to use than ordinary calibrators and require special equipment.

Unless a sufficient calibration quantity is available to be performed, it is not usually practical to maintain such equipment.

An alternative procedure to calibrate a turbine flowmeter for the same service would be to use high pressure air. The following procedures for air calibrations may be used.

(4) Determine the kinematic viscosity of the fluid at the service conditions and compute the pressure required to obtain the same kinematic viscosity in the calibration gas. If the calibration temperature and the service temperature are different, the temperature correction must be applied using the St and Re equations as in procedure 2.

The absolute viscosity of dense phase ethylene gas is 0.04 centipoise at 32°C (90°F) and 1125 psig. The specific volume is given as 0.075 ft^3/lb. It is possible to calibrate this meter on air at 21°C (70°F). The density of the application fluid is calculated to be 13.3 lb/ft^3 or 0.213 gr/cc. Dividing the density into the absolute viscosity yields a kinematic viscosity of 0.188 centistoke at 32°C (90°F) and 1125 psig. To use air as a surrogate fluid, it is necessary to determine what pressure of air at 21°C (70°F) has a kinematic viscosity of 0.188 centistoke. Referring to NBS Circular 564 or other appropriate tables of air properties, it is found that air at 1180 psig and 21°C (70°F) has the desired kinematic viscosity. A calibration at this condition will yield a proper calibration for the desired service. Once the calibration at 21°C (70°F) is obtained, the data must be corrected to the 32°C (90°F) desired operation condition.

If it had been required to operate the meter over a range of temperatures and pressures, then the following additional steps are required. One must first determine the range in absolute viscosity that will occur for the gas conditions desired. The absolute viscosity will change as a function of temperature. For most applications, the change in absolute viscosity as a function of pressure can be ignored. Determine the range in density that will occur in the application fluid (gas) over the range in temperature and pressure of interest. The change in density as a function of pressure *may not* be ignored for this calculation. If the range of pressure and temperature is not too great, the Ideal Gas Law will yield acceptable results for this calculation. Once the range in absolute viscosity and density are determined, the resulting range in kinematic viscosity is determined by dividing the highest absolute viscosity by the lowest density and the lowest absolute viscosity by the highest density (assuming, of course, that the conditions of this cross division are applicable). The resulting range in kinematic viscosity must be addressed during the calibration. Several air pressures may be required to cover the desired range in kinematic viscosity. Keep in mind that a change in kinematic viscosity of the air (surrogate fluid) is achieved at constant temperature (constant absolute viscosity) by changing the pressure (density) and ignoring that a small change in absolute viscosity is occurring with the change in pressure. Once the various separate calibrations are accomplished, the data for each calibration must be corrected to the appropriate temperature for the same kinematic viscosity of the application gas using the same technique as above.

(5) Liquid meters may be calibrated in gas using the same technique of determining the pressure of a calibration gas (usually air) that yields the same kinematic viscosity. Again, the temperature correction must be used if the two conditions are not at the same temperature.

Steam is a fluid that readily adapts to surrogate calibrations for a turbine flowmeter. The kinematic viscosity of steam ranges from 1.0 $(10)^{-5}$ to 1.0 $(10)^{-3}$ ft^2/sec in the normal range of use depending on the temperature, pressure, and quality. The upper half of this range can be simulated with air at pressures between atmospheric and 225 psig. The lower half of the viscosity range can be simulated in a partial vacuum or with a high viscosity gas at low pressures. Helium is a good choice for small flow rates or a vacuum pump with air is a good choice for higher flow rates. As an example, saturated steam at 160°C (320°F) can be simulated with air at 50 psig. Data on the kinematic viscosity of steam at various conditions is available in most steam tables. Simply match the operating viscosity with the appropriate air pressure at room temperature. The temperature correction to Re and St must also be made.

(6) If an alternative fluid of equivalent kinematic viscosity cannot be found or is impractical to use, it is possible to calibrate the meter in a fluid in which the kinematic viscosity is close to the desired viscosity. In this case, the frequency range of the calibration is adjusted to compensate for the difference in the viscosities. Keep in mind that the objective of a proper calibration is to match the Reynolds number or the frequency/kinematic viscosity. If the kinematic viscosity of the calibration fluid is 1.5 times greater than desired, then calibrating the meter at 1.5 times its normal operating range will yield the same Reynolds number. In this technique the meter is overspeed and should not be run for very long, but it can be operated for long enough to get good calibration data. If the calibration temperature and the operating temperature are different, then the data must be corrected using the St and Re equations.

Determine an appropriate surrogate fluid for liquid hydrogen at -259°C (-435°F). The density of liquid hydrogen is determined to be 4.84 lb/ft^3, and the kinematic viscosity is determined to be 0.337 · 10^{-5} ft^2/sec. There is no convenient liquid to use that has such a low kinematic viscosity. Since the liquid hydrogen has a low density as well, air will make a suitable surrogate fluid. Air at 21°C (70°F) and 47.2 atmospheres of pressure has the same kinematic viscosity as the liquid hydrogen. It also will have a density of 3.61 lb/ft^3, which is reasonably close to the liquid hydrogen. The air data, once it is collected, must be corrected to the cryogenic temperature. Once this air calibration and correction are accomplished, the meter will yield accurate flow rate measurements in liquid hydrogen service at the specified conditions.

Water would not make a good surrogate fluid for this calibration, even though it is often used, because the Reynolds number obtained in a water calibration would be over 3.3 times too low. It cannot be assumed that the meter will be linear for that great a range change, and it would not be advisable to overspeed the meter by more than three times its normal speed to achieve the correct Reynolds number range.

It is difficult to find a surrogate fluid for liquid oxygen that will provide good calibration. The density and kinematic viscosity change dramatically with temperature and pressure. It is, therefore, often required to perform a liquid and a gas calibration in order to achieve the desired results. A technique of overranging the meter to achieve the desired Reynolds number may be used in some cases.

Liquid oxygen has a density of 75.1 lb/ft^3 and a kinematic viscosity of 0.232 · 10^{-5} ft^2/sec at -195°C (-320°F). Air at 69.9 atmospheres and 21°C (70°F) has the correct kinematic viscosity, but the density of the air is only 5.3 lb/ft^3. While the theory presented would lead one to the conclusion that an air calibration will yield accurate results, this is one case where a pitfall is encountered. The 14:1 reduction in density may cause the calibration to be in error at the low end of the flow range. There may be insufficient driving force in the greatly lower density of the air than in the liquid oxygen to equally overcome the bearing drag in the meter. In a large meter this may not be a problem, but in a small meter it very well could be. This potential error can be overcome by calibrating the meter in two media: water and air. The Reynolds number for the water calibration is mismatched from the liquid oxygen by over 4:1, since the kinematic viscosity of water is $1.07 \cdot 10^{-5}$ ft^2/sec compared to the $0.232 \cdot 10^{-5}$ ft^2/sec for the liquid oxygen. However, since the turndown on the meter is greater than 10:1, the higher frequency portion of a water calibration will overlap the lower frequency portion of the air calibration. Using this technique, the droop in the air calibration at the lower flow rates (frequencies) can be determined. If the two calibrations are coincident, then no droop will occur; if the air calibration falls below the water calibration, then the water data should be used to obtain the correct St in that portion of the curve. The two calibrations should become contiguous at or below 25% flow rate. Figure 17-13 shows this calibration technique graphically.

If the liquid oxygen were to be measured as -212°C (-350°F) instead of the higher temperature as above, an alternative calibration technique can be considered. The kinematic viscosity of LOx at this temperature is $0.474 \cdot 10^{-5}$ ft^2/sec. This higher viscosity is only 2.3 times less than that of water. The Reynolds number obtained using water as the calibration medium may also be increased to match that of LOx by overspeeding the meter (increasing the velocity, flow rate, and resulting Reynolds number). Running the meter at 2.3 times its normal flow rate is obviously not a good practice, but most good meters will stand that much abuse for long enough to get a calibration point or two. After the calibration, it is a good idea to replace the bearing before putting the meter in service. Replacing

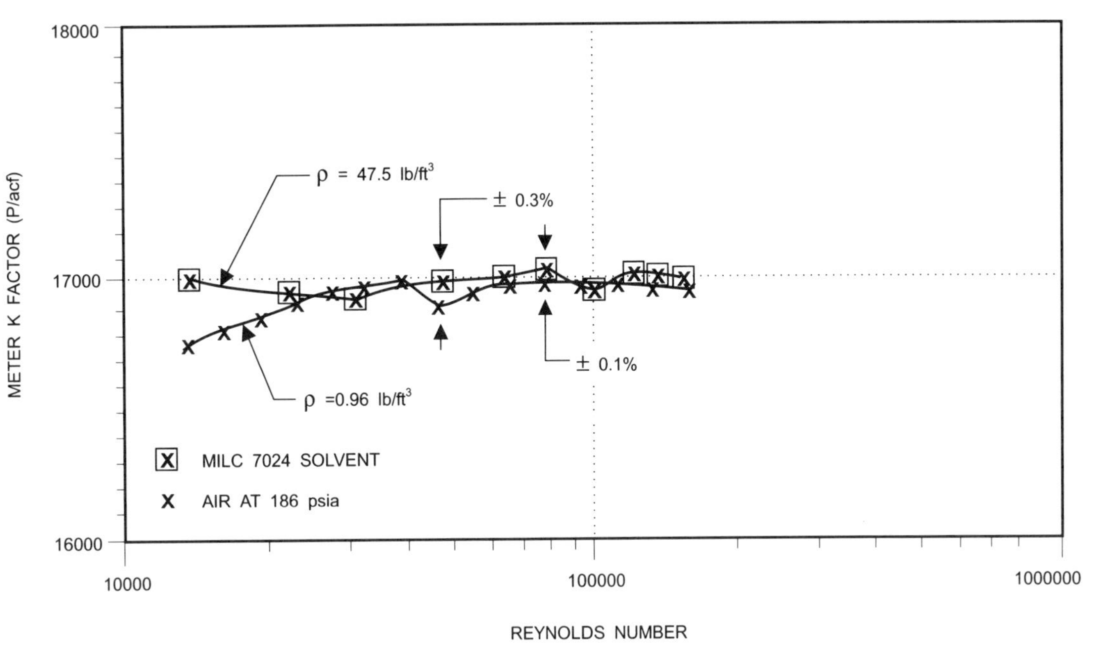

Figure 17-13. Flowmeter Calibration Data for Solvent and Air over the Same Range of Reynolds Number

the bearings after the calibration leads to questions as to whether or not the calibration is affected by this action. However, the effect is probably small, if any, and this technique can be used if no other alternative is available. It is offered as an alternative way to set the desired Reynolds number; in short, any method of matching the Reynolds number is viable. Don't forget that with this technique as well, the data must be corrected for any temperature difference that exists between the calibration temperature and temperature of operation.

Specifications

When purchasing a turbine flowmeter, it is necessary to give as much information as possible to the manufacturer. This is especially important in purchasing an instrument where a large number of options are available. It is not a good idea to order a flowmeter from a manufacturer by a part number or model number unless one is *very* sure what is being ordered. If the user does the meter selection, the manufacturer will usually ship what is ordered, whether it is correct for the application or not. It is best to discuss the application with the manufacturer and jointly select the part number or model number.

The best meter improperly calibrated is nearly worthless.

The more information one gives to the manufacturer, the better the opportunity to get the best meter for the application; and, more important yet, the better the opportunity to get the proper calibration. One of the most important aspects of specifying a meter is to specify directly or have the information available for the manufacturer to select the proper calibration.

A good rule to follow is "Specify the application rather than the meter."

Allow the manufacturer to select the meter to fit the application. The following information is required:

(1) Flow Rate Range

Maximum, minimum, and a nominal rate if applicable. Don't "pad" the numbers; state what is needed. The manufacturer will add whatever allowances are required.

(2) Temperature Range

Maximum and minimum fluid temperatures and ambient external temperature range if other than indoor temperature. If the fluid temperature is above 177°C (350°F), supply more information about the surrounding environment.

(3) Pressure Range

Maximum and minimum pressure in the fluid stream.

Note: For gas applications, the maximum pressure and minimum temperature crossed with the minimum pressure and maximum temperature create the largest change in volume and an accordingly large change in volumetric flow rate. If these crossed conditions are not applicable, be sure to tell the manufacturer, who will most likely do that procedure if not otherwise notified. This may extend the range of the meter beyond its normal range, limit the capability of the system, or erroneously create the need for two different sized flowmeters when one would be sufficient.

(4) Type of Service

The type of fluid media by descriptive name, specification, etc. Specify whether it is a liquid or a gas if not obvious from the other description. Many fluids can be in the liquid or the gas phase—hydrogen, oxygen, petroleum gas, ether, and water (steam), for example. The fluid service is also required to determine the materials of construction and bearing selection. Be specific; don't just say "water service." There are many types of water, and it does make a difference in the material selections. For example, deionized water is a difficult fluid, and river water has particulates in it.

(5) End Fitting Type

Almost any type of fitting is available, so specify what type is desired and the rating of that type if applicable.

(6) Pickoff Type

Several types are available in a number of temperature ranges. This is one subject that it is best to discuss with the manufacturer, unless one is sure what is required. If this decision is left to the manufacturer, the one that best fits the meter for the application will be selected, but it may not interface the outside world as desired.

(7) Power Source

Be sure to tell the manufacturer what type of power is available for the electronics.

(8) Other Information

Explosion-proofing, intrinsically safe, and any special features desired.

(9) Bearing Type

A standard meter is usually supplied with ball bearings. Many other types are available. Here, again, unless one is sure what type to specify, discuss it with the manufacturer. The incorrect type of bearings for an application usually make a disaster of the installation.

(10) Pressure Drop

Unless pressure drop across the meter is a problem, it isn't necessary to specify it. Turbine flowmeters are low pressure drop devices. However, if there is a concern, be sure to let the manufacturer know what the requirements are. Most manufacturers can provide meters with lower than normal pressure drops if required.

(11) Materials of Construction

Most meters are constructed from stainless steel. If other materials or specific forms of stainless steel are required, be sure to specify those materials. Avoid statements such as "All 304 stainless steel," as this is impossible. The rotor must be a magnetic material (usually 430 stainless steel or 17-4 Ph stainless steel or with a magnetic material imbedded in it), and the bearing material cannot be 304 stainless steel; ball bearings are made of 440C stainless steel as described previously.

(12) Fluid Properties

Specify the absolute viscosity and density and/or the kinematic viscosity of the fluid, as well as the range of this parameter. This is very important in determining the proper calibration.

(13) Units of Flow Required

Specify the units in which the flow rate is desired to be read—gpm, cc/sec, scfm, acfm, lb/hr, etc. and the standard pressure and standard temperature (if applicable).

(14) Electronic Output Type

Specify the output expected from the electronics; if visual, what type; analog, 0-5 V DC, 4-20 mA, etc.; totalization or rate only; RS-232; etc.

(15) Electronic Enclosure Type

Specify the type of enclosure desired for the electronics—panel mount, wall mount, meter mount; explosion-proof, weather-proof; etc.

Performance Specifications

If specific performance is of concern, then specify what is desired.

(1) Linearity

± 0.5% of reading is standard with most manufacturers for liquid service and ± 1% of FS is standard for gas over the normal range of the meter. Better linearity is available if required. If the meter is to be connected to microprocessor-based electronics, linearity isn't too important because the microprocessor can compensate for nonlinearity. If the meter is connected to analog electronics, linearity is important.

(2) Repeatability

Most turbine flowmeters possess excellent repeatability, so it isn't usually necessary to specify it. If one desires to specify it, ± 0.05% of reading for liquid and ± 0.10% of reading for gas.

(3) Accuracy

Accuracy has more to do with the calibration than it does with the meter. Once the linearity and repeatability of the meter are specified, only calibration errors are left to contribute to accuracy. If the calibration scheme is not correct for the application, the accuracy has only to do with the accuracy in the calibration scheme. It will be of value and will demand the manufacturer specify the uncertainty of the meter in the application fluid over the range of specified conditions. This technique transfers the burden of providing a proper calibration and makes sure the manufacturer provides the user with what is needed, not merely the cheapest possible meter to be competitive with the "other guy's" equipment and calibration. If large price differences occur from one manufacturer to another, it is time to start asking questions or get some professional help. It probably means that one or more of them don't understand the requirement. Uncertainty of measurement in the application fluid is possible to ± 0.15% of reading for most liquids and ± 0.25% of reading for most gas applications, but these levels are not easily attained. Uncertainties of ± 0.5% of reading for liquid and ± 1.0% of reading for gas are more commonly attained. If the calibration is performed at an inappropriate Reynolds number, the measurement uncertainty can be in error by several percent up to tens of percent for a meter that is repeating to ± 0.05% and is linear in the normal range to ± 0.5%.

Installation

Inspection

The flowmeter and associated electronics should be carefully unpacked and inspected to verify that no damage has occurred to the instrument during shipping and that all packaging materials or other debris are removed from the meter internals. Check the rotation of the rotor by lightly blowing through the meter. It should not be necessary to blow hard through the meter to make it spin rapidly. In fact, it is relatively easy to overspeed a meter by blowing through it. If the meter does not spin freely with normal blowing, it may have bad bearings or other damage. Check with the manufacturer before installing the meter in service.

Inspect the electronics visually for damage and perhaps connect the instrument to a power source and verify that all functions appear to work and the indication reads "zero." If the meter is equipped with a carrier frequency pickoff, the meter must be connected to the electronics to obtain a signal.

Never use an air hose or other high pressure jetting hose to spin a meter! This is probably the easiest way to ruin the bearings in a meter.

Mechanical Connections

The most sensitive external influence on a turbine flowmeter is swirl in the pipeline or other plumbing. Common practice is to specify 10 pipe diameters upstream of a meter; recent work in pipe flow profiles by NIST has shown that swirl can exist in a pipe for more than 100 pipe diameters. Velocity profile disturbances return to normal in 10 pipe diameters in most cases, but the swirl will not.

It is a recommended practice to install flow straighteners upstream of a meter. Disturbances upstream of a meter from pumps, valves, out-of-plane elbows, etc., may require more than the minimum straightener section.

A flow straightener section is not just a straight piece of pipe that is 10 diameters long. It will also contain a series of straightening vanes or tubes. A straightener conforming to industry standards such as ASME, ANSI/ISA-RP31.1, API 2534/ANSI Z11.299, API: RP550, etc., should be used. There are many designs of flow straighteners, and all have advantages and disadvantages in terms of cost, pressure drop, and effectiveness. A good rule of thumb is that the swirl is easiest to eliminate as close to the source as possible.

The flowmeter should be installed in accordance with the directional arrow on the meter. Some meters are symmetrical in design, and the only way to tell which end is upstream is to observe the arrow. Other meters are not symmetrical in design, but the user must be familiar with the meter to physically observe the upstream end.

All meters should have a direction arrow on the meter. OBSERVE IT! This may sound obvious, but many meters are installed backwards and most meters installed backwards will read flow, but not correctly.

For cases in which space is a problem and a meter must be installed close to a large flow disturbance, it may be necessary or advisable to install a screen or perforated plate to help restore the normal turbulent or laminar profile. A number of acceptable designs are available to minimize pressure drop while restoring the profile.

Pulsating flow is difficult to measure accurately. The turbine flowmeter is a very fast responding meter, but it still may not properly respond to pulsations in the flow. Best performance is achieved in a smooth flow stream without pulsations.

Pulsations should be kept to less than 10% of the flow rate to achieve the best performance from the meter.

It is mandatory that all flow lines be purged prior to installing the meter. Purging will remove pipe dope, metal shavings, Teflon tape, slag, dirt, rags, and other debris that will damage or impair the operation of the meter. After the lines are clean and free of loose debris, the meter may be installed.

Installation of a filter or stainer upstream is a good practice, especially when the cleanliness of the system is not certain. For ½-inch diameter meters, a 10-micron nominal rated filter should be used; for ¾- and 1-inch meters, a 20-micron nominal rated filter should be used; and for larger meters, a 50-micron rated filter is recommended. Manufacturers may have different recommendations, so it is advisable to check with the manufacturer before selecting a filter. For clean fluids it is still a good idea to install a strainer in systems where obstacles can get into the system from pipeline maintenance or other means.

The orientation of the meter installation can affect the performance of the meter. The gravitational forces on the rotor will influence the nature of the load on the bearings and, in turn, the drag on the rotor. It is advisable to install the meter in the attitude in which it was calibrated.

Usually the proper orientation is horizontal with the pickoff vertically up. This is not as important in large meters as it is in small meters. Low flow tangential turbine meters are extremely sensitive to orientation and must be installed in the orientation in which they are calibrated. The best orientation is with the rotor axis vertical. Typically, this is with the meter axis horizontal, the pickoff horizontal, and the name plate vertically up; however, it is advisable to check with the manufacturer for this convention.

It is best to install the control valve in a system downstream of the meter. This arrangement allows the pipe to be filled without allowing a slug of fluid to impact the meter or allowing expanding air to escape from the system to overspeed the meter. It also allows the meter to remain full of fluid when not in use in some applications.

When filling a new system, fill it slowly to avoid overspeeding the meter. A large number of needless meter failures arise from this cause.

It is best to keep the meter full of fluid in the line when not in use. A meter that is left in line without fluid when not in use is subject to corrosion of the bearings. If possible and economically feasible, the flowmeter should be removed from the line, cleaned, and stored if the line will be dry for extended periods. Keep in mind that a turbine flowmeter is a precision instrument, and treat it accordingly.

Torque

Flanged flowmeters should be installed using the recommended torque values for the type of flange or fitting on the meter. A precision meter is not just another piece of pipe and should be treated accordingly. It is not advisable to overtorque the bolts to compensate for out of plane and out of flat conditions on the mating pipes. If a flow straightener section is purchased from the manufacturer with the meter, it will be manufactured so that the mating flanges or fittings are square and in plane with the meter. Overtorquing on the other end of the straightener to line up with out of plane pipes and flanges is not as bad as it is on the meter.

A common practice with MS or other screw fittings is to use the meter body to resist the torque. Use the fitting on the other end—not the meter body! This prevents unsightly marks and other damage to the meter body. Also never use a pipe wrench on the meter body. If installing a pipe thread meter, use two wrenches—one on the fittings adjacent to each end of the meter.

Electrical Connections

The only electrical connection to the meter itself is to the pickoff. Meters are supplied with pigtail wires or a variety of connectors. Usually the meter manufacturer supplies the mating connector when a connector type is supplied. Two- and three-wire versions of pickoffs are available. The third wire is usually a ground or shield wire, when it exists.

Connections to the amplifier or other electronics vary greatly from manufacturer to manufacturer. It is advisable to read the instructions for connections supplied with the meter. Amplifiers are available in DC and AC power types and two- or three-wire configurations. Often a temperature or pressure sensor is also part of the meter installation.

It is best practice to keep the amplifier as close to the meter as possible to minimize susceptibility to noise. When it is required to move the amplifier to a remote location, do not exceed 300 ft of wire for magnetic pickoffs and 100 ft of wire for carrier-type pickoffs. If it is necessary to exceed these recommendations, consult the meter manufacturer.

When running the wires to the meter, it is also advisable to keep them separated from power lines or other electromagnetic sources such as electric motors, transformers, welding machines, or high voltage lines. These types of devices transmit electrical noise that can be picked up by the meter electronics. It is advisable to install meter leads in a separate instrument conduit or instrument cable tray.

The shield of a shielded cable or separate ground wire should be grounded only at one point in the system to prevent ground loops. Most instruction manuals will suggest an acceptable grounding method.

Maintenance

Maintenance of a turbine flowmeter consists of periodic inspections to ensure that the internal parts are not fouled with debris or have suffered any corrosion by the fluid in the system. The bearing should be inspected to ensure that it rotates freely.

The internal assembly of the meter may be removed for inspection and cleaning. Most typical meters use a snap ring arrangement to hold the internal assembly into the body. Once removed, the internal assembly may be cleaned with solvent or alcohol. If the meter is to be stored for an extended period, it is advisable to coat the internals with a light oil.

If the flowmeter has been in oxygen service, especially liquid oxygen or other corrosive service, clean the meter thoroughly after removal and again before installation. Special cleaning procedures are applicable to liquid oxygen service.

Contamination in the bearings of a turbine flowmeter is the largest single source of poor meter performance. Many fluids contain residues that can build up on the bearing surfaces over time and retard the bearings' freedom to rotate. Flushing the internal assembly of the meter with an appropriate solvent will remove this buildup. While journal bearings are less susceptible to contamination than ball bearings, debris still can collect in the journal. It is appropriate to periodically clean all types of turbine flowmeters.

Blow lightly through the meter to see how freely the rotor spins and how abruptly it stops spinning. A meter fitted with ball bearings should rotate freely with a slight blowing through the meter and should coast slowly to a stop. If it takes a large blow to start the meter or it coasts quickly to a stop, the bearings are either worn out or contaminated.

Do not use an air hose to attempt to free a rotor that does not spin. If the bearings are not already damaged, they surely will be after the air hose trick!

Journal bearings will take a slightly larger breath to start spinning and will come to a more abrupt stop. It should not take an excessive breath to make the meter spin freely.

Do not force the assembly out of the body! If the internal assembly does not easily come out of the body, send the meter to a repair center.

If, after cleaning, the rotor does not spin freely, the bearings should be replaced. If one is not confident to replace the bearings, send the meter to a competent repair facility or the original manufacturer. Usually a snap ring or other retaining ring at one or both ends of the meter is used to hold the internal assembly into the body. Carefully remove one of these rings and use a blunt rod to push the internal assembly out of the body. If it cannot be pushed out by hand or an *extremely* light tap with a mallet, something is wrong!

Once the internal assembly is out of the housing, usually the end supports will slide off the rotor shaft. Some meters may have a nut or another retaining ring to remove before the supports can be removed from the shaft. In some meters, the bearings will be in the supports, but in most meters the bearings are located in the bore of the rotor. Again, usually a small retaining ring holds the bearings into the rotor.

It is extremely important to note which surface of the rotor was upstream! Many rotors appear symmetrical, and actually are symmetrical, but the calibration will be greatly different depending on which surface is upstream.

The upstream support in some meters may be slightly bent on its downstream end just in front of the rotor. This is normal and is not damage. It is a technique used in trimming the linearity of the meter. The tip of the corners of some rotor blades may also appear to be removed or damaged. This, too, is a technique used by manufacturers to trim in the desired performance of the meter and may not be damage.

Some manufacturers will mark either the upstream or downstream surface with an identifying mark. If one gets confused about how the rotor came out of the meter, call the manufacturer and ask him how the rotor is identified. Also the upstream support must not be exchanged with the downstream support, even if they appear the same.

Replacement bearings are available from the manufacturer or other repair facilities. Some of the journal bearings are not field replaceable. After replacement of a ball bearing, some manufacturers and other repair facilities claim that it is not necessary to recalibrate the meter. However, it is a good idea to recalibrate the meter to assure that it was reassembled correctly and that the new bearings are functioning properly. After replacement of a journal bearing, recalibration of the meter is mandatory.

Pickoff removal is easily accomplished by unscrewing the pickoff from the body.

If a magnetic pickoff is removed from the body, the meter must be recalibrated! A carrier frequency pickoff may be removed and reinstalled without affecting the calibration, since there is no drag force imparted on the rotor by this type of pickoff.

The pickoff should be inserted until it bottoms in the recess, then the jam nut should be tightened lightly. It is not necessary to tighten the pickoff into the body. There is a thin diaphragm at the bottom of the recess and excessive tightening of the pickoff can deform the diaphragm. For meter bodies with an explosion-proof "spud" on the housing, it may be necessary to fabricate a socket to remove the pickoff. This tool is easily fabricated by turning down the O.D. of a deep socket and perhaps putting a hole in the side to allow pigtail wires to protrude if the pickoff has pigtail wires. Figure 17-14 shows an exploded view of a typical turbine meter.

References

1. API Standard 2534, ANSI Z11-2299, "Measurement of Liquid Hydrocarbons by Turbine Meter Systems," Washington, DC.: API, 1971.
2. Beck, Edwin J., "Liquid Gas Measurement and Temperature Compensation," *Cryogenics and Industrial Gases*, pp 21-24, July/August, 1974.

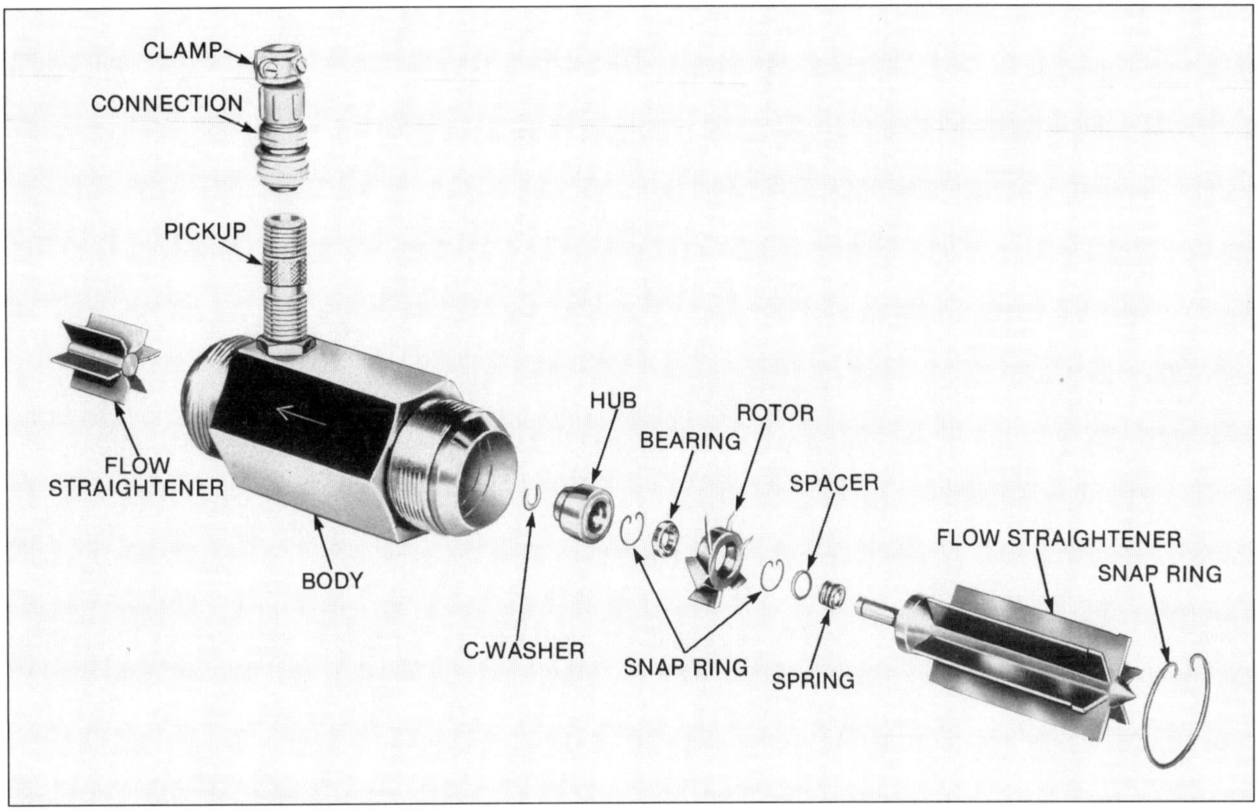

Figure 17-14. Exploded View of Turbine Meter
(Courtesy of Shute & Koerting Div., Ketema: Cox Instruments)

3. Brennan, J. A., and others, "An Evaluation of Several Cryogenic Turbine Flowmeters," *NBS TN 624*.

4. Burgin, J. A., "Turbine Meter Measurement on a Transmission System," *Gas*, Vol. 49, pp 30-32, Feb. 1973.

5. Clayton, C. G., Editor, *Modern Developments in Flow Measurement*, London: Peter Peregrinus, Ltd., pp 305-346.

6. Dijstelbergen, H. H., "Rotameters and Turbine Flowmeters in Pulsating Flow Measurement," *Measurement and Control*, Vol. 3, No. 12: Dec. 1970.

7. Dowdell, R. B., and Liddle, A. H., "Measurement of Pulsating Flow with Propeller and Turbine Type Meters," *Trans ASME*, Vol. 1953, pp 961-968.

8. Francisco, E. E., "Response Characteristics of a Turbine Meter," *Electronic Instrument Digest*, pp 33-35, Feb. 1971.

9. Haalman, A., "Pulsation Errors in Turbine Flowmeters," *Control Engineering*, p. 89, May 1965.

10. Head, V. P., and Hatbord, P. A., "Practical Pulsation Threshold for Turbine Flowmeters," *Trans. of ASME*, p. 1471, Oct. 1956.

11. Higson, D. J., "The Transient Performance of a Turbine Flowmeter in Water," *Journal of Scientific Instrumentation*, Vol. 41, p. 317, May 1964.

12. Hochreiter, H. M., "Dimensionless Correlation of Coefficients of Turbine Type Flowmeters," Paper 57-A-63, *Transactions of the ASME*, Vol. 80, pp 1363-1368, 1958.

13. *Instrumentation Engineers' Handbook*, Vol. 1, Chilton Book Company, pp 488-500.

14. Jepson, P., and Bean, P. G., "Effect of Upstream Velocity Profiles on Turbine Flowmeter Registration," *Journal of Mechanical Engineering Science*, Vol. II, pp 503-510, Nov. 5, 1969.

15. LaNasa, P., "A New Look at Methods of Field Test Turbine Meters," *Pipe Line Industry*, pp 35-38, April 1975.

16. Lee, W. F. Z., and Evans, H. J., "A Field Method of Determining Gas Turbine Meter Performance," *Transactions of the ASME, Journal of Basic Engineering*, pp 724-731, Dec. 1970.

17. Lee, W. F. Z., and Evans, H. J., "Density Effect and Reynolds Number Effect on Gas Turbine Flowmeters," *Transactions of the ASME, Journal of Basic Engineering*, pp 1043-1057, Dec. 1965.

18. Lee, W. F. Z., and Karlby, Henning, "A Study of the Viscosity Effect and Its Compensation on Turbine Type Flowmeters," *Journal of Basic Engineering*, Vol. 82, pp 717-728, 1960.

19. Lee, W. F. Z.; Kirik, M. J.; and Bonner, J. A., "Gas Turbine Measurement of Pulsating Flow," *Journal of Engineering for Power, Transactions of the ASME*, pp 531-539, Oct. 1975.

20. Liptak, B. G., "Flowmetering Accuracy," *Instrumentation Technology*, pp 35-38, July 1971.

21. Mattingly, G. E., "The Characterization of a Piston Displacement-Type Flowmeter Calibration Facility and the Calibration and Use of Pulsed Output Type Flowmeters," Gaithersburg, Md.: NIST, 1991.

22. Minkin, H. L.; Hobart, H. F.; and Warshawsky, I., "Performance of Turbine Type Flowmeters in Liquid Hydrogen," NASA TN D-3770.

23. Olivier, P. D., "The Development of a Primary Standard for Calibrating Flowmeters on Gaseous Media," *First Internal Symposium on Fluid Flow Measurement*, Crystal City, VA: American Gas Association, 1986.

24. Olsin, Lief O., "Introduction to Liquid Flowmetering and Calibration of Liquid Flowmeters," NBS Tech. Note 831, Jan. 1974.

25. Rubin, M.; Miller, R. W.; and Fox, W. G., "Driving Torques in a Theoretical Model of a Turbine Meter," *Transactions of the ASME, Journal of Basic Engineering*, ASME Paper #64-WA/FM-2.

26. Shafer, M. R., "Performance Characteristics of Turbine Flowmeters," *Transactions of the ASME, Journal of Basic Engineering*, Vol. 84, pp 471-485, 1962.

27. Siev, R., "Turbine Flowmeters in Fuel Oil Service," *Instrumentation Technology*, pp 39-41, Jan. 1973.

28. Stevens, G. H., "Dynamic Calibration of Turbine Flowmeters," *Instruments and Control Systems*, pp 109-111, April 1970.

29. Strohneir, W., "Bearings for Turbine Flowmeters," *Instrumentation Technology*, p. 39, April 1972.

30. Thompson, R. E., and Grey, J., "Turbine Flowmeter Performance Model," *Transactions of the ASME, Journal of Basic Engineering*, pp 712-723, Dec. 1970.

31. ISA-RP 31.1-1972, Specification, Installation and Calibration of Turbine Flowmeters, Research Triangle Park, N.C.: ISA.

About the Author

Paul D. Olivier is President of Flow Dynamics, Inc., an independent flowmeter calibration laboratory. His background in precision flow measurement is quite extensive. With roots in the aerospace industry during the 1960s and 1970s as a test engineer and laboratory manager, he developed techniques to improve the current technology in measurement of flow, temperature, and pressure as well as other parameters. In the early '80s he was Engineering Manager at Flow Technology, where he developed primary flow calibration devices for liquid flow calibration and developed correlation techniques for liquid and gas turbine meters. During the late '80s, as General Manager of Flow Measurement Systems, Inc., he continued the industry leadership with further developments in primary flow calibration standards for both liquid and gas service. He also designed and developed the nation's new primary liquid flow calibrator for NIST. In 1990, he formed Flow Dynamics, Inc., to pursue the precision calibration and correlation of flowmeters for all of the metering community.

Mr. Olivier received a Bachelor's degree in Aeronautical Engineering from California State Polytechnic University and advanced degrees in Aeronautical Engineering from the University of Cincinnati and in Business Administration from West Coast University.

18

Ultrasonic Flowmeters

Two types of ultrasonic flowmeters are in general use for closed pipe flow measurement. The first (transit time) usually uses pulse transmission, while the second (Doppler) usually uses continuous wave transmission. These types are more complementary than competitive.

Transit time flowmeters make use of the difference in the time for a sonic pulse to travel a fixed distance, first in the direction of flow and then against the flow. They can measure flow accurately when properly installed and applied.

Doppler instruments make use of the Doppler frequency shift caused by sound scattered or reflected from moving particles in the flow path. Doppler meters are not considered to be as accurate as transit time flowmeters. They are very convenient to use and generally more popular and less expensive than transit time meters.

Principles of Operation

Three separate modes of operation are required to describe these meters: two for transit time and one for doppler.

Transit time flowmeters are divided into those operating in the time domain and those in the frequency domain. Both transmit pulses, from a transmitting transducer, which propagate through the flowing medium to a receiving transducer. The difference between the times of arrival of a pulse propagated in an upstream direction (against the flow) and one propagated downstream (with the flow) is used to calculate flow velocity in both types.

Doppler meters transmit a continuous sonic wave inclined at some angle to the flow. The beat frequency between the transmitted signal and that scattered from moving bodies in the flow provides flow velocity information.

Time Domain Transit Time Meter

Time domain meters, depicted in Figure 18-1, transmit a pulse in a given direction and record the time of arrival of the pulse on the other end of the acoustic path. They then transmit a pulse in the opposite direction and record that time of arrival. The difference between the two time measurements provides information on the motion of the fluid in the flow path. The configuration in Figure 18-1 shows the essential parts of the time domain meter. There are many ways to implement this configuration, but they all contain these essential parts. A typical time domain flow velocity equation is given by:

$$v_t = \frac{C^2 \cdot (T_u - T_d) \cdot \tan\theta}{2 \cdot D} \qquad (18\text{-}1)$$

Note that this expression is sensitive to the speed of sound in the medium. Make sure that the flow calculation algorithm used by the manufacturer takes sound speed into account to avoid significant temperature-dependent error.

where:

v_t = time–based flow velocity
C = sound speed in the fluid
T_u = upstream transit time
T_d = downstream transit time
D = diameter of pipe
θ = angle between sonic path and flow axis

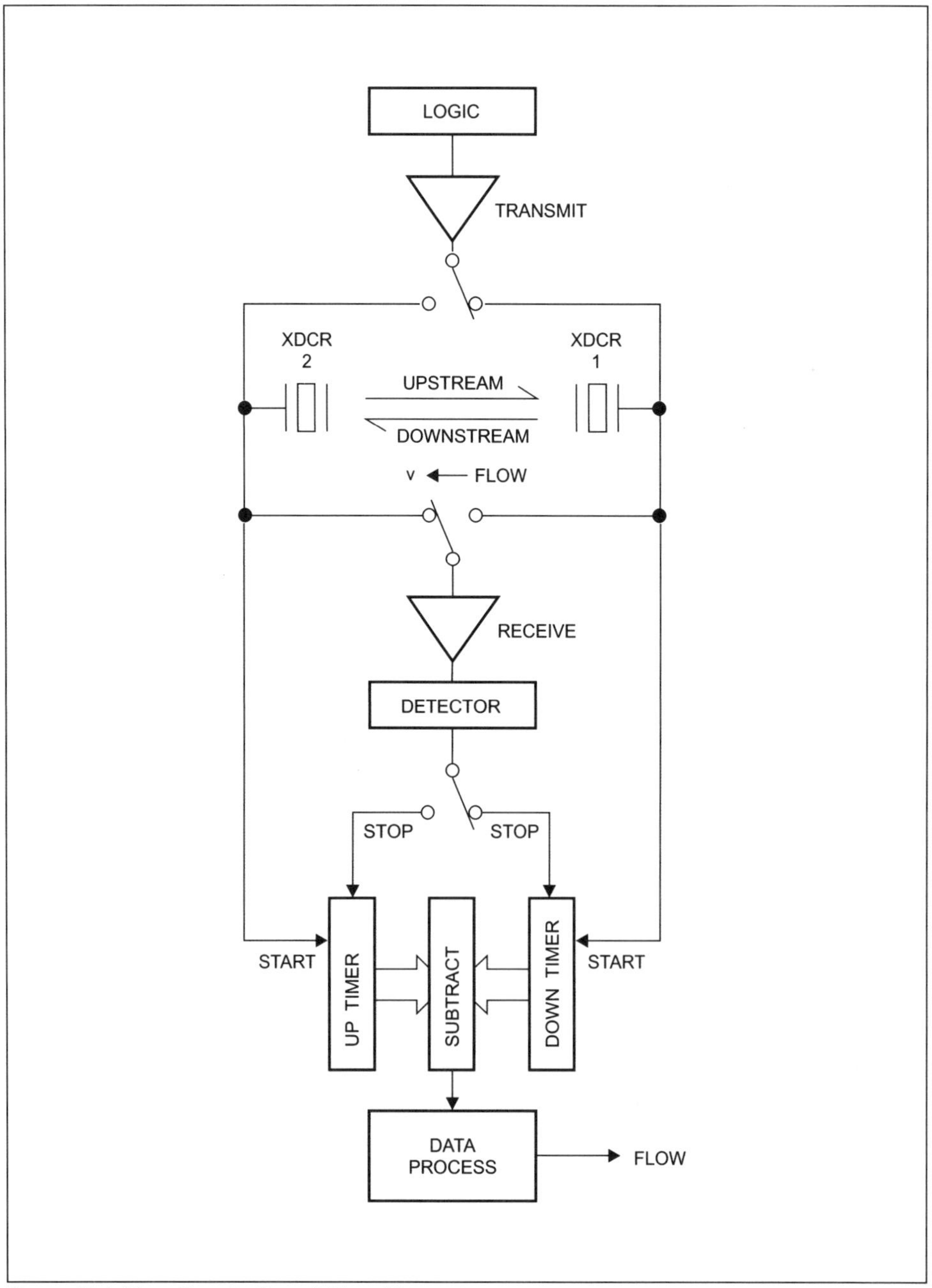

Figure 18-1. Time Domain Transit Time Flowmeter

Practical limits of time domain flowmeters are pipe size and the ability to resolve time differences. Assume time differences can be resolved to 10 nanoseconds (100 MHz clock). If it is desired to measure flow in a 0.5 meter (20 inch) diameter pipe, the minimum measurable change in flow velocity (assuming 1500 m/s (4500 ft/s) sound speed) is about 0.02 meter per second (0.07 foot per second). In a practical measurement, a resolution of about 0.07 meter per second (0.25 foot per second) can be expected, which represents a flow of about 0.14 meter per second (200 gpm). Below this point the ability of the meter to read accurately becomes questionable.

On the other hand, in a three meter (10 foot) diameter pipe the minimum resolvable flow is about 0.003 meter per second (0.01 foot per second). An advantage of time domain meters is that a velocity determination can be made every transmit cycle. This allows for rapid response in large pipes where other meters may have long time constants. Because of the rapid response, time domain meters are the technology of choice for meters above one meter (36 inches) in diameter.

Frequency Domain Transit Time Meter

Transit time meter manufacturers time-share circuit functions between upstream and downstream paths as much as possible. This is not to save costs but to minimize error-producing differential time delays.

Frequency domain meters use the same sensor as time domain meters. The only difference is in the processing of the signals. Rather than read time directly, this implementation converts the time information into a frequency. In the simplest form, as shown in Figure 18-2, as soon as a sonic pulse is received it is immediately retransmitted to form a pulse repetition rate (frequency) proportional to the transit time of the pulse. If two such paths, one in each direction of flow, are used, two frequencies are generated. The difference in these two frequencies is proportional to flow velocity. The configuration of Figure 18-2 is for explanation only. No manufacturer uses independent paths for upstream and downstream measurements. In practice, one time-shared sonic path is used and rather complex phase-locked loops are generally used to obtain the two independent frequencies. Flow velocity for frequency domain meters is given by:

$$\bar{v}_f = \frac{2 \cdot D \cdot (f_{\text{up}} - f_{\text{dn}})}{2 \cdot \sin\theta \cdot \cos\theta} \qquad (18\text{-}2)$$

where:

$\bar{v}_f$ = frequency-based average flow velocity

f_{up} = upstream frequency $\left(\frac{1}{T_u}\right)$

f_{dn} = downstream frequency $\left(\frac{1}{T_d}\right)$

The shortest practical path cannot be calculated as simply as the time domain meter. Transducer diameter, mounting geometry, and other subjective considerations give a practical minimum path length of about 0.075 meter (three inches).

The ability to make simple and accurate readings of frequency gives the frequency domain meter an advantage in resolution. However, bandwidth limitations make these meters sluggish for long path lengths. In general, the frequency domain meter has an advantage for meters with a diameter of from 0.3 to one meter (12 to 36 inches). For pipes under 0.3 meter, frequency domain meters provide superior performance.

Doppler Meter

Industrial Doppler flowmeters generally make use of the continuous transmission of a single frequency rather than pulses. The sound beam is propagated into

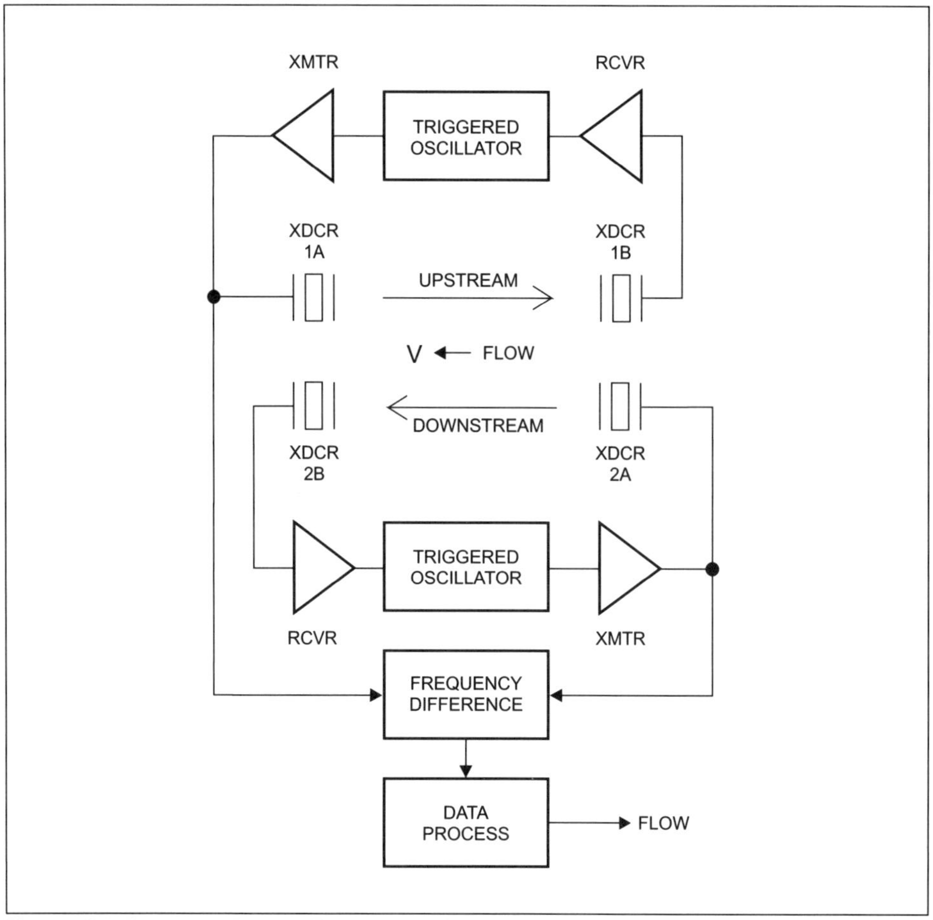

Figure 18-2. Frequency Domain Transit Time Flowmeter

the flowing medium at some angle to the flow. Small inclusions of bubbles, solids, or eddies in the flowing medium reflect or scatter the sound back to a receiver. If there is any motion of these inclusions, there will be a shift in the frequency (Doppler shift) of the returned signal. Each particle or "scatterer" reflects sound while it is in the sonic field of the transmitter. These particles have a random physical distribution and a random distribution of velocities. The reflected composite signal is a random distribution of frequencies that add up to what appears to be a single waveform. The difference between the transmitted and received frequencies is proportional to the motion of the scatterer (flow). Figure 18-3 shows the basic parts of a Doppler meter. The theoretical flow velocity given by the Doppler frequency is:

$$\bar{v}_d = \frac{f_d \cdot C}{(2 \cdot F_o \cdot \cos\theta)} \qquad (18\text{-}3)$$

where:

$\bar{v}_d$ = Doppler-based average flow velocity
f_d = Doppler frequency
f_o = transmitted frequency

Most Doppler systems determine the Doppler frequency by counting each cycle in the Doppler waveform. This technique does not give the theoretical Dop-

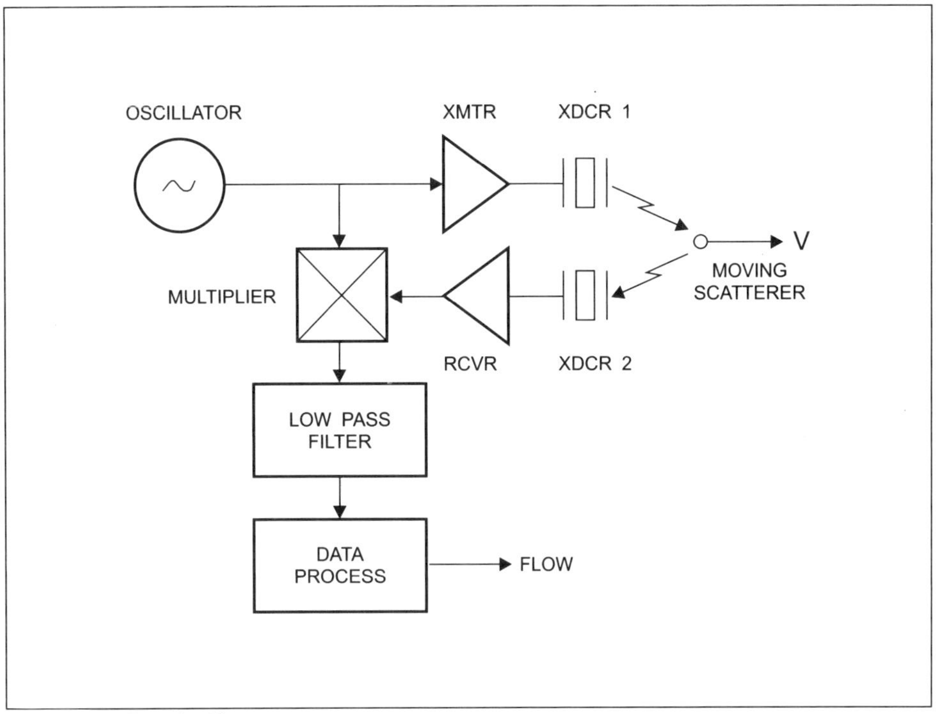

Figure 18-3. Doppler Flowmeter

pler frequency but something close to the mean of the Doppler spectrum. Other systems take the spectral distribution of the signal into account and predict the theoretical frequency. This latter technique gives better performance in the presence of changing concentration of scattering particles.

Transit Time Transmit/Receive Sensors

Because of the wide variety of pipe sizes and flow conditions encountered, there are a number of different sensor configurations for transit time meters. These vary from the small meter shown in Figure 18-4 to the large specially constructed installations of Figure 18-5. Normally, the sensor is in the form of a spool section inserted in the pipe as seen in Figure 18-6.

AXIAL

When small diameter pipes are encountered, it is necessary to pass the sound directly down the axis of the pipe. This is to ensure that there is a sufficient path length over which to measure the transit time. Path lengths of less than about 0.075 meter (3 inches) do not give adequate time difference for an accurate measurement. Sensors of this configuration, shown in Figure 18-4, are generally referred to as "axial" sensors. Figure 18-4(a) is a one-inch axial flow section. Note the mixing chambers at each end. These are to break up refractive velocity gradients in high viscosity material. Such gradients introduce velocity-dependent errors. Figure 18-4(b) is a steam-jacketed, half-inch axial flow section for high temperature materials such as molten sulfur.

RADIAL

Most transit time sensors have the sonic transducers placed on either side of a spool section. They are normally inclined at 45 degrees to the pipe axis to measure a vector component of the flow. A practical lower pipe size limit of this

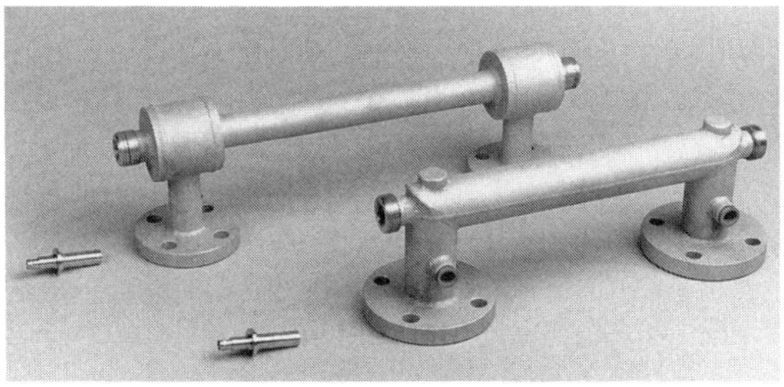

(a) A 1 in. STANDARD AXIAL FLOW SECTION (b) A ½ in. STEAM JACKETED FLOW SECTION

Figure 18-4. Axial Transit Time Flow Sections
(Courtesy of TN Technologies)

configuration is about five centimeters (two inches), with an upper limit in excess of 3 meters (10 feet). Such sensors, shown in Figure 18-6, are generally called "radial" because of the transducer placement.

LARGE DIAMETER

For very large or complex installations, as shown in Figure 18-5, the meter section is often built in place or the transducers mounted in the walls of the conduit. Note the transducers held in place at the pipe wall by a series of expansion rings placed inside the pipe. Each transducer pair forms an independent measuring path. The paths are mathematically combined to provide a flow measurement that is, for practical purposes, independent of velocity profile. In most of these special installations there are several sonic paths. Each such installation is unique and the manufacturer must provide special calibration for that particular installation.

Figure 18-5. Large-Diameter Built-in-Place Flow Section—Multiple Sonic Paths Are Used to Provide Averaging of Asymmetrical Flow Velocity Profiles
(Courtesy of Aqusonic Div., Ferranti O.R.E., Inc.)

Figure 18-6. Radial Transit Time Flow Section
(Courtesy of TN Technologies, Inc.)

CLAMP-ON

In many instances it is desirable to have an installation where the pipe wall is not penetrated by the transducers (non-wetted type transducers). Such installations are very convenient and have accuracies approaching those of direct wetted transducers. While somewhat more complex to calibrate, the convenience of this type of installation makes it very popular. A typical "clamp-on" meter is shown in Figure 18-7. The transducers are mounted on a calibration device and acoustically coupled to the pipe wall with grease or epoxy. In some cases it is possible to improve performance by bouncing the sound from one wall to the other. The normal position of the second transducer is shown as Location 1 in Figure 18-7. To operate on a bounced signal, move the second transducer to Location 2 or 3. While this deteriorates the quality of the received signal (accuracy), it gives a greater path length over which to average the transit time. In small pipe installations this is often necessary.

Unlike direct-wetted transducers, the angle of the sound beam from a the clamp-on installation is determined by the pipe material. Care must be taken in the design so that the critical angle (where no sound enters the fluid) is not exceeded. This angle, controlled by Snell's Law of Refraction, is determined by the sound speed in the pipe material and the sound speed in the fluid.

In order to improve the sonic angle, some clamp-on transducers are designed to use the slower, but less efficient, shear wave propagation. In a steel pipe this will change the critical angle from 14.6 to 27.6 degrees from the normal to the pipe wall. The measured flow component will increase from 25% to 46% of the average velocity.

FIELD MOUNTING

When it is necessary to instrument an existing line, it is usually desirable to make the installation without cutting out a section of the pipe for the insertion of a sensor spool section. This type of installation is accomplished with the use of "field-mountable" transducer bosses. The transducer bosses are attached to the outside of the pipe by strapping, cementing or welding. The pipe is then drilled and the transducers installed. If necessary, the drilling can be done using hot-tap techniques without draining the line. One type of field-mountable transducers is shown in Figure 18-8.

Doppler Transmit/Receive Sensors

In the majority of Doppler installations the transducers are mounted on the outside of the pipe in a temporary manner. The sensor generally consists of two transducers clamped to the pipe wall with straps and acoustically coupled to the metal with grease. Figure 18-9 shows a typical installation.

When the pipe wall is acoustically opaque, it is necessary to penetrate the wall and insert the transducer into the flow. A "probe" type Doppler assembly, shown in Figure 18-10, can be inserted and removed under line pressure. The Doppler probe is a form of insertion meter. By placing the transducer at the correct place along the radius of the pipe, readings of average velocity are obtained. By using the probe as a velocity-determining device, a traverse of the profile can be made and the point of average velocity determined regardless of the flow profile.

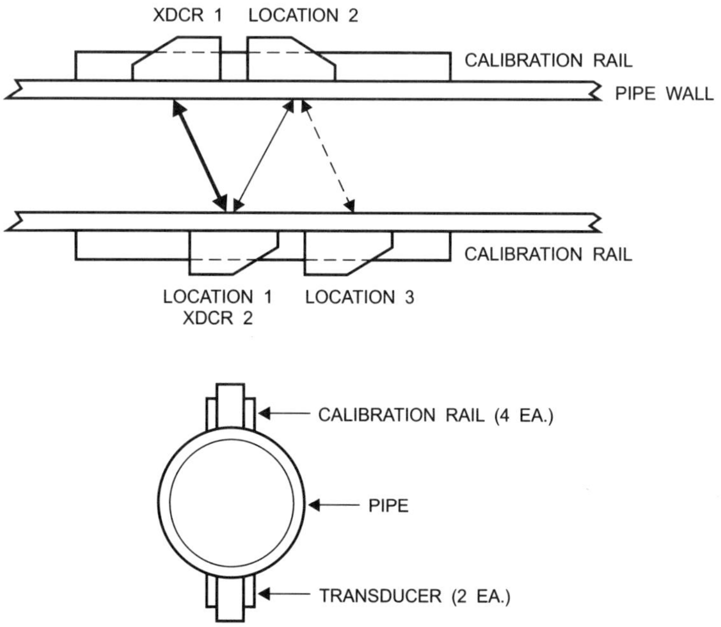

Figure 18-7. Clamp-On Transit Time Flowmeter

Figure 18-8. Field-Mountable Weld-On Transducer Bosses for Pre-Drilled Pipe
(Courtesy of TN Technologies, Inc.)

Velocity Profile

Nothing is more critical to a sonic meter installation than an understanding of the effects of the velocity profile on the fluid within the flowing conduit. Sonic meters do not measure flow volume. The information obtained from both transit time and Doppler meters is an average flow velocity along the sonic path. How velocity relates to flow volume is given by the distribution of velocities across the flowing conduit. In practice, the shape of the velocity profile is different in the laminar and turbulent flow regimes.

In normal piping laminar flow exists as long as the Reynolds number is below about 2,000. The shape of the profile conforms to a parabola, and the velocity of a point on the profile is given by:

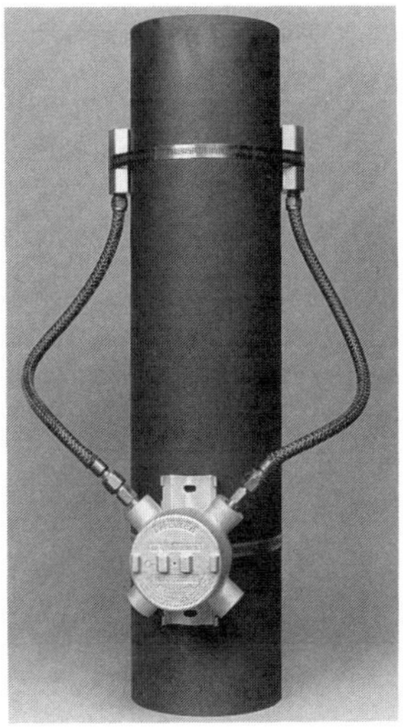

Figure 18-9. Typical Doppler Flowmeter Installation
(Courtesy of TN Technologies, Inc.)

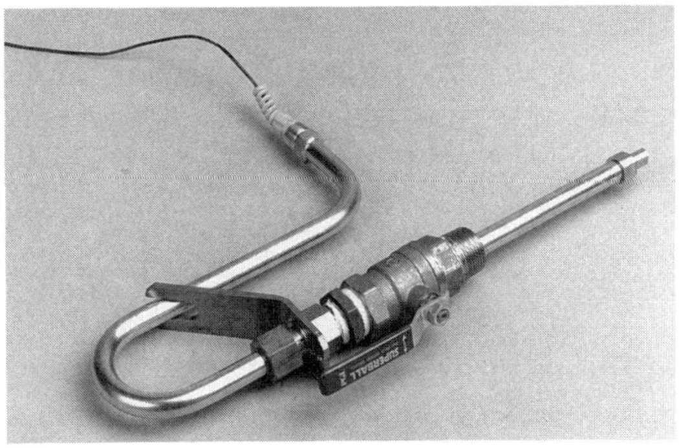

Figure 18-10. Insertion Probe Doppler Sensor
(Courtesy of TN Technologies, Inc.)

$$v_{\text{lam}} = v_{\max} \times \left(1 - \frac{2y^2}{D}\right) \tag{18-4}$$

where:

v_{lam} = instantaneous laminar velocity
$v_{\max}$ = maximum velocity (center-line velocity)
y = radial distance center

When the Reynolds number exceeds 2,000, the fluid starts to "fail in shear." The result is a drop in the velocity and a decrease in the resistance to flow. The lower velocity causes the Reynolds number to drop, and laminar flow is re-established. As soon as this happens the fluid switches back to its turbulent state. This process of switching back and forth, called the critical zone of the transitional flow regime, continues up to a Reynolds number of approximately 4,000.

Above a Reynolds number of about 4,000 the velocity profile enters the turbulent flow regime. The fluid does not switch back to the laminar state because the stress of the fluid is too great. The velocity profile is now given by [Ref. 1]:

$$v = v_{\max}\,(1 + 1.435\sqrt{f}) - 2.15\sqrt{f} \cdot \log\left(\frac{D}{2(D - 2y)}\right) \tag{18-5}$$

where:

v = instantaneous velocity
f = friction factor

The friction factor is a function of the pipe roughness, its diameter, and Reynolds number. As the roughness of the pipe or the velocity increases, the friction effects penetrate farther into the flow. Until this wall effect reaches the center of the pipe, the velocity profile continues to change. For smaller pipes this effect reaches the center much sooner than in a larger pipe. Once the wall effect has reached the center, there is little further change in the shape of the velocity profile. A convenient way of looking at these effects in pipes is by the relative roughness, which is:

$$R = \frac{\varepsilon}{D} \tag{18-6}$$

where:

R = relative roughness
ε = pipe wall roughness

A convenient way to display flow data is to use the "pipe factor" or velocity ratio, which is:

$$P = \frac{v_{\text{ave}}}{v_{\max}} \tag{18-7}$$

where: P = pipe factor

The friction factor, in turn, can be expressed as

$$\frac{1}{\sqrt{f}} = \frac{-2 \cdot \log\left(\frac{R}{3.7} + 2.51\right)}{\mathrm{Re}_D\sqrt{f}} \tag{18-8}$$

The above expressions are an approximation of the profile. The factors involved in determining the profile are so complex that there is no rigorous derivation of its shape. Equation (18-5), based on a rigorous derivation and modified by empirical data, has proved to be one of the best.

The above equations are difficult to manipulate. Transit time meters with microprocessors can usually cope with them. For general use, however, a close approximation [Ref. 2] of the profile is given by the expression:

$$v = v_{max} \cdot \left(1 - \frac{2 \cdot y}{D}\right)^{1/n} \qquad (18\text{-}9)$$

where:

n = a Reynolds number-dependent exponent

For most applications a value of 7 is used for n and the expression is referred to as the 1/7th power law. The value of n will vary from 6 at $Re_D = 4{,}000$ to 10 at Re_D 4,000,000. In the region of $Re_D = 100{,}000$, the equation is quite close.

A plot of the pipe factor as a function of Reynolds number for various relative roughnesses is shown in Figure 18-11. Note that for small values of relative roughnesses the velocities continue to change after leaving the critical zone of the transitional flow regime. While the velocity profile is continuing to change, the flow is in the turbulent transitional flow regime. Finally, when the wall effects extend to the middle of the pipe, the changes stop and the flow is fully developed turbulent flow.

How these various flow regimes are treated by the meter will determine the accuracy of the meter installation. Before selection of a flowmeter, careful consideration should be given to the range of Reynolds numbers an application will cover.

In selecting the meter size, a balance must be made between velocity and pressure drop. As the full scale velocity drops below about one meter per second (three feet per second), the accuracy will deteriorate; as the velocity increases above about 3 meters per second (10 feet per second), the pressure drop (pumping costs) may become excessive.

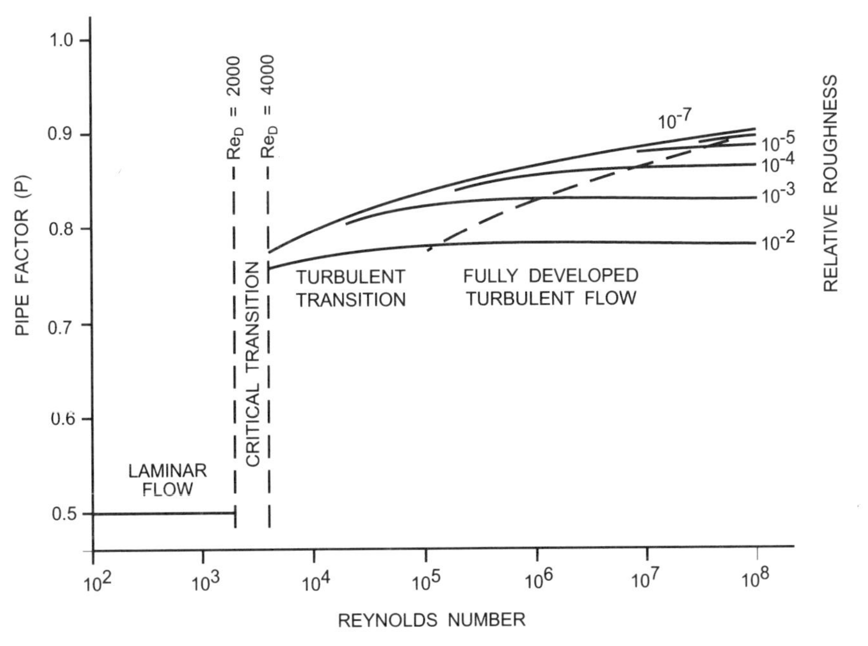

Figure 18-11. Pipe Factor as a Function of Reynolds Number and Relative Roughness

Offset-Path Sensors

An analysis of the above profile equations will show that there is a sonic path, offset from the center, where a velocity measurement will give the same reading of turbulent flow regardless of the pipe factor. This point [Ref. 3] is 0.542 times the radius from the center of the pipe. Additionally, there will be less than 3% difference between turbulent and laminar velocity measurements made along this path. A second location at a distance of 0.493 times the radius from the center is where the laminar velocity profile equals the average of the turbulent profiles. The variations in turbulent profiles as a function of pipe factor is about 1%. Placing a transit time sonic path at one of these locations will provide velocity data that is relatively insensitive to the velocity profile changes. For convenience, the path is usually located at 0.5 times the radius from the center. This sonic path location is different from that used for single-point measuring devices such as insertion probes. The sonic path averages all velocities encountered and takes much more flow into account.

For this concept to work, a very stable velocity profile is required. As a means of offsetting against variations in profile, two parallel paths are generally used, one on each side of the center as shown in Figure 18-12.

Integrating-Path Sensors

An advantage of this technique is that it does not rely on a knowledge of the shape of the velocity profile. Also, it lends itself well to conduits other than round pipes.

Sonic meter designers have made use of classical mathematical techniques to overcome the problems associated with disturbed and asymmetrical profiles. In these meters several readings of velocity are measured at different radial distances of the pipe. The measurements are used to compute a more accurate average flow velocity. Methods of approximating the integration of the area under a curve have been described by Simpson, Gauss, Labato, and others. Most mathematical texts will describe the location of the sample points and the weighting to be assigned to each sample. Figure 18-13 shows a four-path system in which

$$Q = A \times (k_1 v_1 + k_2 v_2 + k_3 v_3 + k_4 v_4) \cos \theta \qquad (18\text{-}10)$$

where:

Q = volumetric flow rate
A = cross-sectional area of pipe
k_i = weighting coefficient of integration technique i
v_i = velocity determined by transducer path i
θ = angle between sonic path and flow axis

Such multiple-path meters usually comprise four paths and are used on larger installations.

In cases where it is not possible to control the velocity profile, multiple paths are installed in more than one plane. This will provide averaging of the various error-producing flow components. Meters of this type are uniquely suited to such difficult flow measurements.

Applications and Practice

For the most part the above discussions have assumed that the velocity profile is ideal and that there are no flow components normal to the pipe wall. In most installations this situation does not exist. Major errors can be introduced depending on the type of sensor and the configuration of the piping in the vicinity of the sensor.

Frictional forces at the pipe wall will eventually cause flow to be symmetrically distributed in the pipe. However, in the presence of a disturbance that causes the flow to change direction, inertial forces are introduced. It is these inertial forces that produce asymmetrical flow. Downstream from an elbow the fluid tends to continue in a straight line, but the pipe will not allow that. Figure 18-14 shows the typical flow profile. Note that the flow can be resolved into vectors. The flow component going directly down the pipe will be referred here to as the "axial component," that normal to the pipe wall will be the "radial component," and, if present, any angular or spiral flow will be the "helical component." Helical components are introduced by causing the flow to change direction in two or more planes.

Profile Effects

AXIAL PATH SENSOR

Sonic paths that pass down the axis of the pipe are the least sensitive to radial and helical components of flow. Usually, the sonic path traverses sufficient diameters of the pipe to average these components to zero. Also, when they are operated through the critical zone of the transitional flow regime (Reynolds numbers of 2000 to 4000), the swings from laminar to turbulent will average, resulting in a usable measurement. There will still be a known change in calibration slope between laminar and turbulent flow, but the output in between will be relatively smooth and stable.

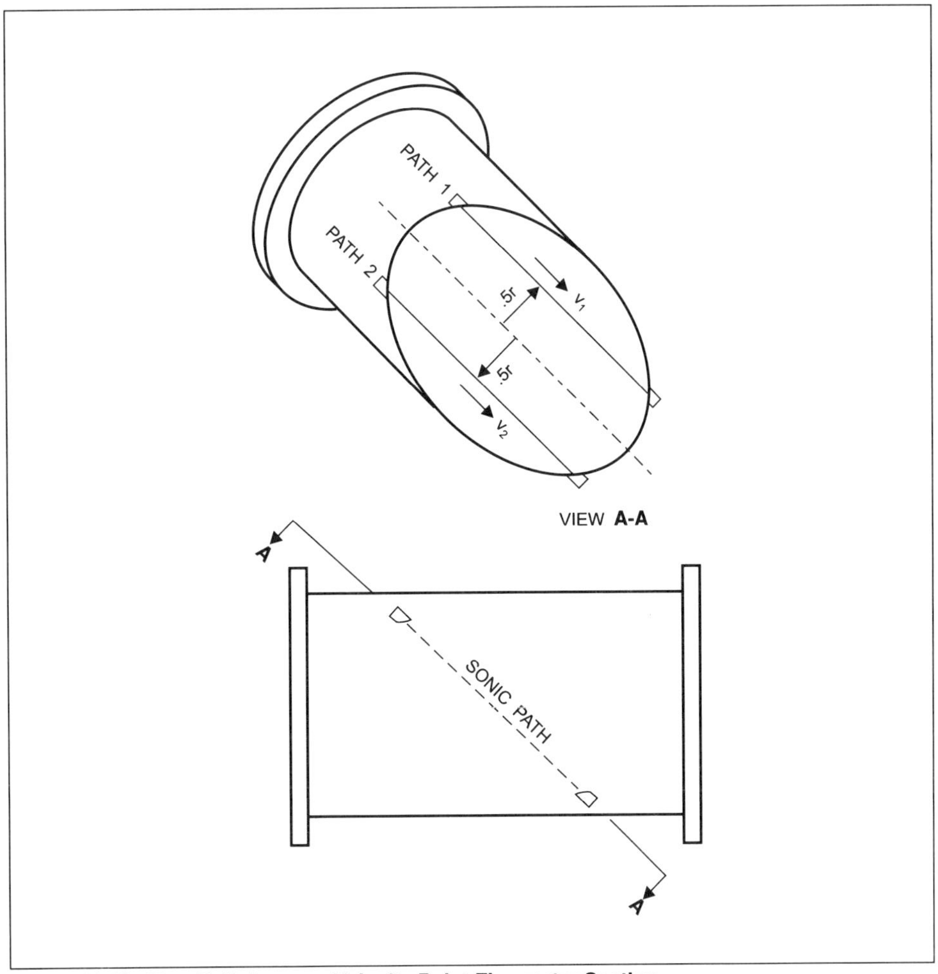

Figure 18-12. Two-Path Average Velocity Point Flowmeter Section

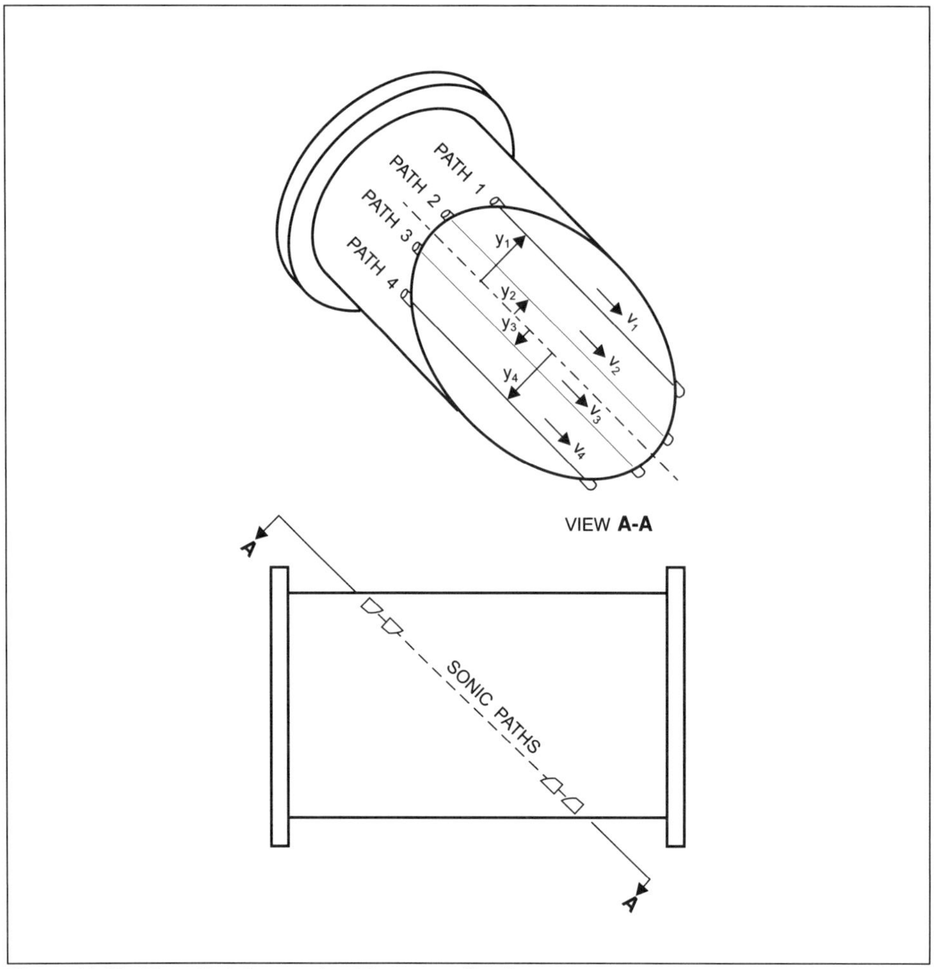

Figure 18-13. Four-Path Integrating Flowmeter Section

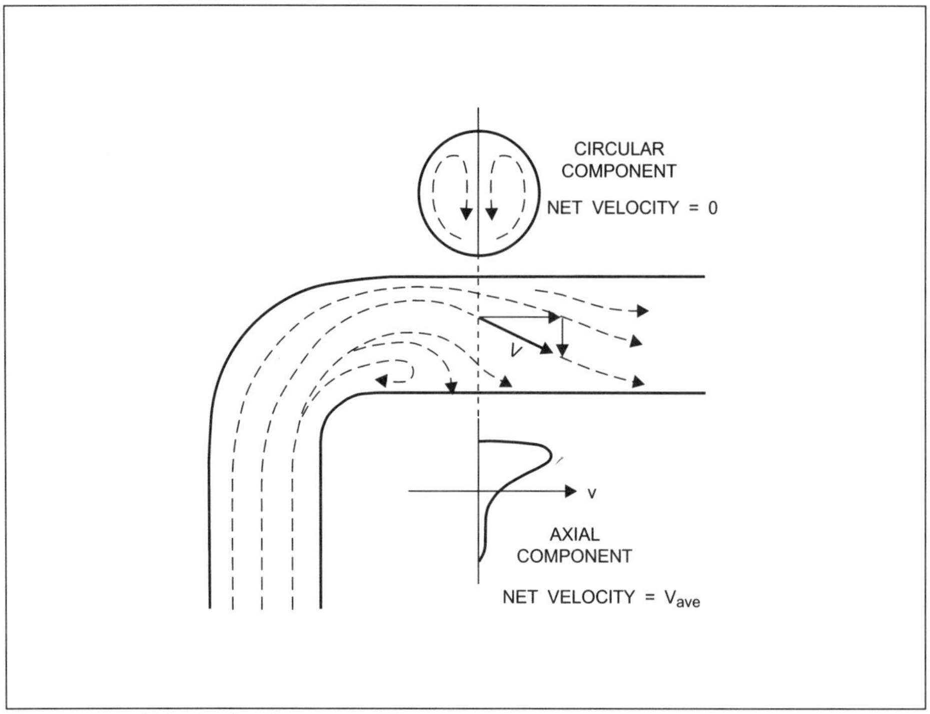

Figure 18-14. Velocity Gradients at an Elbow

Unlike radial meters, there is no requirement for straight diameters upstream and downstream from the sensor. Profile control takes place within the sensor itself.

SINGLE-PATH RADIAL SENSORS

Sensors conforming to the configuration of Figure 18-6 are the most sensitive to secondary (radial and helical) flow components. When the sonic path is inclined at 45 degrees to the flow, the meter is equally sensitive to the desired axial component and the undesired radial component. If, however, clamp-on transducers are used, the sonic path is inclined at about 75 degrees to the axis. In this case the meter is 3.8 times more sensitive to the radial component and magnifies the error introduced by secondary components.

Manufacturers' specifications on the minimum number of straight diameters upstream and downstream of the sensor vary. Most recommend as little as 10 diameters upstream and 5 diameters downstream of the section. Tests have been conducted that indicate the velocity profile effects persists far beyond 10 diameters. If a single path meter is installed 10 diameters downstream from an elbow, one can be assured of a 5 percent error. A better rule of thumb estimate of straight run requirements is 20 and 10 diameters upstream and downstream, respectively. At this distance there will be a 1 percent error.

The ratio of sensitivity of radial flow components to axial flow components is given by the tangent of the angle of inclination of the sonic path to the pipe axis. Care must be taken to minimize these components.

On the other hand, errors introduced by asymmetrical flow profiles tend to be quite stable over a wide range of Reynolds numbers. If an *in situ* (in place) calibration can be conducted, the meter can usually be recalibrated to meet its performance specifications.

MULTI-PATH SENSORS

Meters of this class do not depend on a knowledge of the velocity profile. They make direct readings of time difference and calculate the flow velocity by rigorous mathematical techniques. This makes them insensitive to asymmetrical velocity profiles but not to radial components of flow. If the magnitude of the radial component is known, it can be accounted for in the mathematical algorithm used to compute flow. Otherwise, a set of mirror-image, independent transducers can average out the radial components. Theoretical calibration is usually sufficient, but, when the installation is questionable, an *in situ* calibration is required.

DOPPLER SENSORS

There is no fixed configuration for a clamp-on Doppler installation. As a result, the question of profile sensitivity is complex. It is further complicated by variations in the concentration and size of the sonic scattering material in the flowing medium. Several examples will demonstrate the problems.

Consider an installation on a 0.3 meter (one foot) diameter pipe. The transducers are often mounted with one on each side of the pipe. The sonic patterns of the transducers will overlap as shown in Figure 18-15, giving Doppler information from the center of the pipe where the velocity is maximum. The Doppler emanating from this volume of overlap will be as if there were only one scatterer at a mean volumetric location. This theoretical particle is called the "mean scatterer." If, on the other hand, the transducers are mounted on the same side of the pipe as shown in Figure 18-16, the sonic overlap will be almost complete. This will provide Doppler information from scatterers all the way across the pipe. Now the Doppler frequency will be a function of the average velocity. There is approximately a 15 percent difference between maximum and average velocity.

The next consideration is with changing concentration. If the transducers are mounted on each side of the pipe and the concentration of scatterers increases, the Doppler signal at the receiving transducer will eventually be lost because of absorption and scattering. Since no signal is obtained, the transducers should be relocated to a side-by-side mounting. The Doppler signal will now be satisfactory and will be a function of something close to average velocity. If the concentration continues to increase, the mean scatterer will move closer to the pipe wall. As it does, it provides a measurement of an area of lower velocity, resulting in a lower Doppler reading. This continues until the mean scatterer is confined to the boundary layer at the inside pipe wall. At this point, data from the meter becomes meaningless. Variations of 30 percent can occur as the point of mean scattering approaches the pipe wall.

It is good practice to try the transducers in a number of locations and finally place them at a location that gives a reading of the average of all the locations tested. There is no ideal location. Each installation must be treated individually.

Added to these problems is the effect of asymmetries in the profile on the mean Doppler frequency. Take the case where the transducers are side by side, and the concentration is such that the mean scatterer is half way to the center. Different flow velocities will be read by locating the transducers on the other side of the pipe if the velocity profile is asymmetrical. The extent of this error can be seen if transducers are placed on the outside of an elbow and compared with a reading on the inside of the elbow. Downstream from short elbows the flow is usually flowing backwards in an eddy produced by the inertial effects of the flow.

FLOW LAB CALIBRATION

Engineering specifications often call for flow calibration of transit time meters. As with any calibration, the results reflect the interaction between the meter and the flow lab used for calibration. When the meter is placed in service, the calibration may be totally invalidated by the hydraulic configuration of the installation.

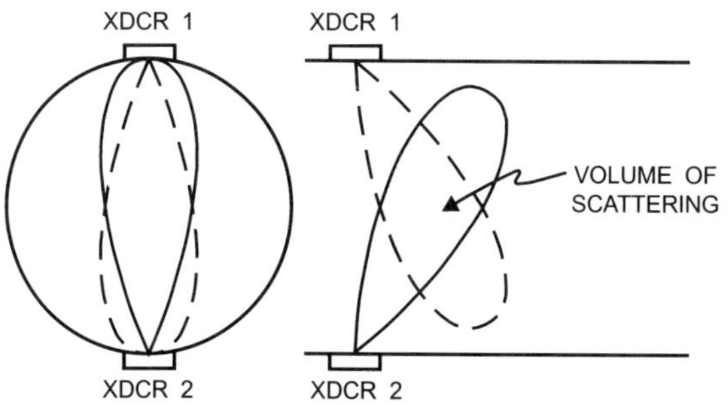

Figure 18-15. Controlled Point of Scattering

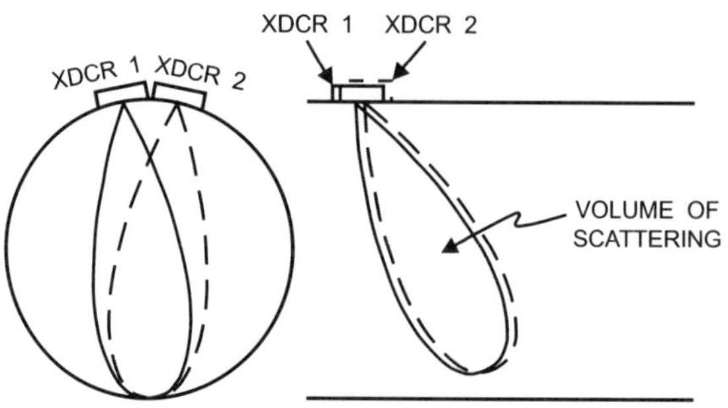

Figure 18-16. Overlapping Fields of Scattering

Great care must be taken to duplicate, as nearly as possible, the conditions of the calibration flow lab. A meter calibrated to within 0.25% can show as much as a 16% error if it is mounted within 5 diameters of an elbow. Partially closed gate valves in close proximity can introduce as much as 30% error. This problem is not unique to ultrasonic meters [Refs. 4, 5]. Meter accuracy is only as good as the installation. If possible, an *in situ* calibration should be conducted whether the meter has been calibrated in a flow lab or not. This will indicate any installation problems and provide a basis for performance history.

On occasion, there is a requirement for flow lab calibration of Doppler meters. This usually occurs because of requirements in standard performance specifications. Given the great variability in Doppler meter performance under varying flow conditions, such calibrations are usually not worth the expense of conducting them.

Temperature Effects

Variations in temperature of the flowing medium produce different effects in the different types of sonic meters. In all cases the effect is seen in changes in sound speed of the fluid and supporting material. For installations where the transducers are in direct contact with the medium and axially aligned and the meter is operated in the frequency domain, there is no temperature sensitivity. However, when operated in the time domain, the temperature effects do cause measurement error. The extent of this error can be calculated from a measurement of the mean transit time and can be accounted for. While early time domain meters had problems with temperature, present microprocessor-based instruments do not.

CLAMP-ON TRANSDUCERS

Many configurations of clamp-on transducers [Ref. 6] are available for both transit time and Doppler meters. They can be operated in various sonic modes of propagation, but all share some common requirements:

(1) The pipe wall must be capable of passing sound.

(2) The mounting surface must be clean and smooth.

(3) An acoustic couplant such as oil, grease, or epoxy must be used.

(4) The inside of the pipe must be free of sound-absorbing material such as scale or dirty grease.

CLAMP-ON TRANSIT TIME TRANSDUCERS

The angle at which sound is refracted into a flowing medium is a function of the angle of incidence and the sound speeds of the materials involved. As temperature (sound speed) changes, the refractive angle changes, which in turn changes the magnitude of the flow component measured. A test fixture, shown in Figure 18-17, simplified to a single refracting interface will have a calibration shift of 0.18 percent for each degree Celsius change (0.11% per degree F). This theoretical change agrees well with test measurements. The measured flow component is a function of the angle of the sonic beam, not the angular location of the transducers.

CLAMP-ON DOPPLER TRANSDUCERS

A different situation exists with clamp-on Doppler transducers. As before, the angle at which the sound is refracted into the flowing medium is a function of sound speed. With reference to Figure 18-18, Snell's Law of Refraction provides:

$$\frac{\cos x}{Cx} = \frac{\cos p}{Cp} = \frac{\cos f}{Cf} \qquad (18\text{-}11)$$

From Equation (18-3),

$$v = \frac{f_d \cdot C}{(2 \cdot f_o \cdot \cos \theta)}$$

which can be rearranged to

$$v = \frac{f_d \cdot Cf}{(2 \cdot f_o \cdot \cos f)}$$

By substitution,

$$v = \frac{f_d \cdot Cx}{(2 \cdot f_o \cdot \cos x)} \qquad (18\text{-}12)$$

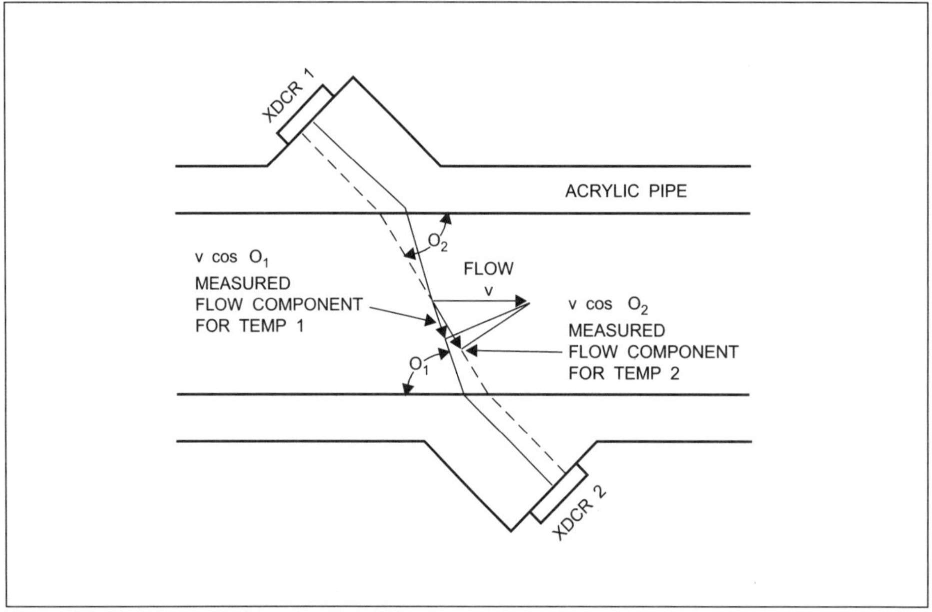

Figure 18-17. Clamp-On Temperature Effect

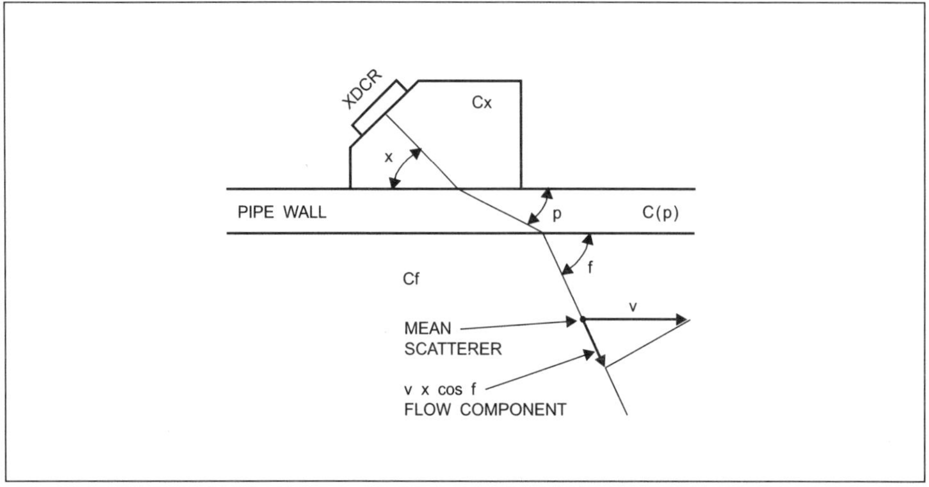

Figure 18-18. Clamp-On Sound Path

From this it is seen that the temperature (sound speed) of the transducer is the element controlling the calibration of clamp-on Doppler installations. Correction for temperature can be made by embedding a temperature sensor in the transducer. If there is no compensation, the calibration will change as a function of transducer temperature. Since various materials have different temperature dependence, the extent of the shift will vary.

Specifications and Installation Guidelines

In general, for each type of flowmeter, meters produced by different manufacturers exhibit similar performance. The difference lies in the optional equipment and features provided by the specific instrument and the availability of service. Overall needs more than the published accuracy specifications should be used as a basis for choice of a given flowmeter.

Specifications are often based on what can be accomplished under laboratory conditions with a carefully controlled flow medium. Such results are not always transferable to the field.

Accuracy

The fundamental measurement of sonic meters is velocity. Volumetric flow is inferred from velocity and, therefore, conditional on the mathematical exactitude of the conversion algorithm.

Transit time meters are "percent of full scale" devices. Since transit time meters are based on differential readings, they do not have a zero output for zero input inherent to the measuring technique. If a specification of percent of rate is given, it must be accompanied by a low velocity limitation.

Doppler meters, however, can be listed as "percent of rate" devices. When flow goes to zero, there is no relative motion within the medium and the Doppler signal goes to zero.

Associated with accuracy is resolution. Both transit time and Doppler meters are inertialess and measure instantaneous conditions. Each measurement is independent of all others, and there is no interaction as a function of time except in the data processing portion of the transmitter. Resolution is determined by the smallest difference of velocity that can be measured. Accuracy is given by the combination of resolution and the ability to determine zero. A third factor is repeatability, which is the ability to come back to the same point each reading under identical flow conditions. Since each measurement in a sonic meter is independent and inertialess, the repeatability of the flowmeter is the same as its resolution.

A simple statement of accuracy is difficult to make due to the large variations of conditions that control accuracy. As a rule of thumb, the following can be used as an approximation of transit time meter accuracy under *controlled flow lab conditions:*

Resolution	0.003 m/s	0.01 ft/s
Zero crror	0.003 m/s	0.01 ft/s

To this must be added other uncertainties such as the velocity profile effect and the accuracy of the conversion algorithm. Based on actual independent flow lab calibrations performed on various manufacturers' instruments, transit time meters with a full scale flow in excess of one meter per second (three feet per second) can be expected to provide between 0.25 and 1.0 percent of full scale if the installation duplicates the flow lab conditions. This, unfortunately, is seldom the case, and the accuracy of the actual installation is primarily a function of profile asymmetry.

It is much harder to estimate the accuracy of Doppler meters. An accuracy statement of one percent full scale is often given by manufacturers, but this applies to their testing conditions. In actual applications, with good attention to the installation detail, and flow in excess of one meter per second (three feet per second), these meters can be expected to provide 3 to 5 percent full scale performance.

Transit Time-Suspended Material

One of the most often asked questions concerns the effect of suspensions in the flow. The most common and also the most troublesome suspension for a transit time meter is air or other gases. If air is injected directly into the flow in the form of large bubbles, most transit time meters will operate satisfactorily with 50 percent by volume of bubbles. If, on the other hand, the gas is in the form of micron-sized bubbles, it is possible to lose signals with as little as 0.1 percent gas by volume.

A gas is much lower in acoustic impedance than a liquid. The small bubbles absorb sound to the point that there is nothing left to receive. Metals, on the other hand, have a much higher acoustic impedance than liquids and act as forward scatterers and, therefore, can pass sound. Micron-sized iron particles (taconite) have been satisfactorily measured at levels of 60% by volume. Conversely, if their size increases, they soon obstruct the sound so that nothing reaches the other side. The effect is the opposite that of a gas.

The practice of sending a sample of the flowing material for tests to see if it is satisfactory for measurement can be misleading. A sample of 6 percent paper pulp is quite satisfactory if it has a chance to out-gas on the way to the lab for tests. On the other hand, if a 2 percent stock has been aerated, it will probably be unreadable.

Following these guidelines will help to ensure a satisfactory installation.

(1) Provide adequate straight sections for radial meters.

(2) Do not attempt to utilize this technology to measure the flow of highly aerated materials such as primary sludge or flowing paper stock.

(3) Do not mount a meter at the discharge of a pump.

(4) Do not meter a fluid immediately after a free fall.

(5) Do not mount a wetted-transducer meter with the transducers in the vertical plane where bubbles can be trapped in the upper transducer well and solids in the lower well.

(6) If possible, install the meter in the highest pressure portion of the system where bubbles will be forced into solution.

(7) If the medium contains solids but no gas, have it tested for sonic transmissivity.

Doppler-Suspended Material

The great appeal of Doppler meters is their ease of installation and operation. Care must be taken, however, in the type of fluid measured. Doppler flowmeters require something to reflect or scatter sound. Published specifications can lead the user into some questionable installations if their meaning is not understood. A meter rated for concentrations of suspensions as low as 50 parts per million may give satisfactory performance at a flow of one meter per second (three feet per second) but quit abruptly at 0.3 meters per second (one foot per second). Raw sewage into a treatment plant will provide good signals during the day. Often in the early morning when most of the flow is from intruding ground water and the velocity is quite low, Doppler meters will stop working. Sewage with an average suspension of 200 ppm may have concentrations as low as 5 ppm under these conditions. However, if the meter is located near the discharge of a pump, there are often enough bubbles present to give adequate readings.

It is not necessary to have suspended scatterers to make a Doppler meter operate. Any flow-related disturbance that can change the volume reverberation of the signal will produce changes in the phase patterns of the receiving transducer. This will be interpreted by the meter as a Doppler signal. Irregularities inside the pipe such as flanges and fittings can cause vortices to be shed. The resulting velocity gradients may provide adequate disturbances to produce Doppler signals. Relatively high flow velocities are required to produce these disturbances, and because they are usually near the pipe wall they may not be traveling at the average flow velocity. For best accuracy, an *in situ* calibration should be conducted.

In situations where there are insufficient scatterers to produce a reliable Doppler signal, injected air can provide the signal source. A small quantity of air is injected upstream of the meter section. This should be done through something such as a ceramic filter that will produce very fine bubbles. When the resulting bubbles pass through the ultrasonic field, they will provide the necessary reflections. In some installations, such a location is provided by the process. Effluent from a water treatment plant, normally very low in suspended solids, can often be read by installing the meter downstream from the chlorine injector.

Applicable Fluids

TRANSIT TIME METERS

A list of fluids that are acceptable for measurement by transit time meters would be prohibitively long. They range from liquid hydrogen to molten sulfur. It is better to describe the required sonic characteristics. To be acceptable the flowing medium must:

(1) support the passage of sound from the transmitting transducer to the receiving transducer;

(2) be in a full conduit;

(3) be continuous, not pulsating flow; and

(4) contain no material to deposit on the pipe wall.

A word of caution concerning grease coatings is in order. A transit time meter with the pipe full of flowing grease will work if the grease is acoustically transparent. As long as the grease is clean and free from any form of entrained contaminants, relatively thick (1 to 2 cm) layers can be accommodated. The form of grease usually encountered in practice is anything but free from acoustically absorbent entrainment. A relatively thin layer of grease and trapped dirt can stop any sonic meter if the sound cannot get through the layer. Also bear in mind that even if the sound can get through a 0.25 centimeter (0.1 inch) layer in a 10 centimeter (4 inch) pipe, there will still be an 11 percent error in flow readings due to the reduction in cross section.

DOPPLER

A listing of the sonic requirements for a Doppler meter are similar to the transit time meter. To be acceptable the fluid must:

(1) support the passage of sound;

(2) contain sufficient scatterers or other disturbances to provide a Doppler reflection;

(3) not contain so many scatterers that the sound cannot penetrate into the flow;

(4) be in a full conduit;

(5) be continuous, not pulsating flow; and

(6) contain no material to deposit on the pipe wall.

Approaches to Problems

Transducer Sensitivity

When a sonic meter fails because of low signal strength, the transducers are often blamed. There are several simple tests to check the condition of the transducers before new ones are installed. If they are clamp-on Doppler transducers, remove them from the pipe. Hold them several inches apart with the radiating surfaces facing each other. Rapidly move them toward and away from each other over a distance of several inches. If a Doppler signal results, the transducers are functioning and the problem is probably in the transmitter or the application. Run this test before the meter is installed to get a feel for the procedure. Transit time transducers can be tested by connecting them to an oscilloscope and tapping the transducer face with a hard substance such as the edge of a coin. A good transducer should produce a pulse of several hundred millivolts when tapped. Try these tests before installing the instrument.

Removable Transit Time Transducers

A removable transducer is one that can be removed from the pipe without needing to drain the line. Such transducers are directly wetted by the fluid. This is done so that in the case of a transducer failure it can be replaced or, more likely, so that the transducer well can be cleaned if the inside of the pipe becomes coated. Figure 18-19 shows the concept of transducer removal. This is implemented in many ways by various manufacturers. Some have permanent valve installations for each transducer, while others have portable extraction tools that can be moved from transducer to transducer as needed. Some manufacturers provide "removable" transducers where the transducer is in an external well that is formed in an insert in the pipe wall. The insert is in contact with the fluid but the transducer is not. Such transducers, shown in Figure 18-20, are a form of clamp-on transducer and not truly a removable transducer. In this type of clamp-on configuration the most common failure is the acoustic coupling medium between the transducer and the insert in the pipe wall. Silicone grease is usually the recommended acoustic couplant, but over a period of time this will dry out and result in an acoustically disconnected transducer.

If decoupling problems are encountered, try using thin neoprene disks the diameter of the transducer. Coat both sides of the disk with a heavy oil. Diffusion pump oil from a vacuum pump is good because of its very low vapor pressure. A good substitute for the above is 20-50W motor oil. Use very little oil. Lightly wipe the disk so that there is a thin oil film left. Place the disk between the transducer and the well and clamp the transducer back in place. Keep the oil handy.

Fluid Path Simulation

When servicing transit time meters, it is desirable to have some means of simulating the flow path. This can be accomplished with a plastic rod. Take a Lucite® rod with a diameter of at least twice that of the transducers. The rod must have the same transit time as the sensor. If water (1480 m/s) is to be simulated by Lucite (700 m/s), there should be 700/1480 centimeters of Lucite for every centimeter of water path. If desired, transducer bosses can be machined in the end of the rods.

Couple the transducers to the plastic rod with any convenient acoustic couplant. Water soluble lubricants available at any drug store (KY Jelly™) are ideal for temporary acoustic couplants. The ends of the rods must be parallel to each other and have a smooth surface. While flow cannot be simulated, it is possible to test the entire meter system under zero flow conditions. This procedure provides a convenient zero-flow simulator, but a meter zero obtained with this

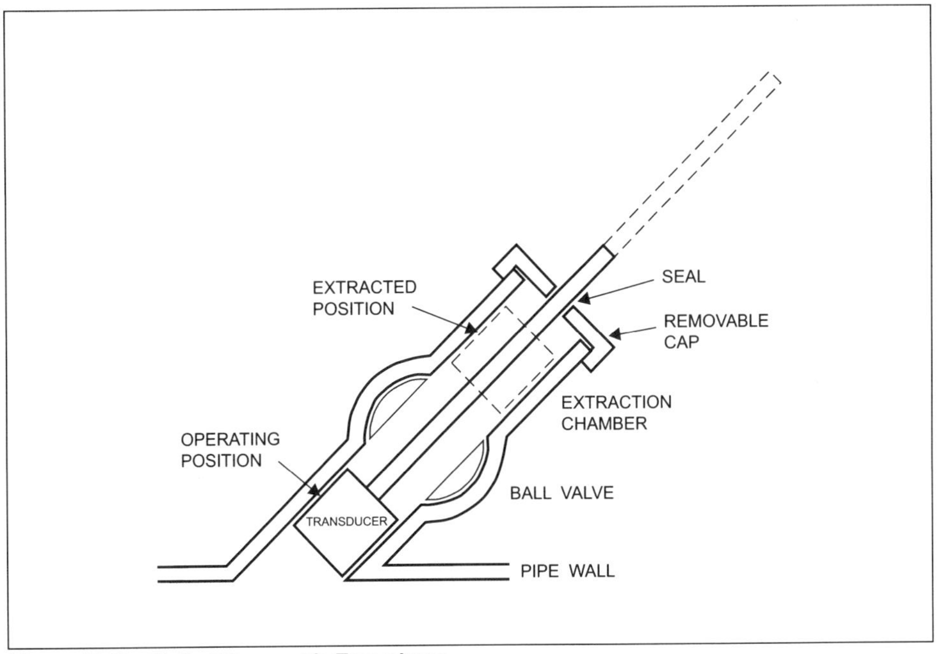

Figure 18-19. Wetted Removable Transducer

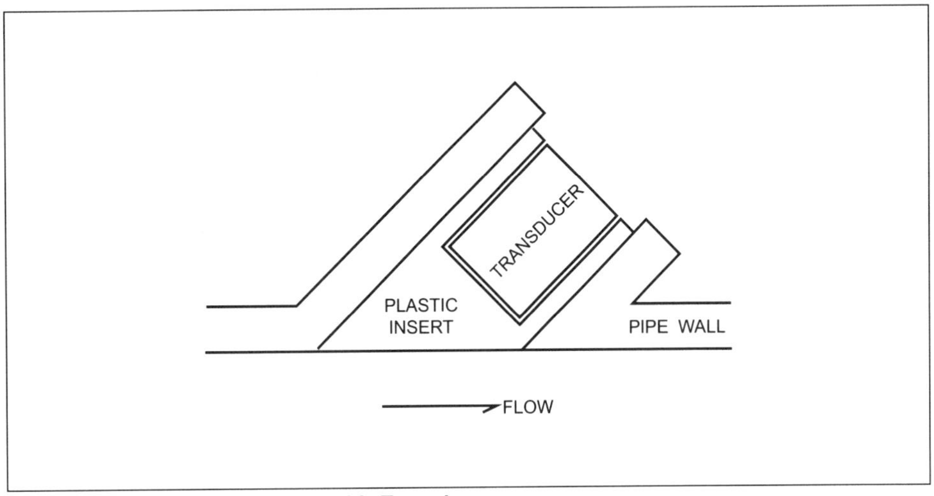

Figure 18-20. Non-Wetted Removable Transducer

device may not be transferable to an actual flow installation. Minor variations in signal phase position resulting from geometric differences between the simulator and the actual transducer installation may produce a slight difference in the meter zero.

Dynamic Zeroing

At times it is required to zero transit time meters when there is no practical means of stopping flow. Some manufacturers provide a means to trick the meter into thinking it has zero flow. This is usually done by transmitting all pulses in the same direction. All portions of the electronics are properly zeroed, but the transducers are not. Any phase difference in the transducers will be interpreted as flow. To determine if the transducers are producing any offset, transpose the transducer cables. The meter should read an identical reverse flow. If there is a difference between the two values, half is because of zero offset. Re-zero the meter until the forward and reverse flow readings are identical.

Externally Induced Doppler

An unusual problem has been known to occur with Doppler meters operating with low concentrations of scatterers. Under these conditions the receiver is running with maximum sensitivity because of the low reflected signal strength. The transducers project sound into the pipe wall, with a small amount being transmitted through the inner pipe wall into the flowing medium. Most of the signal is reflected back into the pipe wall. Some of this internally reflected signal arrives at the outside wall of the pipe. If there is a moving liquid on the outside of the pipe, it will reflect sound just the same as on the inside. Sometimes a single drop of water running down the outside of the pipe can be detected.

Be careful of installations where water can run over the outside of the pipe.

Installations with high concentration of scatterers do not usually have this problem because of the large Doppler signal. If external flow becomes a problem, coat the outside of the pipe with some form of sound-absorbing material or cover the pipe to prevent the liquid contact.

Vibration

Doppler transducers are not sensitive to low frequency vibrations. However, if there are vibrations large enough to shake the pipe, motion can be set up in the particles in the fluid. If these particles move, they will generate a Doppler signal that will be detected. It appears as if the transducers are picking up the vibration, but they are detecting actual Doppler signals generated within the shaking fluid. If this occurs, support the pipe to prevent the vibration or, if necessary, relocate the transducers to a more stable section of pipe.

Radio Interference

Doppler frequencies are chosen as a compromise between penetration into the liquid (low frequency) and flow sensitivity (high frequency). Practical Doppler frequencies are between 500 kHz and 1.5 MHz. Unfortunately, this is also the AM broadcast band. Radio stations can interfere with Doppler meters if the station is close and if the carrier frequency of the station is within several kilohertz of the Doppler carrier frequency. The radio signal usually gets into the transmitter through its wiring to the sensor. Careful attention to conduit connections and grounding can sometimes eliminate problems, but there is no sure cure. Another possibility is to change the frequency of the Doppler transmitter. The bandwidth of the transducers is usually wide enough to allow enough change to avoid the radio interference.

Fouled Transducers

When direct wetted transducers are used, there is a recess in the pipe wall left by the location of the transducer. Under most conditions this recess is kept clean by the action of eddies in the flow. In some cases, however, this recess can become fouled. An option to draining the line and removing the transducers for cleaning is to permanently install a purge port to keep the recesses clean. Figure 18-21 shows a typical installation with a purge port and spray nozzle. A spray pressure of 15 or 20 psi over line pressure will usually be satisfactory. Remember to provide check valves to protect the pressure source from contamination. Liquids are usually used, but gases have also proved satisfactory. Periodic injections (once a day) for a minute or two will usually keep the well free of deposits.

It should be noted that wetted transducers are the only form that can be cleaned in this manner. Internal coatings will render clamp-on and external removable transducer installations inoperative with no means of recovery. For installations with coating problems, wetted transducers with purge ports may be the best choice.

Transitional Flow

If a meter cycles about 20% with a period of one or two seconds at a fixed flow, the meter is probably operating in the critical zone of transitional flow. Figure 18-22 shows the change in calibration as a function of Reynolds number. The problem with the laminar-turbulent transition is rarely a consideration when measuring water flow, but problems do occur with other liquids. In some cases, it is not possible to avoid this type of transitional flow by adjusting pipe diameter.

A means of dealing with this problem is to place a straightening vane bundle in the middle of the meter section. Drill a hole through the bundle to accommodate the sonic path. The hole should be about 1.5 times the diameter of the transducer face (sound beam). The Reynolds number is now a function of the diameter of the tubes in the bundle rather than the entire pipe. The result is a drop in Reynolds number by a factor of approximately 5. This is usually sufficient to avoid operating in the transitional flow regime. This technique is not recommended for clamp-on transducer installations.

Splashing Flow

Transit time meters placed close to a wet well may have trouble from air entrained by the fluid falling into the well. Placing plywood or sheet metal baffles

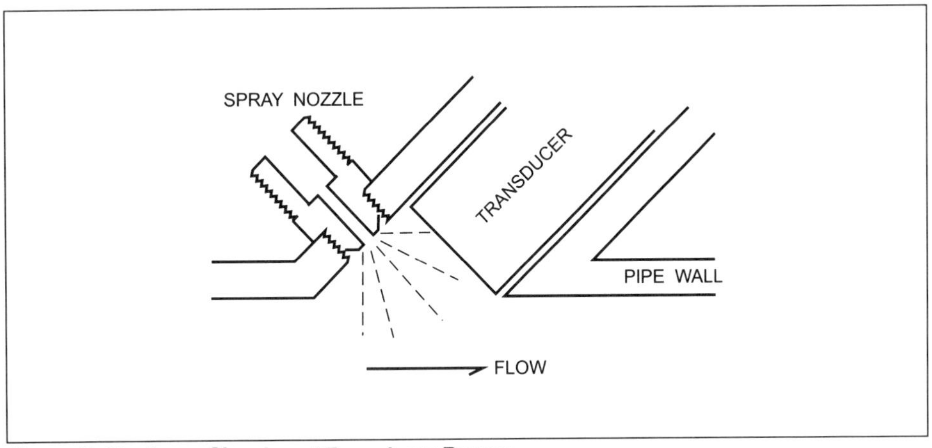

Figure 18-21. Spray Cleaning of Transducer Recess

to control the fall will reduce the entrained air. The baffle should extend close to the bottom of the well.

READINGS AT ZERO FLOW

On occasion, a sonic meter will continue to indicate flow after the flow has apparently stopped. In Doppler meters this is usually caused by the settling of solids or the rising of bubbles after flow has stopped. This is most noticeable in vertical pipe installations. Other than providing low-flow cutoffs, there is no way to prevent this actual movement of scatterers from producing a Doppler signal.

In transit time meters, a similar phenomenon is encountered when flow is stopped. If a valve is rapidly closed and the flow is stopped, inertial forces generate eddy currents in the pipe that can be detected as flow. While there is no net flow, there can be flow components along the sonic path. This eddy effect can persist for several minutes in larger pipes. With pipes over 0.15 or 0.20 meter (6 or 8 inches) in diameter, it is best to wait about five minutes for the eddies to become quiet before zeroing the meter.

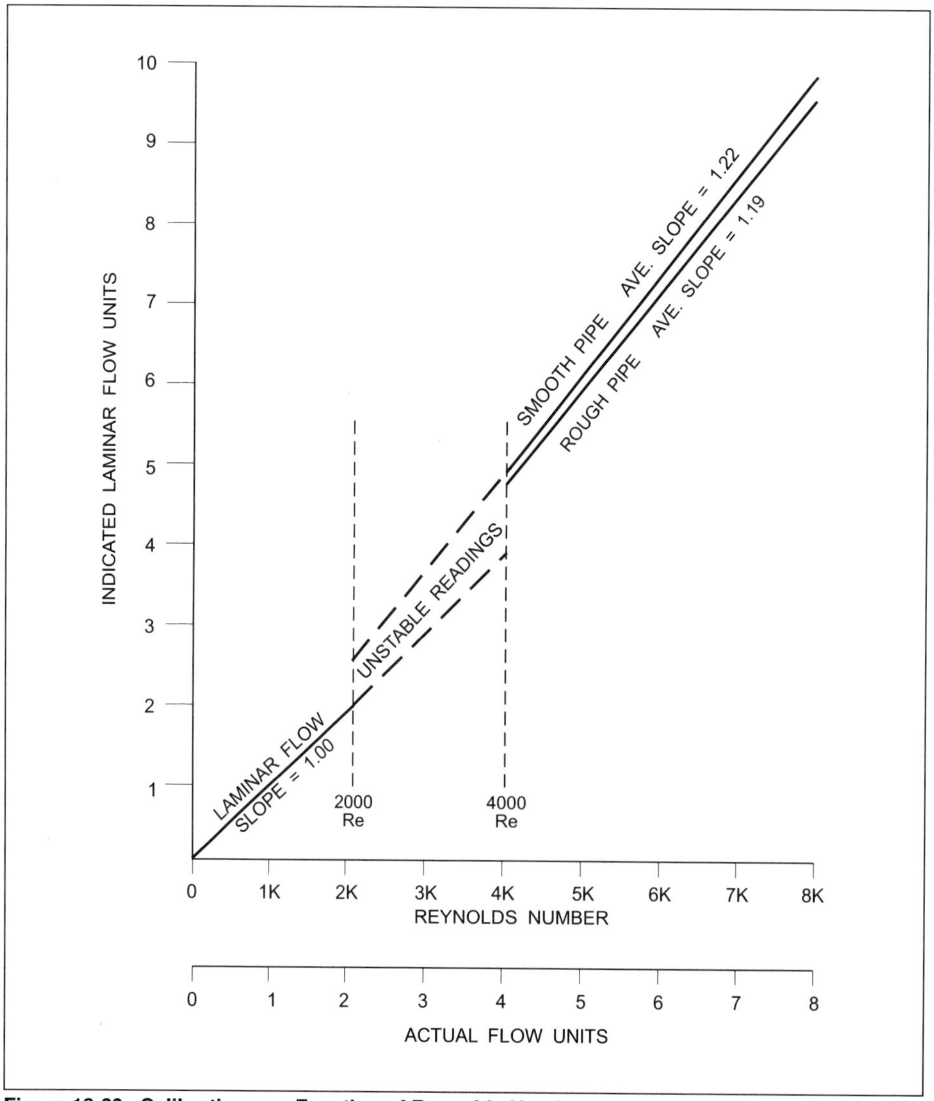

Figure 18-22. Calibration as a Function of Reynolds Number

Conclusion

Sonic flowmeters provide flow measurements not realizable by other means. The following examples are a few of the types of applications that show the versatility of these devices.

(1) Cryogenic flowmeters must have very low pressure drop. The sound-generating material in ultrasonic transducers is not damaged by the extreme low temperature. There are no moving parts to cause flashing in the cryogen and no pressure drop other than the pipe itself.

(2) Molten sulfur is a difficult material to measure. If it freezes, it must be thawed with blowtorches. It is very corrosive. Sonic meters are ideally suited to this service.

(3) Large diameter (0.6 m/2 ft) multi-meter crude oil on-line blending requires fast response and dependable operation. Meters must be capable of measuring a wide range of viscosities and velocities.

(4) Large diameter (3 m/10 ft) water conduits and rivers have few practical means of continuous measurement. Large Venturi meters are prohibitively expensive, and weirs and flumes do not work well in many rivers.

(5) High temperature (300°C) chemicals used in plastic manufacturing are non-conductive and can have no obstructions in the flow.

(6) Hazardous, reactive, and corrosive materials often use exotic materials of construction that are very expensive. A meter that attaches to the existing pipe without requiring modification to the pipe is desirable. The integrity of the pipe is maintained and cost is low.

References

1. Daugherty, R., and Franzini, J., *Fluid Mechanics with Engineering Applications*, New York: McGraw-Hill, 1965.
2. Schlichting, H., *Boundary-Layer Theory*, New York: McGraw-Hill, 1979.
3. Brown, A., "Method and System for Measuring Flow Rate," US Patent No. 4102186, 1978.
4. Shercliff, J., *The Theory of Electromagnetic Flow-Meters*, Cambridge Press, 1962.
5. Scott, R. W. W., "A Practical Assessment of the Performance of Electromagnetic Flowmeters—Limitations and Remedies," *Conf. on Fluid Flow Measurement*, Nat'l Engr. Lab., E. Kilbridge, UK, 1975.
6. Lynnworth, L. C., *Physical Acoustics*, Mason, P. M., and Thurston, R. N., editors, New York: Academic Press, 1979.

About the Author

Alvin E. Brown recently retired as Chief Engineer for TN Technologies, Inc. of Austin, Texas, where his principal duties were in the development of new sonic instrumentation. Beginning as a Laboratory Analyst with Boeing following his graduation from the University of Washington, he has been with Northrop, Lockheed, E. I. DuPont de Nemours and Company, Envirotech, and Manning Technologies, Inc., where he was Vice President, Research. In 1969, he became co-founder of Saratoga Systems, Inc., and developed sonic meters that are used internationally. He continues development work in his own laboratory.

He has presented technical papers at such diverse conferences as the Annual Cryogenic Engineering Conference, the Marine Science Conference at Woods Hole, and the Annual Scientific Meeting of the Aerospace Medical Association. He also was involved in the "Science in Action" Series for PBS.

19

Variable Area Flowmeters

Variable area flowmeters were invented in the nineteenth century but were not introduced for general industrial use until the 1930s. They became very popular and widely used since they offered a low-cost alternative to the differential pressure flowmeter—especially in smaller pipe sizes.

The original designs introduced were available for visual indication of flow rate only. In the 1940s designs became available with secondary functions such as transmission, alarm, recording, and totalizing. Variable area flowmeters are available in pipe sizes from ⅛ inch to 8 inches and full scale flow rates from less than one ml/minute to over 1000 gpm of water and approximately 30 scc/minute to 2500 scfm of air.

Since they are in-line devices, they have been designed for a wide range of line pressure ratings, pipe connection types, and many different materials of construction.

Operating Principle

The variable area flowmeter is, in essence, a special type of differential pressure flowmeter. On differential pressure flowmeters, the area of the opening is fixed and the flow rate is measured as a function of differential pressure across the opening. In the variable area flowmeter, the differential pressure across the opening is constant, and the flow rate is measured as a function of the area of the opening. This area is generally displayed as the position of a "float" or obstruction that is free to move to produce the varying area.

Advantages:
Simple design—one moving part
Requires no power
Relatively low cost

There are three general types of variable area flowmeters.

Rotameters

In this type, a float made of a material of a density greater than that of the fluid is contained in an upright conical tube whose smaller end is at the bottom. The float, free to move vertically in the tube, is lifted to the position of equilibrium between the upward force of the fluid flowing past the float and the downward force of gravity on the float. In its simplest form the tapered tube is made of glass that is graduated, allowing the flow rate to be read directly by observing the position of the float (see Figure 19-1(a)).

Orifice and Tapered Plug Meters

This type is equipped with a fixed orifice inside an upright cylindrical chamber. The float has a tapered body with the small end at the bottom and is free to move vertically through the orifice. The flow rate is indicated by the position of the tapered float (see Figure 19-1(b)).

Piston-Type Meters

In these meters a piston is closely fitted to a cylindrical sleeve with vertical slots or a series of ports that are progressively uncovered as pressure differential across the piston raises it. Flow through the ports exits the meter through an outer chamber. The flow rate is indicated by the position of the piston (see Figure 19-1(c)).

In those meters utilizing the force of gravity to balance the fluid forces, the flowmeter must be mounted in an upright position. Another style uses a spring force to balance the fluid forces. Many of these are virtually independent of the attitude of the flowmeter. All three styles may utilize a transparent tube or cylinder for visual indication of flow rate as a function of float or piston position or may make use of a magnetic coupling to sense float position for flow rate readout. This latter design allows the use of tubes or cylinders made of nonmagnetic metals such as austinitic stainless steel.

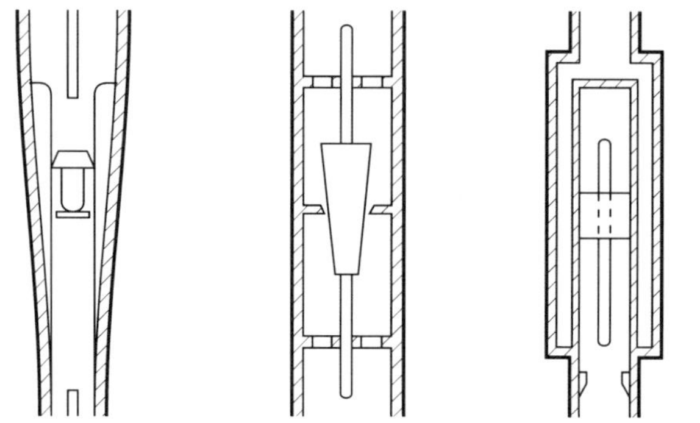

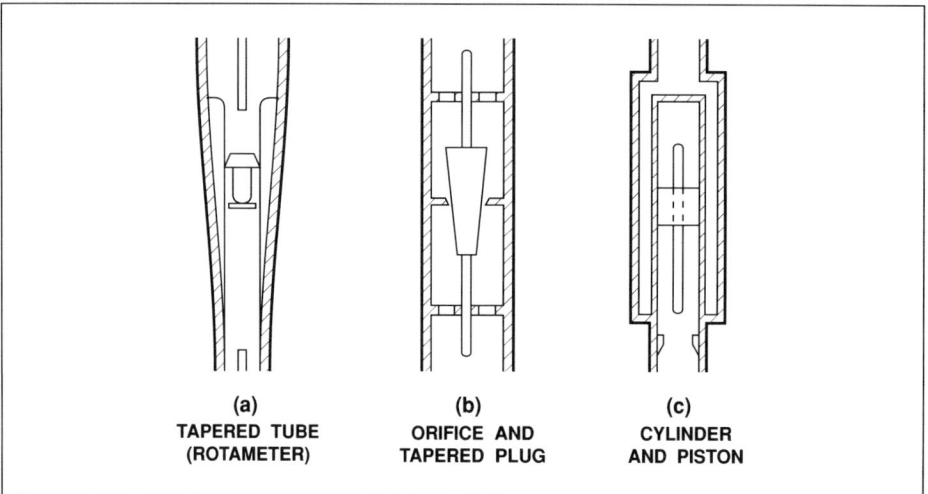

Figure 19-1. Types of Variable Area Flowmeters

Application

Variable area flowmeters can be used to measure the flow rate of almost any liquid or gas.

Variable area flowmeters can be used to measure the flow rate of nearly any liquid or gas. While there is no theoretical limit to the value of viscosity at which they can measure flow rate, practical calibration considerations generally limit their use to fluid viscosities of several hundred centipoise. Some rotameter designs have been used successfully to meter low concentration slurries.

The greatest application of valuable area flowmeters is flow rate indication in relatively small pipe sizes. They represent one of the least expensive methods of measuring rate of flow in pipe size 2 in. and smaller.

The following are some of the major areas of application:

(1) Bubble tube feed for dip tube level measurement

(2) Purge feed to pressure tap lines on differential pressure flowmeters to prevent process fluid from backing up into pressure tap lines

(3) Purge feed to tanks or vessels to blanket a liquid with an inert gas

(4) Purge flow to pump gland seals

(5) Low flow process fluid measurements for batching or continuous control

(6) Flow rate indication for setting flows necessary on various types of chemical analyzers

(7) Low or high flow alarms on cooling water or lubricating oil lines to protect expensive or critical equipment from overheating or lubrication failure

(8) Transmission of flow rate signal for remote indication or control loop operation

(9) Setting and/or monitoring feed rates on filtration systems

(10) Measurement of service fluids such as plant air, plant potable water, cooling water, fuel gas, or steam for purposes of departmental allocations, cost management, and energy conservation

(11) Laboratory flow measurements

(12) Test flowmeters for equipment testing of valves, pumps, prime mover engines, etc.

Basic Equation

The following is a nonrigorous derivation of the basic flow equation for rotameters that allows the development of precise working equations. Reference is made to Figure 19-2. The following are assumptions made to simplify the analysis:

(1) The float has a sharp metering edge.

(2) The fluid velocity at section 1 is negligible compared to that at section 2.

(3) The tube taper is small enough that there is a negligible difference in tube diameter at section 1 and section 2.

(4) The flow lines crossing each section are normal to the sections.

Definition of Terms

v_1 = velocity of fluid at section 1
v_2 = velocity of fluid at section 2
g = acceleration due to gravity
h_1 = hydraulic head at section 1
h_2 = hydraulic head at section 2
P_1 = pressure at section 1
P_2 = pressure at section 2
ρ = density of fluid
Q = flow rate, volumetric units
W = flow rate, weight units
A_2 = area at Section 2
C = coefficient of discharge

According to Bernoulli's theorem:

$$v_2^2 - v_1^2 = 2g(h_1 - h_2) \tag{19-1}$$

The hydraulic head drop can be expressed in terms of pressure drop as:

$$h_1 - h_2 = \frac{P_1 - P_2}{P} \tag{19-2}$$

and the continuity of flow equation may be written as:

$$W = A_2 v_2 \rho \tag{19-3}$$

To allow for factors not included in this analysis, a coefficient of discharge C is introduced. (This factor is normally determined empirically.) Equation (19-3) then becomes:

$$W = CA_2 v_2 \rho \tag{19-4}$$

Neglecting v_1, which is assumed negligible compared to v_2, and combining Equations (19-1) and (19-4), the expression becomes:

$$W = CA_2\sqrt{2g(P_1 - P_2)\rho} \tag{19-5}$$

To express the flow rate in volumetric units:

$$Q = \frac{W}{\rho} \tag{19-6}$$

Substituting $Q\rho$ for W in Equation (19-5):

$$Q = CA_2 \frac{\sqrt{2g\,(P_1 - P_2)}}{\sqrt{\rho}} \tag{19-7}$$

For weight-loaded floats of the rotameter, tapered plug, or piston type, the differential pressure can be assumed to equal the weight of the float with allowance for buoyancy divided by the area of the float head or piston:

$$P_1 - P_2 = \frac{W_f \dfrac{(\rho_f - \rho)}{\rho_f}}{\dfrac{\pi D_f^2}{4}} \tag{19-8}$$

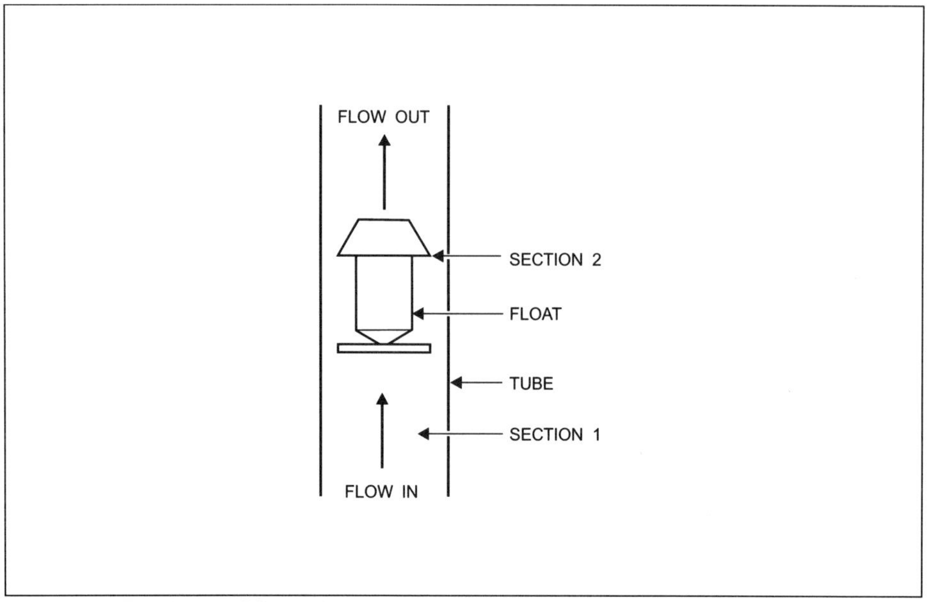

Figure 19-2. Fundamental Operation of a Variable Area Flowmeter

where:

W_f = Float weight
D_f = float diameter
ρ_f = Float material density

Substituting Equation (19-8) into Equations (19-6) and (19-7) and eliminating constants by including them in the coefficient, C:

$$W = CA_2 \frac{\sqrt{2g\rho W_f(\rho_f - \rho)}}{\sqrt{\rho_f}} \tag{19-9}$$

and

$$Q = CA_2 \frac{\sqrt{2gW_f(\rho_f - \rho)}}{\sqrt{\rho\,\rho_f}} \tag{19-10}$$

Flowmeter Design

The greatest asset of the variable area meter is its simplicity. It is a mechanical device not requiring complicated or expensive components; thus, in small pipe sizes it is one of the least expensive means of indicating rate of flow. This advantage disappears, however, in increased pipe sizes. The tube, float, and body become prohibitively expensive in sizes above 3 or 4 in. when compared to other types of flowmeters.

The basic components of variable area flowmeters are the meter tube, float or piston, body, end connections, seals and scale (see Figure 19-3).

Tubes

The resolution and, therefore, the potential accuracy of rotameters or piston meters is a function of their length. There is a trade-off between length (and cost) and accuracy. Typical glass tube rotameter scale lengths are 3 in., 5 or 6 in., 10 in. and, in a few cases, 24 in. The original glass tubes developed industrially in the 1930s were plain conical sections, and the floats were kept centered by spin stabilization. This was accomplished by using floats shaped like a carpenter's

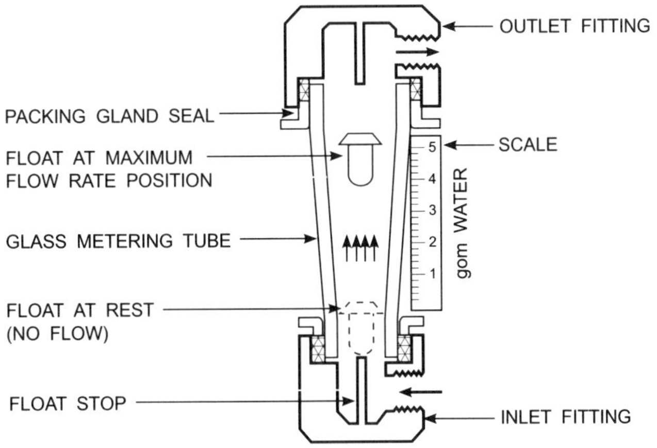

Figure 19-3. Basic Components of a Rotameter

plumb bob with slanted notches in the float head diameter to impart spin—hence, the name rotameter. Current practice for most suppliers is to center the float with ribs or beads molded into the glass tube (see Figure 19-4).

Due to the friable nature of glass, extra precautions must be taken in the selection, installation, and operation of flowmeters using glass tubes or cylinders as part of the pressure boundary. Carefully review and comply with all data in the manufacturer's installation and operating manual.

Floats

Floats can be made from a wide array of materials but must have density greater than that of the fluid if the meter is a gravity dependent design. The flow rate in any given tube is dependent on both the weight of the float and its shape. Suppliers optimize the float design for one or more of the following variables:

(1) Flow rate

(2) Viscosity immunity

(3) Pressure drop

(4) Simplicity (low cost)

(5) Durability

Figure 19-5 shows some common float shapes.

For glass tube rotameters it is very important to check the operating manual to ensure that the float is installed right side up and that the proper edge of the float is used to read the scale.

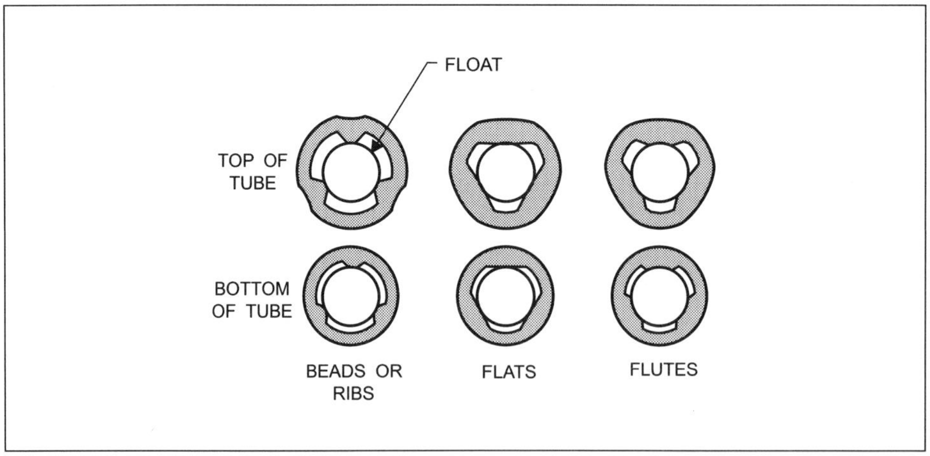

Figure 19-4. Rotameter Tubes with Ribs or Beads for Float Guides

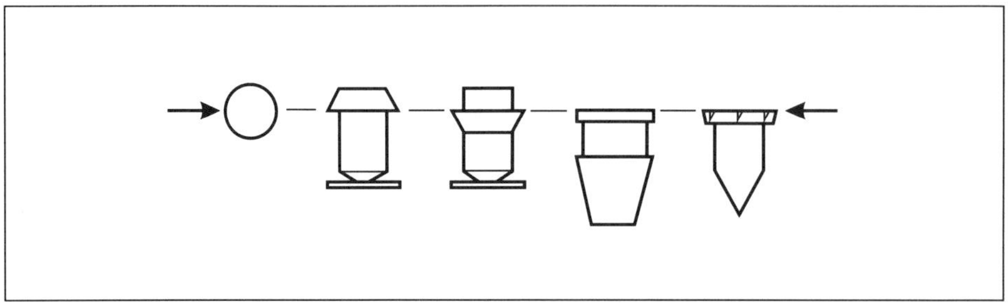

Figure 19-5. Partial Representation of Typical Rotameter Float Shapes

Bodies

The function of the body is to hold the metering components together and to provide a means of connecting them to the pipe line. Its design must have adequate strength to meet the forces created by the internal pressure as well as external forces such as pipe strain or vibration. In some styles, the body is part of the pressure boundary and is wetted by the process fluid. In those cases the body must be made of a material that is compatible with the process fluid. Bodies that house glass tubes should be equipped with a clear plastic or safety glass shield that will allow clear observation of the float and scale yet provide protection for the viewer in the event of accidental rupture of the glass tube under pressure.

End Fittings/Connections

Since variable area flowmeters are in-line devices, they must be equipped with end connections that allow insertion into a pipe line. The connection must seal against line pressure and generally must afford the meter support, since most variable area flowmeters are supported solely by the pipe. Glass tube rotameters require a method of sealing the tube to the metal or plastic end fittings. In many metal tube rotameters and piston-type meters, the metal tube and end fittings are an integrally welded assembly. The most common type of pipe connection for small sizes is pipe threads, the majority of which are female national pipe threads (FNPT). Metal meters, and those of larger sizes (2 in. and larger), are often available with flanged connections, ANSI class 150 and class 300 being the most common. Special flange facings such as tongue and groove or ring joint are also available from some manufacturers.

Other connection types available are hose connection, various compression seal connectors with straight threads, solder joint on brass or bronze bodies, and socket weld on steel or alloy steel bodies.

Do not assume that by specifying ANSI Class 300 flanges you can increase the safe working pressure of a glass tube rotameter. In many cases the rated pressure is dictated by the glass tube regardless of the rating of the flanges.

Seals

The seal between glass rotameter tubes and fittings can be a static O-ring seal or a packing gland seal. O-ring seals are generally used where there will be no corrosion of the material of the end fitting. Packing glands are somewhat more forgiving of slight corrosion of the sealing surface of the fitting and also allow the use of Teflon™ seals.

Materials of Construction

Tubes

Glass is the most commonly used material for the rotameter tube. Tubes of stainless steel or other nonmagnetic metals are also available as well as Teflon lined tubes and tubes of polyvinylchloride (PVC). Metering tubes are also available fabricated of transparent plastic such as acrylic, polycarbonate, or polysulfon.

Floats

Since the shape, diameter, and weight of the float are all critical to the accuracy of the flowmeter readings, it is important that the float exhibit no corrosive attack by the process fluid. Stainless steel is by far the most commonly used material, but manufacturers generally offer many other materials to handle those applications for which stainless steel is not entirely impervious to corrosive attack. These other materials include glass or sapphire for ball-type floats, PVC, Teflon, Hastelloy-C™, Monel™, tantalum, and titanium.

A rotameter was to be selected for indicating the flow rate of 93% sulphuric acid at room temperature. Generally, carbon steel fittings and a stainless steel float are deemed satisfactory for this application. If the concentration falls below 93% for short periods of time, the stainless steel float could suffer some corrosive attack. In this case, it might be prudent to use a float made of one of the high nickel-chromium alloys, which are satisfactory for all concentrations of sulphuric acid.

End Fittings/Bodies

End fittings and bodies are generally made of brass, carbon steel, or stainless steel but are also available fabricated from a wide range of structural materials that range from cast iron to high nickel-chrome alloys. Glass tube rotameters are also available with PVC and Kynar™ end fittings.

Seals

Seals are typically Neoprine™ or Teflon for packing glands or nitrile or Viton™ for O-ring construction.

As a rule of thumb, materials deemed satisfactory for valves give a good indication of suitability for use in variable area flowmeters. The material used for the valve body can be used for the flowmeter body wetted parts and end fittings. The material used for valve trim can be used for the flowmeter float or piston as well as the tube if the meter is of the metal tube type. The seal material used in the valve packing can generally be used as the O-ring or packing gland for a glass tube flowmeter.

Performance Data

Accuracy

The rotameter is inherently repeatable since it has only one moving part with virtually no friction or hysteresis. Since the indicator scale is nearly linear, the resolution or readability and accuracy are rated in terms of percent of full scale and are a function of the scale length. Typical rated accuracies are shown in Table 19-1.

Table 19-1. Typical Rotameter Accuracies

	Scale Length			
	1 ½ to 3 in.	5 to 6 in.	9 to 12 in.	20 to 24 in. (Logarithmic scale)
Repeatability	±2% FS	±½% FS	±¼% FS	±¼% Rate
Composite accuracy using prediction data	±5%-10% FS	±2% FS	±2% FS	±1%-2% Rate
Composite accuracy with custom fluid calibration	±2%-5% FS	±1% FS	±1% FS	±½%-1% Rate

Linearity

A conical rotameter tube or a piston meter with a constant area slot will produce a curve of flow rate versus float travel that is nearly linear. These flowmeters typically deviate from linearity by approximately 5 percent. When equipped with the transmitting function, linearity adjustments are usually supplied to guarantee linearity of the output versus flow rate. The linearity on transmitters is usually rated at ± 1% of full scale.

When selecting a variable area flowmeter with an accuracy of 2 percent of full scale, remember that the accuracy in terms of percent of rate deteriorates at lower flow rates. For example, the percent of rate accuracy at half scale (50%) is 4% of rate and at 25% of scale is 8% of rate. It is prudent to specify a maximum flow rate that is not more than about 30% higher than the normal flow.

Flow Ranges

The factors that determine flow range in a given size meter are (1) area of the orifice, (2) shape of the float, and (3) weight of the float. Typical maximum flows of nominal rotameter tube sizes are shown in Table 19-2. Nearly all variable area flowmeters are supplied with scales having at least a 10:1 turndown.

Maximum Pressure Rating

The pressure rating of glass tube meters is a function of the nominal tube diameter. On flanged meters, the pressure rating cannot exceed the ANSI code rating regardless of the strength of the glass or metal tube. Typical pressure ratings are shown in Table 19-3.

Table 19-2. Typical Range of Maximum Flow Rates

Nominal tube size, in.	Water	Air
⅛	0.5-200 cc/min	50-7500 scc/min
¼	100-2000 cc/min	4000-34000 scc/min
⅜	0.13-0.55 gpm	075-2.4 scfm
½	0.25-4.0 gpm	1-20 scfm
¾	1.9-5.0 gpm	8-20 scfm
1	4.0-20 gpm	20-45 scfm
1½	9.0-50 gpm	38-112 scfm
2	20.0-100 gpm	80-200 scfm

Table 19-3. Typical Pressure Ratings for Glass Tube Meters with Pipe Thread Connections

Tube Size, in.	Typical Maximum Rated Pressure	
	psig	kPa
1/16-¼	250-500	1724-3448
½	300	2069
¾	200	1379
1	180	1241
1½	130	896
2	100	690
3	70	483

Note: 1/16 in. to ½ in. glass tube meters with ANSI class 150 flanged connections would be limited to a rating of 270 psig (1826 kPa) @ 100°F by the ANSI code rating.

Temperature Rating

The minimum process fluid temperature of most variable area flowmeters is limited by the formation of frost on the exterior surfaces, which limits the visibility of the float in glass tube meters or inhibits the motion of the magnetic follower in metal tube meters. Typical low temperature ratings for the flowing fluid are between -4°F(-20°C) and 32°F(0°C).

The maximum temperature rating is generally limited by the elastomeric seal material for glass tube meters or the electronic components in electronic transmitters or alarm-type meters. Typical maximum temperature ratings are shown in Table 19-4.

Table 19-4. Typical Maximum Temperature Ratings

Glass tube meters	250°F-350°F (121°F-177°C)
Metal tube, indicating only	400°F-500°F (204°C-260°C)
Metal tube, transmitters or alarms	350°F-450°F (177°C-232°C)

Reynolds Numbers

When specifying a variable area flowmeter, it is important to give the viscosity at operating temperature. A viscosity value at 100°F is of no value to a flowmeter manufacturer for sizing and calibrating a flowmeter that will be operated at 200°F.

Because the viscosity effect on variable area floats or pistons varies so much from one float design to another, manufacturers do not use Reynolds number to rate their flowmeters. Generally, the viscosity "rating" of variable area meters is expressed as an upper limit in centipoise. Sharp edged floats weighing several grams or more (most ½ in. and larger floats) exhibit negligible viscosity effects on their calibration at viscosities of 1 centipoise or less. This means that standard prediction scale data can be used for water, water-like solutions, and all gas service applications. However, the very small floats, usually spherical, used in 1/16 in., ⅛ in., and ¼ in. rotameter tubes are so low in mass that they are sensitive to viscosity effects even in the very low viscosities exhibited by gases. These low-flow meters and larger meters operated at viscosities over their limit must have empirically determined prediction curve data established in order to supply scales with accurate flow data. Table 19-5 shows the typical range of viscosity limits for rotameters designed for optimum insensitivity to viscosity. Figure 19-6 shows the effect of viscosity on the full scale flow range of a typical rotameter float.

Table 19-5. Typical Viscosity Limits with Floats Designed for Optimum Insensitivity to Viscosity (Glass Tube Rotameters)

Size, in.	Viscosity Limit, Centipoise
½	2 cP-7.5 cP
¾	7.5 cP-15 cP
1	15 cP-20 cP
1½	20 cP-35 cP
2	35 cP-50 cP

Pressure Drop

Total pressure loss in a variable area flowmeter is the sum of the fixed loss across the float and the variable frictional loss through the fittings and body. The pressure loss shown by manufacturers is usually the total loss across the meter at 100% of flow. In the lower flow ranges on a given nominal size,the pressure loss

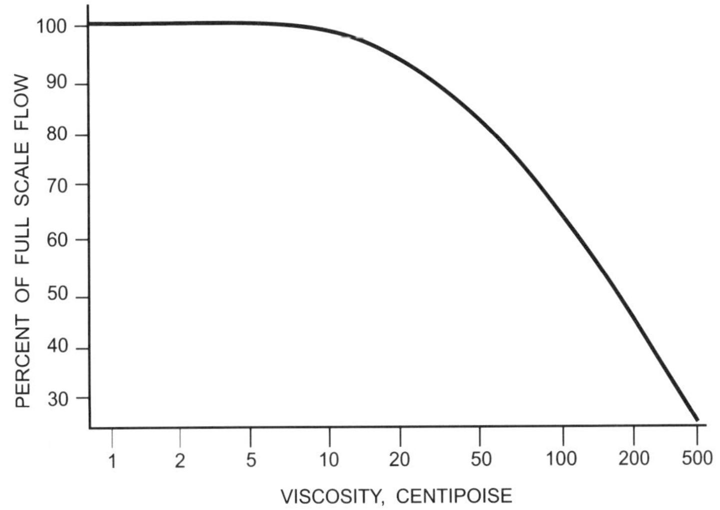

Figure 19-6. Typical Viscosity Effect Curve for Sharp Edged Rotameter Float

is quite low, usually under 0.5 psi. However, for higher flow rates the pressure loss can be as high as several psi at full scale flow. Manufacturers can supply meters with extremely light floats that will have very low pressure losses. The price paid for these low losses is that the nominal tube size is much larger than normally used for the given flow rate.

The pressure loss across the float for liquid flows can be calculated using the expression:

$$h_l = \frac{(0.078)\, W_f\, (\rho_f - \rho)}{D^2_f\, \rho_f \rho} \tag{19-11}$$

For gas flows the expression becomes:

$$hg = \frac{(0.078)\, W_f}{D_f^2} \tag{19-12}$$

where:

h_l = pressure drop across float, inches of the flowing liquid
hg = pressure drop across float, inches of water
W_f = float weight, grams
D_f = float diameter, inches
ρ_f = float density

A glass tube meter was ordered to indicate a flow rate of 3 gpm of a water-like liquid whose source is a head tank with a level of 24 inches above the outlet of the flowmeter. The user did not specify an operating pressure or a pressure loss limit. When installed, the float would not rise from the bottom with the metering valve wide open. The flowmeter turned out to have a pressure loss of 50 inches of water at 3 gpm, and 25 inches of water at 0.3 gpm. Because there was some loss in the ½ in. pipe and the metering valve, only 10 inches of usable head was available for the rotameter. By replacing the ½ in. meter with a 1 in. meter sized for the same 3 gpm, the pressure drop through the meter was reduced to 7 inches of head, allowing the feed and metering of the desired flow rate of 3 gpm.

Power Requirements

Variable area flowmeters used only for indication need no power for operation. Most electronic transmitters, alarm units, and totalizers supplied for the U.S. market require 115 V 60 Hz power or a 24-48 V 2-wire DC power supply.

Sizing Calculations

For general-purpose variable area flowmeters, suppliers usually show flow ranges in terms of water flow or air flow metered at 14.7 psia and 70°F.

To determine meter size for an application other than water or air at 14.7 psia and 70°F, it is necessary to first determine the equivalent flow of water or air. The working equations for calculating water and air equivalents can be developed from Equations (19-9) and (19-10).

Gravimetric Units

Starting with Equation (19-9) and setting up a ratio between the flow rates for fluid density 1 and density 2,

$$\frac{W_2}{W_1} = \frac{CA_2 \dfrac{\sqrt{2g\rho_2 W_{f2}(\rho_{f2}-\rho_2)}}{\sqrt{\rho_{f2}}}}{CA_2 \dfrac{\sqrt{2g\rho_2 W_{f1}(\rho_{f1}-\rho_1)}}{\sqrt{\rho_{f1}}}} \qquad (19\text{-}13)$$

Assuming that the coefficient of discharge C does not change with density, it and all rotameter physical constants will cancel, giving the gravimetric correction formula:

$$W_2 = W_1 \frac{\sqrt{\rho_2(\rho_{f2}-\rho_2)}}{\sqrt{\rho_1(\rho_{f1}-\rho_1)}} \qquad (19\text{-}14)$$

Volumetric Units

Similarly, starting with Equation (19-10), the volumetric correction formula is:

$$\frac{Q_2}{Q_1} = \frac{CA_2 \dfrac{\sqrt{2gW_{f2}(\rho_{f2}-\rho_2)}}{\sqrt{\rho_2\,\rho_{f2}}}}{CA_2 \dfrac{\sqrt{2gW_{f1}(\rho_{f1}-\rho_1)}}{\sqrt{\rho_1\,\rho_{f1}}}}$$

$$Q_2 = Q_1 \frac{\sqrt{\rho_1(\rho_{f2}-\rho_2)}}{\sqrt{\rho_2(\rho_{f1}-\rho_1)}} \quad \text{(Volumetric Correction Formula)} \qquad (19\text{-}15)$$

Standard Volume Units

Many gases used in variable area flowmeters follow the ideal gas laws very closely. In such cases it is often convenient for users to use "standard" or "normal" volume flow units such as standard cubic feed per minute (scfm) or normal cubic meters per hour (m^3/hr). Standard conditions used by the majority of the U.S. manufacturers of flowmeters are 14.7 psia and 70°F in English units and 1 bar and 0°C in metric units, but user standard conditions can be different. A convenient correction formula for standard volume flow of perfect gases can be derived from the gravimetric correction formula, Equation (19-14).

$$Q_{s2} = Q_{s1}\frac{\sqrt{SG_1 \times P_2 \times T_1}}{\sqrt{SG_2 \times P_1 \times T_2}} \tag{19-16}$$

where:

Q_{s1} = standard volumetric flow at fluid condition 1
Q_{s2} = standard volumetric flow at fluid condition 2
SG_1 = specific gravity of fluid condition 1
SG_2 = specific gravity of fluid condition 2
P_1 = absolute pressure at fluid condition 1
P_2 = absolute pressure at fluid condition 2
T_1 = absolute temperature at fluid condition 1
T_2 = absolute temperature at fluid condition 2

Find the water equivalent of 10 gpm of liquid with an operating density of 1.52 g/ml. Using Equation (19-15) and assigning subscript 1 to the liquid with density of 1.52 and subscript 2 to water and maintaining consistent units:

$$Q_2 = Q_1\frac{\sqrt{\rho_1(\rho_{f2} - \rho_2)}}{\sqrt{\rho_2(\rho_{f1} - \rho_1)}}$$

Assume a float of stainless steel, $\rho_{f1} = \rho_{f2} = 7.92$ g/ml

$$\text{water equivalent} = 10\frac{\sqrt{1.52(7.92 - 1)}}{\sqrt{1.0(7.92 - 1.52)}}\text{ gpm}$$

$$= 12.8\text{ gpm}$$

For determination of water equivalent for liquid flow rotameters, it is essential to use the float density upon which the manufacturer's capacity table is based. If the specifications call for a float other than that upon which the capacity table is based, it is necessary to use the density of the float specified as ρ_f in the denominator, e.g., if specifications call for a Hastelloy-C™ float $\rho_{f1} = 8.94$ (Hastelloy C) and $\rho_{f2} = 7.92$ (stainless steel).

Find the air equivalent of 100 scfm of helium metered at 50 psig and 120°F (specific gravity of helium = 138). Using Equation (19-16), where subscript 2 = air at standard conditions, subscript 1 = helium at operating conditions, and P_2 = 14.7 psia, P_1 = 50 + 14.7 = 64.7 psia, T_2 = 460 + 70 = 530°R, T_1 = 460 + 120 = 580°R:

$$\text{air equivalent} = 100\frac{\sqrt{1.52 \times 14.7 \times 580}}{\sqrt{1.0 \times 64.7 \times 530}}\text{ scfm}$$

$$= 61.5\text{ scfm}$$

When using Equation (19-16) for determining air equivalent, it is essential that the absolute pressures and temperatures be used. It is also important that the standard conditions upon which "standard" volume is based are equal. If the capacity table standard conditions are different from those specified, one set must be corrected.

Find the air equivalent of 10 m³/hr of hydrogen metered at 3 kg/cm² gage and 25°C. Assume that normal conditions are 1 atmosphere (14.696 psia) and 0°C (32°F).

Assuming the supplier's capacity table is based on scfm where standard conditions are 14.7 psia and 70°F, the differing standard conditions and units must be corrected.

$$\text{scfm} = \text{m}^3/\text{hr} \times \frac{\text{ft}^3}{\text{m}^3} \times \frac{1}{\text{min/hr}} \times \frac{P_{\text{normal}}}{P_{\text{standard}}} \times \frac{T_{\text{standard}}}{T_{\text{normal}}}$$

$$= 10 \times \frac{35.31}{60} \times \frac{14.68}{14.7} \times \frac{530}{492} = 6.33$$

Now that the normal cubic meters per hour value has been converted to scfm, the standard variable area flowmeter density correction must be made using Equation (19-16):

$$P_1 = 3 \times 14.22 + 14.7 = 57.36 \text{ psia}$$

$$T_1 = 25 \times \frac{9}{5} + 32 + 460 = 537°\text{R}$$

$$\text{air equivalent} = 6.33 \frac{\sqrt{0.0695 \times 14.7 \times 537}}{\sqrt{1.0 \times 57.36 \times 530}} = 0.850 \text{ scfm}$$

Find the air equivalent for 1000 pounds per hour (lb/hr) of a vapor whose density at operating temperature and pressure 0.198 pounds per cubic foot.

Use Equation (19-14):

$$W_2 = 1000 \times \frac{1}{60 \text{ min/hr}} \frac{\sqrt{\rho_2(\rho_{f2} - \rho_2)}}{\sqrt{\rho_1(\rho_{f1} - \rho_1)}}$$

But for gases, $\frac{\rho_{f2} - \rho_2}{\rho_{f1} - \rho_1}$ is very nearly equal to unity, when $\rho_{f2} - \rho_{f1}$ (usually, for stainless steel: $\rho_{f2} = \rho_{f1} = 7.92$

Therefore:

$$W_2 = \frac{1000}{60} \frac{\sqrt{\rho_2}}{\sqrt{\rho_1}}$$

The density of air at standard temperature and pressure (70°F and 14.7 psia) is 0.075 pound per cubic foot.

Therefore:

$$W_2 = \frac{100}{60} \frac{\sqrt{0.075}}{\sqrt{0.198}} = 10.26 \text{ pounds per minute}$$

$$\text{air equivalent} = 10.26 \times \frac{1}{0.075} = 136.8 \text{ scfm}$$

Safety

Since variable area flowmeters are in-line devices, they contain a pressure boundary that must withstand the static line pressure. It is incumbent upon the system designer to ensure that the pressure and temperature rating of the flowmeter specified is adequate for the maximum possible pressure to be encountered in that line. Glass tube meters are available with polycarbonate shields designed and tested to prevent broken glass from exiting forward in the event of accidental tube rupture. Notwithstanding this safety feature, it is recommended that all lines containing a glass tube rotameter be equipped with a pressure relief device (valve or rupture disc) set to open at a pressure somewhat under the pressure rating of the flowmeter.

Glass tube rotameters equipped with shields should never be operated without the shield in place.

Care must also be taken not to exceed the pressure rating of flanged flowmeters, bearing in mind that the ANSI flange pressure ratings decrease with increased operating temperature.

Calibration

Most rotameters are originally supplied with a scale for a specific application. The scale can be (1) direct reading in engineering units, (2) percentage with a 100% factor for a specific fluid, or (3) a reference scale used with a curve plotting the flow rate of a specific fluid versus the reference scale graduations.

It is important to maintain records to ensure that correct scale data is used with each flowmeter.

Because the variable area flowmeter readings are affected by fluid density, it is necessary to either order a new scale or make the necessary corrections when changing the liquid density or gas specific gravity, pressure, or temperature. Flow corrections can be calculated using Equations (19-14), (19-15), or (19-16).

Calculate liquid flow correction for liquid flow in volumetric units using the following:

Liquid specific gravity: 0.86
New liquid specific gravity: 0.95
Float: stainless steel (specific gravity 7.92)

Using Equation (19-15):

$$Q_2 = Q_1 \frac{\sqrt{(0.86)(7.92 - 0.95)}}{\sqrt{(0.95)(7.92 - 0.86)}}$$

$$Q_2 = Q_1 \times 0.945$$

Calculate liquid flow correction in gravimetric units using the following:

Liquid specific gravity: 1.86
New liquid specific gravity: 1.52
Float: Hastelloy C (specific gravity 8.94)

Using Equation (19-14):

$$W_2 = W_1 \times \frac{\sqrt{1.52(8.94 - 1.52)}}{\sqrt{1.86\,(8.94 - 1.86)}}$$

$$W_2 = W_1 \times 0.925$$

Gas Flow - Standard Volume Units

Calculate gas flow correction in standard volume units using the following:

Nitrogen specific gravity: 0.867 metered at 50 psig and 80°F
New flow: oxygen specific gravity 1.105 metered at 100 psig and 70°F

Using Equation (19-16):

$$\text{scfm}_2 = \text{scfm}_1 \frac{\sqrt{(0.867)(114.7)(540)}}{\sqrt{(1.105)(64.7)(530)}}$$

$$\text{scfm}_2 = \text{scfm}_1 \times 1.190$$

Calculate gas flow correction in gravimetric units using the following:

Original scale: vapor with operating density of 0.270 pounds per cubic foot.
New operating density: 0.168 pounds per cubic foot.

Using Equation (19-14):

$$W_2 = W_1 \times \frac{\sqrt{\rho_2(\rho_{f2} - \rho_2)}}{\sqrt{\rho_1(\rho_{f1} - \rho_1)}}$$

But $\frac{\rho_{f2} - \rho_2}{\rho_{f1} - \rho_1}$ **very nearly equals 1.0. Therefore,**

$$W_2 = W_1 \frac{\sqrt{\rho_2}}{\sqrt{\rho_1}} = W_1 \times \frac{\sqrt{0.168}}{\sqrt{0.270}} = W_1 \times 0.789$$

Recalibration/Recertification of Accuracy

The cycle time between recalibrations for certification will depend on the application, the type of meter, and the accuracy requirement. Most variable area flowmeters can operate for years with very little calibration shift. The typical time between calibrations is one year in those applications that require periodic recertification. Recalibration can be accomplished by the user with either primary standard or secondary standard flow calibrators. Most flowmeter manufacturers also have calibration services available, many times traceable to NIST (formerly NBS).

Specification

A lack of understanding of the effects of operating density and viscosity on the sizing and calibration of variable area flowmeters often results in incomplete specifications or incorrectly ordered flowmeters.

The use of the ISA specification sheets or a similar document will ensure that all the necessary data is supplied. Refer to Figure 19-7.

Installation

Variable area flowmeters generally have fewer installation restrictions than do other types of flowmeters. With the exception of the spring-loaded type, they must be mounted in a vertical position (see Figure 19-8).

Guidelines

(1) Most variable area flowmeters can be supported by the pipe, although some metal tube rotameters and piston-type meters are quite heavy and should be supported directly or quite near the pipe connection.

(2) Flowmeters in critical lines in a continuous process should be equipped with a three-valve bypass (see Figure 19-9).

(3) Avoid quick opening valves (ball valves, solenoid valves, etc.) for start-up on compressed gas service. The sudden propulsion of the float to the top of the meter is likely to cause damage.

(4) Always install some safety device to ensure that line pressure cannot exceed the pressure rating of the flowmeter.

(5) The pressure gage used for gas pressure control for correction calculations should be located as close to the outlet of the flowmeter as possible, with no valve or other restrictions between the meter and the pressure gage (see Figure 19-10).

(6) On gas service applications, a throttling device (valve, back pressure regulator, or pressure regulator) should be mounted as close to the flowmeter outlet as possible (see Figure 19-10).

(7) Piston-type meters should be equipped with a strainer upstream.

(8) Straight-through flowmeters with a magnetic coupling type of readout should be equipped with a magnetic trap upstream if there is any wetted iron or steel in the piping.

(9) If the float must be inserted at the time of installation, be sure to follow the manufacturer's instructions as to which end is up.

©ISA S20

Specification Forms for Process Measurement and Control Instruments, Primary Elements and Control Valves

	ROTAMETERS (VARIABLE AREA FLOWMETERS)	SHEET ____ OF____
	NO / BY / DATE / REVISION	SPEC. NO. / REV.
		CONTRACT / DATE
		REQ. - P.O.
		BY / CHK'D / APPR.

GENERAL	1	Tag Number				
	2	Service				
	3	Line No./Vessel No.				
	4	Function				
	5	Mounting				
	6	Power Supply				
	7	Conn. Size / Type				
	8	Inlet Dir. / Outlet Dir.				
	9	Fitting Material				
	10	Packing or O-Ring Mtl.				
	11	Enclosure Type				
METER	12	Size / Float Guide				
	13	Tube Mtl. / Float Mtl.				
	14	Meter Scale: Length & Type				
	15	Meter Scale Range				
	16	Meter Factor				
	17	Rated Accuracy				
	18	Hydraulic Calib. Required				
FLUID DATA	19	Fluid				
	20	Color or Transparency				
	21	Maximum Flow Rate				
	22	Norm Flow / Min Flow				
	23	Oper. Specific Gravity (Liq)				
	24	Max Oper. Viscosity				
	25	Oper. Press. / Oper. Temp.				
	26	Oper. Density (Gases)				
	27	Std. Density / Mol. Wgt.				
	28	Max. Allowable Press. Drop				
	29					
EXT	30	Extension Well Mtl.				
	31	Gasket Mtl.				
XMTR	32	Transmitter Output				
	33	Trans. Enclosure Class				
	34	Scale Range				
ALARM	35	No. of Contacts / Form				
	36	Rating / Housing				
	37	Action				
	38					
OPTIONS	39	Valve Size & Material				
	40	Valve Location				
	41	Const. Diff. Relay Mtl.				
	42	Purge Meter Tubing				
	43	Airset				
	43a					
	44	Manufacturer				
	45	Model Number				
	46	Tube Number				
	47	Float Number				

Notes:

ISA FORM S20.22

Figure 19-7.

Maintenance Guidelines

(1) Rotameter glass tubes require cleaning in many applications. If the tube is removed from the body for cleaning, the seals (O-rings or packing rigs) should be inspected and replaced if worn, corroded, swollen, or have taken a permanent set.

(2) Glass tubes should be visually inspected periodically for signs of wear, chemical attack, chips, or cracks. The burst pressure of glass tubes is decreased dramatically if the surface is chipped, scratched, or cracked, so, if such damage is visible, the tube should be replaced.

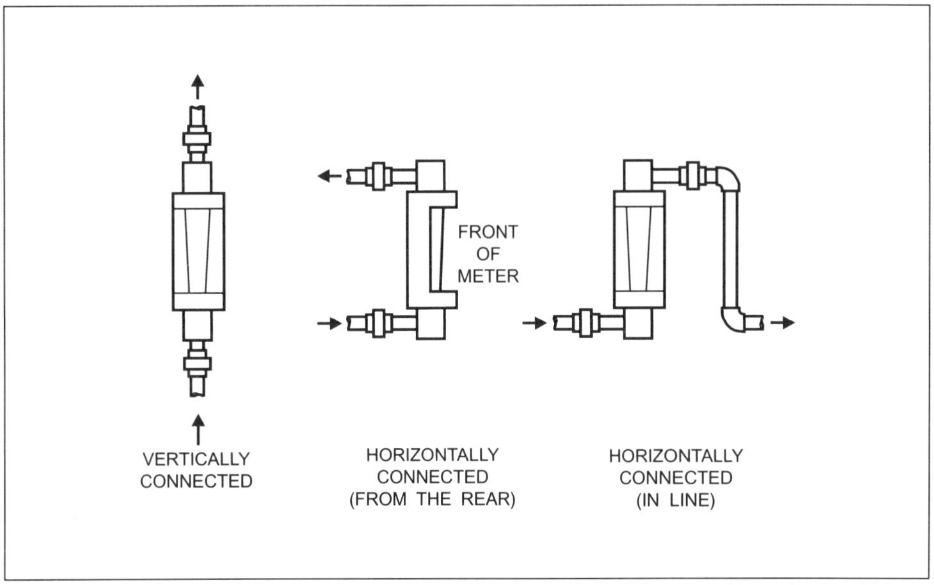

Figure 19-8. Typical Piping Arrangements for Variable Area Flowmeters (without Bypass)

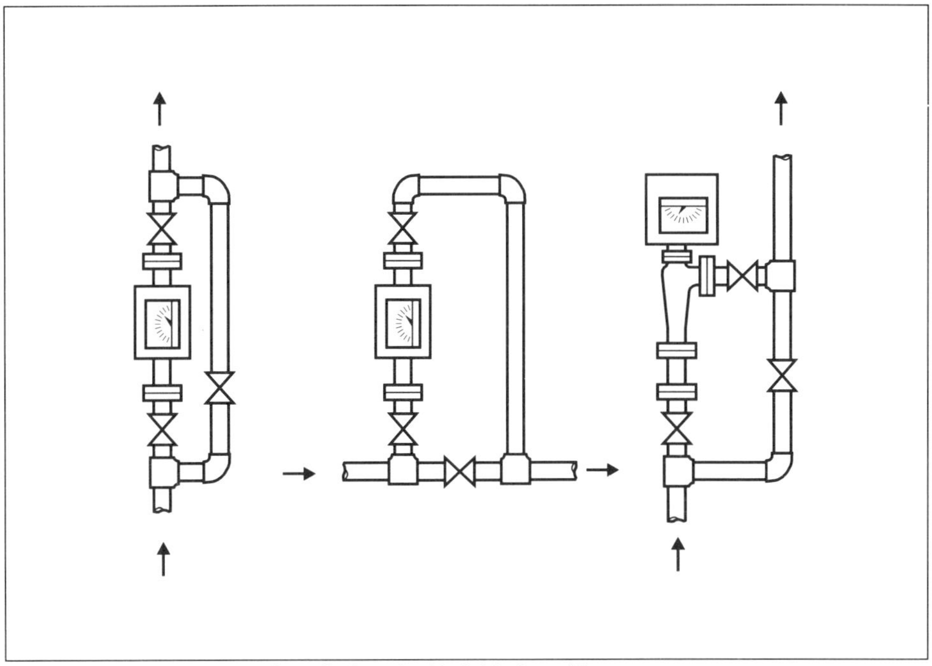

Figure 19-9. Bypass Piping Installations for Variable Area Flowmeters

(3) If the scale is engraved on a separate plate, make sure it is aligned with the reference mark on the rotameter tube.

(4) Magnetically coupled transmitters should be periodically checked for pointer position versus output signal. On most designs this can be done by removing the instrument cover and manually moving the pointer to various scale positions, while monitoring the output value. Since only a magnetic coupling is being "broken," no damage will result, and the pointer assembly will return to its magnetically bound alignment with the float when released.

Special Applications

Flow Rate Measurement with Limited Supply Pressure

Applications for liquid flow where the inlet pressure is limited, such as a small head tank or reflux from a small distillation column, have very limited pressure to give up in the form of a pressure loss across the flowmeter. Include the maximum allowable pressure loss into the specification and have the manufacturer(s) quote accordingly. For example, 3 gpm can be measured with a ½ inch rotameter with a pressure loss of approximately 30 inches of water column. If the maximum allowable loss is 6 inches of water column, a 1 inch rotameter should be specified.

The same scenario would apply to gas applications with very low available pressures.

Float Instability on Low Pressure Gas Applications

Rotameter floats at the high end of the weight range in any given size tend to be unstable on low pressure gas service, which leads to cycling over a wide scale band at frequencies exceeding 1 Hz. This renders indicating meters unreadable and transmitters useless in a control, record, or totalization application. It is imperative to indicate the lowest operating pressure as well as the pressure for which the meter is sized and the scale graduated.

If "float bounce" should occur, contact the manufacturer for specific suggestions. Some of the corrective strategies include the following:

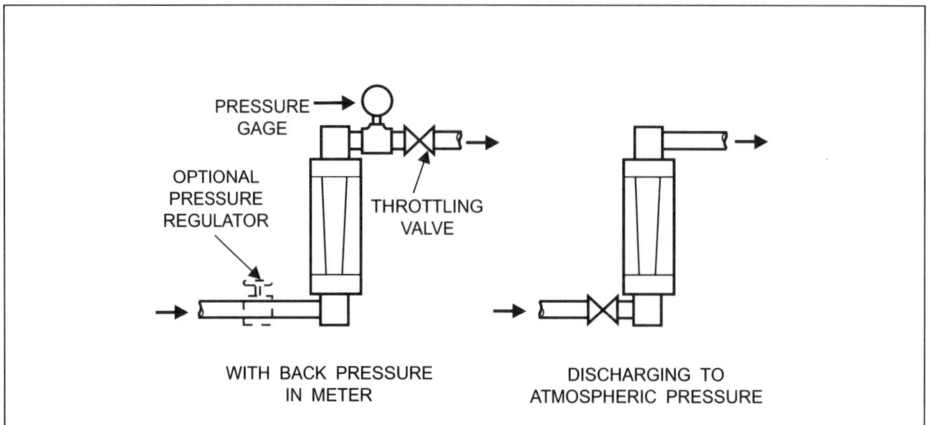

Figure 19-10. Typical Piping Arrangements for Variable Area Flowmeters on Gas Service

(1) Use a lighter float and new scale if normal flow is well below the full scale flow.

(2) Increase tube size with an attendant decrease in float weight to maintain approximately the same flow range.

(3) Operate in a higher pressure portion of the flow loop by moving a throttling valve downstream of the flowmeter or by setting the pressure source to a higher value.

Many models are equipped with "snubbers" or damping devices designed to inhibit float instability with heavy floats.

Inadequate Accuracy at Flow Rates near Bottom of Range

Wherever possible variable area flowmeters should be sized so that the normal flow is above mid scale. Many users forget that a 2 percent rotameter is rated at plus or minus 2% of full scale. When operating at 15% of scale, they are then dissatisfied when their meter exhibits an error of over 10 percent of rate.

A meter with a scale range of 10-100 gph and rated at ± 2% of full scale is warranted to be within ± 2 gph at any point on scale. A reading of 15 gph has a tolerance of 15 ± 2 or ± 13.3% of rate.

If the normal flow rate is 15 gph and the highest flow rate for which a reading is required is 25 gph, a meter sized for 25 gph would be a much better choice. Plus or minus 2% of 25 is ± 0.5 gph, and at 15 gph the accuracy would be 15 ± 0.5 gph, or ± 3.33% of rate.

Material Selection Problems with Ammonia

There are numerous applications for variable area flowmeters for measurement of liquid or gaseous anhydrous ammonia or gaseous disassociated ammonia. Rotameter suppliers typically offer meters with brass fittings and nitrile (Buna-N) O-rings for noncorrosive fluids and stainless steel fittings with Viton O-rings for corrosive fluids. Ammonia attacks brass and causes Viton to swell up to four times its original volume. For all ammonia applications, it is imperative to specify stainless steel with nitrile (Buna-N) seals (or Teflon if available).

References

1. Ruppel, G., and Umpfenbach, K. J., "Measurement of Flow Rate and Viscosity of Fluids with Resistance Bodies," *Zeitschift fur Technische Physik*, No. 12, pp 647-651, 1929.

2. Schoenborn, Jr., E. M., and Colburn, A. P., "The Flow Mechanism and Performance of the Rotameter," *Transactions of the American Institute of Chemical Engineers*, Vol. 35, No. 3, June 1939.

3. Fischer, K., "How to Predict Calibration of Variable Area Flowmeters," *Chemical Engineering*, pp 180-184, June 1952.

4. Coleman, *Transactions Institute of Chemical Engineers*, No. 34, pp 339-350, 1956.

5. Perry, R. H., and Chilton, C. H., *Chemical Engineers' Handbook*, Fifth Edition, pp 5-15, 5-16, New York: McGraw-Hill, 1973.

6. Salisbury, J. Kenneth, Editor, *Kent's Mechanical Engineer's Handbook*, Power Volume, 12th Edition, pp 18-22, John Wiley & Sons, 1950.

7. Considine, Douglas M., Editor-in-chief, *Process Instruments and Controls Handbook, Third Edition*, pp 4.32-4.50, New York: McGraw-Hill, 1985.

8. McCabe and Smith, *Unit Operations of Chemical Engineering-Second Edition*, pp 233-238, New York: McGraw-Hill, 1967.

About the Author

Charles E. Fees, B.S. in Mechanical Engineering at Drexel University, Philadelphia, PA, has been employed at Fischer & Porter Company in various positions related to variable area flowmeters since 1954. In 1980 he became Product Manager for Variable Area Flowmeters.

20

Insertion (Sampling) Flow Measurement

Most flow measurements are made by meters that are installed in-line such that all of the fluid passes through the meter body. Insertion (sampling) flow measurement is a different technique because the meter is inserted into the pipe and the flow rate is inferred from the measurement at a point in the flow stream.

There are both similarities and differences between in-line and insertion techniques. On the one hand, fluid properties, pipe flow phenomena, and the basic principles of flow sensing are the same. There are significant differences, however, in operating principles, installation, and applications.

What Is an Insertion Flowmeter?

An insertion flowmeter consists of three basic components: a sensor, a probe assembly, and electronic circuitry. The sensor provides a signal proportional to the velocity or flow rate at a point in the flowstream from which total flow rate is inferred. The circuitry properly conditions the raw signal from the sensor and performs required computational functions. The probe assembly rigidly positions the sensor within the flow and provides a process seal.

An insertion meter is *inserted* into the line. An in-line meter is *installed* in the line.

Where Can an Insertion Meter Be Used?

An insertion meter can be used in most flow measurement applications. Insertion meters are available that utilize most of the traditional flow sensing technologies (i.e., turbine, vortex, magnetic, dP). The range of fluid conditions (i.e., pressure, temperature, fluid type) under which insertion meters operate is similar to that of in-line meters. The performance of the sensing techniques is likewise similar.

Why Are Insertion Meters Used?

The applications in which an insertion meter may be a better choice than an in-line meter can be divided into three categories:

(1) Applications in which an insertion meter is the only option. This is the case of large line sizes for which in-line meters are unavailable. Another example is an application where the process cannot be shut down for meter installation.

(2) Applications in which the insertion meter is a more economical option. Installations in intermediate pipe sizes are suited to either in-line or insertion meters, but the insertion meter may be more economical.

(3) The unique operation of an insertion meter can sometimes provide a better solution. For example, the velocity range of an insertion meter can be changed by installing a different sensor.

Principles of Operation

An insertion flowmeter operates by inferring the total flow rate in a pipe or duct from a measurement(s) of the velocity or flow rate at specific location(s) within the pipe or duct.

Pipe Flow Concepts

An *ideal* fluid (zero viscosity) travels along a pipe with a uniform average velocity. The volumetric flow rate (Q) is the product of average velocity ($\bar{v}$) and flow area (A):

$$Q = A\bar{v} \tag{20-1}$$

The flow of an ideal fluid is illustrated in Figure 20-1. The volumetric flow rate can be determined from a velocity measurement made anywhere in the pipe because the velocity is uniform.

A *real* fluid possesses viscosity that creates shear between individual fluid particles and between the pipe wall and the adjacent fluid. The result of the shear is a non-uniform distribution of local velocity. A well established equation [Ref. 1] describes the theoretical velocity distribution in a fully developed flow:

$$v_y = v_o\left(1 - \frac{y}{R}\right)^{1/n} \tag{20-2}$$

where:

v_y = velocity at a point y
y = distance from centerline
v_o = centerline velocity
R = pipe radius

The velocity v_y is also called the local velocity.

The coefficient n can be expressed as a function of the pipe Reynolds number ($\mathrm{Re_D}$):

$$n = 3.299 + 0.3257 \ln \mathrm{Re_D} \quad \text{for } \mathrm{Re_D} < 400{,}000 \tag{20-3a}$$

$$n = 5.5365 + (5.498 \times 10^{-6}) (\ln \mathrm{Re_D})^5 \quad \text{for } \mathrm{Re_D} > 400{,}000 \tag{20-3b}$$

The velocity distribution in a pipe is three dimensional. It can be thought of as a series of concentric rings, each representing a line of constant velocity. This visualization is illustrated in Figure 20-3.

The velocity distribution, also called the velocity profile, is shown in Figure 20-2 for several values of $\mathrm{Re_D}$. The velocity profile provides the theoretical basis for insertion flowmetering in a circular conduit by relating the total flow rate to the velocity at any point in the flow.

The practical application of Equation (20-2) requires a term called the profile factor. The profile factor $F_p(y)$ is defined as the ratio of average velocity ($\bar{v}$) to the velocity (v_y) at a point in the flow (y):

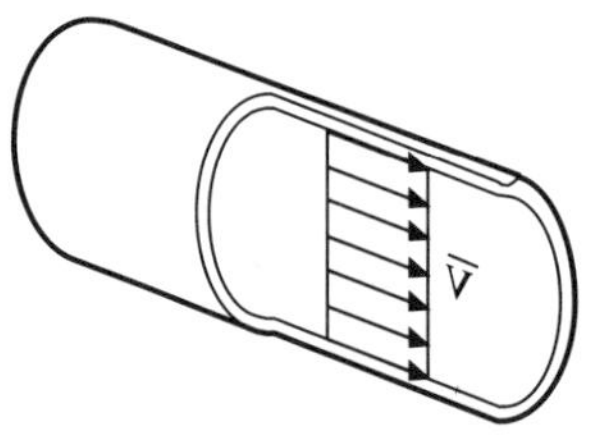

Figure 20-1. Ideal Fluid Flow

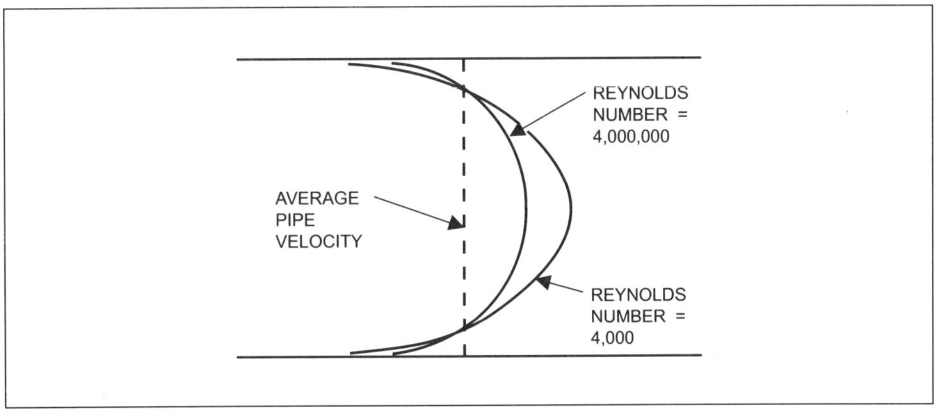

Figure 20-2. Theoretical Velocity Profile

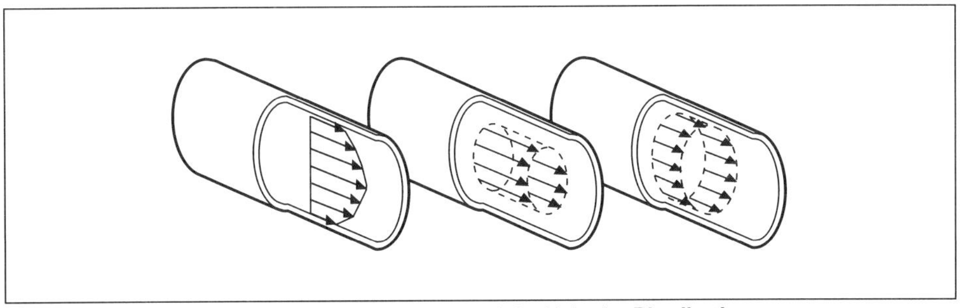

Figure 20-3. Three Dimensional Representation of the Velocity Distribution

$$F_p(y) = \frac{\bar{v}}{v_y} \qquad (20\text{-}4)$$

Combining an appropriate formulation of $F_p(y)$ with Equation (20-1), and adding the obscuration factor ($F_o(y)$) to take blockage into account, provides a means to determine the volumetric flow rate from the velocity at any position (y):

$$Q = A\bar{v} = AF_p(y)\,F_o(y)\,v_y \qquad (20\text{-}5)$$

Positioning The Sensor

Two locations within the pipe are traditionally used to sense local velocity: the pipe centerline and the critical position.

An equation relating the average velocity to that at the pipe centerline ($y = y_o$) can be determined by integrating Equation (20-2) over the entire pipe area:

$$\bar{v} = \frac{Q}{A} \qquad (20\text{-}6a)$$

$$= \frac{1}{A}\int v_y\,dA = \frac{1}{A}\int\left(1 - \frac{y}{R}\right)^{1/n} y\,dy$$

$$\bar{v} = v_o\left(\frac{2n^2}{(2n+1)(n+1)}\right) \qquad (20\text{-}6b)$$

With a centrally positioned sensor, the profile factor is defined by Equation (20-6b):

$$F_p(y_o) = \frac{\bar{v}}{v_o} = \left(\frac{2n^2}{(2n+1)(n+1)}\right) \qquad (20\text{-}6c)$$

Figure 20-4 shows the profile factor above as a function of Reynolds number.

There is a critical position where the local velocity is equal to the average velocity. It is determined by setting Equation (20-2) equal to Equation (20-6b) and solving for the position (y_{crit}):

$$\left(1 - \frac{y_{crit}}{R}\right)^{1/n} = \left(\frac{2n^2}{(2n+1)(n+1)}\right) \qquad (20\text{-}7a)$$

$$\left(\frac{y_{crit}}{R}\right) = 1 - \left(\frac{2n^2}{(2n+1)(n+1)}\right)^n \qquad (20\text{-}7b)$$

The location of the critical position (y_{crit}/R) is shown plotted in Figure 20-5. The location is nearly constant over a wide range of Reynolds number; the total variation is less than one percent for Reynolds numbers between several thousand and two million.

The profile factor for a critically positioned sensor is defined as unity because the local velocity is the average pipe velocity:

$$F_p\,(y_{crit}) = 1.0 \qquad (20\text{-}8)$$

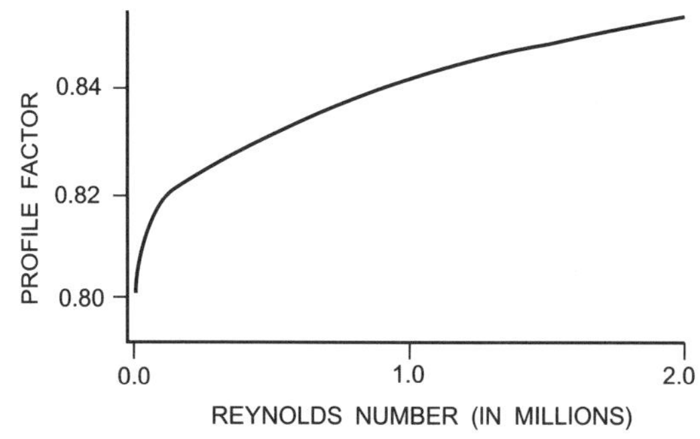

Figure 20-4. Profile Factor

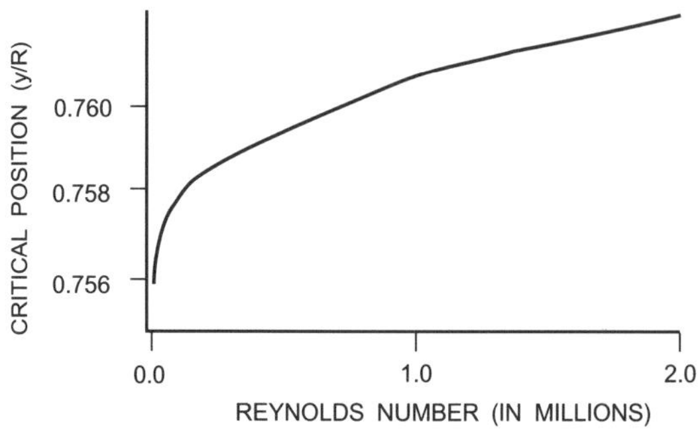

Figure 20-5. Critical Position

To derive a general profile factor, Equation (20-4) can be expanded into the product of two velocity ratios:

$$F_p(y) = \frac{\bar{v}}{v_y} \tag{20-4}$$

$$F_p(y) = \left(\frac{\bar{v}}{v_o}\right)\left(\frac{v_o}{v_y}\right) \tag{20-9a}$$

Substituting Equation (20-2) ($\bar{v}_y/v_o$) and Equation (20-6b) ($\bar{v}/v_o$) results in:

$$F_p(y) = \left(\frac{2n^2}{(2n+1)(n+1)}\right)\left(\frac{1}{(1-y/R)^{1/n}}\right) \tag{20-9b}$$

There are advantages and disadvantages to consider when selecting the sensor location:

In general, the centerline position is recommended whenever possible. The critical position should be used only when the pipe is large enough to accurately position the sensor well away from the wall.

(1) The centerline position is less sensitive to errors in positioning the sensor because the velocity profile is flatter; this is illustrated in Figure 20-6.

(2) The theory discussed above assumes that the local velocity is measured at a point. An actual sensor measures velocity over a region, resulting in a difference from the theoretical case. The deviation from theory is less if a sensor is positioned centrally because the velocity profile is flatter.

(3) The critical position does not require a profile factor correction (F_p is always unity). This simplifies the electronics because the output signal is always proportional to average velocity. On the other hand, the location of critical position varies with Reynolds number (as shown in Figure 20-5).

(4) The centerline position provides the most uniform velocity distribution. This situation will typically result in the best performance from a given velocity sensor.

(5) When a meter is installed, the pipe wall is cut, and the adjacent velocity distribution will be distorted. A critically positioned sensor is more likely to be located within the region of velocity distortion, especially with smaller pipe sizes.

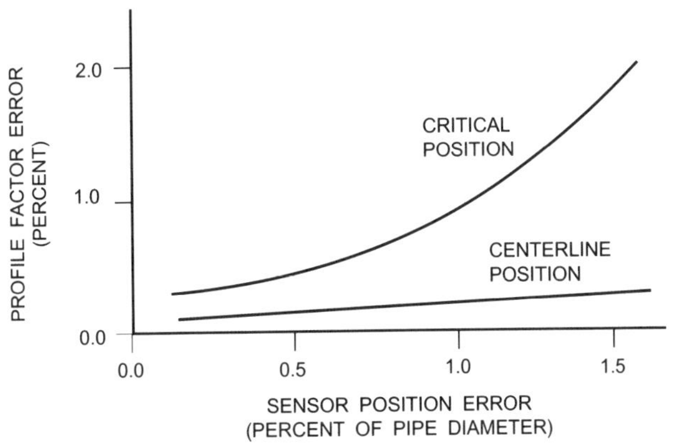

Figure 20-6. Sensitivity to Positioning Error

(6) The smaller the velocity sensor, the less likely it will be affected by the error sources noted above.

(7) With a knowledge of the flow profile and appropriate correction factors, the sensor can be positioned anywhere within the pipe. The optimum location may be a function of the performance of a particular sensing technique. Consult the meter manufacturer regarding the best location for their sensor. Some manufacturers may recommend sensor locations other than the centerline or critical positions.

The Blockage Effect

When the meter is inserted into the pipe, the flow area is decreased due to the presence of the meter body. This results in an increase in velocity for a constant flow rate. A correction factor is defined based on the projected area of the meter and the pipe area:

$$F_o(y) = \frac{\text{(Projected area of meter)}}{\text{(Pipe cross–sectional area)}} \tag{20-10}$$

which is commonly referred to as the blockage or obscuration factor. Note that the obscuration factor is dependent on the position of the velocity sensor. In general, the smaller the velocity sensor in relation to the pipe cross-sectional area, the less concern there must be about the blockage effect.

The volumetric flow rate is determined by combining Equation (20-5) with the obscuration factor:

$$Q = A\,F_p(y)\,F_o(y)\,v_y \tag{20-11}$$

where $F_p(y)$ is a general profile factor defined at a position y.

Exact values of $F_O(y)$ are dependent on the particular flowmeter; consult the manufacturer for this information.

Other Flow Geometries

The above development assumed the measurement to be in a circular pipe running full, the most common industrial application. Other geometries include:

(1) Square and rectangular ducts. A typical application is flow rate measurement in HVAC systems. ISO 3354 and ISO 3966 provide details regarding insertion flow measurement in rectangular ducts.

(2) Partially filled pipes, for example, the flow in sewer pipes. The flow measurement can be combined with a level detector to yield the flow area and volumetric flow rate.

(3) Open channels and basins. Examples include river and tidal flow measurement. ISO 3455 discusses techniques for measurement in open channels.

Accuracy of the Method and How To Improve It

Basic Accuracy of the Method

The accuracy of a flow measurement is made up of the sum of two uncertainties: bias (or systematic) uncertainty and precision (or random) uncertainty. The precision uncertainty of an insertion meter measurement is comparable to that resulting from a measurement with an in-line meter employing the same technique. The bias uncertainty of an insertion meter measurement is typically greater than that of an in-line meter because the flow rate is inferred from a single point measurement. Typical insertion meter accuracies of 1 to 5% can be expected.

Three factors contribute to the bias uncertainty of an insertion meter measurement:

(1) The obscuration factor assumes that the velocity everywhere increases in proportion to the blocked area. In reality, the intrusion of the meter results in flow disturbances that change the velocity profile. Furthermore, the pipe wall geometry has been altered by installation of the meter, creating an additional flow disturbance.

(2) It is assumed that the velocity sensor measures the velocity at a point in the flow. In reality, the sensor is of significant size that an *average* value of velocity over the sensor area is obtained.

(3) It is required that the flow area be known.

In-line meter measurement accuracy is not affected by these three factors. A fourth factor, the actual shape of the flow profile, will affect the accuracy of any flow measurement.

The repeatability of an insertion flowmeter is as good as that of an in-line flowmeter using the same sensing technique.

Improving Accuracy by Calibration

Calibration means testing the sensor or meter in a flowing fluid and determining its meter factor. Dry calibration generally means simulating the effect of the flowing fluid and inferring the calibration constants. Dry calibration decreases the basic accuracy of the measurement.

Wet calibrations are performed by one of two methods: free stream calibration or fixed installation calibration. Free stream calibration involves subjecting the sensor to a uniform velocity field. This is generally accomplished by moving the sensor through a stagnant fluid, for example, a liquid towing tank. Fixed installation calibration involves mounting the sensor within a pipe under known flow conditions.

It is unlikely that a flowmeter (either in-line or insertion) can be calibrated under conditions identical to those found in the field. The fluid composition and pipe configuration can rarely be duplicated in a calibration facility. Calibration includes knowledge of how a meter performs under different conditions. The major distinction between in-line and insertion meters is the pipe diameter: An in-line meter is always calibrated in the same line size as the application; an insertion meter will likely not be.

The level of accuracy achieved by calibration will not necessarily be achieved in a particular application.

Care should be taken in the interpretation of accuracy statements based on sales literature.

An insertion meter (or several meters) can be installed in a fixed position within a pipe and calibrated as a *unit* in the same manner as an in-line meter. This will result in accuracy that rivals an in-line meter because the pipe section and meter together effectively form an in-line meter.

Improving Accuracy with Proprietary Correction Factors

Insertion meter manufacturers will often apply a correction factor to the flow rate determined by Equation (20-11). Such a factor is based on prototype testing and meter performance histories maintained by the manufacturer. It is an empirical method to correct for the sources of bias uncertainty.

The correction factor is often used in conjunction with calibration. Sensor calibrations are done with certain fluid properties in a certain pipe size (or in a tow tank). The correction factor is defined based on the calibration results and the conditions (pipe size, fluid properties) found in the customer application.

To experimentally determine a correction factor is costly for a manufacturer because of the testing involved. The results are, therefore, typically not available. When a meter is purchased, the correction factor(s) will be applied for specific applications.

Improving Accuracy with Pressure and Temperature Compensation

The theoretical profile factors are dependent on Reynolds number, which varies with pressure, temperature, and viscosity. If the temperature and/or pressure fluctuate, the volumetric flow, the profile factor, and, therefore, the measurement accuracy can be altered. Proprietary correction factors will also likely depend on pressure and temperature.

Pressure and temperature measurements in the same location as the flowmeter will allow compensation for such fluctuations. Some insertion meters can be purchased with integral pressure and temperature transmitters to eliminate the need for extra installation.

Improving Accuracy with a Flow Computer

A flow computer accepts input signals from flow, pressure, and temperature transducers. It calculates flow-related parameters that can be displayed, totalized, or converted into electrical signals. A flow computer can significantly improve the accuracy of any flow measurement but especially an insertion flow measurement. Examples can be found in [Refs. 2, 3, 4].

A flow computer is the central component of a measurement system that may include pressure and temperature sensors as well as several flowmeters. Due to advances in electronics the cost of a good flow computer is not a significant part of the entire system cost.

Some typical capabilities of a flow computer include the following:

(1) Calculation of profile and obscuration factors

(2) Corrections for nonlinear behavior of the velocity sensor

(3) Application of proprietary correction factors

(4) Density calculations from temperature and/or pressure data

(5) Mass flow rate calculations from density and volumetric flow rate data

(6) Energy flow calculations from flow rate and temperature data

(7) Engineering unit conversion

(8) Totalization

(9) Alarm outputs

(10) Multiple channel units that accept inputs from several measurement locations

(11) Interface to larger computer-based control systems

Improving Accuracy by Flow Conditioning

An insertion flow measurement requires a specific velocity profile. If the profile at the sensor is distorted, the measurement accuracy can be reduced. Profile distortion is caused by a variety of piping components. One of the worst distorting elements is a pair of out of plane elbows; this is illustrated in Figure 20-7. Experimental data are contained in [Ref. 5].

The following are some other sources of profile distortion:

(1) Partially open valves

(2) Valve-elbow combinations

(3) Pipe area changes

(4) A tee section with a small diameter, high velocity flow entering a large diameter, low velocity flow

(5) Spiral welded pipe

The degree of distortion in a particular application is difficult to predict; the resulting reduction in accuracy is also difficult to predict. Insertion meters are unique because they can be used to measure the degree of profile distortion; a technique is described in the next section. The flow conditions in a particular location can be evaluated for suitability of an insertion flowmeter installation.

The use of a flow conditioner is recommended if an accurate measurement is required in a location where a distorted profile is suspected. A flow conditioner is a device that, when installed in the flow, will produce a predictable flow profile at its outlet. A number of different configurations are used; some typical units are shown in Figure 20-8. Installation guidelines are available from ASME (the American Society of Mechanical Engineers) and AGA (the American Gas Association).

Consult the manufacturer regarding the best flow conditioner and installation technique for a particular application.

The Effect of Sensor Positioning and Installation on Accuracy

The accuracy of an insertion flowmeter measurement can be significantly reduced by sloppy installation. The following are some basic installation guidelines to maintain accuracy:

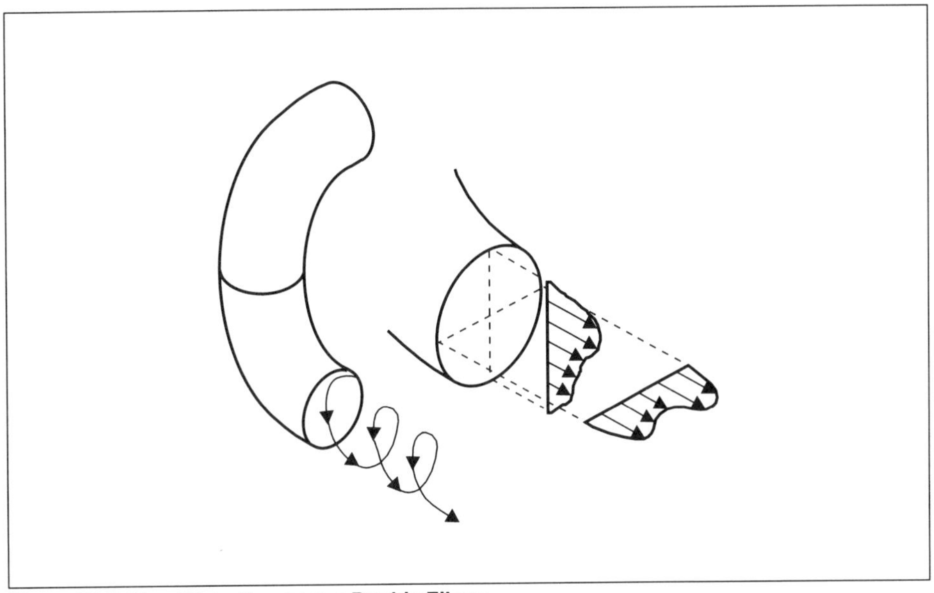

Figure 20-7. Flow Distortion from a Double Elbow

(1) Select an installation site as far downstream as possible from sources of velocity profile distortion. A good rule of thumb is to install the meter 20 diameters of straight run upstream and 10 diameters downstream.

(2) The probe must be perpendicular to the pipe axis and aligned with the pipe center.

(3) The sensor must be oriented parallel to the pipe axis.

(4) Make sure that the sensor is located at the proper position within the pipe.

(5) Welded fittings should be installed to minimize any protrusions into the flow.

To achieve the highest possible accuracy from a particular meter the manufacturer's installation procedure should be carefully followed.

Meter Design and Construction

A typical insertion meter consists of three basic components:

(1) The *sensor*, mounted on a stem, produces a signal proportional to the local velocity at the sensor.

(2) The *preamplifier* converts the sensor signal to a useable form (for example, 4-20 mA or pulse).

(3) The *retractor* holds the stem such that the sensor can be positioned within the pipe.

Measuring the Local Velocity—Sensing Techniques

The primary component of an insertion flowmeter is the velocity sensor. Five traditional sensing techniques employed in in-line flowmeters (differential pressure, turbine, thermal, vortex and magnetic) are also used in insertion flowmeters. Discussed below are the theory, features, and limitations of each of the five sensing techniques.

DIFFERENTIAL PRESSURE

The oldest and most commonly used flow sensing technique is differential pressure, which is the principle behind the orifice plate and the Venturi. The flow velocity is sensed based on the Bernoulli principle:

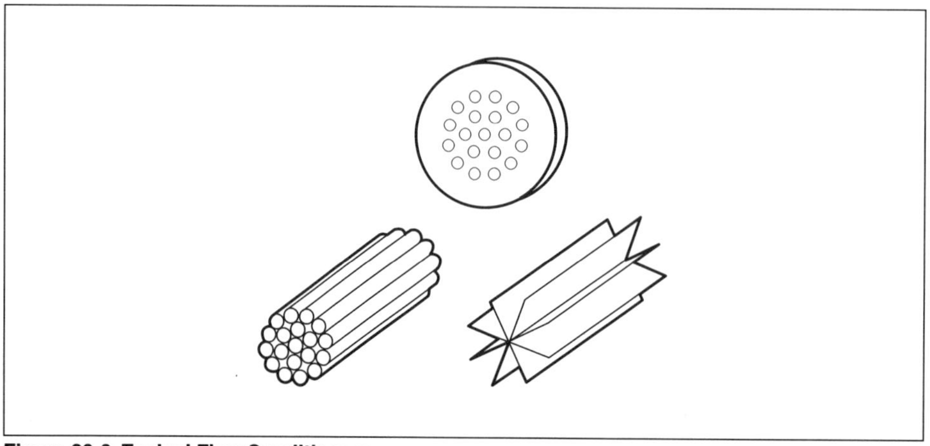

Figure 20-8. Typical Flow Conditioners

$$\Delta P = (\text{constant})\rho\bar{v}^2 \tag{20-12}$$

The differential pressure measurement is proportional to fluid density and the square of the velocity. The basic principle of differential pressure measurement has two inherent disadvantages. First, the velocity range is limited because of the square law behavior; for example, a 10:1 turndown in the differential pressure measurement is required to measure a 3.5:1 range of velocity. Second, for low density (gas) measurements, the differential pressure signals can be very low.

The Pitot static probe is the traditional differential pressure insertion measurement. Physically, the Pitot static probe (also called a Pitot tube) consists of a cylindrical tip fastened to a stem. The probe, shown in Figure 20-9, has pressure tapping holes that sense the static and total pressure (also called stagnation pressure). The difference between static and total pressure is the ΔP term in Equation (20-12). The total pressure is sensed at the probe tip, and the static pressure is sensed by a series of holes around the perimeter. The probe tip is streamlined to minimize flow disturbances. The static tapping holes are located well back from the tip to minimize the effect of any flow disturbances that do occur.

The differential pressure signal is detected external to the flow. Therefore, the stem must accommodate transmission of the pressure signals. This is accomplished by means of two concentric tubes—two passages that transmit the pressure signals. At the end of the probe the pressure signals are connected to a differential pressure sensor.

The following are features of a Pitot tube:

(1) A Pitot tube can be used in liquids or gases.

(2) The operation is simple, the design rugged, and there are no moving parts to fail. The electronic components are limited to those required to detect the differential pressure.

(3) The performance is well established because the device has been used for many years. Considerable reference and design literature are available.

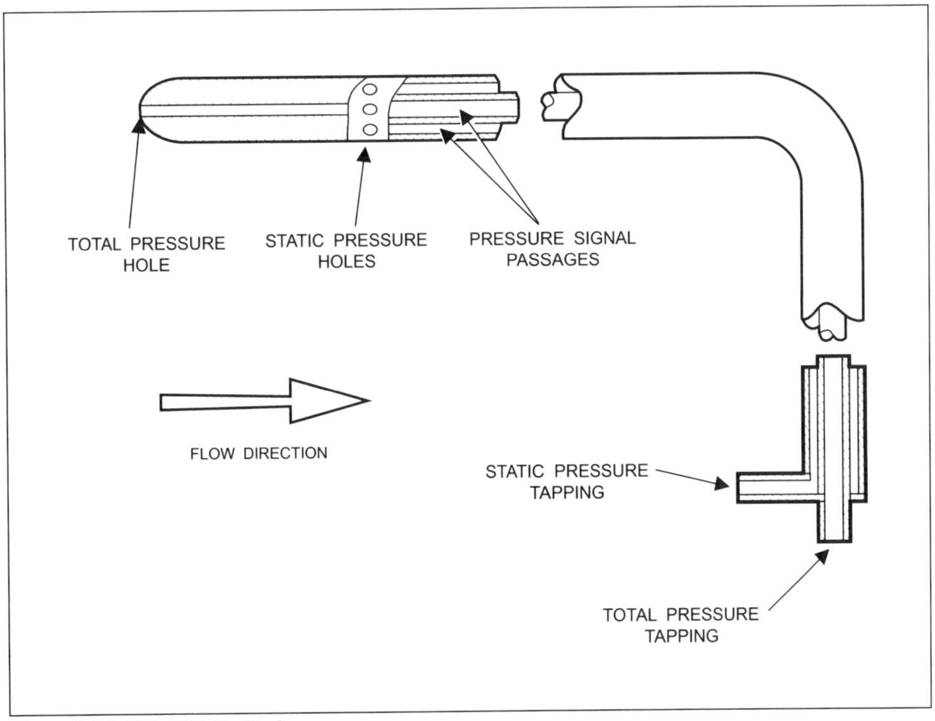

Figure 20-9. Pitot Static Probe

The standards ISO 3966, ISO 7145, and ISO 7194 provide detailed information. Additionally, manufacturers of Pitot tubes can provide extensive bibliographies of reference literature.

(4) Since the tube can be very small in diameter, the technique is less intrusive into the flow than other sensing techniques. Therefore, the accuracy of the measurement is less dependent on a correction for blockage effects.

(5) The operating temperature range is limited only by the differential pressure sensor. The Pitot tube itself is limited only by the material from which it is fabricated.

(6) The primary failure mode is via blockage of the pressure transmitting passages. This type of failure can be readily detected and easily fixed by removing the tube and cleaning it.

(7) A Pitot tube cannot be damaged by overranged flow conditions. The differential pressure sensor can, however.

The following are limitations of the Pitot tube:

(1) The velocity range is limited by the square root of the differential pressure sensor range because of the Bernoulli principle. A pressure sensor with 10:1 turndown will result in a velocity range of 3.5:1.

(2) Accuracy decreases with flow rate. The accuracy of a differential pressure sensor is expressed as a percentage of the full scale. Therefore, the accuracy as a percent of rate decreases as the sensed pressure and the flow decreases.

(3) The output is dependent on the fluid density, so density must be measured or inferred to ensure accuracy.

(4) A Pitot tube is not suited for "dirty" fluids because the pressure transmitting passages are subject to clogging.

(5) Care must be taken when connecting the pressure signals from the Pitot tube to the differential pressure sensor. With liquid measurements the lines must be purged of gases. With steam measurements the installation must account for condensation. ASME/ANSI MFC-8M contains details regarding different situations.

(6) The output is nonlinear.

The use of multiple sensors can reduce the sensitivity of the measurement to profile distortions by averaging the velocity profile in several locations. Differential pressure sensing is unique in that a number of multiple-sensor configurations have been developed. This is not true for most other sensing techniques. A multiport differential pressure measurement is made by connecting several total pressure sensing holes together, thus averaging the readings. Appropriate corrections need to be made to determine the total flow rate. A photograph of a typical multiported sensor is shown in Figure 20-10. Further information can be found in References 6 and 7.

The operation of an insertion target flowmeter is similar to the in-line model. An object, usually a disk, is positioned in the flow such as shown in Figure 20-11. The velocity streamlines are distorted by the presence of the disk in a manner similar to the distortion caused by an orifice plate. The Bernoulli principle states that the pressure drop across the disk ($P_u - P_d$) will be proportional to the fluid density (ρ) and the square of the velocity (v^2). The pressure drop results in a force (F_t) that is the product of the drag object area (A_t) and differential pressure ($\Delta P = P_u - P_d$):

$$F_t \propto A_t \Delta P = A_t (P_1 - P_2) \qquad (20\text{-}13a)$$

$$\propto A_t \rho \bar{v}^2 \qquad (20\text{-}13b)$$

The drag force is measured by strain gages or a force balance feedback system connected to the mounting arm.

The target meter shares many performance characteristics with the Pitot tube because it operates based on the Bernoulli principle. The following are some differences:

(1) A target meter is tolerant of dirty fluids because there are no small passages to block.

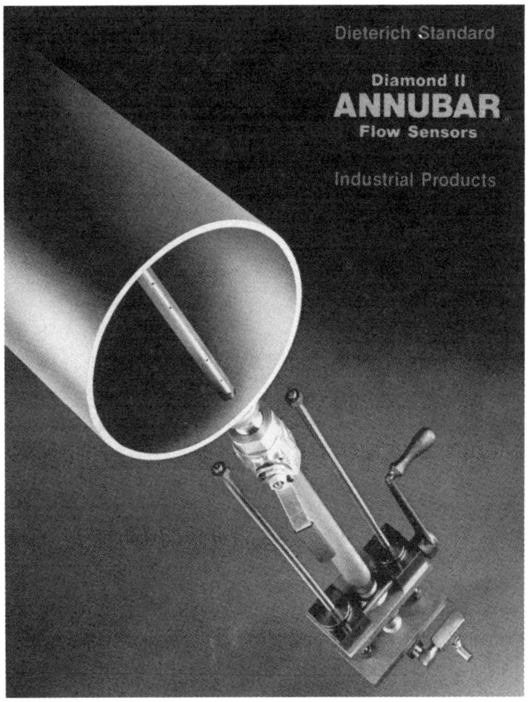

Figure 20-10. Typical Multiport Differential Pressure Insertion Meter
(Courtesy of Dieterich)

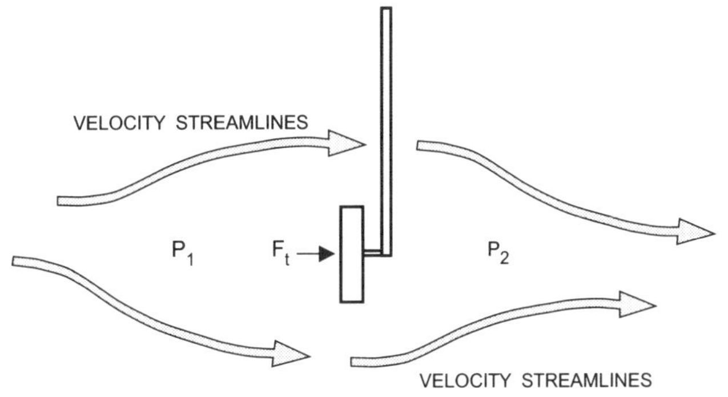

Figure 20-11. Target Sensor Operation

(2) A target meter is more intrusive to the flow than the Pitot tube. Not only is the meter physically larger, but the differential pressure variation is larger due to velocity averaged over a larger area.

(3) The maximum operating temperature range is dependent on the sensing technique. A force balance feedback system will operate at a higher temperature than a strain gage system because there is no need for electronic components to be mounted near or in the flow stream.

(4) A target meter is less susceptible to damage from overranging because there is no differential pressure sensor.

(5) Accuracy is dependent on the target size; if the target wears, the calibration factor will change. This change in performance will be difficult to detect because it occurs gradually.

TURBINE METER

The insertion turbine sensor is second only to the Pitot static tube in popularity. It consists of a turbine blade assembly and electrically based detector. Typical units are shown in Figure 20-12. A rotor blade assembly held between two bearings spins in response to velocity. A variety of bearing techniques are used in the industry. A turbine blade spins due to the kinetic energy in the flow. The kinetic energy is proportional to density and velocity squared:

$$E_k \propto \rho\, v^2 \tag{20-14}$$

Opposing the spin are forces due to friction, both in the bearings and due to fluid viscosity. This is illustrated in Figure 20-13.

The performance of a typical turbine meter is shown in Figure 20-14. The meter factor is the ratio of frequency to velocity. In the flow rate range where the meter factor is constant, the meter is linear. The upper limit of the linear range is the flow rate where the bearings will fail prematurely due to overspeeding. At low flows, nonlinear performance is dependent on the friction forces. The total operating turndown (linear plus nonlinear) of an insertion turbine is typically 25:1 to 35:1. The linear turndown is typically 7:1 to 15:1.

The meter factor of a turbine sensor is proportional to the blade pitch: the greater the blade pitch, the faster the rotor will spin at a given fluid velocity. The performance curves in Figure 20-14 are applicable to a variety of flow rate ranges, each corresponding to a different blade pitch.

Figure 20-12. Typical Turbine Sensors
(Courtesy of EMCO)

The following are some features of a turbine sensor:

(1) A flow rangeability of 30:1 (10:1 linear) can be achieved with a single sensor. Variations in blade pitch allow a rangeability of 100:1 to be achieved with a pair of sensors.

(2) The output is linear.

(3) The accuracy is a percent of rate.

(4) A turbine meter can be applied to liquids, gases, and vapors.

(5) The output is electrical, as opposed to mechanical; thus, interfacing to instrumentation and control systems is straightforward.

(6) A lightweight rotor assembly can measure low gas flow rates and densities.

(7) The turbine meter can be configured to operate bidirectionally.

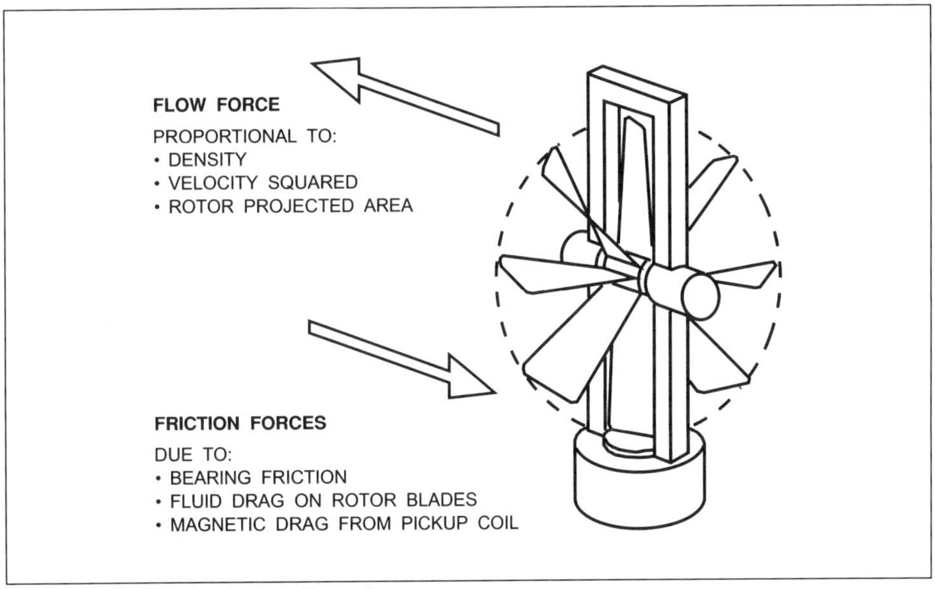

Figure 20-13. Turbine Sensor Operation

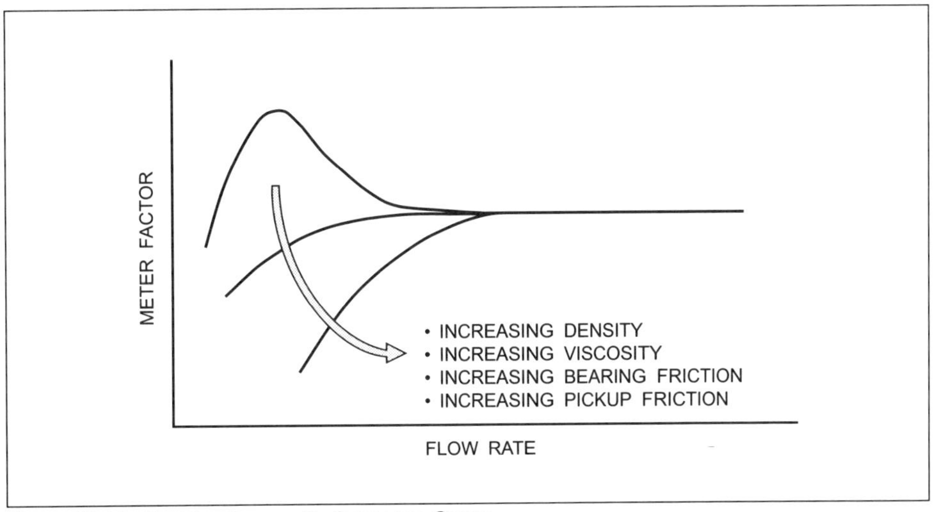

Figure 20-14. Turbine Sensor Performance Curve

(8) The sensor is removable and replaceable.

(9) The range can be adjusted by installing a new sensor with a different pitch.

(10) Turbine meters are typically immune to flow noise and vibration. The signal generated due the blades spinning is generally much larger than that caused by noise sources.

(11) The high temperature limitation in operating a turbine meter is usually the blade pulse detection sensor. A turbine meter can operate at temperatures as high as 750°F.

The following are some limitations:

(1) The turbine meter utilizes a moving part. A turbine sensor therefore has a finite life, after which it must eventually be replaced.

(2) The bearings can fail gradually, causing increasing drag and a slow change in calibration factor. This failure mode is difficult to detect.

(3) A turbine meter is limited to relatively clean fluids because of its bearings. Care must be taken to ensure that the bearing materials are compatible with the fluid.

(4) Turbine rotors are not as rugged other sensor techniques and are, therefore, more susceptible to damage. This includes handling during installation, overrange, and fluid contamination.

THERMAL RESISTANCE

This technique is more commonly known as thermal anemometry or constant temperature anemometry. A wire or film exposed to the fluid flow acts as the sensor. The basic principle of operation involves heating the wire with an electric current and allowing the flowing fluid to cool it, as shown in Figure 20-15. The heat delivered is:

$$q_{in} = i^2 R \qquad (20\text{-}15)$$

The flowing fluid cools the wire by convection:

$$q_{out} = c_0 + c_1 \sqrt{\rho\, v} \qquad (20\text{-}16)$$

where:

q_{in} = electrical heating of the sensor
q_{out} = convective cooling of the sensor
i = electrical current
R = electrical resistance
v = fluid velocity
ρ = fluid density
c_0, c_1 = constants that depend on the fluid thermal properties, sensor geometry, fluid temperature, and pressure

The product ρv is referred to as the mass velocity.

A thermal anemometer can be operated in two ways—in the constant current mode or the constant temperature mode. Applying a constant current to the wire results in an equilibrium temperature difference between the wire and fluid (ΔT) that is proportional to the square root of the mass velocity:

$$\rho v = c_2 + c_3\, \Delta T^2 \qquad (20\text{-}17)$$

The temperature can be detected by using a pair of temperature sensors or the heated wire itself. Typically, the resistance of the sensor wire varies linearly with temperature; hence, the mass velocity is proportional to the square of the resistancc:

$$\rho v = c_4 + c_5 R^2 \qquad (20\text{-}18)$$

The constants c_2 through c_5 are derived from those in Equations (20-15) and (20-16). The constant temperature mode involves a resistance bridge as shown in Figure 20-16. The feedback current is maintained by the feedback amplifier to keep the resistance (and temperature) of the wire constant. The current required to maintain equilibrium is proportional to the fourth root of the velocity. The mass velocity is, therefore, proportional to the current raised to the fourth power:

$$\rho v = c_6 + c_7 i^4 \qquad (20\text{-}19)$$

The constants c_6 and c_7 are derived from those in Equations (20-15) and (20-16).

A thermal sensor is unique in that the output signal is dependent on both density and velocity. If the total mass flow rate is to be measured, the output signal needs only to be corrected for profile factor, obscuration factor, and flow area. Independent measurement of density is not required. If the total volumetric flow rate is required, the density must be measured.

The traditional advantage of the thermal technique is a sensor that can be small and fast (in terms of dynamic response). Sensor wires can be as small as 0.00015 inch in diameter with a frequency response as high as 600 kHz. A probe can be as small as 0.125 inch in diameter and, therefore, virtually non-intrusive. In some applications the price of such extreme performance can be a very delicate sensor. For this reason, anemometers of this type are traditionally used in laboratory environments.

For industrial flow measurement applications, the thermal anemometer principle has been ruggedized, and the sensor is larger and better protected from the environment. A photograph of a typical industrial unit is shown in Figure 20-17.

The following are some features of thermal sensors:

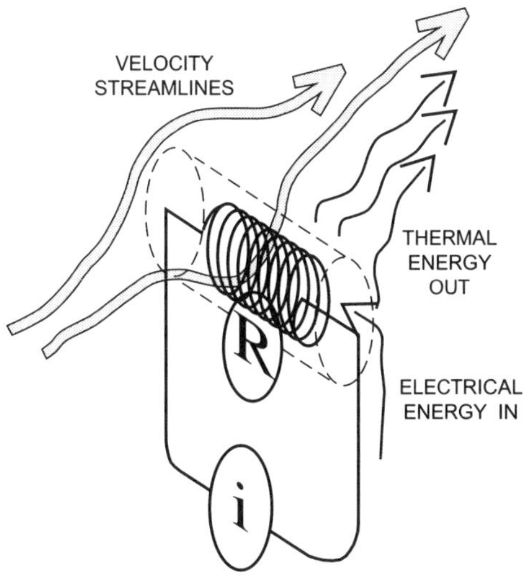

Figure 20-15. Thermal Sensor Operation

(1) Extreme sensitivity to low gas velocities—lower than other insertion techniques (For example, a thermal sensor can measure the velocity of human breathing.)

(2) No moving parts to fail

(3) Large turndown, up to 500:1, due to the electronic compensating circuitry

(4) Can be configured to operate bidirectionally

(5) Immunity to flow noise and vibrations

(6) Improved accuracy when multiple thermal sensors are mounted on the same probe.

The following are some limitations:

(1) The output is nonlinear. The electronics are more complex than those required for some of the other techniques.

(2) The accuracy is expressed as a percent of full scale. This results in low accuracy (as a percent of rate) at low flow rates.

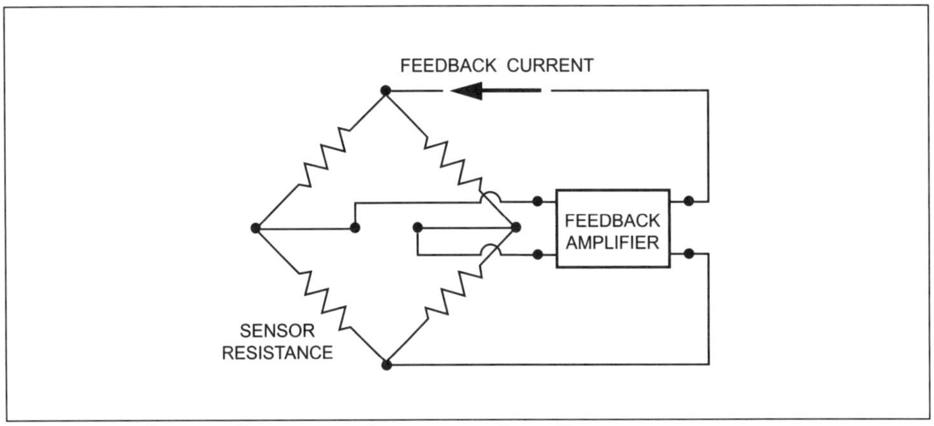

Figure 20-16. Constant Temperature Anemometer Operation

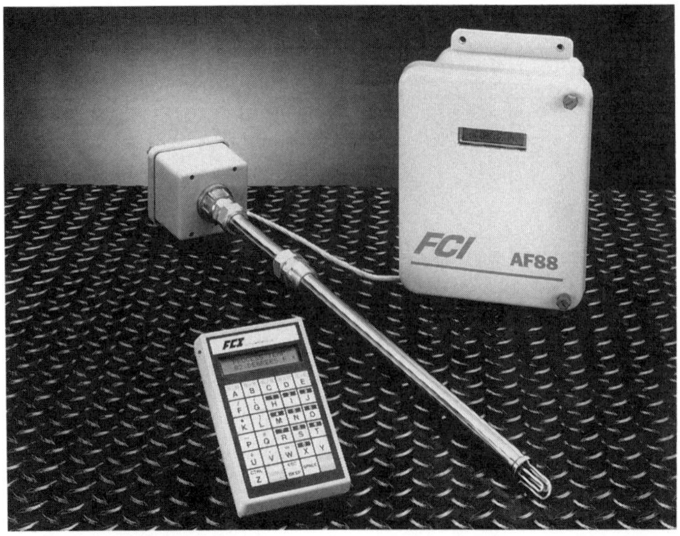

Figure 20-17. Typical Insertion Thermal Meter
(Courtesy of Fluid Components, Inc.)

(3) Industrial thermal anemometers are typically not applicable to liquid flow measurement.

(4) The various constants depend on fluid properties; the output signal must, therefore, be corrected if these properties vary significantly.

(5) Measurement error can occur if liquid is present in the flow stream because of the resulting increase in thermal conductivity.

(6) Material buildup on the sensor will result in loss of sensitivity due to the insulating effect. This failure mode is slow; therefore, it may not be detected. Applications where buildup could occur should be avoided, or the sensor should be periodically cleaned.

VORTEX SHEDDING

The vortex shedding sensor principle is based on the von Karman effect, which states that a bluff body within a flowing fluid will exhibit a periodic series of vortices in the flow downstream of the body. This effect can be observed in the waving of a flag; vortices shed by the pole cause the flutter in the flag. Another common example is the whistling of telephone wires in the wind, when vortices shed by the wire are high enough in frequency to be audible.

The implementation of the von Karman effect as a flowmeter is illustrated in Figure 20-18. The flow separates at the bluff body into a pair of shear layers. At a high enough Reynolds number the shear layers are unstable and “roll up” into the region behind the bluff body. A series of vortices, called the von Karman vortex street, extends for a significant distance downstream of the bluff body. A typical insertion vortex meter is shown in Figure 20-19.

Flow measurement applications are described by the dimensionless Strouhal number (St), which is a function of the fluid velocity (v), shedding frequency (f) and bluff body width (d):

$$\mathrm{St} = \frac{fd}{\bar{v}} \tag{20-20}$$

The Strouhal number is related to Reynolds number. A typical curve is shown in Figure 20-20. The linear region is the Reynolds number range over which the Strouhal number is constant. The minimum linear Reynolds number is typically between 10,000 and 30,000. Within the linear region a constant factor (K_f) can be defined:

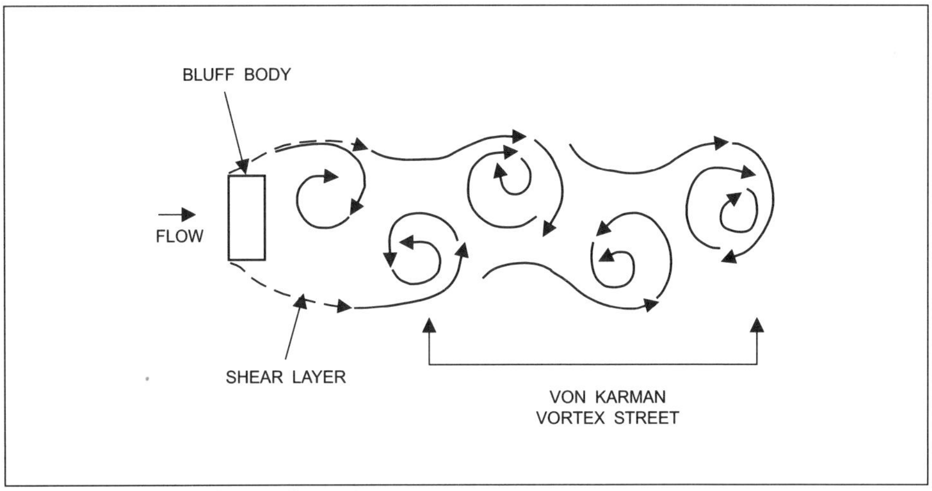

Figure 20-18. Vortex Sensor Operation

$$K_f = \frac{\mathrm{St}}{\mathrm{d}} = \frac{f}{\bar{v}} \tag{20-21}$$

Considerable study has been made of the shape of the bluff body and its effect on performance. A variety of shapes are used by the different manufacturers. The performance details are beyond the scope of this text.

A variety of different methods are used by manufacturers to detect the vortices. These include detection of pressure and temperature and the distortion of an ultrasonic beam. Some methods involve a second body installed downstream of the shedder. The method used to detect vortices will affect the performance of the meter, especially the maximum operating temperature.

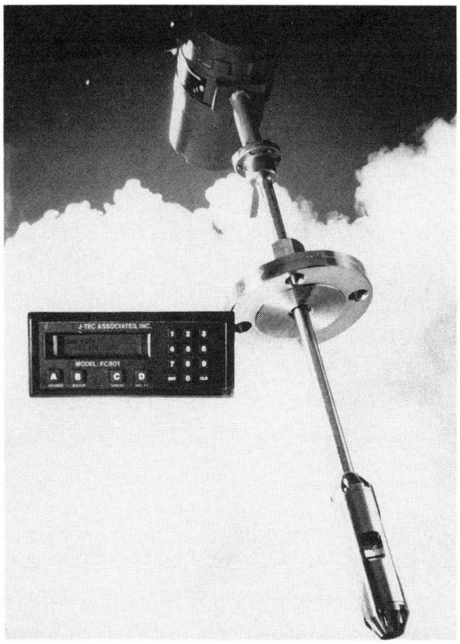

Figure 20-19. Typical Insertion Vortex Meter
(Courtesy of J-Tec)

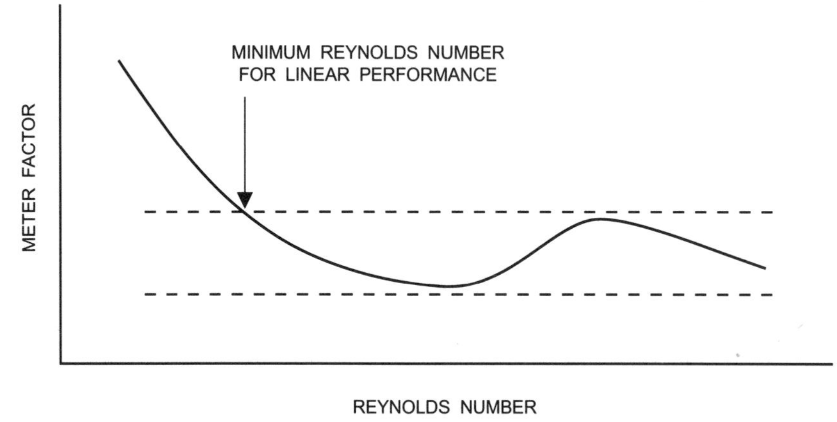

Figure 20-20. Vortex Sensor Performance Curve

The following are some features of vortex shedding sensors:

(1) With no moving parts, the sensor will theoretically never wear out. Little or no maintenance is required.

(2) A vortex shedder is rugged. Damage during installation or due to adverse flow conditions is unlikely, as is damage due to overrange.

(3) A vortex shedder will tolerate significant dirt and particulates in the flow. Some calibration shift can occur if the shedding strut size changes due to wear. This is usually a detectable failure; the output signal will degrade significantly before enough wear has occurred to affect accuracy.

(4) The output signal is linearly proportional to flow rate.

(5) The accuracy is expressed as a constant percent of rate. Accuracy is maintained at low flow rates.

The following are some limitations:

(1) The shedding phenomenon is very weak (for example, as compared to a turbine meter). A vortex shedder is sensitive to mechanically based noise sources. This includes pipe vibration, turbulence, and upstream noise sources (for example, partially open valves, pumps, and compressors). Noise is particularly troublesome if it occurs at a frequency within the operating range of the sensor.

(2) A particular sensor operates (with a constant Strouhal number) over a fixed Reynolds number range. This range cannot be adjusted.

(3) Most practical applications are limited by the low end rather than the high end of the linear range.

MAGNETIC INDUCTION

The magnetic sensor is based on Faraday's Law of Induction. This law states that a conductive medium passing through a magnetic field will induce an electromotive force (voltage). The induced voltage ($\overline{E}$) is proportional to the velocity of the conductive medium ($\overline{v}$) the magnetic field strength ($\overline{B}$) and the length between detecting electrodes (L):

$$\overline{E} = L(\overline{v} \times \overline{B}) \tag{20-22}$$

The equation is written in this manner because $\overline{E}$ (a vector) is the cross product of $\overline{v}$ and $\overline{B}$ (two other vectors). The induced voltage is, therefore, perpendicular to both the magnetic field and the velocity of the conductive medium.

The application of Faraday's Law to a magnetic flow sensor is illustrated in Figure 20-21. The magnetic field ($\overline{B}$) is produced by current flowing through a coil within the sensor. The fluid whose velocity ($\overline{v}$) is being measure is the conductive medium. The induced voltage ($\overline{E}$) is measured across the two electrodes (that are separated by a distance L). A typical insertion mag meter is shown in Figure 20-22.

The magnetic field cannot be driven by a constant current or electrolytic polarization of the electrodes will occur. To prevent this, the magnetic field is driven by an AC current. The flow is then proportional to the amplitude of the induced voltage.

Magnetic sensor electronics need to be "zeroed" with process fluid at temperature after the meter has been installed. This is to compensate for process fluid conductivity and electrical noise sources that are present in the application. These effects are not present in laboratory calibration; adjustments must be made after the meter has been installed, with the pipe filled and the flow rate zero.

To eliminate the need for field zero adjustment, a pulsed DC excitation technique can be used. The magnetic field is turned on with a DC current and the resulting induced voltage is measured. The drive current is then turned off and the voltage is again measured. The voltage measured when the magnetic field is off is due to noise. It is subtracted from the "magnet on" voltage to yield the signal proportional to the flow rate. A pulsed DC excitation scheme will continuously compensate for variations in the background noise level. The limitation is a slower response time than with AC excitation.

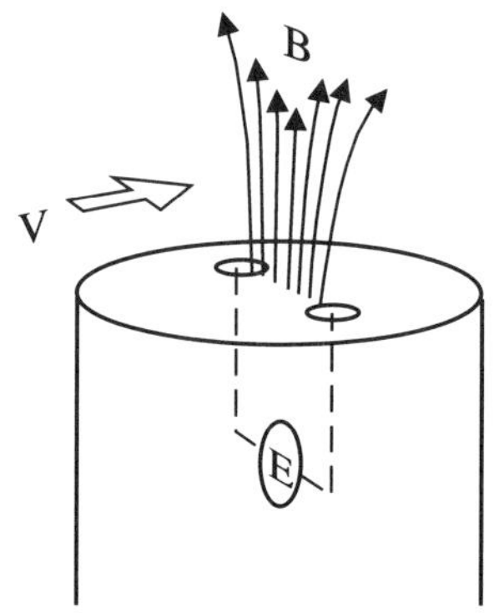

Figure 20-21. Magnetic Sensor Operation

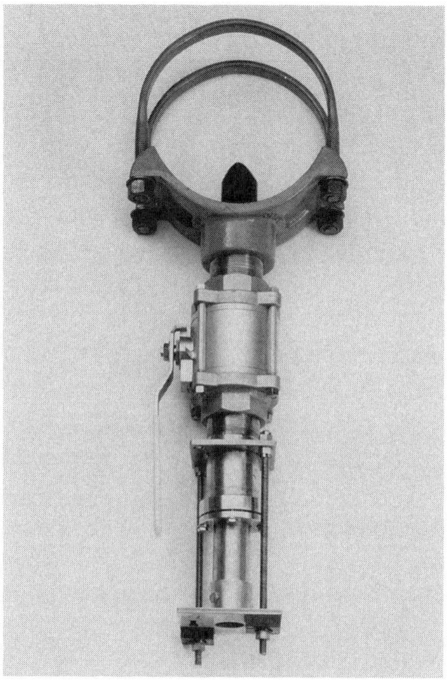

Figure 20-22. Typical Insertion Magnetic Meter

The following are some features of magnetic flow sensors:

(1) With no moving parts, a sensor can theoretically last forever.

(2) The sensor is rugged. Maintenance requirements and breakage during handling are, therefore, reduced.

(3) DC excited sensors are percent of rate devices.

(4) The magnetic sensor is unaffected by suspended solids and dirty fluids. For this reason, magnetic flowmeters are found in pulp and paper and wastewater applications.

(5) The output signal is linearly proportional to velocity.

The following are some limitations:

(1) AC excited sensors are percent of full scale devices.

(2) The electronics are more complex than those used with other sensing techniques.

(3) The induced voltage level is extremely low, and the electrode signal is sensitive to a variety of electrical noise sources. The meter must be installed to minimize the possibility of electrical noise affecting the flow signal.

(4) The sensor can measure only conductive fluids. Conductivity limits vary from one manufacturer to another.

(5) In some applications the electrodes can become coated; the coating acts as an insulation, and the induced voltage decreases. This is a failure mode that cannot be readily detected. An insertion meter has the advantage over an in-line flowmeter that it can be periodically removed and cleaned.

(6) The magnetic field can become distorted if the sensor is mounted too close to the pipe wall. This distortion will affect the accuracy of the measurement.

Applications

The decision of where to use an insertion meter comes down to a comparison of performance features between in-line and insertion metering techniques. For line sizes greater than one inch, an insertion meter can be used virtually anywhere that an in-line meter is used. Described below are a number of applications where an insertion meter is a practical solution. More application examples can be found in the references listed at the end of the chapter.

Large Line Sizes

The traditional application for insertion metering is large line sizes. In-line meters are generally unavailable for line sizes greater than eight inches. Insertion meters can accommodate virtually any line size.

The cost of an insertion meter is nearly independent of line size, and the only extra costs are those associated with extending the retractor stem. The cost of an in-line meter rises dramatically with line size because of the meter size. The cost comparison between an in-line and an insertion meter is illustrated in Figure 20-23. The total meter cost includes purchase, installation, and operation. An insertion meter is a more economical solution above a certain line size (the intersection point of the two curves). The pipe diameter of this point varies between 4 and 12 inches, depending on the sensing technology.

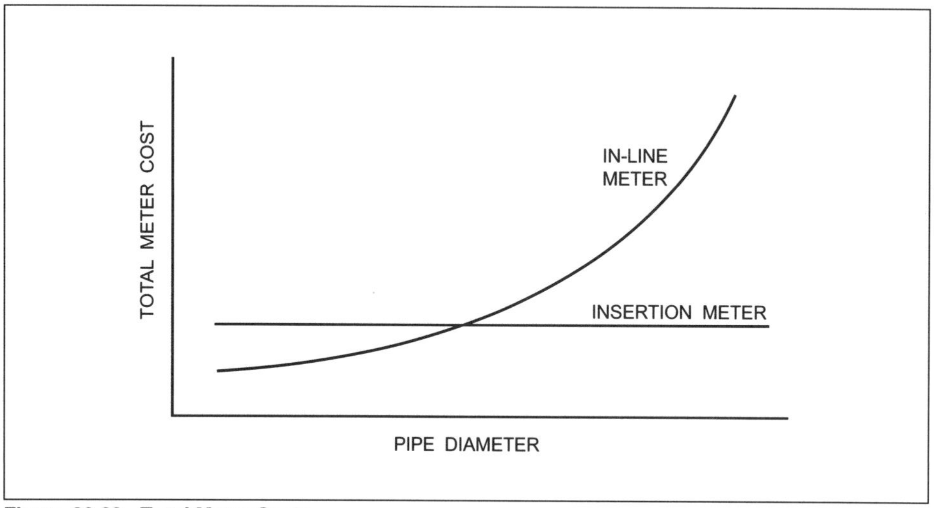

Figure 20-23. Total Meter Costs

Survey Metering

Some applications require only occasional measurement of a flow rate. An example would be a process that requires a flow measurement only during start-up. A meter can be inserted only when required and used elsewhere the remainder of the time. This application is termed a "survey meter" because the meter is used to periodically survey the flow rate at several locations. To perform the same task with in-line meters would require one flowmeter at each location of interest; in most applications a far more costly alternative.

A survey meter can be used as a flow measurement standard. Periodic measurements made with one flowmeter at multiple locations within a facility can be used to correlate the readings from one area to another. Recording performance over time creates a record that can be used to detect variations in process or flow changes. Such information can be useful in early detection of equipment problems. Examples describing the use of a survey meter are contained in References 20 and 21.

Measurement of Distorted Velocity Profiles

Most insertion flowmeters are sensitive to velocity profile distortions. The traditional solution to this problem is to install the meter as far downstream as possible from known sources of velocity profile distortion. This means long, straight runs of pipe that can be expensive and take up valuable space within a facility. If adequate space is unavailable, a flow conditioner is installed upstream of the meter.

It is very difficult to predict the degree of profile distortion due to a particular pipe layout. The accuracy of a flowmeter installed in a location with suspected distortion cannot, therefore, be predicted. An insertion meter can be used to measure the profile by taking multiple velocity readings across the pipe.

Profile measurement is based on Equation (20-1):

$$Q = A\,\overline{v}$$

which states that the volumetric flow rate (Q) is the product of average velocity ($\overline{v}$) and flow area (A). The objective is to measure the average velocity at a number of locations. In theory, each location has a corresponding flow area; thus, the volumetric flow rate can be determined. The total volumetric flow rate is the sum of the individual flow rates.

An example of the procedure is illustrated in Figure 20-24. The velocity has been measured at twelve equally spaced locations across the diameter of the pipe, where the velocity profile is less than ideal. For each velocity measurement, the volumetric flow rate is determined based on the appropriate flow area. For example, the flow rate Q_2 is the product of the flow areas A_2 and the average velocity $\bar{v}_2$; Q_9 is the product of A_9 and $\bar{v}_9$. The total flow rate is the sum of Q_1 through Q_{12}.

The average pipe velocity can be determined from the calculated total flow rate. The sensor can be positioned at the critical position, which is determined from the velocity measurements. Alternatively, the profile factor can be determined at any point along the diameter.

The flow must be as steady as possible while the velocities are being measured. A reference measurement should be periodically checked for flow variation. A reference measurement can be another flowmeter or the rotational speed of a pump or fan. If a reference measurement is not available, the insertion meter can be used in a reference position. The reference position is a location (for example, the pipe centerline) to which meter sensor is returned periodically.

If the flow is found to be unsteady, it is recommended that each velocity measurement be accompanied by a reference measurement. An individual velocity reading is then adjusted, depending on how much the reference reading varies from the mean.

The following are some general comments regarding flow profile measurements:

(1) When the sensor is repositioned, the obscuration factor changes; this must be taken into account at each measurement location.

(2) The accuracy of the procedure improves if the velocity is measured at more locations. A minimum of ten measurements is recommended.

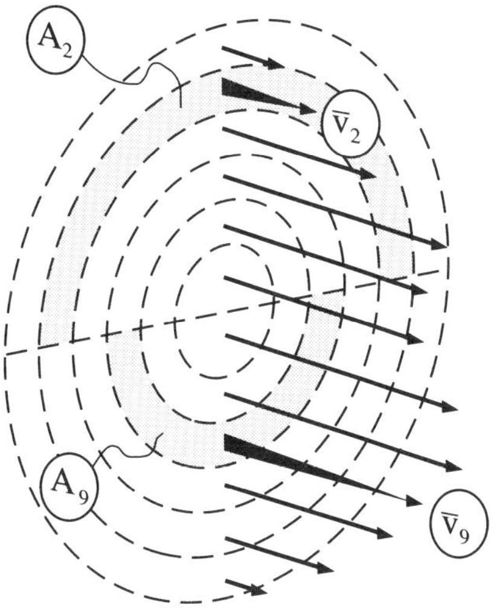

Figure 20-24. Velocity Profile Measurement

(3) The accuracy of the procedure also improves if the velocity measurements are repeated. It is recommended that the velocity be measured four times at each location.

(4) It is recommended that the velocity data be plotted and a smooth curve drawn through the data points. The velocity values obtained from the curve, rather than the actual data, should be used to calculate the flow rate.

(5) The example shown in Figure 20-24 is based on equally spaced velocity measurement locations. Another approach involves velocity measurements that are not equally spaced but results in equal flow rates through each flow area.

(6) This technique is applicable to any flow geometry, not just pipes.

(7) The ISO standards ISO 3354 and ISO 3966 provide detailed descriptions of velocity profile measurements with insertion meters. Meter manufacturers may have techniques that are designed to be used with their equipment.

Examples of the method are given in References 12, 14, and 16.

Multiple Point Measurements

A multiple sensor configuration is less sensitive to profile distortion than a single point measurement. An increase in flow rate at one sensor is offset by a decrease detected elsewhere. A multiple point insertion meter contains several sensors that are mounted on the same probe. A measure of average flow rate is determined by averaging the signals from all the velocity sensors.

An example of the multiple sensor technique is contained in Figure 20-25. Three possible velocity profiles of the same flow rate are shown. The first is representative of a fully developed profile; the other two are distorted. Four velocity sensors are positioned as shown (numbered 1 through 4): two near the top of the pipe and two near the bottom. Local velocity readings for the four sensors are tabulated for each of the three profiles. For each profile an average is made of the four local velocity measurements. Also tabulated for the three profiles are velocity measurements made at the centerline and two (upper and lower) critical positions. The multiple point measurement is less sensitive to profile distortion than single point measurements.

Two of the sensing techniques are suited to multiple sensor configurations: Pitot-static (see Figure 20-10) and thermal resistance. Further information can be found in References 7, 8, and 22.

Two limitations are imposed by multiple sensor configurations:

(1) A sensor assembly is designed to be used on a single pipe size only. The meter cannot be used on different line sizes.

(2) For hot tap installation, a multiple sensor retractor assembly can be considerably larger than a corresponding single sensor unit. The multiple sensor assembly is the same size as the pipe diameter, while a single sensor is typically only a few inches in length.

Higher Accuracy Configurations

An insertion meter can be considered an in-line meter by mounting it into a section of pipe. The pipe is mounted as if it were an in-line meter. The pipe/meter assembly can be calibrated as an in-line flowmeter. Accuracy is dependent on the calibration accuracy and the degree to which the calibration conditions match the application conditions.

Control Applications

When a flow measurement is required for control purposes, repeatability is important. A bias error can be eliminated when the control loop set point is adjusted. The repeatability of an insertion meter is typically comparable to an in-line meter that uses the same sensing technique. If the velocity profile is repeatable, an insertion meter will perform similarly to an in-line meter for control applications.

Energy Monitoring Applications

Historically, insertion flowmeters have been used in three applications:

(1) The Pitot tube and the thermal anemometer are the velocity sensors most commonly used in the research of fluid dynamic behavior; see, for example, Reference 11.

(2) The industrialized thermal anemometer is used extensively in HVAC applications. It is also used for low pressure gas flow applications in process industries, typically stack gas. See, for example, Reference 22.

(3) Magnetic flowmeters have been applied in water treatment and wastewater applications. A mag meter will often be used in conjunction with a level detector in a partially filled pipe. See, for example, Reference 12.

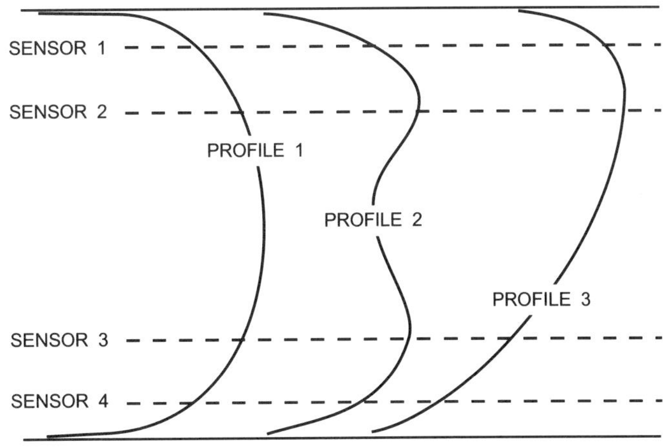

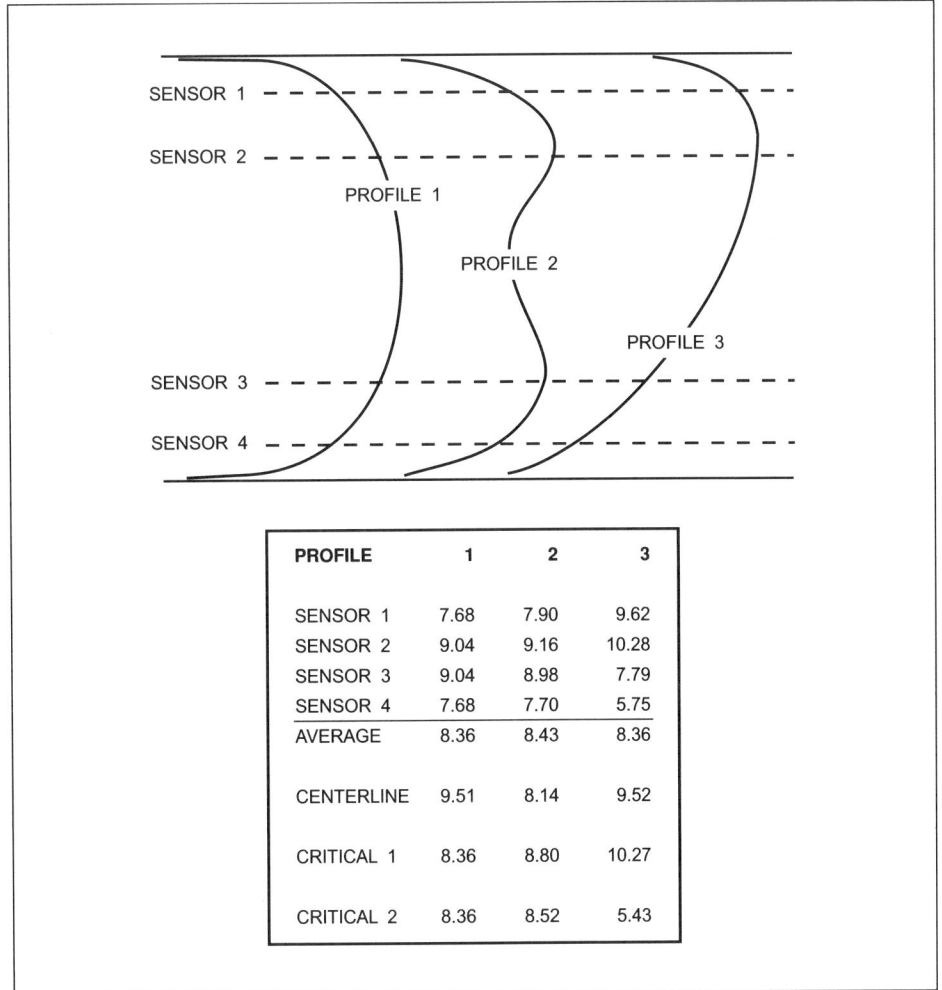

PROFILE	1	2	3
SENSOR 1	7.68	7.90	9.62
SENSOR 2	9.04	9.16	10.28
SENSOR 3	9.04	8.98	7.79
SENSOR 4	7.68	7.70	5.75
AVERAGE	8.36	8.43	8.36
CENTERLINE	9.51	8.14	9.52
CRITICAL 1	8.36	8.80	10.27
CRITICAL 2	8.36	8.52	5.43

Figure 20-25. Multiple Sensor Averaging Technique

A fourth application evolved during the first energy crisis in the early 1970s. Large multiple-building facilities often use a centralized steam plant and pipe distribution system to provide heat energy to the different buildings. Typical facilities include hospitals, universities, military installations, and central heating districts. When the cost of energy increased significantly, the monitoring of energy flow became an economic necessity. Such facilities typically utilize pipes that are too large to be economically metered by an in-line meter. The insertion meter has proven to be an economical solution to the energy monitoring needs of large steam distribution systems.

Many older installations use large orifice plates. The insertion meter is a better solution for two reasons:

(1) An insertion meter is more economical.

(2) Steam distribution systems typically exhibit a wide range in flow rates between summer and winter because of changing heating load. Flow rangeability of 100:1 is not uncommon. Orifice plates could be switched every six months to accommodate the change. However, the changeover is accomplished much more easily with insertion meters. Additionally, line shutdown is not required.

Energy monitoring applications also spurred the use of increasingly sophisticated flow computer-based monitoring systems. A typical energy measurement consists of flow rate and temperature measurements at several locations. A significant number of sensor inputs and mathematical calculations are required. The insertion meter is only one component in a measurement system. Some typical applications are described in References 23 through 26.

Installation

Installation begins with the selection of an appropriate location. A number of factors that should be considered:

(1) Adequate space must be available to accommodate the meter when fully retracted from the pipe. Any new equipment or construction performed after installation should take this requirement into account.

(2) Access by service personnel must be considered. Maintenance costs are proportional to the difficulty in accessing the equipment. Think of human factors: heat, cold, ventilation, light, access.

(3) The meter may operate best if it is installed in a particular orientation (consult the manufacturer). Most meters will operate at any orientation.

(4) Installation at the bottom of a pipe is not recommended. In liquid flow, particulate matter tends to collect at the bottom of the pipe. In gas or vapor flow, liquid will collect at the bottom. Whatever is found in the pipe bottom will tend to foul the isolation valve and retractor mechanism seals.

(5) The electronic components must be tolerant of the ambient environment. Be sure that the environmental limitations (typically temperature and humidity) are not exceeded. If the pipe is hot, the meter will be exposed to more heat if mounted vertically than if mounted horizontally.

(6) The meter is sensitive to the flow profile; hence, upstream pipe disturbances should be avoided or appropriate flow conditioning applied. A good rule of thumb is to position the meter with 20 diameters of straight run upstream and 10 diameters downstream.

An insertion meter can be installed without a process shutdown by hot tapping the line and installing the appropriate valving. An in-line meter cannot be installed without shutdown unless the pipe layout was installed with a bypass line. Similarly, maintenance can be performed on an insertion meter without process shutdown, while many in-line meters require a shutdown for maintenance. If a process cannot be shut down, the insertion meter may be the only viable alternative.

(7) In liquid applications the pipe must be full.

(8) Be sure that the wetted parts of the meter are compatible with the process fluid.

Installation of an in-line flowmeter requires access to the inside of the pipe. An in-line flowmeter consists of a spool piece or wafer section that mounts between flanges, and installation is much like installation of a section pipe. An insertion meter is installed from the outside of the pipe.

One common installation of an insertion meter is into an existing piping system without process shutdown. The following five-step procedure is illustrated in Figure 20-26:

(1) A nipple is welded to the outside surface of the pipe. The welding surface of the nipple is sized for the curvature of the pipe. Nipples are available with different pressure ratings.

(2) A flange that may be welded or threaded is attached to the nipple.

(3) An isolation valve (either a ball valve or gate valve) is bolted to the flange. The bore must be large enough to accommodate the velocity sensor.

(4) A hot tap drill is mounted to the valve. This drill is designed to operate under process pressure and will drill a hole through the pipe wall without dropping the disc (called a coupon) into the flow.

(5) The hot tap drill is removed, the valve is closed, and the meter is mounted.

Details regarding the hot tap procedure can be found in Reference 27.

The hot tap procedure should be performed only by trained personal. Local regulations may require a hot-tap permit.

The same basic procedure can be followed if the process is shut down. A tee section or saddle can be utilized if a pipe system is designed to accommodate an insertion meter. During installation, a number of precautions need to be followed:

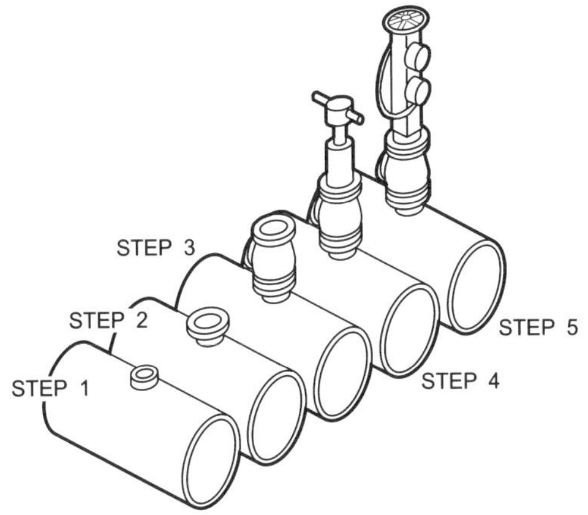

Figure 20-26. Hot Tap Procedure

(1) The hole and isolation valve diameters must allow for clearance of the sensor and stem. Alignment of all components should be checked carefully before installation. A "check tool" is a good way to verify clearance. This can be done using a disc of appropriate diameter mounted to the retractor stem. If the disc fits, so will the sensor.

(2) The hole should be smooth and free of burrs and welding slag. Protrusion of any material into the pipe can reduce accuracy. A shell saw or fly cutter leaves a cleaner hole than does a twist drill.

(3) The welds should be static pressure checked before cutting pipe with the hot-tap drill.

(4) The accuracy of the measurement is dependent on the meter being aligned perpendicular to the pipe axis. Consult the meter manufacturer regarding specific tolerances.

(5) Be sure to install a pressure relief valve in the line to protect the flowmeter, if necessary.

(6) Put an anti-seize compound on all threaded parts.

Mounting the Sensor—Retractor Mechanisms

The sensor of an in-line meter is held in place by means of the meter body. The meter body becomes an integral part of the piping system. The sensor of an insertion meter is mounted to a retractor mechanism that positions the sensor within the flow.

The selection of a retractor mechanism is based on three basic constraints:

(1) The process pressure and temperature. The selection of the process seal as well as the retraction mechanism will be affected.

(2) The anticipated need to reposition or remove the sensor. Does the probe need to be retractable?

(3) Whether the process can be shut down for installation and maintenance. A retractable probe is required if shutdown is not possible.

Three basic retractor mechanisms are available for different applications. Each is described below. Pressure ratings are typical values. Consult the manufacturer for specific performance of the particular product.

(1) The simplest retractor mechanism consists of a clamp to hold the stem in position and a process seal made up of O-rings. The pressure rating is low, typically below 50 psi. The stem can be repositioned by hand under process pressure because the reaction force is low. The operating temperature is defined by the materials and the design of the sealing mechanism.

(2) A high pressure version of the same retractor involves a compression fitting over the stem. A typical value of operating pressure is up to 5000 psi. Once tightened, the fitting seal *cannot* be reused; hence, this type of retractor is not used where the stem needs to be repositioned frequently. The stem cannot be repositioned under process pressure; line shutdown is required.

(3) A more sophisticated retractor moves the stem by means of a threaded rod or similar mechanism. Process seals are provided that can accommodate stem positioning under process pressure. Pressure ratings are typically several hundred psi. The sensor can be repositioned without process shutdown because of the retraction mechanism. The operating temperature is

defined by the materials and design of the sealing mechanism. A retractor of this type can be configured with an automatic positioner to enable remote controlled repositioning of the velocity sensor.

Typical retractor mechanisms are shown in Figures 20-27 and 20-28 that illustrate:

(1) fixed position and retractable units,

(2) threaded and flanged fittings,

(3) integral pressure transmitter, and

(4) isolation valve.

Installation of an insertion meter requires two sets of measurements; the first is cross-sectional pipe area. Ideally, the pipe inside diameter should be measured in several directions and the results averaged. A more practical solution is to measure the outside diameter and subtract the wall thickness. Corrections may have to be made for external corrosion and pipe quality.

After the coupon is removed, attach it to the retractor with a piece of wire. An indication of pipe wall thickness will be available at the meter site.

The second measurement is the insertion depth, which is the distance from the inside pipe wall to the sensor. The probe assembly is usually marked to indicate the insertion depth from the flange face. The sizes of the isolation valve, weld assembly, and pipe wall thickness must be taken into account to determine the sensor position.

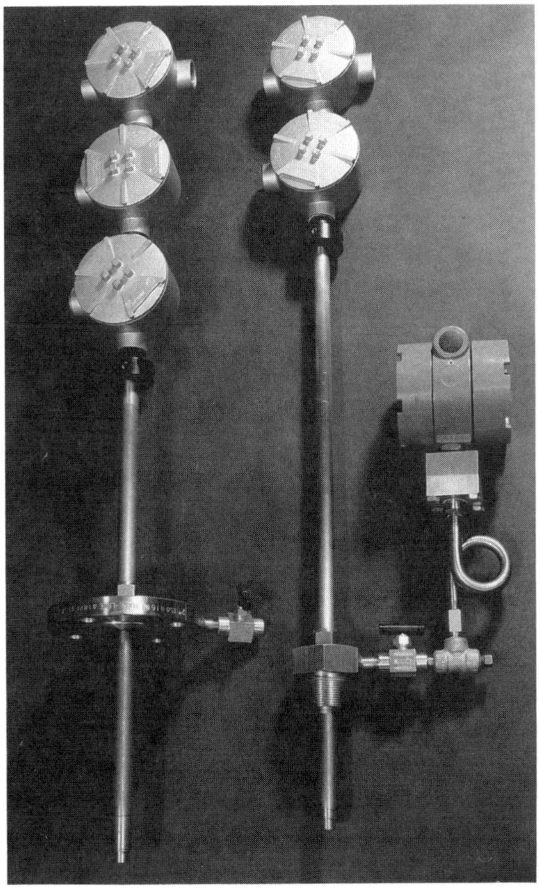

Figure 20-27. Typical Retractor Mechanisms
(Courtesy of EMCO)

The opposite pipe wall can be located by extending the probe until the sensor makes contact. Be careful to not damage the sensor.

Maintenance

The major difference between maintenance of an insertion meter and an in-line meter is that the insertion meter can be removed without process shutdown. Beyond that, the maintenance requirements of a particular insertion sensor will be similar to those of an in-line unit.

The following are some general guidelines regarding the removal and installation of an insertion meter:

(1) Do not insert the sensor so far that it is forced against the opposite pipe wall.

(2) Do not close the isolation valve on the sensor or stem.

(3) An insertion meter can be heavy. Be aware of the weight when the assembly is removed. Care must be taken to avoid damaging the sensor during removal. Be aware of connecting wires before lifting the weight of the meter.

(4) Be careful to not damage the velocity sensor when removing the meter.

(5) Be sure to purge the pressure in the retractor before removing the meter.

(6) When reinstalling the meter, always verify that a sensor will clear by using a hot-tap tool.

Each of the sensor types has unique maintenance requirements. Described below are some features.

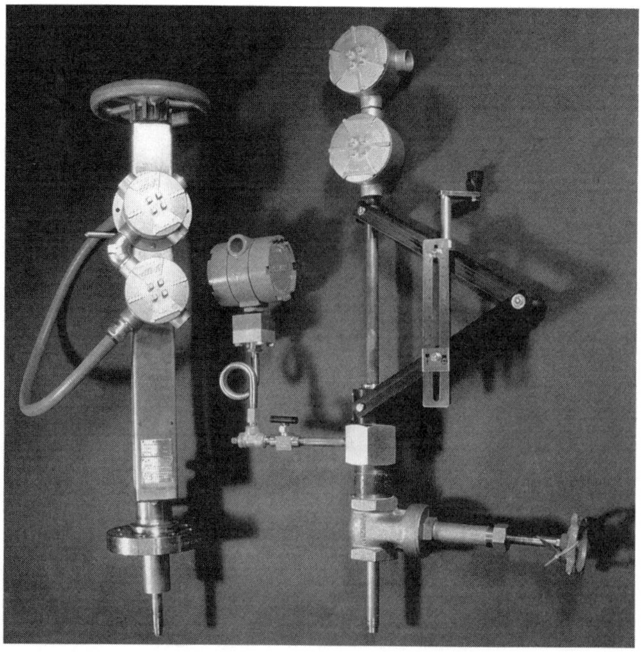

Figure 20-28. Typical Retractor Mechanisms
(Courtesy of EMCO)

The turbine is more likely to wear out than any of the other sensors because it has moving mechanical parts. Depending on flow conditions and service, a turbine may last for ten years or several months. Difficult environments include high temperature steam, non-lubricating fluids, corrosives, and unsteady flow. Consult the manufacturer for a suggested maintenance schedule for your application.

The following are some maintenance aspects of turbine meters:

(1) A turbine sensor is usually field replaceable; often the bearings are also field replaceable.

(2) Turbine rotor bearings can exhibit either wear or buildup. This cannot be detected without removing the sensor.

(3) Turbine rotor blades can become bent or can break. This can usually be detected without removing the sensor by looking at the signal on an oscilloscope. Bent or broken blades will affect the calibration factor.

(4) If a system is restarted, it is recommended that flowmeters subject to damage be retracted. During start-up, high fluid velocities can occur when valves are opened. Additionally, in gas applications, the gas may contain entrained liquid that has collected in low spots throughout the piping system. The combination of high gas velocity steam and high density entrained liquid can damage many flow sensors.

Both magnetic and thermal sensors may build up a coating that acts as an (electrical or thermal) insulator. This cannot be detected without removing the sensor, and it will affect the accuracy of the measurement. Usually, the coating can be cleaned and the sensor reinstalled. The required maintenance schedule depends on how fast the fluid coats the sensor.

Differential pressure sensors may plug with particulates. The shape of the sensor can provide a cleaning action. The most certain method for cleaning accumulation is removal from the process and mechanical cleaning. An alternative is high pressure purging from the differential pressure sensing ports. Some models are fitted with suitable ports. Be careful not to damage the pressure sensor. The frequency of cleaning depends on the contamination level in the process fluid.

Cost Savings

Typical Installed Cost Savings

In this section the typical installed costs of insertion and in-line meters are compared. Installation of an insertion meter is similar for all line sizes: A nipple is welded to the pipe, the pipe wall is cut, and the meter is mounted to the nipple. This typically takes one or two hours. An in-line meter requires cutting the line and welding flanges onto the ends. Eight or more hours may be required.

Figure 20-29 shows typical savings as a function of line size, the saving being the difference between installed costs of an in-line meter and insertion meter. The installed cost includes purchase of the meter and labor for installation. This example is based on a comparison between an orifice plate and a multiport Pitot sensor. Details are contained in Reference 17.

Typical Operating Cost Savings

The major cost of operating a flowmeter is the energy required to overcome the permanent pressure loss. In large line sizes the difference in pressure loss between in-line and insertion meters becomes significant. In the example, a comparison is made between an in-line (orifice plate) and an insertion meter. The orifice plate is used because it is the most common, large line size in-line meter that exhibits a significant pressure loss. Magnetic flowmeters are available in large line sizes but exhibit no pressure drop.

Hydraulic energy loss calculations can be made from equations found in Reference 17:

For liquids:

$$hp = \frac{Q\,\Delta p}{47{,}500\,\eta}$$

For gases:

$$hp = \frac{W\,\Delta p}{3.8 \times 10^5\,\rho\,\eta}$$

where:

hp = horsepower
Q = liquid volumetric flow rate (gal/min)
W = gas mass flow rate (lb/hour)
ρ = gas density (lb/cu ft)
η = efficiency of pump or compressor
Δp = permanent pressure loss (inches of water)

The energy cost associated with these energy losses can be calculated from:

energy cost ($/year) = 0.746 (hp)(operating hours/year)($/kWh) (20-22)

Contained in Figures 20-30(a) and 20-30(b) are examples of potential energy cost savings resulting from the use of an insertion meter instead on an orifice plate. Significant operating cost savings can be achieved by the use of the insertion meter if the piping system is controlled in such a manner as to *not* generate

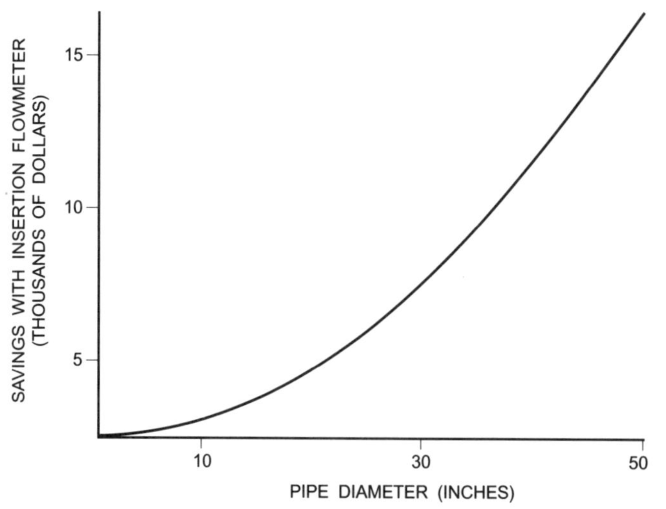

Figure 20-29. Installed Cost Savings (Insertion Meter Compared to Orifice Plate—1984 Dollars)

the hydraulic energy wasted across the orifice plate. Therefore, the equations above describe the *potential* energy cost savings that may be realized. Energy cost savings will not be realized if the hydraulic system is designed, for example, to increase the pressure drop across a control valve if the selection of an insertion flowmeter *reduces* the pressure drop across the flowmeter.

The figure examples assume a "typical" orifice plate with a permanent pressure loss of 65 inches of water. Such an orifice plate has a beta ratio of 0.6 and a full scale output of 100 inches of water. Further information regarding the pressure loss of different flowmeters can be found in Reference 18. It is also assumed that the insertion meter has no permanent pressure loss. This is a reasonable assumption for larger line sizes.

More details regarding cost savings can be found in Reference 19.

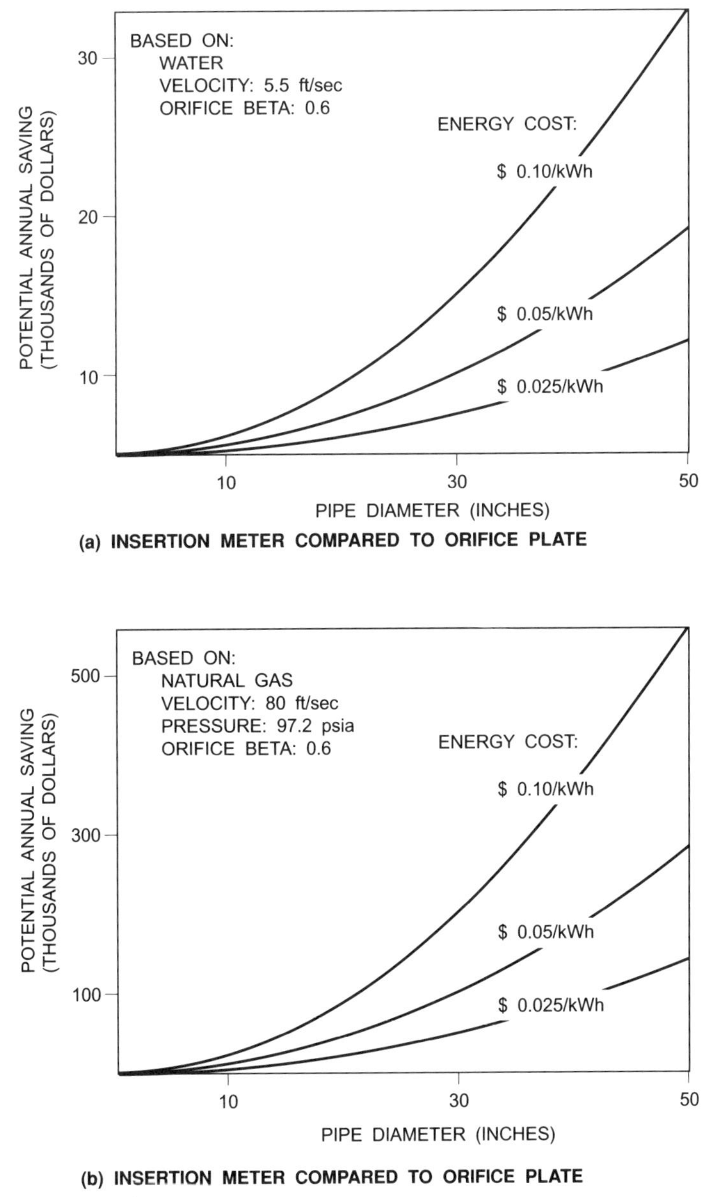

Figure 20-30. Potential Cost Savings

Glossary

Profile factor—The ratio of local velocity to average velocity.

Obscuration factor—The ratio of the projected area of the wetted part of the meter to the cross-sectional pipe area.

Velocity profile—The distribution of local velocity across a pipe diameter.

Local velocity—The velocity at a point in the flow stream.

Pitot probe—A velocity sensor that operates based on detecting a differential pressure.

Coupon—A piece of the pipe wall removed during the hot tap procedure.

Hot tap—A procedure to install a mounting section on a pipe without stopping down the flow.

Centerline position—Positioning the sensor in the center of the pipe.

Critical position—The point in the flow stream where the local velocity is equal to the average velocity.

Retractor mechanism—The mechanism that positions the sensor within the pipe and allows for sensor repositioning and removal.

References

1. Schlichting, H., *Boundary Layer Theory*, (University of Braunschweig, Germany), New York: McGraw-Hill Book Company, 1979.
2. Rusnak, J. J., et al., "Energy Monitoring—The State of the Art," Annual Industrial Energy Technology Conference, 1985.
3. Bailey, S. J., "Computer Enhanced Metering Results in Better Flow Control," *Control Engineering*, Feb., 1983.
4. Mesnard, D., "Flow Measurement Using Field Flow Computers," *Measurement and Control*, Dec., 1986.
5. Mattingly, G. E., and Yeh, T. T., "Mixing Motions Produced by Pipe Elbows," National Institute of Standards and Technology, NISTIR 89-4029, Jan., 1989.
6. *Annubar Flow Handbook*, Boulder, Colorado: Dieterich Standard Corporation, April 1986.
7. Britton, C., and Mesnard, D., "A Performance Summary of Round and Diamond-Shaped Averaging Pitot-Type Primaries," *Measurement and Control*, Sept. 1982.
8. Miller, C. E., "On the Application and Performance of Insertion Turbine Meters for Steam Flow," International District Heating Association Annual Conference, 1979.
9. Clarke, W. R., "Thermal Mass Flowmeters Join the Workhorse Class," *Measurement and Control*, Sept. 1987.
10. Kurz, J. L, "Thermal Flow Sensors," *Measurement and Control*, June 1983.
11. Freymuth, P., "A Bibliography of Thermal Anemometry," TSI Incorporated, 1982.

12. Powers, J., "Insertable Magnetic Flowmeter Fights Debris, Handles Difficult Lines," *Chemical Processing*, April 1986.

13. "Flow Sensors Can Also Make Dramatic Energy Savings," *Fluids Handling*, October 1981.

14. Rusnak, J. J., "Why Insertion Turbine Meters Are Replacing Orifice Plates for Steam Flow Measurement," Conference on Industrial Energy Conservation Technology, 1983.

15. Ginesi, D., and Grebe, D., "Flowmeters, A Performance Review," *Chemical Engineering*, June 22, 1987.

16. Ginesi, D., "Try Insertion Flowmeters for Low-Cost, Hot-Tap Installations," *Power*, March 1987.

17. "Installed Cost Savings Estimator," Boulder, Colorado: Dieterich Standard Corporation, DS-7110, Nov. 1984.

18. Miller, R. W., *Flow Measurement Engineering Handbook*, New York: McGraw-Hill Publishing Company, 1989.

19. "Energy Cost Savings Estimator," Boulder, Colorado: Dieterich Standard Corporation, 71110, Nov. 1984.

20. Andersen, S. A., and Dieckert, J. C., "On-Site Chiller Testing," *ASHRAE Journal*, April 1990.

21. Rodencal, T. E., and Atkins, J. W., "Practical Aspects of Conducting a Paper Machine Steam Survey," *TAPPI Journal*, Sept. 1980.

22. Muller, M. R., "On the Use of Self-Heated Thermistor Arrays for the Measurement of Air Flow in Ducts," ASME FED-17, Mass Flow Measurements Symposium, 1984 Winter Annual Meeting.

23. Hayes, J. W., Rusnak, J. J., "Use Meters To Put Teeth into Your Conservation Programs," Conference on Industrial Energy Conservation Technology, 1983.

24. Porter, P. E., "Steam Metering Is Essential for Energy Conservation Decision Making," American Textile Manufacturer's Institute - Energy Conservation Forum, 1983.

25. Jaehne, H., and Smith, K., "The St. Paul Metering Experience," International District Heating Association Annual Conference, 1984.

26. Rusnak, J. J., "You Can Justify Meters for Your Energy Conservation Program," Conference on Industrial Conservation Technology, 1984.

27. Tinsley, R. J., "Hot Tapping of Live Steam Lines," Civil Engineering Laboratory, Naval Construction Battalion Center, Technical Memorandum TM 53-79-4, 1979.

The following standards are applicable to insertion flow measurement:

ISO 3354-1988(E), *Measurement of clean water flow in closed conduits: Velocity-area method using current-meters in full conduits and under regular flow conditions.*

ISO 3455-1976(E), *Liquid flow measurement in open channels: Calibration of rotating-element current meters in straight open tanks.*

ISO 3966-1077(E), *Measurement of fluid flow in closed conduits: Velocity area method using Pitot static tubes.*

ISO 7145-1982(E), *Determination of flow rate of fluids in closed conduits of circular cross section: Method of velocity measurement at one point of the cross section.*

ISO 7194-1983(E), *Measurement of fluid flow in closed conduits: Velocity-area methods of flow measurement in swirling or asymmetric flow conditions in circular ducts by means of current meters or Pitot static tubes.*

About the Author

Mr. Thomas Kegel received his Bachelor's and Master's degrees from the University of New Hampshire in 1980 and 1982, respectively. Graduate studies focused on fluid mechanics and the modeling and control of dynamic systems. He was employed by The Foxboro Company as a Research Engineer from 1982 until 1987. This work involved research and development projects leading to new fluid measurement products. Considerable investigations were made into the nature of multiphase flow. From 1988 to 1991 Mr. Kegel was employed by Engineering Measurements Company as a Principal Engineer. This role included responsibility for the flow calibration facilities and engineering responsibility for the insertion flow products. Mr. Kegel is currently the Senior Staff Engineer with Colorado Engineering Equipment Station, Inc. (CEESI) and is a member of the ASME Flow Measurement Standards Committee.

21

Custody Transfer Measurement

Why should a separate chapter be written about custody transfer measurement since flow measurement is flow measurement? This is true, except that, when money is to be exchanged, achieving the best accuracy becomes important. The desired limit on accuracy for flow measurement is that it must be 100% correct. However, for reasons discussed elsewhere, no measurement is absolutely accurate, but it is accurate to some limit.

For some flow measurements (such as control applications), a repeatable signal may be more important than an accurate signal. Once a process is under control, repeatability is of prime importance.

On the other hand, if money is going to change hands, accurate flow measurement is required so that the parties to the transaction are treated fairly. The accuracy limit for control signals may be several percent; for custody transfer it should be tenths of a percent.

An example of accuracy and the relationship of dollars to flow accuracy can be shown in a large crude-oil measuring facility (see Figure 21-1). This facility was delivering 12-million barrels of oil per day. Calculating on a basis of $30 per barrel, this meter station was handling $360 million of oil per day. A 0.1% error in flow represents $360,000 per day.

In the example, a 0.1% accuracy represents a significant quantity of dollars, but it is the limit of accuracy that can be obtained. Operational measurement may require no more than ± 5%. Control measurement may be acceptable at ± 2%. But custody transfer measurement converted to dollars is not ± anything, since accountants send out bills to the penny. For this reason, quantities for custody transfer are treated as absolute. The goal for this measurement is to reduce all inaccuracies to a minimum so that a measured quantity can be the agreed upon quantity for exchanging custody.

It is important in custody transfer metering to be constantly aware that flow measurement is dollars, and the perspective on the measurement changes accordingly.

Measurement Contract Requirements

Measurement becomes more complex when two or more parties must agree to the quantity of product that is measured and agree to pay money based on this quantity. To protect each party's interest, a contract is normally written that specifies all requirements for the measurement of the delivered material such as:

(1) definitions used in the contract,

(2) quantity of material,

(3) point of delivery,

(4) properties of the material,
(5) measurement station design,
(6) measurements to be made,
(7) quality of material,
(8) price,
(9) billing,
(10) force majeure,
(11) default or termination,
(12) term,
(13) warranty of title,
(14) government requirements,
(15) arbitration, and
(16) miscellaneous.

Figure 21-1. Crude-Oil Measuring Facility

These are all items of interest that are settled prior to commencing measurement for custody transfer purposes. A number of these items are typical information of any contract, but it is of value to expand on the ones that impact the measurement equipment.

Quantity of Material

This will specify not only the quantity of the material to be measured by the Seller but also any rights the Seller may have to quantities of material above or below the agreed upon amount. This requires the responsible measurement personnel to be aware of these values, to see that the contract limits are being met, and to ensure that the Seller has the capability to meet them.

Point of Delivery

The contract sets forth the point of custody transfer. If the measurement point and the point of delivery are not the same, an agreement must be reached between the Buyer and the Seller for the responsibilities for the material between the two points.

Properties of the Material

Limits are specified for certain basic properties (such as pressure and temperature) and actions to be taken are stated if the material is outside the limits.

Measurement Station Design

The ownership and responsibility for design, installation, maintenance, and operation of the meter station of both the Buyer and Seller are spelled out. For metering stations covered by standards, specific references to the standards are made.

These standards may be governmental requirements, industry practices, or individual company guidelines and are usually combinations of them. They detail the kinds of meters to be used and their attendant correcting and readout systems. Details of access by both parties to the equipment and the requirements of frequency of testing and/or reports are spelled out. For large dollar-exchange quantities, there may be an allowance for a check station with the same provisions listed as above, stating how any discrepancies between the two measurements will be handled.

Some provision is made for estimating deliveries during times when the meter is out of service or registering inaccurately, and the procedure for resolving quantities during these periods is included. Some limits of accuracy are set; and, if these are exceeded as determined by test or check meters, then provisions of settlement are implemented.

A time period during which a correction can be made is stated if it is not possible to determine the error source and the time of change. Requirements for retaining records and reports is spelled out for both parties. This relates to the specified time period allowed for the quantity measurement to be questioned.

Measurements

This specifies in nonconfusing terms the unit of quantity that is to be delivered.

In a weight measurement, only the unit of weight need be specified. For most commercial purposcs, the terms "weight" and "mass" are used interchangeability without concern for the effects of the attraction of gravity on the weight being measured.

In a volumetric measurement, the base conditions of temperature and pressure are spelled out for the volume, which, in essence, affects a weight or mass measurement.

Requirements are specified for all related equipment (beyond the basic meter) and how these secondary measurements will be used to correct basic meter readings. These requirements are particularly important since it is possible that the quantities may be kept in different ways by interested government parties and by the parties to the contract (with their plant quantity reports). Major confusion can arise if all of these requirements are not spelled out and understood.

Quality of Material

Any natural or manufactured product can have small and varying amounts of foreign material that is not desirable, but, at least, whose quantities must be limited. The quality section defines the rights of the Buyer and the Seller if such limits are exceeded. These specifications may also include separate pricing for mixed streams, so quantities must be delineated for proper payment of the mix. If there are too many unwanted contaminants, a price reduction may be allowed rather than a delivery curtailment. These details are spelled out in the quality requirement section.

Billing

This section sets a deadline for the computation of quantity with a provision for correcting of errors. It specifies the procedure for billing, the payment period, and penalties for late payment.

A properly written contract protects the interests of the Buyer and the Seller. That fair and equitable billings can be made for an exchange of the quantity of material is a basic requirement for establishing custody transfer.

Summary of Contract Requirements

The ultimate definition of measurement accuracy is when the Seller sends a bill, the Buyer pays the bill, and both parties are satisfied with the results. The last place that flow measurement should take place is in a court room, so all possible misunderstandings and means of their solution should be defined by the contract.

Metering System Design Concerns

To minimize flow errors in a metering system as required in custody transfer situations, a full study and understanding of the effects of the various influences on accuracy is required before the design is set.

FLOW MEASUREMENT IS MORE THAN BUYING A METER.

For purposes of looking at the problem, three concerns should be studied: the fluids, the flow, and the meter.

Fluid Concerns

The purpose of a custody transfer flow measurement is to determine the quantity of a given fluid delivered so that ownership may be transferred. Fluid characteristics can be of little consequence, or they can be the most important concern in arriving at proper measurement.

The first concern is that volumetric fluid measurements at flowing conditions must be converted to base conditions. Physical properties are available for clean, pure fluids, but accurate data for mixtures may not be readily available. The mixture law calculations are a means of approximation, but with some fluids these calculations can introduce significant errors in custody transfer metering.

Not all flows are single-phase, and multi-phase flows cause significant errors in most flowmeters. Some fluids are handled near their critical points. Because of the wide variations in their densities with small changes in temperature and/or pressure, measurement accuracy requirements are difficult to achieve. Fluids that are corrosive or erosive may require special materials of construction and/or filtering to obtain accurate measurement.

When a fluid characteristic is not conducive to accurate measurement, consideration should be given to upgrading the fluid by filtering, cooling, heating, separating, or whatever is required to prepare it for measurement. Otherwise, custody transfer measurement accuracy will be compromised. This may include moving the meter to another location in the process where fluid conditions are more appropriate.

Flow Concerns

A meter should be given the best chance for performance by proper preparation of the flow it measures. Most meters operate with minimum error if the flow is at a fairly constant rate (at least at a frequency change slower than the frequency response of the meter being used). Trying to measure downstream of flow fluctuation sources (such as pumps, compressors, fluctuating control valves, or fluctuation-inducing piping) will create questions of accuracy. Therefore, a proper choice for location of the meter with reference to these variations is pertinent (see Figure 21-2).

A custody transfer flowmeter should be operated only within the limits of its performance curve. No meter used for custody transfer should be used at the extremes of its performance curve. This may require the use of multiple meters with control systems if wide ranges of flows are to be encountered. Such an installation will be a valuable investment in metering confidence.

Minimum initial cost of an installation is of little importance if the measurement accuracy obtained cannot be proven. The most important design parameter is the accuracy obtained on the custody transfer station as a total system. Such stations are more complex and more expensive than process control stations.

The choice of meter may be dictated by flow variations. Meters with low rangeability should not be chosen for wide-ranging flows.

Consideration should be given to the "real" normal flow rates so that meters are operating most of the time within their most accurate ranges. There is a tendency for designers to concern themselves overly with maximum and minimum flows (which may exist only for short durations) while ignoring the fact that the normal flow represents the major portion of total flow to be measured. Bills for custody transfer are based on total flow, not instantaneous flow rates; whereas, in control applications, it is more important to control the flow rate.

Meter Concerns

Selection of a meter to accomplish the measurement should be chosen based on its ability to do the job required. After establishing fluid and flow concerns, the meter best able to accomplish the job should be chosen.

Valuable sources of information for successful meter use are various company and industry standards (see the Bibliography). If these are not available, then a person in another industry doing the same type of flow measurements may be contacted. Manufacturers also may have experiencc in solving all kinds of measurement problems. A caution: Ask specific questions to define all of the

In evaluating meters, the source of the information should be qualified to supply correct information based on experience rather than advertising.

A.G.A. FIGURE A

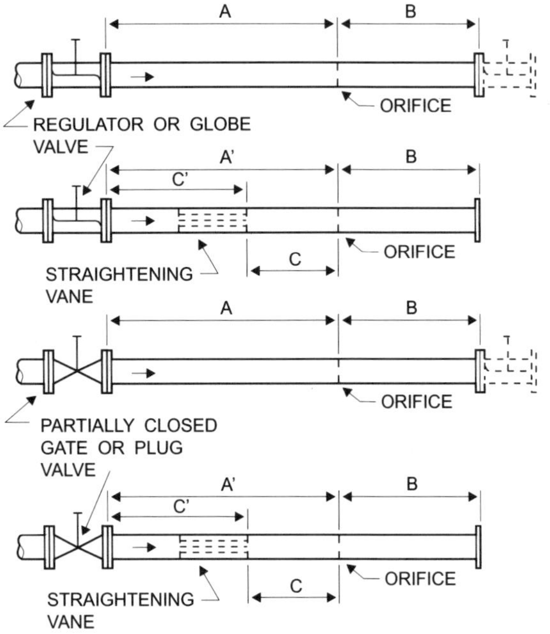

Illustration shows regulator or partially closed gate, globe, or plug valve preceding the meter tube. If the valve is wide open, length requirements should be based on the fitting immediately preceding it, but in no case should length be less than permitted by Figure B, C, or E. Where there are no fittings preceding the valve, dimensional data for Figure C shall apply.

A.G.A. FIGURE B

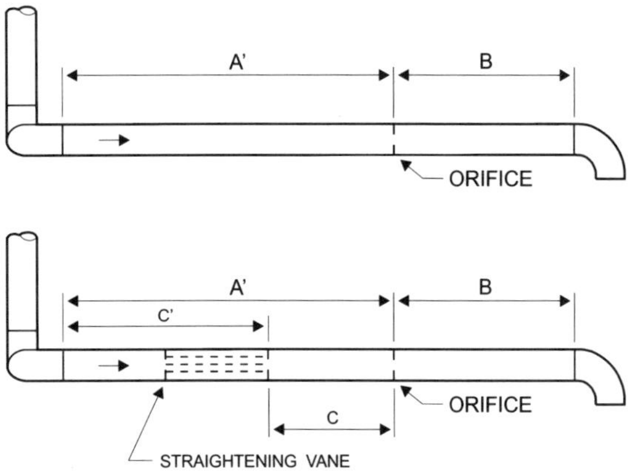

Two ells or bends not in the same plane preceding the meter tube are illustrated in Figure B. This type of installation is common in small metering stations similar to Figure C, including power plant stations. Note: If the two ells illustrated are closely (less than three pipe diameters) preceded by a third ell in a different plane, the piping requirements shown for dimension "A", should be doubled.

A.G.A. FIGURE C

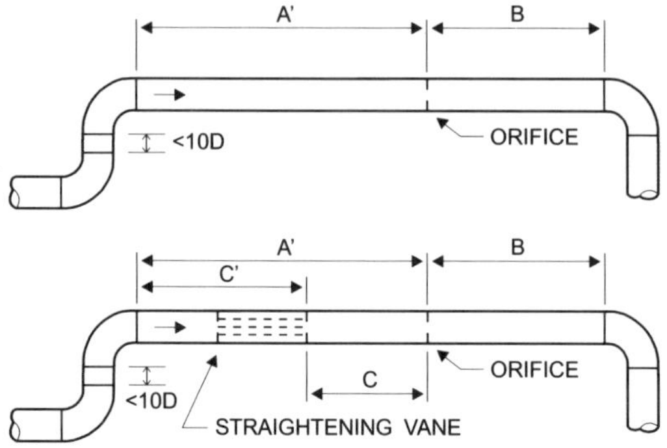

A.G.A. FIGURE D

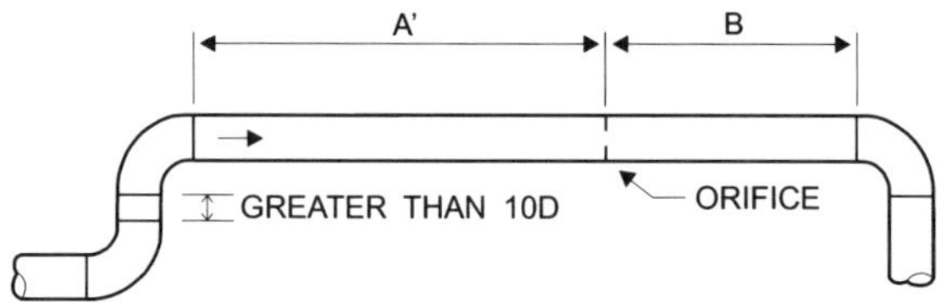

Illustrated are meter tubes preceded by two ells or bends in the same plane. These types of installations are commonly found in small metering stations such as well-head measurement, transmission purchase stations and town border stations. Where there are no other fittings, dimensional data for Figure C applies when meter tube is preceded by a wide open gate, plug or globe valve.

*NOTE: The meter tube lengths given in this catalog are Daniel interpretations of the charts from the 1985 printing of ANSI/API 2530 (A.G.A.#3) orifice metering of natural gas, section 4.4 entitled "Length of Pipe Preceding and Following Orifice".

Figure 21-2. A.G.A. Piping Installation Classifications

A.G.A. FIGURE E

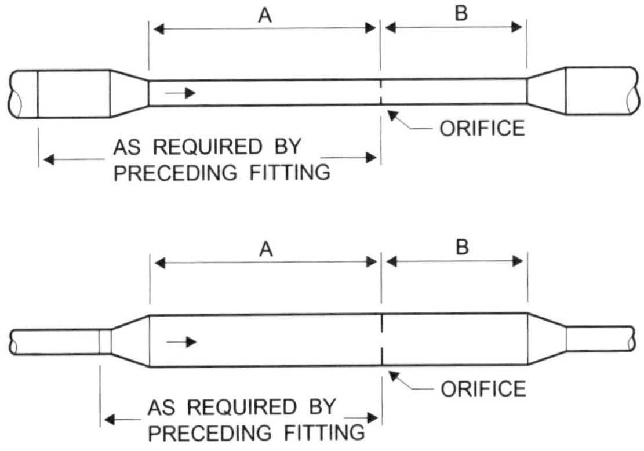

This figure illustrates a concentric reducer preceding the meter tube. This type of installation is used to increase or decrease internal diameter of meter tube from that of flow line. Although not commonly used, this type of installation is sometimes employed in power and chemical plants or other applications to increase or decrease the pressure drop across the plate so that proper Beta ratio can be used. For installations using eccentric reducers, refer to A.G.A. Figure A.

Note: Straightening vanes will not reduce required lengths of straight pipe A. Straightening vanes are not required because of the reducers. They may be required because of other fittings which precede the reducer. Length A is to be increased by an amount equal to the length of the straightening vanes whenever they are used.

For those installations not explicitly covered in A.G.A. installation Figures B through E, it is recommended that the meter tube lengths shown in installation sketch Figure A be used.

MINIMUM METER TUBE LENGTHS IN TERMS OF PIPE DIAMETERS AND BETA RATIO
Use for All Pipe Sizes*

INSTALLATION FIGURE	DIMENSION	BETA RATIO				
		.5	.6	.67	.70	.75
A.G.A. Fig. A	A	25.2	30.2	36.0	38.8	44.5
	A′	10.3	12.2	14.2	15.3	17.5
	B	3.7	4.0	4.2	4.3	4.5
	C	5.0	5.5	6.2	6.3	7.0
	C′	5.3	6.7	8.0	9.0	10.5
A.G.A Fig. B	A	21.0	25.0	28.8	31.3	35.3
	A′	10.0	11.3	12.8	13.5	15.0
	B	3.7	4.0	4.2	4.3	4.5
	C	5.0	5.5	6.2	6.7	7.2
	C′	5.0	5.8	6.6	7.0	7.8
A.G.A Fig. C	A	10.1	13.7	17.4	18.9	21.5
	A′	9.2	10.3	11.7	12.0	13.5
	B	3.7	4.0	4.2	4.3	4.5
	C	5.1	5.6	6.2	6.4	7.1
	C′	4.1	4.7	5.5	5.6	6.4
A.G.A Fig. D	A	6.9	9.3	12.3	13.9	16.6
	B	3.7	4.0	4.2	4.3	4.5
A.G.A Fig. E	A	7.2	9.8	11.6	12.1	13.4
	B	3.7	4.0	4.2	4.3	4.5

*Above table based on flange taps. For pipe taps add 2 diameters to A, A′ and C and 8 pipe diameters to B.

Figure 21-2. A.G.A. Piping Installation Classifications
(continued)

parameters of the measurement being investigated. The ability to ask these specific questions requires that the user be knowledgeable in the subject of flow measurement. Simply asking, "What is a good steam meter?" is not meaningful. Steam meters have many specifics that need to be considered, such as (1) is the steam superheated, saturated, or quality, and (2) are flow rate variations and the effects of ambient conditions surrounding the meter significant? Without such specifics, the answer to the initial question may be worse than useless, since you can be mistakenly led to believe that success is assured.

ACCURACY

A term used frequently in flow measurement is "accuracy." Accuracy is more abused than correctly used. Unfortunately, it is a sales tool used commercially by both suppliers and users of metering equipment. The supplier with the "best number" wins the bid. The user will sometimes require accuracy beyond the capabilities of any meter available. In either case, the definition of accuracy may serve a purpose for the user or supplier but has little relevance.

In custody transfer measurement, accuracy is usually defined as "the difference between the measured value and the true value expressed as a percentage." The problem with this definition is that the indicated value is read from the meter, but the method of obtaining the true value cannot be specified. Therefore, the true value is not precisely known. For this reason it is the subject of many arguments. The term "uncertainty," with a specified procedure, is a statistical statement with at least a comparative meaning when examining various meter capabilities.

UNCERTAINTY

An evaluation of the performance of the measurement under flowing conditions can be made by making an uncertainty calculation. Many uncertainty calculation procedures are available in the standards and flow measurement literature, one of which is ANSI/ASME MFC-2M-1983, "Measurement Uncertainty for Fluid Flow in Closed Conduits." The value of this calculation is not so much in the number obtained but in examining the significance of each variable that impacts the flow calculation and in relating these to the flow measurement job in question.

*These calculations must consider the **particular operating conditions for a specific application** in order to be most useful in getting the most accurate measurement.*

Equation (21-1) shows the calculation of uncertainty applied to an orifice meter; it is, therefore, representative of this device in specifics, but it also shows the manner in which the calculations can be made for any device. The equation for the calculation is divided into two types of error — bias (B) and precision (S) — then combined as the square root of the sum of the squares.

$$U = \sqrt{(B_1^2 + B_2^2 + ...) + (S_1^2 + S_2^2 + ...)} \qquad (21\text{-}1)$$

This equation will calculate the uncertainty of a flow measurement, assuming the variables are measured over a long period of time. The uncertainty will represent the deviation from the true value for 95 percent of the time. The value of the calculation will depend on the accuracy of the values used for the bias and precision errors. The errors are weighted, based on the manner in which they enter the flow equation. In the simplified orifice equation:

$$Q = Kd^2 \cdot (\Delta p \cdot P)^{1/2} \qquad (21\text{-}2)$$

where:

Q = rate of flow in specified units
K = a coefficient based on the mechanical installation and minor flow variables
d = orifice diameter in appropriate units
Δp = differential pressure in appropriate units
P = absolute static pressure in appropriate units

the values are either direct multipliers, squared, or square root values. The weighing values have a power factor of 1 for the direct multiplied variables, a power factor of 2 for the squared value, and a power factor of ½ for the square-root value multiplied by the errors. Thus, in Equation (21-2), K is multiplied by 1, d is multiplied by 2 and Δp and P are multiplied by ½. Thus, the uncertainty in the orifice flow rate would be:

$$U = \sqrt{B_k^2 + (2B_d)^2 + (\tfrac{1}{2}B_{\Delta p})^2 + (\tfrac{1}{2}B_P)^2 + S_k^2 + (2S_d)^2 + (\tfrac{1}{2}S_{\Delta p})^2 + \tfrac{1}{2}(S_P)^2} \qquad (21\text{-}3)$$

The values substituted for the bias in the equation may be obtained from manufacturers of the respective pieces of equipment, provided that the values are adjusted for actual flowing conditions. The precision error must be determined by comparison tests.

A differential device with a range of 100 inches of water has an error based on full scale. If the transmitter will operate at an average of 25 inches of water (50% of the full scale flow), then the full scale value should be applied for the 25-inch measurement — not full scale. Thus, for a transducer with an error of ± 0.1% of full scale (or 0.1 inch), this percent at flowing conditions of 25 inches would be ± 0.2% of flow (differential error of ± 0.4% times a sensitivity multiplier of ½ = ± 0.2%). Similar adjustments should be made for the calculation of operating error of other components of the measurement system. This procedure assumes that the differential device has been *properly* calibrated by an appropriate standardized device and the standardized device uncertainties taken into account in the error determinations.

The foregoing is a somewhat detailed description of the influence of only one of the factors on the determination of uncertainty from the equation. Many other values should be similarly examined. Calculation of the uncertainty associated with the variables in the flow equation is not the only concern for a complete uncertainty determination. Allowance must be made for human interpretation errors, recorder or computer errors, installation errors, and fluid characteristic errors. Most of these errors are minimized provided industry standard requirements are met and ***properly trained personnel*** *are responsible for the operation and maintenance of the station. Since these effects cannot be quantified, they are minimized by recognizing their potential existence and properly controlling the meter station design, operation, and maintenance. Without proper attention to the total problems, a simple calculation of the equation variables may mislead a user into believing measurement is better than it actually is.*

The uncertainty calculation assumes that the meter has been properly installed, operated, and maintained. If maintenance is neglected and the meter has deposits on it that change its flow characteristics, then the calculation is meaningless until the meter is cleaned up.

MAINTENANCE OF METER EQUIPMENT

Both the supplier and the customer must have confidence that a custody transfer meter is measuring the proper delivery volumes. Equipment calibration may change over time, so both parties should take an active part in periodic testing of the meter system. Without tests to reconfirm the original accuracies of the metering system, a statement of accuracy is not complete.

The most significant difference in custody transfer metering and in plant or operative metering is the frequency of the maintenance/testing of equipment.

Maintenance tests will depend on contractual requirements for type and frequency and may be as often as weekly. They may be a calibration of the readout equipment only, a complete mechanical inspection of the entire system, or an actual throughput test against some agreed upon correct volume. In any case, equipment used to test the meter must be approved and agreed upon. Such test devices include certified thermometers for temperature, certified dead-weight testers or test gages for pressure, certified differential testers for differential meters, certified chromatographs for component analysis, and certified provers for throughput tests. Many models of each are available and can be supplied with accuracy certification. Certification is important to both parties to minimize concern about the equipment.

Operators who have had experience with similar metering systems will also increase confidence in the calibration equipment and test procedures. The test equipment itself should be recertified on a timely basis by the agency or manufacturer that originally certified the equipment.

The first step in testing any meter is visual inspection for any signs of improper operation such as leakage and unstable flow. This includes a review of all of the attendant equipment and their indications or recordings. If the station appears to be operating properly, the individual elements of the station, such as the meter and the correctors for pressure, temperature, density, and composition, should be individually calibrated with the assumption that if all parts are in calibration, the system will be in calibration to the limits calculated by the uncertainty equation. This procedure is commonly used for industrial flowmetering.

If there is a desire to reduce tolerances on this measurement, then a throughput can be run against a "master meter" or a prover system.

The master meter is calibrated and certified to some accuracy limit by a testing facility of a government agency, a private laboratory, a manufacturer, or the user using agreed upon flow standards. Periodically, the master meter has to be sent back to the laboratory for recertification. The frequency of this retesting depends on the fluids being tested and the treatment of the master meter between tests. The best throughput test is one that can be run directly in series with a "prover." The prover can come in many forms, but essentially involves a basic volume that has been certified by a government or industry group (see Figure 21-3). Being one step closer to a basic calibration, this is probably the most accurate test of a meter's throughput. Such provers for liquid may be calibrated seraphin cans (for fluids with no vapor pressure at flowing temperature), pressurized volume tanks (for fluids with vapor pressure at the flowing temperature), or pipe provers (formerly called mechanical displacement provers as described in API's Manual of Petroleum Measurement Standards). These pipe provers are permanently installed in large-dollar-volume meter stations but there are portable units for testing multiple meter stations. Gas provers may be master meters with computer controls so that testing requires little or no calculation or critical flow nozzles (where accurate thermodynamic properties of the gas are available). Critical nozzles require a permanent pressure drop of some 15 to 20% of the upstream static pressure and cannot be run at static pressure below approximately 30 pounds per square inch absolute.

Great care must be exercised in using such equipment as detailed by the standards or the manufacturer's instructions, to ensure accurate testing. Since these tests are subject to errors, they should be conducted only by *qualified technicians.* Run correctly, these tests ensure the best measurement and provide proof of accuracy.

Figure 21-3. Prover

Proper maintenance and calibration of a billing meter is essential to accurate custody transfer metering.

Testing requires participation by both the supplier and the customer. Diagnostics and evaluation with proper test equipment ensure that recorded volumes are correct. Any proving must be documented and signed by both parties so that contract provisions can be implemented on any corrections required.

The study below is based on natural gas measured with an orifice meter. The exact magnitude of errors and dollars is not as important as the realization that significant dollars are at stake and proper testing is vital. A similar study could be performed for other custody transfer applications.

Natural gas costs have risen from 50 cents per 1000 cubic feet (Mcf) to a range as high $4.00 to $8.00. Tables 21-1, 21-2, and 21-3 show volumes using correct values and volumes calculated with small errors. Although the examples are based on errors of equipment reading low, similar calculations can be made for equipment reading high. The calculations are based on one eight-inch meter tube using a 4.000 inch orifice plate and a gas with a specific gravity of 0.580 with zero percent carbon dioxide and nitrogen. Differential pressure, static pressure, and temperature are calculated with the small errors to show the monetary results of calibration errors. A value of $4.00/Mcf was used to show the daily and yearly errors in revenue.

Table 21-1. Error Magnitude - Differential Pressure

Differential, in.	Flow, Mcf	Loss/Day	Loss/Year
2.0	3837		
1.8	3641		
-0.2	196	$784.00	$286,160.00
25.0	13553		
24.8	13499		
-0.2	54	$216.00	$78,840.00
90.0	25678		
89.8	25650		
-0.2	28	$112.00	$40,880.00

Reflects an error in differential pressure measurement of -0.2 in.

Table 21-2. Error Magnitude - Pressure

Pressure, lb	Flow, Mcf	Loss/Day	Loss/Year
(Differential of 2.0")			
600.00	3837		
598.00	3831		
-2	6	$24.00	$8760.00
(Differential of 25.0")			
600.00	13553		
598.00	13528		
-2	25	$100.00	$36500.00
(Differential of 90.0")			
600.00	25678		
598.00	25632		
-2	46	$184.00	$67160.00

Reflects an error in pressure measurement of -2 lb at differential flow rates.

Table 21-3. Error Magnitude - Temperature

Temperature, °F	Flow, Mcf	Loss/Day	Loss/Year
(Differential of 2.0")			
60.0	3837		
62.0	3828		
+2	9	$36.00	$13140.00
(Differential of 25.0")			
60.0	13553		
62.0	13519		
+2	34	$136.00	$49640.00
(Differential of 90.0")			
60.0	25678		
62.0	25615		
+2	63	$252.00	$91980.00

Reflects an error in temperature of +2 °F at differential flow rates.

The study shows that a small error can cause a considerable revenue loss over a period of time. It is very important that the people who maintain and test metering equipment understand the amount of revenue for which they are responsible and maintain the equipment in accurate working condition.

It is equally important that test reports be accurately filled out in their entirety so that office personnel can make adjustments for any errors that are found.

Properly trained personnel who understand the importance of the equipment they maintain are the key to accurate measurement. With proper test procedures, accurate test equipment, and a good maintenance procedure, any company should have an acceptable loss or "unaccounted for" record.

OPERATIONAL CONSIDERATIONS

Choosing the correct meter type is the first step to achieving measurement with a minimum uncertainty. However, recognition of the meter's limitations is required to achieve the possible accuracy. Most meters operate within stated limits and should not be used in the extremes of ranges for custody transfer metering. The first choice for custody transfer is to minimize flow variations by obtaining better control of the flow rate. At times, this may not be possible, and the need for a wide flow range will necessitate further consideration of the flowmeter choice. If there is no one meter with the range required to operate in the accurate part of its range, the use of multiple meters with some type of flow switching control to put meters in and out of service is required (see Figure 21-4).

For example, consider the fuel to a process with three heat exchangers. The range of fuel flow required may be from the pilot load to all three exchangers in full load service. This could require a measurement flow range of over 100 to 1. The metering used could be a combination of a positive displacement meter for the low flows and several turbine or orifice meters for the high flows. At one time, implementation of this type of complex system could have been a problem, but computers and electronic controllers have made implementation practical. Such a system is easily managed, and volumes can be accurately measured with the total flow for the whole station reported as a single measurement.

In addition to basic meter problems at the extremes, the secondary equipment that measures the pressure, temperature, differential pressure, density, or specific gravity and composition of the flow can also have shortcomings. Typical specifications for these devices are stated as a percent of full scale. Selecting an instrument with the wrong range for the parameters to be measured introduces errors.

If the gas pressure to be measured is 75 pounds per square inch gage (psig) and a 1000-psig range spring that has a ± 0.5% accuracy based on full scale is used to measure the gas pressure, then the pressure measurement error could be as high as 5 pounds out of 75 or ± 6.7% for linear meters and ± 3.5% for differential meters where the square root of the value is taken. (Note: The difference in the two values is that the pressure enters the linear meter as a direct multiplier but as a square-root in the differential meter).

The best operating range for metering systems is within the 25% to 90% range of the meter. If operation changes do not overrange the meter, the meter should be selected to operate near its full scale reading.

A system properly chosen, designed, and installed may still fail to meet performance expectations if the meter is not designed to operate in its most accurate range.

Figure 21-4. Multiple Meters with Flow Switching Control

Custody Transfer Auditing

When money is exchanged for measured material, the agreement will include a means of auditing the volumes obtained. Sufficient operation and maintenance records are made available to all parties so that the calculated volumes can be arrived at independently. At least a check of the values used by the other party should be made to see that agreement is reached on the volume. This procedure is an important aspect of custody transfer metering and is usually completed within 30 to 60 days after a bill is submitted. It will keep both parties involved in the measurement and prevent disagreements about procedures and volumes at some later date. With the data still current, any disagreements can be settled while knowledge of the measurement is fresh in the minds of both parties. A complete file of any disagreements must be kept, including their resolutions. Records can be reviewed to see if a particular station or particular errors have recurring problems that need to be addressed by an equipment or maintenance upgrade.

Summary

Custody transfer measurement begins with a contract between two parties that specifies the data needed to choose a metering system. To get the most accurate measurement required to minimize settlement problems, *maintenance and operation* of the system must be controlled so that the accuracy capabilities of the meter may be realized in service. Information in this chapter should be supplemented by reference to other chapters for a complete understanding of the advantages and limitations of individual meters. If all precautions are taken, proof of the station's performance will be when bills are submitted, audited, and paid, and the custody of the fluid has been successfully transferred.

Bibliography

1. ANSI/API 2530 (R1985), "Orifice Metering of Natural Gas and Other Related Hydrocarbon Fluids"
2. ANSI/ASME MFC-4M-(1986), "Measurement of Gas Flow by Turbine Meter"
3. ASME MFC-3M-(1989), "Measurement of Fluid Flows in Pipes Using Orifice, Nozzle, or Venturi"
4. ANSI/ASME MFC-2M-(1983), "Measurement Uncertainty for Fluid in Closed Conduits"
5. *API Manual of Petroleum Measurement Standards*
   Chapter 4, Proving Systems (Parts 1-7)
   Chapter 5, Metering (Parts 1-5)
   Chapter 6, Metering Assemblies (Parts 1-7)
   Chapter 7, Temperature Measurement
   Chapter 8, Sampling
   Chapter 9, Density Determination
   Chapter 11, Physical Properties Data (Parts 1-3)
   Chapter 12, Calculation of Petroleum Quantities (Parts 1, 2)
   Chapter 13, Statistical Aspects of Measuring and Sampling (Part 1)
   Chapter 14, Natural Gas Fluids Measurement (Parts 3, 5, 6, 8)

About the Author

E. L. Upp graduated from Louisiana State University with a Bachelor of Science in Chemical Engineering and went to work for Tennessee Gas Pipline Company as Assistant Measurement Engineer.

Joining Daniel Industries in 1967, initially as Manager of Fluid Mechanics and later as Vice President of Flow Technology, he is now Director of Technology.

Mr. Upp has been associated with the AGA Transmission Measurement Committee for over 27 years, having served the ASME Fluid Meters Committee for 18 years and the U.S. Advisory Committee for International Standards Organization Technical Committee #30 (Measurement of Fluid Flow in Closed Conduits). He served as Chairman of the AGA Committee in 1966, is a Past Chairman of the ASME Standards Committee for Measurement of Fluid Flow in Closed Circuits, and Past Chairman of the U.S. Advisory Committee for ISO-TC-30. He is a member of the American Petroleum Institute Natural Gas Fluid Measurement Subcommittee of the Petroleum Measurement Committee.

He is a contributing author of the American Gas Measurement Committee Report No. 3 - Orifice Metering of Natural Gas; the American Gas Association Gas Measurement Manual; the American Society of Mechanical Engineering Fluid Meters Sixth Edition; and the Natural Gasoline Processors Supplies Association Engineering Data Book.

A member of API and the Gulf Coast Gas Measurement Society, Mr. Upp has received awards of merit from the Gulf Coast Measurement Society, the Appalachian Gas Measurement Short Course, and the American Gas Association. He received a Citation for Service from the Committee on Petroleum Measurement of the API in April 1977, and on March 21, 1984 he was the recipient of a GPA Recognition Award. On May 21, 1991, Mr. Upp received the Laurence Reid Award at the International School of Hydrocarbon Measurement for his contributions to hydrocarbon measurement and control.

22

Sanitary Flowmeters

Many industries require equipment that meets specific design and construction criteria to ensure "sanitary" conditions. Here, "sanitary" does not refer to waste treatment applications but to highly "clean" conditions. The most common usage of sanitary flowmeters is in industries where bacteria growth and product contamination are a critical concern, such as in the production of food, dairy, and pharmaceutical products. Although this is the prime consideration in selecting a sanitary-type flowmeter, applications in a variety of non-food industries also require sanitary design flowmeters for reasons other than prevention of bacterial growth or product contamination.

Design requirements and specifications for sanitary flowmeters originated in the dairy industry. The handling and packaging of a perishable product such as fluid milk required components that did not compound the problem of product spoilage and bacterial growth. Equipment of all types, including flowmeters, had to be designed to ensure that milk residue did not become trapped or be caught and left to spoil. These pockets of spoiled product could harbor areas of harmful bacterial growth and later contaminate fresh product. Flowmeters were designed to eliminate cracks, crevices, and dead ends where residue could collect or pockets of bacteria could form. In addition, flowmeters had to be easily disassembled for hand cleaning and periodic visual inspection.

The same standards and specifications developed by the dairy industry began to be adapted by the food industry as well. Processed food products such as fruits, vegetables, sauces, gravies, pudding, dressings, and candy often required the same sanitary conditions to prevent bacterial contamination and product spoilage. In the last few years, soft drink bottlers have also become more concerned with product spoilage, particularly with the high juice content found in many fruit juice-based soft drinks. Although baked goods and cereals do not require the same rigid requirements as some food products, these industries also specify sanitary flowmeters on such ingredients as color additives, flavoring ingredients, liquid eggs, liquid vitamin slurries, edible oils, and liquid sugars.

Outside the food industry, sanitary flowmeters are used in the production of products such as liquid ointments, shampoo, hand lotion, and cosmetics. Because these products are for use on the human body, similar sanitary practices and guidelines are followed during the manufacturing process.

Probably the most stringent sanitary requirements can be found in the pharmaceutical and biotechnology fields. Bacteria-free environments and the prevention of product contamination are essential in the production of such products as oral and injectable drugs, purified buffer solutions, water for injection, and in the separation of blood serums. Highly purified products must not only be free of bacteria but must remain free of all foreign contaminants to ensure proper processing. In the biotech area, sanitary flowmeters are used to add nutrients such as aqua-ammonia solutions, glucose sugars, and buffer solutions to highly sensitive bioreactors for the cultivation of cells.

Within many industries, multiple uses and applications for sanitary flowmeters can be found. Although the products, system requirements, and the degree of sanitation may vary, a few applications are common to these industries.

Typical Sanitary Applications

Bulk Receiving

Use of sanitary flowmeters can be found in bulk receiving areas. By metering bulk product deliveries into a plant, delivery receipts can be verified and monitored for shortages. Sometimes the receiving systems are used not only to verify receipts but the measurement also provides a legally accepted means for custody transfer.

During farm collection, raw milk is picked up from each individual producer and measured at the farm by inserting a dipstick into a calibrated storage tank. Using the reading from the dipstick markings, a calibration chart estimates the approximate weight of the milk received. Although milk is still commonly purchased in bulk by the pound, volumetric flowmeters are used to provide a check on the total quantity received. Because of the crude method for collecting and measuring milk at the farm, the plant relies on receiving meters to verify the load receipts and help spot potential shortages. In the dairy industry, positive displacement, magnetic, and turbine flowmeters are used to receive raw milk off over-the-road tank trucks.

With recent technical improvements, magnetic and mass flowmeters are frequently used in bulk receiving applications because multiple products of differing viscosities and densities, such as raw milk, condensed milk, and heavy cream, can be received using the same equipment with no affect on accuracy.

With the improvement in the accuracy and reliability of flowmeters, many states now calibrate, test, and approve flowmeters for use in trade. In dairy plants, for example, certified metering systems are used for buying and selling product based on the reading of the flowmeter. Rotary piston flowmeters are most common in trade-type applications because of their excellent repeatability. Figure 22-1 shows a certified plant milk receiving system used for legal custody transfer. This system has approval from the National Institute of Standards and Technology (NIST) under Handbook 44 requirements. These receiving systems must meet an acceptance tolerance of ± 0.15%.

Other bulk receiving systems for liquid sugars and corn sweeteners, such as those found in soft drink plants, use positive displacement or mass flowmeters to

Figure 22-1. Certified Plant Milk Receiving System for Legal Custody Transfer
(Courtesy of Accurate Metering Systems, Inc.)

verify the delivery receipts. Processed food plants meter incoming quantities of edible oils and liquid eggs. Virtually any liquid received in bulk from tankers or rail cars can be metered with the proper flowmetering system.

Batch Processing

Flowmeters are used extensively in batching operations of liquid products. A batch process involves delivering one or more liquid ingredients into a mixing tank or vat. The entire batch is held in the tank until the exact quantity of all ingredients has been delivered. The order of addition of the ingredients may be critical in some processes, but most often all ingredients are added simultaneously in the batch tank. Meter-based batching can significantly decrease production times compared to scales or load cells, where each ingredient must be delivered sequentially with a tare weight taken after each product.

In the dairy industry, magnetic flowmeters are the most common meters for batching dairy products. Not only are all milk-based products highly conductive, but the volumetric reading of a magnetic flowmeter is not affected by the different viscosities and densities of products from skim milk to cottage cheese. In the production of ice cream, for example, magnetic flowmeters measure the various liquid ingredients such as condensed milk, cream, and milk into batch tanks. Rotary piston or mass flowmeters deliver the required amounts of liquid sugars. Stabilizers or milk solids in the form of dry powders are then dissolved into the batch tank. After proper agitation, the ice cream mix is pasteurized, flavored, frozen, and packaged.

Batching is a common use of flowmeters within the food and beverage industries. Positive displacement flowmeters are used on liquid corn sweeteners and treated water systems for the manufacture of soft drink syrups, while the production of mayonnaise, salad dressings, sauces, gravies, ketchup, beer, and vinegar use sanitary flowmeters in similar processes.

Continuous Ratio Blending

In addition to batch operations, meter-based ratio control systems are frequently used in continuous blending processes. Unlike batch operations, ratio blending systems meter products in the exact proportions directly in line to the next stage of the process, to surge tanks, or to filling and packaging equipment through a single header or pipe. No batch tank or intermediate handling is required.

Orange juice concentrate is often reconstituted with a two-stream meter-based blending system. Typically, the concentrated juice is diluted with water in a ratio ranging from three to five parts water to one part orange juice concentrate. By precisely controlling the addition of water, the final juice dilution can be carefully controlled. With a flowmeter in the total blended juice line and a second flowmeter in the juice concentrate line, a ratio set point of concentrate to juice can be maintained by controlling the flow rates of the metered streams. No batch tank or double pumping is necessary. The final juice blend can be pumped directly to the filler.

In recent years, low fat milk containing two percent butterfat or less has become very popular. Flowmeters are used to blend continuously the proper volumetric ratios of two milk streams to standardize a butterfat content. The milk standardization process involves blending a small amount of cream back into skim milk or blending skim milk into whole fat milk to reach a desired butterfat level. Once the exact butterfat contents for both the cream and the skim milk are determined by lab analysis, an exact ratio of the two streams can be calculated. This ratio is then maintained by allowing the skim milk stream (the master stream) to run uncontrolled and controlling the flow rate addition of cream (the slave stream). Other systems achieve the same result by adding skim milk to raw whole milk to obtain the desired butterfat level.

Flavored drinks are commonly manufactured with continuous blending systems. Typically, flavored drinks will be made from water, a liquid corn sweetener, and a prepackaged flavoring component. Three flowmeters are required: one in the total combined line for the finished drink, one in the liquid sugar line, and the third in the flavor line. Commonly, a positive displacement or mass flowmeter will be used on the liquid sugar line. Smaller capacity positive displacement or magnetic meters are used in the flavor line. In the final beverage line, depending upon the characteristics of the product (i.e., conductivity, flow rate, fruit pulps in the product, etc.), a magnetic or mass flowmeter is applied.

The water stream runs essentially uncontrolled and unmonitored. A "recipe" ratio is established for the correct proportions. In this case, the recipe would be the percent of liquid sugar in the final beverage and the percent of flavor in the final beverage based on volumetric additions of each. The water makes up the remaining proportion.

Once these percentages are entered into the system controller, AC variable frequency drives control the flow rates of the two slave streams (liquid sugar and flavoring) to maintain proper proportions. Should the uncontrolled water stream flow rate change or should some downstream conditions cause a system upset, it will be detected by the flowmeter located in the finished beverage line. The controller, in turn, drives the sugar and flavor pumps to a different flow rate to maintain the prescribed volumetric ratios.

The flowmeters provide not only the flow rate signal to the controller to maintain the ratio but also usage totals of the ingredients (liquid sugar and flavor) and the total amount of drink prepared for the day.

Product Transfer and Monitoring

Flowmeters provide accurate usage and inventory information for a variety of applications such as transferring product from a batching area to storage and collecting data on raw ingredient inventories used in production.

In ice cream plants, ice cream mix is batched into multiple batch tanks. A continuous cycle of filling one tank while emptying another is established. Sequential timing allows one tank to empty, while a second batch is ready for transfer, as a third batch is initiated. Liquid ingredients such as condensed milk, cream, water, and liquid sugar are all metered into the batch tanks where dry components such as stabilizers are added. Once the ice cream mix is batched and properly agitated, flowmeters are used to meter the mix out of the batch tank to a pasteurizing system or raw product surge tanks. The meters not only provide production totals of ice cream mix batched, but, by metering into and out of the batch tanks, real-time inventory information is continuously available. This tells the operator how much mix is left in each tank, allowing the batch tank to be monitored as it empties for preparation of the next batch in that tank.

Inventory levels of raw ingredients also can be monitored in a similar manner. In a ketchup batching system, for example, raw products such as liquid sugar are metered from a bulk truck into raw storage silos. The raw ingredients are also metered out of the storage silos during the batching operation. By adding quantities received and subtracting quantities batched, real-time inventory information tells the production manager how much product remains in each raw storage silo and allows better production schedule planning and subsequent raw ingredient ordering.

In simple product transfer processes, flowmeters provide useful data. Breweries install flowmeters to monitor transfer to mash kettles, lauter tuns, brew kettles, fermenters, and filling lines. Flowmeters are used by large processors when selling bulk product to smaller plants. The meter is used to accurately load out product onto trucks. This procedure is common with bulk item transfers such as ice cream mix, orange juice concentrate, liquid sugars, high gravity beer, liquid eggs, and edible oils.

Product Yield and Loss Control

As plant operations become larger and more automated, flowmeters are playing an increasing role in providing product yield and loss information. This involves accounting for how much product was actually transferred from a batching or production area to the filling lines. In large soft drink operations, for example, ten thousand to fifteen thousand gallons of soft drink syrup are batched into a single batch tank. This one batch tank, however, may be serving several filling lines at once. Flowmeters are commonly installed prior to each filler line to provide exact totals on product usage of each line. By knowing the final quantity of bottles, cans, or cases produced from each line and the amount of syrup used, efficiencies can be determined for each line to help pinpoint where product losses occur.

With individual product usage figures for each fill line, material balances and yield data be calculated not only for each line but for the entire batch as well. By separating the plant into a production area and a packaging area, the flowmeters can help locate areas where losses occur.

The proper placement of flowmeters not only provides loss control information, but, since ingredients are metered into batch tanks and finished syrup is metered out, real-time inventory data can be provided for all the batch tanks. Since some plants may have twenty or more batch tanks running more than fifty different recipes and formulations, the inventory data tells the operator how much product remains in each batch tank. Real-time inventory allows timed shutdown procedures for packaging lines and control systems interlocks for automatic cleaning cycles.

Flow Rate Control and Critical Alarming

In critical operations involving flow control, it is necessary to assure a product or system operates within a specified flow range. Flowmeters are commonly used to provide the flow rate signal for an alarming device. In the pasteurization of milk products, specific time and temperature relationships must be maintained to ensure all harmful bacteria are destroyed. Using meter-based timing systems on high temperature short time (HTST), the flow rate control and, hence, the minimum legal time and temperature requirements are based on a signal from the meter. A maximum flow is determined at which the system may operate and still maintain the required holding time for the product during pasteurization. Should any upset condition occur, causing the flow rate to exceed this maximum flow rate limit, the pasteurizing system discharge is immediately diverted back to the raw ingredient balance tank. A magnetic flowmeter provides an analog flow rate signal that the flow alarm monitors for proper flow range requirements (see Figure 22-2).

Besides flow rate alarm functions, it may be important to maintain a single stream at a specific flow rate. Using the flow rate signal from the flowmeter, a set point controller will drive a variable speed pump or position a control valve to maintain a preselected flow rate. One example is automated cleaning systems. These systems often use vortex shedding flowmeters to monitor and control the flow rate of the cleaning solution through the system, thus ensuring adequate flow rates for proper cleaning. Additionally, the system monitors the return flow rate of the recirculated cleaning solution. If the flow rate is too low, the system is alarmed to indicate inadequate cleaning of the system.

Automation of Processes

In automating processes with programmable controllers, flowmeters are used as process sensing elements to control and monitor a variety of events such as sequencing production steps or detecting product changeovers. In the production of cheese, cottage cheese, and other cultured products, processing vats are filled with milk and cultures are added to begin a fermenting process. Flowmeters not only ensure that an exact quantity of milk is delivered to each vat, they also provide the feedback in the process to control the automatic sequential filling of multiple vats.

During the pasteurization of milk, the high temperature short time (HTST) pasteurization system is started on water until proper temperature and flow rate conditions are met (see Figure 22-2). Once the HTST unit reaches legal operating conditions, the unit switches over to the milk product. Flowmeters help to automate this changeover sequence by metering a known quantity of water out of the system to the drain, ensuring a complete product changeover while minimizing any loss of milk.

A similar changeover sequence is used in the soft drink industry. Near the end of a production run, the remaining cola syrup in the lines is recovered by pushing it with water to the carbonation equipment and filler. By knowing the precise amount of syrup in the lines between the source batch tank and the filler, a meter allows only the amount of flush water required to empty the lines. This helps to minimize product dilution by creating a repeatable flush cycle.

In statistical product sampling devices, flowmeters are often used as the basis for taking periodic samples for a product. For example, when receiving raw milk, it is important for the processor to know how much butterfat, antibiotic, etc., is in the load. Statistical product samplers are driven off the signal from a flowmeter.

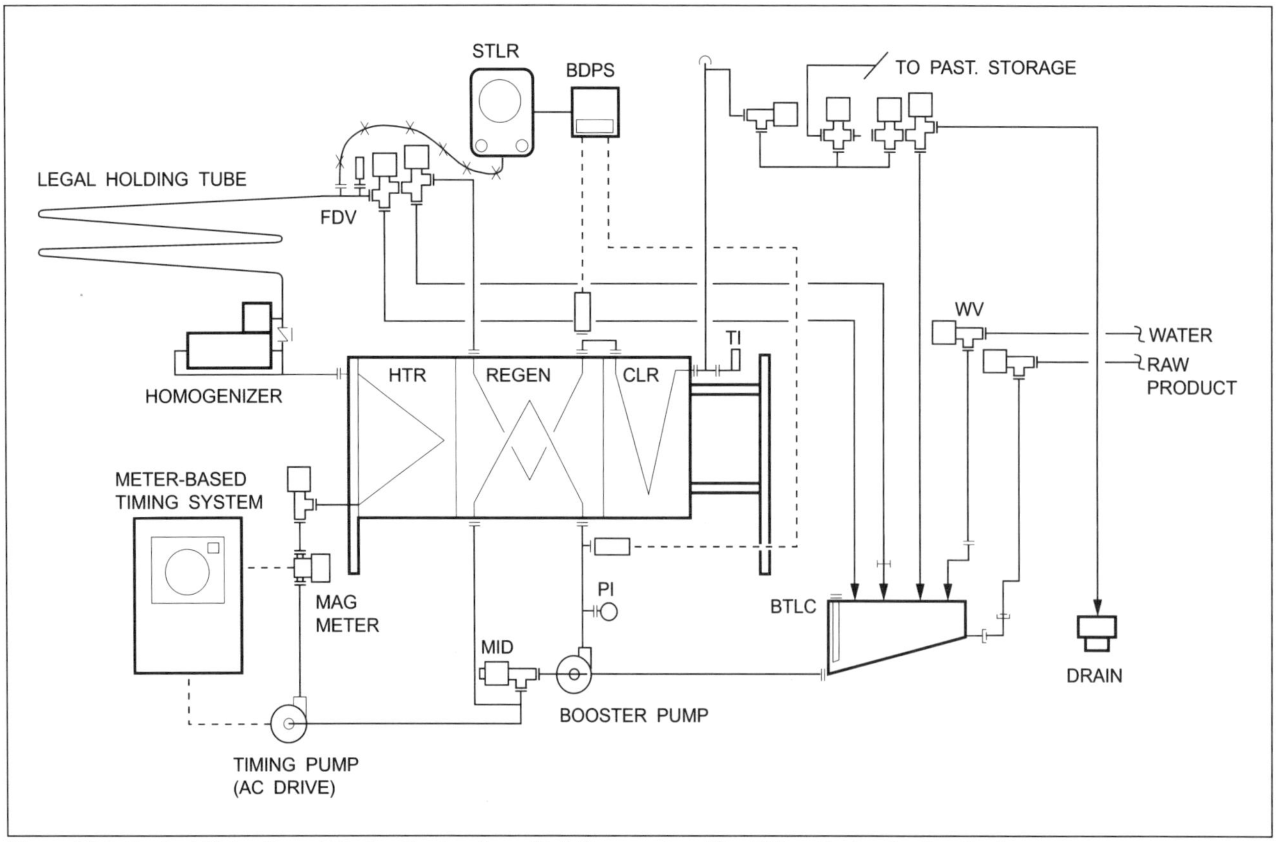

Figure 22-2. Typical Meter-Based Timing System for a High Temperature Short Time Pasteurization System

The sampling device will take a known size sample quantity for every few gallons of product through the flowmeter. Should the butterfat rise to the top of the load and stratification occur, a statistical sampling device provides a representative composite sample over the entire load. The sampling rate is adjustable for varying size loads received.

Other Industry Applications

PRODUCT CARRY-OVER

Food, beverage, dairy, and pharmaceutical applications must be concerned with the sanitary aspects of bacterial contamination and product spoilage. But sanitary flowmeters are often used for other reasons. In the production of some carbonated soft drinks, product spoilage is often not a problem because of the high acidity and the large amount of sugars. A soft drink bottler is, however, concerned with cross-contaminations. For example, after running a cola product, followed by proper rinsing, the bottler may want to run a decaffeinated cola. The same design considerations of the sanitary flowmeter that prevent bacterial cross-contamination also prevent carry-over of caffeine, colorings, flavoring additives, or other components.

Even in areas outside the food and beverage industries, the prevention of product carry-over has opened up new applications for sanitary flowmeters. In automotive paint booths, for instance, sanitary flowmeters are used on the paint supply lines. The sanitary flowmeter ensures that no color traces or residuals will be left behind that may taint the next color. Other applications where there is a chance of carry-over of paints, pigments, or coloring agents also incorporate sanitary design flowmeters.

CORROSIVE APPLICATIONS

Another important design feature of sanitary flowmeters is the material of construction. Highly polished stainless steel is required in equipment used in the manufacture of food products. Because of the stainless steel construction, sanitary flowmeters are resistant to many corrosive compounds. Other food grade materials such as Teflon™ and ceramic are also used in flowmeter construction since these materials will not break down or leave particulate traces in the products. Chemical plants use stainless steel flowmeters in the production of acidic and caustic cleaning agents and compounds where other nonstainless steel metals may be corroded or oxidized. The semiconductor industry uses sanitary flowmeters on highly corrosive purified water solutions used in the production of silicon chips.

Flowmeter Usage

Because of these design requirements, the sanitary industries do not use as many types of flowmeters as the industrial market. The most common types of sanitary flowmeters include positive displacement, magnetic, turbine, vortex shedding, and mass flowmeters. Although the ultrasonic flowmeter is not usually considered in the sanitary category, occasional uses can be found for this meter, which externally clamps onto the pipe line. Since no direct product contact is made, the unit may be used in sanitary applications. Other types of flowmeters will not typically meet the criteria necessary for most food, dairy, and pharmaceutical applications.

Some flowmeter manufacturers have simply taken an existing design of an industrial-type flowmeter and added a sanitary end connection.

Let the buyer beware, because some manufacturers offer optional sanitary connections for their industrial flowmeters that do not meet sanitary requirements.

Having sanitary connections does not ensure the internal design is sanitary.

To be sanitary, the meter itself must be free of internal cracks, crevices, and steps to prevent the trapping of product particles. Most manufacturers of sanitary equipment have been careful to make necessary modifications or to design the flowmeter specifically for the sanitary market.

Flowmeter Design

3-A Sanitary Standards

Because of the rigid requirements necessary to ensure equipment is suitable for use in food and pharmaceutical products, design standards were developed in 1920. The most commonly accepted set of standards in the United States is the 3-A Sanitary Standards. The 3-A Sanitary Standards outline specific criteria for all types of sanitary process equipment and accepted sanitary practices. For process equipment like flowmeters, the standards outline design, material, fabrication, and construction criteria.

Originally, the 3-A Standards were developed for equipment and systems used in the production of milk and milk products. Today, however, most industries requiring sanitary equipment have adopted the 3-A Standards or require equipment to carry the 3-A symbol.

Organizationally, the 3-A Standards were developed and are periodically reviewed by the three separate branches that make up the 3-A Standard Committees. These three branches, representing several special interest groups, include an equipment manufacturing group (represented by the Dairy and Food Industry Suppliers Association—a non-profit trade organization for equipment manufacturers), a user group (represented by the Dairy Industry Council—a non-profit trade organization for dairy processors), and the sanitarians (represented by the International Association of Milk, Food, and Environmental Sanitarians and by the United States Public Health Service). A lengthy and detailed review process is encountered for each new standard or when existing standards are revised or changed. Each branch is then allowed input into the creation or change of any standard.

The 3-A Standards are a voluntary, self-policing set of criteria. Manufacturers wishing to sell equipment for sanitary applications should design or modify their equipment to meet these standards. Although compliance to the standards is not mandatory, most buyers and users of sanitary equipment will not consider its use without 3-A compliance. To obtain 3-A compliance, an application and a signed affidavit stating that the equipment meets the specific standards are submitted by the manufacturer to the 3-A Council for review. Along with the application, a complete set of drawings and photographs are included. The 3-A Council reviews and evaluates whether the equipment meets the published standards. They do not "approve or certify" the equipment. The burden of proof and responsibility remains with the manufacturer to ensure that the equipment meets all standards. An application must be resubmitted annually. Any design change in a particular piece of equipment also requires a submittal of an Application for Amendment.

Figure 22-3. The 3-A Symbol Designating Compliance with the 3-A Standards
(Courtesy of the 3-A Sanitary Standards Symbol Council)

For equipment found to be in compliance with the 3-A Standards, a manufacturer is issued a certificate by the 3-A Symbol Council indicating compliance with the standards and is given permission to display the 3-A symbol (see Figure 22-3). Only then is a manufacturer allowed to display the 3-A symbol on a specific piece of equipment. Most flowmeters will display the 3-A symbol on the meter itself. Figure 22-4 shows a mass flowmeter with the 3-A symbol and the number of the applicable standard (number 2800) immediately below the symbol.

The specific standards for flowmeters are found in 3-A Standard Number 28-01, "The 3-A Sanitary Standards for Flow Meters for Milk and Liquid Milk Products." Some manufacturers, however, choose not to display the 3-A symbol on the flowmeter even though permission has been granted for its use and the equipment does meet all requirements.

Should a question arise regarding a specific piece of equipment, the manufacturer will be able to produce upon request a current certificate from the 3-A Council verifying the equipment compliance.

Sanitary Requirements

To ensure that sanitary construction criteria are met, the design of a sanitary flowmeter differs from an industrial type in several areas. The broad and basic guidelines that drive the design and construction of sanitary flowmeters is to prevent the contamination of the product by ensuring that all direct product contact surfaces can be effectively cleaned. If this is not possible, the flowmeter must be easily disassembled for hand cleaning and for visible inspection.

DEAD ENDS

Sanitary flowmeters must be free of dead ends, cracks, and crevices where product residue may become trapped or caught within the meter. The concern is not only small crevices where liquid residue could be trapped but also with obstructions within the flowmeter that can catch or trap particulates or solids that pass through the flowmeter. No sharp corners, edges, steps, or breaks are allowed where particulates like fruit juice pulps or suspended solids could become lodged.

To prevent product residue and to facilitate constant flushing within the flowmeter, all internal angles within the flowmeter or on product contact surfaces must be at least 135°. This allows for easy drainage and provides a wide enough angle so particulates can easily pass through the meter, particularly during a mechanical cleaning cycle. Angles of less than 135° must have minimum radii of 0.25 inch. The only exception to the 0.25-inch minimum radius is in the main body of the flowmeter or in the base of any teeth in gear-type meters. Rotary piston flowmeters, for example, are allowed to maintain 90° corners in the main measuring chamber. The main body of the meter provides a wide surface area leading into the corners to allow product to be properly flushed. Positive displace-

Figure 22-4. A Sanitary Mass Flowmeter Displaying the 3-A Symbol
(Courtesy of Micro Motion, Inc.)

ment flowmeters that use gear-type rotors are also allowed to have smaller radii at the base of the gear. Most gear-type flowmeters must, however, be cleaned by hand.

If any internal radii are less than 1/32 inch, the flowmeter must be easily disassembled for inspection and hand cleaning.

Other considerations to prevent dead ends or areas where product may stagnate must be weighed in the design of the flowmeter. Some types of positive displacement flowmeters use a rotating piston in the measuring chamber. The piston may have a pin extending out the back of the piston. This pin inserts into a magnetic coupling or a clutch device that transfers the rotations to an external magnetic assembly. This magnetic assembly, in turn, drives a pulse transmitter or roller register. Figure 22-5 shows a disassembled rotary piston meter with the magnetic coupling. The magnetic clutch eliminates direct product contact by any external components and, thus, maintains a sanitary seal. The design must be such that product is not trapped within or around the magnetic coupling assembly. The pin and shaft must have sufficient clearance to allow for continuous flushing. Although no specific criteria outline the construction of this assembly, a thorough visual inspection will suggest whether this may be a problem. The point is, however, that product must flow freely throughout the entire meter to prevent small pockets of stagnation.

Another example involves the turbine flowmeter. Figure 22-6(b) shows a cross-sectional view of a sanitary turbine flowmeter. Notice that a bearing flush hole has been added to allow product and cleaning solution to properly flush the thrust ball and bearing area. This prevents product stagnation. Similarly, it is critical that the flow straightening vanes at the inlet and outlet be secure enough to prevent rotation during operation but loose fitting enough to allow product to flow easily between the vanes and the outside wall to keep the area flushed.

In any sanitary application, absolutely no threaded surfaces are allowed to have contact with the product. This includes cover plates, end connections, bearing assemblies, and so on. Several high pressure sanitary connections are

As specific as the 3-A Standards are, they do not cover all possible design configurations of all flowmeters. If the buyer is unfamiliar with a particular flowmeter of a specific manufacturer, a thorough visual inspection of all internal product contact surfaces should be done. A common sense review will answer many questions and concerns not specifically addressed within the 3-A Standards.

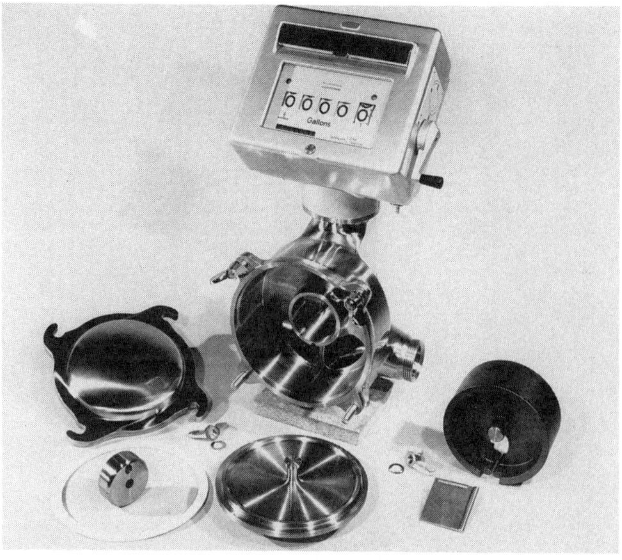

Figure 22-5. Rotary Piston Flowmeter with Magnetic Coupling and Roller Counter Register
(Courtesy of Accurate Metering Systems, Inc.)

(a) SHOWING 3-A COMPLIANCE

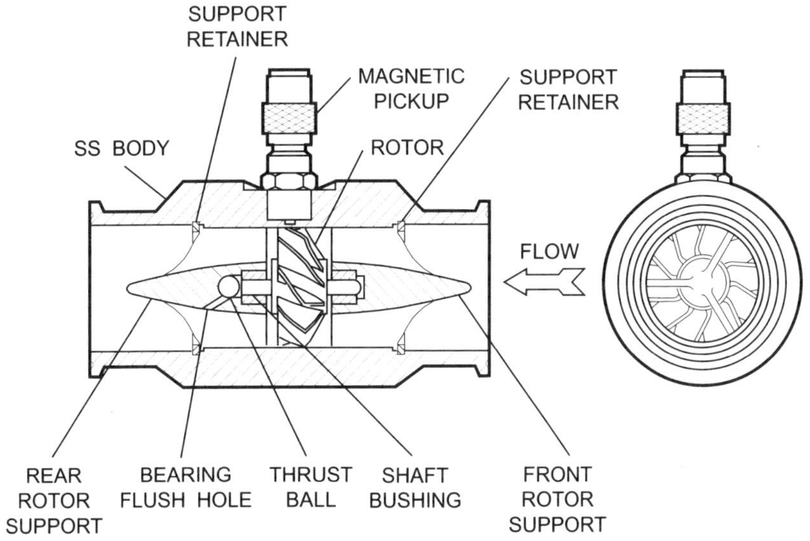

(b) CROSS-SECTIONAL VIEW OF A SANITARY TURBINE FLOWMETER

Figure 22-6. A Sanitary Turbine Flowmeter
(Courtesy of Combustion Engineering)

threaded, but the threads themselves are external and have no direct contact with the product.

Any flowmeter using internal coil springs that have contact with product must have at least a 3⁄32 inch opening between coils on the spring. It is rare, however, to find any springs or snap rings of any type in a sanitary flowmeter. Even with the required 3⁄32 inch opening, suspended pulps, fruit skins, or suspended solids in some products may be trapped in the spring.

SELF-DRAINING

All product contact surfaces in the flowmeter must be totally self-draining except for normal product clingage. Pockets or traps must not exist. This is especially critical with the use of Coriolis mass flowmeters. The double tube designs must be such that the unit drains completely without any product being trapped in one of the bends or legs. In the U-tube design, for example, this means installing the unit so the U-tube is upside down to ensure complete drainage. Mass flowmeters using tubes with 360° bends must have both ports installed at the lowest point for drainage. Oftentimes, this means installing the unit upside down or on its side, which may differ from the manufacturer's suggested installation practice and may or may not affect meter performance. With some mass flowmeters, this may mean a trade-off between a sanitary installation or following the manufacturer's installation guidelines. Also in positive displacement flowmeters, the inlet and outlet ports must be designed and connected to the main body so the measuring chamber drains completely with no puddles left standing.

Often, proper installation ensures the units are self-draining. Ideal installation for other types of sanitary flowmeters such as magnetic, turbine, and vortex shedding should be in a vertical line with flow from bottom to top. This not only ensures proper operation of the flowmeter by maintaining a full flow tube, it also allows the unit to drain completely by gravity.

GASKETING

Any gaskets or O-rings used within sanitary flowmeters must be removable. Grooves holding the gasket in place can be no deeper than their width. The minimum radius of any internal angle on a gasket retaining groove must be more than 1⁄8 inch. For standard size O-rings, smaller radii are permitted. For example, with a 1⁄4 inch O-ring, a 3⁄32 inch radius may be used. A 1⁄8 inch O-ring can use a 1⁄32 inch radius.

The practical side of these guidelines means that, once the gasket is properly installed, it must not protrude past the adjacent surfaces it is sealing. Figure 22-9(c) shows a typical sanitary connection; notice that the gasket and connecting tubing form a smooth, flat, stepless surface.

DISASSEMBLY

Sanitary flowmeters must be easily disassembled for periodic visual inspection or hand cleaning. "Easily disassembled" implies that the flowmeter can be taken apart by hand with no tools or with very little difficulty. This is most noticeable on positive displacement meters where the cover of the meter is secured with wing nuts that can be removed by hand. A family of various size positive displacement flowmeters is shown with the wing nut secured front covers in Figure 22-7.

Magnetic flowmeters usually have adapting connectors that can be removed for inspection. Figure 22-8 shows adapting connectors on a magnetic flowmeter that can be easily removed by hand for inspection of the flow tube liner for cuts or damage.

Materials of Construction

MATERIALS

All metallic surfaces that have direct product contact must be constructed of a type 300 series stainless steel as described by the Iron and Steel Society. Typically, this will be 304, 316, or 316L stainless steel. Other equally corrosion-resistant metals can be used provided they are nontoxic and nonabsorbent. Magnetic flowmeters, for example, are available with a variety of metallic electrode materials such as Hastelloy™ or platinum.

The only exception still permitted is the use of a 400 series stainless steel for the rotors of turbine flowmeters. The 400 series stainless steel rotor contains enough ferrous material that it can be sensed magnetically. Originally, the only pickup probes available for turbine meters were magnetic pickup probes that could sense the slightly magnetic vanes of the rotor as product flowed through the meter. Not only is there enough ferrous material in the rotor to magnetize it, but, after prolonged use and cleaning, the ferrous material in the rotor can begin to oxidize. With current technology, however, manufacturers have developed electronic pickup probes that are capable of sensing a 300 series stainless steel nonmagnetic rotor through a 300 series stainless steel housing. This has provided the industry with a turbine meter constructed completely from 300 series stainless steel. New applications have opened up for turbine meters on corrosive products such as deionized water, high grain vinegars, brine solutions, and so on, without the fear of oxidation or deterioration of the rotor.

Other materials used in shafts, sleeves, gaskets, linings, pistons, bearings, etc., of sanitary flowmeters can be made of rubber, rubber-like materials, or other nontoxic and nonabsorbent components. The nontoxic requirement is obvious for food and pharmaceutical applications. The material also must be nonabsorbent to prevent product or cleaning agents from being absorbed by the material and later released, causing cross-contamination. Common materials include Teflon™ and ceramics in the liners of magnetic flowmeters. Turbine flowmeters use a variety of bearing surfaces, including ceramic, Rulon™, and tungsten carbide. The pistons of positive displacement meters are made of carbon, polypropylene, Kynar™ polysulphone. Viton™, Buna™, neoprene, butyl, and Teflon are all com-

Figure 22-7. Sanitary Positive Displacement Flowmeter with Wing Nut Secured Covers for Easy Disassembly
(Courtesy of Badger Meter, Inc.)

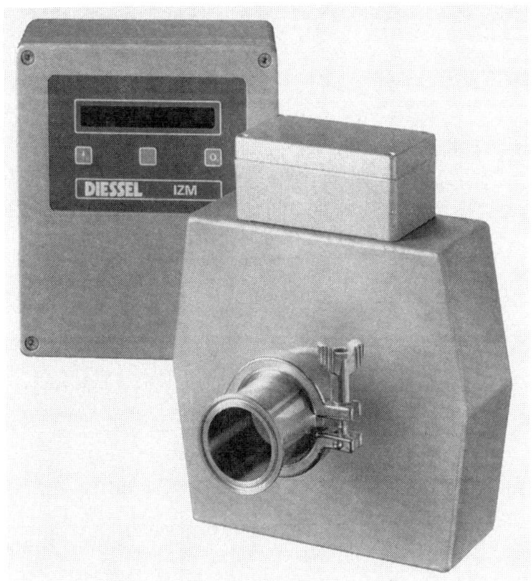

Figure 22-8. Sanitary Magnetic Flowmeter with Adapting Connections Terminating with Standard Sanitary Clamps.
(Courtesy of Accurate Metering Systems, Inc.)

mon gasket materials. These rubber-type materials and synthetic polymers must receive an approval for their use from the Food and Drug Administration under the Food, Drug, and Cosmetic Act. Listings of all acceptable materials are available from the Food and Drug Administration (FDA), the United States Department of Agriculture (USDA), and the 3-A Sanitary Committee. From this list of approved food grade materials, manufacturers may provide a proprietary blend of various materials to create new materials with special temperature, strength, or chemical resistivity characteristics. All components going into a proprietary blend still must meet the requirements of the Code of Federal Regulations, Title 21, Part 177—Indirect Food Additives: Polymers.

The material of construction should be chosen so it does not flake, chip, dissolve, break down, or leave trace elements in the product. The material also must be able to withstand any caustic or acidic cleaning solutions to which it may be exposed as well as the high temperatures of the cleaning or sterilization process. Often, specific users will perform other tests to verify that the material will not affect the color or taste of the product.

Materials should be carefully chosen for the specific application and product. Some materials may not be compatible for all uses depending on products run, operating temperatures, or the cleaning agents involved.

SURFACE FINISHES

All product contact surfaces must be smooth and free of any pits or blemishes. The finish on metallic surfaces that have direct contact with product must be at least a number four (#4) mill finish. All product contact surfaces, metallic and nonmetallic, must have at least an equivalent of 150 grit finish as obtained with silicone carbide. Standard measurement of a surface finish is a root-mean-square average (rms) deviation of the surface roughness centerline measured in microinches. A 32 rms surface finish is typical for sanitary equipment. In some highly pure environments, such as in the production of purified water solutions or intravenous solutions, flowmeters can require up to a 4 rms finish followed by electropolishing.

Electropolishing is a process by which the roughness of the surface can be further smoothed by an electrochemical procedure of "leveling" the peaks. The process is a form of reverse plating in which the "peaks" of the surface are removed. This not only results in a highly polished finishing surface on the equipment but virtually levels and removes any place for particles to become trapped.

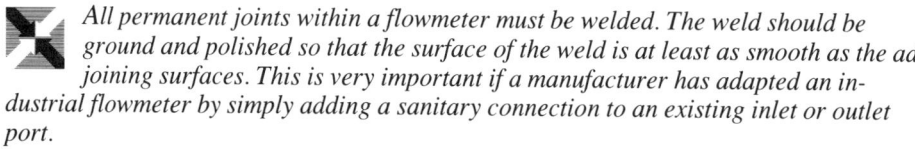

All permanent joints within a flowmeter must be welded. The weld should be ground and polished so that the surface of the weld is at least as smooth as the adjoining surfaces. This is very important if a manufacturer has adapted an industrial flowmeter by simply adding a sanitary connection to an existing inlet or outlet port.

End Connections

One difference between an industrial-type flowmeter and a sanitary flowmeter is the end connections. Sanitary connections must be designed so that, when tightened, there exist no cracks, crevices, pits, folds, steps, sharp edges, or corners where product may become trapped. The most common sanitary connection is the sanitary clamp.

SANITARY CLAMP

The sanitary clamp connection has two identical, mating ferrules separated by a sanitary gasket. The ferrule has a slight taper, making it wider at the base and narrower approaching the edge. Figure 22-9(a) shows the two mating ferrules. A V-groove in the clamp (Figure 22-9(b)) fits over the two joining, tapered ferrules. As the clamp is tightened, it forces the ferrules tightly together. The wing nut on the clamp allows the connection to be put together and taken apart completely by hand and without the aid of tools. This connection may be taken apart daily for a manual cleaning step or for periodic visual inspection of the connection. When fully tightened, the gasket is compressed to form a smooth flush surface with the connecting tubing. As seen in Figure 22-9(c) and 22-9(d), no steps or crevices

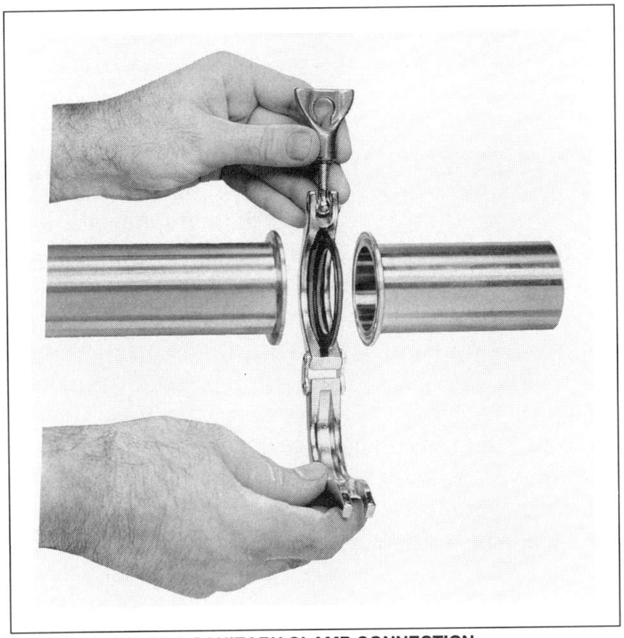

(a) ASSEMBLY OF A SANITARY CLAMP CONNECTION
(Courtesy of Alfa-Laval, Tri-Clover Division)

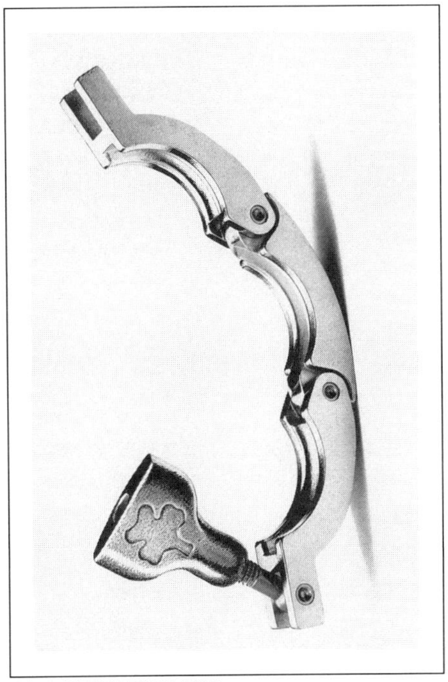

(b) A SANITARY CLAMP
(Courtesy of Alfa-Laval, Tri-Clover Division)

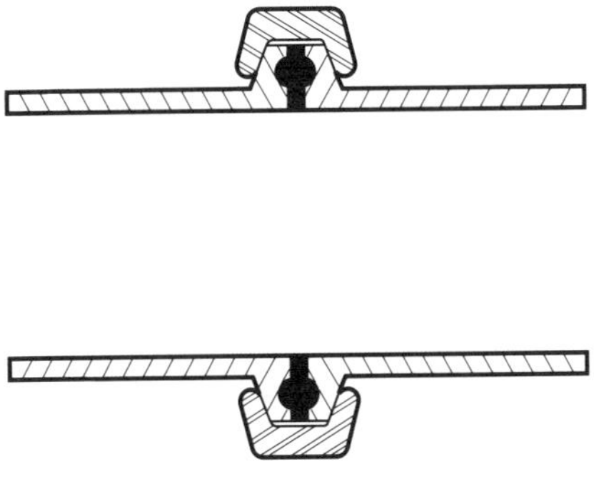

(c) CROSS SECTION OF A TIGHTENED SANITARY CLAMP CONNECTION
(Courtesy Accurate Metering Systems, Inc.

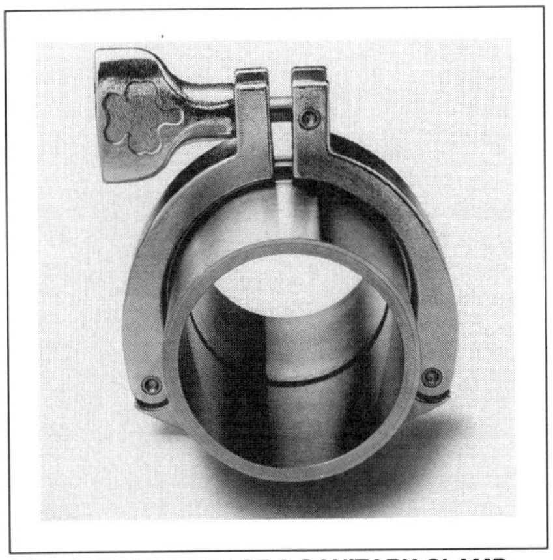

(d) INTERNAL VIEW OF A SANITARY CLAMP CONNECTION
(Courtesy of Alfa-Laval, Tri-Clover Division)

Figure 22-9. Sanitary Clamps

exist. Often this connection is called a "Tri-Clamp" connection. Tri-Clamp™, however, is a brand name from a specific manufacturer. The proper name is "sanitary clamp connection."

Sanitary clamp connections are available for pipe line sizes ranging from ½ inch to twelve inches with four inches being about the largest practical size. The size of sanitary connections typically refers to the outside diameter of the tubing. Thus, on a two-inch connection, the outside diameter is two inches and the inside diameter is 1.856 inches. Pipe schedules are also available but are not common in sanitary applications.

Maximum pressure limitations for the clamp connection are determined by the clamp size and the temperature of the product being run. With higher temperatures, the gasket will begin to soften. Because of this, the pressure rating of the clamp is reduced as the temperature increases. According to some manufacturers' specifications, a four-inch clamp at 21°C (70°F) will handle pressures of up to 200 pounds per square inch (psi) but only 125 psi at 121°C (250°F). Similarly, a one-inch clamp at 21°C (70°F) is rated up to 500 psi but is reduced to 250 psi as the temperature reaches 121°C (250°F). When manufacturers specify a maximum pressure rating for a flowmeter, it may be the maximum pressure rating for the meter body itself or it may be the pressure rating of the connection. Typically, the lower of the pressures is stated in the technical specifications.

Using the same mating ferrules as the clamp connection, a variety of heavy duty clamps are available from various manufacturers. These usually require tools to tighten and are safe to much higher pressure ratings.

BEVEL SEAT

Some high pressure applications use connecting ends that have distinct male and female ends. The bevel seat connection (see Figure 22-10) has a male end with a protruding ferrule and a captured hex nut over it. The female end of the connection is threaded on the outside with a smooth tapered lip on the inside. The gasket is formed to the same shape as the bevel in the mating connections and, when properly placed between the male and female ferrules, forms a smooth, stepless connection. Although the connection is threaded externally, there is no product contact with the threads. As the nut is tightened, the mating ferrules help to self-center the gasket and ensure a smooth internal connection. A hex wrench is required to tighten the nut. Bevel seat connections are common on the larger line sizes found on bulk delivery tankers and rail cars. Both the clamp and the bevel seat connections meet 3-A requirements.

DIN STANDARD

Another type of connection standard used in most European countries is one that meets the DIN 18651 (Deutsches Institut für Normung) standard. This Ger-

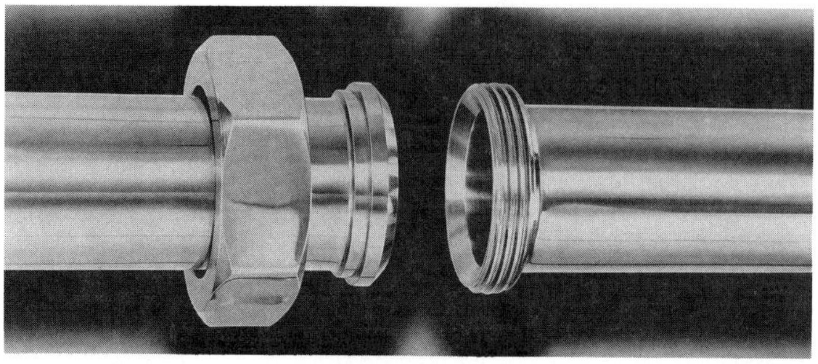

Figure 22-10. Sanitary Bevel Seat Connection
(Courtesy of Alfa-Laval, Tri-Clover Division)

man standard is common throughout the world and is often found on imported equipment. Similar to the bevel seat connection, the DIN fitting has a male and a female end and requires its own special gasket. A special spanner wrench is required to tighten the captured nut. Figure 22-11 shows the DIN standard connection on a magnetic flowmeter.

Although flange connections are usually not used significantly in sanitary applications, a few do comply with 3-A Standards. Some sanitary magnetic flowmeters are designed for insertion between two ANSI (American National Standards Institute) raised-face flanges. Figure 22-12 shows a magnetic flowmeter inserted between two raised-face flange connections and terminated with the sanitary clamp connection. This flowmeter is used in applications when higher temperature and pressure requirements are needed. It may even be welded into the line to eliminate the clamp connections. Flange connections are also common when line sizes get up above six inches. Since the meter body is inserted between the two flanges and bolted together with a binding post, this is considered a permanent connection because it not easily inspected. This permanent connection must still be free of any crevices. Custom-cut gaskets of food grade materials are used to ensure smooth internal surfaces when tightened.

ASEPTIC CONNECTION

Aseptic processing is a sterile process in which food products are processed under high temperatures and packaged so that no bacteria are allowed into the product. Food products such as single-serve pudding, fruit drinks, apple sauce, and even milk can be aseptically processed. Aseptic packaging allows the product to maintain a longer shelf life without spoilage and does not require refrigeration.

To obtain this highly clean environment, additional precautions are taken to eliminate bacteria. Aseptic connections require steam tracing to prevent airborne bacteria from getting into the lines. The turbine meter in Figure 22-13 is specially designed with aseptic connections with steam inlet and outlet ports. A small channel or groove surrounds the entire connecting point (see Figure 22-14). The two small ports allow saturated culinary steam to be continuously injected into the channel to form a sterile seal.

Figure 22-11. Magnetic Flowmeter with the German DIN Standard Connection—A Special Spanner Wrench Is Required to Tighten the Nut
(Courtesy of Accurate Metering Systems, Inc.)

Figure 22-12. Sanitary Magnetic Flowmeter Inserted between Two Raised-Face Flanges and Terminated with Standard Sanitary Clamp Connections
(Courtesy of Accurate Metering Systems, Inc.)

Installation

CIP/Mechanical Cleaning

Most modern sanitary processing plants have some type of automated cleaning process. This mechanical cleaning process is referred to as clean in place (CIP). CIP is a series of rinses and washes of product contact surfaces of designated circuits and lines within a process. With proper combinations of flow velocity, temperature, concentration of cleaners, and time, all organic and inorganic soils can be removed under automatic control. Within a particular circuit or line, the system may be cleaning process piping, valves, pumps, tanks, fillers, flowmeters, and so on.

A typical CIP cycle for a process line, including the flowmeter, will start with one or more steps of a fresh water rinse to remove any product clingage. Following the rinse cycle, the system will be washed for approximately ten minutes with a hot, caustic cleaning solution of 140°F (60°C) to 150°F (66°C) to remove organic contaminates within the line. A postrinse step of fresh water flushes out any remaining caustic. Next, the system is filled with an acidic rinse to neutralize the remaining caustic and to remove any inorganic contaminates. Another fresh water postrinse removes the acid left in the lines. Finally, a diluted chlorine or iodine sanitizer fills the system. This will either remain in the line until the next production run, or, if it is drained, the line is not flushed until the next run. The sanitizer is allowed to remain in or to coat the line until the equipment is used again, preventing future contamination.

For proper mechanical cleaning and rinsing of lines, the solution needs to create enough turbulence to remove clingage and to "scrub" the product contact surfaces. In straight pipe, this requires a flow velocity of at least five feet per second. Not only must flowmeters be able to handle the higher flow rates, the design must be such that enough velocity and turbulence are generated to properly clean the surfaces.

Most sanitary flowmeters are designed to be cleaned in place; that is, the meter can be left in line completely assembled for the CIP cycle. Magnetic and vortex shedding flowmeters are essentially straight pipe and clean very well at five feet per second. There are no chambers or cavities that may affect their cleanability.

Most mass flowmeters are fully CIP. One concern for proper CIP of mass meters is on models with split tube design. Proper turbulence must be generated to ensure the "Y" connection for the split flow at the inlet and outlet is cleaned.

Figure 22-13. Aseptic Turbine Flowmeter with Steam Seal Connections
(Courtesy of Accurate Metering Systems, Inc.)

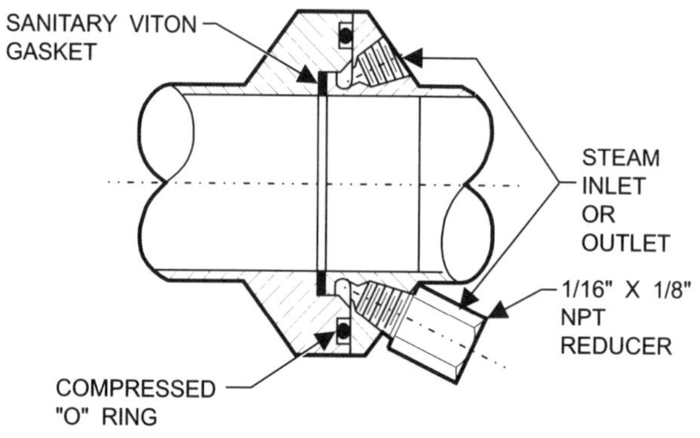

Figure 22-14. Cross-Sectional View of an Aseptic Connection with Two Steam Ports for a Sterile Seal
(Courtesy of Accurate Metering Systems, Inc.)

The bearing surfaces of turbine flowmeters are another area where turbulent flow must be created to remove any product clingage. Recall Figure 22-6(b), which shows the bearing flush hole to ensure proper flushing and cleaning.

A more difficult meter to clean in place is the positive displacement flowmeter. Because the measuring chamber is so much larger than the incoming line size, the piston itself must create turbulence and a scrubbing action. In designs that incorporate a magnetic coupling assembly or transmission unit, turbulence is difficult to create in the small area where the transmission occurs. These units must often be cleaned out of place (COP) by hand. The entire meter does not need to be removed, however, to clean the unit. The body itself stays in the line and is cleaned by the standard mechanical cleaning process. The internals, however, such as the piston, the coupling, and the dividing wall, are removed and washed by hand. Since the meter must be frequently disassembled and cleaned by hand, non-CIP meters tend to wear faster and have higher maintenance requirements. Most popular designs have limited the number of moving parts. By using an electronic transducer rather than a magnetic coupling, the piston is often the only moving part, allowing fully clean-in-place operation.

Sanitary flowmeters meeting 3-A Standards and displaying the 3-A symbol may not be fully compatible for clean-in-place practices. Some disassembly and hand cleaning may be required. Although most sanitary meters can be left fully assembled when cleaned in place, this is a separate issue that must be considered when selecting a flowmeter that is best suited for a particular application. The meter must be able to withstand the flow rates, pressures associated with the increased flow, the corrosive cleaners, and the temperature, and should require minimal hand cleaning, if any.

Steam Sterilization

In many aseptic and pharmaceutical processes that must maintain a very high degree of sanitation, one further step of cleaning is done. This involves sterilizing the system with culinary saturated steam. As the saturated steam is pumped through the circuit into the flowmeter, temperatures can rise to as high as 121°C (250°F). This temperature must often be maintained for at least thirty minutes to sterilize product contact surfaces. This additional step can influence which flowmeter is best for these conditions.

Magnetic and vortex shedding flowmeters with integral electronic transmitters may be affected by these temperatures. Usually, when steam sterilizing turbine flowmeters, the pickup probes are removed to prevent damage.

To steam sterilize a flowmeter, it must be of a CIP design. Once the meter has been steam sterilized, it cannot be opened for reassembly.

References

1. 3-A #08-17, As amended September 1988, Part One and Part Two of the 3-A Sanitary Standards for Fittings Used on Milk and Milk Products Equipment and Used on Sanitary Lines Conducting Milk and Milk Products.
2. 3-A #20-14, As amended March 1989, 3-A Sanitary Standards for Multiple Use Plastic Materials Used as Product Contact Surfaces for Dairy Equipment.
3. 3-A #28-01, September 1988, 3-A Sanitary Standards for Flow Meters for Milk and Liquid Milk Products.

4. "3-A: The Mark of Compliance," 3-A Sanitary Symbol Council, Waukehsa, WI.

5. "The 3-A Story," 3-A Sanitary Symbol Council, Waukehsa, WI.

6. *Accepted Meat and Poultry Equipment*, MPI-2, United States Department of Agriculture, Food Safety and Inspection Service, Washington, D.C., July 1990.

7. Atherton, Henry V., and Gilmore, Thomas M., "Cleanability Requirements of Dairy Processing Equipment Meeting 3-A Standards," *Dairy, Food and Environmental Sanitation*, Vol. 9, No. 2, pp 75-76, 1989.

8. Clem, Lyle W., and Seiberling, Dale A., "CIP Concepts in Process and Equipment Design," Presented to the American Institute of Chemical Engineers, August 1984.

9. Code of Federal Regulation, Title 21 - Food and Drugs, Part 110—Current Good Manufacturing, Processing, Packing, or Holding of Human Food, Superintendent of Documents, U.S. Government Printing Office, Washington, D.C. 21 CFR 110.

10. Code of Federal Regulation, Title 21 - Food and Drugs, Part 177—Indirect Food Additives: Polymers, Superintendent of Documents, U.S. Government Printing Office, Washington, D.C. 21 CFR 177.

11. *Grade A Pasteurized Milk Ordinance, 1978 Recommendations, 1989 Revision*, Public Health Service Publication No. 229, Public Health Service, Food and Drug Administration, Washington, D.C.

12. *Steel Products Manual: Stainless and Heat Resisting Steels*, Iron and Steel Society, Warrendale, PA, 1974.

Acknowledgments

The author would like to thank Lyle W. Clem, Engineering Manager, of Accurate Metering Systems, Inc., and Roger Dickerson, Former Chief, Food Engineering Branch of the Food And Drug Administration, for their technical review.

About the Author

Michael A. Lucas serves as the National Sales and Marketing Manager for Accurate Metering Systems, Inc., of Schaumburg, IL, which is a manufacturer of sanitary flowmeters for the food, dairy, beverage, and pharmaceutical industries. Joining Accurate Metering Systems, Inc., in 1985, Mr. Lucas has been extensively involved in product development and training as well as in areas of the sales and marketing program. He has a B.S. in Electrical Engineering from Bradley University and an MBA from the University of Chicago.

23

Flowmeter Selection

A large number of many types of flowmeters are available today. The most commonly used is the variable differential pressure type, which includes "the orifice meter," an engineering expression for the thin plate, concentric hole, sharp-edged orifice plate with its associated transmitter. The use of this flowmeter is so well established that within some organizations other types of flowmeters will be used only "when and if the orifice meter is proven *not* to work." That's one way to select a flowmeter; however, it is not a method that easily optimizes flowmeter selection. The variable differential pressure flowmeter is only one type, even though it is available in many forms:

(1) Concentric, thin, sharp edged orifice

(2) Eccentric orifice

(3) Segmental orifice, and the wedge

(4) Quadrant and/or conical orifice

(5) Venturi

(6) Flow nozzle

(7) Flow tube

(8) Pitot

(9) Averaging Pitot

(10) Target (annular orifice)

(11) V-cone

(12) 90 degree elbow

Other types of flowmeters, without all of their various formats, include the following:

(1) Constant differential pressure (variable area)

(2) Displacement (positive and inferential)

(3) Sonic

(4) Electromagnetic

(5) Coriolis

(6) Thermal types

(7) Oscillatory (fluidic, vortex precession, and vortex shedding)

No one type of flowmeter is suitable and certainly cannot be the optimum choice for every application. A selection process is needed to find which meter is the best solution for a specific measurement problem.

Initial Approaches

There are at least three (3) initial approaches to selecting a meter:

(1) Choose the meter that is most familiar. This may mean the type that is most readily understood. It may be the type that has been used for the longest period of time. It may be familiarity based on a large quantity of meters in use.

(2) Choose the same meter that has been used on prior, similar applications. This is one of the simplest approaches and is not necessarily bad. However, blindly following what has been done before will not always lead to the best meter for the application. It would be unfortunate if such a selection process perpetuated previous problems!

(3) Review and consider all the factors that will influence the selection. This is a time-consuming process and can be justified only for the more critical flow applications. A variation of this approach can be justified and can lead to the optimum flowmeter for the application.

The Large Number of Selection Factors

A number of factors require at least some level of investigation when determining which flowmeter to select for an application.

An analytical approach has been proposed by Sam Bailey of *Control Engineering* magazine (see Figure 23-1). Another analytical version appeared in *Plant Engineering* magazine (see Figure 23-2). Each analysis is thought provoking and helpful, but, for engineers, the need is for more direct solutions. A typical list of selection factors would include the following:

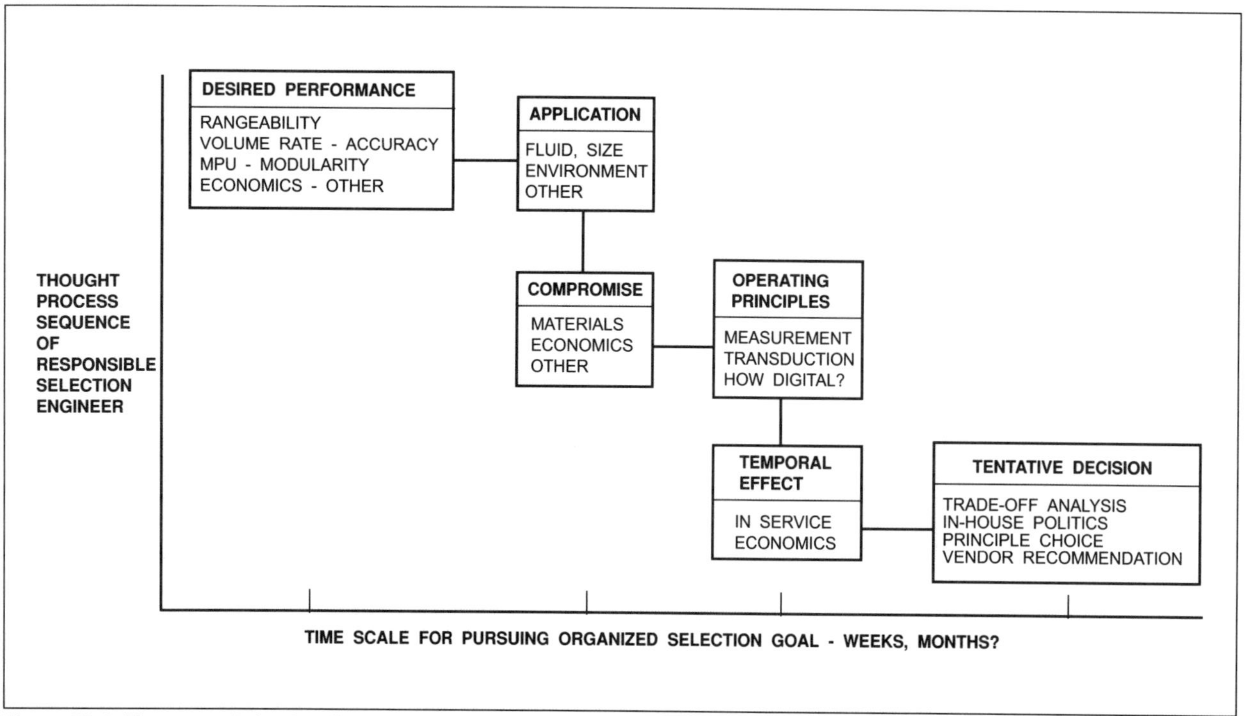

Figure 23-1. Flowmeter Selection Problem Statement [Ref. 2]
(Courtesy of Control Engineering)

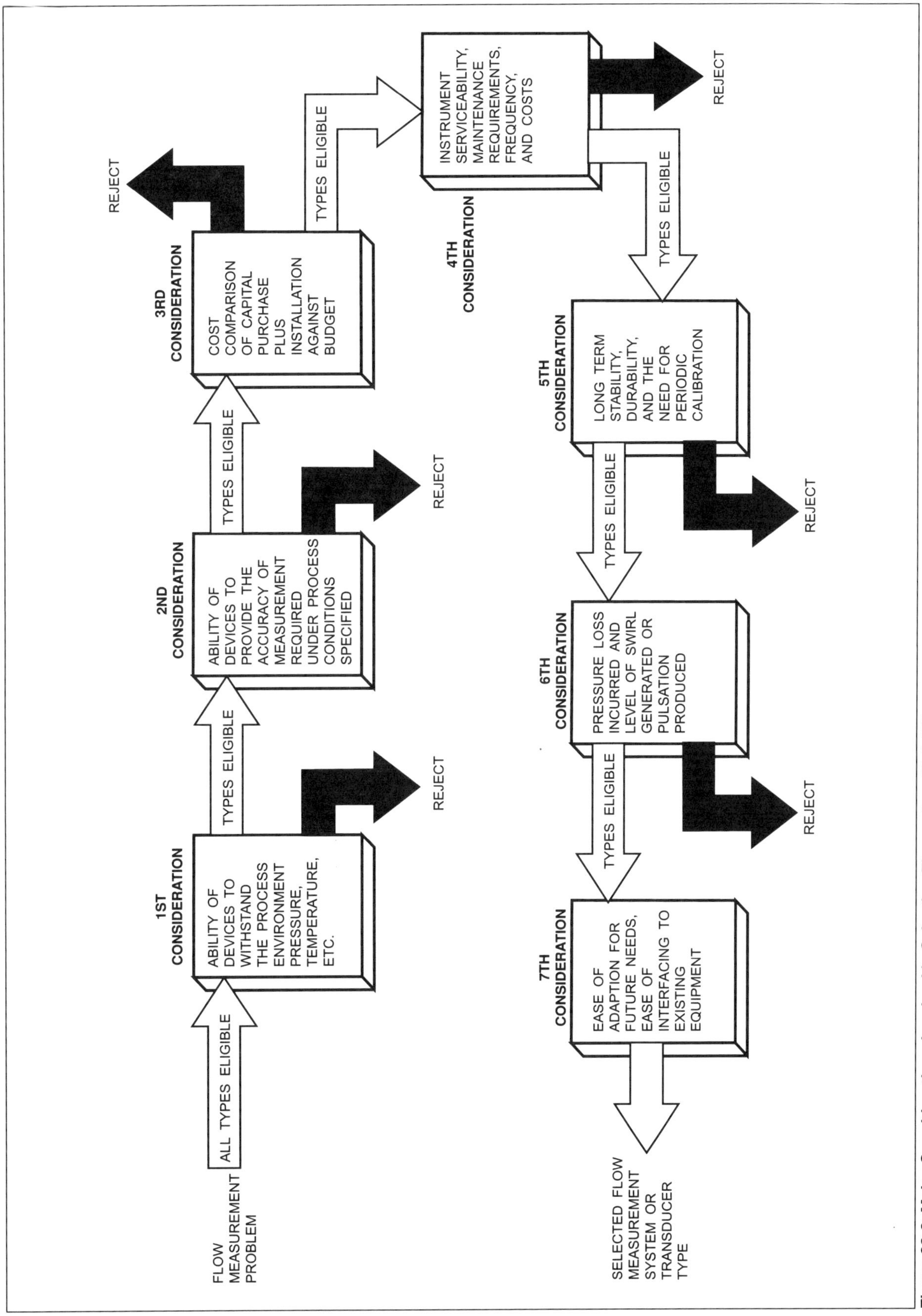

Figure 23-2. Major Considerations Involved in Selecting Flowmeters [Ref. 20]
(Courtesy Plant Engineering Magazine)

Is the measurement to be mass or volume?

Is the measurement of liquid, gas, or vapor, or are there mixtures, or are they in combination with solids?

Is the flow Newtonian, or is the fluid a non-Newtonian?

Is rate of flow or totalization required?

What signal is required?

What display is needed?

Is the fluid corrosive or passive?

What are the environmental constraints?

Is the fluid clean or dirty?

What power is needed?

What range is required; that is, what is the ratio of the maximum flow rate to the minimum flow rate?

What performance is needed? (This is generally stated as accuracy.)

What is the cost? (Hardware cost and/or total cost of ownership.)

What maintenance is required and who has to do it?

What are the operating temperatures and pressures (normal and extremes)?

What pressure drop is permissible; that is, what energy consumption will there be?

What fluid properties must be considered? (These usually include viscosity, viscosity index, density, compressibility, electrical conductivity, lubricating qualities, opacity, etc.)

Such a list is only typical; it is not comprehensive. The problem becomes more difficult because each application places different priorities on the individual factors, adding further complexity to making the optimum flowmeter selection for the application. A computer solution to the problem would be ideal. Development work in this area is reported by J. Baker-Counsell in *Process Engineering* magazine [Ref. 3].

> Items such as pumps, control valves, and instrumentation cover an extremely wide range of techniques, types, models, and manufacturers, and, although it may be fairly easy for an experienced engineer to pick a suitable component fairly quickly, the choice of the *optimum flowmeter*, for instance, in any particular case, is time consuming and far more difficult. If, however, all the information and selection rules used intuitively by the engineer are embodied in an expert system, the selection can be made more accurately and much more speedily.

The main selection criteria, suggested by Baker-Counsell in the magazine, are shown in Table 23-1.

Decide what factors are fundamental to making a successful measurement and start the selection process there.

Until expert systems are available, it will be necessary to manually sort through the choices. Ted Higham has provided a sorting of the factors into four groups (see Figure 23-3). At first glance this appears to be a simplification that would facilitate a solution. It certainly helps, but Higham's replotting of the factors (see Figure 23-4) provides an interesting insight into the benefits of expert systems. It dramatically shows the interrelation of the factors and the high level of interdependence that exists between them. No wonder so few meters are

Table 23-1. Constructing an expert system reveals just how many criteria have to be taken into account in arriving at a choice of flowmeter. Obviously, many factors are almost subconsciously dealt with by the engineer as he works through the problem in his mind, but many more require deep understanding and detailed consideration. The main criteria are outlined below. [Ref. 3]

Company standards	Flowmeter installation—flanged, etc.
Legal requirements	Space limitations
Phase(s) of principal fluid(s)	Upstream pipe run
Solids content	Nearest upstream component
Calibration factors	Level of vibration
Type of fluid—air, methane, water . . .	Accuracy of reading
Nature of fluid—corrosive, conducting . . .	Repeatability of
reading	
Size and shape of piping	Can a microprocessor be used for linearization?
Will a flow indicator suffice?	Instrument response time
Are we measuring flow rate or fluid quantity?	Flow obstruction—pressure loss
Is the flow likely to reverse direction?	Type of control output required
Max. and min. flow rates	Time between inspection/maintenance
Max. and min. pressures	Reliability
Max. and min. temperatures	Intrinsic safety
Fluid density	Calibration drift
Fluid viscosity	External conditions—moisture, heat . . .
Is flow pulsating?	Cost
Is the flow transparent?	Suppliers
Material of pipework	

selected by a thorough review of all the factors. Figure 23-4 also shows the way to a solution.

Categorizing Flowmeters by Process-Dominated Factors

The most frequent way in which flowmeters are categorized is by the measurement technology they use. However, flowmeter selection is superficial if it is based only on the measurement method the flowmeter uses. The measurement technology, however, provides access to other flowmeter characteristics that are worthy of investigation. As Higham suggests, it is these additional characteristics that form the basis for categorizing flowmeters and facilitate the selection process. Some of the basic characteristics would include:

(1) the type of measurement, mass or volume;

(2) the information provided, rate of flow or total; and

(3) the fluid state the flowmeter can handle—liquid, gas, steam, or slurries.

Mass or Volume Measurement

Some processes require mass flow information, while others will require volumetric flow information.

Categorizing flowmeters based on whether they provide mass or volume measurement reveals that:

(1) some meters categorized traditionally as the volumetric type are, in fact, velocity detectors, and

(2) some meters provide neither mass nor volume measurement.

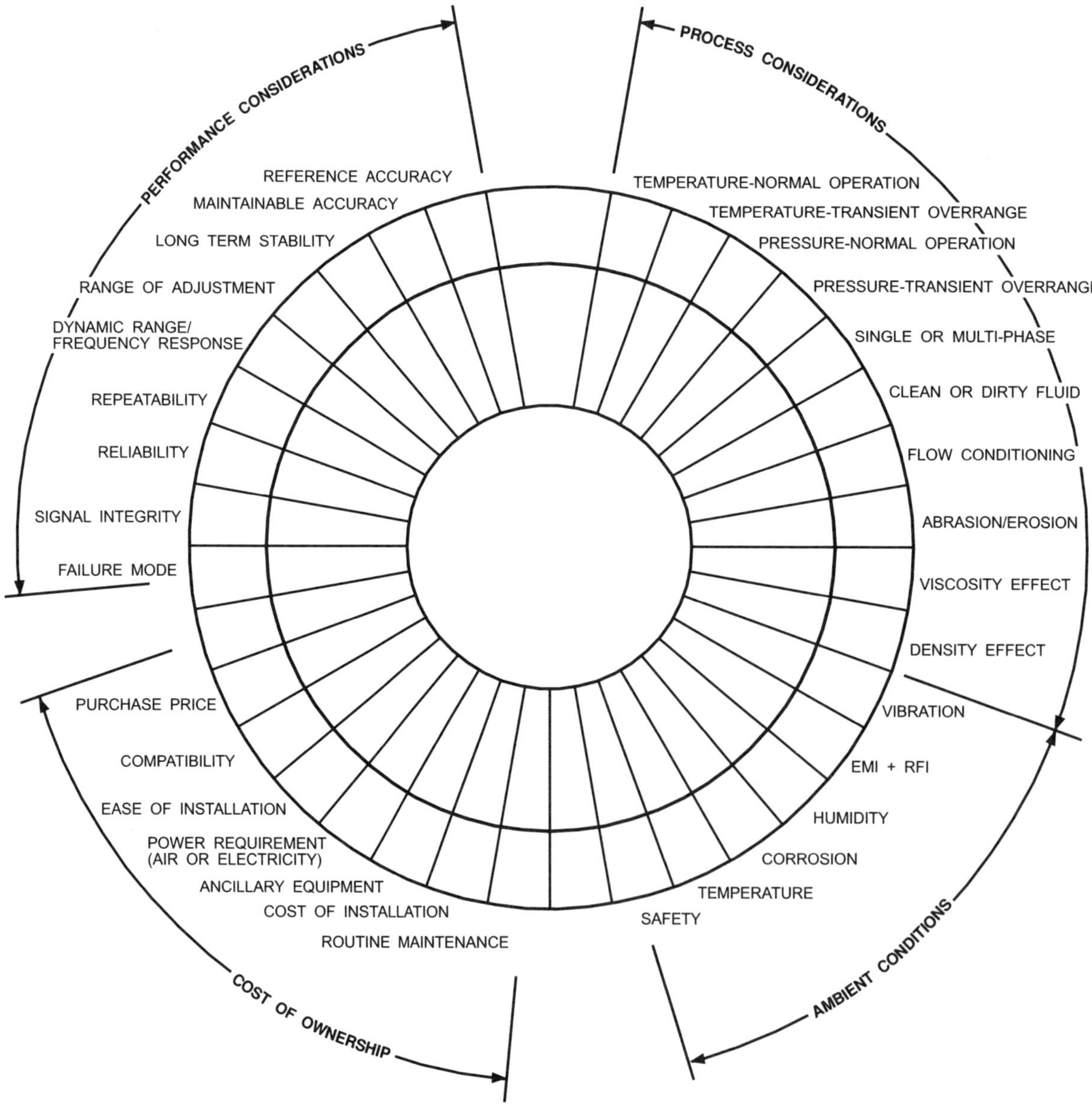

Figure 23-3. Selection Considerations [Ref. 12]
(Courtesy of The Foxboro Company)

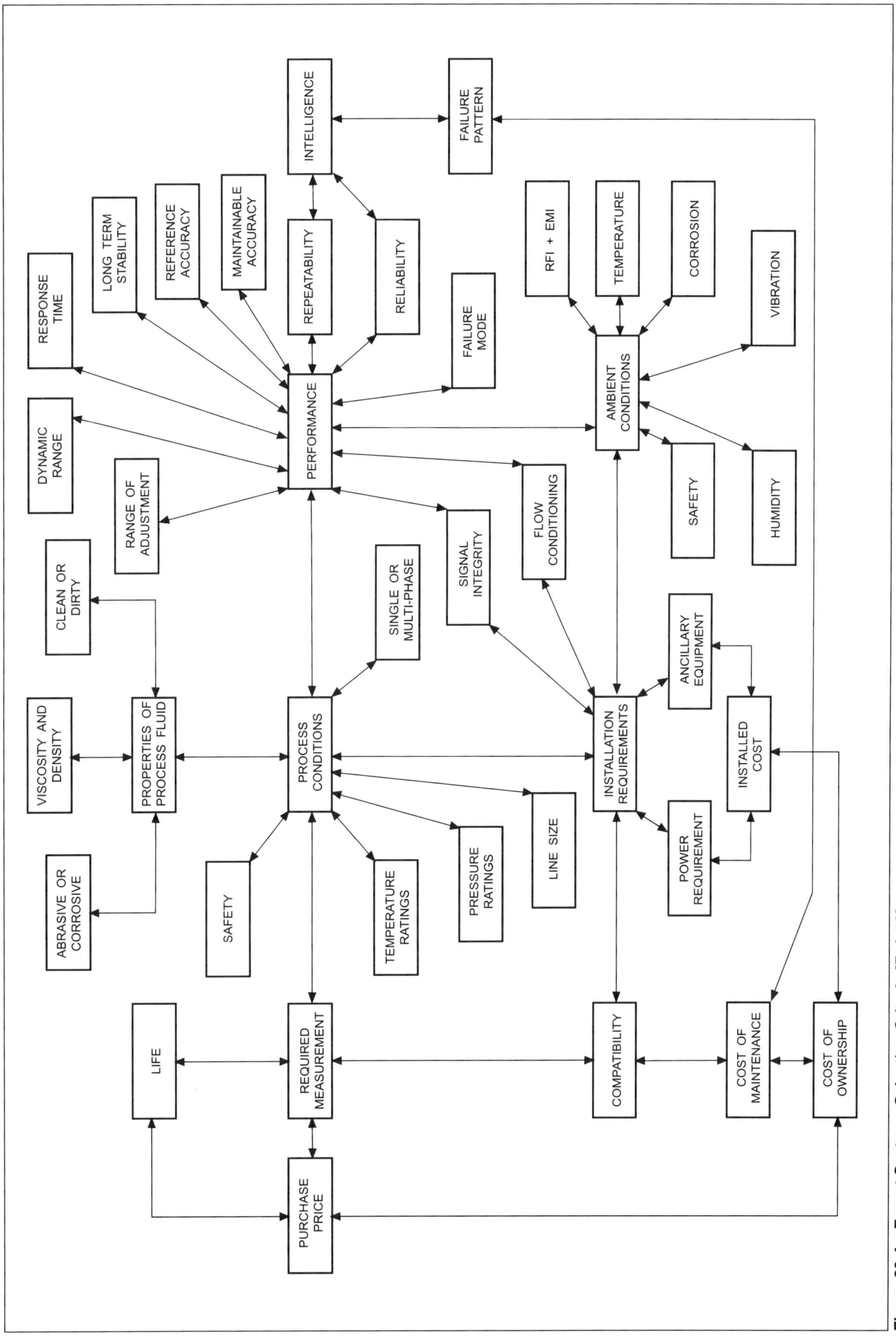

Figure 23-4. Expert Systems Selection Criteria [Ref. 12]
(Courtesy of The Foxboro Company)

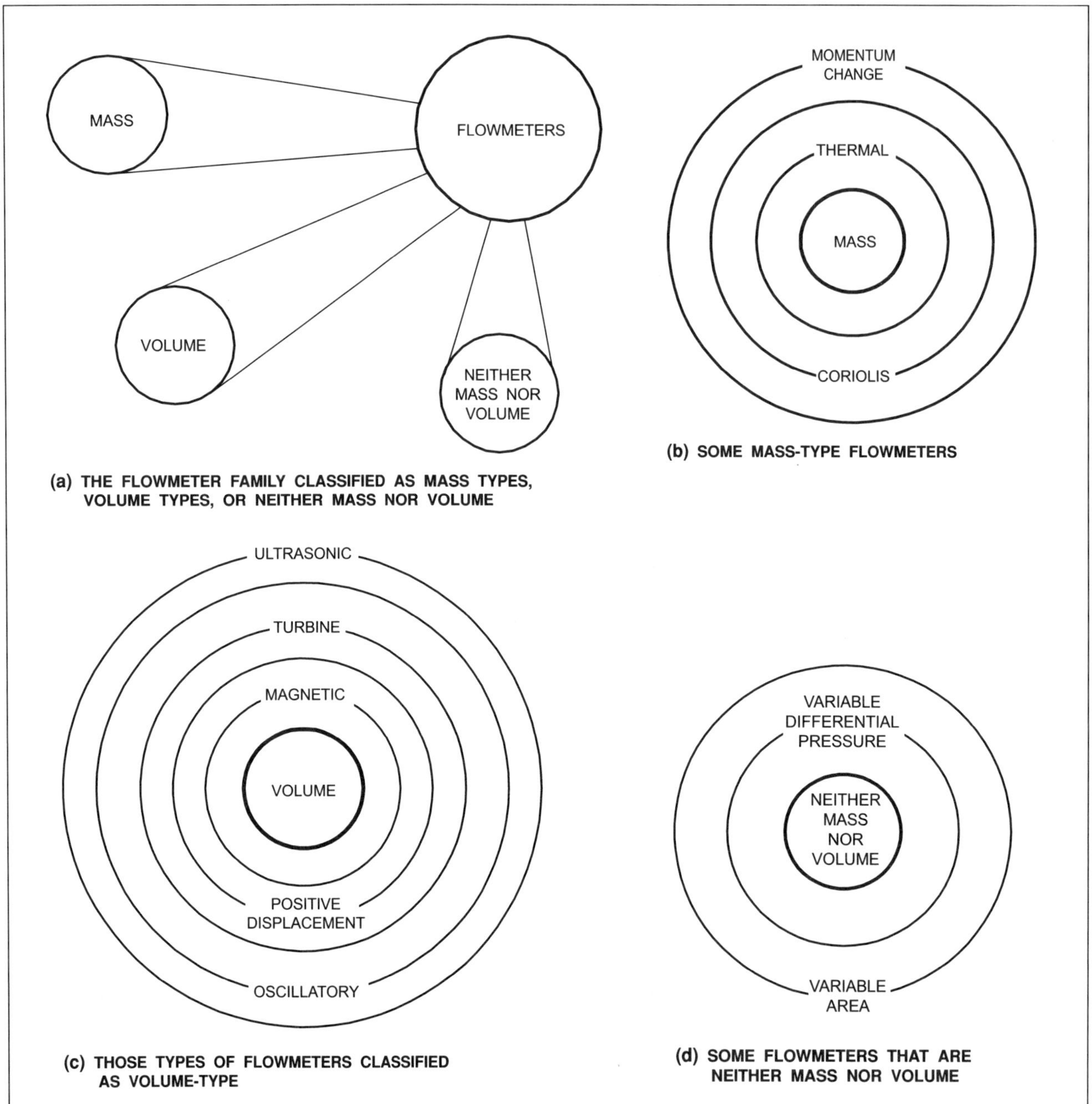

Figure 23-5. Types of Flowmeters [Ref. 13]
(Courtesy of The Foxboro Company)

Figure 23-5 shows the meters that would be included in each category. Of the meters categorized as the volume type, only the positive displacement truly measures volume. All other meters in this group sense flow velocity (hopefully, average fluid velocity).[1] This is fine as long as the cross-sectional area of the flowmeter, at the point where the measurement is made, is known and constant. When the meter is calibrated, this area is determined precisely and, hence, is

---

1 Flow rate, Q, equals the cross section of the pipe (flowmeter), A, times the *average velocity*, $\bar{v}$, in that area; $Q = A\,\bar{v}$.

known. As long as there is neither coating of the pipe (flowmeter) interior nor corrosion of the meter, the cross-sectional area will remain constant and the relationship is valid. Velocity-sensing flowmeters can be legitimately used to infer volumetric flow when these conditions are met:

(1) The velocity sensed is the average flow velocity.

(2) The cross-sectional area, where the average velocity is measured, is known and constant.

Processes that include chemical reactions and heat transfer will usually require mass flow information. For these applications meters that provide mass flow data would be preferred. There are, however, only two types of mass flowmeters with broad industrial acceptance:

(1) The Coriolis mass flowmeter provides a flow signal that is directly proportional to the mass rate of flow for liquids and many slurries. Gases at specific operating conditions can also be measured.

(2) The thermal type provides a mass flow signal for gases of known and constant thermal conductivity. They have been principally used for lower mass flow rates of industrial gases, with limited application of the principle to liquids.

Although mass meters are available, mass flow rate, for many reasons, is still widely inferred by using volume measurement multiplied by the fluid density. Fluid density can be measured with the appropriate densitometer, which provides a density signal regardless of the fluid or its composition. These are slowly gaining recognition and acceptance. In most inferred mass flow measurements, density is inferred by measuring temperature and/or pressure of the fluid.

The pressure-temperature-density relationship applies to one fluid. Variations in fluid composition invalidate the relationship.

Categorization of flowmeters based on whether they provide mass or volume information is shown in Figure 23-6.

Rate or Total Measurement

Process needs may be for rate of flow information or for the total accumulated flow; in some cases, both forms of information will be required. If only rate information or only total information is required, the flowmeters can be categorized as those that are by nature rate of flow devices and those that are totalizing devices, as shown in Figure 23-7. It is possible to transduce from rate of flow to totals, and vice versa.

Remember, when taking any additional transducing step, a price must be paid, in terms of transducer performance, that must be considered in the overall system accuracy.

All else being equal, if only total accumulated flow is needed, superior performance is available from those meters that do not need a transducing step to provide totalization. If only rate of flow information is required, preference should be given to those meters that naturally provide such information (see Figure 23-8).

Some integrators can provide total information using either linear or square root analog flow rate signals. With frequency signals, scalers (electronic dividers) are used to provide the selected volumetric units; usually they can be easily reprogrammed in the field. Gravimetric units can also be totalized, when the density or specific gravity is known and does not change with time, by including the appropriate scaling factor. Where there is changing density, an additional measurement or inference of density is required so that the compensated signal can be totalized.

If density is inferred by temperature and/or pressure measurement, it applies only to a given fluid or composition.

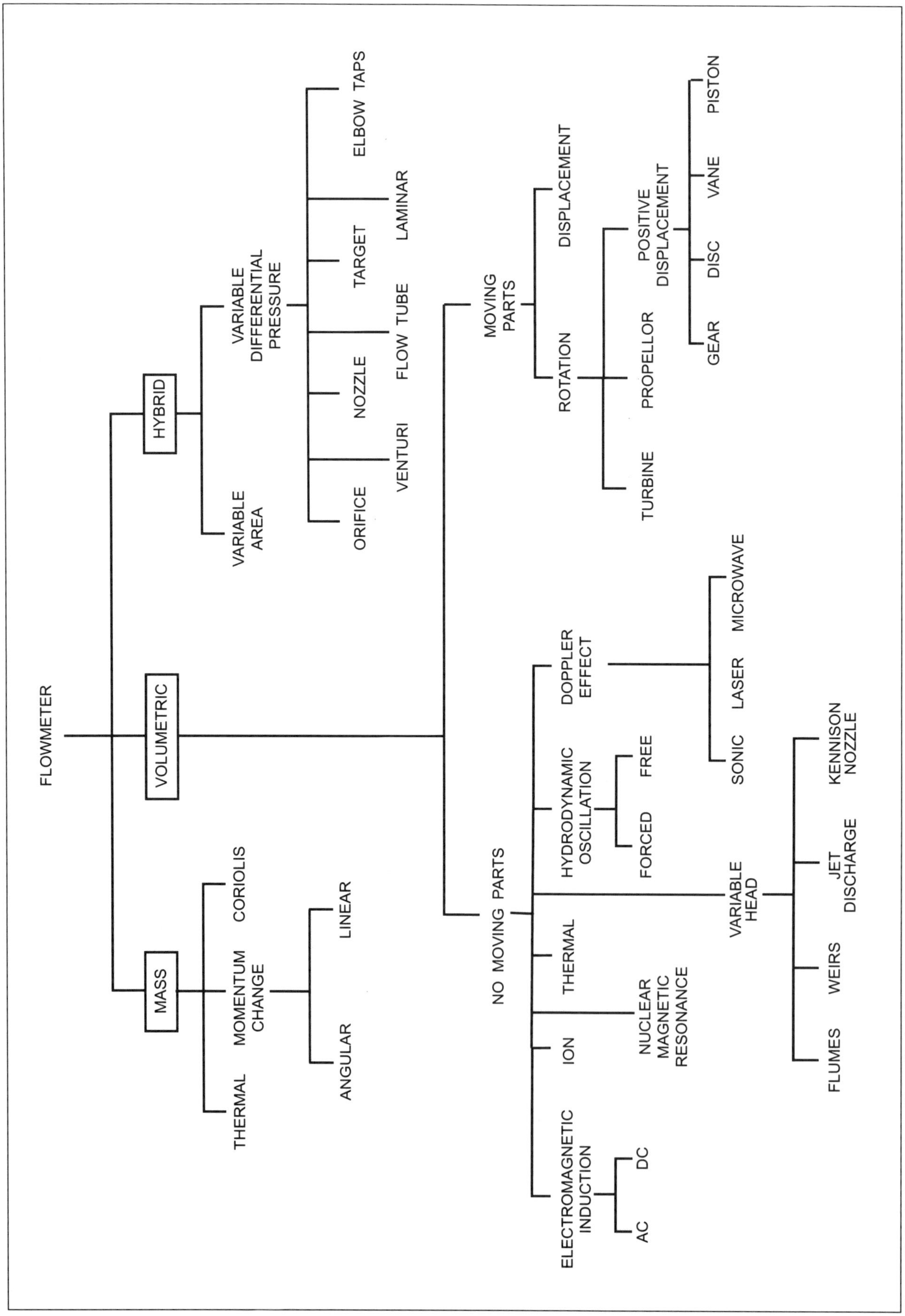

Figure 23-6. Volumetric or Mass [Ref. 15]

Fluid State

Flowmeters can be categorized by their ability to handle various fluid types. An excellent guide to making a selection on this basis is shown in Figure 23-9, "Guide to Flow Measurement," which appeared in *Instruments and Control Systems (I&CS)* magazine [Ref. 10]. In this approach John Hall provides a relative rating for the more commonly used flowmeter types for a wide range of fluids.

Beyond the fluid types given in Figure 23-9, there are the problem applications where gas is entrained in the liquid and where liquid phase is carried along with the gaseous phase. Volume meters handling liquids with entrained gas will be in error by the percent volume of gas present. Gaseous phase is normally transported through piping systems at fluid velocities at least ten times higher than liquid phase. If liquid droplets are present with the gaseous phase, they too are transported at high velocity, and the flow can be very erosive.

Other selection guidelines using fluid state as the starting point are shown in Figures 23-10 and 23-11.

Fluid Cleanliness

There is an associated consideration when making the selection of a flowmeter for specific fluid applications (see Figure 23-12). If the meter has moving parts, fluid cleanliness (or lack of it) is an issue. In this case, "cleanliness" means the presence of solid particles.

A knowledge of the particles is imperative. Their concentration and distribution in the fluid is important, but equally important is their nature:

(1) Are they soft and amorphous?

(2) Are they sticky?

(3) Are they sharp and abrasive?

(4) Are they fibrous?

(5) Are the fibers long or short?

Meter construction must be taken into account when considering fluid cleanliness. This is shown in Figure 23-13 where the central questions is, "Is the liquid clean?"

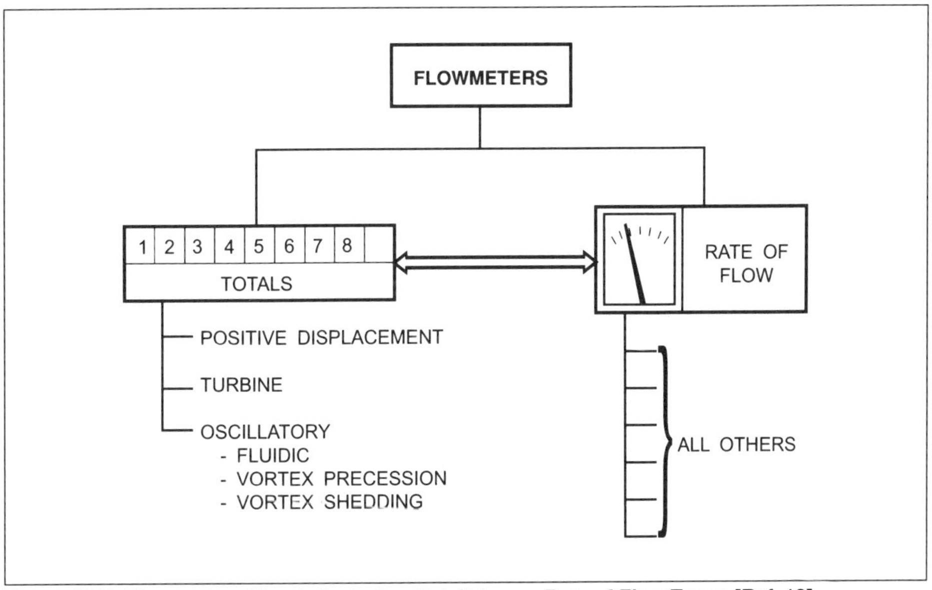

Figure 23-7. Flowmeters Discriminated as Totalizing or Rate of Flow Types [Ref. 13]

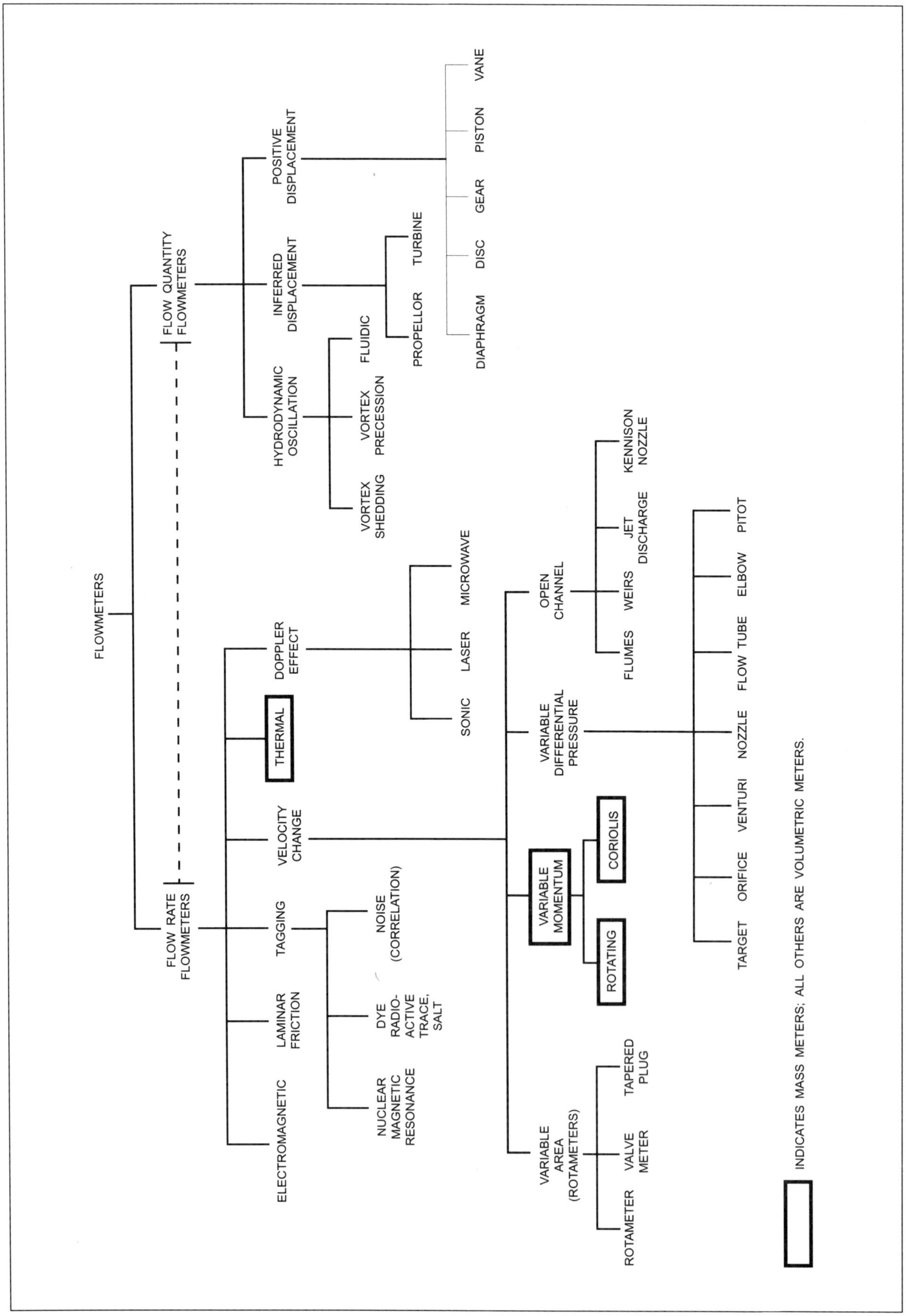

Figure 23-8. Flowmeter Family Tree [Ref. 16]

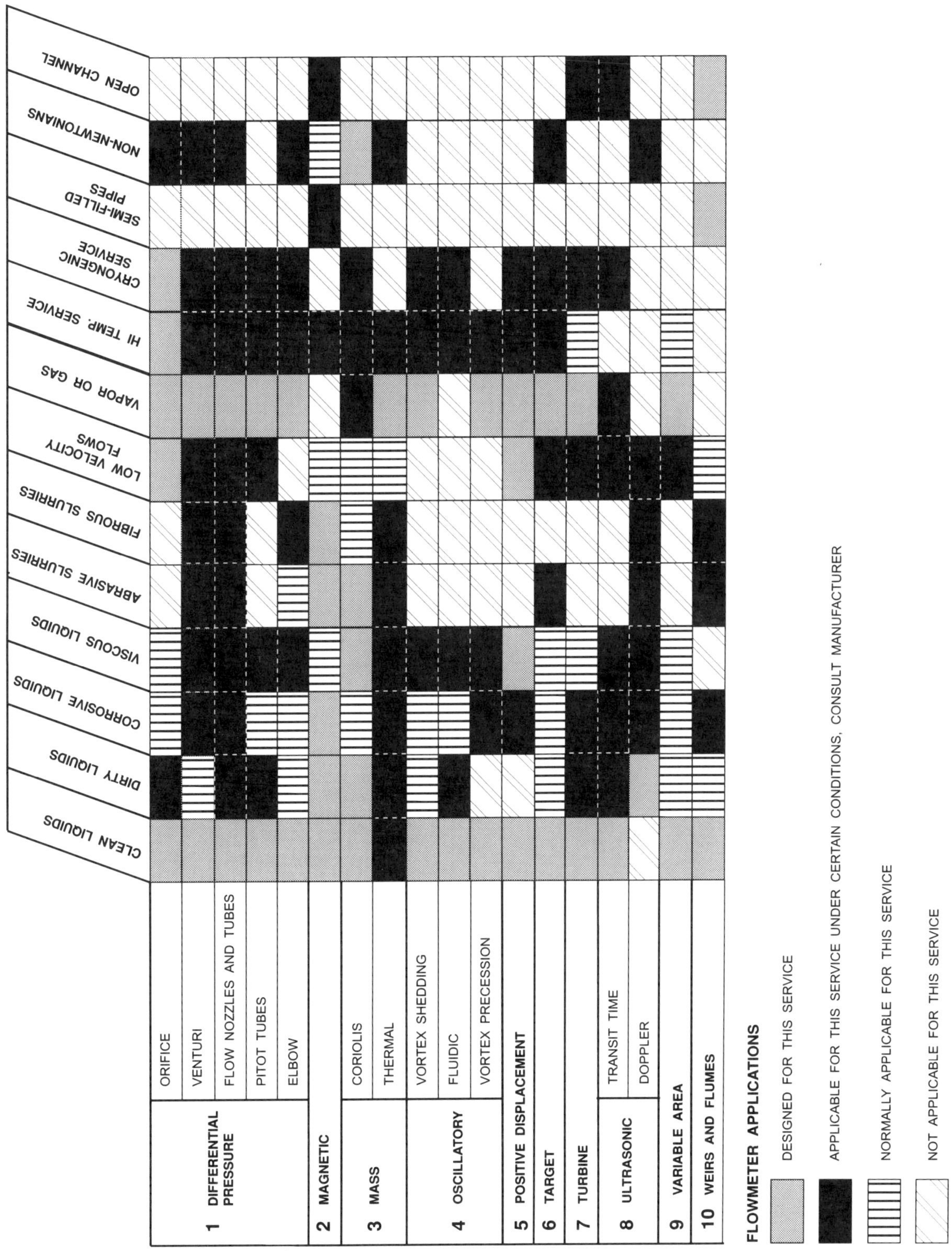

Figure 23-9. Guide to Flow Measurement [Ref. 10]
(Courtesy Instruments and Control Systems Magazine)

1 DIFFERENTIAL PRESSURE

	Orifice	Venturi	Flow Nozzles and Tubes	Pitot	Elbow
Service:	Liquids and gases including steam	Liquids and gases including steam	Liquids and gases includng steam	Liquids and gases	Liquids and gases
Design Pressure:	Determined by transmitter	Determined by transmitter	Determined by transmitter	Determined by transmitter	Determined by transmitter
Design Temperature:	Determined by materials	Determined by materials	Determined by materials	Determined by materials	Determined by materials
Flow Range:	From 0.1 cc/min up* or gas equivalent	From 5 gpm liquid; 20 scfm gas; determined by pipe size	From 5 gpm liquid; from 20 scfm gas equivalent	Determined by pipe size	Determined by pipe size
Scale:	Square root	Square root	Square root	Square root	Square root
Signal:	Analog electronic or pneumatic	Analog electronic or pneumatic	Analog electronic or pneumatic	Analog electronic or pneumatic	Analog electronic or pneumatic
Accuracy:	±0.6% of max flow uncalibrated and includes transmitter; sizes smaller than 2" usually calibrated	±1% of max flow or better; uncalibrated and includes transmitter	±1% full scale includng transmitter; flow calibration recommended	±5% full scale or better including transmitter	±5% to ±10% full scale including transmitter
Rangeability:	4:1 for given transmitter span setting	4:1 for given transmitter span setting	4:1 for given transmitter span setting	4:1 for given transmitter span setting	4:1 for given transmitter span setting
End Connections:	Mounts between flanges	Flanged	Flanged or mounted between flanges	Insert probe	Mounts in 90° pipe elbow
Sizes:	Determined by pipe size	To 72" and larger	3" to 48"	Unlimited probe length	Determined by pipe size
Advantages:	Easy-to-install; uses one transmitter regardless of pipe size; low cost; wide variety of types and materials available; easy-to-change capacity. Versions available that do not require power.	Low permanent loss; good for slurries and dirty fluids; uses one transmitter regardless of pipe size.	Economical, low permanent loss; uses one transmitter regardless of pipe size; nozzle commonly used for steam and has higher capacity for same generated ΔP.	Very low cost; uses one transmitter regardless of pipe size. Averaging types available.	Very economical; easy-to-install; uses one transmitter regardless of pipe size; can be bi-directional by using 45° tap location; very low pressure loss. Minimum upstream piping required.
Limitations:	Use eccentric orifices or segmental plates for very dirty fluids or slurries; quadrant orifice for viscous liquids; venturi, flow tube, pitot, or elbow taps to reduce energy consumption; straight run of upstream and downstream piping required. Some fluid must leave pipe except when chemical seal protectors are used.	Most expensive ΔP producer; generally limited to air and water; big and heavy especially in larger pipe sizes.	Flow tubes lack extensive background data compared to orifice plates; application on viscous liquids limited. Calibration recommended for optimum performance.	Doesn't sample full stream; limited accuracy. Low differential pressure for given flow rate.	Not good for low flow velocity; not as accurate as other ΔP types. Low differential pressure for given flow rate.

*Note: For less than 1" pipe size use integral orifice ΔP transmitter.

Figure 23-9. Guide to Flow Measurement (continued)

2 MAGNETIC

Service: Electrically conductive liquids or slurries

Design Pressure: Up to 740 psig

Design Temperature: Up to 360°F

Flow Range: 0.01 through 500,000 gpm

Scale: Linear

Signal: Analog electronic; digital

Accuracy: ±0.5% of rate to ±1.0% of full scale; factory calibrated

Rangeability: 10:1 for any span*

End Connections: Flanged, sanitary, or wafer; Dresser and Victaulic ends available in larger sizes.

Sizes: 0.1" to 96" (sampling type available); also used as by-pass meter around mainline orifice.

Advantages: Unaffected by change in fluid density, viscosity; zero head loss; bi-directional; no flow obstruction; easy to respan; versions for dc power.

Limitations: Moderate to expensive cost; liquids or slurries only; required minimum electrical conductivity varies with manufacturer; larger sizes big and heavy.

*30:1 of span adjustment plus 10:1 range or greater for any given span setting.

3 MASS

	Coriolis Effect	Thermal
Service:	Liquids and slurries (limited gas service)	Gas (some designs for liquid)
Design Pressure:	Up to 2800 psig	500 psig and higher
Design Temperature:	Up to 400°F	Up to 150°F and higher
Flow Range:	Up to 23,000 lb/min	Up to 4000 gpm liquid. Up to 1500 scfm of gas.
Scale:	Linear	Exponential
Accuracy:	±0.5% of rate or better	±1% of full scale
Rangeability:	10:1 or better	10:1 or better
End Connections:	Threaded or flanged	Threaded, flanged, hose
Sizes:	1/16" to 6"	1/8" to 10"; Sampling types available; by-pass type available.
Advantages:	Measures mass flow directly. Can handle very difficult applications.	Measures mass flow directly. Very low pressure loss. Good for low velocity gas measurement.
Limitations:	Moderate to expensive cost. There are specific installation requirements. Head loss may be high. Be careful with 2-phase flows.	Affected by coatings. Some designs are fragile.

4 OSCILLATORY

	Vortex Shedding (Bluff body)	Fluidic (Coanda Effect)	Vortex Precession
Service:	Liquids and gases including steam	Liquids	Liquids and gases
Design Pressure:	Up to 3600 psig	Up to 600 psig	Up to 1400 psig
Design Temperature:	Up to 750°F	0 to 250°F	-100°F to 350°F
Flow Range:	3 to 5000 gpm liquid; 10,000,000 scfh gases	1 to 1000 gpm	1.8 to 3082 gpm liquid; 10,000,000 scfh gases
Scale:	Linear at high Reynolds No.	Linear at high Reynolds No.	Linear at high Reynolds No.
Signal:	Frequency or analog electronic	Analog electronic or pneumatic; pulse	Frequency or analog electronic
Accuracy:	±1% of rate or better on liquid; factory calibrated; ±2% of rate on gas	±1% of rate or better; factory calibrated	±1% of rate or better; factory calibrated
Rangeability:	8:1 to 15:1	Up to 30:1	8:1 to 25:1 (deter. by size, application)
End Connections:	Flanged, threaded, wafer or insert; also can be used as by-pass meter around mainline orifice.	Mounts between flanges	Flanged
Sizes:	1/2" thru 8"; larger sizes available (sampling and by-pass types available)	1" thru 4"; by-pass types available	1/2" thru 12"
Advantages:	No moving parts; suitable for wide variety of fluids; excellent combination of price and performance.	No moving parts; suitable for wide variety of liquids; excellent combination of price and performance.	No moving parts; ideal for corrosive and difficult gases.
Limitations:	Straight piping required; sensitive to increasing viscosity below a given Reynolds number.	Straight piping required; sensitive to increasing viscosity below a given Reynolds number.	Expensive; has operating minimum dependent on flow rate and gas density.

Figure 23-9. Guide to Flow Measurement (continued)

5 POSITIVE DISPLACEMENT

Service: Clean liquids and gases

Design Pressure: Up to 1400 psig for liquid or gas

Design Temperature: Up to 600°F liquids; up to 250°F gas

Flow Range: 0.1* to 9000 gpm liquid; 0 to 100,000 scfh gas.

Scale: Linear

Signal: Pulse or analog electronic

Accuracy: ±½% of rate on liquid; ±1% of full scale on gas; factory calibrated

Rangeability: Typically 10:1

End Connections: Flanged or threaded

Sizes: Up to 12"

Advantages: Ideal for viscous liquids; good for custody transfer, batching, blending; simplest versions do not require electrical power; very little straight upstream pipe required.

Limitations: Subject to mechanical wear; requires periodic proving; sensitive to dirt and may require upstream filters; larger sizes are excessive in size and weight, may require special installation care.

*Ultra-low flow rate versions also available.

6 TARGET

Service: Liquids and gases including steam

Design Pressure: Up to 10,000 psig

Design Temperature: Up to 750°F

Flow Range: 0.07 gpm and up liquid; 0.3 scfm and up gas

Scale: Square root

Signal: Analog electronic or pneumatic

Accuracy: ±½% to ±5% of full scale; factory calibrated

Rangeability: 3:1*

End Connections: Flanged, threaded, flared tube

Sizes: Up to 8" (sampling types available)

Advantages: No moving parts; relatively inexpensive; good for hot, tarry and sediment bearing fluids.

Limitations: Need 20 diameters upstream and 10 diameters downstream of straight pipe to maintain accuracy; reading is per cent of scale; limited range.

*10:1 span adjustment plus 3:1 range for any given span setting.

7 TURBINE

Service: Clean liquids and gases including steam

Design Pressure: Up to 3000 psig

Design Temperature: -450°F to 500°F

Flow Range: 0.001 through 40,000 gpm liquids; to 10,000,000 scfh gases

Scale: Linear when Reynolds number is 10,000 or higher

Signal: Frequency or analog electronic

Accuracy: ±0.25% of rate liquids; ±1% of rate gas; factory calibration should simulate operating viscosity and lubricity for liquids.

Rangeability: 10:1 to 50:1

End Connections: Flanged or threaded

Sizes: Up to 24" (sampling types available); also used as by-pass meter around mainline orifice.

Advantages: One of the most accurate iquid meters; good operating range; easy-to-install and maintain; very low flow rate designs available; small in size; lightweight. Versions optimized for gas; sampling types for steam. Some versions do not require external power.

Limitations: Sensitive to increasing viscosity; avoid use where state may change from liquid to gas; gas versions require care when used in varying flow rate applications; straight upstream pipe is required; flow straighteners may be recommended.

8 ULTRASONIC

	Transit Time (Pulsed type)	Doppler (Frequency shift)
Service:	Relatively clean liquids (some designs for gas)	Liquids with entrained gas or suspended solids
Design Pressure:	Wetted transducers; 1000 psig up, Clamp-on; pipe rating	Wetted transducers; 1000 psig up, Clamp-on; pipe rating
Design Temperature:	-300°F tp +500°F	-300°F to 500°F
Flow Velocity:	Typically to 40 ft/sec	Typically to 40 ft/sec
Scale:	Linear	Linear
Signal:	Analog electronic or digital	Analog electronic or digital
Accuracy:	±1% of rate to ±5% of full scale depending on type and calibration	±5% of full scale or better
Rangeability:	Up to 40:1	Typically 10:1
End Connections:	Flanged (clamp-on design available)	Clamp-on; body versions available (Sampling type available)
Sizes:	⅜" up	¼" up
Advantages:	No flow obstruction; can be bidirectional; use with practically any relatively clean liquid. Versions for gas. Clamp-on versions available.	Can handle inorganic slurries and aerated liquids; clamp-on version can be installed without process shut-down.
Limitations:	Straight upstream piping required to provide uniform flow profile; clean liquids only.	Not suitable for clean liquids; requires straight upstream piping.

Figure 23-9. Guide to Flow Measurement (continued)

9 VARIABLE AREA (Rotameter)

Service: Liquids and gases including steam

Design Pressure: Up to 350 psig (glass tube); to 720 psig (metal tube)

Design Temperature: Up to 400°F (glass tube); to 1000°F (metal tube)

Flow Range: Liquids 0.01 cc/min to 300 gpm; gases 0.3 cc/min to 1500 scfm at 10 psi

Scale: Linear or logarithmic

Signal: Visual; electronic or pneumatic analog

Accuracy:±0.5% of rate to ±10% of full scale depending on type, size, and calibration

Rangeability: 5:1 to 12:1

End Connections: Female pipe threaded or flanged

Sizes: Up to 3"; also used as a by-pass meter around a mainline orifice for larger pipe sizes.

Advantages: Inexpensive; somewhat self-cleaning; insensitive to viscosity variations below a given threshold; direct indicating; no power required; can be direct mass device; minimum piping requirements. Versions available with plastic liners.

Limitations: Requires accessories for data transmission; must be vertically mounted; gas use requires minimum backpressure.

10 WEIRS AND FLUMES

Service: Liquids in open channels

Flow Range: From ½ gpm up

Scale: Proportional to the measured head to the $3/2$ power for rectangular and trapezoidal weirs and Parshall flumes; proportional to the measured head to the $5/2$ power for V-notch weirs.

Signal: Analog electronic or pneumatic

Accuracy: 2% to 5% full scale

Rangeability: 75:1 rectangular, trapezoidal weirs, Parshall flumes; 500:1 V-notch weirs; Palmer-Bowlus flumes 10:1

Sizes: From 1" up

Advantages: Ideal for water and waste flows; flumes have low head loss, low cost

Limitations: Weirs are more accurate than flumes but require cleaning; flumes are self-cleaning.

Figure 23-9. Guide to Flow Measurement (continued)

Cleanliness or dirtiness of fluids is a relative consideration. Some fluids, because of process needs, are scrupulously clean. Most fluids are relatively clean and may have solids present because of open storage, as corrosion products, or as catalyst carry-over, none of which are desirable from a process perspective. Regardless, the degree of cleanliness of the fluid has an effect on selection of the optimum flowmeter for the application. As a general rule, meters with moving parts have the lowest tolerance of solids, particularly abrasive particles and fibers. Meters that represent an obstruction in the flow stream, even though parts do not move, require specific consideration. Target flowmeters and vortex shedding flowmeters tolerate particles in the flow stream more than conventional orifice plates (thin, concentric, sharp-edge) where edge condition is absolutely critical to good measurement, but fibrous particles do present a potential problem. The flow nozzle and the V-cone have a high degree of particle tolerance. The flow nozzle is usually the only flow element that survives in wet steam (steam with quality). Where fluids are very dirty, or where the liquid or gas is a vehicle to convey solids (slurries), flowmeter parts in the flow stream are a detriment, whether they are moving or not. The meters of choice for slurries or very dirty streams are those that are obstructionless or are non-invasive.

For some food, beverage, pharmaceutical, and even some chemical applications, the design of the meter must not contain cavities that allow stagnation of the fluid and/or growth of bacteria.

Other charts that use fluid state to initiate the selection process are shown in Figures 23-14 and 23-15.

Fluid Compatibility

Compatibility, however, is a two-way consideration. It includes potential corrosion of the flowmeter parts but also includes the potential contamination of the process fluid.

Another area of important consideration is the compatibility of the fluid with the materials of construction of the meter. For the more corrosive applications, correct material selection is vital. Some meters traditionally are available in very corrosion-resistant materials of construction and, hence, should be given priority for these applications. Magnetic flowmeters are widely selected for this reason.

LIQUID	GAS	STEAM	SLURRY
VARIABLE AREA	VARIABLE AREA	VARIABLE AREA	MAGNETIC
VARIABLE DIFFERENTIAL PRESSURE	VARIABLE DIFFERENTIAL PRESSURE	VARIABLE DIFFERENTIAL PRESSURE	CORIOLIS
POSITIVE DISPLACEMENT	POSITIVE DISPLACEMENT	TURBINE	ULTRASONIC (DOPPLER)
TURBINE	TURBINE	OSCILLATORY	VARIABLE DIFFERENTIAL PRESSURE (ECCENTRIC, SEGMENTAL, VENTURI)
MAGNETIC	OSCILLATORY		
ULTRASONIC	THERMAL		
THERMAL	CORIOLIS		
OSCILLATORY			
CORIOLIS			

Figure 23-10. Flowmeters Categorized by Application [Ref. 13]

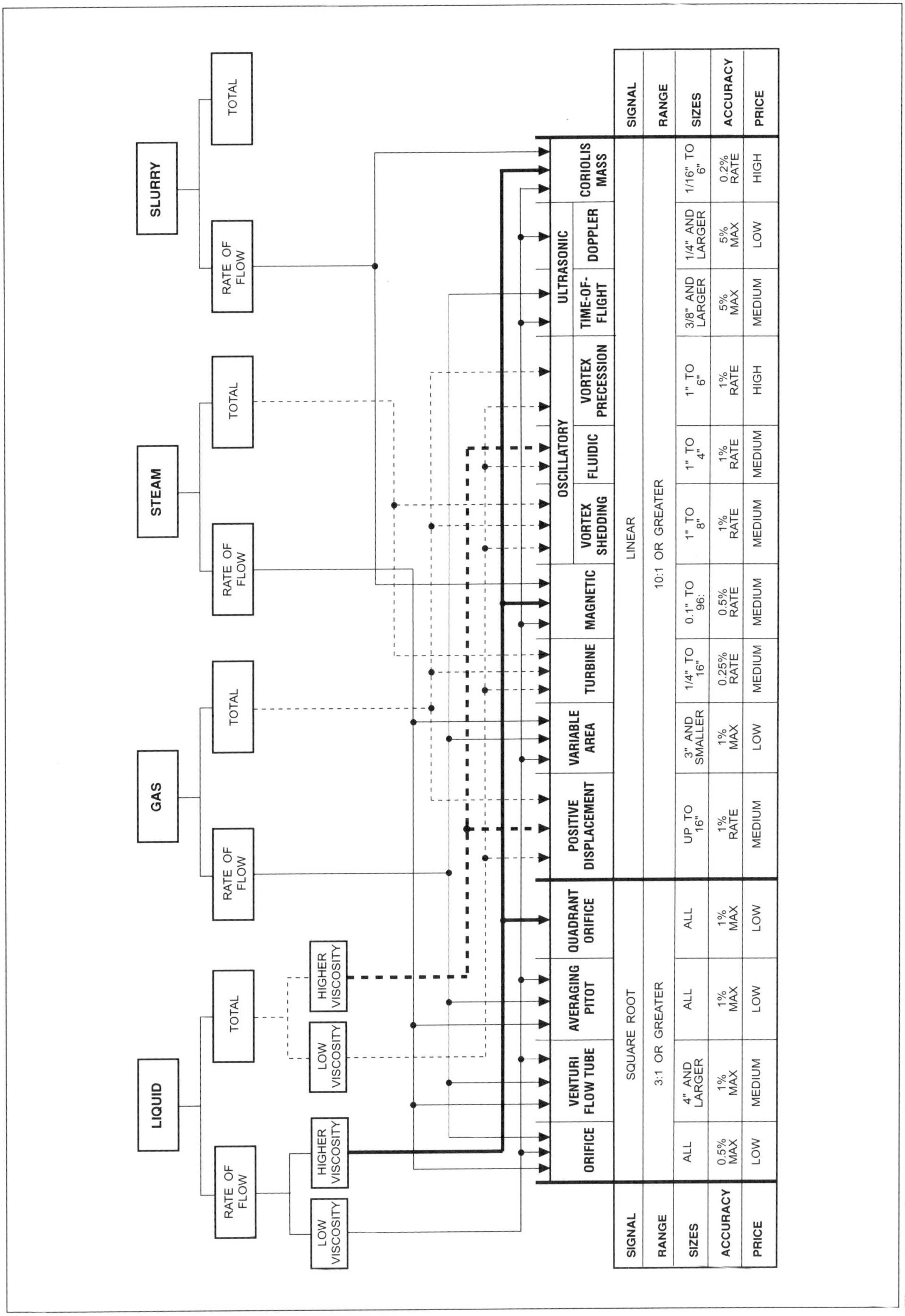

Figure 23-11. Flowmeters Categorized by Fluid State

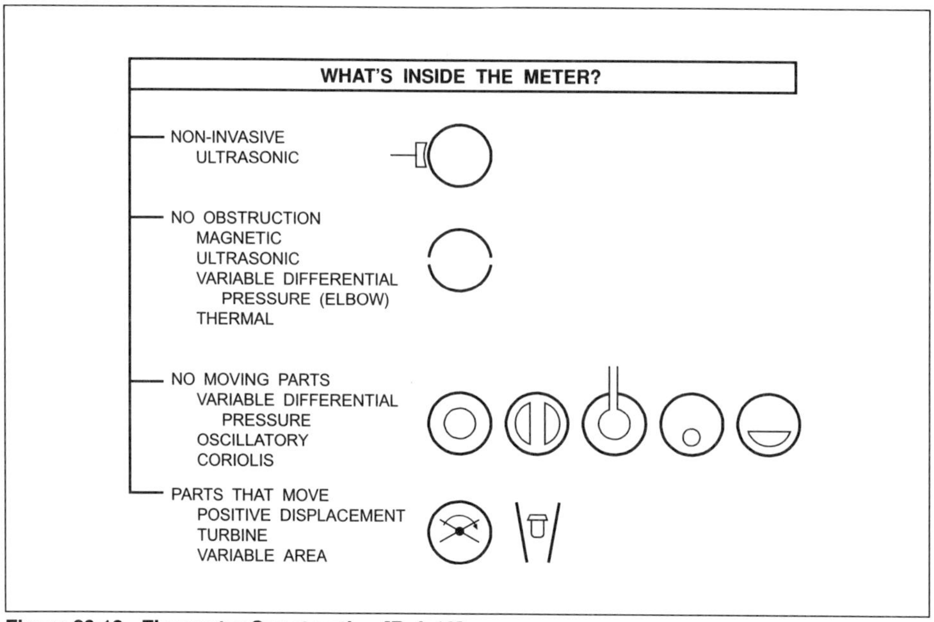

Figure 23-12. Flowmeter Construction [Ref. 13]

Certain meters can be eliminated from consideration and the choice narrowed down to a reasonable number by following this flowchart. Start in the middle, at "Is the liquid clean?"
Abbreviations used are:

EM = electromagnetic		PD = positive displacement	
DT = Dall tube		VA = variable area or rotameter	
TM = turbine meter		OPE = orifice plate, eccentric	
VM = vortex meter		OPC = orifice plate, concentric	

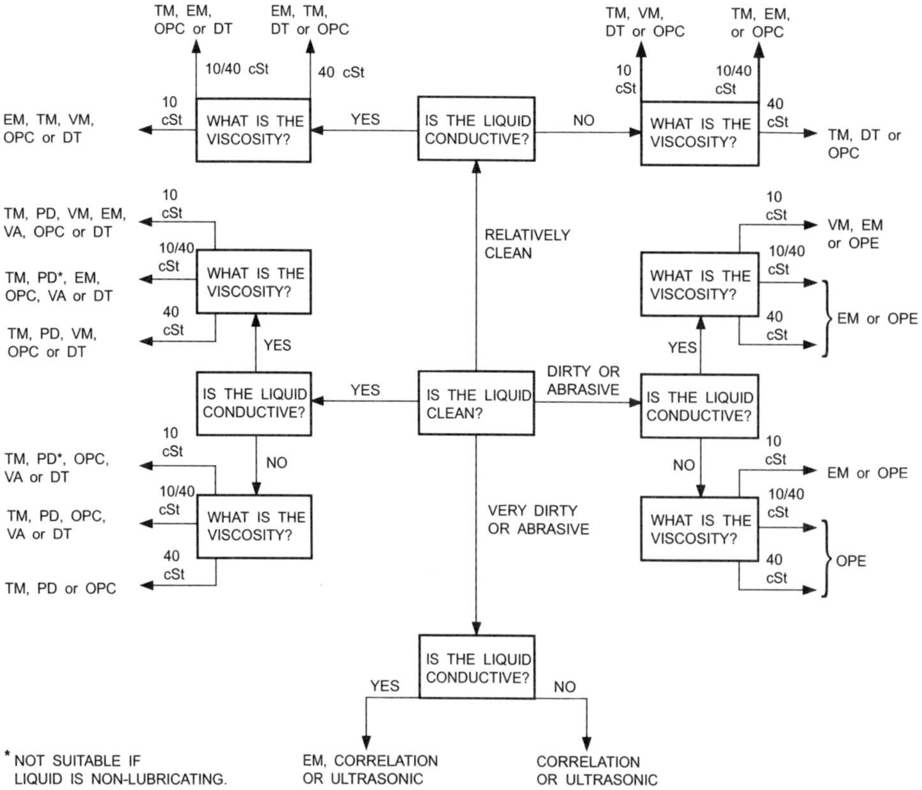

* NOT SUITABLE IF LIQUID IS NON-LUBRICATING.

Figure 23-13. One Scheme for Making Flowmeter Selections [Ref. 18]
(Courtesy of InTech)

CORIOLIS — 1/16" - 6"
MAGNETIC — 1/16" - 72"
ULTRASONIC — ALL
VORTEX — 1/2" - 8"
DIFFERENTIAL PRESSURE — ALL
POSITIVE DISPLACEMENT — 1/16" - 12"
TURBINE — 1/16" - 24"
VARIABLE AREA — 1/16" - 3"

APPLICATION	CORIOLIS	MAGNETIC	ULTRASONIC	VORTEX	DIFFERENTIAL PRESSURE	POSITIVE DISPLACEMENT	TURBINE	VARIABLE AREA
LIQUID - CLEAN OR DIRTY, LOW VISCOSITY	4	5	5	5	5	3	5	5
- CLEAN OR DIRTY, VISCOUS	5	5	1	2	2	5	3	3
- CLEAN OR DIRTY, CORROSIVE	3	5	5	3	3	2	1	2
SLURRY - ABRASIVE	2	5	4	2	1	0	0	1
- AMORPHOUS	5	5	4	3	2	3	1	2
- FIBROUS	3	5	4	0	0	0	0	0
GAS	2	0	1	5	5	2	2	2
STEAM	0	0	0	5	5	0	1	2
TEMP > 400°F	0	0	0	4	4	0	1	2
PRESS > 1500 psi	1	1	4	2	4	1	4	1

LEGEND: "0" NOT USED TO "5" MOST FREQUENTLY USED

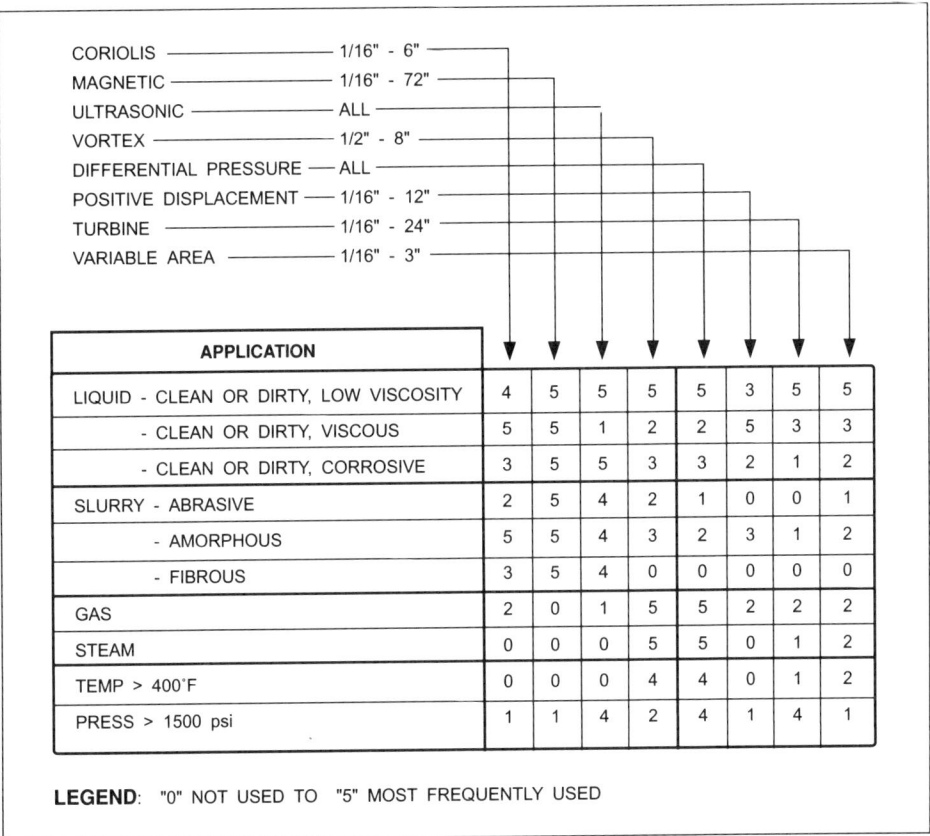

Figure 23-14. Where Flowmeters Are Applied
(Courtesy The Foxboro Company)

Fluid Properties

Other fluid properties will be important to specific types of meters. These properties are beyond the ones that are almost inevitably considered, such as temperature, pressure, density, and viscosity. Magnetic flowmeters are dependent upon a given threshold of electrical conductivity. Thermal meters are dependent on specific thermal properties. Glass tube variable area flowmeters are dependent on the fluid being transparent or at least translucent. Ultrasonic flowmeters are dependent on the ability of the fluid to transmit sound waves; but those that are non-invasive in design are also dependent on the properties of the pipe.

Flow Profile Dependency

Whole families of flowmeters can be categorized as viscosity sensitive or, more precisely, Reynolds number dependent. All meters are so classified except Coriolis, magnetic, and some thermal types. The Reynolds number dependent meters have performance characteristics that will change when a given threshold is crossed. They generally have linear performance (a relatively constant coefficient) when the Reynolds number is greater than a given threshold. Performance will change to a less linear or a nonlinear characteristic below that threshold.

Associated with Reynolds number change is a flow profile change from the parabolic laminar profile at low Reynolds numbers to the "squared-up" turbulent profile at higher Reynolds numbers.

	APPLICATIONS								FEATURES										
	CLEAN LIQUID	DIRTY LIQUID	VISCOUS LIQUID	CORROSIVE	SLURRIES	CLEAN GAS	DIRTY GAS	STEAM	SIZES AVAILABLE	ACCURACY	RANGEABILITY	REYNOLDS NUMBERS	VISCOSITY EFFECTS	PRESSURE LOSS	RELATIVE COST	PIPING REQUIRED UPSTREAM	INSTALLATION	MAINTENANCE	TYPE OF OUTPUT
ORIFICE PLATE	●					●		●	> 1"	3/4%	3/1	> 30,000	H	H	L	10-30D	M-H	M-H	√
INTEGRAL ORIFICE	●					●		●	1", 1.5"	2%	3/1	> 10,000	H	H	L	10-30D	L	M-H	√
WEDGE™	●	●	●	○	●	●	●	●	> .5"	1/2%	3/1	> 500	H	M	H	10-30D	L	L	√
FLOW NOZZLE	●	○	○	○		●	○	○	> 2"	1-1/2%	3/1	> 75,000	H	M	M	10-30D	M	L	√
VENTURI TUBE	●	○	○	○	○	●	○	○	> 2"	1%	3/1	> 75,000	H	L	H	5-10D	M	L	√
ELBOW	●	○		○		●	○		> 2"	5%	3/1	> 10,000	H	L	L	NONE	M	L	√
PITOT-VENTURI	○					●			> 6"	3%	3/1	> 100,000	H	L	L	20-30D	M	L	√
PITOT	●		○	○		●			> 3"	3%	3/1	> 100,000	H	L	L	20-30D	M	L	√
FLOBAR™	●	○		○		●	○	●	> .5"	1%	3/1	> 40,000	H	L	L	10-20D	L	L	√
MAGNETIC	●	●	●	●	●				> .1"	1/2%	10/1	NONE	N	L	H	5D	H	M	LINEAR
ROTAMETER	●			○		●		○	≤ 3"	2%	10/1	NONE	M	M	L	NONE	L	L	LINEAR
TURBINE	●			○		●		●	> .25"	1/2%	10/1 TO 50/1	≤ 2-15 cSt	H	H	M	10-20D	L	M-H	LINEAR
POSITIVE DISPLACEMENT	●	○	●	○		●			< 12"	1%	20/1	≤ 8000 cSt	N	H	H	NONE	H	H	LINEAR
VORTEX SHEDDING	●	○		○		●	○	●	> 1"	1%	20/1	> 10,000	N	M	M	15-25D	M	M	LINEAR
DOPPLER	○	●	○	●	●				> .5"	2-5%	10/1	NONE	N	L	M	5-20D	L	L	LINEAR
TRANSIT TIME	●		○	●					> .5"	2-5%	10/1	NONE	N	L	M	5-20D	L	L	LINEAR
MASS	●	●	●	○	●	●	○		< 6"	1/4%	25/1	NONE	N	M	H	NONE	H	L-M	LINEAR
TARGET	●	●	●	○		●	●		> .5"- 4"	1-1/2 - 5%	10/1	> 100	M	M	L	10-20D	L	M	√

● RECOMMENDED
○ LIMITED APPLICABILITY
(blank) NOT RECOMMENDED

N — NONE
L — LOW
M — MEDIUM
H — HIGH

Figure 23-15. Flowmeter Selection Guide [Ref. 7]

Between the laminar and turbulent regions is the transition zone where there is repeated and random change of profile from laminar to turbulent and vice versa. Attempting to operate Reynolds number dependent flowmeters in this unstable zone will give poor measurement results. Where possible, select a meter size, usually smaller than pipe size, so that the Reynolds number for the meter does not fall in the transition zone.

Flow profile is not only a function of Reynolds number but is also influenced by the piping. All meters classified as Reynolds number dependent, except positive displacement and variable area types, will have specific piping installation requirements to assure a known and stable flow profile, which is essential to making a good flow measurement.

Signal

The signal characteristic of the meter (see Figure 23-16) may be a deciding factor in some applications. The preferred signal, in which there is a small variation in flow rate, is square root, which provides a large change in signal amplitude for a small change in flow rate. The linear signal is preferred where there is a wide operating range and where totalization accuracy is important. The logarithmic characteristic is found in some types of variable area and thermal flowmeters. Incidentally, the 5/2 characteristic is for triangular or V-notch weirs and the approximate 3/2 characteristic is for Cipolletti and rectangular weirs and Parshall flumes. Table 23-2 is a method for selecting among flowmeters where signal characteristic is a primary criterion.

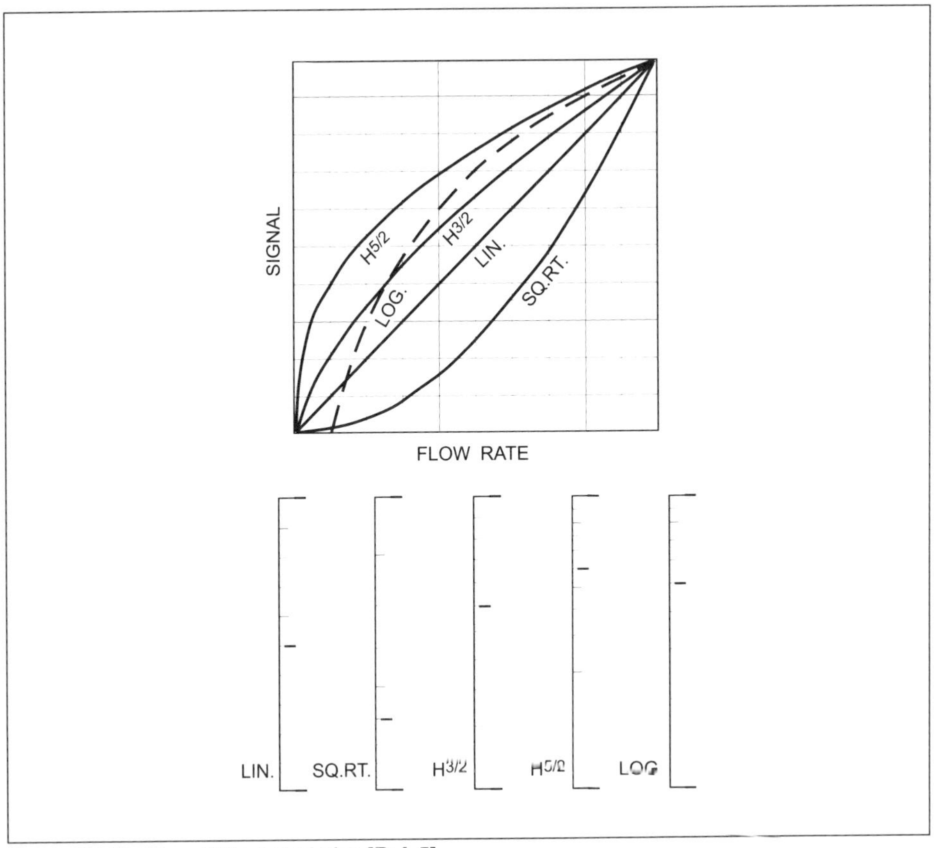

Figure 23-16. Signal Characteristics [Ref. 5]

Table 23-2. Flowmeter Selection Table [Ref. 21]
(Courtesy of The Foxboro Company)

FLOWMETER	PIPE SIZE, in (mm)	GASES (VAPORS)		LIQUIDS				SLURRIES		TEMPERATURE, °F (°C)	PRESSURE, psig (kPa)	ACCURACY, UNCALIBRATED (INCLUDING TRANSMITTER)	REYNOLDS NUMBER
		CLEAN	DIRTY	CLEAN	VISCOUS	DIRTY	CORROSIVE	FIBROUS	ABRASIVE				
SQUARE ROOT SCALE: MAXIMUM SINGLE RANGE 4:1													
ORIFICE		■	□	■	□	□	□	□	□	PROCESS TEMPERATURE TO 1000°F (540°C); TRANSMITTER LIMITED TO –20-250°F (–30-120°C)	TO 6000 PSIG (41,000 kPa)		
SQUARE-EDGED	>1.5 (40)	■	□	■	□	▒	▒	□	□			±1-2% URV*	R_D>2000
HONED METER RUN FOXBORO IFOA	0.5-1.5 (12-40)	■	□	■	▒	□	▒	□	□			±1% URV	R_D>1000
INTEGRAL	<0.5 (12)	■	□	■	■	□	▒	□	□			±2-5% URV	R_D>100
QUADRANT/CONIC EDGE	>1.5 (40)	□	□	■	■	□	▒	□	□			±2% URV	R_D>200
ECCENTRIC	>2 (50)	▒	■	▒	□	■	▒	□	□			±2% URV	R_D>10,000
SEGMENTAL	>4 (100)	▒	■	▒	□	■	▒	□	□			±2% URV	R_D>10,000
ANNULAR	>4 (100)	▒	■	▒	□	■	▒	□	□			±2% URV	R_D>10,000
TARGET	>0.5-4 (12-100)	■	■	■	■	■	▒	□	□			±1.5-5% URV	R_D>100
VENTURI	>2 (50)	■	▒	■	▒	▒	▒	▒	▒			±1-±2% URV	R_D>75,000
FLOW NOZZLE	>2 (50)	■	▒	■	▒	▒	▒	□	□			±1-±2% URV	R_D>10,000
LO-LOSS	>3 (75)	■	□	■	□	□	■	□	□			±1.25% URV	R_D>12,500
PITOT	>3 (75)	■	□	■	▒	□	▒	□	□			±5% URV	NO LIMIT
MULTIPORT AVERAGING	>1 (25)	■	▒	■	▒	□	▒	□	□			±1.25% URV	R_D>10,000
ELBOW	>2 (50)	■	▒	■	□	▒	▒	▒	▒			±4.25% URV	R_D>10,000
LINEAR SCALE: TYPICAL RANGE 10:1													
MAGNETIC	0.1-72 (25-1800)	□	□	■	■	■	■	■	■	360 (180)	≤ 1500 (10,500)	±0.5% OF RATE TO ±1% URV	NO LIMIT
MASS FLOWMETER CORIOLIS		▒	□	■	■	■	▒	□	□	<570 (300)	<2000	±0.2% TO ±1% OF RATE	NO LIMIT
POSITIVE DISPLACEMENT	<12 (300)	■	□	■	□	□	▒	□	□	GASES: 250 (120) LIQUIDS: 600 (315)	≤ 1400 (10,000)	GASES: ±1% URV LIQUIDS: ±0.5% OF RATE	≤ 8000 cSt
TURBINE	0.25-24 (6-600)	■	□	■	□	□	▒	□	□	–450-500 (–268-260)	≤ 3000 (21,000)	GASES: ±0.5% OF RATE LIQUIDS: ±1% OF RATE	≤ 2-15 cSt
ULTRASONIC		□	□	□	□	□	□	□	□	–450-500 (–268-260)			
TIME-OF-FLIGHT	>0.5 (12)	□	□	■	▒	□	■	□	□	–300-500 (180-260)	PIPE RATING	±1% OF RATE TO ±5% URV	NO LIMIT
DOPPLER	>0.5 (12)	□	□	□	▒	■	■	■	■	–300-250 (–180-120)	PIPE RATING	±5% URV	NO LIMIT
VARIABLE AREA	≥ 3 (75)	■	□	■	■	□	□ / ▒	□	□	GLASS: ≤ 400 (200) METAL: ≤ 1000 (540)	GLASS: 350 (2400) METAL: 720 (5000)	±0.5% OF RATE TO ±1% URV	TO HIGHLY VISCOUS FLUIDS
VORTEX	0.5-16 (12-400)	■	▒	■	□	▒	▒	□	□	≤ 750 (400)	≤ 1500 (10,500)	±0.5–1.5% OF RATE	<10,000

■ = DESIGNED FOR THIS APPLICATION;
▒ = NORMALLY APPLICABLE;
□ = NOT DESIGNED FOR THIS APPLICATION

*URV = UPPER RANGE VALUE OF THE FLOW RATE: FORMERLY FULL SCALE FLOW RATE.

Accuracy

The quality (accuracy) of the flow measurement is an important element in flowmeter selection. The selection chart in Figure 23-17 uses accuracy as the primary selection factor. It discriminates between applications where highest accuracy is needed (called "test accuracy") and applications where less accurate measurement (called "industrial") is needed. The terminology would be different today, but of real consequence is the discrimination between percent of rate measurement and percent of full scale accuracy.

In today's competitive world, the only statement of consequence is *percent of rate.*

Table 23-3 provides a quick and easy way to convert the various accuracy statements that are used to the equivalent percent of rate accuracy that is needed.

It is important to restate that the quality of the achieved measurement is a function not only of the capability of the meter selected but also a function of the care with which it is installed.

Installation

Flowmeter installation requirements constitute important selection criteria, particularly where facilities, units, or processes are being modernized. These are shown in Figure 23-18 and reflect the traditional installation requirements such as straight runs of upstream piping. They also include those factors being studied, such as piping-induced stresses and their effect on the operation of the meter. There is an increasing awareness of the interaction of installation and the performance of the meter.

It is only recently that installation considerations have been included in the selection of a flowmeter. The considerations go beyond those included when a "hot tap" type of insertion meter is selected rather than an in-line flowmeter; in this case, the dominating factor is the need for process interruption. The awakening awareness that accuracy and installation are intimately intertwined is forcing consideration of the degree of installation care that must be used to assure the desired level of acceptable performance. As L. McCarthy has said [Ref. 19],

Table 23-2. Comparison of Accuracy Statements [Ref. 4]

Flow Rate (% of Max.)	Equivalent "% of Actual Flow Rate" Accuracies				
	X % Actual Flow Rate	X % Full Scale Flow Rate	X % Max. Differential	X % Full Scale Level, 3/2 Power Flumes, and Weirs	X % Full Scale Level, 5/2 Power, and Weirs
100	X	X	0.50X	1.5X	2.5X
90	X	1.11X	0.62X	1.6X	2.6X
80	X	1.25X	0.78X	1.7X	2.7X
70	X	1.43X	1.02X	1.9X	2.9X
60	X	1.67X	1.39X	2.1X	3.1X
50	X	2.00X	2.00X	2.4X	3.3X
40	X	2.5X	3.1X	2.8X	3.6X
30	X	3.33X	5.6X	3.4X	4.0X
20	X	5.00X	12.5X	4.4X	4.8X
10	X	10.00X	50.0X	7.0X	6.3X

All too often flowmeters are selected on the basis of accuracy or cost without considering installation and maintenance requirements. Flowmeters with complex installation requirements often will not meet their accuracy rating unless those procedures are followed exactly.

In addition, flowmeters that are used in marginal applications (such as using a magnetic flowmeter for a fluid that is only slightly conductive) must undergo continual repair or adjustment. More important than the purchase price and rated accuracy of a flowmeter is a total system cost that factors in the complexity of installation and maintenance requirements.

Summary

Selecting the optimum flowmeter requires an assessment of what is of consummate value in making the measurement. This is the starting point in making the "best selection" or to reduce the number of candidates to a manageable level. It is a progressive prioritizing of distinctive qualities that leads to "the correct choice of flowmeters for the application" and, most importantly, has the greatest potential of providing results that fulfill the expectations for the measurement.

The selection process is not simple. It is complex because there is interaction and interdependence of process-related factors and the flowmeter characteristics. Good measurement results are dependent not only on making the best meter choice but also on installing and using the meter properly. The meter selected will truly satisfy the measurement requirement only when total cost of ownership permeates the entire decision process.

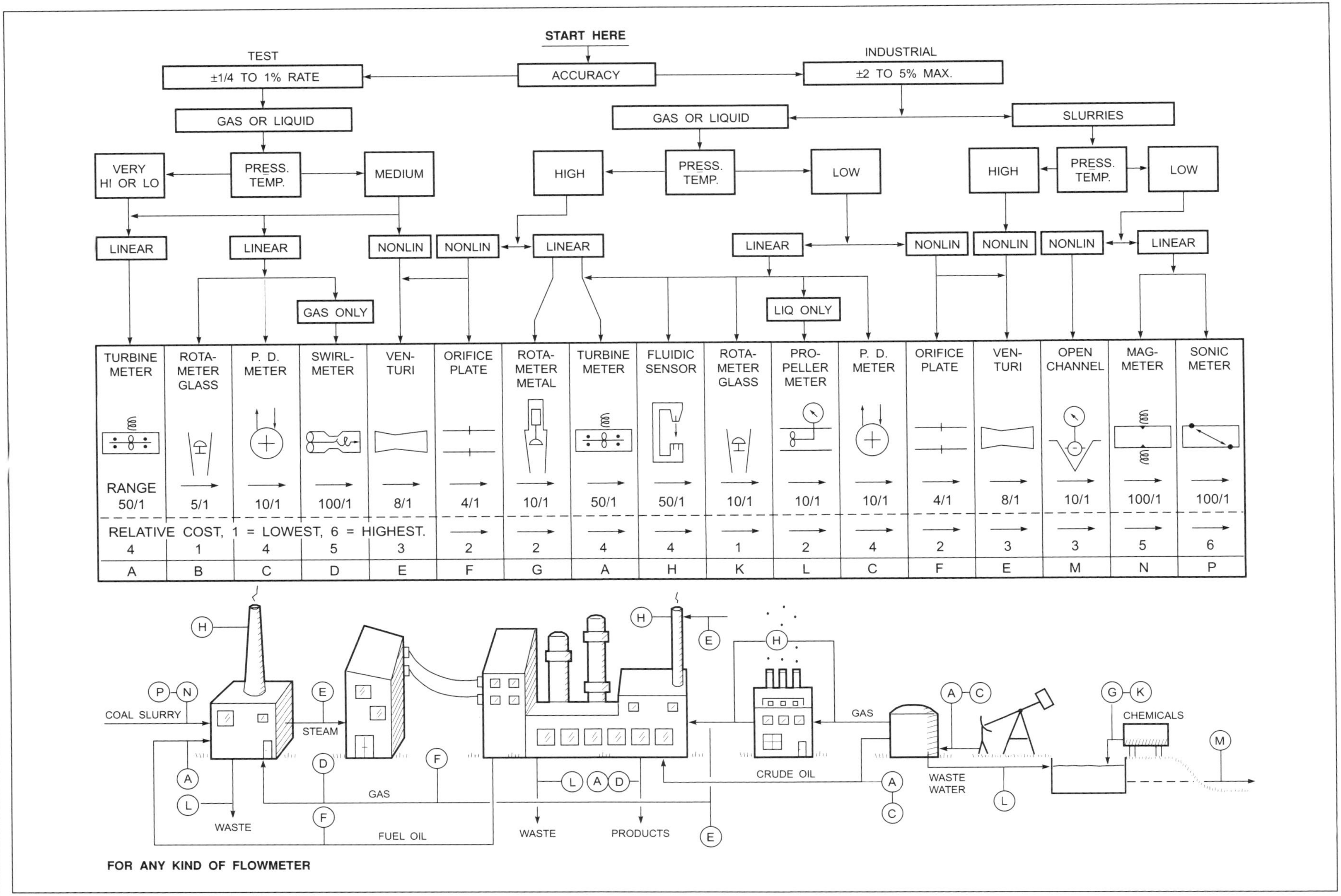

Figure 23-17. Flowmeter Finder [Ref. 22]
(Courtesy of Flowmetrics)

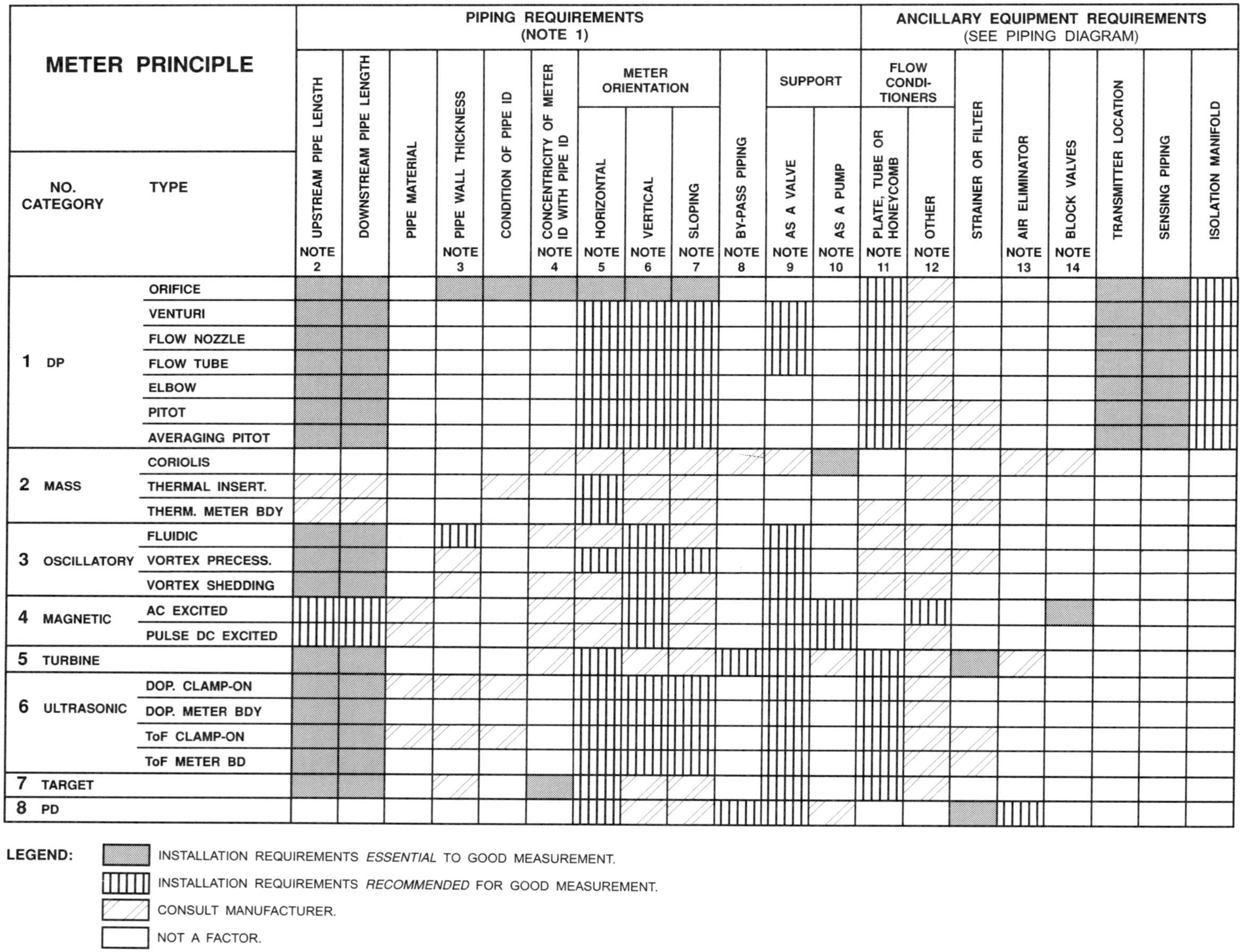

Figure 23-18. Guide to Flowmeter Installation [Ref. 11]
(Courtesy Instruments and Control Systems Magazine)

1 DIFFERENTIAL PRESSURE (DP)

	Orifice	Venturi/Flow Nozzle	Flow Tube	Elbow	Pitot/Averaging Pitot
Minimum pipe length (diameters):	Concentric, eccentric, and segmental types: Before plate, 6; after plate, 2. See Note 15. Conical and quadrant types best installed after a disturbance such as a tee, then: Before plate, 6 to 8; after plate, 3.5.	Before meter, 6; After meter, 2. Note 15.	Before meter, 4; After meter, none.	Usually the same as a concentric orifice.	Before meter, 7; After meter, 3. Note 15.
Power requirements:	DC or pneumatic	DC or pneumatic	DC or pneumatic	DC or pneumatic	DC or pneumatic
Piping connections:	Flanged, threaded, welded	Flanged, threaded, welded, insert	Flanged, insert	Flanged, welded	Insertion, hot tap
Type of measurement:	Whole body. Concentric can be in a bypass configuration.	Whole body	Whole body	Whole body	Sampling
Installation requirements:	Upstream pipe ID, circularity and roughness are important factors. Upstream pipe joints must be smooth and gaskets must not project into the flow.	The average diameter of the pipe at the meter inlet should be within ±1% of the meter diameter and the pipe out of roundness should not exceed 2% of nominal. Gaskets must not project into the flow.	Types with manifolds which average local pressure are recommended if less than desirable flow profiles are anticipated.	Bidirectional when pressure taps are at 45°. Flow must enter end nearest pressure taps when using 22½° taps. Inside diameter of elbow at taps and the radius of the bend must be measured for best results: Pipe ID and elbow ID must agree within 1%. Gaskets must not project into the flow.	Some designs bidirectional. Locate to assure the equivalent of fully developed turbulent flow.
Orientation requirements:	Orifice plate inlet must face upstream. The centerline of concentric orifice plates must be the same centerline as that of the pipe. Plane of orifice plate must be perpendicular to the pipe centerline. Pressure tap orientation depends upon pipe orientation and the type of fluid being metered, see piping diagram.	Pressure tap orientation depends upon pipe orientation and the fluid measured.	None	None	Dependent on pipe orientation and type of fluid. Tube must be within ±3% of flow direction (yaw), ±5% of transverse plane (pitch), and ±3% of pipe centerline perpendicular to flow (roll).
Ancillary equipment requirements:	Flow conditioners are recommended for some applications. Drain, vent, and blow-off valves may be needed. Shut-off valves are usually used at each pressure tap location.	Flow conditioners are recommended for some applications. Shut-off valves are usually used at every pressure tap location.	Flow conditioners are recommended for some applications.	Flow conditioners are recommended for some applications. Low range differential pressure transmitter required.	Flow conditioners are recommended for some applications. Low range differential pressure transmitter required.
Standards or recommended practices:	AGA 3, ANSI/API 2530, ANSI/ASME MFC 3M, ASME Fluid Meters, ISO 5167, Shell Flowmeter Engineering Handbook.	ANSI/API 2530, ANSI/ASME MFC 3M, ASME Fluid Meters, ISO 5167	None	None	ISO 7145 (Pitot)

Figure 23-18. Guide to Flowmeter Installation (continued)

2 MASS

	Coriolis	Thermal
Minimum pipe length (diameters):	Before meter, none; After meter, none	Body type: before meter, 10; After meter, 0. Insertion type: before and after meter, 10
Power requirements:	AC/DC	AC/DC
Piping connections:	Flanged, threaded, sanitary, flaretube	Flanged, threaded, flaretube
Type of measurement:	Whole body	Whole body, by-pass, sampling
Installation requirements:	Pipe supports must be located on adjoining pipe and go to a common reference to minimize transfer of pipe stress to the meter. By-pass piping is recommended by some manufacturers. Meters in series must be separated by 15 pipe diameters. Some designs are bidirectional.	In some insertion types, the sensor head must not touch the opposite pipe wall.
Orientation requirements:	Must be oriented so that gas bubbles or sediment do not collect in the measuring region of the meter. Specific orientations allow self-draining and vary with the meter design.	Some types require installation with the same orientation used during calibration. Insertion types will have orientation requirements that vary with their design. Some designs require insertion plane to be parallel with flow plane to within ±2°.
Ancillary equipment requirements:	Block valves are required for some designs to set zero with no flow.	None
Standards or recommended practices:	California Weights & Standards Bureau, PTB	None

3 OSCILLATORY

	Fluidic	Vortex Shedding	Vortex Precession
Minimum pipe length (diameters):	Follow recommendations for 0.7 Beta orifice meter	Before meter, 10; After meter, 5. Varies with manufacturer. Note 15	Before meter 10; After meter, 5
Power requirements:	AC/DC, 2-wire DC available	AC/DC, 2-wire DC, battery power	AC/DC
Piping connections:	Flanged, flangeless (wafer style) Note 4	Flanged, threaded, sanitary, weld ends, flangeless (wafer style) Note 4	Flanged
Type of measurement:	Whole body; by-pass larger than 4 inches	Whole body; insertion type is sampling.	Whole body
Installation requirements:	Pipe wall thickness should be specified. Gaskets must not project into the flow stream.	Gaskets should not protrude into flow stream; some designs require specific pipe inside diameter at meter entrance.	Avoid pipe size increases upstream of the meter.
Orientation requirements:	Horizontal plane is preferred.	Some designs require specific orientation based on the type of application or operating temperature. Check manufacturer.	None
Ancillary equipment requirements:	Flow conditioners are recommended for persistent swirl profiles or jet flow profiles.	Flow conditioners are recommended for persistent swirl profiles or jet flow profiles.	None
Standards or recommended practices:	None	ASME/ANSI MFC 6M	None

Figure 23-18. Guide to Flowmeter Installation (continued)

4 MAGNETIC

Minimum pipe length (diameters): Before meter, 5; After meter, 2.

Power requirements: AC/DC

Piping connections: Flanged, Victaulic, Dresser, sanitary, flangeless (wafer-style) Note 4.

Type of measurement: Whole body; insertion type is sampling.

Installation requirements: Some designs are bidirectional. Grounding is a function of the adjoining pipe material for most designs. Gaskets must not project into the flow. There are specific pipe support requirements for some designs. Note 10. Do not inject additives immediately upstream of the meter.

Orientation requirements: Electrodes must be in horizontal plane. Larger sizes have an integral base support for installation on a slab at ground level.

Ancillary equipment requirements: Flow conditioners are recommended for severely distorted profiles and severe swirl. Block valves are required for ac excitation type meters to set transmitter zero.

Standards or recommended practices: ISO 6817

5 TURBINE

Minimum pipe length (diameters): Before and after meter, 5; Note 15.

Power requirements: None for most versions. AC/DC or battery used for RF signal systems and accessories.

Piping connections: Flanged, threaded, sanitary, flaretube, flangeless (wafer style) Note 4, and insertion

Type of measurement: Whole body; insertion type is sampling.

Installation requirements: Locate meter as far as practical from flow disturbance. Strainers and flow conditioners must be specifically located; varies with standards, recommended practices, and manufacturers. By-pass piping frequently recommended; it is essential for cryogenic applications. Clean piping before installing meters. Abnormal pipe stresses at meter connections not allowed with some designs. Some deigns are bidirectional.

Orientation requirements: Some designs must be oriented as calibrated.

Ancillary equipment requirements: Flow conditioners and strainers or filters are usually required. A separator for condensate removal is recommended for gas flows.

Standards or recommended practices: AGA7, API 2534, API Manual for Petroleum Measurement Standards, Chapter 5, Section 3, ISO 2715, ASME Fluid Meters

6 ULTRASONIC

	Doppler (Dop.)	Time of Flight (ToF)
Minimum pipe length (diameters):	Follow recommendations for a 0.7 Beta ratio orifice meter installation. Swirling and jet flows must be avoided.	Before meter, 10; After meter, 2.
Power requirements:	AC/DC	AC/DC
Piping connections:	Usually non-invasive (clamp-on); meter body types are also available.	Flanged; non-invasive (clamp-on) also available
Type of measurement:	Sampling	Sampling
Installation requirements:	Allow adequate upstream pipe length following a disturbance to assure a fully developed turbulent profile. Avoid severely vibrating pipe sections. Deposits on pipe ID may affect the meter's ability to make measurements. Clamp-on: Pipe material and or type of pipe lining may affect the measurement. Particle (bubble) velocity being sensed, so particles must be of uniform size, uniformly distributed, be of uniform materials, and have a velocity the same as the liquid. Meter locations must be selected so there is adequate velocity to prevent particles from rising or settling which introduces measurement errors.	Allow adequate upstream pipe length to assure a fully developed turbulent profile. Clamp-on types are dependent on pipe material, pipe wall thickness, and condition of pipe interior.
Orientation requirements:	Transducers must be in the horizontal plane. Locate transducers so they look away from flow disturbances.	Transducers must be in the horizontal plane.
Ancillary equipment requirements:	None	Flow conditioners recommended.
Standards or recommended practices:	None	ANSI/ASME MFC-YY

Figure 23-18. Guide to Flowmeter Installation (continued)

7 TARGET

Minimum pipe length (diameters): Before meter, 6; After meter, 3.5. Note 15.

Power requirements: AC/DC or pneumatic

Piping connections: Flanged, threaded, flaretube, flangeless (wafer style). Note 4, and insertion

Type of measurement: Whole body; insertion type is sampling.

Installation requirements: Some designs require specific pipe IDs at meter entrance. Gaskets should not protrude into flow stream.

Ancillary equipment requirements: Flow conditioners recommended for some flow profiles to reduce upstream piping requirements. Strainers usually recommended.

8 POSITIVE DISPLACEMENT (PD)

Minimum pipe length (diameters): Before or after meter, none.

Power requirements: Basic versions do not require power. AC/DC or battery used for accessories.

Piping connections: Flanged, threaded, sanitary

Type of measurement: Whole body (True volumetric)

Installation requirements: Larger and heavier designs require base mounting. Housing should not be subjected to undue pipe stress. By-pass piping may be required, especially where blocked flow due to meter failure is not acceptable. Drip traps for wet gas applications are recommended. Meters should be kept from draining to minimize deposits forming on meter internals. Piping interior should be flushed before meter is installed.

Orientation requirements: Meters should not be at low point in the piping where particulate matter would accumulate. Meter should be level.

Ancillary equipment requirements: Strainers and air eliminators are generally recommended. Flow restricting orifices or flow limting valves used where transient high flow rates are experienced. Air chamber or shock arrestor may also be recommended.

Standards or recommended practices: API—Chapter 5, Section 2, PD Meters; RP2535, Recommended Practice for Viscuous Hydrocarbons; ASME Fluid Meters; ANSI: B109.1 and B109.2 for diaphragm type PD meters; AGA 6; API 1101

Notes:

1. When measuring liquids and slurries, pipe must be **full**. Flow through meters must be in the direction marked on the meter body.

2. Includes straight length of pipe before a flow conditioner as well as that at the meter entrance; see piping diagram.

3. Most meters prefer that the inside diameter of upstream pipe be slightly greater than the meter inside diameter. For some meters, such as orifice, the inside diameter should be precisely known.

4. Concentricity is important to many meters including orifice types. It is essential to all wafer-style meters regardless of the operating principle. Wafer-style meters are either furnished with or recommend centering devices which must be used. Gaskets must not protrude into the flow stream.

5. Horizontal is the most common orientation, but special care must be taken when measuring liquids which have entrained gas or particles, and gases in which liquids are present. Some meters calibrated in the horizontal will require installation in that orientation.

6. Preferred flow direction is up to assure a full pipe; this is mandatory in gravity feed systems.

7. Sloping lines are generally handled as horizontal lines.

8. The criteria are whether flow will be able to pass through the meter if there is a meter failure and whether the measurement is critical to the process so that access to the meter, at any time, is important.

9. Many smaller size meters are essentially another piece of pipe and can accept piping stresses. They are considered in the piping system as a concentrated mass.

10. Some meters require that there are low (no) piping stresses transferred to the meter body (housing). Performance may be impaired if this criteria is not followed.

11. Plate, tube, honeycomb, and similar conditioners (flow straighteners) are beneficial for swirling flows. They may be detrimental if used when distorted profiles are present.

12. If profiles are severely distorted, or swirling flow is persistent, flow conditioners such as Mitsubishi, Sprenkle, Vortab, Zanker, and so on should be used. Pressure drops should be checked.

13. Entrained gas will affect the accuracy of volumetric liquid measurement and calibration; signal output is related to total volume, not just liquid phase. Air eliminators are an important consideration in custody transfer and billing applications. Some liquid mass meters have specific limits on percent gas they can tolerate (void fraction).

14. Block valves must be leak-tight so that a true no-flow condition is established. Similar requirements exist for by-pass piping arrangements.

15. The required length of straight upstream pipe increases with the need for accuracy. It also increases with an increasing Beta-ratio, where that is a factor. It also varies with the type of upstream disturbance and whether or not the correct flow conditioner is used.

Figure 23-18. Guide to Flowmeter Installation (continued)

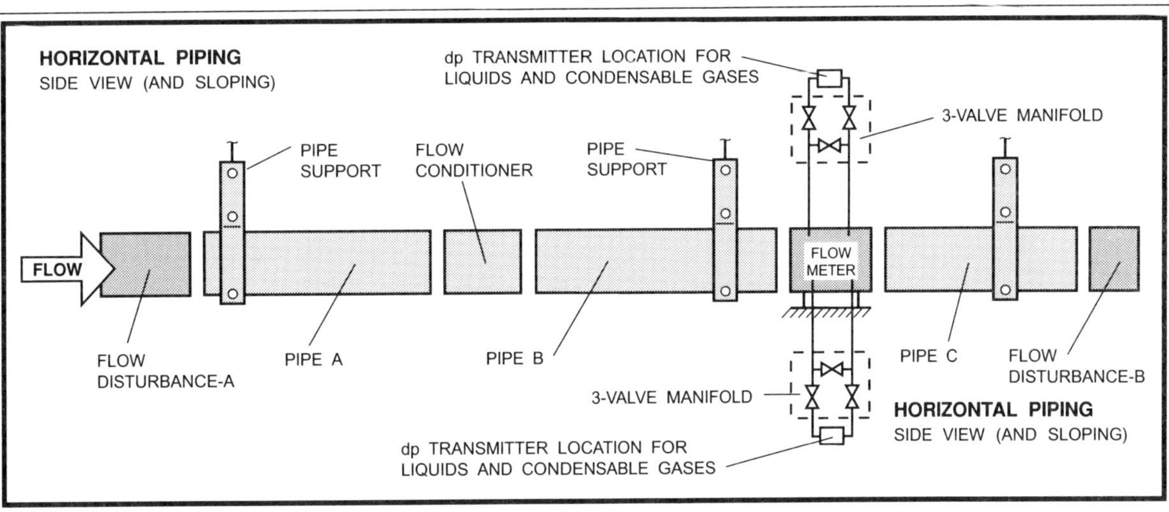

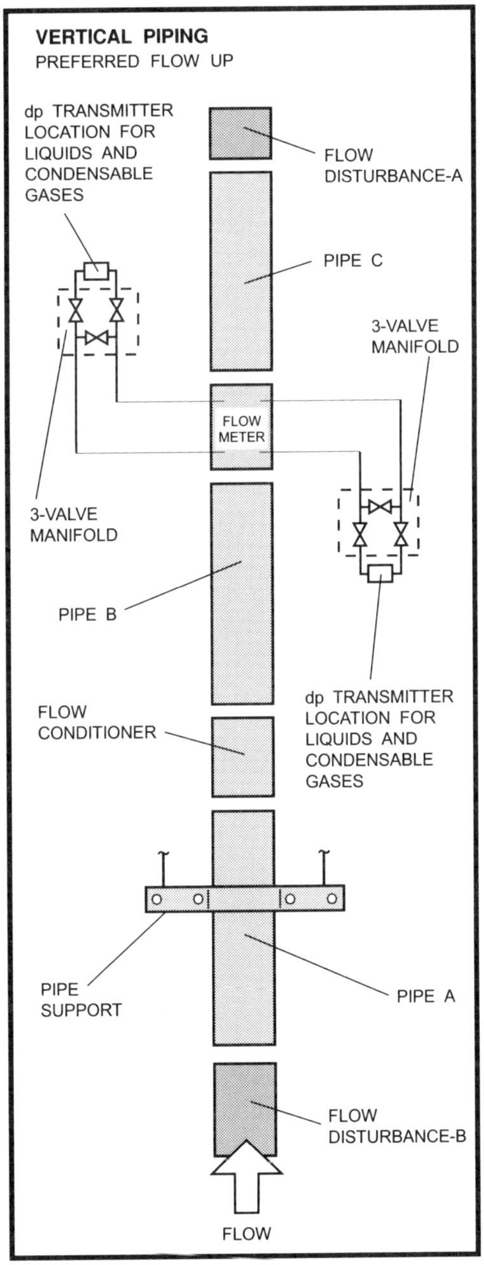

PIPING NOTES

Flow conditioner: Includes conventional flow straighteners such as tube-bundles, plate or honeycomb types for swirl. For severely distorted flow profiles, jets, or persistent swirl patterns, proprietary flow conditioner designs such as Sprenke, Zanker, Mitsubishi, Vortab, and so on are recommended.

Flow Disturbance:

(A) Includes pipe fittings such as elbows, tees, and so on, as well as fabricated pipe bends or other fabricated pipe configurations. It also includes equipment such as pumps, strainers, air eliminators, and so on. Shut-off valves must be fully opened or closed. Throttling valves must be located downstream from the meter (see Flow Disturbance B).

(B) Includes elbows, tees, and other pipe fittings as well as fabricated pipe bends or other fabricated pipe configurations. Also included are throttling or shut-off valves.

Flowmeter: Includes volumetric, velocity, and mass type flowmeters. Variable differential pressure types include all orifice variations, venturis, flow nozzles, flow tubes, pitots, averaging pitots, and elbows, all of which require associated sensing piping (lead lines), a 3-valve manifold, and a dp transmitter. The transmitter location relative to the dp producer, such as plate, tube, and so on, is determined by the type of fluid being measured (liquid, gas, or steam). Orientation is important for some types of meters. Some types of turbine meters, for example, are intended for installation in horizontal piping; installation in sloping or vertical piping must be reviewed with the manufacturer.

Pipe:

(A) Included in the upstream piping requirements when a flow straightener or flow conditioner is used. Pipe size should be the same as the meter size. Length is determined by Standards, Recommended Practices, or the Manufacturer's Specifications.

(B) Pipe size should be the same as the meter size. Pipe length is established by Standards, Recommended Practices, or Manufacturer's Specifications. The inside diameter of the pipe, and its condition may be important; concentricity of the pipe with the meter's inside diameter may also be important. Gaskets must not protrude into the flow stream.

(C) Pipe size should be the same as the meter size. Pipe length is defined by Standards, Recommended Practices, or Manufacturer's Specifications.

Pipe Support: Type hanger (from above), or support (from below) is influenced by the type of flowmeter being used. Some flowmeter types can accept piping induced line stresses, and are handled as just another concentrated mass from a pipe/hanger point of view. Some types of flowmeters must be isolated from external stresses, including piping induced stresses. These meters require specific care in the type and location of hangers/supports which are used. Although large and heavy meters include integral base supports, associated hanger/support requirements must be considered to minimize (eliminate) transfer of stresses to the flowmeter housing.

Figure 23-18. Guide to Flowmeter Installation (continued)

References

1. American Society of Mechanical Engineers (ASME): *Fluid Meters*, 6th Edition, 1971.
2. Bailey, S. J., "Trade-offs Complicate Decisions in Selecting Flowmeters," *Control Engineering*, April 1980.
3. Baker-Counsell, J., "Flowmeter Selection: Expert Help Is on Its Way," *Process Engineering*, pp 71-73, October, 1985.
4. Buzzard, W. S., "Predicting Flow Rate System Accuracy," *ISA Symposium*, Newark, Delaware, June 1979.
5. Buzzard, W. S., "Flowmetering Now, and Future," ISA, South Jersey Section, September 1981.
6. DeCarlo, J. P., *Fundamentals of Flow Measurement*, ISA, 1984.
7. Ginesi, D., and Grebe, G., "Flowmeter Selection, A Comparison of Performance Features vs. Economic Costs," Paper 85-0835, Instrument Society of America International Conference and Exhibit, 1985.
8. Grebe, G., "Flowmeter Selection: Broadening—But Still a Challenge," *InTech*, pp 65-69, October 1984.
9. Grebe, G., "Application Considerations for Selecting Flowmeters," Paper 84-765, ISA/84 International Conference and Exhibit.
10. Hall, J. T., "Guide to Flow Measurement," *Instruments and Control Systems*, p. 58, February 1990.
11. Hall, J. T., "Guide to Flowmeter Installation," *Instruments and Control Systems*, p. 104, Sept. 1989.
12. Higham, E. H., Foxboro Great Britain Ltd., Internal Communication, April 1985.
13. Kopp, J. G., "Primary Flow Element Applications—Which One and Why," ISA/82 International Conference and Exhibit, Oct. 1982.
14. Kopp, J. G., "How to Get the Accuracy You Expect from Flowmeters," *Instruments and Control Systems*, pp 65-68, Sept. 1989.
15. Kopp, J. G., Memphis Section, ISA, February 1971 address.
16. Kopp, J. G., ISA Flow Measurement Course, University of Delaware, April 1979.
17. Liptak, Bela G., *Instrument Engineers' Handbook—Process Measurement*, Chilton Book Company, Revised Edition, 1982.
18. Lomas, D. J., "Selecting the Right Flowmeter," *Instrumentation Technology*; Part I, pp 55-62, May 1977, and Part II, pp 71-77, June 1977.
19. McCarthy, L., "Flowmeter Installation and Maintenance," *Chemical Processing*, Sept. 1989.
20. Meinhold, T. F., "Liquid Flowmeters—An Overview of Types and Capabilities plus Guidelines on Selection, Installation, and Maintenance," *Plant Engineering*, pp 46-60, November 1984.
21. Miller, R. W., *Flow Measurement Engineering Handbook*, Second Edition, McGraw-Hill Publishing Company, 1989.

22. Nelson, P., and Woodring, E., "Flowmetrics Flow Finder," c. 1973.

23. Owen, R. E., "Selecting Flowmeters for Viscous Fluids," *Plant Engineering*, pp 137-140, Aug. 1980.

24. *Shell Flowmeter Engineering Handbook*, Royal Dutch/Shell Group, Waltman Publishing Company, Delft, The Netherlands, 1968.

25. Slomiana, M., "Selecting Pressure and Velocity Head Primary Elements for Flow Measurement," *InTech*, pp 40-49, Nov. 1979.

26. Spitzer, D. W., *Industrial Flow Measurement*, ISA, 1990.

About the Author

John G. Kopp is a graduate of Drexel University with a Bachelor of Science degree in Mechanical Engineering and is a licensed Professional Engineer in Pennsylvania.

He began his career as an application engineer for flow products and associated equipment including magnetic flowmeters, oscillatory-type flowmeters, and turbine flowmeters and has been a product manager for various flow products and product marketing manager for magnetic flowmeters and vortex shedding flowmeters.

A contributor to *Instrument Engineers' Handbook* (Chilton), Mr. Kopp is now a consultant for product planning, recommending content, enhancements, or extensions for flow products including magnetic, target, turbine, vortex shedding, and Coriolis mass flowmeters.

24

Flow Metrology: Standards, Calibrations, and Traceabilities

As increased concerns for improved fluid flow rate measurement drive the fluid metering community—meter manufacturers and users alike—to search for increased flow measurement accuracy and precision, better verification and documentation are needed to substantiate fluid meter performance. These concerns affect both domestic and international market places; they permeate instrumentation and control technologies—aerospace, chemical processes, automotive, bioengineering, etc. They involve public health and safety, and they impact our national defense. These concerns are based upon the rising value of fluid resources and products and the importance of critical material accountability. These values directly impact the increased accuracy needs of fluid buyers and sellers in custody transfers. These concerns impact the designers and operators of chemical process systems where increased control and productivity optimization depend critically upon measurement precision. Public health and safety depend upon advancing the quality of numerous pollutant measurements—both liquid and gaseous. The performance testing of engines—both automotive and aircraft—are critically based upon accurate fuel measurements—both liquid and oxidizer streams. For all these reasons, flow metrology, its standards, its calibrations, and its traceabilities need to be understood, well established, and properly used to document and validate fluid quantity and flow rate measurements.

Fluid flow rate measurements are established differently from counterparts in length and mass measurement systems because these have the benefits of "identity" standards. For rate measurement systems, the metrology is based upon "derived standards." These use facilities and transfer standards that are designed, built, characterized, and used to constitute basic measurement capabilities and quantify performance (accuracy and precision). Because "identity standards" do not exist for flow measurements, facsimiles or equivalents must be concocted and used to quantify the systematic errors that might exist between or among measurement facilities for fluid flow rate or air speed, etc. This is the purpose of this chapter: to describe the ways that flow measurement facilities can be characterized and how traceability of these facilities can be established. Examples of the performance assessment for flow rate measurement facilities are given using typical values prevailing at the National Institute of Standards and Technology (NIST, formerly the National Bureau of Standards).

Standards

Fluid flow rate standards could be significantly simplified if the fundamental bases of these measurements were as simple as those for mass, length, and so on.

These systems of measurement are based upon discrete standards[1] or artifacts. For examples, the platinum kilogram known as "K-20" is the ultimate artifact to provide the fundamental basis for mass measurement in the U.S., and the platinum meter bar (or its modern-day wavelength equivalent) is the ultimate artifact to provide the fundamental basis for length measurement. These artifacts can be considered "identity" standards.

Identity Standards

These mass and length artifacts can be considered "identity" standards because under the appropriate conditions of use they define the basic quantity in their respective measurement systems. However, for flow rate measurements of fluids, i.e., liquids or gases, there does not exist an identity standard such as a gallon per minute, a liter per second, or a kilogram per hour. To supply the fundamental basis upon which to establish a flow measurement system, a "derived" standard is needed.

Derived Standards for Flow

For fluid flow rate measurements, as needed to form the basis of a national reference system, calibration facilities spanning a range of fluid and flow conditions are maintained by NIST for use by industry and others, [Refs. 1-6]. These facilities consist usually of:

(1) a source of flow, generally a compressor or a pump, and a supply of the fluid with appropriate auxiliary equipment such as a regulated, pressurized tank of gas or a reservoir of liquid;

(2) a test section into which the meter and its adjacent piping can be installed so that the flow and fluid conditions into it duplicate those expected where the meter will actually be used; and

(3) a flow determination system having a specified level of performance and appropriate proof of this to specify and assure the desired metering performance of the devices in question. Calibration systems are generally categorized according to the type of flow determination scheme used.

Flow Determination Systems

The heart of the fluid flowmeter calibration facility is the flow determination system, [Refs. 1-6]. This generally uses a timed collection of the fluid that flows through the meter being calibrated. The amount of fluid collected is determined by gravimetric or volumetric techniques. This collected fluid is converted to flow rate using the collection time; the volumetric flow rate through the meter can be determined via conservation of mass principles using the pertinent thermodynamic properties measured at the meter. This system can be made to perform at a high level of performance to determine the bulk flow rate of the fluid.

---

1 The term "standard" has many meanings. It is used to refer to "paper" standards, which are documents; it is also used to refer to reference facilities and equipment; it is also used to refer to the specific materials needed to transfer measurement quality from or between facilities. These specific materials are referred to in what follows as "artifacts."

Levels of Performance

Measurement systems can be characterized through their accuracy and precision. These terms are briefly defined as follows:

Accuracy—The degree, generally expressed as a percent, to which a measured result approximates the true value of the quantity being measured.

Precision—The degree, generally expressed as a percent, to which successive determinations of the same quantity duplicate each other. Precision is sometimes further subdivided into:
"reproducibility," which involves "how closely will successive determinations duplicate each other," or
"repeatability," which involves "how closely can successive determinations be made to duplicate each other" (i.e., when conditions are the same and there is only a short time between measurements).

These characteristics apply to measurements made by flowmeters and to measurements made using calibration facilities, [Refs. 7-11].

Facility Performance

For fluid flow calibration facilities, the precision can be theoretically evaluated from the appropriate error budget and from the precision of the component measurements that constitute the system. This evaluation technique is often referred to as the propagation of error approach, [Refs. 9-11]. It should be stressed that this approach can lead to serious underestimates of the actual conditions. This is because the physical model for the actual process may not conform to that used for the propagation of errors. Furthermore, difficulty is encountered when facility accuracy is to be quantified, because the true value of the fluid flow rate is not easily obtained. To estimate possible systematic offsets from true value, approximations, generally very conservative, are frequently used. Alternatively, and more preferably, a realistic and highly defensible traceability scheme either should be used or can be generated and then appropriately used to quantify the systematic offset of a calibration facility. This quantification should be done on a continuing basis to assure traceability to national standards.

Traceability

The concept of measurement traceability is based upon the need to check measurement results. As such, traceability has come to mean many things to many persons. There are a number of definitions for traceability [Ref. 12]. For example, a prevalently used definition for traceability "is to calibrate into a hierarchical scheme of measurements that leads, ultimately, to the national references for the respective measurements." For flow rate measurement systems that are based upon timed gravimetric or volumetric collection schemes, this definition could be implemented by checking, individually, the weighing or volumetric technique in addition to checking the timing device. However, limitations to this type of traceability for fluid flow rate measurement can be that errors can occur in the other components that contribute to the end result. Examples would be the associated temperature, pressure, or humidity measurements that may be influential. Equivalently, the mechanism that starts and stops the timing device can be in error so that even if the timing device itself is accurate and traceable, the timing can be wrong due to faulty activation. Many other errors can affect flow measurement processes.

Conventional Calibration Procedures

With conventional calibration procedures, a testing laboratory or a meter manufacturer might own and routinely use a master meter technique to assess the flow rate measurement performance of the laboratory with a report on its performance in an NIST facility. The meter would be placed into the respective facility in the laboratory and then calibrated. The relative performance of these calibrations would hopefully compare very favorably and, thereby, document the closeness of agreement between the laboratory's facility and NIST. This procedure, while widely used at the present time, can leave a considerable number of factors affecting measurement completely unassessed.

Traceability might also be established for a flowmeter calibration laboratory in the following manner. If calibrated weights (for example, from a state office of weights and measures) were used to check the scale system and if a timing standard were used to check the lab's timing system, then traceability could be asserted for the weigh-time system. However, the overall ability of the lab to calibrate a flowmeter can be quite incomplete. For such reasons, it is widely believed that more complete assessment of the measurement capabilities of a flow measurement laboratory is preferred. This type of traceability can be established and maintained via flow measurement assurance programs, i.e., flow MAPs.

Flow Measurement Assurance Programs (MAPs)

In the case of flow MAPs, a procedure is used that is different from conventional calibrations [Refs. 13 and 14]. This involves NIST (or an initiating laboratory) sending a very reliable and well characterized artifact package (i.e., tandem meter arrangements consisting of two meters in series) to the laboratory in question with the request for a calibration of the arrangement according to tightly specified and prearranged conditions [Refs. 15 and 16]. The results, which contain the effects of all the lab's routine calibration procedures, its facilities, its operating conditions, its personnel, and its techniques for calculating final results from raw data, are then sent to NIST. These can be objectively (and informedly) compared to similar results from a number of other labs that have performed the same tests in a "round robin" set of these calibrations. In these comparisons, NIST results are also incorporated as one of the participants. The results show, quantitatively, the agreement (or disagreement) among the participants' results. Algorithms have been developed to handle these results [Refs. 17 and 18]. Figure 24-1 shows a comparison of conventional calibration procedures and those that can occur with MAPs. The comparison shows that the crucial advantages of the MAP program are that: (1) all aspects of the laboratory's measurement processes are checked, and (2) there is a "feedback" and, if necessary, a "follow-up" activity that can make improvements, etc. These follow-up activities can be directed either at the lab's procedures or at its calibration procedures and facilities, depending upon the results obtained from previous rounds of testing.

Analysis and Results

Conservation of Mass Equation

Flow calibrations are usually performed using a system that includes a source of flow, the instrument being calibrated, connecting conduits, and a scheme for determining the fluid flow rate. When the calibration is based upon the bulk flow rate, i.e., either volumetric or gravimetric, the scheme for determining the fluid flow is based on conservation of mass considerations, [Refs. 1-4]. When the calibration is based upon the local fluid velocity, as in the case for air speed calibrations, the scheme for determining fluid velocity is generally a transfer standard such as an accurate Pitot static tube, an anemometer, or a laser Doppler velocimeter (LDV). For each of these schemes, the ideal error budget should be known and maintained so that overall performance levels are as quoted [Refs. 1-3].

Figure 24-2 is a sketch of a calibration arrangement with labelled components. The meter and its downstream piping are considered as a part of the meter and volume a. Depending on the type of calibrator, control surface 4 of volume c may be a moving piston, the stationary end of a tank, etc. Conservation of mass principles applied to an arbitrary, stationary control volume, V, which is surrounded by the control surface S, can be written:

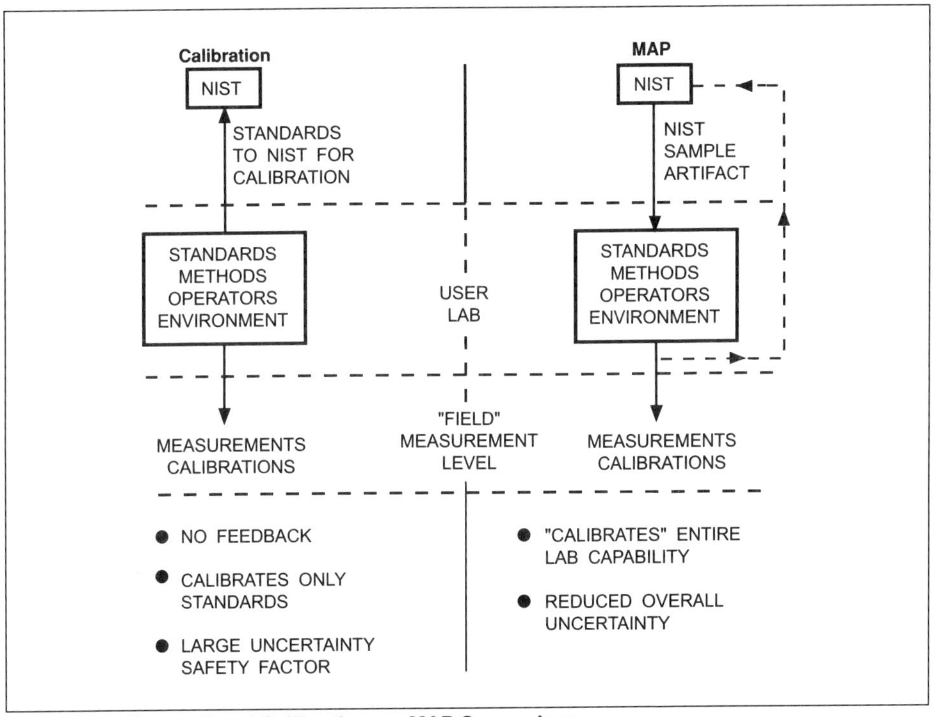

Figure 24-1. Conventional Calibration vs. MAP Comparison

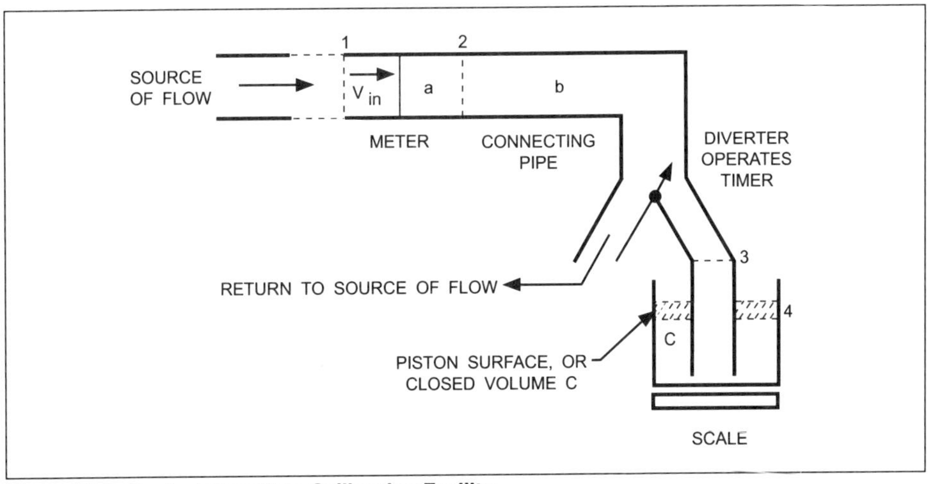

Figure 24-2. Typical Flow Rate Calibration Facility

$$0 = \frac{\partial}{\partial t}\int_V \rho dV - \int_S \rho \bar{n} \cdot \bar{v} dS \qquad (24\text{-}1)$$

where ρ is the fluid density, $\partial/\partial t$ is the partial derivative with time, V is the control volume, which is composed of all the subvolumes in Figure 24-2. The quantity $\bar{v}$ is the vector velocity of the fluid and $\bar{n}dS$ is the vectorial control surface element of area with direction taken inward and normal to the surface. Applications of Equation (24-1) to the control volume and surfaces shown in Figure 24-2 gives:

$$\dot{M} = \int_{S_1} \rho_1 v_{1n} dS_1 = \frac{\partial M_c}{\partial t} + \int_{S_4} \rho_4 v_{4n} + \int_{V_a} \frac{\partial}{\partial t} \rho_a \, dV_a + \int_{V_b} \frac{\partial}{\partial t} \rho_b \, dV_b \qquad (24\text{-}2)$$

where $\dot{M}$ is the mass flow rate through the 1 surface and $\frac{\partial M_c}{\partial t}$ is the rate of fluid mass collected in volume c. Subscripts n refer to vector components normal to the numbered surfaces; integer subscripts refer to surfaces; lettered subscripts refer to volumes.

Performance levels for bulk flow rate calibration facilities can be assessed using the above principles. These principles have been used to produce the quantifications of the uncertainties of the NIST flow facilities, [Refs. 1-4].

Fluid Meter Calibration Facilities

To attain the improved flow measurement accuracy needed to calibrate fluid meters, a range of techniques are used. These generally consist of systems that are based upon timed collections of the fluid passing through the meter being calibrated. The collected fluid quantities are assessed, using either volumetric or gravimetric methods. These calibration facilities are arranged and used so that the uncertainty of the unit can be determined and quantified as described below.

Because of the importance of the precision and accuracy levels of fluid quantity and flow rate measurement, calibration facilities are needed frequently to validate results. For these reasons, a range of calibrators and provers are manufactured by several makers of fluid meters. These are either laboratory based or are mobile and used to perform calibration tests in situ on installed flowmeters.

An alternative approach to using mobile calibrators or provers to calibrate meters in situ is to use other flowmeters as transfer standards. These should be designed to match, or preferably to exceed, the levels of performance of the installed units. By properly controlling the test conditions and the associated uncertainties, the flow measurement results from the installed units can attain the desired validity or credibility or both.

Uncertainty Assessment

The performance of a calibration facility can be assessed in several ways. Before the facility is designed and built, performance can be assessed on the basis of the operational equation for the facility and the specifications of the component measurements. For example, a static gravimetric facility for measuring liquid volumetric flow rate can operate with the equation:

$$\dot{V} = \frac{M_N}{\rho t} \qquad (24\text{-}3)$$

where, in compatible units, $\dot{V}$ is the volumetric flow rate, M_N is the net mass of liquid collected (i.e., the difference between the gross mass collected, M_G, and the tare mass of the collection tank, M_T), ρ is the appropriate liquid density, and t is the collection time. Based upon this model, the uncertainty in the determination of $\dot{V}$ can be specified in terms of the uncertainties in the values of M_N, ρ, and t. Assessment of the magnitudes of these results can be estimated by several techniques for combining component uncertainties. Two such examples are:

$$\frac{\Delta \dot{V}}{\dot{V}} \le \left[\left(\frac{\Delta M_N}{M_N}\right)^2 + \left(\frac{\Delta \rho}{\rho}\right)^2 + \left(\frac{\Delta t}{t}\right)^2\right]^{1/2} \qquad (24\text{-}4)$$

and

$$\frac{\Delta \dot{V}}{\dot{V}} \leq \left|\frac{\Delta M_N}{M_N}\right| + \left|\frac{\Delta \rho}{\rho}\right| + \left|\frac{\Delta t}{t}\right| \tag{24-5}$$

By inserting values for the precisions (percents of rate, *not* percents of full scale) for the respective components in the right-hand sides of Equations (24-4) and (24-5) one can obtain an initial estimate for the precision that can be expected in the determination of the volumetric flow rate, $\dot{V}$. These determinations are based upon a number of important assumptions such as: (a) Equation (24-3) is the proper model of the process, (b) an adequate data base is used to form the component uncertainties in Equations (24-4) and (24-5), and (c) no other factors are involved. To varying degrees, a number of other factors can be involved, and, for these reasons, further assessments are needed.

After the facility is built, improved assessment of performance is possible, and this should be done in several stages. In the first stage, the components should be checked individually against the respective standards for each respective measurement. These can be considered "static" checks. They could consist of checking weigh systems with mass standards, checking timing and density measuring systems against appropriate standards, and so on.

For liquid flow rate measurement using static gravimetric techniques at NIST-Gaithersburg, the uncertainties (3 standard deviations) for the component measurements have nominal values, as follows:

Item	Uncertainty (%)
Net mass determination	0.02
Liquid density	0.02
Collection time	0.01

These can be combined using Equations (24-4) or (24-5) to produce liquid flow rate precision levels of ± 0.03% or ± 0.05%, respectively.

For gas flow rate measurement using piston-volumetric displacement techniques at NIST-Gaithersburg, the uncertainties (3 standard deviations) for the component measurements have nominal values as follows:

Item	Uncertainty (%)
Net mass determination	
1. Volume	0.04
2. Density	
a. Pressure effects	0.13
b. Temperature effects	0.05
Collection time	
1. Device	0.01
2. Switching	0.02

These can be combined using Equations (24-4) or (24-5) to produce gas flow rate precision levels of ± 0.15% or ± 0.25%, respectively. It should be noted that these performance levels are those obtained after the respective instruments have been calibrated.

When these static checks of instrument performance give satisfactory results, one should proceed to the next phase of checking. The facility should be operated over its pertinent parameter ranges and data should be obtained for all the measurable quantities under realistic ("dynamic") conditions. This data quantifies the precision of the volumetric flow rate determined "dynamically." These values quantify the left-hand side of Equations (24-4) or (24-5). Additionally, these data should be compared to that obtained statically for the right-hand sides of Equations (24-4) and (24-5). Satisfactory agreement should be achieved for these precision assessments before the third stage of assessment is started.

The third stage of assessment should be directed at the systematic errors that may be present in the facility's measurement processes. This is properly done by conducting appropriate interlaboratory or "round robin" tests, thereby establishing its traceability. In this way the performance of the laboratory is quantified using its normal, routine materials, procedures, and personnel, and in its environmental conditions. Such quantifications are based upon the test results produced using transfer standards or "artifacts." These artifacts are comprised of flowmeters; the type of flowmeter, its size, the fluid used, and the other test conditions should be selected according to the routine types of flow testing that occur in the laboratory. These artifact meters are tested, i.e., calibrated, according to strictly controlled algorithms as described above. These algorithms are arranged to precisely stipulate all the details of the artifact testing procedures, complete with "go" and "no-go" check points to ensure the validity of the meters and the techniques for analyzing and presenting the data. Done properly and on a continuing basis, the third stage of quantifying flow measurement facility performance provides and maintains realistic traceability for the facility and, in turn, for the measurement products (i.e., calibration data produced by the facility). When this data is properly processed and analyzed to demonstrate that the facility's performance is satisfactory, considerable assurance can be placed in this facility. For this reason, these round robin activities have been named flow MAPs (measurement assurance programs). When these programs include or closely connect to the national reference systems (NIST), strong traceability links are produced.

NIST has initiated a number of round robin flowmeter testing programs as described below. Based upon these tests, NIST uses an estimated systematic uncertainty of ± 0.1% for both its liquid and gas measurement facilities. If this estimated systematic uncertainty is root-sum-squared with the precisions described previously, the total accuracy quotes for liquids and gases would be ± 0.10% and ± 0.18%, respectively. However, because the systematic error is estimated, it is generally preferred to use the more conservative addition method to produce the accuracy quote. This produces the total accuracy quotes for liquid flow measurement of ± 0.13% and for gas flow measurement of ± 0.25%.

Flow Measurement Traceability

To establish the realistic traceability described above, a test program must be devised so that:

(1) high confidence can be placed in the artifact package—the meters assembled and the specifics of the procedures, checkpoints, responses to anticipated anomalies, etc.;

(2) the data base produced is adequate to the task of clearly evaluating the significant components of the systems that participate; and

(3) the algorithm for processing the data producing the results is an unbiased and clear procedure that is adequate to this task.

Artifact confidence is established via calibration testing over an extended period of time for the kind of conditions that will be used in the round robin. This testing should occur in the initiating laboratory and it should establish a credible background data base for the units being tested. Specifically, high confidence can be attained both in meter performance and in facility operation by calibrating two (2) meters in series according to tightly specified conditions. This type of configuration is shown in Figure 24-3. Pretesting of these configurations gives expected values for the respective meter factors as well as for the relative performance of the meters, i.e., the ratio of their outputs.

Adequacy of the data base is established by specifying the number of repeat calibrations done for each flow rate and meter configuration. These results should produce sufficient data so that statistical significance can be generated to exhibit the quality of measurement performance: (1) how this varies for successive calibrations done for the same conditions over short periods of time, i.e., repeatability; and (2) how this varies from day to day for conditions that may vary slightly, i.e., reproducibility. It is recommended here that the data base be generated efficiently and for the expressed purpose of testing laboratory performance. To do this, a minimum number of flow rates are used and sufficient tests at each are done.

An alternative approach might be to use numerous flow rates and minimal replications at each. However, this alternative approach tends to place an undesirable emphasis on meter characteristics as opposed to test laboratory characteristics.

The algorithm for data processing should be well established. This attribute is achieved when it is (has been) used for a number of MAPs for other measurement systems, i.e., the procedures produced by W. J. Youden and co-workers [Ref. 19].

By testing in both configurations shown in Figure 24-3, the upstream data and the downstream data, individually, have the statistical independence requirement that is needed to apply the Youden procedure, etc. The SFC unit shown in Figure 24-3 is a "super flow conditioner" placed between the tandem meters [Refs. 15-17]. Here, it is intended to isolate the downstream meter from flow profile (or other anomalies) that might exist in the laboratory pipeline that connects to the upstream meter. Thus, the tandem meter configuration affords one the opportunity of generating data both without and with pipeflow profile effects, because

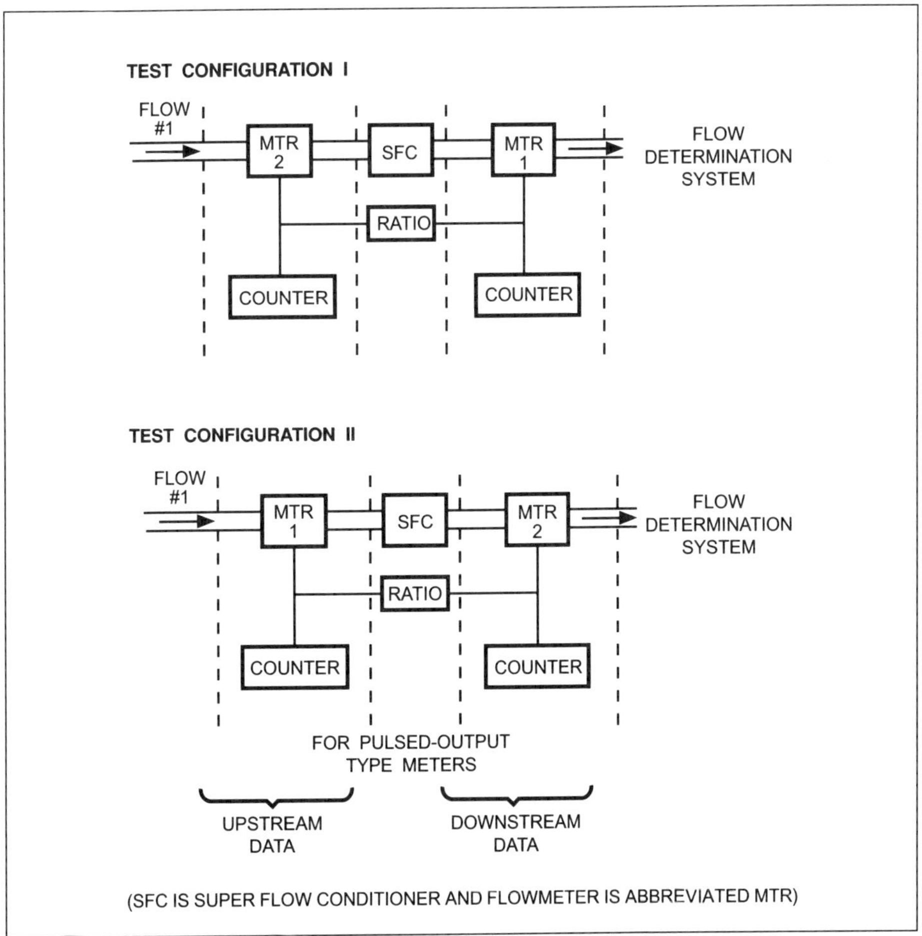

Figure 24-3. Sketch of Tandem Meter Test Configuration for Each Test Flow Rate

downstream meter and upstream meter performances can be analyzed separately. Comparisons can give unique global insights into laboratory pipeflow phenomena without having to measure these distributions.

The types of flowmeters for this type of laboratory testing should be selected according to the experiences of the participating laboratories. This consensus selection should produce the type of meter, the size, manufacture, associated instrumentation, etc. This selection process should be extended to include the fluid conditions, the flow rates, etc., as well as the tolerances to be used in arranging these.

The data generated via the round robin testing program is analyzed for each of the flow rates selected and for each of the meter positions. For each of these conditions, plots are produced of the respective meter performance characteristics (i.e., meter performance characteristics such as meter factor, discharge coefficient, etc. [Refs. 15-18]). Individual results, or averages thereof, can be plotted (see Figure 24-4). Each point represents the combined results for both meters when they were tested in each position in each laboratory.

The data processing procedures consist of determining median values for the respective sets of data for the meters. In this plot, thirteen (13) data points are shown, each representing one of the participating laboratories. Similar plots should be made for each of the other flow rates. Similar plots should be made for the meter results obtained when the meters were in the downstream position. By drawing horizontal and vertical lines through the median points for each meter, the plot is divided into four Cartesian quadrants, as shown by the dashed lines. The origin of this Cartesian system is, according to the available data, the best estimate of the true values of the meter factors for the two meters tested according to the specified conditions. In the northeast Cartesian quadrant, the data can be considered systematically inaccurate in that points are each higher than those of the origin. Similarly, in the southwest quadrant points are lower. Thus, the degree to which data is distributed in these quadrants is a measure of the systematic offsets prevailing in the laboratory data.

In the northwest and southeast quadrants the data can be considered inconsistent or random, in that one value is low while the other is high. Therefore, the degree to which the data is distributed in a northwest to southeast manner about the median intersection is a measure of the random variation in the data.

The preferred result, indicating good control, would be to find that the measurement of systematic distribution (northeast to southwest) is equal to the random

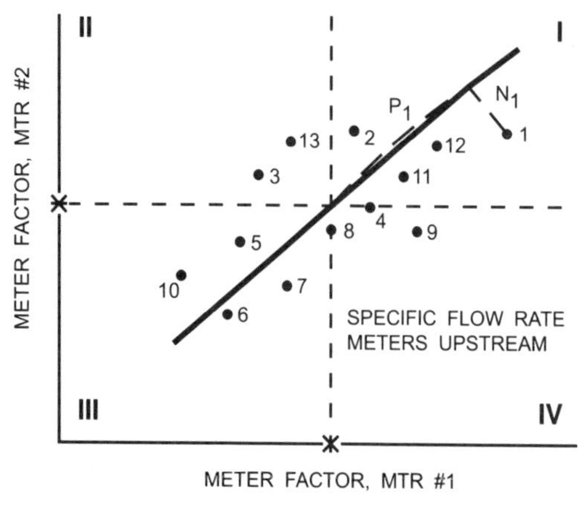

Figure 24-4. Sketch of Youden Plot for Round Robin Test Results for Each Flow Rate and for Each Meter Position

distribution (northwest to southeast) and that these measures are acceptably small. The respective levels of uncertainty can be quantified.

Where, as is usually the case, the two meters are identical, a procedure for quantifying the respective random and systematic levels of the data can be used as follows, [Refs. 15-17]. A line of slope + 1 is drawn through the intersection of medians on Figure 24-4. The data is then projected perpendicular and parallel to this diagonal line. The respective projections are then used to produce standard deviations:

$$\sigma_r = \left[\frac{1}{N-1}\sum_{i=1}^{N} N_i^2\right]^{1/2} \tag{24-6}$$

$$\sigma_s = \left[\frac{1}{N-1}\sum_{i=1}^{N} P_i^2\right]^{1/2} \tag{24-7}$$

where N_i and P_i are the normal and parallel components of the data projected to the diagonal line. The ratio of these quantities produces the degree of ellipticity of the data:

$$e = \frac{\sigma_s}{\sigma_r} \tag{24-8}$$

When this ratio is larger than unity, the interpretation is that systematic variations prevail among the labs; this is quantified by magnitude of e. Analogous conclusions can be drawn for $e < 1$.

Depending upon the results obtained for ellipticity, a number of reactions can occur. If e is large and this is produced by one or more laboratories, then the reaction should be to examine the components of their flow measurement processes to find systematic causes, etc. If e is small and this is produced by one or more laboratories, the reaction should be to examine the components of their processes with respect to their precision. If e is near unity but the levels of uncertainty are considered too large, then the appropriate response would be for the labs responsible to search and repair the pertinent components' systematic and random errors.

When such search and repair efforts are completed, the round of tests should be repeated for the same conditions so that improvements can be quantified. Even when such search and repair efforts are not needed, repeat testing is needed to produce the continuous data record desired to substantiate that the realistic traceability established has not diminished in time.

Conclusion

The standard philosophies for flow rate measurements have been presented. Uncertainty analyses are given for successive stages of flow rate measurement laboratory assessment. The techniques used for fluid flow rate calibration facilities have been described briefly. Nominal levels of performance have been given for typical facilities at NIST-Gaithersburg, MD.

The NIST flow rate measurement accuracy quotes of $\pm$ 0.13% for liquids and of $\pm$ 0.25% for air are described, where precisions are produced by the root-sum-square method and systematic errors are added to the random errors. An alternative way of combining systematic and random errors could be by root-sum-square. However, because systematic errors in flow laboratory assessments are generally estimated, it is felt that the more conservative method of combination is preferred. The systematic portion of these quotes is estimated to be ±0.1% on the basis of round robin tests.

Techniques for establishing and maintaining flow rate measurement traceability have been presented. A specific scheme has been described in some detail so that realistic data, produced on a continuing basis, can be generated so that a laboratory's entire flow rate measurement process can be assessed.

It is concluded that once these types of traceability chains are produced so that flow measurement laboratories are linked within and across national borders and boundaries, satisfactory fluid measurements can be achieved at specified levels. In this manner, the increasingly critical and costly measurements of valuable fluid resources and products can occur satisfactorily for the widely varying conditions and reasons for making flow rate measurements.

References

1. Ruegg, F. W., and Shafer, M. R., "Flow Measurement: Procedures and Facilities at NBS," Procs. Semi-Annual ASHRAE Meeting, San Francisco, CA, 1970.
2. Benson, K. R., et al., "NBS Primary Calibration Facilities for Air Flow Rate, Air Speed, and Slurry Flow," Procs Amer. Gas. Assoc. Symposium on Fluid Measurement, Crystal City, VA, No. 1986.
3. Mattingly, G. E., "Gas Flow Measurement: Calibration Facilities and Fluid Metering Traceability at the National Bureau of Standards," Procs. Inst. for Gas Technology Conference on Natural Gas Energy Measurement Chicago, 1986.
4. Uriano, G. A. (editor), *NIST Calibration Services—Users Guide*, NIST Special Publication 250, January 1989.
5. Mattingly, G. E., "Primary Calibrations, Reference and Transfer Standards," in *Developments in Flow Measurement-I*, edited by R. W. W. Scott, Englewood, N. J.: Applied Science Publishers, pp 31-73, 1981.
6. Shafer, M. R., and Ruegg, F. W., "Liquid-Flowmeter Calibration Techniques," Trans. *ASME 80*, 1369, 1958.
7. ANSI/ASME Standard—Glossary of Terms Used in the Measurement of Fluid Flow in Pipes, ANSI/ASME-MFC-1M-1979.
8. ISO International Vocabulary of Basic and General Terms in Metrology, Int'l Ingam. for Standardization, Geneva, Switz., 1984.
9. ANSI/ASME Standard—Uncertainties in Flow Measurement, ANSI/ASME-MFC-2M.
10. ISO Standard 5168—Measurement of Fluid FLow—Estimation of Uncertainty of a Flow rate Measurement, Int'l Organ. for Standardization, Geneva, Switz., 1982.
11. Abernethy, R. B., and Thompson, J. W., *Measurement Uncertainty Handbook* (Revised 1980), Aerospace Industries Div. of the Instrument Society of America (ISA).
12. Belanger, B. C., *Traceability—An Evolving Concept*, ASTM Standardization, Feb. 1979.
13. Cameron, J. M., "Measurement Assurance," NBSIR No. 77-1240, Apr. 1977.

14. Croarkin, M. C., *Measurement Assurance Programs*, NBS Special Publications 676-I and II, 1984.

15. Mattingly, G. E., and Spencer, E. A., "Steps Toward an Ideal Flow Transfer Standard," FLOMEKO Symposium, Groningen, The Netherlands, 1978.

16. Mattingly, G. E., "Dynamic Traceability of Flow Measurements," Invited Lecture, IMEKO Tokyo Flow Symposium, Society of Instrument and Control Engineers, Tokyo, Japan, 1979.

17. Mattingly, G. E., "An Interlaboratory Round Robin Flowmeter Test Using Turbine Meters in Flowing Water," in preparation.

18. Youden, W. J., "Graphical Diagrams of Interlaboratory Test Results," *Journ. of Industrial Quality Control*, Vol. 15, No. 11, pp 133-137, May 1959.

About the Author

Dr. George E. Mattingly has been at NIST since 1975 and is the leader of the NIST Fluid Flow Group. In this position, he is responsible for the maintenance and dissemination of the flow measurement standards that use water, air, and hydrocarbon liquids. Additionally, he is responsible for air speed measurements and liquid volume and density standards. In these capacities, he is involved in numerous committee activities—both national and international—to upgrade old or to generate new paper standards on these and related topics.

Dr. Mattingly is also involved in a wide range of fluid mechanics and flow measurement research projects. These include flowmeter installation effects; establishing realistic traceability for fluid meter calibration facilities; devising improved accuracies for fluid density and flow calibration systems; studying fluid-structure interaction phenomena; and characterizing and improving liquid metal atomization processes. Dr. Mattingly has authored or co-authored almost 100 publications or reports on a wide range of flow topics.

25

Standards in Flow Measurement

What is a standard? In the dictionary the word can mean a reference weight, length, time, etc., such as those found at the National Institute of Standards and Technology (NIST). It also means a document that provides a basis for technical communication regarding a device or procedure. A distinction is made between a *recommended practice* (describing a particular use) and a *standard* (with a technical foundation). This latter is the type of standard discussed in this chapter. In the IEC and ISO, the term "Technical Report" covers Recommended Practices.

Good standards set down on paper contain a vast store of human knowledge; they are a product of past experience and present knowledge and a guide for the future. They are a measure of human progress. Standards are intended to make life easier for everyone—not just a few.

There is no attempt in this chapter to cover in-house standards (mostly standard practices specific to that organization). Lists of U.S. and international standards are included.

How Does One Interface with Standards?

Standards for flow measurement fill the same function as the everyday standards that assure the stability of specifications for clothing, food, traffic lights, automobiles, etc. Terminology, technical features, dimensional interchangeability, etc., ensure that the purchaser gets what was ordered and enable one to design a system.

Certification ensures performance under specified conditions so the user need not test every meter received.

This chapter is intended to increase the awareness and understanding of the importance and benefits in the development of standards. Lack of this appreciation has, at times, resulted in inefficiencies and lack of competitiveness, particularly in the world market.

A good standard will be helpful to the product designer, the product manufacturer, the sales and marketing group, and, of course, the user of the product.

A properly developed standard will establish terminology to identify the device and the effects of installation and ambient conditions. It will identify test procedures, and it supplies required equations for the user to calculate flow rate and uncertainty for a specific application.

The Flow Measurement User

Standards ensure that one will get what was ordered, even when ordered from another country with a different language. Performance and quality are assured by certification.

The standard terminology will identify the principle and features of the device and the type of effects on the performance. The standard will describe (or refer to other standards that do) the test procedures used to determine the magnitudes of the environmental, application, and installation effects, and it will enable the potential user to understand the application limits listed by the supplier. It will describe the mechanical and electrical interfaces that need to be met. Certification will ensure the performance; it assists in dealing with new vendors.

Standards enable purchasing from multiple sources. By comparing the literature from several competitors, using the standard as a reference, the user can select between interchangeable devices.

Standards minimize training of personnel. With common terminology and test procedures, the instrument people will have an easier job shifting from one manufacturer to another and using the next generation devices to obtain more useful features.

Standards enable updating system parts with advancing technologies. With common terminology and standard interfaces, the newer devices or system parts will be understood more easily and with less retraining required.

Standards enable lower system and product costs (fewer custom designs). The process of writing the standard by many experts will standardize on more common designs and avoid the proliferation found in a disorganized market when each supplier uses his own way to describe products and their features. Product manufacturers will find a need for fewer product variations and, thus, lower stocking costs, better delivery, and fewer special designs.

Standards give guidance for application and installation to achieve the performance required. For mature technologies standards can state what effects there will be on accuracy due to variations of application, environment, and installation conditions. For new technologies they will refer the user to the manufacturer with specified test procedures. Instructions and precautions are given to enable the user to obtain the performance needed for a particular application.

Standards provide instructions so that the flowmeter may be tested. The test procedures and a list of environmental and installation conditions will enable the user to test the performance under expected application conditions or have a third-party testing laboratory perform the testing.

Standards provide the precautions for the safety of people and protection of equipment under hazardous conditions. Standards will provide design requirements and test procedures to establish the maximum internal pressure/temperature that the device will safely withstand. They will specify the designs and test procedures governing installation in hazardous locations and the definitions of the various classes of hazardous locations. They will specify the power source standard voltages and tolerances, and, for AC, the harmonic content allowable for the given performance.

Some standards describe information the manufacturer should be prepared to supply, some of which may be mandatory. Some standards require specified information on labels (such as flow direction, material of construction, etc.) and for the manufacturer to be prepared to answer user questions regarding effects of environmental variations (such as corrosive atmospheres, vibration, motion, etc.). Some provide guidance as to what items should be included in the instruction manual (such as installation and troubleshooting instructions). Some standards will include a list of the information that a manufacturer can expect users to request.

Standards ensure interchangeability and compatibility. An example of the value of a compatibility standard is the classic Baltimore fire. Fire equipment from as far as 100 miles came, but their hose couplings weren't compatible so all they had were men without water. At present, there is a national standard for fire hose couplings.

A specialist will gain from participating in writing standards. From participating in the writing of standards comes assurance that: (1) application needs are recognized, (2) the factors that affect the performance are defined and quantified if the meter is sufficiently mature, and (3) the environmental conditions that must be met are identified.

The Party to an Agreement for the Custody Transfer of Fluids

In addition to the above items, standards are the backbone of the contracts for ownership transfer. Standards intended to be used in custody transfer:

(1) establish, in a legally clear manner, a complete and very specific detailed description of the procedures and methods for installation and calculation, often using tables rather than equations;

(2) give details of the measurement procedures and practices and the expected accuracies that both parties agree to in their contract; and

(3) are best accepted if historically based on extensive past experience. Standards also establish the basis for taxation on the transfer. Custody transfer is a sale in which there are sales taxes to be paid. Thus, the standard must be accepted by the taxing authorities as well.

Standards establish the tests needed in the certification of performance, as required in many contracts.

The Vendor of Flow Measurement Equipment

Standards enable a clear understanding of what has been contracted to be supplied. Standards provide the vendor with an assurance of what has been contracted for delivery. They give a supplier confidence in the product, the knowledge of the market, and the competition. The terminology supplies the description of the functions the equipment is expected to perform. Terminology and test procedures identify the performance required, such as effects of application, environmental, and installation variations. Input and output interface requirements will be established. Power supply availability and hazardous location, if any, will be given by the user.

Standards result in fewer custom designed systems. In the development of system standards it is necessary to establish standard system functions and elements to provide the functions. This reduces the need for many special designs. Interfaces should be standardized and inter-element communication codes, formats, protocols, and common commands established.

Standards enable a reduced inventory. With the reduced need for so many special designs, standards will enable a reduced inventory and faster delivery.

Standards enable the manufacturer to evaluate a product in the same manner the user does. To the manufacturer, this will indicate best directions for development and marketing, whereas the user can more fairly evaluate flowmeter performance.

Helping to write standards will keep a company current. Through participation in writing standards the supplier can be sure of technically sound standards and that the technology will not be left out, and will gain insight into what competition is up to.

Standards will support the manufacturer's assurance of quality products, especially if the manufacturer participates in a certification program.

Standards Related to Flow Measurement

A large number of standards have been written and more are now being written covering methods and devices for the measurement of fluid flow. In addition, there are many related standards, such as those for pipe and flange pressure/temperature rating and dimensions; threads and flange bolting; mechanical, pressure, and electrical safety; enclosures; electrical and pneumatic signal transmission; physical measurements; and even format and procedural standards for writing standards.

Mature Methods and Devices for Measuring Fluid Flow

When the method or product has become public property and there are no controlling patents or proprietary designs, standards can give detailed designs and closely specified installation requirements. Sufficient test data allows specific performance to be stated in the standard. Uncertainty limits can be given for various application conditions. Certification can be performed; that is, a sample of the device has been tested to prove that it will perform within the limits specified in the standard. With enough data available, statistical analysis can be used to develop equations to eliminate individual calibrations, except when extreme accuracy is required or when the application conditions are not covered by the standard. Examples of mature methods and devices are the square edge orifice, Venturi, nozzles, Pitot tubes, and tracer methods. Sometimes two or more classes can be identified by different performance limits. Performance classification is more acceptable when a technology is mature to the point that there is enough data available and there are no proprietary designs.

Generic Standards

Many of the fluid flow measurements being standardized use new principles and technologies in sensors and computation. Almost all are represented by proprietary devices that are very attractive, promising, or offer performance superior to the "classical" flowmeter technologies.

When a developing technology is being used, there normally will be several suppliers, each with a proprietary version of the product. No design details or application limitations can be covered; each manufacturer must supply these for each device. In a standard, it is essential to describe the principle in a manner that does not favor any one of the versions. Thus, standards for new technology products can attempt only to identify the important effects (such as installation, environment, and application) on the performance, with no magnitudes. The standard should indicate the test procedures needed to obtain the magnitudes of the effects.

At this stage of development, the need is for communication standards. It is necessary for a professional organization (such as ASME or ISA) to assemble experts from the developers, users, and educators to establish a stabilizing reference document. This may be called a Technical Report or a Generic Standard, but it must not be delayed by argument over the name.

There seldom is enough public data available to support the development of flow equations. Each manufacturer has proprietary equations for each design. A few examples of new technologies are vortex shedders, Coriolis mass flowmeters, and ultrasonic flowmeters (transit time and Doppler).

Test Procedures

Testing is needed to establish the relation between the fluid flow and the output signal from the flow measuring device. Standardized testing facilities and procedures are an essential part of the standardizing of flow measuring devices and their installation. An accuracy claim must be based on actual tests to have credibility, and the user must have confidence that independent tests will yield the same results.

There are a number of standard testing arrangements and procedures. The least controversial require a well-equipped flow laboratory with a well-established flow profile. Some can be used for *in situ* calibrations, which are often required. If there is space, the most common *in situ* method is to install in series a calibrated meter or transfer standard (probably employing a different principle and possibly more limited in rangeability and application) for comparison. Some device standards allow for approximating the *in situ* procedure by building a special installation that matches the flowing conditions as closely as practical. When the flow profile is expected to be distorted (velocity profile and/or swirl) because of upstream piping conditions, the *in situ* comparison procedure is recommended if the transfer meter has been proven to be less affected by these conditions.

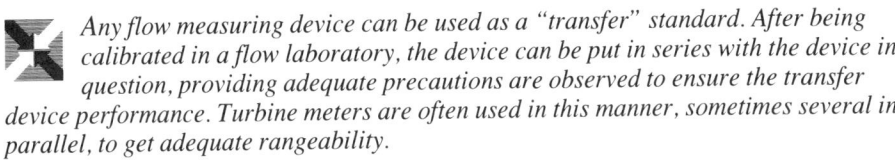

Any flow measuring device can be used as a "transfer" standard. After being calibrated in a flow laboratory, the device can be put in series with the device in question, providing adequate precautions are observed to ensure the transfer device performance. Turbine meters are often used in this manner, sometimes several in parallel, to get adequate rangeability.

In the case of mature technologies, the tests establish the fundamental design characteristics, but for developing technologies each design must be tested.

Certification of Design

In Europe there is increasing pressure to require certification for all products covered by standards. For industrial processes, flow measurement third-party testing labs periodically check all of the performance claims on representative production samples and certify the performance of the design. These labs are financed primarily by users.

The argument is that marketing people can tout the excellence of their product, but there is more assurance with third-party certification that the product will perform in accordance with a particular standard. The user can be confident that the delivered meter is the one ordered; thus, it assists in dealing with new suppliers.

There has been reluctance for U.S. manufacturers to participate in certification for non-safety requirements, claiming that their normal quality control and testing practices make certification unnecessary. Products intended to meet standards developed for the protection of life or property (safety standards) require certification to ensure the compliance. This certification can be by recognized third parties (who have often written the testing standard) or by self-certification such as the ASME Boiler and Pressure Codes or the Japanese JIS system, where unscheduled visits by inspectors "keep the suppliers honest." It covers industry and most consumer products as well.

The use of the ASME boiler and pressure vessel stamp is a monitored self-certification arrangement. The manufacturer is expected to ensure compliance, and random inspections by ASME agents check the manufacturers and give approval for continued use of the stamp.

In the U.S., insurance companies employ third-party testing laboratories to ensure compliance. Underwriters Laboratories (UL), National Fire Protection Association (NFPA), Factory Mutual (FM), and the Canadian Standards Association (CSA) are the most active examples of third-party testing. Manufacturers and users rely on the laboratories to check designs as well.

Europeans rely more on certification by government laboratories such as DIN in Germany. Usually, certification in one is accepted by all European countries. Certification presumes well-established terminology, and national standards on performance testing and evaluation procedures have been developed.

Manufacturers and user companies in Europe may utilize laboratories such as WIB in Holland and SIRA in the UK to verify compliance with claimed performance.

Pressure Rating

Flowmeters are mounted in pipes, normally bolted between flanges. The pressure rating of the meter has no required relation to the flange rating, since the flanges have a standard pressure/temperature rating for high temperature where the metal will be weaker. The pressure and temperature ratings of the meter should be for the maximum operating conditions and will normally be limited by the mechanical parts or electronics. These ratings must be displayed on the outside of the meter or the user may mistakenly assume that the flange ratings apply.

Installation in a Hazardous Location

If the flowmeter is to be located in a hazardous location or is to measure the flow of an explosive fluid, there are safety standards that specify the precautions to be taken.

In the U.S., protection against possible explosions caused by electrical equipment can be accomplished through the use of explosion-proof housings and metal conduit protection of wiring that contain an explosion. Hazardous atmospheres are classified in the NFPA National Electronic Code (NEC) articles 500 through 503 and indicate the type of protection and precautions required. UL (Underwriters Laboratories), FM (Factory Mutual), and CSA (Canadian Standards Association) standards and their certification of the designs assure the user of equipment approval.

Explosion-proof housings and wiring requirements are heavy, clumsy, and expensive. Another approach (intrinsic safety) became popular in Europe, whereby the potential energy level in any circuit is kept low enough that ignition is not possible. Therefore, the housings need not be explosion-proof and the equipment can be maintained in the hazardous atmosphere. The German PTB laboratory established test procedures while manufacturers developed the equipment and circuits. FM (Factory Mutual) and UL also test for this feature, and many manufacturers design the feature into their flowmeters due to its increased popularity.

Products designed to be used in hazardous locations need to be examined and tested by appropriate authorities and given a formal approval or certification.

These products are checked by labs such as Underwriter's Laboratory (UL) and Factory Mutual (FM) and government laboratories such as PTB in Germany and JIS in Japan. There is currently little acceptance of U.S. designs in international markets. The big challenge to U.S. companies is to make designs that will meet both the U.S. and European requirements, perhaps necessitating a change in the U.S. design requirements and standards.

Safety Codes

Safety standards for life and property are normally written to be legislatively adopted as codes for inspectors to enforce, such as the NFPA National Electric Code. These standards limit the design and application of equipment. Boilers, elevators, bridges, and electrical equipment and installation are but a few examples. Minute technical details must be spelled out, and inspectors must be educated to enforce them. Maximum flexibility should be written into them to minimize these limitations. Certification tests should be based on physical fundamentals as much as possible, rather than on arbitrary dimensions or other absolute limits. Certification in one European country is usually accepted by the others, but this has virtually no effect on approval for use in the U.S.

Enclosures

Standards have been written to give guidance in the design of the protective enclosures for the flowmeter and for the electrical or mechanical associated equipment. NEMA 250 (National Electrical Manufacturers Association) describes the design details for various classes of environments, up to hosing with high pressure water, and the protection required. Originally intended for electrical switchgear, these standards have been used satisfactorily for the protection of all types of electrical and electronic equipment.

Under IEC procedures, IEC 529 has been developed to identify the environment and degree of protection required with a two-digit IP (ingress protection) number. These numbers (e.g., IP 65) identify the testing procedures to be used to evaluate an enclosure, with water (from drip to fire hose) or with a dust box. Another test includes a dummy finger to determine if one can get into trouble by being too inquisitive.

Interfaces

For satisfactory use, a flowmeter must match several standardized interface conditions, such as installation, piping, compatibility of output signals, and environmental protection.

If the meter is designed to be supported by flanges in the flow line, the flanges must match the physical dimensions. The flowmeter need not match the full flange pressure rating but must be rated for the process pressure and temperature. Standard bolts are used for centering unless some other provision is made. Pipe dimensions are given in standards and appropriate handbooks. Care must be taken not to confuse metric and U.S. dimensions.

Each flowmeter standard specifies the requirements for the upstream and downstream piping, and they must be followed. Some flowmeters have a requirement for matching the inside diameters and permissible offset; others claim to tolerate a misfit, but this should be avoided unless specifically approved by the vendor.

Piping

Low pressure piping in ISO metric and U.S. inch dimensions generally are comparable, with rounded-off nominal sizes (e.g., 4 inch is same as 100 mm). Flanges and their bolts, however, are different, so care must be taken to ensure that the proper flanges and bolts are used.

ISO and U.S. pressure/temperature ratings are different, so it is important to follow the correct standards when designing for installation of a flowmeter in a pipeline, being careful to use the correct schedule designation.

Metric tubing has close tolerances and therefore these fittings are not compatible with the U.S. tubing.

Output and Transmission Signals

Standards have been written to specify pneumatic (SI metric, or U.S.), analog DC current, analog DC voltage, and several digital signals. Most specify supply pressure, voltage, or power supply required for the performance claimed by the vendor. Some devices may have non-standard outputs as part of proprietary systems.

The IEC standard for analog DC transmission signal is IEC 382-1 (4 to 20 Ma), and IEC 382-2 (DC voltages, several). It also specifies a power supply of 24 V maximum.

The IEC standard for analog pneumatic transmission signal is IEC 381 (20 to 100 kPa).

ISA-S7.4 is the standard for analog pneumatic transmission signal (20 to 100 kPa [3 to 15 psi]). It includes pneumatic supply pressure limits of 0 and 140 kPa (0 and 22 psi).

Standard digital transmissions are covered in EIA RS-232, RS-485, and RS-422.

Measurement Units—SI Metric vs. U.S.

Measurement units today are in two sets—U.S. (formerly called "English units") and SI (Standards International or metric) covered by ISO 1000. The U.S. stands alone in the technical world in its refusal to recognize the SI metric units.

ISO 1000 teaches technology by making a very clear distinction between force (newton) and mass (kilogram) in contrast with the thorough confusion caused in the U.S. by the dual role of the pound and the associated slugs and poundals. ISO 1000 also establishes a rational relation between all physical units in the metric system. For these reasons SI units are mandatory in almost all of the countries in the world.

Many U.S. companies have had difficulty selling products overseas because they have not been designed to meet the standards of the international marketplace; thus, standards have often become nontariff trade barriers. One such standard that must be recognized by U.S. companies is ISO 1000. It is now mandatory in the European Community (EC) and many other countries that products be described in SI units. There is growing pressure that the products be designed in SI units as well.

Engineering societies such as ASME, ISA, IEEE, ASTM, etc., are insisting that their papers, publications, and standards show SI units; some call for both sets of units.

Although called "English," the U.S. units for length and mass are based on metric primary standards:

(1) The inch is based on the meter and is defined as 25.4 mm exactly.

(2) The mass standard is a 1-kg cylinder of platinum-iridium. The U.S. lbm is a decimal (to 10 places) times that value.

(3) The U.S. temperature unit, °F, is based on the same physical phenomenon as the SI unit, K (kelvin).

(4) Physical references for measurements (such as temperature, length, time, mass, resistance, voltages, etc.) are kept in the U.S. at The National Institute for Standards and Technology (NIST).

(5) Time is the same in both sets of units, based on astronomical observations, with the reference being an atomic clock.

(6) Electrical units (ohm, volt, ampere, and watt) are based on the basic units of time, length, and mass and are derived using accepted electromagnetic theory. For practical reasons they are translated into a standard ohm and a standard ampere by a series of international congresses, settling in 1948 on a system of absolute values, with the primary references being resistors and Weston standard cells at NIST and derived from time, length, and mass standard units through the use of carefully constructed inductance coils.

Traceability Concept

Testing laboratories have their own reference standards for length, mass, volume, temperature, resistance, pressure, etc. Periodically they are sent to NIST for comparison with the standards there to maintain traceability confidence. For general use, laboratories will also have instruments or working standards that they compare periodically with the reference standards. Good practice in a test laboratory is to have measuring instruments with an accuracy of $\frac{1}{3}$ to $\frac{1}{10}$ that of the required accuracy of the devices being tested. Generally there will be, for flow measurement, two or more instruments, the accuracy of which must be combined to determine the overall uncertainty of the laboratory.

Converting Units

When it is necessary to convert from U.S. units to SI units or reverse, guides can be obtained from ASME, IEEE, ANSI, etc.; some flowmetering texts provide the equations in both sets of units, and all ASME flowmetering standards present both sets of units.

Standards for Special Industry Needs

A contrast is found between standards written for general industry and those written for special industry needs. The following are some examples.

The Power Test Code (ASME PTC) for fluid flow describes measurements under laboratory conditions to determine the performance of steam turbines, etc., in contrast with the less controlled environment of industrial flow measurement. Potable water flow measurement standards are written by the AWWA to address their unique testing and application conditions (including the fact that they are in the custody transfer business and must expect that their standards may be written into codes for public authority supervision). Likewise, heating and air conditioning people have found that other standards do not adequately cover their practices, so they have developed their own flow measuring standards. The natural gas and hydrocarbon custody transfer industry, for apparently the same reasons, has also chosen to promote their own standards.

Custody Transfer

A very large percentage of present day industry is based on flowing fluids (gasoline, natural gas, fuel oil, water, chemicals). Many of these fluids need to be metered, either continuously or in batches, in order to bill the customer; thus, custody transfer is a major function in our economy.

The most important feature of custody transfer standards is the need for unchanging uniformity throughout the industry. Standards are written into custody transfer contracts, and the seller and buyer agree to accept these standards. Unchanging uniformity is seemingly more important than high accuracy because it results in a stable, competitive marketplace. If different standards were available, sellers and buyers would each demand the standard most favorable to them.

Custody transfer standards for flow measurement devices or systems cover details on product performance, strict application details, testing procedures, and methods of calculation that are common in an industry or application but may not be satisfactory elsewhere. The supplier-customer contract establishes the details of which standards are applicable for a given application.

To support the contract, the flowmeter standard has to specify the uncertainty (accuracy) under reference conditions and the additional uncertainties (bias and random) due to the installation, the fluid conditions and properties, and the ambient environment. The standard must also show how to combine these uncertainties or refer to an uncertainty standard.

Why Write Standards?

Individuals and groups exert the effort to write standards for many reasons:

(1) To improve marketing by enabling one to see more closely the user's demands. Since all standards writing committees operating under ANSI rules have a balance of users, suppliers, and general interest participants, there is an opportunity for users to express their needs and desires on features and performance.

(2) To influence the standard to be technically sound, rather than repeating practices based on old entrenched and often outdated technology. The tendency was to write a standard that described what had been done for years because it is so comfortable, often without regard to either technical soundness or new developments.

(3) To eliminate *de facto* "standards" with their shortcomings (there is no clear responsibility for updating, and often the "sponsor" has a narrow view of the field).

(4) To avoid unnecessary limitations to the development of new designs. Often new technology makes it possible to improve a measurement in a manner not compatible with the "traditional" way of doing it. A new or revised standard should allow the improvement.

(5) To gain some insight into competitors' thinking by watching the emphasis they put on their input. It is possible to get some new ideas from the requirements competitors ask for. This is a two-way street, so be extremely careful until there is adequate patent coverage.

(6) To teach and learn new technology. Every new product design will have some interesting aspects. Writing standards for products using new technology will provide opportunities to learn.

(7) To make certain that products will not be left out of the mainstream of the flowmeter technologies.

Ten Commandments for Standards Committee Meetings

1. Standardization means sacrifice. It is unlikely that all of your ideals will be realized.
2. "This is our standard practice" is not a valid argument. The practice in other countries may be as good or better.
3. Each proposal must be judged on its own merits. Experience and unbiased judgment must be taken fully into account.
4. If the ideal solution cannot be reached at this moment, the best compromise must be adopted. This is better than no decision at all.
5. If the national delegate cannot make a decision, the opinions of others must not be condemned. The international standard may be in conflict with a national standard but still be useful and may be adopted nationally at a later date.
6. Do not insist on discussing items of minor importance. There is no time to spare.
7. Do not try to rearrange the sequence of paragraphs or argue about editorial details. The editing committee can take care of these.
8. Standardization means cooperation. With success, all parties will have great moral and material profits.
9. Standardization is based on consensus. Find a path between weak compromise and heavy-handed overruling of minority views. Avoid voting as much as possible. Obtaining a consensus is the best way.
10. Enjoy the meetings and the hospitality of the hosts.

Proprietary devices can be included in a design standard only if the patent owner has agreed to release the use of the design details, even to competitors, at a reasonable fee. A generic standard can be written with no reference to designs or performance limits, making it a communication document rather than a design or performance limit type of standard.

"Don't fight, make it right," says the voice of experience.

Teaching Technology

The most successful standards depend on adequate understanding of the underlying technology. When this is lacking, there should be a background of experience that supports an empirical standard.

Standards development requires the extensive teaching of technology to bring all of the committee together to produce a document that is current and doesn't give one manufacturer a proprietary advantage over others. It is, of course, important that a standard recognize the latest technology and support it to the same extent that it supports the older technology. This requires that the new technology be taught in the standard to the extent possible at that stage of development. If this is not possible (because of too many proprietary devices), the standard should be generic and cover terminology, test procedures, principles, and application conditions. In this case it is important that all suppliers of the new technology participate in the development of the standard to ensure that their designs will not be "left out."

The Boiler Codes the ASME are good examples of standards with a good understanding of the technology. However, the technology was not always known. An example of the advantage of teaching technology by standards is the following classic: In 1884 a number of empirical standards were being followed in the design and construction of steam boilers. More than 10,000 boilers and pressure vessels exploded that year. In 1984, with the understanding of the fundamentals as shown in the Boiler and Pressure Vessel Code, there were no such explosions. Standards teach technology through the strict requirements in the design in order to maintain safety.

How to Respond to Proposed ISO or IEC Draft Standards

The U.S. should have an expert on each working group for a flowmeter standard development (or revision of an existing standard). This should avoid surprises in drafts that are circulated for vote as well as maximize the U.S. technical input.

When the proposal is received for comment, it must be remembered that the objectives of ISO and IEC are to develop standards. One nation's negative vote will not stop the process regardless of the intensity of the vote.

The comment that "It doesn't match current U.S. practice" or that "It doesn't match past practice" is of no help. Calling attention to a technical weakness with suggested acceptable alternative wording is conducive of progress. Every comment should be supported by an explanation, and proposed wording should be offered in each case.

Suggested editorial improvements will be appreciated, but the final editorial treatment will be given in Geneva, and the main influence by the U.S. is to call attention to unfortunate differences between various forms of the English language (example: French-English for "orifice" is "diaphragm").

Government Procurement

In many countries, government standards are mandatory if one wants to be a government supplier. Government purchasing is based on detailed product descriptions. Recently, commercial standards have been used whenever possible. This puts pressure on the standards-writing committees to make the standards very specific, including accuracy classifications and limited ambient effects on performance.

When the technology is mature, there should be enough data for support, but, at an early stage of the technology, care must be taken to avoid favoring proprietary designs.

Strategy for Writing Standards

The place to best influence the development of a standard is at the meetings of the Working Group. If the technology is sound and it can be made demonstrably better, acceptance by the working group is most likely. If possible, use an ANSI standard as a starter in international standardization.

Be sure that the committee is balanced with experts from users, suppliers, and academia. Follow ANSI procedures carefully if the objective is to develop an American National Standard.

Emphasize the objective of being helpful and useful for communication worldwide, even if you are developing a U.S. standard.

Develop standards based on sound technology, with experimental data whenever possible. The best data is that which has been collected under controlled (laboratory) conditions. To be practical, however, very suitable data is often obtained by comparison with a reference meter (a standard flowmeter using a different principle) and staying within the limitations of the reference meter.

If others have better technology, accept it and help build a good standard that will also support present products. This may mean some temporary pain, but in the long run it will definitely be better.

Develop generic standards when the technology is too young for design standards.

Never try to push commercial aspects as a reason for certain parts of a draft standard. Many Europeans and some Americans get very irritated.

A new or revised U.S. standard should be consistent with an existing ISO or IEC standard, if there is one. It must not be more restrictive lest it be judged a nontariff trade barrier. After the U.S. has approved an ISO or IEC standard, it is not necessary to adopt it verbatim, but the U.S. version must not be more restrictive.

The U.S. should start with the ISO or IEC standard and revise as necessary to "Americanize" it (such as adding U.S. units).

The best standards can still be improved. Prepare to introduce improvements by revising ISO standards in their 5-year cycle. The latest revision of ISO 5167 is a good example of this. Many of the changes have come from ASME MFC-3M.

Participate energetically in the continuing process of preparing for the mandated five-year reaffirmation or revision.

If the above practices are followed, there should be no surprises when it comes to voting and commenting on the draft. In some cases, the U.S. must disapprove, but that will normally be due to a misunderstanding that can be resolved.

U.S. Participation in Developing International Standards

Before the IEC and ISO were the international standards organizations, every nation had its own standards, and international trade was a nightmare with the variation between standards providing effective trade barriers—sometimes deliberately. International standards are the best assurance of reduction of technical barriers to trade.

The IEC and ISO, with ANSI as a member of both, have provided an interface between the standards-writing bodies of the world. Almost every technically or commercially important nation is a member of one or both of these organizations. In general, their representation has been by government-supported bodies, while industry-supported ANSI is the U.S. member.

IEC and ISO made it possible for all nations to develop internationally accepted standards, but there still were differences in practices between nations as they separately wrote standards that agreed more or less with the ISO/IEC. As new technology developed, standardization lagged behind the industry demand, even in ISO and IEC.

U.S. expertise has been very effectively incorporated in many of the IEC and ISO standards when the U.S. has been ahead in the technology and the experts have taught it.

In recent years, as the European Community (EC) has steadily become more consolidated, their industry organized CEN (European Committee for Standardization in the nonelectrotechnical field), CENELEC (the same for electrotechnical standards), and CEPT/ETSI (European Telecommunications Standards Institute). These bodies have been writing standards for the EC and EFTA, the companion group of nations, and also have been selecting which of the IEC and ISO standards are to be adopted throughout Europe. They have given assurance that

IEC and ISO standards will be used whenever possible. With the European Community planning to achieve a single market by the end of 1992, they are planning to use a single set of standards rather than individual standards for each country.

In most European countries the approval of an international standard means that the national standard will be the same as the international one, sometimes verbatim. ANSI has encouraged U.S. standards writers to make U.S. standards compatible with the IEC and ISO standards; but, more often than not, U.S. standards follow U.S. past practices.

European companies working together on standards could hurt U.S. industries. If they develop standards faster than the U.S., they could dictate standards to U.S. companies for international trade.

One of the major objectives in the EC92 (1992 true common market) is to establish a single set of standards for all of the European nations—the EC and also the EFTA—making a single market similar in size to the entire U.S. by removing technical barriers to trade.

ANSI can participate in the ISO and IEC, but not in the EC activities, so it behooves American companies who will sell into Europe to work aggressively with ISO and IEC committees to write standards that recognize the latest technologies and avoid parochial, antique, and other practices that might exclude American products from the EC markets.

To reduce this problem, ANSI has recently announced the opening of an office in Brussels, the EC government center, to observe the European standardization activities and report back to the U.S. public.

ANSI's president, Manuel Peralta, commented in 1991, "The EC standards effort will, by definition, have major consequences for U.S. producers and suppliers. Most American companies are only beginning to awaken to the strategic significance of standards technology to competitiveness in global trade and commerce—a lesson the Japanese and Europeans learned some time ago." The development of standards is of vital importance because standards establish acceptable criteria for goods and thereby offer the benefits of economy of scale and simplicity of promotion and distribution. Without skilled experts in attendance, the U.S. cannot insist on the inclusion of technical fundamentals to ensure the soundness of a standard and cannot participate to achieve an intelligent compromise.

To be effective, U.S. participation must have continuity of experts in the technology. The practice of having someone attend because it is convenient should not be acceptable. It takes many meetings before the experts can learn of the sincerity and knowledge of the others. Since most international standards are developed in meetings in Europe, the Europeans usually have strict continuity of expert participation, and it is very difficult for an American, however knowledgeable, to contribute at the first few meetings. The "regulars" are reluctant to go back over the many previous discussions.

For many years, the U.S., because of its market dominance, was able to drive the development of international standards, but, with the development of a global economy, that advantage no longer exists. The 1992 European Community will erode the U.S. trade position even further, making it essential that the U.S. rely on better technology, experienced experts, and aggressive involvement to lead in standards development.

Performance and Classification

Originally, standards were written to confirm previous practices, and most still do. The result has been standards for devices with performance classes that relate uncertainty to the application conditions (ASME MFC-3M and API 2530 are examples). This kind of standard may limit designs and opportunities for future innovation. If, for instance, an inventor could show superior reliability with a differently shaped orifice, it would not be covered by the standards—not, at least, until revision dates.

Standards intended to be used in custody transfer often classify the flowmeter by the accuracy to be expected under certain installations. Thus, a meter might be labeled a 5% overall device, considering the effects of ambient temperature variations, fluid properties changes, vibration in the mounting, etc., when in normal use it could be relied upon to have a 2% maximum uncertainty. The same meter might be a 1% device in calibration in a flow laboratory. For industrial applications, the user needs to know the separate effects on performance and what performance to expect in a particular application.

When the meter is used in custody transfer, the classification approach is much in order, since a stated performance must be mutually accepted for billing and by the government for tax purposes.

As new technology is applied to flowmeters, performance must be specified by the vendor. The standard can only indicate what performance statements the vendor must supply for certain application conditions. Thus, there can be no classification supported by data.

Legal Aspects of Standards Writing

Two areas of legal implications have been raised regarding participating in the development of product standards: product liability and antitrust.

Product Liability

Counsel for the American National Standards Institute (ANSI) have concluded: "We do not believe that members of voluntary standards committees operating under ANSI procedures incur significant risks of product liability for damages based on deficient standards." There are no reported cases to indicate that such a suit has ever been instituted. Negligence is the most likely basis to be used, and that is very difficult to prove.

A member of a standards-writing committee would be vulnerable if he or she had deliberately intended to harm one or more companies. The development of a standard that deliberately excludes one or more competitive product designs is likely to cause committee members and the standards-writing organization to be vulnerable.

Product liability suits are escalating at an ever-increasing rate. One cannot escape the fact that badly designed or unsafe products do exist, but following appropriate standards is an efficient, cost-effective way to take advantage of the cumulative experiences of thousands of expert man hours. It is unfortunate that, in too many instances, this information is unknown or ignored.

It should be remembered that in all but three states the provable use of the appropriate industry standard is acceptable defense in a product liability suit.

Antitrust Implications

To avoid Federal Trade Commission (FTC) action, following standards must be voluntary, particularly in the case of trade associations. The writing of industry standards is one of the main activities of trade associations. However, it is important that nothing can be construed as limiting competition. Thus, all standards must be written on the basis that everyone has the right to voluntarily follow or not to do so.

Terminology Standards

Every flowmeter product standard has a number of definitions for special terms, but two comprehensive standards specifically cover flow measurement terminology:

ASME/ANSI MFC-1M, 1987, Glossary of Terms Used in the Measurement of Flow in Pipes

ISO DIS 4006, 1988, Measurement of Fluid Flow in Closed Conduits—Vocabulary and Symbols

Other relevant industry terminology standards are:

ANSI/ISA-S51.1-1979, Process Instrumentation Terminology (being revised, 1991)

IEC Pub 359, Expression of the functional performance of electronic measuring equipment

IFAC, Multilingual Terminology of Automatic Control Terminology

INFOTERM (International Center for Terminology) (UNESCO)

OIML, Vocabulary of Legal Metrology, Fundamental Terms

Acronyms for Standards Organizations Involved in Flowmeter Standards

Technical Societies

ANSI	American National Standards Institute
ASME	American Society for Mechanical Engineers
IEEE	Institute of Electrical and Electronic Engineers
ISA	Instrument Society of America
MFFCC	ASME Standards Committee for Measurement of Fluid Flow in Closed Conduits
VDE/VDI	German engineering societies

Trade Associations

AGA	American Gas Association
API	American Petroleum Institute
ASHRAE	American Society of Heating and Air Conditioning Engineers
AWWA	American Water Works Association
NAMUR	German chemical industry standards association
NFPA	National Fire Protection Association
SAMA	Scientific Apparatus Manufacturers' Association

Testing Laboratories

ASTM	American Society for Testing and Materials
CSA	Canadian Standards Association
FM	Factory Mutual Research, commercial

NIST	National Institute of Standards and Technology (formerly NBS)
PTB	German, Governmental Physical-Technical Institution
SIRA	British Scientific Instrument Research Association (commercial)
UL	Underwriters Laboratory, commercial
WIB	Dutch Association of Users of Measurements and Control (commercial)

International Organizations

CCITT	International Consulting Committee on Telegraph and Telephone
CEN	European Committee for Standardization
CENELEC	European Committee for Electrotechnical Standardization
EC	European (Economic) Community
EFTA	European Free Trade Association
IEC	International Electrotechnical Commission
IFIP	International Federation for Information Processing
ISO	International Standards Organization
OIML	International Office for Legal Metrology

National Standards Organizations

AFNOR	Association Francaise de Normalisation
BSI	British Standards Institution
DIN	Deutsches Institut für Normung
DS	Dansk Stanardiseringsraad
GOST	USSR State Committee for Standards
IBN	Belgium Institute for Standardization
JISC	Japanese Industrial Standards Committee
NNI	Netherland Standards Institute
NSF	Norwegian Standards Organization
SCC	Standards Council of Canada

Procedures, Policies, and Guides

ANSI—Operating Procedures of the Board of Standards Review

ANSI—Procedures for the Development and Coordination of American National Standards

ANSI—Procedures of the Executive Standards Council

ANSI—Style Manual for the Preparation of Proposed American National Standards

ANSI/IEEE Std 268, Metric Practice Guide

API—Policies and Procedures for the Committee on Petroleum Measurement

ASME—A Guide to Writing ASME Codes and Standards

ASME—Text Booklet: SI Units in Fluid Mechanics

ASME—The Codes and Standards Policy

ISA—A Guide for (new) S&P Chairmen

ISA—Standards and Practices Department Manual of Procedures

ISA—Style Manual for Standards and Recommended Practices

IEC: Guiding documents of a general nature
Guiding documents for national committees
Guiding documents for secretaries on general matters
Guiding documents for secretaries on specific matters
Guiding documents relating to meetings
Joint ISO/IEC Guides

ISO—Directives for the Technical Work of ISO
Part 1, Procedure and Working Methods
Part 2, Methodology
Part 3, Presentation of International Standards

U.S. Flowmeter Standards and Active Developments

AGA 7, Turbine Meters

AGA TMC Report No.8, Properties of Natural Gas

ANSI/API 2530-1985, Manual of Petroleum Measurement Standards, Chapter 14—Natural Gas Fluids Measurement; Section 3—Orifice Metering of Natural Gas and Other Related Hydrocarbon Fluids (1989 being revised)

ANSI/ASME MFC-1M 1986R, Glossary of Terms Used in the Measurement of Fluid Flow in Pipes (being revised)

ANSI/ASME MFC-2M 1988R Measurement Uncertainty for Fluid Flow in Closed Conduits

ANSI/ASME MFC-4M-1990, Measurement of Gas Flow by Turbine Meters (being revised)

ANSI/ASME MFC-5M-1989R, Measurement of Fluid Flow in Closed Conduits Using Transit-Time Ultrasonic Flowmeters

ANSI/ASME MFC-6M-1987, Measurement of Fluid Flow in Pipes Using Vortex Flowmeters

ANSI/ASME MFC-7M-1987, Measurement of Gas Flow by Means of Critical Flow Venturi Nozzles (being reaffirmed)

ANSI/ASME MFC-8M-1988, Fluid Flow in Closed Conduits—Connections for Pressure Signal Transmissions between Primary and Secondary Devices

ANSI/ASME MFC-9M-1988, Measurement of Liquid Flow in Closed Conduits by Weighing Method

ANSI/ASME PTC-19.5-1972, Orifice Meters for Power Test Code Testing

ANSI/AWWA-C701-1988, Cold Water, Turbine Type

ANSI/ISA-RP31.1-1977, Specification, Installation, and Calibration of Turbine Flowmeters

ASHRAE-41.8-1978R, Standard Methods of Measurement of Flow of Fluid in Pipes Using Orifice Flowmeters (1989 submitted to ANSI)

ASME MFC-3M-R1989 Measurement of Fluid Flow in Closed Conduits using Orifice, Venturi, Nozzle

ASME MFC-10M-1988, Method for Establishing Installation Effects on Flowmeters

ASME MFC-11M-1990, Measurement of Fluid Flow by Means of Coriolis Flowmeters

ASME MFC-12M, Measurement of Fluid Flow in Closed Conduits Running Full Using Multiport Averaging Pitot-Primary Meters (in draft form)

ASME MFC-14M, Measurement of Fluid Flow in Closed Conduits—Velocity Area Method Using Insertion Meters (in draft form)

ASME MFC-?M, Flow Conditioners (1990 Active development, SC19)

ISA-RP16.5-1961, Glass Tube Variable Area Meters (Rotameters)

ISA-RP16.6-1961, Calibration of Variable Area Meters (Rotameters)

Related Standards

3-A Sanitary Piping Standards, available from the *Journal of Food Protection*, Box 701, Ames IA 50010

ANSI/ASME B31 series, Piping Sizes and Ratings

ANSI B16 series, Piping Flanges

ANSI/ISA-RP12.6-1987, Installation of Intrinsically Safe Systems for Hazardous (Classified) Locations

ANSI/ISA-S7.4-1981, Air Pressures for Pneumatic Controllers, Transmitters, and Transmission Systems

ANSI/ISA-S12.10-1988, Area Classification in Hazardous (Classified) Dust Locations

ANSI/ISA-S12.12-1984, Electrical Equipment for Use in Class 1, Division 2 Hazardous (Classified) Locations

ANSI/ISA-S50.1-1982, Compatibility of Analog signals for Electronic Industrial Process Instruments

ANSI/ISA-S51.1-1979, Process Instrumentation Terminology

ANSI/ISA-S75.01-1986, Flow Equations for Sizing Control Valves

ANSI/NEMA 250-1985, Enclosures for Electrical Equipment (1000 V maximum)

ANSI/NFPA 496-1989, Purged Enclosures for Electrical Equipment in Hazardous Locations

ANSI/NFPA 497A-1986, Classification of Class I Hazardous (Classified) Locations for Electrical Installations in Chemical Process Areas

ANSI/NFPA 497M-1986, Classification of Gasses, Vapors, and Dusts for Electrical Equipment in Hazardous (Classified) Locations

ANSI/UL 1203-1988, Explosion-proof and Dust-ignition-proof Electrical Equipment for Use in Hazardous (Classified) Locations

ANSI/UL/NFPA 4913 1988, Intrinsically Safe Apparatus and Associated Apparatus for Use in Class I, II, and III, Div. I Hazardous (Classified) Locations

ASME/ANSI B16 series standards for valves, fittings, flanges, and gaskets

ASME/ANSI B31 series codes for pressure piping

ASME/ANSI PTC-19.1-1987, Measurement Uncertainty

EIA 422-A, Electrical Characteristics of Balanced Voltage Digital Interface Circuits

EIA 485, Electrical Characteristics of Generators and Receivers for Use in Balanced Digital Multipoint Systems

ISA-RP2.1-1978, Manometer Tables

ISA-RP12.1-1960, Electrical Instruments in Hazardous Atmospheres, General

ISA-RP12.4-1970, Instrument Purging for Reduction of Hazardous Area

ISA-S12.11-1973, Electrical Instruments in Hazardous Dust Locations

International Standards and Drafts

IEC 381-1 1982, DC Current Signal Transmission (for process industry use)

IEC 381-2 1971, DC Voltage Signal Transmission (for process industry use)

IEC 382 1990R, Air Pressures for Pneumatic Analog Signal Transmission

IEC 529, Ingress Protection Classification and Testing Procedures

ISO 2186-1988R, Lead lines, Reconfirmed

ISO 2975/1 '74, Tracer method, General

ISO 2975/2 '75, Tracer method, Constant rate injection, non-radioactive

ISO 2975/3 '76, Tracer method, Constant rate injection, radioactive

ISO 2975/6 '77, Tracer method, Transit time, non-radioactive

ISO 2975/7 '77, Tracer method, Transit time, radioactive

ISO DIS 3313 '89, Measurement of pulsating flow in a pipe by means of orifice plates, nozzles or Venturi tubes, in particular in the case of sinusoidal or square wave intermittent periodic-type fluctuations (revision of DTR)

ISO 3354 1988, Measurement of clean water flow in closed conduits—Velocity area method using current meters in full conduits and under regular flow conditions

ISO 3966 '77, Velocity/area with Pitot static tubes

ISO DIS 4006-1988R, Measurement of fluid flow in closed conduits—Vocabulary and symbols (being revised)

ISO 4053-1 '77, Gas flow by tracer, General

ISO 4053-4, 1988, Tracer, radioactive, (Reconfirmed)

ISO DP4064/1 (revision), Cold water meter specifications

ISO 4064-2, 1988R, Potable water meters, installation

ISO 4064-3, 1988, Potable water meters, materials, methods (Reconfirmed)

ISO 4185 '80, Measurement of liquid flow—weighing method

ISO 5167-1, 1991, Measurement of fluid flow by means of orifice plates, nozzles and Venturi tubes inserted in circular cross-section conduits running full

ISO DP 5167-2, Measurement of fluid flow by means of differential pressure devices - Part 2 - Orifice plate or nozzle at the inlet of a pipe

ISO 5168 '78, Measurement of fluid flow - Estimation of uncertainty of a flow-rate measurement (being revised)

ISO DIS 6817 '80, Measurement of conductive fluid flow rate in closed conduits—Method using electromagnetic flowmeters (being revised as a standard)

ISO DIS 7066/1 1989R, Assessment of uncertainty in the calibration and use of flow measurement devices - Part 1: Linear calibration relationships

ISO 7066-2 1988, Assessment of uncertainty in the calibration and use of flow measurement devices, Part 2—Nonlinear relationships

ISO DIS 7066/1.2, Editorial revision of 7066/1

ISO 7145 '82, Determination of flow rate of fluids in closed conduits of circular cross section—Method of velocity measurement at one point of the cross section

ISO 7194 1988R, Measurement of fluid flow in closed conduits—Velocity-area methods of flow measurement in swirling or asymmetric flow conditions in circular ducts by means of current meters or Pitot static tubes

ISO 7278-2 1988, TC 28, Proving systems for volumetric meters, pipe provers

ISO 7858/1 '85, Measurement of water flow in closed conduits—Meters for cold potable water, combination meters, Part 1: Specifications

ISO 7858/2 '88, Measurement of water flow in closed conduits—Meters for cold potable water, combination meters, Part 2: Installation requirements

ISO DIS7858/3 '88, Measurement of water flow in closed conduits—Meters for cold potable water, combination meters, Part 3: Test methods and equipment

ISO 8316 1987, Measurement of liquid flow in closed conduits—Method by collection of the liquid in a volumetric tank

ISO DP 8959/2, Measurement of gas flow—Volumetric measurement—Part 2: Bell provers

ISO 9104, Methods of evaluating the performance of electromagnetic flowmeters for incompressible liquids in closed conduits for use in industrial process control systems

ISO DIS 9300, Measurement of gas flow by means of critical flow Venturi nozzles (being revised)

ISO DIS 9368-1, Installations for liquid flow rate measurement by the weighing method—Test methods—Part 1: Static weighing systems

ISO DP 9464, Code of practice for ISO 5167

ISO DP 9695/1, Measurement of gas flow rate, general

ISO DP 9695/2, Measurement of gas flow rate with bell prover

DIS 9951, Turbine meter specifications for gas flow rate measurement

ISO Draft Proposals

02.10	Conditions differing from specified in 5167, study
(8959)	Part 3 Gas, volumetric, Liquid displacement, study Part 4 Gas, volumetric, Forced piston, study Part 5 Gas, volumetric, Tube provers with free piston displacement

15.02	Turbine meter use, Gas, study
16.01	WG16-Straighteners, conditioners, study
20.02	WG19-Vortex, active draft development (4th Draft)
20.03	WG20-Ultrasonic, study (First Draft Follows U.S. and German)
20.1	SC12-Dynamic mass flow measurement, active draft development
07.05.03	Combination meters, cold potable water, Part 3, test methods, study
07.06	Hot water meter specifications, Part 1, study, DIS 10385-1
17.01	Variable area, gravity type, study

Flowmeter Testing Standards

ANSI/ASME MFC-7M, 1987, Measurement of Gas Flow by Means of Critical Flow Venturi Nozzles

ANSI/ASME MFC-9M, 1988, Measurement of Liquid Flow in Closed Conduits by Weighing Method

ASME MFC-3M, 1985, Measurement of Fluid Flow in Closed Conduits Using Orifice, Venturi, and Nozzle

ISO 4185, 1980, Flow measurement by weighing method

ISO 7278, 1988, Proving systems for volumetric meters, pipe provers

ISO 8316, 1987, Method by collection of the liquid in a volumetric tank

ISO DIS 9368, Test by static weighing

ISO DP 8959/2, Volumetric method by collection of gas in a bell prover

Where to get Copies of Standards and Procedures

3-A	3-A Sanitary Standards Committee (available from the *Journal of Food Protection*, Box 701, Ames, IA 50010)
ANSI	American National Standards Institute, 1430 Broadway, New York, NY 10018
API	American Petroleum Institute, 1220 L Street NW, Washington, DC 20005
ASHRAE	American Society of Heating and Air Conditioning Engineers, 1791 Tullie Circle, N.E. Atlanta, GA 30329
ASME	American Society of Mechanical Engineers, 345 East 47th Street, New York, NY 10017
AWWA	American Water Works Association, 6666 West Quincy Avenue, Denver, CO 80235
EIA	Electronic Industries Association, 2001 Eye Street, NW, Washington, D.C. 20006
FM	Factory Mutual Engineering and Research, 1151 Providence Highway, Norwood, MA 02062
IEC	International Electrotechnical Commission (order from ANSI)
ISA	Instrument Society of America, 67 Alexander Drive, P. O. Box 12277, Research Triangle Park, NC 27709

ISO	International Standards Organization (order from ANSI) Handbook #15 includes all ISO standards for fluid flow in closed conduits
NEMA	National Electrical Manufacturers' Association, 2101 L Street NW, Washington DC 20037
NFPA	National Fire Protection Association, Batterymarch Park, Quincy, MA 02269
SCC	Standards Council of Canada, 2000 Argentia Rd., Suite 2-401, Mississauga, Ontario L5N 1V8
UL	Underwriters Laboratories, 333 Pfingsten Road, Northbrook, IL 60062

About the Author

Mead Bradner has been involved in using and writing standards for 45 years. He was employed by The Foxboro Company for 39 years as Research Director, Product Marketing Manager, International Engineering Coordinator, Technical Director for Ideas, etc. During this time he worked with the American Society of Mechanical Engineers (ASME), the Scientific Apparatus Manufacturing Association (SAMA), the Instrument Society of America (ISA), the International Electrotechnical Commission (IEC), the International Standards Organization (ISO), and others. Since retiring, Bradner has continued to work as an independent consultant with ASME, ISA, ISO, and IEC on industrial process measurement and control standards, with major emphasis on the measurement of fluid flow in pipes. Mr. Bradner is a Fellow in ISA, a Fellow in ASME, and has a degree from MIT.

26

Flowmeter Specifications

Various methods are used to specify and purchase flowmeters. When flowmeters are purchased by specifying only the model number, the manufacturer is excluded from providing any expertise that may improve the measurement system, providing information that may deem the flowmeter be suitable for the application, or reducing costs.

Composing the flowmeter description may exhibit the same pitfalls as those of specifying the model number, depending upon the type and amount of descriptive information that is provided to the manufacturer. It is truly amazing that the correct instrument is purchased, given the wide diversity of literary styles and varying amounts of technical substance used to specify flowmeters.

Instrument specification forms, such as those presented herein, allow the user to virtually avoid varying literary styles and inconsistent technical content. The concise fill-in-the-blank specification format allows the user to generate a complete technical document by completing the form and perhaps adding some notes. This approach allows the manufacturer to better evaluate the suitability of the instrument in a specific application.

The Instrument Society of America (ISA) has developed several specification forms for flow measurement instruments (ISA-S20-1981, Specification Forms for Process Measurement and Control Instrumentation, Primary Elements and Control Valves). When ISA specification forms are not available for a particular type of flowmeter, specification forms from sources other than ISA have been included.

Specification Forms for Process Measurement and Control Instruments, Primary Elements and Control Valves

ORIFICE PLATES and FLANGES

SHEET ____ OF ____

NO	BY	DATE	REVISION

SPEC. NO.	REV.
CONTRACT	DATE
REQ. - P.O.	

BY	CHK'D	APPR.

ORIFICE PLATES

1. Concentric ☐ Other ________
2. ISA Standard ☐ Other ________
3. Bore: Maximum Rate ☐ Nearest 1/8 in. ☐
4. Material: 304SS ☐ 316SS ☐ Other ________
5. Ring Material & Type ________
6. MFR. & Model No. ________

ORIFICE FLANGES

7. Taps: Flange ☐ Vena Contracta ☐ Pipe ☐ Other ________
8. Tap Size: 1/2 in. ☐ Other ________
9. Type: Weld Neck ☐ Slip On ☐ Threaded ☐
10. Material: Steel ☐ Other ________
11. Flanges included ☐ By others ☐
12. Flange Rating ________

FLUID DATA	13	Tag Number				
	14	Service				
	15	Line Number				
	16	Fluid				
	17	Fluid State				
	18	Maximum Flow				
	19	Normal Flow				
	20	Pressure				
	21	Temperature				
	22	Specific Gravity at Base				
	23	Operating Spec. Gravity				
	24	Supercomp. Factor				
	25	Mol. Weight / Cp/Cv				
	26	Operating Viscosity				
	27	Quality % or °Superheat				
	28	Base Press. / Base Temp.				
METER	29	Type of Meter				
	30	Diff. Range – Dry				
	31	Seal sp. gr. at 60° F				
	32	Static Press. Range				
	33	Chart or Scale Range				
	34	Chart Multiplier				
PLATE & FLANGE	35	Beta=d/D				
	36	Orifice Bore Diameter				
	37	Line I.D.				
	38	Flange Rating				
	39	Vent or Drain Hole				
	40	Plate Thickness				

Notes:

ISA FORM S20-21

©ISA S20

Specification Forms for Process Measurement and Control Instruments, Primary Elements and Control Valves

ORIFICE PLATES AND FLANGES

Instructions for ISA Form S20.21

Refer to ISA Recommended Practice RP3.2, "Flanged Mounted, Sharp Edged Orifice Plates for Flow Measurement."

1. Check if concentric bore, or write in eccentric, segmental, etc.
2. ISA Standard reference given above. This also conforms to AGA-ASME requirements.
3. Check whether plate is to be bored odd size for exact maximum rate, or to nearest 1/8 in. for approximate maximum rate.
4. Select plate material.
5. If ring joint assembly is used, give ring material and configurations.
6. Refers to plate, not flanges.
7. Select one of the standard tap locations or write in other.
8. Select tap size.
9. Select flange construction.
10. Select flange material. If stainless steel, show type; such as, "304 SS."
11. Indicate whether orifice flanges are to be included with the plate, or furnished by others.
12. Note Flange Rating.
13. Tag number or other identification No.
14. Process service.
15. Line number. Include line size.
16. List fluid, unless classified.
17. Liquid, gas, or vapor.
18. Maximum flow assumed to be meter maximum. Give flow units.
19. Figure only if units given above.
20. Upstream operating pressure and units. This is also the contract figure unless otherwise noted.
21. Operating temperature, °F or °C. See comment in 20 above.
22. Specific gravity at Base Temperature.
23. Liquid specific gravity at operating temperature given on Line 21.
24. Applies to gas, at operating pressure. Supercompressibility factor normally required for gases over 100 psig because the gas at this pressure and above does not follow the ideal gas laws.
25. Applies to vapor or gas. C_p specific heat at constant pressure, C_v specific heat a constant volumes — Ratio = K at the operating temperature.
26. Viscosity and units, at operating temperature given on line 21.
27. Applies to vapor or steam. Write "SAT" if saturated; otherwise give % quality or degrees superheat, in F or C.
28. Contract base conditions. Pressure must be given in absolute units.
29. Bellows, diaphragm, mercury, etc.
30. Set range and units.
31. Applies to wet meters.
32. Fill in if applicable.
33. Full scale range and units. See comment under 18 above.
34. Fill in if required.
35. Fill in for final records after approved bore calculation is available.
36. For final records, see comment on 35.
37. In inches; or give line size and Schedule.
38. ANSI Flange Rating, i.e., 4 in. 300 lb RF
39. If desired, state whether top or bottom.
40. Give plate thickness.

ROTAMETERS
(VARIABLE AREA FLOWMETERS)

NO	BY	DATE	REVISION

SHEET ____ OF ____		
SPEC. NO.		REV.
CONTRACT		DATE
REQ. - P.O.		
BY	CHK'D	APPR.

GENERAL	1	Tag Number					
	2	Service					
	3	Line No./Vessel No.					
	4	Function					
	5	Mounting					
	6	Power Supply					
	7	Conn. Size	Type				
	8	Inlet Dir.	Outlet Dir.				
	9	Fitting Material					
	10	Packing or O-Ring Mtl.					
	11	Enclosure Type					
METER	12	Size	Float Guide				
	13	Tube Mtl.	Float Mtl.				
	14	Meter Scale: Length & Type					
	15	Meter Scale Range					
	16	Meter Factor					
	17	Rated Accuracy					
	18	Hydraulic Calib. Required					
FLUID DATA	19	Fluid					
	20	Color or Transparency					
	21	Maximum Flow Rate					
	22	Norm Flow	Min Flow				
	23	Oper. Specific Gravity (Liq)					
	24	Max Oper. Viscosity					
	25	Oper. Press.	Oper. Temp.				
	26	Oper. Density (Gases)					
	27	Std. Density	Mol. Wgt.				
	28	Max. Allowable Press. Drop					
	29						
EXT	30	Extension Well Mtl.					
	31	Gasket Mtl.					
XMTR	32	Transmitter Output					
	33	Trans. Enclosure Class					
	34	Scale Range					
ALARM	35	No. of Contacts	Form				
	36	Rating	Housing				
	37	Action					
	38						
OPTIONS	39	Valve Size & Material					
	40	Valve Location					
	41	Const. Diff. Relay Mtl.					
	42	Purge Meter Tubing					
	43	Airset					
	43a						
	44	Manufacturer					
	45	Model Number					
	46	Tube Number					
	47	Float Number					

Notes:

ISA FORM S20.22

Specification Forms for Process Measurement and Control Instruments, Primary Elements and Control Valves

ROTAMETERS

Instructions for ISA Form S20.22 (Refer to ISA RP16.1,2,3,4)

1. List tag number.

2. Refers to process applications.

3. Show line number, vessel number, or line specification.

4. Give functions such as INDICATE RECORD, CONTROL TRANSMIT, INTEGRATE, etc.

5. FLUSH PANEL, FRONT PANEL, PIPE, etc.

6. Give voltage, dc or ac, and ac frequency.

7. Give nominal connection size and type such as SCREWED, 150 lb FLANGED, etc.

8. Select orientation of inlet and outlet and designated as RIGHT, LEFT, VERTICAL or REAR.

9. Select material of end fittings. Note if lining is required.

10. Select either packing or "O" ring design and note material.

11. Select type of enclosure, if any, such as SIDE PLATE, SAFETY GLASS, etc.

12. Give meter size. Note that this is not the same as connection size but refers to the nominal size of the tube and float combination.

Give the method of float guiding such as NONE, FLUTES, POLE, EXTENSIONS.

13. Select tube and float material.

14. Select type meter scale: NONE, ON GLASS, METAL STRIP. Select meter scale length.

15. Select meter scale range and flow units. Remember that rotameters' scales cannot start at zero but typically have rangeability of 10:1 or 12:1.

16. Meter factor if not direct reading.

17. Accuracy statement does not imply any specific calibration.

18. Note if hydraulic calibration is required and state required accuracy.

19. If fluid cannot be identified, state if liquid or gas.

20. Give fluid color or transparency which will affect float visibility in glass tube meters.

21. List maximum operating flow rate and units, usually the same as maximum of meter scale.

22. Show normal and minimum flow rates expected.

23. Give operating specific gravity of liquid. (Numerically equal to density in gm/cm^3).

24. Give maximum expected viscosity and units.

25. Give operating pressure and temperature, with units.

26. For gases give operating density and units, unless molecular weight is given on Line 27.

27. For gases give density at standard conditions (14.7 psia and 60°F unless stated otherwise, and/or molecular weight if known.

28. State maximum allowable pressure drop at full flow, if applicable.

30. If meter has an extension well, state material of well.

31. Select material of gasket on extension.

32. If meter transmits, state pneumatic or electronic output such as 21-103 kPa (3-15 psig), 4-20 mA, etc.

33. Give transmitter electrical classification such as General Purpose, Class 1, Group D, etc.

34. Give transmitter scale size and range. Note that this is not the meter scale but the scale of the attached instrument.

35. Number of alarm contacts in case.

Form of contacts: SPDT, SPST, DPDT, etc.

36. Contact electrical load rating. Contact housing — GP, Class I, GR.D, etc. Use NEMA identification.

37. HIGH, LOW, DEVIATION.

39. Specify needle valve if required.

40. Valve may be on the inlet, outlet or separately mounted. Do not list here if valve is to be furnished by others.

41. This relay may be used on purge assemblies.

44 - 47. When manufacturer is selected fill in exact model and part numbers.

Specification Forms for Process Measurement and Control Instruments, Primary Elements and Control Valves

MAGNETIC FLOWMETERS				SHEET ____ OF____
				SPEC. NO. / REV.
NO	BY	DATE	REVISION	CONTRACT / DATE
				REQ. - P.O.
				BY / CHK'D / APPR.

	1		Meter Tag No.
	2		Service
	3		Location
METERING ELEMENT	4	CONN'S.	Line Size, Sched.
	5		Line Material
	6		Connection Type
	7		Connection Mat'ls.
	8	METER	Tube Material
	9		Liner Material
	10		Electrode Type
	11		Electrode Matl.
	12		Meter Casing
	13		Power Supply / Elect. Code
	14		Grounding, Type & Matl.
	15		Enclosure Class
	16		
	17	FLUID	Fluid
	18		Max. Flow, Units
	19		Max. Velocity, Units
	20		Norm. Flow / Min. Flow
	21		Max. Temp. / Min. Temp.
	22		Max. Press. / Min. Press.
	23		Min. Fluid Conductivity
	24		Vacuum Possibility
	25		
ASSOCIATED INSTRUMENT	26		Instrument Tag Number
	27		Function
	28		Mounting
	29		Enclosure Class
	30		Length Signal Cable
	31		Type Span Adjustment
	32		Power Supply
	33	TRANS.	Transmitter Output
	34		
	35	DISPLAY	Scale Size / Range
	36		Chart Drive / Speed
	37		Chart Range / Chart No.
	38		Integrator
	39	CONTR.	Modes / Output
	40		Action / Auto-Man.
	41		
	42	ALARM	Contact No. / Form
	43		Rating / Elec. Code
	44		Action
	45		Manufacturer
	46		Meter Model Number
	47		Instrument Model Number

Notes:

ISA FORM S20.23

Specification Forms for Process Measurement and Control Instruments, Primary Elements and Control Valves

MAGNETIC FLOWMETERS

Instructions for ISA Form S20.23

1. Tag number of meter only.
2. Refers to process application.
3. Show line number or identify associated vessel.
4. Give pipeline size and schedule. If reducers are used, so state.
5. Give material of pipe. If lined, plastic or otherwise non-conductive, so state.
6. Give connection type: FLANGED, DRESSER COUPLINGS, ETC.
7. Specify material of meter connections.
8. Select tube material. (Non-permeable material required if coils are outside tube).
9. Specify material of line.
10. Select electrode type: STD., BULLET NOSED, ULTRASONIC CLEANED, BURN OFF, etc.
11. Specify electrode material.
12. Describe casing: STD., SPASH PROOF, SUBMERSIBLE, SUBMERGED OPERATION, etc.
13. Give ac voltage and frequency, along with application NEMA identification of the electrical enclosure.
14. State means for grounding to fluid: GROUNDING RINGS, STRAPS, etc.
15. State power supply and enclosure class to meet area electrical requirements.
16.
17. State fluid by name or description.
18. Give maximum operating flow and units; usually same as maximum of instrument scale.
19. Give maximum operating velocity, usually in ft/s.
20. List normal and minimum flow rates.
21. List maximum and minimum fluid temperature °F.
22. List maximum and minimum fluid pressure.
23. List minimum (at lowest temp.) conductivity of fluid.
24. If a possibility of vacuum exists at meter, so state and give greatest value. (highest vacuum).
25.
26. List tag number of instrument used directly with meter.
27. Control loop function such as INDICATE, RECORD CONTROL, etc.
28. Mounting: FLUSH PANEL, SURFACE INTEGRAL WITH METER, etc.
29. Give NEMA identification of case type.
30. State cable length required between meter and instrument.
31. Span adjust: BLIND, ft/s DIAL, OTHER.
32. Give ac supply voltage and frequency.
33. If a transmitter, state analog output electrical or pneumatic range, or pulse train frequency for digital outputs, i.e., pulses per gallon.
34. List scale size and range.
35. List Scale Size and Range for indicating transmitter
36. Recorder chart drive – ELECT. HANDWIND, etc. and chart speed in time per revolution or inch per hour.
37. List chart range and number.
38. If integrator is used, state counts per hour, or value of smallest count; such as "10 GAL UNITS".
39. For control modes: (Per ANSI C85.1-1963, "Terminology for Automatic Control.") Write-in PI_f, I_f, PI_s, PI_fD_f, etc.

P = proportional (gain)
I = integral (auto reset)
D = derivative (rate)

Subscripts:

f = fast
s = slow
n = narrow

State output signal range, pneumatic or electronic.

40. Controller action in response to an increase in flowrate – INC. or DEC.

State auto-man. switch as NONE, SWITCH ONLY, BUMPLESS, etc.

42. Number of alarm lights in case. Give form of contacts; SPDT, SPST, etc.
43. Contact electrical load rating. Contact housing General Purpose, Class I, Group D, etc., if not in the same enclosure described in line 29.
44. Action of alarms: HIGH, LOW, DEVIATION, etc.
45. Fill in manufacturer and model numbers for meters
46. and
47. instrument after selection.

Specification Forms for Process Measurement and Control Instruments, Primary Elements and Control Valves

TURBINE FLOWMETERS				SHEET ____ OF____
NO	BY	DATE	REVISION	SPEC. NO. / REV.
				CONTRACT / DATE
				REQ. - P.O.
				BY / CHK'D / APPR.

METER	1	Tag Number				
	2	Service				
	3	Meter Location				
	4	Line Size				
	5	End Connections				
	6	Body Rating				
	7	Nominal Flow Range				
	8	Accuracy				
	9	Linearity				
	10	K Factor, Cycles per Vol. Unit				
	11	Excitation				
	12	Materials: Body				
	13	Support				
	14	Shaft				
	15	Flanges				
	16	Rotor				
	17	Bearings: Type				
	18	Bearing Material				
	19	Max. Speed				
	20	Min. Output Voltage				
	21	Pickoff Type				
	22	Enclosure Class				
	23					
FLUID DATA	24	Fluid				
	25	Flow Rate: Min. / Max.				
	26	Normal Flow				
	27	Operating Pressure				
	28	Back Pressure				
	29	Operating Temp. Max. / Min.				
	30	Operating Specific Gravity				
	31	Viscosity Range				
	32	Percent Solids & Type				
	33					
SECONDARY INSTR.	34	Secondary Instr. Tag No.				
	35	Preamplifier				
	36	Function				
	37	Mounting				
	38	Power Supply				
	39	Scale Range				
	40	Output Range				
OPTIONS	41	Totalizer Type				
	42	Compensation				
	43	Preset Counter				
	44	Enclosure Class				
	45	Strainer Size & Mesh				
	46					
	47					
	48					
	49	Manufacturer				
	50	Meter Model No.				
	51	Secondary Instr. Model No.				

Notes:

ISA Form S20.24

Specification Forms for Process Measurement and Control Instruments, Primary Elements and Control Valves

TURBINE FLOWMETERS

Instructions for ISA Form S20.24
Refer to ISA Standard S31, "Specification, Installation, and Calibration of Turbine Flowmeters"

1. Show meter tag number. Quantity is assumed to be one unless otherwise noted.

2. Refers to process service or applications.

3. Give line number or process area.

5. Specify size and style of connections, such as "1 in. NPT", "2 in. 150 lb ANSI", etc.

6. Pressure and temperature design rating required.

7. Nominal flow range is obtained from manufacturer's data. This usually defines linear range of selected meter.

8. Turbine meter accuracy figures are in terms of percent of instantaneous flow rate.

9. Degree of linearity over nominal flow range.

10. K factor relates cycles per second to volume units. Enter this figure after selection is made.

11. Excitation modulating type only expressed as volts_____ at_____ hertz.

12 to 16. Specify materials of construction or write in "MFR STD."

17. Specify sleeve or ball bearings, or none if floating rotor design.

18. Bearing material — will be MFG STD if not stated otherwise.

19. Maximum speed or frequency which the meter can produce without physical damage.

21. Pickoff may be standard hi-temp., radio-frequency type (RF) or explosion proof. Minimum output voltage _______ volts peak to peak.

22. Specify electrical classification of enclosure such as General Purpose, Weather Proof, Class 1, Group D, etc.

23. Specify fluid data as indicated, using line 28 for additional item if required.

34. Give Tag No. of secondary instrument if different from meter Tag No.

35. Pre-amplifier if used.

36. Specify function of instrument, such as rate indicator, totalizer, or batch control.

37. Flush, surface or rack.

38. Power Supply, i.e., 117 Vac.

39. Applies to rate indicator

40. Give output range such as "40-20 mA", 21-103 kPa (3-15 psig), etc.

41. May be used for number of digits, and to state whether counter is reset or non-reset type.

42. Specify range of compensation, if required, in pressure and/or temperature units or viscosity units.

43. Pre-set counter.

44. Specify NEMA classification of enclosure.

45. Specify strainer size and mesh size. Request vendor's recommendation if not known.

50. Fill in after selection is made.

51. Fill in after selection is made.

52. Fill in after selection is made.

Specification Forms for Process Measurement and Control Instruments, Primary Elements and Control Valves

(ISA logo)	POSITIVE DISPLACEMENT METERS	SHEET ____ OF ____
	NO / BY / DATE / REVISION	SPEC. NO. / REV.
		CONTRACT / DATE
		REQ. - P.O.
		BY / CHK'D / APPR.

	1	Tag Number				
	2	Service				
	3	Line No./Vessel No.				
METER	4	Type of Element				
	5	Size				
	6	End Connections				
	7	Temp. & Press. Rating				
	8	Flow Rate Range				
	9	Totalized Units				
	10	Enclosure Class				
	11	Power Supply				
	12	Materials: Outer Housing				
	13	Main Body Cover				
	14	Rotating Element				
	15	Shaft				
	16	Blades				
	17	Bearings: Type & Material				
	18	Packing				
	19	Type of Coupling				
	20					
COUNTER	21	Register Type				
	22	Totalizer				
	23	Reset				
	24	Capacity				
	25	Set-Stop				
	26					
FLUID DATA	27	Fluid				
	28	Flow Rate: Min. / Max.				
	29	Normal Flow				
	30	Oper. Press. / Oper. Temp.				
	31	Oper. Specific Gravity				
	32	Oper. Viscosity				
	33	Coef. of Expansion				
OPTIONS	34	Flow Units				
	35	Shut-Off Valve				
	36	Switch: Single or 2-Stage				
	37	Temp. Compensator				
	38	Transmitter Type				
	39	Transmitter Output				
	40	Air Eliminator				
	41	Strainer: Size & Mesh				
	42					
	43					
	44					
	45	Manufacturer				
	46	Model Number				

Notes:

ISA FORM S20.25

©ISA S20

Specification Forms for Process Measurement and Control Instruments, Primary Elements and Control Valves

POSITIVE DISPLACEMENT METERS

Instructions for ISA Form S20.25.

1. Tag No. of instrument.

2. Process service.

3. Pipe line or vessel identification.

4. Write in type of rotating element, such as, disc, piston, vane, helical, rotors, etc.

5. Show connection pipe size.

6. Specify end connections type and ANSI rating such as 300 lb R.F.

7. Specify the manufacturer's recommended body pressure and temperature rating, such as 250 psi at 190°F.

8. Write in manufacturer's recommended normal operating range.

9. Specify smallest totalized unit, such as "Tens of Gallons", "Pounds", "Barrels".

10. Specify enclosure electrical classification, if applicable, such as "Class 1, Group D., Div. 2", "General Purpose", etc.

11. Specify power supply, if applicable.

12. Specify materials of construction. If no preference, write in, MFR. STD. (Manufacturer's Standard).

13-18. Specify materials of construction, if no preference, write in, Manufacturer's Standard (MFG-STD)

19. Specify type of coupling

20. Specify coupling such as "Magnetic", or MFR. STD.

21. Specify register type such as horizontal, vertical, inclined, inline reading, dial reading, print, etc.

22. Specify number of figures such as 6 digit, 5 digit, or 0-99, 999, etc.

23. If totalizer reset required, write in type. If reset is not required, write in "none".

24. Write in number of figures or maximum quantity (in flow units) that can be held in counter.

25. Specify by writing in "yes" if a set-stop is required to operate shutoff valve, switch, etc.

27-34. Specify fluid data as completely as possible, note at operating conditions. Be sure to note if liquid is at saturation conditions.

35. Specify by writing in "yes" if a shut-off valve is required. Valve to be manufacturer's standard construction unless otherwise noted.

36. Specify by writing in "yes" if a switch is required. Two switches are required for 2-stage shut-off control.

37. Write in "yes" if manufacturer's standard temperature compensator is required. Write in "no" if not required.

38. Specify, if transmitter is required, by writing in type such as pulse, rate of flow, etc.

39. Give transmitter output in pulse per gallon, 4-20 mA, etc.

40. Write in "yes" if air eliminator is required, otherwise write in "no".

41. Specify, if strainer is required, by writing in type such as "Y", "Basket", etc. Strainer to have same pressure and temperature rating, end connections and material as meter body unless otherwise noted.

45-46. Identify manufacturer's name and model number after selection is made.

MASS FLOWMETERS

ENGINEERING SPECIFICATION

Sheet Of

Location	No. Req.	Spec. No.
TAG NO.		
QUANTITY		
SERVICE		
MANUFACTURER		
MODEL NO.		
SERVICE CONDITIONS		
Fluid		
State		
Flow Min/Nom/Max		
Temperature		
Pressure		
Specific Gravity or Density		
Viscosity		
Area Electrical Classification		
FLOWMETER ELEMENT - TAG NO.		
Size & End Connections		
Wetted Parts		
Mounting		
Pressure Drop & Max Flow		
Face to Face Dimension		
FLOWMETER TRANSMITTER - TAG. NO.		
Input Power		
Calibration Range		
Temperature Compensation		
Output Signal		
Accuracy		
Housing		
Cable Length		

Revision						
	5			Spec. By		
	4			Design Approval		
	3			Process Approval		
	2			Purchase Order No.		
	1			Vendor		

OPEN CHANNEL FLOWMETERS

Sheet Of

ENGINEERING SPECIFICATION

Location	No. Req.	Spec. No.

Tag No.
Quantity
Service
Manufacturer
Model No.

Operating Conditions
- Fluid Name
- Fluid State
- Pipe Size and Shape
- Operating Density
- Min/Max Flow
- Temperature
- Operating Viscosity

Flowmeter Characteristics
- Size and Connections
- Face to Face Dimension
- Materials of Construction
- Transmitter Mounting
- Input Power
- Output Signal
- Calibration Range

$$R_D = \frac{3160\ Q_{gpm}\ SG}{\mu cp^D\ in.} = \frac{379\ Q_{acfm}\ \rho\ lb/ft^3}{\mu cp^D\ in.}$$

Revision						
	5			Spec. By		
	4			Design Approval		
	3			Process Approval		
	2			Purchase Order No.		
	1			Vendor		

THERMAL FLOWMETERS

ENGINEERING SPECIFICATION

Sheet Of

Location	No. Req.	Spec. No.

Tag No.

Quantity

Service

Manufacturer

Model No.

Operating Conditions

- Fluid Name
- Fluid State
- Pipe Size and Shape
- Specific Gravity
- Min/Max Flow
- Temperature
- Operating Viscosity
- Pressure

Flowmeter Characteristics

- Size and Connections
- Insertion or Face to Face Dimension
- Materials of Construction
- Transmitter Mounting
- Input Power
- Output Signal
- Calibration Range
- Accuracy

$$R_D = \frac{3160\ Q_{gpm}\ SG}{\mu cp^D\ in.} = \frac{379\ Q_{acfm}\ \rho\ lb/ft^3}{\mu cp^D\ in.}$$

Revision						
	5			Spec. By		
	4			Design Approval		
	3			Process Approval		
	2			Purchase Order No.		
	1			Vendor		

Index

A

D

E

F

G

H

I

J

K

Q

R

S

T

U

V

W

Y

Z

Special Characters